The Power of Functional Resins in Organic Synthesis

Edited by
Judit Tulla-Puche and
Fernando Albericio

The Power of Functional Resins in Organic Synthesis

Edited by
Judit Tulla-Puche and Fernando Albericio

WILEY-VCH Verlag GmbH & Co. KGaA

The Editors

Dr. Judit Tulla-Puche
Institute for Research in Biomedicine.
University of Barcelona
Baldiri Reixac 10
08028 Barcelona
Spain

Prof. Dr. Fernando Albericio
Institute for Research in Biomedicine.
University of Barcelona
Baldiri Reixac 10
08028 Barcelona
Spain

All books published by Wiley-VCH are carefully produced. Nevertheless, authors, editors, and publisher do not warrant the information contained in these books, including this book, to be free of errors. Readers are advised to keep in mind that statements, data, illustrations, procedural details or other items may inadvertently be inaccurate.

Library of Congress Card No.: applied for

British Library Cataloguing-in-Publication Data
A catalogue record for this book is available from the British Library.

Bibliographic information published by the Deutsche Nationalbibliothek
Die Deutsche Nationalbibliothek lists this publication in the Deutsche Nationalbibliografie; detailed bibliographic data are available in the Internet at <http://dnb.d-nb.de>.

© 2008 WILEY-VCH Verlag GmbH & Co. KGaA, Weinheim

All rights reserved (including those of translation into other languages). No part of this book may be reproduced in any form – by photoprinting, microfilm, or any other means – nor transmitted or translated into a machine language without written permission from the publishers. Registered names, trademarks, etc. used in this book, even when not specifically marked as such, are not to be considered unprotected by law.

Typesetting SNP Best-set Typesetter Ltd., Hong Kong
Printing betz-druck GmbH, Darmstadt
Binding Litges & Dopf Buchbinderei GmbH, Heppenheim

Printed in the Federal Republic of Germany
Printed on acid-free paper

ISBN: 978-3-527-31936-7

Contents

Preface *XIX*

Part One Introduction

1 The (Classic Concept of) Solid Support *3*
Fernando Albericio and Judit Tulla-Puche
1.1 Introduction *3*
1.2 Linkers/Handles *4*
1.3 Solid Supports *7*
1.3.1 Gel-Type Support *7*
1.3.1.1 Polystyrene (PS) Resins *7*
1.3.1.2 Poly(Ethylene Glycol)–Polystyrene (PEG-PS) Resins *8*
1.3.1.3 Hydrophilic PEG-Based Resins *8*
1.3.2 Modified Surface Type Supports *10*
1.3.2.1 Cellulose Membranes *10*
1.3.2.2 Polyolefinic Membranes *11*
1.3.2.3 Pellicular Solid Supports *11*
Acknowledgments *12*
References *12*

2 Molecularly Imprinted Polymers *15*
Henrik Kempe and Maria Kempe
2.1 Introduction *15*
2.2 The Concept of Molecular Imprinting *16*
2.2.1 Non-covalent Molecular Imprinting *17*
2.2.2 Covalent Molecular Imprinting *17*
2.2.3 Semi-covalent Molecular Imprinting *19*
2.2.4 Metal Ion Mediated Molecular Imprinting *19*
2.3 Formats of Molecularly Imprinted Polymers *19*
2.3.1 Irregularly Shaped Particles *21*
2.3.2 Beads *22*
2.3.2.1 Homogeneous Polymerization *22*
2.3.2.2 Heterogeneous Polymerization *22*

2.3.2.3	Two-Step Swelling Polymerization	24
2.3.2.4	Core–Shell Polymerization	24
2.3.2.5	Silica Composite Beads	24
2.3.3	Films and Membranes	24
2.4	Design of MIPs	25
2.4.1	Functional Monomers	25
2.4.2	Cross-linking Monomers	28
2.4.3	The Porogen	28
2.4.4	Initiation of Polymerization	30
2.4.5	Optimization of Imprinting Conditions	30
2.5	Characterization of Molecularly Imprinted Polymers	31
2.5.1	Characterization of Binding Properties of MIPs	31
2.5.2	Characterization of Chemical and Physical Properties of MIPs	34
2.6	Applications of Molecularly Imprinted Polymers	34
2.6.1	Liquid Chromatography	34
2.6.2	Solid-Phase Extraction	35
2.6.3	Solid-Phase Binding Assay	35
2.6.4	Sensors	35
2.6.4.1	Optical Sensors	35
2.6.4.2	Mass Sensitive Sensors	36
2.6.4.3	Electrochemical Sensors	36
2.6.5	Synthetic Enzymes	36
2.7	Conclusions	37
	References	37

3 Nanoparticles Functionalized with Bioactive Molecules: Biomedical Applications 45

Ivonne Olmedo, Ariel R. Guerrero, Eyleen Araya and Marcelo J. Kogan

3.1	Introduction	45
3.2	MNPs	46
3.2.1	Gold Nanoparticles	46
3.2.1.1	Synthesis and Properties	46
3.2.1.2	Functionalization of *GNPs* with Bioactive Compounds and Biomedical Applications of Functionalized *GNPs*	47
3.2.2	Nanoshells and Metal Heterodimers	55
3.2.3	Iron Oxide NPs	57
3.2.3.1	Synthesis and Properties	57
3.2.3.2	Functionalization of IONPs	57
3.2.4	Silver NPs	61
3.2.5	Quantum Dots	62
3.2.6	Nanowires	65
3.3	CNTs	67
3.4	Organic Nanoparticles (ONPs)	68
3.4.1	Synthesis and Properties of ONPs	68
3.4.2	Functionalization Strategies	70
3.4.3	ONPs Types and Applications	73

3.4.3.1	Fluorescent ONPs	73
3.4.3.2	Cancer-Aimed ONPs	73
3.4.3.3	Delivery of ONPs through the Blood–Brain Barrier (BBB)	74
3.4.3.4	Nucleic Acids/Gene Delivery	74
3.4.3.5	Other Biomedical Uses of ONPs	74
3.5	Conclusions	75
	Acknowledgments	75
	List of Abbreviations	76
	References	77

Part Two Solid-Supported Reagents and Scavengers

4	**Oxidizing and Reducing Agents** *83*	
	Samuel Beligny and Jörg Rademann	
4.1	Introduction	83
4.2	Considerations Concerning the Nature of the Solid Support Used for Polymer-Supported Redox Reagents	84
4.3	Oxidizing Resins	84
4.3.1	Novel Oxidative Resins	84
4.3.1.1	Solid-Supported Hypervalent Iodine Reagents	84
4.3.1.2	Supported TEMPO	85
4.3.1.3	Supported Co-oxidants	86
4.3.1.4	Asymmetric Oxidation	88
4.3.1.5	Oxidation with Multi Supported Reagents	88
4.3.1.6	Nanoparticles and Polymer Incarcerated Oxidants	89
4.3.1.7	High-Loading Resins	90
4.3.2	Applications of Functional Resins for Oxidation	90
4.3.2.1	Polymer-Supported Chromic Acid	90
4.3.2.2	Applications of Supported TPAP	90
4.3.2.3	Aerobic Oxidation with Polymer-Supported Catalysts	91
4.3.2.4	Osmium Tetroxide	91
4.4	Polymer-Supported Reducing Reagents	92
4.4.1	Novel Reagents	92
4.4.1.1	Chiral Boranes	92
4.4.1.2	Polymer-Supported Ru-TsDPEN	94
4.4.1.3	Polymer-Supported Hydrazine Hydrate	95
4.4.1.4	Nanoparticles and Polymer Incarcerated Reagents	95
4.4.1.5	Polycationic Ultra-borohydride	95
4.4.2	Solid–Supported Reducing Agents and Their Applications	96
4.5	Conclusion	96
	References	96
5	**Base and Acid Reagents** *101*	
	Tadashi Aoyama	
5.1	Base Reagents	101
5.1.1	KF/Al_2O_3	101

5.1.1.1	Generation of Dichlorocarbene *102*
5.1.1.2	Acceleration of the Metal-Catalyzed Reaction *102*
5.1.1.3	With Microwave *103*
5.1.2	Silica-Gel Bound Organic Bases *104*
5.1.2.1	For Electroorganic Synthesis *106*
5.1.3	Others *106*
5.2	Acid Reagents *108*
5.2.1	SSA *108*
5.2.2	$HClO_4/SiO_2$ *110*
5.2.3	HBF_4/SiO_2 *111*
5.2.4	Silica-Gel Supported Heteropoly Acid *112*
5.2.5	Others *112*
5.3	Applications *113*
5.4	Conclusions *115*
	List of Abbreviations *116*
	References *116*

6 Nucleophilic, Electrophilic and Radical Reactions *121*
Magnus Johansson and Nina Kann

6.1	Introduction *121*
6.2	Polymer-Supported Nucleophilic Reagents *121*
6.2.1	Polymer-Supported Secondary Amines *121*
6.2.2	Reactions with Polymer-Bound Phosphanes *123*
6.2.3	The Wittig Reaction and Related Olefination Reactions *126*
6.2.4	Nucleophiles Anchored to Anion Exchange Resins *127*
6.3	Reagents for Electrophilic Reactions *129*
6.3.1	Reagents for Alkylation and Silylation *129*
6.3.2	Reagents for Acylation and Sulfonylation *131*
6.3.3	Nitrogen Electrophiles on Solid Phase *132*
6.3.4	Electrophilic Bromination Reagents *133*
6.4	Radical Reactions Using Supported Reagents *134*
6.5	Conclusions *136*
	List of Abbreviations *136*
	References *137*

7 Coupling and Introducing Building Block Reagents *141*
Rafael Chinchilla and Carmen Nájera

7.1	Introduction *141*
7.2	Supported Carbodiimide and Isourea Reagents *142*
7.3	Supported Aminium and Uronium Reagents *146*
7.4	Supported Phosphorous-Containing Reagents *150*
7.5	Other Supported Coupling Reagents *152*
7.6	Active Ester-Forming Polymeric Reagents *158*
7.6.1	Phenol-Derived Polymers *159*
7.6.2	*N*-Hydroxysuccinimide-Derived Polymers *163*

7.6.3	*N*-Hydroxybenzotriazole-Derived Polymers	*168*
7.6.4	Other Active Ester-Forming Polymers	*174*
7.7	Conclusions	*176*
	References	*177*

8 Supported/Tagged Scavengers for Facilitating High-Throughput Chemistry *183*
Paul R. Hanson, Thiwanka Samarakoon and Alan Rolfe

8.1	Introduction	*183*
8.2	Supported/Tagged Reactive Functionalities for the Sequestration of Nucleophiles (Nucleophile/Electrophilic Scavengers)	*185*
8.2.1	Polystyrene Resin-Supported Scavengers	*185*
8.2.1.1	Resin-Supported Isocyanates	*185*
8.2.1.2	Resin-Supported Anhydrides and Acid Chlorides	*188*
8.2.1.3	Resin-Supported Aldehydes	*188*
8.2.1.4	Resin-Supported Diketones	*189*
8.2.1.5	Resin-Supported Maleimide	*189*
8.2.1.6	Resin-Supported Diazonium Salts	*192*
8.2.1.7	Resin-Supported Halide	*192*
8.2.2	PEG-Supported Scavengers	*193*
8.2.2.1	PEG-Supported Dichlorotriazine (DCT)	*193*
8.2.3	Silica-Supported Scavengers	*194*
8.2.3.1	Silica-Supported DCT	*194*
8.2.4	Fluorous-Tagged Scavengers	*194*
8.2.4.1	Fluorous-Tagged Anhydrides and Isocyanates	*196*
8.2.4.2	Fluorous-Tagged Sulfonyl Chlorides and Acid Chlorides and Epoxides	*197*
8.2.4.3	Fluorous-Tagged Aldehyde	*197*
8.2.4.4	Fluorous-Tagged DCT	*198*
8.2.5	ROMP-Derived Electrophilic Scavengers	*200*
8.2.6	ROMP-Gel Anhydride	*202*
8.2.6.1	Oligomeric Bis-acid Chloride	*203*
8.2.6.2	Oligomeric Sulfonyl Chloride (OSC)	*204*
8.2.6.3	Oligomeric Phosphonyl Chloride	*205*
8.3	Supported/Tagged Reactive Functionalities for the Sequestration of Electrophiles (Electrophile/Nucleophilic Scavengers)	*205*
8.3.1	Polystyrene Resin-Based Scavengers	*206*
8.3.1.1	Supported Amine	*206*
8.3.1.2	Supported Thiols	*207*
8.3.1.3	Supported Phosphines	*207*
8.3.1.4	Supported Anthracene for Dienophile Scavenging	*209*
8.3.1.5	Supported Hydrazines	*209*
8.3.1.6	Supported Acids/Bases for Scavenging	*209*
8.3.1.7	Scavengers in "Tablets"	*211*
8.3.1.8	Magnetically-Tagged Nanoparticle for Impurity Removal	*211*

8.3.1.9 Phase Switching–Bipyridyl Tagging Approaches to Scavengers *212*
8.3.1.10 PEG-Supported Amine Scavenger Microgels *212*
8.3.1.11 Application of Polystyrene Resin-Supported Scavengers towards the Synthesis of Natural Products *213*
8.3.1.12 Application of Polystyrene Resin-Supported Scavengers towards Library Synthesis *214*
8.3.1.13 Application of Polystyrene Resin-Supported Scavengers in a Flow-through Platform towards Combinatorial Libraries and Natural Product Synthesis *215*
8.3.2 Silica-Supported Scavengers *217*
8.3.2.1 Silica-Supported Amines *217*
8.3.2.2 Silica-Supported Thiols *218*
8.3.3 Fluorous-Tagged Scavengers *219*
8.3.3.1 Fluorous-Tagged Thiols *219*
8.3.3.2 Fluorous-Tagged Dienophile *220*
8.3.4 Supported Ionic Platforms for Scavenging Electrophiles *220*
8.3.4.1 Task Specific Ionic Liquid Amine *220*
8.3.5 ROMP-Derived Nucleophilic Scavengers *222*
8.4 Conclusion *223*
References *223*

9 Metal Scavengers *227*
Aubrey Mendonca
9.1 Background *227*
9.2 Medicinal Chemistry Uses of Metals *228*
9.2.1 New Chemistries *228*
9.2.2 Examples of Drugs Involving Use of Metals as Catalysts *229*
9.3 Process Chemistry Uses of Metals *230*
9.3.1 Large-Scale Removal of Metals in Chemistry *230*
9.3.2 Alternatives to Functional Resins *230*
9.4 Players in the Field *231*
9.4.1 ChemRoutes *231*
9.4.2 Silicycle *231*
9.4.3 Aldrich *232*
9.4.4 Reaxa *233*
9.4.5 Polymer Laboratories/Varian *233*
9.4.6 Degussa *235*
9.4.7 Johnson Matthey Chemicals *236*
9.5 Silica Manufacture of the Metal Scavenger Resins *236*
9.5.1 Irregular Silica *237*
9.5.2 Uniform Silica *237*
9.6 Polymer Manufacture of the Metal Scavenger Resins *237*
9.6.1 Polystyrene Based Resins *237*
9.6.2 Macroporous Based Resin *238*
9.6.3 Gel-Type Resin *238*

9.7	Palladium Removal in Organic Chemistry	*239*
9.8	Platinum Removal in Organic Chemistry	*240*
9.9	Ruthenium and Tin Removal in Organic Chemistry	*240*
9.10	Other Metals Removal	*241*
9.11	Conclusion	*241*
	List of Abbreviations	*242*
	References	*242*

Part Three   Resin-Bound Catalysts

10   Polymer-Supported Organocatalysts   *247*
Belén Altava, M. Isabel Burguete and Santiago V. Luis

10.1	Introduction	*247*
10.2	Polymer-Supported Acidic Catalysts	*250*
10.2.1	Catalysis by Strongly Acidic Ion Exchange Resins	*250*
10.2.1.1	Conventional Polystyrene Sulfonic Resins	*250*
10.2.1.2	Modification of Conventional Sulfonated Polystyrene Resins	*254*
10.2.2	Catalysis by Perfluorinated Sulfonic Acid Polymers	*256*
10.2.3	Catalysis by Functional Polymers Containing Specific Acidic Sites	*260*
10.3	Polymer-Supported Basic Catalysts	*261*
10.3.1	Catalysis by Anion Exchange Resins	*261*
10.3.2	Catalysis by Non-chiral Polymer-Supported Amines and Phosphines	*263*
10.3.3	Polymer-Supported Chiral Basic Catalysts	*270*
10.4	Polymer-Supported Phase Transfer Catalysts	*273*
10.4.1	Non-chiral PTC Using Insoluble Supports	*273*
10.4.2	Chiral PTC Using Insoluble Supports	*276*
10.4.3	PTC Using Soluble Supports	*278*
10.5	Polymer-Supported Oxidation Catalysts	*280*
10.5.1	Non-chiral Polymer-Supported Oxidation Catalysts	*280*
10.5.2	Chiral Polymer-Supported Oxidation Catalysts	*283*
10.6	Polymer-Supported Organocatalysts Based on Amino Acids	*285*
10.6.1	Polymer-Supported Proline and Related Catalysts	*286*
10.6.2	Polymer-Supported Peptides and Related Catalysts	*290*
10.7	Polymer-Supported Imidazolium, Thiazolium and Related Structures	*293*
10.8	Miscellaneous Organocatalysts	*294*
10.9	Conclusions	*296*
	List of Abbreviations	*297*
	References	*298*

11   Transition Metal Catalysts   *309*
Rajiv Banavali, Martin J. Deetz and Alfred K. Schultz

11.1	Introduction	*309*
11.2	Synthetic Avenues for Producing Metal Loaded Organic Resins	*309*

11.2.1	Covalent Bonding *310*
11.2.2	Coordination Complexes *310*
11.2.3	Precipitation *311*
11.2.4	Microstructural Aspects of Organic Functional Resins *313*
11.3	Reactions Utilizing Transition Metal Catalysts Supported on Organic Resins *316*
11.3.1	Hydrogenation *316*
11.3.2	Bi-functional Catalysis *319*
11.3.3	Isomerization of Olefins *319*
11.3.4	Hydrosilylation Reaction *320*
11.3.5	Aldol Reactions *320*
11.3.6	The C–O Coupling Reaction *321*
11.4	Commercial Interest in Transition Metal Catalysts *322*
11.4.1	Methyl Isobutyl Ketone (MIBK) Synthesis *322*
11.4.2	Conversion of Glycerol into 1,2-Propanediol *323*
11.4.3	Oxidation–Synthesis of H_2O_2 *324*
11.4.4	Hydrogenation of N–O Bonds *324*
11.4.5	Removal of Oxygen from Water *325*
11.5	Conclusions and Future Outlook *326*
	Acknowledgments *326*
	References *326*

12 Chiral Auxiliaries on Solid Support *329*
Peter Gaertner and Amitava Kundu
12.1	Introduction *329*
12.2	Carbohydrate Derived Auxiliaries *330*
12.3	Alcohols as Chiral Auxiliaries *332*
12.4	Amine Derived Auxiliaries *336*
12.5	Oxazolidinones, Oxazolidines and Oxazolines as Auxiliaries *341*
12.6	Sulfoxide, Sulfinamide and Sulfoximine Auxiliaries *348*
12.7	Cyclohexanone as a Chiral Auxiliary *352*
12.8	An Enone as a Chiral Auxiliary *352*
12.9	Hydrobenzoin Derived Auxiliaries *355*
12.10	Conclusion *360*
	References *361*

13 Immobilized Enzymes in Organic Synthesis *365*
Jesper Brask
13.1	Introduction *365*
13.1.1	Enzyme Classification and Availability *366*
13.1.2	Popular Enzymes for Biocatalysis *366*
13.1.2.1	Lipases *366*
13.1.2.2	Oxidoreductases *367*
13.1.2.3	Nitrile-Converting Enzymes *368*
13.1.3	Solvents and Stability *368*

13.1.4	Predicting Hydrolase Enantiopreference	369
13.2	Immobilized Enzymes	370
13.2.1	Immobilization without Carrier	371
13.2.1.1	Cross-linking	371
13.2.1.2	Entrapment	372
13.2.2	Immobilization on a Carrier	372
13.2.2.1	Adsorption	373
13.2.2.2	Covalent Binding	374
13.3	Applications	375
13.3.1	Novozym 435 – A Versatile Immobilized Biocatalyst	375
13.3.2	Immobilized Enzymes in Industrial-Scale Biocatalysis	377
13.4	Concluding Remarks	377
	References	378

Part Four Resins for Solid Phase Synthesis

14   Acid-Labile Resins   *383*
Peter D. White

14.1	Introduction: Linker Design	383
14.2	Linker Types	387
14.2.1	Benzyl-Based Linkers	387
14.2.2	Benzhydryl-Based Linkers	397
14.2.3	Trityl-Based Linkers	400
14.2.4	Cyclic Linkers	402
14.2.5	Silyl-Based Linkers	403
14.2.6	Acetal/Aminal-Type Linkers	404
14.3	The Cleavage Reaction	405
14.4	Functional Group and Linker Combinations	408
	List of Abbreviations	412
	References	413

15   Base/Nucleophile-Labile Resins   *417*
Francesc Rabanal

15.1	Introduction	417
15.2	Nucleophile-Labile Resins	417
15.2.1	Intermolecular Nucleophilic Displacement	420
15.2.1.1	Hydroxy Functionalized Linkers and Resins and Related Ones	420
15.2.1.2	Carboxy- and Sulfonic-Based Linkers and Resins	424
15.2.2	Intramolecular Nucleophilic Displacement: Cleavage by Cyclization	426
15.2.2.1	Cyclic Peptides	426
15.2.2.2	Diketopiperazines	427
15.2.2.3	Hydantoins	427
15.2.2.4	Benzodiazepinones	428
15.3	Base-Labile Linkers and Resins	429

Acknowledgments *432*
List of Abbreviations *432*
References *433*

16 Safety-Catch and Traceless Linkers in Solid Phase Organic Synthesis *437*
Matthias Sebastian Wiehn, Nicole Jung, and Stefan Bräse
16.1 Introduction *437*
16.2 Safety-Catch Linkers *437*
16.3 Traceless Linkers *450*
16.4 Conclusions *460*
List of Abbreviations *460*
References *462*

17 Photolabile and Miscellaneous Linkers/Resins *467*
Soo Sung Kang and Mark A. Lipton
17.1 Introduction *467*
17.2 Types of Photolabile Linker *467*
17.2.1 o-Nitrobenzyl as a Photolabile Group *467*
17.2.1.1 o-Nitrobenzyl Linker *468*
17.2.1.2 4-Bromomethyl-3-Nitrobenzoic Acid Derived Linker *469*
17.2.1.3 Hydroxynitrobenzyl and Nitroveratryl Linkers *471*
17.2.1.4 α-Substituted Nitrobenzyl Linkers *474*
17.2.1.5 Comparison of o-Nitrobenzyl Linkers *476*
17.2.2 Functionalized Phenylacyl Linkers *477*
17.2.2.1 Benzoin Linker *478*
17.2.2.2 Phenacyl Linker *479*
17.2.2.3 o-Methylphenacyl Linker *482*
17.2.3 Pivaloyl Linker *484*
17.2.4 Miscellaneous Photolabile Linkers *485*
17.3 Miscellaneous Linkers *487*
17.3.1 1,3-Dithiane Linker *487*
17.3.2 p-Alkoxybenzyl Linker *488*
17.3.3 Alkene-Functionalized Linker *488*
17.4 Conclusion *489*
List of Abbreviations *490*
References *490*

Part Five Solid Phase Synthesis of Biomolecules. The State of the Art (from the Resin Point of View)

18 Peptides *495*
Judit Tulla-Puche and Fernando Albericio
18.1 Introduction *495*
18.2 Methodological Remarks *496*
18.2.1 Solid Supports *497*

18.2.2	Protecting Groups 497
18.3	Coupling Reagents 498
18.4	Synthesis of Long Peptides 499
18.4.1	Convergent Approaches 500
18.4.1.1	Hybrid Approaches: The Synthesis of Enfuvirtide 500
18.4.1.2	Solid Phase Fragment Coupling: Synthesis of Hirudin 502
18.4.2	Stepwise Approaches 503
18.4.2.1	PEG Resins 504
18.4.2.2	Pseudoprolines 504
18.4.2.3	O-Acyl Isopeptide 506
18.5	Native Chemical Ligation 508
18.5.1	Example: Synthesis of a 203 Amino Acid Covalent Dimer, HIV-1 Protease [146] 515
18.6	Cyclic Peptides 516
18.6.1	Example: Solid Phase Synthesis of Argadin by a Side-Anchoring Approach [169] 517
18.7	Depsipeptides 519
18.7.1	Example: Synthesis of H-L(1)EAKLKELEAKλ(12)AALEAKLKELEAKL-OH (L12λ) 519
18.8	Click Chemistry 520
	List of Abbreviations 521
	References 522

19 Oligonucleotides and Their Derivatives   529
Dmitry A. Stetsenko

19.1	Polymer Supported Oligonucleotide Synthesis: An Overview and History 529
19.1.1	Introduction to Nucleic Acid Synthesis: Never-Ending Quest for Excellence 529
19.1.2	Dawn of the Oligonucleotide Synthesis 530
19.1.3	Phosphodiester and Phosphotriester Methods: A Slow Maturity 530
19.1.4	Phosphite Triester and Phosphoramidite Methods: A Breakthrough 534
19.1.5	H-Phosphonate Method: Encouraging Diversity 536
19.2	Supports for the Polymer-Supported Synthesis of Oligonucleotides 538
19.2.1	Polystyrene (PS) Resins 538
19.2.1.1	"Popcorn" Polystyrene 538
19.2.1.2	Low Cross-linked Polystyrene 538
19.2.1.3	Highly Cross-linked (Macroporous) Polystyrene 540
19.2.1.4	Polystyrene–Poly(ethylene glycol) (PEG-PS) Composite Supports 540
19.2.1.5	Linear Polystyrene Grafted onto Other Polymer 540
19.2.2	Polyacrylamide (PA) Resins 541
19.2.3	Silica Gel Supports 542

19.2.4 Miscellaneous Polymers, Surface-Modified Materials and Composite Supports *542*
19.3 Linkers and Anchor Groups for Solid Phase Oligonucleotide Synthesis *545*
19.3.1 Linkers for the Synthesis of 3'- or 5'-Unmodified Oligonucleotides *545*
19.3.1.1 Acid-Labile Trityl Linkers *545*
19.3.1.2 Base-Labile Acyl Linkers *546*
19.3.1.3 Carbonic Acid Linkers, Carbamate and Carbonate *549*
19.3.1.4 Silyl, Silanediyl and Disiloxanediyl Linkers *551*
19.3.2 Linkers for the Synthesis of Phosphorylated, Thiophosphorylated or Other Related Oligonucleotides *553*
19.3.2.1 Phosphoramidate and Phosphorothiolate Linkages *553*
19.3.2.2 Base-Labile Linkers Based on β-Elimination *554*
19.3.2.3 Reduction Linkers Based on Disulfide Bond *556*
19.3.2.4 Miscellaneous Phosphate Linkers *557*
19.3.3 Linkers for the Synthesis of Oligonucleotides 3'-Functionalized with Other Chemical Groups *559*
19.3.3.1 Amino Group *559*
19.3.3.2 Thiol Group *561*
19.3.3.3 Carboxyl Group *563*
19.3.4 Speciality Linkers *564*
19.4 Oligonucleotide Synthesis on Soluble Polymer Supports *565*
19.5 Oligonucleotide Synthesis with Polymer-Supported Reagents *568*
19.6 Conclusions *570*
List of Abbreviations *570*
References *572*

20 Oligonucleotides *585*
Peter H. Seeberger and Harald Wippo
20.1 Introduction *585*
20.2 Insoluble Resins *586*
20.2.1 Controlled-Pore Glass (CPG) *586*
20.2.2 Polystyrene Resins *586*
20.2.2.1 Polystyrene Resin Cross-linked with Divinylbenzene *586*
20.2.2.2 PEG Grafted Polystyrenes *588*
20.2.3 Magnetic Particles *588*
20.3 Soluble Resins *590*
20.3.1 MPEG Resins *590*
20.3.2 Hyperbranched Soluble Resins *591*
20.3.2.1 Hyperbranched Polyester *591*
20.3.2.2 PAMAM *592*
20.3.3 Ionic Liquids *594*
20.4 Linkers *594*
20.4.1 Acid- and Base-Labile Linkers *594*

20.4.1.1	Acid-Labile Linkers	*594*
20.4.1.2	Base-Labile Linkers	*596*
20.4.2	Linkers Cleaved by Olefin Metathesis	*597*
20.4.3	Photocleavable Linkers	*597*
20.4.4	Silyl Ether Linkers	*598*
20.4.5	Boronate Linkers	*599*
20.4.6	Thiol-Group Containing Linkers	*600*
20.4.7	Linkers Cleaved Under Oxidative Conditions	*601*
20.4.8	Linkers Cleaved by Reduction	*602*
20.4.9	Cleavage by Hydrogenolysis	*602*
20.4.10	Enzymatically Cleavable Linkers	*603*
20.5	Capture and Release Techniques	*604*
20.6	Conclusions	*607*
	List of Abbreviations	*607*
	References	*609*

21 High-Throughput Synthesis of Natural Products *613*
Nicolas Winssinger, Sofia Barluenga and Pierre-Yves Dakas

21.1	Introduction	*613*
21.2	Solid Phase Elaboration of Natural Product Scaffolds	*614*
21.3	Solid Phase Synthesis of Natural Products	*616*
21.4	Synthesis of Natural Products Using Immobilized Reagents	*623*
21.5	Synthesis of Natural Products Using Isolation Tags	*627*
21.6	Combinatorial Synthesis of Libraries Based on Important Natural Product Motifs	*630*
21.6.1	Structure-Based Libraries Targeting Kinases and Other Purine-Dependent Enzymes	*631*
21.6.2	Libraries Based on a Privileged Scaffold – Discovery of Fexaramine	*633*
21.6.3	Inhibitors of Histone Deacetylases (HDAC)	*634*
21.6.4	Secramine	*634*
21.6.5	Carpanone	*635*
21.6.6	Natural-Product Inspired Synthesis of αβ-Unsaturated-δ-Lactones	*635*
21.7	Conclusion	*636*
	List of Abbreviations	*636*
	References	*637*

Index *641*

Preface

The Nobel Prize is usually awarded to scientists who have contributed to our understanding of the processes of nature. However, to improve scientific knowledge, it is essential to have access to strategies and scientific equipment that increase the productivity of the scientific process. A paradigm of this concept is the solid-phase methodology proposed by R. Bruce Merrifield in the early 1960s. In his Nobel lecture Merrifield said: "One day I had an idea about how the goal of a more efficient synthesis might be achieved" (R.B. Merrifield, Solid Phase Synthesis, Nobel lecture, 8 December 1984), and on that day not only was a new method for peptide synthesis born but also a new synthetic concept that revolutionized the life sciences and organic chemistry.

Starting with the synthesis of peptides, the solid-phase approach moved rapidly on to other biomolecules, such as oligonucleotides and, to a lesser extent, oligosaccharides, as well as small organic molecules. Thanks to the solid-phase approach, new polymers, such as foldamers, peptoids, peptide nucleic acids (PNAs) and chimeras of several molecules, comprising unnatural building blocks were discovered and found applications in diverse fields of biomedicine and biomaterials. Furthermore, combinatorial sciences have been fueled by solid-phase techniques, as shown by the pioneering work by Furka, Lam, and Houghten.

The initial concept derived from Merrifield's work of associating the resin to a protecting group has also been adapted for other parts of the synthetic process: supported reagents and catalysts, and supported scavengers and tags for improving and speeding up the work-up and/or purification processes. This book was conceived with the idea of reflecting all the applications of solid-phase methodology, taking the resin as cornerstone. Thus, not only resins for solid-phase synthesis classified by cleavage mechanism (Part IV), but also solid-supported reagents and scavengers (Part II), and resin-bound catalyst (Part III) are discussed in depth. As a colophon, Part V addresses the state of the art of biomolecules such as peptides, oligonucleotides, and oligosaccharides as well as other natural products. Furthermore, the Introduction (Part I) combines the classic concept of the solid support with two emerging fields related to resins and polymers, such as imprinted polymers and nanoparticles functionalized with bioactive molecules.

We close these words with special thanks and a dedication. Thanks go to all those who have made this project possible. Their dedication, promptness and

The Power of Functional Resins in Organic Synthesis. Judit Tulla-Puche and Fernando Albericio
Copyright © 2008 WILEY-VCH Verlag GmbH & Co. KGaA, Weinheim
ISBN: 978-3-527-31936-7

excellent work are truly appreciated. Finally, we dedicate this volume to all young investigators, mainly PhD students and post-doctoral fellows, in the hope that they will learn how to use resins and the solid-phase approach for the advancement of science.

June 2008, Barcelona

Fernando Albericio
and Judit Tulla-Puche

Part One   Introduction

1
The (Classic Concept of) Solid Support

Fernando Albericio[1,2] and Judit Tulla-Puche[1]

[1]Institute for Research in Biomedicine, Barcelona Science Park, 08028 Barcelona, Spain
[2]University of Barcelona, Department of Organic Chemistry, 08028-Barcelona, Spain

1.1
Introduction

The use of a solid support in a chemical process can be traced back to the birth of chromatography. However, it was at the beginning of the 1960s with the publication by the Nobel Prize winner R.B. Merrifield of his seminal paper "Solid phase peptide synthesis. I. The synthesis of a tetrapeptide" [1] that solid-supported or solid-phase chemistry was truly born. The idea of using a solid support as protecting group (carboxylic-protecting group in the case of the Merrifield tetrapeptide) was rapidly and enthusiastically adapted by Letsinger to the synthesis of oligonucleotides (protection of the exocyclic amino function of the 5′-O-trityldeoxycytidine) [2]. Although at the end of the 1960s Merrifield himself predicted that: "... it seems quite clear that a gold mine is awaiting the organic chemist who would look to solid supports for controlling and directing his synthetic reactions" [3], only a few examples of the extension of the solid supported protecting group concept to other areas of synthetic organic chemistry can be found in the literature at that time. These examples involved the monoalkylation and monoacylation of carboxylic acids [4, 5], the synthesis of aldehydes and ketones [6] and the use of resin-bound dienes or dienophiles to trap reactive intermediates [7]. A new milestone in this long journey was reached at the beginning of the 1990s when Ellman published the synthesis of 1,4-benzodiazepine derivatives on a solid phase [8]. This achievement can be considered the advent of small molecule combinatorial chemistry strategies for drug discovery. This field had its roots in the work of Houghten [9] and Lam [10], who examined the solid-phase synthesis of peptide libraries. In contrast to what happens in other disciplines, where knowledge and technology transfer from academia to industry is slow, in this case the pharmaceutical companies set up their own facilities and departments and sponsored the establishment of small- and medium-sized companies dedicated to this branch of chemistry.

The Power of Functional Resins in Organic Synthesis. Judit Tulla-Puche and Fernando Albericio
Copyright © 2008 WILEY-VCH Verlag GmbH & Co. KGaA, Weinheim
ISBN: 978-3-527-31936-7

Simultaneously, numerous academic groups started and even reoriented their activities towards the implantation of solid-phase strategies in their laboratories. Although progress towards fulfillment of the initial great expectations has slowed in recent years, solid-supported chemistry is now a very useful tool in all modern organic chemistry laboratories [11] and, of course, is the method of choice for both research and industrial synthesis of peptides [12] and oligonucleotides [13].

The solid support in synthetic chemistry has amplified its range of applications. Thus, from the protecting group concept applied for the synthesis of peptides, oligonucleotides, oligosaccharides and other biomolecules, it has evolved to solid-supported reagents and resins as scavengers for high-throughput purification.

Usually, chemical processes take longer in solid-phase mode than in solution. Furthermore, the solid support should be considered analogous to a co-solvent [14]. Thus, this support is an integral part of the process and very often each new support requires optimization. Consequently, the translation of organic processes from solution- to solid-phase commonly calls for some work to optimize the overall process. In general, reactions that tolerate excesses of reagent can be transferred well to the solid phase. In contrast, those reactions that require stoichiometric amounts of reagents rarely work well because the solid-phase approach is based on the use of large excesses of reagents to drive the processes to completion [15].

1.2
Linkers/Handles

There is a tendency to confuse the terms "solid support/resin" and "handle/linker" [16]. As pointed out by Bradley and coworkers, resins/solid supports are inert matrices, which are passive to chemistry [17]. In contrast, linkers/handles are immobilized protecting groups [16, 17], which can be classified into two types, integral and non-integral [17]. Integral linkers/handles are characterized by the fact that part of the solid support forms part or all of the linker/handle, as shown in some polystyrene-based linkers. Non-integral linkers/handles are attached permanently to the solid support (Figure 1.1). In all cases, the linker/handle should contain a functional group that will link it temporarily to a growing molecule. The final step in all solid-phase synthetic processes is the release of the final compound.

Although non-integrated linkers/handles can be attached to the solid support through an ether (Wang resin) or even a C–C bond, the most convenient strategy is through an amide bond. Thus, the carboxylic acid-containing linker/handle is prepared and characterized in solution and then incorporated into an amino solid support. In this strategy, the amide that links the linker/handle to the solid support should be totally stable to all synthetic processes, including the final treatment that will detach the target compound from the solid support. Very often the linker/handle is attached to a MBHA resin, commonly used for the preparation of peptide amides using a *tert*-butoxycarbonyl (Boc)/benzyl strategy. In this case, the bond

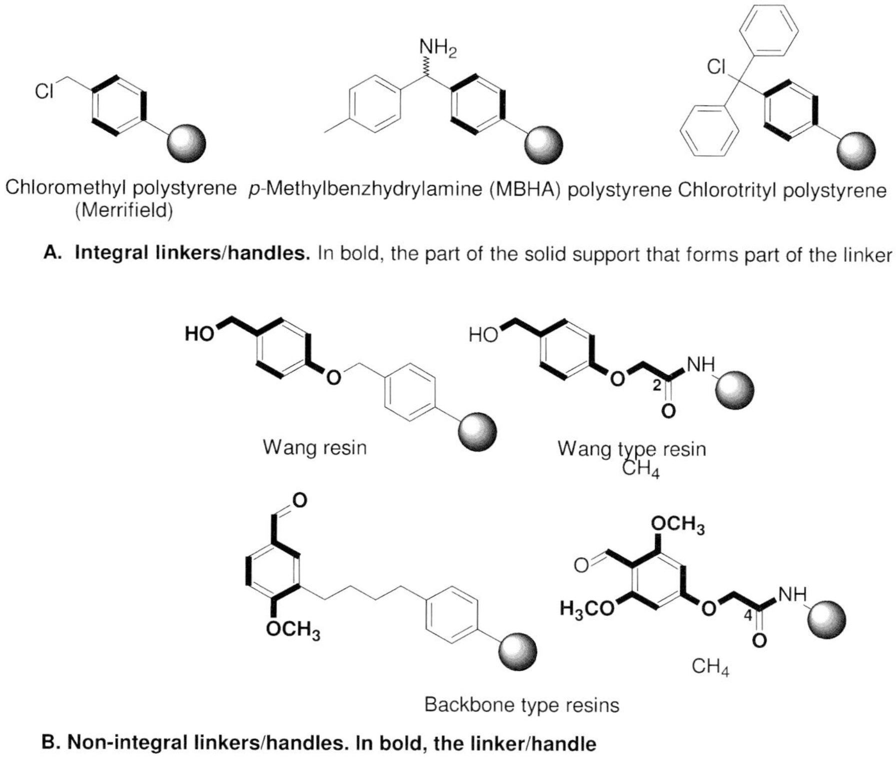

1.2 Linkers/Handles | 5

Chloromethyl polystyrene   p-Methylbenzhydrylamine (MBHA) polystyrene   Chlorotrityl polystyrene
(Merrifield)

A. Integral linkers/handles. In bold, the part of the solid support that forms part of the linker

Wang resin   Wang type resin
                  CH_4

Backbone type resins

B. Non-integral linkers/handles. In bold, the linker/handle

Figure 1.1 Examples of integral (a) and non-integral
(b) linkers/handles. Adapted from reference [16, 17].

formed between the linker/handle and the BHA resin is not totally stable to the acid conditions (TFA–scavengers–CH_2Cl_2 mixtures at 25–60 °C) and therefore the carbocation-containing linker is detached from the solid support. This can add impurities to the final crude or, what is even more damaging, can cause back-alkylation of the target compound [18, 19]. To overcome this side-reaction, the use of aminobenzyl polystyrene or aminoalkyl resins, which form a more acid stable bond, is recommended [19].

A similar undesired cleavage can take place with (poly)alkoxybenzyl [Wang, backbone amide linker (BAL), Rink-type resins] [19, 20] (Figure 1.1B).

Thus in peptide synthesis, Tsikaris *et al.* [21] have described the incorporation of the *p*-hydroxybenzyl moiety cleaved from the Wang resin into the N- of the C-terminal amide of a peptide during TFA cleavage (Figure 1.2). Similarly, Martinez *et al.* [22] have reported the alkylation of the indol ring of Trp-containing peptides by the *p*-hydroxybenzyl moiety. Furthermore, Stanger and Krchnak have demonstrated the formation of O-(4-hydroxy)benzyl derivatives [23]. The use of the Wang resin for the solid-phase preparation of small molecules has led to the introduction

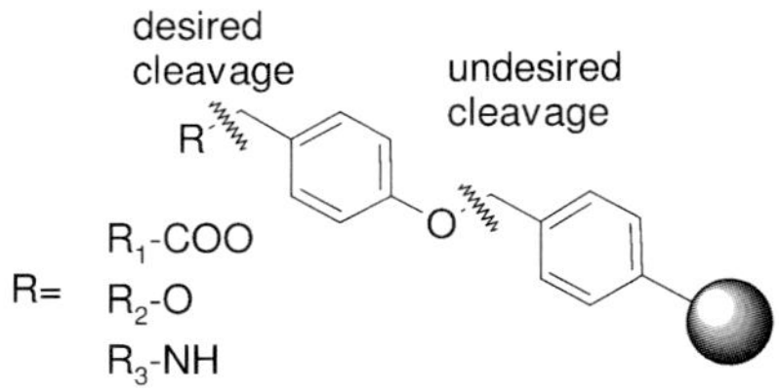

Figure 1.2 Dual cleavage on a Wang resin as an example of this side-reaction in other (poly)alkoxybenzyl resins. Adapted from reference [20].

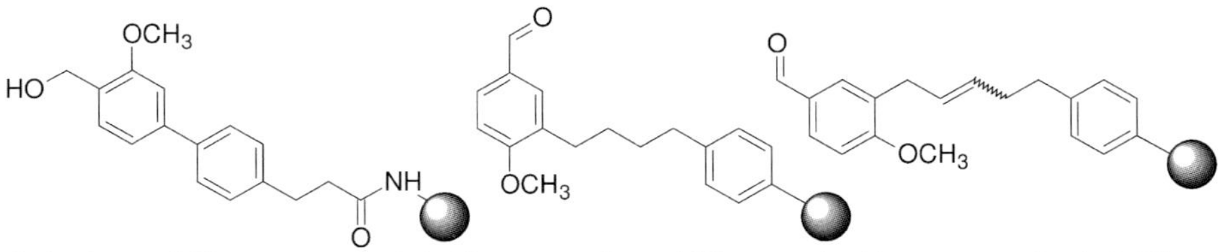

Figure 1.3 Non-acid degradable linkers.

of impurities due to the undesired cleavage from the resin (no cleavage at the benzyl position) or from a back-alkylation of the *p*-hydroxybenzyl cation in the case of furopyridine and furoquinoline target derivatives [24]. To overcome these problems, two resins have been developed based on the activation of the benzyl position by a MeO group, a non-cleavable electron-donating group, in either the ortho or para position. Thus, Gu and Silverman [25] incorporated the precursor of their backbone linker to the resin through a metal-catalyzed coupling reaction, and Colombo *et al.* incorporated the precursor of their Wang type resin through an amide bond (Figure 1.3) [20].

The use of non-integral linkers/handles is usually more recommendable because they provide control and flexibility for the synthetic process [26]. Thus, (i) any functionalized solid support can be used; (ii) when the linker/handle is attached through an amide bond, the quality of the starting resin can be assured by controlling the purity of the initial linker/handle; (iii) loadings can be easily fine-tuned; (iv) it allows the introduction of internal reference amino acids (IRaas) [27–29] between the linker/handle and the solid support, which can facilitate monitoring of the synthetic process; and (v) when the first building block (BB) is incorporated to the linker/handle through a more demanding ester bond, it can be attached to the linker/handle in solution and, after characterization and, if required, purification, incorporated to the solid support (preformed linker/handle, Figure 1.4) [29–30].

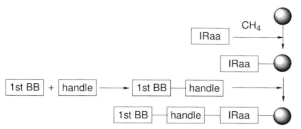

Figure 1.4 Optimized solid-phase strategy using a preformed linker/handle and IRaa.

1.3
Solid Supports

An optimal solid support should (i) be mechanically robust; (ii) be stable to variation in temperatures; (iii) have reagent-accessible sites; (iv) show acceptable loadings; (v) present acceptable bead sizes, if applicable, when required, facilitating filtration; (vi) be stable in diverse media, and in the case of being used in biochemical assays (vii) show biocompatibility and swelling in aqueous buffers.

In contrast to most chemical reagents, which are usually of the same quality regardless of the supplier, the source of solid support is extremely important. Thus, distinct solid supports from different manufacturers or even from the same one, but from separate batches, may differ in performance. Consequently, the choice of the solid support source can have a significant influence on the result of the chemical process. If the large-scale synthesis of a molecule or the production of a library is preceded by an optimization step, both steps should be carried out with the same batch of solid support.

Several classifications of solid supports have been proposed on the basis of their physical properties [31–32]. However, the solid supports most widely used today can be classified into just two groups: gel type and modified surface type.

1.3.1
Gel-Type Support

These types of supports are the most commonly used. The polymer network is flexible and can expand or exclude solvent to accommodate the growing molecule within the gel. Thus, the chemistry occurs within the well-solvated gel that contains mobile and reagent-accessible chains [33]. Three types of gel resin are commercially available:

1.3.1.1 Polystyrene (PS) Resins
These are, without doubt, the most widely used solid supports. PS systems used for the synthesis of peptides and small molecules consist of 1% cross-linked hydrophobic resins obtained by suspension polymerization from styrene and divinylbenzene. For other uses, PS with 2% cross-linking, which is mechanically more stable than those with less cross-linking, is also employed. This 2% PS was used

in the early years of solid-phase peptide synthesis (SPPS). However, due to problems observed during the synthesis of difficult sequences, the cross-linking of this resin was then reduced to 1%. Although full derivatization of PS would give a substitution of approximately 10 mmol g^{-1}, practical loadings can vary from 0.3 to 3 mmol g^{-1}. For SPPS, the loading should be kept at around 0.5 mmol g^{-1} to ensure a successful synthetic outcome [34]. PS as hydrophobic beads swell well in nonpolar solvents like toluene or CH_2Cl_2 [35]. However, these systems can also be used in combination with other more polar solvents such as N,N-dimethylformamide (DMF), dioxane, and tetrahydrofuran. In SPPS, the use of acetonitrile is incompatible with a good quality final product [36].

1.3.1.2 Poly(Ethylene Glycol)–Polystyrene (PEG-PS) Resins

Based on the early work of Mutter [37], the first commercially available PEG-PS resins were developed independently in the mid-1980s by Zalipsky, Albericio, and Barany (PEG-PS) [38–39] and Bayer and Rapp (TentaGel) [40–41]. These systems can be synthesized either by reaction of preformed oligooxyethylenes with aminomethylated polystyrene beads (PEG-PS) or by graft polymerization on polystyrene beads (TentaGel).

The first PEG-PS resin was developed with the idea of combining a hydrophobic PS core with hydrophilic PEG chains on the same support. Furthermore, in some of the aforementioned PEG-PS resins, the PEG unit might act as a spacer, separating the starting point of the solid-phase synthesis from the PS core [40]. This early hypothesis has not been corroborated in any other formulations of PEG-PS resins and it is, therefore, possible to conclude that the environment effect of the PEG is far more critical than any spacer effect [15, 42]. Given the unique conformational flexibility of PEG chains, PEG-PS resins are compatible with both polar and nonpolar solvents [43]. The high degree of swelling provides the beads with a firmer and flow-stable character and makes these resins physically stable in flow systems. The content of PEG varies significantly between distinct resins and therefore their swelling properties can differ markedly. Furthermore, several of these systems can lose PEG during treatment with TFA. PEG-PS resins usually show lower loadings (0.15–10.3 mmol g^{-1}) than PS ones.

1.3.1.3 Hydrophilic PEG-Based Resins

These resins have their roots in the polyacrylamide resins introduced in the 1980s for SPPS using the fluorenylmethoxycarbonyl (Fmoc)-tert-butyl (tBu) strategy. They were developed following the concept that the insoluble support and peptide backbone should have comparable polarities [44].

Kempe and Barany [45] developed the CLEAR family, which is based on the copolymerization of branched PEG-containing cross-linkers such as trimethylolpropane ethoxylate triacrylate, which contain various ethylene oxide units, with amino-functionalized monomers such as allylamine or 2-aminoethylmethacrylate. These amino groups constitute the starting points for the solid-phase synthesis. The loadings of these solid supports are affected by the amount of functionalized monomer used for polymerization. Typical loadings are in the range

$0.1–0.3\,\text{mmol}\,\text{g}^{-1}$. The CLEAR family supports swell in a broad range of hydrophobic and hydrophilic solvents and has excellent physical and mechanical properties for both batch-wise and continuous-flow systems.

Meldal designed a series of PEG-based resins, first of all combining them with a small amount of PS or polyamide and, finally, using neat PEG [31]. The first member of this family, and the most widely used, was the PEGA resin [46], obtained by inverse suspension radical polymerization of various sizes of linear bis- and branched tris-2-aminopropyl-PEG samples with acryloyl chloride. The uniform beads swelled in all solvents, ranging from toluene to aqueous buffers. To prevent the presence of secondary amide bonds, which can interfere with reactions involving carbon and carbenium ions, the POEPS, a new family of PEG-based resins, was developed by partial derivatization of linear PEG with vinylphenylmethyl chloride or vinylphenylpropyl chloride, followed by inverse suspension radical polymerization [47–48]. The methyl version of this resin is not stable to Lewis acids because of the benzylic linkage between the PEG and the PS. In addition, the preparation of vinylphenylpropyl chloride is not straightforward – even the presence of a small amount of PS in the polymer leads to a slight decrease in the favorable swelling observed for PEGA resins in polar solvents. Alternatively, polyoxyethylene cross-linked polyoxypropylene (POEPOP), which contains only ether bonds, was developed from a polymerization of PEG that was partially derivatized with chloromethyloxirane [49]. Although the POEPOP resin is mechanically robust, shows relatively high loading (primary and secondary alcohols) and good performance for organic transformations, the presence of secondary ether bonds implies that this solid support is not totally stable to strong Lewis acids [50]. To overcome this problem, the SPOCC resin, in which all ether bonds and functional alcohol groups are primary, was developed [51].

In a parallel manner, Côté [52] developed the ChemMatrix resin, a total PEG-based resin consisting of primary ether bonds. Because of its highly cross-linked matrix, ChemMatrix has greater mechanical stability than other PEG resins. This resin swells well in all of the most common solvents and is, therefore, useful for a broad range of organic chemistries. ChemMatrix resin performs extremely well compared to PS resins in the solid-phase synthesis of hydrophobic, highly structured, poly-Arg peptide, β-amyloid (1-42), RANTES – a complex aggregated chemokine – and HIV protease [53–56], showing that the presence of PFG chains impairs the aggregation of the growing peptide chain, facilitating the solid-phase synthesis of complex peptides. Furthermore, ChemMatrix is convenient for the synthesis of oligonucleotides and oligonucleotide peptide conjugates and oligonucleotide hybrids by Cu^+-catalyzed cycloaddition reactions [57].

The compatibility of all these PEG-based resins with aqueous buffers allows their use for biochemical applications such as on-resin screening of chemical libraries and in the development of affinity chromatography [58–61]. All these families of PEG-based resins, except POEPS, are free of aromatic rings. This feature makes these solid supports highly suitable for a broad range of applications where such rings can react with reagents or/and jeopardize the solid-phase NMR control of the reactions [62].

Regarding the functional site distribution in these gel-type supports, there is some controversy. Thus, using fluorescence optical analysis, McAlpine and Schreiber determined that for PS and TentaGel resins there is a higher effective concentration of functional sites at the surface relative to the macroporous resin ArgoPore and controlled pore glass (CPG) [63–64]. In contrast, Bradley and coworkers [65], using confocal Raman and fluorescence microscopy, showed that for both resins there is a uniform distribution of reactive sites throughout the beads but that the spatial distribution of reacted sites depends on the polymer type, with a fine balance between reaction and diffusion rate. On PS resin beads, there is a uniform distribution of sites, with the reaction rate being slower than diffusion. However, on TentaGel this is not the case because the reaction is diffusion-controlled. These findings were confirmed by Rademann and coworkers for PS resin, using confocal and non-confocal fluorescence microscopy as well as by FT-IR microscopy [66].

1.3.2
Modified Surface Type Supports

Although numerous materials have been used for surface functionalization for solid-phase synthesis [15, 31, 32], the most common, which are commercially available, are membranes and pellicular solid supports, where a mobile polymer is grafted to a rigid and inert plastic.

1.3.2.1 Cellulose Membranes

These systems are the basis of the SPOT concept, a highly parallel and technically simple arrangement that was first developed by Frank [67–69]. SPOT is very flexible and inexpensive compared to other multiple solid-phase procedures, especially regarding miniaturization and array geometries. Cellulose paper was first used for the solid-phase synthesis of oligonucleotides [70] and peptides [71]. Cellulose, usually Whatman 50 or 540, can be easily modified by O-acylation with protected amino acids to form amino acid esters [72]. However, although these functionalized membranes have been applied for peptide synthesis and screening of peptide libraries, their use in a broader range of chemical reactions is jeopardized by several factors. The lability of the ester bond limits their use. To circumvent this problem, the hydroxy functions of cellulose have been alkylated with epoxides containing a protected amino function [73]. However, the preparation of the N-protected epoxypropylamines is laborious and therefore this method is impractical for the functionalization of cellulose membranes. Optimized methods have been developed that involve the incubation of cellulose membranes with epibromohydrin in the presence of perchloric acid, followed by reaction with 4,7,10-trioxa-1,13-tridecanediamine or diaminopropane [74–75]. Furthermore, chemical degradation of cellulose can occur under the conditions used for more conventional organic reactions [68]. For instance, the glycosidic bonds of cellulose can be labile under nucleophilic and acidic conditions. Finally, the large number of hydroxy functionalities can cause instability and interfere with subsequent reactions.

1.3.2.2 Polyolefinic Membranes

Polyalkanes (polyethylene and polypropylene) and their fluorinated derivatives [poly(vinylidene fluoride) and polytetrafluoroethylene (PTFE)] have been successfully used for solid-phase synthesis in reactors [76–78] as well as in the SPOT technique [68]. Polypropylene membranes are highly stable, show low expansion in the presence of most common solvents, and display acceptable loadings. Functionalization can be performed by photo-induced graft copolymerization with functional acrylates by the following sequence: (i) coating with photo-initiator (benzophenone) and selective UV-excitation, causing H-abstraction and radical formation; (ii) application of an acrylate monomer solution, resulting in addition of radical sites to the double bond of the monomer; and (iii) free radical polymerization. If acrylic acid and its methyl ester are used as monomers, the final functionalization in the form of amine groups can be achieved by amidation with 4,7,10-trioxa-1,13-tridecanediamine after activation of the carboxylic acids with $SOCl_2$, $(COCl)_2$, PCl_3 or PCl_5.

1.3.2.3 Pellicular Solid Supports

In these systems a mobile polymer is grafted onto a rigid and inert plastic. This idea was first developed by Geysen and initiated the "Multipin concept" [79, 80]. This design has several handling advantages in that it can be made to take any particular shape/form and can be adapted to any desired array. Furthermore, the size of the inert plastic, together with the efficiency of the grafting, will determine the amount of product that can be synthesized. Thus, unlike gel-type supports, it is the surface area and not the volume that determines the loading capacity. A further advantage of this kind of support is the consistency of the kinetics between grafted supports of different shapes and sizes. Such a correlation is not possible with gel-type supports because the rate of diffusion changes with size and, therefore, the reaction rates are modified. However, comparison studies carried out in our laboratory between pellicular supports and PS resins have demonstrated that kinetic reactions are usually slower with the pellicular solid supports.

The original pin support was poly(acrylic acid) grafted onto inert polyethylene, where the carboxylic acid moieties were capped with mono Boc-protected ethylenediamine [79]. PS and the extremely hydrophilic polyhydroxyethyl methacrylate and poly(methacrylic acid/dimethylacrylamide) were also grafted on [81]. The last grafted support in the "Multipin concept" is the so-called Lanterns [80], where the actual solid support is PS, which can be functionalized in a similar way to PS gel-type supports.

In conclusion, the concept of solid-supported chemistry has been extended from the preparation of peptides and other biomolecules to any organic molecule. PS- and PEG-based resins are the most widely used; however, syntheses can be carried out in any solid support. The development of new solid supports and linkers/handles is crucial to fulfill the new requirements of modern drug discovery programs.

Acknowledgments

The work carried out in the laboratory of the authors was partially supported by CICYT (CTQ2006-03794/BQU), the *Generalitat de Catalunya* (2005SGR 00662), ISCIII (CIBER, nanomedicine 0074), the Institute for Research in Biomedicine, and the Barcelona Science Park.

References

1 Merrifield, R.B. (1963) *Journal of the American Chemical Society*, **85**, 2149–54.
2 Letsinger, R.L. and Mahadevan, V. (1965) *Journal of the American Chemical Society*, **87**, 3526–7.
3 Merrifield, R.B. (1969) *Advances in Enzymology and Related Areas of Molecular Biology*, **32**, 221–96.
4 Patchornik, A. and Kraus, M.A. (1970) *Journal of the American Chemical Society*, **92**, 7587–9.
5 Camps, F., Castells, J., Ferrando, M.J. and Font, J. (1971) *Tetrahedron Letters*, 1713–14.
6 Leznoff, C.C. and Sywanyk, W. (1977) *Journal of Organic Chemistry*, **42**, 3203–5.
7 Rebek, J. and Gavina, F. (1974) *Journal of the American Chemical Society*, **96**, 7112–14.
8 Bunin, B.A. and Ellman, J.A. (1992) *Journal of the American Chemical Society*, **114**, 10997–8.
9 Houghten, R.A., Pinilla, C., Blondelle, S.E., et al. (1991) *Nature*, **354**, 84–6.
10 Lam, K.S., Salmon, S.E., Hersh, E.M., et al. (1991) *Nature*, **354**, 82–4.
11 Cironi, P., Álvarez, M. and Albericio, F. (2006) *Mini Reviews in Medicinal Chemistry*, **6**, 11–25.
12 Verlander, M. (2007) *International Journal of Peptide Research and Therapeutics*, **13**, 75–82.
13 Pon, R.T., Yu, S., Prabhavalkar, T., Mishra, T., et al. (2005) *Nucleosides, Nucleotides and Nucleic Acids*, **24**, 777–81.
14 Czarnik, A.W. (1998) *Biotechnology and Bioengineering*, **61**, 77–9.
15 Hudson, D. (1999) *Journal of Combinatorial Chemistry*, **1**, 333–60.
16 Van den Nest, W. and Albericio, F. (2001) *Optimization of Solid-Phase Combinatorial Synthesis* (eds B. Yan and A.W. Czarnik), Marcel Dekker, New York, pp. 91–107.
17 Guillier, F., Orain, D. and Bradley, M. (2000) *Chemical Reviews*, **100**, 2091–157.
18 Albericio, F., Kneib-Cordonier, N., Biancalana, S., et al. (1990) *Journal of Organic Chemistry*, **55**, 3730–43.
19 Yraola, F., Ventura, R., Vendrell, M., et al. (2004) *QSAR and Combinatorial Science*, **23**, 145–52.
20 Colombo, A., De la Figuera, N., Fernandez, J.C., et al. (2007) *Organic Letters*, **9**, 4319–22.
21 Stathopoulos, P., Papas, S. and Tsikaris, V. (2006) *Journal of Peptide Science*, **16**, 227–32.
22 Giraud, M., Cavalier, F. and Martinez, J. (1999) *Journal of Peptide Science*, **5**, 457–61.
23 Stanger, K.J. and Krchnack, V. (2006) *Journal of Combinatorial Chemistry*, **8**, 652–4.
24 Cironi, P., Tulla-Puche, J., Barany, F., et al. (2004) *Organic Letters*, **6**, 1405–8.
25 Gu, W. and Silverman, R.B. (2003) *Organic Letters*, **5**, 415–18.
26 Songster, M.F. and Barany, G. (1997) *Methods in Enzymology*, **289**, 126–74.
27 Atherton, E., Clive, D.L.J. and Sheppard, R.C. (1975) *Journal of the American Chemical Society*, **97**, 6584–5.
28 Matsueda, G.R. and Haber, E. (1980) *Analytical Biochemistry*, **104**, 215–27.
29 Albericio, F. and Barany, G. (1984) *International Journal of Peptide and Protein Research*, **23**, 342–9.
30 Mitchell, A.R., Kent, S.B.H., Engelhard, M., Merrifield, R.B. (1978) *J. Org. Chem.*, **43**, 2845–52.
31 Meldal, M. (1997) *Methods in Enzymology*, **289**, 83–104.

32 Forns, P. and Fields, G.B. (2000) *Solid-Phase Synthesis. A Practical Guide* (eds S. A. Kates and F. Albericio), Marcel Dekker, New York, pp. 1–77.

33 Sarin, V.K., Kent, S.B.H. and Merrifield, R.B. (1980) *Journal of the American Chemical Society*, **102**, 5463–70.

34 Chiva, C., Vilaseca, M., Giralt, E. and Albericio, F. (1999) *Journal of Peptide Science*, **5**, 131–40.

35 Pugh, K.C., York, E.J. and Stewart, J.M. (1992) *International Journal of Peptide and Protein Research*, **40**, 208–13.

36 Zalipsky, S., Chang, J.L., Albericio, F. and Barany, G. (1994) *Reactive Polymers*, **22**, 243–58.

37 Becker, H., Lucas, H.W., Maul, J., et al. (1982) *Makromolekulare Chemie. Rapid Communications*, **3**, 217–23.

38 Zalipsky, S., Albericio, F. and Barany, G. (1986) Peptides 1985, in *Proceedings of the Ninth American Peptide Symposium* (eds C.M. Deber, V.J. Hruby and K.D. Kopple), Pierce, Rockford, Illinois, pp. 257–60.

39 Kates, S.A., McGuinness, B.F., Blackburn, et al. (1998) *Biopolymers (Peptide Science)*, **47**, 365–80.

40 Bayer, E., Hemmasi, B., Albert, K., Rapp, W. and Dengler, M. (1983) Peptides, in *Proceedings of the Eighth American Peptide Symposium* (eds V.J. Hruby and D.H. Rich), Pierce, Rockford, Illinois, pp. 87–90.

41 Rapp, W. (1996) *Combinatorial Peptide and Nonpeptide Libraries, A Handbook* (ed. G. Jung), Wiley-VCH Verlag GmbH, Weinheim, pp. 425–64.

42 Barany, G., Albericio, F., Kates, S.A. and Kempe, M. (1997) Poly(ethylene glycol)-Containing Supports for Solid-Phase Synthesis of Peptides and Combinatorial Organic Libraries. In 'ACS Symposium Series 680. Poly(ethylene glycol): Chemistry and Biological Applications' (eds J.M. Harris and S. Zalipsky), American Chemical Society, Washington, DC, pp. 239–64.

43 Rademan, J., Grötli, M., Meldal, M. and Bock, K. (1999) *Journal of the American Chemical Society*, **121**, 5459–66.

44 Atherton, E. and Sheppard, R.C. (1989) *Solid Phase Peptide Synthesis: A Practical Approach*, IRL Press, Oxford.

45 Kempe, M. and Barany, G. (1996) *Journal of the American Chemical Society*, **118**, 7083–93.

46 Meldal, M. (1992) *Tetrahedron Letters*, **33**, 3077–80.

47 Renil, M. and Meldal, M. (1996) *Tetrahedron Letters*, **37**, 6185–8.

48 Buchardt, J. and Meldal, M. (1998) *Tetrahedron Letters*, **38**, 8695–8.

49 Rademan, J., Meldal, M. and Bock, K. (1999) *Chemistry – A European Journal*, **5**, 1088–95.

50 Rademan, J., Meldal, M. and Bock, K. (1999) *Proceedings of the 25th European Peptide Symposium* (eds S. Bajusz and F. Hudecz), Akadémiai Kiadó, Budapest, pp. 38–9.

51 Rademan, J., Grötli, M., Meldal, M. and Bock, K. (1999) *Journal of the American Chemical Society*, **121**, 5459–66.

52 Côté, S. (2005) PCT Int. Appl. WO 2005012277.

53 García-Martín, F., Quintanar-Audelo, M., García-Ramos, et al. (2006) *Journal of Combinatorial Chemistry*, **8**, 213–20.

54 García-Martín, F., White, P., Steinauer, R., et al. (2006) *Biopolymers (Peptide Science)*, **84**, 566–75.

55 De la Torre, B.G., Jakab, A. and Andreum, D. (2007) *International Journal of Peptide Research and Therapeutics*, **13**, 265–70.

56 Frutos, S., Tulla-Puche, J., Albericio, F. and Giralt, E. (2007) *International Journal of Peptide Research and Therapeutics*, **13**, 221–7.

57 Mazzini, S., García-Martin, F., Alvira, M., et al. (2008) *Chemical Biodiversity*, **5**, 209–218.

58 Meldal, M. and Svendsen, I. (1995) *Journal of the Chemical Society, Perkin Transactions 1*, 1591–6.

59 Meldal, M., Svendsen, I., Juliano, L., et al. (1997) *Journal of Peptide Science*, **4**, 83–91.

60 Camperi, S.A., Marani, M.A., Iannucci, et al. (2005) *Tetrahedron Letters*, **46**, 1561–4.

61 Meldal, M. (2005) *QSAR and Combinatorial Science*, **24**, 1125–6.

62 Kempe, M., Keifer, P.A. and Barany, G. (1998) In 'Peptides 1996: Proceedings of the Twenty-Fourth European Peptide Symposium' (eds R. Ramage and R. Epton), Mayflower Scientific, Kingswinford, UK, pp. 533–4.

63 McAlpine, S.R. and Schreiber, S.L. (1999) *Chemistry – A European Journal*, **5**, 3528–32.
64 McAlpine, S.R., Lindsley, C.W., Hodges, J.C., et al. (2001) *Journal of Combinatorial Chemistry*, **3**, 1–5.
65 Kress, J., Zanaletti, R., Rose, A., et al. (2003) *Journal of Combinatorial Chemistry*, **5**, 28–32.
66 Rademann, J., Barth, M., Brock, R., et al. (2001) *Chemistry – A European Journal*, **7**, 3884–9.
67 Frank, R. (1992) *Tetrahedron*, **48**, 9217–32.
68 Wenschuh, H., Volkmer-Engert, R., Schmidt, M., et al. (2000) *Biopolymers (Peptide Science)*, **55**, 188–206.
69 Frank, R. (2007) *International Journal of Peptide Research and Therapeutics*, **13**, 45–52.
70 Frank, R., Heikens, W., Heisterberg-Moutsis, G. and Blöcker, H. (1983) *Nucleic Acids Research*, **11**, 4365–77.
71 Frank, R. and Döring, R. (1988) *Tetrahedron*, **44**, 6031–40.
72 Blankemeyer-Menge, B., Nimtz, M. and Frank, R. (1990) *Tetrahedron Letters*, **31**, 1701–4.
73 Volkmer-Engert, R., Hoffman, B. and Schneider-Mergener, J. (1997) *Tetrahedron Letters*, **38**, 1029–32.
74 Ast, T., Heine, N., Germeroth, L., et al. (1999) *Tetrahedron Letters*, **40**, 4317–18.
75 Licha, K., Bhargava, S., Rheinländer, C., et al. (2000) *Tetrahedron Letters*, **41**, 1711–15.
76 Berg, R.H., Almdal, K., Pederson, W., et al. (1989) *Journal of the American Chemical Society*, **111**, 8024–6.
77 Daniels, S.B., Bernatowicz, M.S., Coull, J.M. and Köster, H. (1989) *Tetrahedron Letters*, **30**, 4345–8.
78 Wang, Z. and Laursen, R.A. (1992) *Peptide Research*, **5**, 275–80.
79 Geysen, H.M., Rodda, S.J., Mason, T.J., et al. (1987) *Journal of Immunological Methods*, **102**, 259–74.
80 Rasoul, F., Ercole, F., Pham, Y., et al. (2000) *Biopolymers (Peptide Science)*, **55**, 207–16.
81 Maeji, N.J., Valerio, R.M., Bray, A.M., et al. (1994) *Reactive Polymers*, **22**, 203–12.

2
Molecularly Imprinted Polymers*

Henrik Kempe and Maria Kempe

Lund University, Biomedical Center, Department of Experimental Medical Science, 22184 Lund, Sweden

2.1
Introduction

One of the corner-stones of life is recognition. This phenomenon occurs on a macroscopic level as well as on a microscopic one. For example, the ability to recognize a familiar face is practical in our everyday life and the recognition of a transmitter substance by its receptor is essential for the function of the nervous system. Molecular recognition is the creation of a complex between a host molecule and a guest molecule and often involves non-covalent interactions such as hydrogen bonds, hydrophobic interactions, metal coordination, van der Waals interactions and ionic interactions.

A sophisticated example of molecular recognition is the ability of the immune system to distinguish between self and non-self. Antibodies play a crucial role in this process and are produced as a result of stimuli in the form of antigens, such as bacteria, viruses, pollens or foreign molecules. A theory of the formation of antibodies was postulated by Linus Pauling in 1940 [1]. He suggested that the antigen acts as a template during the folding of the polypeptide chain constituting the antibody. Although the theory of Pauling was incorrect, his ideas are strikingly similar to those behind a technique that was later termed molecular imprinting. Pauling's work inspired his student Frank Dickey to polymerize sodium silicate in the presence of dye molecules. After removing the dyes, Dickey showed that the silica materials bound preferentially the respective dye present during the polymerization [2]. This work, together with work by Polyakov [3], is considered to be the first demonstrations of molecular imprinting.

Molecular recognition elements are useful in a broad range of applications where selective binding is advantageous, for example, chromatographic and batchwise separation and purification, trace enrichment by solid-phase extraction,

---

* This chapter is dedicated to Professor Klaus Mosbach on the occasion of his seventy-fifth 75th birthday.

The Power of Functional Resins in Organic Synthesis. Judit Tulla-Puche and Fernando Albericio
Copyright © 2008 WILEY-VCH Verlag GmbH & Co. KGaA, Weinheim
ISBN: 978-3-527-31936-7

analysis and monitoring of clinical, environmental, and food samples, sustained drug release, and cell targeting for diagnostic and therapeutic purposes. The need for molecular recognition elements can to a certain degree be satisfied by naturally occurring species such as enzymes, antibodies and receptors. Sometimes these biomolecules are, however, difficult and/or expensive to isolate or generate, suffer from limited stability, or simply not exist for the desired target molecule. Since the papers by Pauling and Dickey, large efforts have been put into the development of synthetic receptors and molecular recognition elements. These efforts have resulted in recognition elements based on supramolecular assemblies, such as crown ethers, cyclophanes, and molecular clefts and cavities [4–10], de novo designed peptides and proteins [11–13], aptamers [14–16], ligands derived by combinatorial methods [17–20] and molecularly imprinted polymers (MIPs) [21, 22], the subject of this chapter.

2.2
The Concept of Molecular Imprinting

Molecular imprinting is a technique to synthesize polymers capable of selective molecular recognition and binding. The polymers are synthesized in the presence of a template, sometimes referred to as the print molecule. Figure 2.1 shows the procedure, starting with dissolution of template, functional monomers, cross-linking monomers and initiator in a porogenic solvent. The functional monomers are chosen so that they can interact with the template molecule. The monomers will arrange spatially around the template (step 1 in Figure 2.1). The positions of the monomers are fixed by copolymerization with cross-linking monomers (step 2). The polymerization is most often carried out as a free-radical polymerization initiated by photolytic or thermolytic homolysis of peroxide or azo compounds.

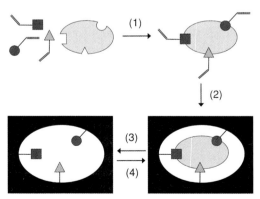

Figure 2.1 Concept of non-covalent molecular imprinting:
(1) arrangement of monomers around the template;
(2) cross-linking of monomers; (3) removal of template; and
(4) rebinding of template.

The polymerization runs through the chain-reaction steps, that is, initiation, propagation and termination. After completion of the polymerization, the template is removed from the polymer by extraction (step 3). The resulting polymer is then able to selectively rebind the template molecule (step 4).

Traditionally, molecular imprinting is classified according to the nature of the interactions between the monomers and the template during the polymerization (i.e. non-covalent, covalent or metal coordinated). Semi-covalent molecular imprinting is a special case of covalent imprinting where the interactions during the imprinting procedure are covalent while the interactions during the rebinding are of non-covalent nature.

The different classes of molecular imprinting are detailed in this section and various formats of MIPs are described in Section 2.3. In Section 2.4, questions on how to formulate and design MIPs are addressed and Section 2.5 describes various methods for characterizing MIPs. Finally, Section 2.6 exemplifies application areas of MIPs.

2.2.1
Non-covalent Molecular Imprinting

Non-covalent MIPs [23] utilize the same kind of interactions as biological recognition systems do, that is, relatively weak interactions such as hydrogen bonding, ion-pairing, hydrophobic interactions and dipole–dipole interactions. The interactions between monomers and templates are of non-covalent nature during the synthesis of the polymer as well as during the subsequent rebinding. The monomers can be acidic, basic or neutral. The most widely applied monomer in non-covalent imprinting so far is methacrylic acid. An overview of other commonly used monomers is given in Section 2.4.

The non-covalent molecular imprinting approach has been applied to a broad range of templates, including free amino acids [24], protected amino acids [25–31], herbicides [32–37], pesticides [38], fungicides [39], narcotics [40], antibiotics [41–47], barbiturates [48] and steroids [49–52]. The non-covalent approach is not restricted to small molecules but has also been applied to the imprinting of larger molecules such as proteins [53–55] and even bacteria [56].

Since the interactions between the monomers and the templates are of weak nature, several combinations of monomer-template complexes exist in the prepolymerization mixture. In addition, the monomers are often present in excess, which results in randomly distributed non-specific binding points. After polymerization, this plurality will exist also in the binding sites of the polymer; the sites are heterogeneous. The heterogeneity of the recognition sites is reflected in a distribution in affinity for the template.

2.2.2
Covalent Molecular Imprinting

In the covalent approach of molecular imprinting, one or more polymerizable group is coupled covalently to a functionality on the template to form a polymerizable template–monomer complex. Upon completion of the polymerization, the template is cleaved off from the resulting polymer and extracted. This leaves a polymer with positioned functional groups capable of re-forming the covalent bonds under appropriate conditions. Theoretically, this approach will give recognition sites with very similar affinity for the template throughout the polymer. The recognition sites are thus more homogeneous than those formed during non-covalent molecular imprinting. The kinetics of rebinding is often quite slow. For this to be a practical method, the cleavage and condensation reactions should occur under rather mild conditions.

The most successful approach in covalent molecular imprinting is probably the coupling of polymerizable boronic acids to hydroxyl groups present on the template to form boronate esters (Figure 2.2). This approach has, for example, been applied to the imprinting of sugars [57–59].

Templates containing diols can be reacted with monomers carrying a carbonyl functionality to form acetals or ketals. In the same way, carbonyl containing templates can be derivatized with polymerizable groups by reaction with diol containing monomers. This approach has, for example, been applied to the imprinting of mono- and di-ketones [60–63] and alcohols [64].

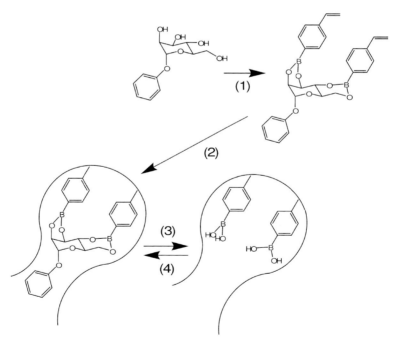

Figure 2.2 Imprinting of phenyl α-D-mannopyranoside using (4-vinylphenyl)boronic acid. Formation of monomer-template complex (1); polymerization (2); cleavage and extraction of template (3); and rebinding (4).

Schiff's bases are formed between primary amines and carbonyl compounds. The amide bond formed is readily reversible and has been utilized in the imprinting of, for example, amino acid derivatives [65, 66] and mono- and di-aldehydes [67, 68].

2.2.3
Semi-covalent Molecular Imprinting

In semi-covalent molecular imprinting, the imprinting step is carried out in the same way as in the covalent approach. The rebinding, however, relies on non-covalent interactions. When the covalently bound template moiety is cleaved from the MIP, a functional group capable of interacting non-covalently with the target molecule is left behind in the polymer at the cleavage site. The semi-covalent approach has been used successfully for the preparation of MIPs selective for *p*-aminophenylalanine ethyl ester [69], phenol [70] and triazines [71] among others.

The cavity left behind after cleaving off the covalently bound template moiety is often too small to accommodate the target molecule during non-covalent rebinding. This problem was addressed by Whitcombe in 1994 when the sacrificial spacer approach was introduced [72]. In this approach, the template and the monomer is joined by a spacer, which is cleaved off when the template is cleaved from the polymer (Figure 2.3). The sacrificial spacer approach has been applied to the imprinting of cholesterol [73], DDT [74] and heterocyclic aromatic compounds [75].

2.2.4
Metal Ion Mediated Molecular Imprinting

Most molecular imprinting polymerizations described so far has been carried out in organic solvents and the resulting polymers have shown best performance in organic solvents. Template molecules insoluble in organic solvents are not readily imprinted by conventional methods. A feasible method for such templates is the application of metal coordinating (chelating) monomers. Figure 2.4 shows the use of the monomer [*N*-(4-vinyl-benzyl)imino]-diacetic acid for the imprinting of histidine containing templates [76]. The approach has been applied also for the imprinting of proteins [77].

2.3
Formats of Molecularly Imprinted Polymers

Most MIPs are synthesized as three-dimensional polymer networks in which functional monomers, arranged around a template, are fixed in position by co-polymerization with a cross-linker. The spatial arrangement of the monomers creates complementary binding sites capable of rebinding the template.

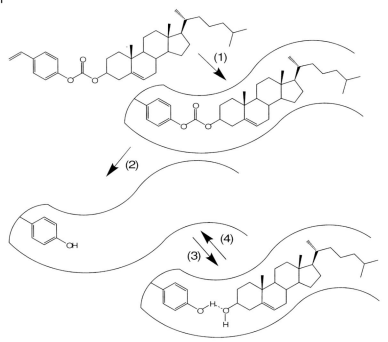

Figure 2.3 Semi-covalent molecular imprinting of cholesterol using the sacrificial spacer approach. Polymerization (1); cleavage and extraction (2); rebinding (association) (3); and dissociation (4).

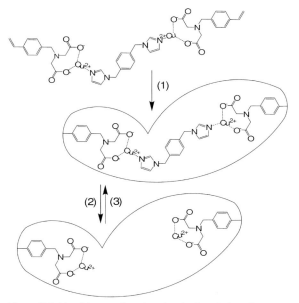

Figure 2.4 Metal ion mediated molecular imprinting. Cross-linking and polymerization (1); extraction of template (2); and rebinding of template (3).

Monolithic polymers, which are crushed and ground into smaller entities, as well as spherical beads are prepared this way. For this method to work, the template must be small enough to be able to diffuse within the polymer matrix. If the template has restricted mobility in the network, a surface imprinting approach is preferred. This approach has been demonstrated with macromolecules such as ribonuclease A [77] and lysozyme [78]. Even though imprinting in the bulk polymer network normally is preferred for low molecular weight templates, the surface imprinting approach can be applied also for such templates. The strategy can also be used to confer new properties to polymer surfaces, as exemplified by the imprinting of the β-lactam antibiotic ampicillin [79]. The imprinted surface retained the antibiotic and inhibited adhesion of bacteria.

2.3.1
Irregularly Shaped Particles

The traditional way to synthesize MIPs is referred to as "bulk polymerization". The terminology is slightly misleading, but has gained acceptance within the MIP community (in the strictest polymer chemistry sense this is not a real bulk polymerization but rather a solution polymerization since the pre-polymerization mixture contains a solvent, the porogen). Wulff [57, 80], Mosbach [23] and Shea [60] introduced the method in their pioneering works on MIPs. The template, the functional monomer(s), the cross-linker and the initiator are dissolved in a solvent (porogen). This mixture is hereafter referred to as the "pre-polymerization mixture." The polymerization is carried out in a sealed container. After completion of the polymerization, the container is broken and the polymer monolith is coarsely crushed, then ground and sieved to obtain particles of appropriate size range. The template molecules are extracted from the polymer. Finally, the polymer particles are dried. This protocol is time- and labor-intensive and produces large amounts of particles of unwanted size fractions. In addition, the shape of the particles is unpredictable and usually highly irregular (Figure 2.5).

Figure 2.5 Scanning electron micrograph of irregular MIP particles. Accelerating voltage: 5 kV; working magnification 150×. Reproduced with permission from (2006) *Analytical Chemistry*, **78**, 3659–66. Copyright 2006 American Chemical Society.

2.3.2
Beads

Spherical beads possess better hydrodynamic and diffusion properties than irregularly shaped particles. It is, hence, desirable to apply MIPs in a spherical bead format, especially for flow-through applications. Methods to synthesize spherical polymer beads are often classified according to the initial state of the polymerization mixture: (i) homogeneous (i.e. precipitation polymerization and dispersion polymerization) or (ii) heterogeneous (i.e. emulsion polymerization and suspension polymerization). In addition, several other techniques have been applied for the preparation of spherical MIP beads. The techniques of two-step swelling polymerization, core–shell polymerization, and synthesis of composite beads will be detailed here.

2.3.2.1 Homogeneous Polymerization
When beads are synthesized by polymerization from a homogeneous mixture, a pre-polymerization mixture, composed of functional monomer(s), cross-linker(s), initiator and template dissolved in a solvent, is prepared. This procedure is similar to that of "bulk polymerization" discussed above. The pre-polymerization mixture is, however, in this case further diluted with solvent so that polymer beads are formed rather than a monolithic polymer. In a sense, it seems more logical to use the term "polymerization medium" rather than "porogen" when referring to the solvent in a homogeneous polymerization.

Polymerization of beads from homogeneous mixtures can be carried out by either dispersion polymerization or precipitation polymerization. In both methods, the initiation and nucleation take place in the polymerization medium. The difference between the methods emanates from how good a solvent the polymerization medium is for the polymer nuclei formed. In precipitation polymerization, the nuclei do not swell in the polymerization medium. The particle growth proceeds via coagulation of nuclei into larger particle aggregates. The resulting particles will thus be irregularly shaped and polydisperse. MIP beads prepared by precipitation polymerization were pioneered by the Mosbach group [81]. The nuclei produced during dispersion polymerization swell in the polymerization medium and the particle growth takes place to a large extent in the swollen particles. The resulting particles are spherical and can be made monodisperse by the addition of stabilizers to the polymerization medium. Criteria for stabilizers used in dispersion polymerization are that they should have low solubility in the polymerization medium and moderate affinity for the polymer particles.

2.3.2.2 Heterogeneous Polymerization
Two liquids that are immiscible or nearly immiscible will form an emulsion upon mixing. The liquid forming the droplets is called the dispersed phase and the surrounding liquid is referred to as the continuous phase. Emulsions can be of the "oil-in-water" or "water-in-oil" types. Emulsions are by nature often

unstable and hence often need to be stabilized by the use of an emulsifier. The emulsifier forms a protective layer around the droplets and decreases the rate of droplet coalescence. Other ways to stabilize emulsions are by increasing the viscosity of either the continuous or the dispersed phase. Increased viscosity of the continuous phase reduces the movement of the droplets and thus lowers the rate of coalescence. This can, in general, be achieved by addition of viscosity modifiers such as xanthan gums, clays or gelatin. A higher viscosity of the dispersed phase results in more rigid droplets that are less prone to coalesce.

Heterogeneous polymerization can be further divided into emulsion polymerization and suspension polymerization. In both types of polymerization, the monomers are dissolved in the dispersed phase. In suspension polymerization, the initiator is dissolved in the dispersed phase as well and nucleation and growth of the beads take place in the droplets. In emulsion polymerization, on the other hand, the initiator is dissolved in the continuous phase, leading to nucleation and bead growth from the continuous phase.

Suspension polymerization was described already in 1909 in a patent by Hoffman and Delbruch. The technique has been used for the preparation of MIPs using both aqueous [45, 82–85] and non-aqueous continuous phases [86–90]. The aqueous approach requires either covalent interactions or strong non-covalent interactions between monomers and templates. Liquid perfluorocarbon, mineral oil and silicon oil have been used as the continuous phase to avoid water that can disturb hydrogen bonds and other polar interactions between functional monomers and templates. An added advantage with suspension in oil is that, due to the high viscosity of the oil, no addition of stabilizing agents or emulsifiers is needed to form stable droplets. Figure 2.6 shows a scanning electron micrograph of MIP beads prepared by polymerization in mineral oil.

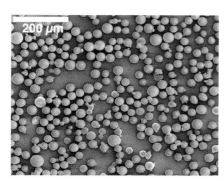

Figure 2.6 Scanning electron micrograph of beads prepared by suspension polymerization in mineral oil. Accelerating voltage: 5 kV; working magnification 150×. Reproduced with permission from (2006) *Analytical Chemistry*, **78**, 3659–66. Copyright 2006 American Chemical Society.

2.3.2.3 Two-Step Swelling Polymerization

Two-step swelling polymerization starts with the preparation of sub-micron sized non cross-linked seed particles by emulsion polymerization. The seed particles are then added to a pre-polymerization mixture. The particles swell in the mixture and the polymerization takes place within the volume of the swollen particles [91]. The procedure, which results in monodisperse particles, has been used for the preparation of MIPs imprinted with a range of templates [92–97].

2.3.2.4 Core–Shell Polymerization

Core–shell polymerization is a seed particle polymerization variation of emulsion polymerization. The seed particles are suspended in the continuous phase. The pre-polymerization mixture of monomer, cross-linker, template and initiator is added to the particle suspension as an emulsion prepared in the continuous phase. The mixture is stirred until the polymerization has completed. The addition of pre-polymerization mixture is repeated several times until the spheres reach the desired size range. The beads formed are composed of a core (i.e. the seed particle) and a shell of MIP [98, 99].

2.3.2.5 Silica Composite Beads

MIPs can be synthesized in the pores and on the surface of pre-made porous particles. Porous silica particles have been applied for this purpose. To ensure that the imprinted polymer is attached firmly to the particle, the particles are often chemically modified by coupling of polymerizable groups or initiator molecules to the particle surface prior to the MIP polymerization [100–102]. The use of immobilized initiators is often referred to as the iniferter (initiator–transfer agent–terminator) approach [103]. The method has been applied to the imprinting of a range of templates [104–107].

The silica can be removed from a MIP-silica composite bead by dissolution with hydrofluoric acid, leaving a MIP bead that is a replica of the pore structure of the original silica bead [108, 109]. MIP composites have also been made using, for example, chitosan as the support [110, 111].

2.3.3
Films and Membranes

MIP films are polymerized on flat surfaces. Surfaces derivatized with polymerizable groups are preferred since they allow covalent attachment to the surface. MIP films are attractive as sensing elements in MIP based sensors [112].

Molecularly imprinted membranes can be prepared either as thick films or as composites in the pores of base-membranes. In composite membranes, the selective properties of the imprinted material are combined with the properties of the base-membrane. Membranes can also be prepared by phase inversion polymerization. The selective nature of MIPs makes it possible to prepare membranes with selective permeability [113, 114].

2.4 Design of MIPs

The development of a MIP normally starts with a given template. The first task is to select the components of the MIP, that is, the monomer(s), the cross-linker(s), the porogen and the initiator. The components can be added in various ratios and the polymerization can be carried out under a range of conditions. The design of a MIP is often quite a complex task and requires some kind of optimization strategy.

2.4.1 Functional Monomers

Much of the selectivity of MIPs arises from the interactions between the templates and the functional monomers. The functional monomers are chosen so that their functionalities complement the functionalities of the template molecules. A wide variety of functional monomers, including acidic, basic, neutral and hydrophobic ones, have been tested in MIP synthesis (Table 2.1). Methacrylic acid is the most widely used monomer and has been applied in the synthesis of MIPs selective for a wide range of templates.

Table 2.1 Functional monomers used in non-covalent molecular imprinting.

Name	Structure	References
Acrylic acid		[25, 115]
Methacrylic acid		[26, 34, 35, 49, 116, 117]
2-(Trifluoromethyl)acrylic acid		[35, 118]
Methyl methacrylate		[119]
2-Hydroxyethyl methacrylate		[50]

Table 2.1 Continued.

Name	Structure	References
2-(Methacryloyloxy)ethyl phosphate		[52]
N,N-diethylaminoethyl methacrylate		[41]
Acrylamide		[120–122]
2-Acrylamido-2-methyl-1-propane-sulfonic acid		[123]
N-(2-aminoethyl)methacrylamide		[124]
2-Methyl-N-(6-methyl-pyridine-2-yl)-acrylamide		[125]
Allyl alcohol		[126]
Itaconic acid		[127]
1-Vinylimidazole		[128]
Vinyl pyrrolidone		[129]

Table 2.1 Continued.

Name	Structure	References
2-Vinylpyridine		[130, 131]
4-Vinylpyridine		[128]
Styrene		[132]
4-Ethyl styrene		[133]
4-Amino styrene		[125, 134]
p-Vinyl benzoic acid		[133]
[N-(4-vinylbenzyl)imino]-diacetic acid		[76, 77]
1-(3,5-Bis-trifluoromethyl-phenyl)-3-(4-vinyl-phenyl)-urea		[47, 135]

2.4.2
Cross-linking Monomers

Several molecules containing two or more polymerizable groups have been applied as cross-linkers in molecular imprinting. The cross-linking monomers lock the position of the functional monomers relative to the template to form recognition sites in a rigid polymer network. The cross-linker is normally chosen so that it does not interact with the template, although some exceptions exist. The cross-linkers N,α-bismethacryloyl glycine and N,O-bismethacryloyl ethanolamine have been shown to possess interesting qualities in forming interactions with templates [136–138]. Table 2.2 summarizes some commonly used cross-linking monomers. Ethylene glycol-dimethacrylate is the most common cross-linker in MIP synthesis.

2.4.3
The Porogen

The porogen is the solvent used for dissolving the monomers, the template and the initiator. The porogen is also responsible for the creation of the pores in the polymer, and thus also affects the surface area. In non-covalent molecular imprinting, the porogen is usually chosen so that it promotes the interactions between

Table 2.2 Cross-linking monomers.

Name	Structure	Reference
Divinylbenzene		[133]
Ethylene glycoldimethacrylate		[25–28, 116]
N,N'-1,4-phenylenediacrylamide		[100]

Table 2.2 Continued.

Name	Structure	Reference
2,6-Bis(acrylamido)pyridine		[‹8]
N,N'-methylenediacrylamide		[100]
N,O-bisacryloyl-L-phenylaninol		[115]
Pentaerythritol triacrylate		[139]
Trimethylolpropane trimethacrylate		[139, 140]
Pentaerythritol tetraacrylate		[139]
N,α-bismethacryloyl glycine		[136, 137]
N,O-bismethacryloyl ethanol-amine		[136–138]

the functional monomer(s) and the template. Aprotic organic solvents, such as acetonitrile, chloroform, and toluene, are often chosen as the porogen. In covalent molecular imprinting, the template has been coupled covalently to the monomer beforehand to form a template–monomer complex that is co-polymerized with the cross-linker during the imprinting. The nature of the porogen is therefore less important in this case, as long as the monomer–template complex and the cross-linker are soluble and the porogen is inert to the polymerization reactions.

2.4.4
Initiation of Polymerization

Most functional monomers and cross-linkers contain one or more vinyl functionalities. Polymerization of this type of compound for the preparation of MIPs is traditionally performed as a free-radical polymerization, initiated via either thermolytic or photolytic homolysis of an initiator. One of the most commonly used free radical initiators for this purpose is 2,2′-azobis (isobutyronitrile) (AIBN). Other examples of free-radical polymerization initiators are phenyl-azo-triphenyl-methane, tert-butyl peroxide (TBP), acetyl peroxide, benzoyl peroxide (BPO), lauroyl peroxide, *tert*-butyl hydroperoxide and *tert*-butyl perbenzoate.

2.4.5
Optimization of Imprinting Conditions

For a newcomer to the field of molecular imprinting, the easiest way to approach the task of preparing a MIP is to search the literature and simply copy an existing protocol developed for a similar or non-similar template molecule. The resulting MIP will probably bind the template, but it will not necessarily be optimal in binding capacity and selectivity.

In the beginning of the molecular imprinting era, polymers were mainly developed by a trial-and-error approach. Optimization studies were often limited to variations of one parameter at a time. Later, the empirical knowledge gained formed the basis of several rules of thumb. For example, it is advisable that the content of the cross-linking monomer in the monomer mixture is above 50 mol % [117, 141, 142]. Several promising methods to approach the optimization problem in a more systematic way have been demonstrated, for example, by using statistical methods to perform variations and evaluations (the chemometrical approach) or by investigating the molecular interactions either experimentally or by computational methods. Combinatorial synthesis approaches combined with high-throughput screening have also been applied. MIP libraries have been synthesized in several different formats, for example, on filter membranes in single use modules [143] and on the bottom of HPLC sample vials [144–146] and the wells of 96-well micro titer plates [147].

In the experimental approach, the complex formation between the template and the functional monomer(s) is studied by spectroscopic methods such as NMR [116, 148–150] and UV [151]. The experiments are often performed as titrations where the template is titrated with the functional monomer and the complex formation is followed by recording the spectra.

The computational approach involves evaluation of the interactions between the template and the functional monomer(s) using molecular modeling software. The procedure often starts with a screening of a virtual library of monomers against the template molecule to decide the appropriate monomer. An iterative routine places a monomer molecule at different positions around the template molecule and the interactions are calculated. During a later stage of the procedure, the number of functional monomer molecules needed per template molecule for optimal interaction is determined. This approach has been used successfully by several investigators [152–162].

The chemometrical approach utilizes statistical experimental design and multivariate data analysis. The number of experiments needed to draw statistically significant conclusions on a problem is minimized by co-variation of parameters in the model. The complexity of the model chosen depends on the nature of the problem to be solved. Models can be chosen for screening, optimization or robustness testing purposes. The investigator chooses the model, the factors (i.e. the parameters that will be varied), the range of the variation of the factors and the responses to be included in the model. The design tool then creates an appropriate experimental design containing the minimal number of experiments needed. Most chemometrics tools also include an evaluation tool based on multivariate data analysis. Different combinations of the factors are tested against the measured responses. Chemometrics has proven useful in the design and evaluation of MIPs [88, 163–167].

2.5
Characterization of Molecularly Imprinted Polymers

2.5.1
Characterization of Binding Properties of MIPs

After designing and preparing a novel MIP, the next step is to elucidate whether the polymer possesses the expected binding properties; in other words, the imprinting effect should be verified. The binding to a MIP often originates from both specific interactions and non-specific ones. The specific interactions take place in the recognition sites created during the molecular imprinting procedure while the non-specific binding is due to random interactions with the polymer. The non-specific binding is normally estimated from the binding to a control polymer (CP). The CP is prepared in the same way as the MIP, but in the absence of template or with a different template present during polymerization.

The binding to a MIP and the corresponding CP is usually evaluated either by batch incubations with polymer and analyte or by packing the polymer into a column and applying it as the stationary phase in liquid chromatography. The final use of the MIP often determines the preferred method. In batch-wise binding studies, the analyte is incubated with the MIP and the CP, respectively, in a suitable solvent in a closed vessel. To assure reproducible results it is advisable to allow the binding to reach equilibrium. The fraction bound analyte (B/T, where B is the bound amount and T is the total initial amount added), the percentage bound or the bound amount (B) are often used as measures of the binding efficiency. Although this strategy is useful, especially for the screening of large numbers of MIPs, the parameters should be interpreted with caution since single-point binding investigations give only limited information on the binding characteristics. The equilibrium concentrations (i.e. the free concentrations) are not the same for two polymers that are compared even if the initial concentration of the analyte was the same in both cases. The comparison is thus made at different points on the binding isotherms of the two polymers. An alternative parameter, which to some degree considers the concentration differences, is the imprinting factor, which is calculated as the amount template bound per gram MIP divided by the free concentration of the template.

The best way to characterize the binding to MIPs and CPs is to determine the complete binding isotherms [168, 169]. The binding of the analyte to the MIP and the CP, respectively, is determined after batch incubations of known amounts of polymer with analyte at a range of concentrations during a period of time required to reach equilibrium. An isotherm model is fitted to the experimental data and the parameters of the model can be used to describe the binding (Figure 2.7). For example, the Langmuir isotherm gives the dissociation constant and the number of binding sites. Since the binding sites are heterogeneous, more complex models have also been applied, such as the bi-Langmuir and the Freundlich isotherms [170].

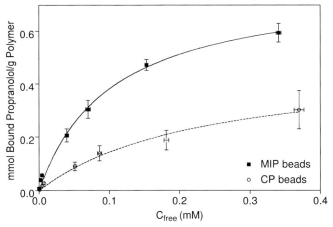

Figure 2.7 One-site Langmuir binding isotherms of propranolol to MIP and CP beads (Ø 25–50 µm). Data points are the mean values of three replicates and standard deviations are indicated with error bars (data from reference [89]). Reproduced with permission from (2006) *Analytical Chemistry*, **78**, 3659–66. Copyright 2006 American Chemical Society.

Isotherms can also be obtained by frontal chromatography [171–173]. The MIP is then packed into a column and used as the stationary phase in liquid chromatography. A solution of known concentration of analyte is applied continuously to the column until a breakthrough curve is obtained. After washing the column, the procedure is repeated with increasing concentrations of analyte. The amount analyte bound for each concentration is calculated from the breakthrough curves.

The cross-reactivity of a MIP is a measure of the selectivity and is calculated as the ratio of the EC_{50} value (effective concentration 50%) of the template to that of a competitor. The EC_{50} value is determined by incubating the MIP with labeled template and increasing concentrations of a competitor. The EC_{50} value is the required concentration of competitor to displace half of the labeled template bound to the MIP [174]. Typical displacement curves are shown in Figure 2.8.

Parameters of interest to determine for MIPs used as stationary phases in liquid chromatography are the retention factors [$k = (t - t_0)/t_0$, where t is the retention of the analyte and t_0 is the void], the separation factors ($\alpha = k_1/k_2$, where k_1 and k_2 are the retention factors of compound 1 and 2) and the resolution (R_s) [175, 176]. For MIPs used as stationary phases in solid-phase extraction, the recovery is also of interest.

A range of physical methods has been applied to study the binding to MIPs. The methods include NMR [62, 177], FT-IR [62, 177], microcalorimetry [178–180],

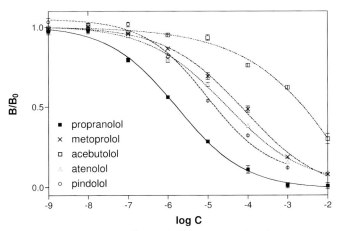

Figure 2.8 Displacement of ^{3}H-labeled propranolol (1 nM) binding to propranolol MIP beads in acetonitrile–acetic acid (199:1) by increasing concentrations of competing ligands. B/B_0 is the ratio of the amount of ^{3}H-labeled propranolol bound in the presence of displacing ligand to the amount bound in the absence of displacing ligand. Samples were prepared in triplicate and data were fitted by GraphPad Prism software (San Diego, CA) to a sigmoidal dose–response curve with variable slope model. Data points are the mean values and standard deviations are indicated with error bars (data from reference [89]). Reproduced with permission from (2006) *Analytical Chemistry*, **78**, 3659–66. Copyright 2006 American Chemical Society.

cyclic voltammetry [181] and Raman spectroscopy [182]. In addition, various sensors can be used to study the binding properties of MIPs. Optical sensors, mass sensitive sensors and electrochemical sensors are discussed further in Section 2.6.4.

2.5.2
Characterization of Chemical and Physical Properties of MIPs

An elemental analysis of a MIP ascertains that the composition of the polymer is the expected one. Comparison of the analytically determined amounts of the elements with the theoretical values gives an indication of the outcome of the polymerization. An estimation of the success of the removal of the template can also be obtained by elemental analysis if the template contains elements not present in the monomers. Other characterization methods that give information on the composition of a polymer include FTIR and NMR.

In flow-through applications, such as liquid chromatography, stationary phases consisting of spherical monodisperse beads give better flow properties than irregularly shaped particles. Spherical beads are in general preferred over irregular particles also for other applications due to their easier handling. The synthesis of MIP beads was discussed in Section 2.3.2. The particle size distribution of a particulate sample can be determined in several ways, for example, by (i) fractionation of the sample by sieving and gravimetric determination of the fractions; (ii) laser diffraction; (iii) dynamic light scattering; (iv) electrical zone sensing (the Coulter principle); and (v) microscopy and image analysis [183].

The binding capacity and the site availability of MIPs depend on parameters such as surface area, pore diameter and pore size distribution. These parameters are therefore often determined, by gas sorption or mercury penetration, when MIPs are characterized.

2.6
Applications of Molecularly Imprinted Polymers

2.6.1
Liquid Chromatography

In the beginning of the era of molecular imprinting, the method of choice for analysis and characterization of the binding to the MIP was to pack the polymer into a column and apply it as the stationary phase in liquid chromatography. Even though an immense number of studies on MIP stationary phases have been carried out, most have aimed at characterizing the molecular recognition event and not many applications have found practical use outside academia. This is partly because MIP stationary phases generally suffer from severe tailing effects due to nonlinear binding isotherms [168]. The nonlinearity of the isotherms originates from the heterogeneity of the binding sites as previously discussed.

The high selectivity of MIPs is demonstrated when an optically active compound is imprinted; the resulting MIP will normally resolve the racemate. Numerous reports on MIP chiral stationary phases have appeared [184–188]. Chiral templates studied include amino acids [26, 29, 120, 139, 189–192], peptides [139, 192, 193], carbohydrates [58, 194, 195] and drugs [127, 196].

2.6.2
Solid-Phase Extraction

The application of MIPs as the stationary phase in solid-phase extraction (SPE), often referred to as molecularly imprinted polymer solid-phase extraction (MISPE), is a rapidly growing area [197–199]. With MISPE, highly specific enrichment of substances present at trace levels is possible. The technique has been applied to the analysis of drugs, for example, caffeine [200], scopolamine [201], naproxen [202], tetracycline [203], cholesterol [204] and local anesthetics [205], as well as environmental pollutants, exemplified by organophosphate flame retardants [206–208], triazines in soil and vegetable samples [71] and naphthalene sulfonates in river water [209].

2.6.3
Solid-Phase Binding Assay

Solid-phase binding assays with MIPs, also know as molecularly imprinted sorbent assays (MIA), are equivalents to immunoassays employing antibodies as the recognition elements [198, 210, 211]. The strategy is to use a reporter molecule that is either displaced from the MIP or is competing with the analyte during binding. The reporter molecule can be radioactively labeled or possess optical properties that allow easy detection. An alternative route is to include a fluorescent moiety in the polymer network that either fluoresces or is quenched upon binding of the analyte. MIA has been developed for a broad range of compounds present in various matrices [89, 198, 212–214].

2.6.4
Sensors

2.6.4.1 Optical Sensors
Most MIP optical sensors reported in the literature utilize fluorescence as the detectable signal [215, 216]. With fluorescent analytes, the fluorescence of the MIP will increase upon binding [217, 218]. With non-fluorescent analytes, a fluorescent reporter molecule can be applied. The reporter molecule competes with or is displaced by the analyte [219–221]. Alternatively, a fluorescent monomer can be incorporated in the polymer network [222]. Upon binding of the analyte, the fluorescence of the MIP is altered.

Another approach for MIP sensors is detection based on surface plasmon resonance (SPR) [223–225]. The strategy is to synthesize the MIP on a gold or silver surface. The reflective properties of the surface change when the MIP binds the analyte.

2.6.4.2 Mass Sensitive Sensors

The quartz crystal microbalance (QCM) is a mass sensitive sensor [226]. The sensor is based on a piezoelectric crystal covered with an adsorbing agent. The crystal oscillates with a characteristic frequency, usually in the range 5–20 MHz, when an electric field is applied. When an analyte adsorbs to the crystal surface a decrease in the frequency, ideally proportional to the oscillating mass, is detected. By attaching a thin MIP film on the crystal surface the rebinding of the template can be studied. MIP-QCMs have been applied to a wide range of analytes, for example, aromatic and halogenated hydrocarbons [227], amino acids [122, 228], peptides [229], propranolol [230], enzymes [231] and cells and viruses [231].

2.6.4.3 Electrochemical Sensors

The first report on an electrochemical MIP based sensor was published in 1993 [232]. The sensing device was a field-effect capacitor with a thin polymer membrane imprinted with L-phenylalanine anilide. Binding of the template gave a shift in the capacitance-voltage. Other early demonstrations of electrochemical MIP-sensors include a morphine sensitive amperometric sensor based on competitive binding of codeine [233] and a conductometric sensor targeted for benzyltriphenylphosphonium ions [234]. In 1999, a capacitive sensor consisting of a thin layer of phenylalanine imprinted polyphenol electropolymerized on a gold surface was reported [235]. The same year, a cholesterol sensitive amperometric sensor fabricated by spontaneous self-assembly of hexadecyl mercaptan on gold surfaces [236] as well as atrazine sensitive conductometric sensors utilizing imprinted membranes were presented [237, 238]. Several investigators have reported on glucose sensitive sensors [239–241] and an amperometric sensor for fructosylamines has been prepared using a catalytic MIP [242].

2.6.5
Synthetic Enzymes

The possibility to tailor-make MIPs towards a desired selectivity in combination with the high stability of the materials under a broad range of conditions has rendered MIPs attractive for the development of synthetic enzymes [243, 244]. A popular strategy has been to imprint a transition state analog to obtain a polymer that reduces the activation energy of the reaction. Catalytically active groups are often included in the polymer network. This approach has been applied towards ester and amide hydrolysis reactions [245, 246]. Examples of other reactions where MIPs have been utilized as enzyme mimics are isomerization [247], transamination [248], Diels–Alder reaction [249], β-elimination [250] and regioselective cycloaddition [251].

2.7 Conclusions

Molecular imprinting offers an attractive method for the preparation of tailor-made recognition elements. There has been a rapid growth in the field and many significant advances have been achieved. MIPs have been explored for several applications and more are expected in the future. The industry needs selective separation media and the regulatory authorities demand methods that can guarantee safe products for society. Therefore, MIPs should be useful for highly selective chromatographic separations as well as solid-phase extractions. Other intriguing applications for MIPs are as sensing elements in sensors, as recognition elements in solid-phase binding assays, and as synthetic enzymes. To improve the technique, however, it is vital that the methods to design and synthesize MIPs are improved and the underlying mechanisms of the imprinting procedure are further explored.

References

1. Pauling, L. (1940) *Journal of the American Chemical Society*, **62**, 2643–57.
2. Dickey, F.H. (1949) *Proceedings of the National Academy of Sciences of the United States of America*, **5**, 227–9.
3. Polyakov, M.V., Kuleshina, L.P. and Neimark, I.E. (1937) *Journal of Physical Chemistry, U.S.S.R.*, **10**, 100–12.
4. Lehn, J.-M. (1988) *Angewandte Chemie – International Edition in English*, **27**, 89–112.
5. Rebek, J. Jr (1990) *Angewandte Chemie – International Edition in English*, **29**, 245–55.
6. Cram, D.J. (1992) *Nature*, **356**, 29–36.
7. Webb, T.H. and Wilcox, C.S. (1993) *Chemical Society Reviews*, **22**, 383–95.
8. Conn, M.M. and Rebek, J. Jr (1997) *Chemical Reviews*, **97**, 1647–68.
9. Bell, T.W. and Hext, N.M. (2004) *Chemical Society Reviews*, **33**, 589–98.
10. Anslyn, E.V. (2007) *Journal of Organic Chemistry*, **72**, 687–99.
11. Baltzer, L. (1998) *Current Opinion in Structural Biology*, **8**, 466–70.
12. Cooper, W.J. and Waters, M.L. (2005) *Current Opinion in Chemical Biology*, **9**, 627–31.
13. Kriplani, U. and Kay, B.K. (2005) *Current Opinion in Biotechnology*, **16**, 470–5.
14. Celia, M.H. (2004) *Chemical and Engineering News*, **82** (13), 32–3.
15. Brody, E.N. and Gold, L. (2000) *Reviews in Molecular Biotechnology*, **74**, 5–13.
16. Osborne, S.E., Matsumura, I. and Ellington, A.D. (1997) *Current Opinion in Biotechnology*, **1**, 5–9.
17. Lowe, C.R., Lowe, A.R. and Gupta, G. (2001) *Journal of Biochemical and Biophysical Methods*, **49**, 561–74.
18. Labrou, N.E. (2003) *Journal of Chromatography B*, **790**, 67–78.
19. Srinivasan, N. and Kilburn, J.D. (2004) *Current Opinion in Chemical Biology*, **8**, 305–10.
20. Schmuck, C. and Wich, P. (2006) *New Journal of Chemistry*, **30**, 1377–85.
21. Sellergren, B. (ed.) (2001) *Molecularly Imprinted Polymers. Man Made Mimics of Antibodies and Their Application in Analytical Chemistry*, 1st edn, Elsevier Science, Amsterdam.
22. Komiyama, M., Takeuchi, T., Mukawa, T. and Asanuma, H. (2003) *Molecular Imprinting from Fundamentals to Applications*, 1st edn, Wiley-VCH Verlag GmbH, Weinheim.
23. Arshady, R. and Mosbach, K. (1981) *Makromolekulare Chemie – Macromolecular Chemistry and Physics*, **182**, 687–92.

24 Vidyasankar, S., Ru, M. and Arnold, F.H. (1997) *Journal of Chromatography A*, **775**, 51–63.
25 Sellergren, B., Ekberg, B. and Mosbach, K. (1985) *Journal of Chromatography A*, **347**, 1–10.
26 O'Shannessy, D.J., Ekberg, B., Andersson, L.I. and Mosbach, K. (1989) *Journal of Chromatography A*, **470**, 391–9.
27 O'Shannessy, D.J., Ekberg, B. and Mosbach, K. (1989) *Analytical Biochemistry*, **177**, 144–9.
28 O'Shannessy, D.J., Andersson, L.I. and Mosbach, K. (1989) *Journal of Molecular Recognition*, **2**, 1–5.
29 Sellergren, B. (1989) *Chirality*, **1**, 63–8.
30 Andersson, L.I., O'Shannessy, D.J. and Mosbach, K. (1990) *Journal of Chromatography A*, **513**, 167–79.
31 Andersson, L.I. and Mosbach, K. (1990) *Journal of Chromatography A*, **516**, 313–22.
32 Piletsky, S.A., Parhometz, Y.P., Lavryk, N.V., et al. (1994) *Sensors and Actuators B*, **19**, 629–31.
33 Muldoon, M.T. and Stanker, L.H. (1995) *Journal of Agricultural and Food Chemistry*, **43**, 1424–7.
34 Matsui, J., Doblhoff-Dier, O. and Takeuchi, T. (1995) *Chemistry Letters*, **24**, 489.
35 Matsui, J., Miyoshi, Y. and Takeuchi, T. (1995) *Chemistry Letters*, **24**, 1007–8.
36 Siemann, M., Andersson, L.I. and Mosbach, K. (1996) *Journal of Agricultural and Food Chemistry*, **44**, 141–5.
37 Baggiani, C., Giraudi, G., Giovannoli, C., et al. (2000) *Journal of Chromatography A*, **883**, 119–26.
38 Baggiani, C., Trotta, F., Giraudi, G., et al. (1999) *Analytical Communications*, **36**, 263–6.
39 Liu, H., Yang, G., Liu, S. and Wang, M. (2005) *Journal of Liquid Chromatography and Related Technologies*, **28**, 2315–23.
40 Andersson, L.I., Müller, R., Vlatakis, G. and Mosbach, K. (1995) *Proceedings of the National Academy of Sciences*, **92**, 4788–92.
41 Levi, R., McNiven, S., Piletsky, S.A., Cheong, S.-H., Yano, K. and Karube, I. (1997) *Analytical Chemistry*, **69**, 2017–21.
42 Senholdt, M., Siemann, M., Mosbach, K. and Andersson, L.I. (1997) *Analytical Letters*, **30**, 1809–21.
43 Siemann, M. and Andersson, L.I. (1997) *Journal of Antibiotics*, **50**, 89–91.
44 Skudar, K., Brüggeman, O., Wittelsberger, A. and Ramström, O. (1999) *Analytical Communications*, **36**, 327–31.
45 Lai, J.P., Cao, X.F., Wang, X.L. and He, X.W. (2002) *Analytical and Bioanalytical Chemistry*, **372**, 391–6.
46 Cederfur, J., Pei, Y., Zihui, M. and Kempe, M. (2003) *Journal of Combinatorial Chemistry*, **5**, 67–72.
47 Urraca, J.L., Hall, A.J., Moreno-Bondi, M.C. and Sellergren, B. (2006) *Angewandte Chemie – International Edition in English*, **45**, 5158–61.
48 Tanabe, K., Takeuchi, T., Matsui, J., et al. (1995) *Journal of the Chemical Society D – Chemical Communications*, 2303–4.
49 Ramström, O., Lei, Y. and Mosbach, K. (1996) *Chemistry and Biology*, **3**, 471–7.
50 Sreenivisan, K. (1998) *Journal of Applied Polymer Science*, **68**, 1863–6.
51 Baggiani, C., Giraudi, G., Trotta, F., et al. (2000) *Talanta*, **51**, 71–5.
52 Kugimiya, A., Kuwada, Y. and Takeuchi, T. (2001) *Journal of Chromatography A*, **938**, 131–5.
53 Hjertén, S., Liao, J.-L., Nakazato, K., et al. (1997) *Chromatographia*, **44**, 227–34.
54 Liao, J.-L., Wang, Y. and Hjertén, S. (1996) *Chromatographia*, **42**, 259–62.
55 Shi, H., Tsai, W.-B., Garrison, M.D., et al. (1999) *Nature*, **398**, 593–7.
56 Aherne, A., Alexander, C., Payne, M.J., et al. (1996) *Journal of the American Chemical Society*, **118**, 8771–2.
57 Wulff, G., Sarhan, A. and Zabrocki, K. (1973) *Tetrahedron Letters*, **14**, 4329–32.
58 Wulff, G., Vesper, W., Grobe-Einsler, R. and Sarhan, A. (1977) *Makromolekulare Chemie – Macromolecular Chemistry and Physics*, **178**, 2799–816.
59 Wulff, G. and Schauhoff, S. (1991) *Journal of Organic Chemistry*, **56**, 395–400.
60 Shea, K.J. and Dougherty, T.K. (1986) *Journal of the American Chemical Society*, **108**, 1091–3.
61 Shea, K.J. and Sasaki, D.Y. (1989) *Journal of the American Chemical Society*, **111**, 3442–4.

62. Shea, K.J. and Sasaki, D.Y. (1991) *Journal of the American Chemical Society*, **113**, 4109–20.
63. Marty, J.D., Tizra, M., Mauzac, M., et al. (1999) *Macromolecules*, **32**, 8674–7.
64. Wulff, G. and Wolf, G. (1986) *Chemische Berichte*, **119**, 1876–89.
65. Wulff, G., Best, W. and Akelah, A. (1984) *Reactive Polymers*, **2**, 167–74.
66. Wulff, G. and Vietmeier, J. (1989) *Makromolekulare Chemie – Macromolecular Chemistry and Physics*, **190**, 1727–35.
67. Wulff, G., Heide, B. and Helfmeier, G. (1986) *Journal of the American Chemical Society*, **108**, 1089–91.
68. Shea, K.J., Stoddard, G.J., Shavelle, D.M., et al. (1990) *Macromolecules*, **23**, 4497–507.
69. Sellergren, B. and Andersson, L. (1990) *Journal of Organic Chemistry*, **55**, 3381–3.
70. Joshi, V.P., Karode, S.K., Kulkarni, M.G. and Mashelkar, R.A. (1998) *Chemical Engineering Science*, **53**, 2271–84.
71. Cacho, C., Turiel, E., Martín-Esteban, A., et al. (2006) *Journal of Chromatography A*, **1114**, 255–62.
72. Whitcombe, M.J., Rodriguez, M.E. and Vulfson, E.N. (1994) in *Separation for Biotechnology 3* (ed. D.L. Pyle), Royal Society of Chemistry, Cambridge, pp. 565–71.
73. Whitcombe, M.J., Rodriguez, M.E., Villar, P. and Vulfson, E.N. (1995) *Journal of the American Chemical Society*, **117**, 7105–11.
74. Graham, A.L., Carlson, C.A. and Edmiston, P.L. (2002) *Analytical Chemistry*, **74**, 458–67.
75. Kirsch, N., Alexander, C., Davies, S. and Whitcombe, M.J. (2004) *Analytica Chimica Acta*, **504**, 63–71.
76. Dhal, P.K. and Arnold, F.H. (1992) *Macromolecules*, **25**, 7051–9.
77. Kempe, M., Glad, M. and Mosbach, K. (1995) *Journal of Molecular Recognition*, **8**, 35–9.
78. Hirayama, K., Sakai, Y. and Kameoka, K. (2001) *Journal of Applied Polymer Science*, **81**, 3378–87.
79. Sreenivisan, K. (2005) *Macromolecular Bioscience*, **5**, 187–91.
80. Wulff, G. and Sarhan, A. (1972) *Angewandte Chemie – International Edition in English*, **11**, 341.
81. Ye, L., Cormack, P.A.G. and Mosbach, K. (1999) *Analytical Communications*, **36**, 35–8.
82. Strikovsky, A., Hradil, J. and Wulff, G. (2003) *Reactive and Functional Polymers*, **54**, 49–61.
83. Flores, A., Cunliffe, D., Whitcombe, M.J. and Vulfson, E.N. (2000) *Journal of Applied Polymer Science*, **77**, 1841–50.
84. Strikovsky, A.G., Kasper, D., Grun, M., et al. (2000) *Journal of the American Chemical Society*, **122**, 6295–6.
85. Zhang, L.Y., Cheng, G.X. and Fu, C. (2003) *Reactive and Functional Polymers*, **56**, 167–73.
86. Mayes, A.G. and Mosbach, K. (1996) *Analytical Chemistry*, **68**, 3769–74.
87. Suedee, R., Srichana, T. and Martin, G.P. (2000) *Journal of Controlled Release*, **66**, 135–47.
88. Kempe, H. and Kempe, M. (2004) *Macromolecular Rapid Communications*, **25**, 315–20.
89. Kempe, H. and Kempe, M. (2006) *Analytical Chemistry*, **78**, 3659–66.
90. Wang, X., Ding, X., Zheng, Z., Hu, et al. (2006) *Macromolecular Rapid Communications*, **27**, 1180–4.
91. Hosoya, K., Yoshizako, K., Tanaka, N., et al. (1994) *Chemistry Letters*, **24**, 1437–8.
92. Haginaka, J., Takehira, H., Hosoya, K. and Tanaka, N. (1997) *Chemistry Letters*, **26**, 555–6.
93. Haginaka, J., Sakai, Y. and Narimatsu, S. (1998) *Analytical Sciences*, **14**, 823–6.
94. Haginaka, J. and Sanbe, H. (1998) *Chemistry Letters*, 1089–90.
95. Sanbe, H., Hosoya, K. and Haginaka, J. (2003) *Analytical Sciences*, **19**, 715–9.
96. Masci, G., Aulenta, F. and Crescenzi, V. (2002) *Journal of Applied Polymer Science*, **83**, 2660–8.
97. Zhang, L.Y., Cheng, G.X. and Fu, C. (2002) *Polymer International*, **51**, 687–92.
98. Pérez, N., Whitcombe, M.J. and Vulfson, E. (2000) *Journal of Applied Polymer Science*, **77**, 1851–9.
99. Pérez, N., Whitcombe, M.J. and Vulfson, E. (2001) *Macromolecules*, **34**, 830–6.

100 Norrlöw, O., Glad, M. and Mosbach, K. (1984) *Journal of Chromatography A*, **299**, 29–41.
101 Otsu, T., Yamashita, K. and Tsuda, K. (1986) *Macromolecules*, **19**, 287–90.
102 Sulitzky, C., Rückert, B., Hall, A.J., et al. (2002) *Macromolecules*, **35**, 79–91.
103 Rückert, B., Hall, A.J. and Sellergren, B. (2002) *Journal of Materials Chemistry*, **12**, 2275–80.
104 Sellergren, B., Rückert, B. and Hall, A.J. (2002) *Advanced Materials*, **14**, 1204–8.
105 Hattori, K., Hiwatari, M., Iiyama, C., et al. (2004) *Journal of Membrane Science*, **233**, 169–73.
106 Baggiani, C., Baravalle, P., Anfossi, L. and Tozzi, C. (2005) *Analytica Chimica Acta*, **542**, 125–34.
107 Tamayo, F.G., Titirici, M.M., Martin-Esteban, A. and Sellergren, B. (2005) *Analytica Chimica Acta*, **542**, 38–46.
108 Yilmaz, E., Ramström, O., Möller, P., et al. (2002) *Journal of Materials Chemistry*, **12**, 1577–81.
109 Titirici, M.M., Hall, A.J. and Sellergren, B. (2002) *Chemistry of Materials*, **14**, 21–3.
110 Guo, T.Y., Xia, Y.Q., Hao, G.J., et al. (2004) *Biomaterials*, **25**, 5905–12.
111 Guo, T.Y., Xia, Y.Q., Wang, J., et al. (2005) *Biomaterials*, **26**, 5737–45.
112 Jakusch, M., Janotta, M., Mizaikoff, B., et al. (1999) *Analytical Chemistry*, **71**, 4786–91.
113 Silvestri, D., Cristallini, C., Ciardelli, G., et al. (2004) *Journal of Biomaterials Science – Polymer Edition*, **15**, 255–78.
114 Wang, H.Y., Xia, S.L., Sun, H., et al. (2004) *Journal of Chromatography B*, **804**, 127–34.
115 Andersson, L., Ekberg, B. and Mosbach, K. (1985) *Tetrahedron Letters*, **26**, 3623–4.
116 Sellergren, B., Lepistö, M. and Mosbach, K. (1988) *Journal of the American Chemical Society*, **110**, 5853–60.
117 Sellergren, B. (1989) *Makromolekulare Chemie – Macromolecular Chemistry and Physics*, **190**, 2703–11.
118 Matsui, J., Doblhoff-Dier, O. and Takeuchi, T. (1997) *Analytica Chimica Acta*, **343**, 1–4.
119 Andersson, L.I. and Mosbach, K. (1989) *Makromolekulare Chemie – Rapid Communications*, **10**, 491–5.
120 Yu, C. and Mosbach, K. (1997) *Journal of Organic Chemistry*, **62**, 4057–64.
121 Yu, C., Ramström, O. and Mosbach, K. (1997) *Analytical Letters*, **30**, 2123–40.
122 Liu, F., Liu, X., Ng, S.C. and Chan, H.S.O. (2006) *Sensors and Actuators B*, **113**, 234–40.
123 Dunkin, I.R., Lenfeld, J. and Sherrington, D.C. (1993) *Polymer*, **34**, 77–84.
124 Spivak, D. and Shea, K. (1999) *Journal of Organic Chemistry*, **64**, 4627–34.
125 Ju, J.-Y., Shin, C.S., Whitcombe, M.J. and Vulfson, E.N. (1999) *Biotechnology and Bioengineering*, **64**, 232–9.
126 Joshi, V.P., Kulkarni, M.G. and Mashelkar, R.A. (1999) *Journal of Chromatography A*, **849**, 319–30.
127 Fischer, L., Müller, R., Ekberg, B. and Mosbach, K. (1991) *Journal of the American Chemical Society*, **113**, 9358–60.
128 Kempe, M., Fischer, L. and Mosbach, K. (1993) *Journal of Molecular Recognition*, **6**, 25–9.
129 Takagishi, T., Hayashi, A. and Kuroki, N. (1982) *Journal of Polymer Science Part A: Polymer Chemistry*, **20**, 1533–47.
130 Sarhan, A. and El-Zahab, M.A. (1987) *Makromolekulare Chemie – Rapid Communications*, **8**, 555–61.
131 Ramström, O., Andersson, L.I. and Mosbach, K. (1993) *Journal of Organic Chemistry*, **58**, 7562–4.
132 Andersson, L. (1988) *Reactive Polymers*, **9**, 29–41.
133 Andersson, L., Sellergren, B. and Mosbach, K. (1984) *Tetrahedron Letters*, **25**, 5211–4.
134 Ju, J.-Y., Shin, C.S., Whitcombe, M.J. and Vulfson, E.N. (1999) *Biotechnology Techniques*, **13**, 665–9.
135 Hall, A.J., Manesiotis, P., Emgenbroich, M., et al. (2005) *Journal of Organic Chemistry*, **70**, 1732–6.
136 Sibrian-Vazquez, M. and Spivak, D. (2003) *Macromoleucles*, **36**, 5105–13.
137 Sibrian-Vazquez, M. and Spivak, D.J. (2004) *Journal of Polymer Science, Part A: Polymer Chemistry*, **42**, 3668–75.
138 Sibrian-Vazquez, M. and Spivak, D. (2004) *Journal of the American Chemical Society*, **126**, 7827–33.

139 Kempe, M. (1996) *Analytical Chemistry*, **68**, 1948–53.
140 Kempe, M. and Mosbach, K. (1995) *Tetrahedron Letters*, **36**, 3563–6.
141 Wulff, G., Kemmerer, R., Vietmeier, J. and Poll, H.G. (1982) *Nouveau Journal de Chimie – New Journal of Chemistry*, **6**, 681–7.
142 Wulff, G. (1986) Polymer reagents and catalysts; ACS Symposium Series **308** (ed. W.T. Ford), American Chemical Society, Washington DC, pp. 186–230.
143 El-Toufaili, F.A., Visnjevski, A. and Brüggemann, O. (2004) *Journal of Chromatography B*, **804**, 135–9.
144 Takeuchi, T., Fukuma, D. and Matsui, J. (1999) *Analytical Chemistry*, **71**, 285–90.
145 Lanza, F. and Sellergren, B. (1999) *Analytical Chemistry*, **71**, 2092–6.
146 Lanza, F., Hall, A.J., Sellergren, B., et al. (2001) *Analytica Chimica Acta*, **435**, 91–206.
147 Chassaing, C., Stokes, J., Venn, R.F., et al. (2004) *Journal of Chromatography B*, **804**, 71–81.
148 Karlsson, J.G., Karlsson, B., Andersson, L.I. and Nicholls, I.A. (2004) *Analyst*, **129**, 456–62.
149 Farrington, K., Magner, E. and Regan, F. (2006) *Analytica Chimica Acta*, **566**, 60–8.
150 O'Mahony, J., Molinelli, A., Nolan, K., et al. (2006) *Biotechnology and Bioengineering*, **21**, 1383–92.
151 Andersson, H.S. and Nicholls, I.A. (1997) *Bioorganic Chemistry*, **25**, 203–211.
152 Piletsky, S.A., Karim, K., Piletska, E.V., et al. (2001) *Analyst*, **126**, 1826–30.
153 Subrahmanyam, S., Piletsky, S.A., Piletska, E.V., et al. (2001) *Biosensors and Bioelectronics*, **16**, 631–7.
154 Chianella, I., Lotierzo, M., Piletsky, S.A., et al. (2002) *Analytical Chemistry*, **74**, 1288–93.
155 Wu, L., Sun, B., Li, Y. and Chang, W. (2003) *Analyst*, **128**, 944–9.
156 Meng, Z., Yamazaki, T. and Sode, K. (2004) *Biosensors and Bioelectronics*, **20**, 1068–75.
157 Piletsky, S., Piletska, E., Karim, K., et al. (2004) *Analytica Chimica Acta*, **504**, 123–30.
158 Dong, W., Yan, M., Zhang, M., et al. (2005) *Analytica Chimica Acta*, **542**, 186–92.
159 Diñeiro, Y., Menéndez, I.M., Blanco-López, M.C., et al. (2005) *Analytical Chemistry*, **77**, 6741–6.
160 Pavel, D. and Lagowski, J. (2005) *Polymer*, **46**, 7528–42.
161 Pavel, D. and Lagowski, J. (2005) *Polymer*, **46**, 7543–56.
162 Chianella, I., Karim, K., Piletska, E.V., et al. (2006) *Analytica Chimica Acta*, **559**, 73–8.
163 Vicente, B., Navarro Villoslada, F. and Moreno-Bondi, M.C. (2004) *Analytical and Bioanalytical Chemistry*, **380**, 115–22.
164 Navarro Villoslada, F., Vicente, B. and Moreno-Bondi, M.C. (2004) *Analytica Chimica Acta*, **504**, 149–62.
165 Navarro Villoslada, F. and Takeuchi, T. (2005) *Bulletin of the Chemical Society of Japan*, **78**, 1354–61.
166 Rosengren, A.M., Karlsson, J.G., Andersson, P.O. and Nicholls, I.A. (2005) *Analytical Chemistry*, **77**, 5700–5.
167 Mijangos, I., Navarro Villoslada, F., Guerreiro, A., et al. (2006) *Biosensors and Bioelectronics*, **22**, 381–7.
168 Tóth, B., Pap, T., Horvath, V. and Horvai, G. (2006) *Journal of Chromatography A*, **1119**, 29–33.
169 Garcia-Calzon, J.A. and Diaz-Garcia, M.E. (2007) *Sensors and Actuators B*, **123**, 1180–94.
170 Umpleby, R.J.II, Baxter, S.C., Rampey, A.M., et al. (2004) *Journal of Chromatography B*, **804**, 141–9.
171 Kempe, M. and Mosbach, K. (1991) *Analytical Letters*, **24**, 1137–45.
172 Ye, L., Ramström, O., Ansell, R.J., et al. (1999) *Biotechnology and Bioengineering*, **64**, 650–5.
173 Szabelski, P., Kaczmarski, K., Cavazzini, A., et al. (2002) *Journal of Chromatography A*, **964**, 99–111.
174 Motulzki, H.J. (1999) *Analyzing Data with GraphPad Prism*, GraphPad Software, Inc., San Diego, CA, p. 307.
175 Wulff, G., Poll, H.G. and Minarik, M. (1986) *Journal of Liquid Chromatography*, **9**, 385–405.

176 Meyer, V.R. (1987) *Chromatographia*, **24**, 639–45.
177 Katz, A. and Davis, M.E. (2000) *Nature*, **403**, 286–9.
178 Kirchner, R., Seidel, J., Wolf, G. and Wulff, G. (2002) *Journal of Inclusion Phenomena and Macrocyclic Chemistry*, **43**, 279–83.
179 Chen, W.Y., Chen, C.S. and Lin, F.Y. (2001) *Journal of Chromatography A*, **923**, 1–6.
180 Weber, A., Dettling, M., Brunner, H. and Tovar, G.E. (2002) *Macromolecular Rapid Communications*. **23**, 824–8.
181 Zeng, Y.N., Zheng, N., Osborne, P.G., et al. (2002) *Journal of Molecular Recognition*, **15**, 204–8.
182 Kim, J.H., Cotton, T.M. and Uphaus, R.A. (1988) *Journal of Physical Chemistry*, **92**, 5575–8.
183 Shekunov, B.Y., Chattopadhyay, P., Tong, H.Y.H. and Chow, A.H.L. (2007) *Pharmaceutical Research*, **24**, 203–27.
184 Sellergren, B.A. (1994) *Practical Approach to Chiral Separations by Liquid Chromatography* (ed. G. Subramanian), Wiley-VCH Verlag GmbH, Weinheim, pp. 69–93.
185 Kempe, M. and Mosbach, K. (1995) *Journal of Chromatography A*, **694**, 3–13.
186 Kempe, M. (2000) *Molecular Imprints as Stationary Phases*, in *Encyclopedia of Separation Science* (eds I.D. Wilson, T.R. Adlard, C.F. Poole and M. Cook), Academic Press, London, pp. 2387–97.
187 Kempe, M. (2001) Techniques and instrumentation in analytical chemistry, in *Molecularly Imprinted Polymers. Man-made Mimics of Antibodies and Their Application in Analytical Chemistry* (ed. B. Sellergren), Elsevier Science, Amsterdam, Vol. **23**, pp. 395–415.
188 Sellergren, B. (2001) *Chiral Separation Techniques: A Practical Approach* (ed. G. Subramanian), Wiley-VCH Verlag, Weinheim, pp. 151–84.
189 Kempe, M., Fischer, L. and Mosbach, K. (1993) *Journal of Molecular Recognition*, **6**, 25–9.
190 Kempe, M. and Mosbach, K. (1994) *International Journal of Peptide and Protein Research*, **44**, 603–6.
191 Allender, C.J., Brain, K.R. and Heard, C.M. (1997) *Chirality*, **9**, 233–7.
192 Kempe, M. and Mosbach, K. (1995) *Journal of Chromatography A*, **691**, 317–23.
193 Ramström, O., Nicholls, I.A. and Mosbach, K. (1994) *Tetrahedron: Asymmetry*, **5**, 649–56.
194 Wulff, G. and Vesper, R. (1978) *Journal of Chromatography*, **167**, 171–86.
195 Wulff, G. and Minarik, M. (1990) *Journal of Liquid Chromatography*, **13**, 2987–3000.
196 Kempe, M. and Mosbach, K. (1994) *Journal of Chromatography A*, **664**, 276–9.
197 Olsen, J., Martin, P. and Wilson, I.D. (1998) *Analytical Communications*, **35**, H13–4.
198 Andersson, L.I. (2000) *Journal of Chromatography B*, **739**, 163–73.
199 Martín-Esteban, A. (2001) *Fresenius' Journal of Analytical Chemistry*, **370**, 795–802.
200 Theodoridis, G. and Manesiotis, P. (2002) *Journal of Chromatography A*, **948**, 163–9.
201 Theodoridis, G., Kantifes, A., Manesiotis, P., et al. (2003) *Journal of Chromatography A*, **987**, 103–9.
202 Caro, E., Marcé, R.M., Cormack, P.A.G., et al. (2004) *Journal of Chromatography B*, **813**, 137–43.
203 Caro, E., Marcé, R.M., Cormack, P.A.G., et al. (2005) *Analytica Chimica Acta*, **552**, 81–6.
204 Shi, Y., Zhang, J.H., Shi, D., et al. (2006) *Journal of Pharmaceutical and Biomedical Analysis*, **42**, 549–55.
205 Andersson, L.I. (2000) *Analyst*, **125**, 1515–7.
206 Möller, K., Nilsson, U. and Crescenzi, C. (2001) *Journal of Chromatography A*, **938**, 121–30.
207 Möller, K., Crescenzi, C. and Nilsson, U. (2004) *Analytical and Bioanalytical Chemistry*, **378**, 197–204.
208 Möller, K., Nilsson, U. and Crescenzi, C. (2004) *Journal of Chromatography B*, **811**, 171–6.
209 Caro, E., Marcé, R.M., Cormack, P.A.G., et al. (2004) *Journal of Chromatography A*, **1047**, 175–80.
210 Ansell, R.J., Ramström, O. and Mosbach, K. (1996) *Clinical Chemistry*, **42**, 1506–12.
211 Haupt, K. and Mosbach, K. (1998) *Trends in Biotechnology*, **16**, 468–75.

212 Vlatakis, G., Andersson, L.I., Müller, R. and Mosbach, K. (1993) *Nature*, **361**, 645–7.
213 Ansell, R.J. (2001) *Bioseparation*, **10**, 365–77.
214 Benito-Peña, E., Moreno-Bondi, M.-C., Aparicio, S., et al. (2006) *Analytical Chemistry*, **78**, 2019–27.
215 Al-Kindy, S., Badía, R., Suárez-Rodríguez, J.L. and Díaz-García, M.E. (2000) *CRC Critical Reviews in Analytical Chemistry*, **30**, 291–309.
216 Gao, S., Wang, W. and Wang, B. (2005) *Molecularly Imprinted Materials: Science and Technology* (eds M. Yan and O. Ramström), Marcel Dekker, New York, pp. 701–26.
217 Matsui, J., Kubo, H. and Takeuchi, T. (2000) *Analytical Chemistry*, **72**, 3286–90.
218 Suárez-Rodríguez, J.L. and Díaz-García, M.E. (2000) *Analytica Chimica Acta*, **405**, 67–76.
219 Piletsky, S.A., Piletskaya, E.V., El'skaya, A.V., et al. (1997) *Analytical Letters*, **30**, 445–55.
220 Suárez-Rodríguez, J.L. and Díaz-García, M.E. (2001) *Biosensors and Bioelectronics*, **16**, 955–61.
221 Rachkov, A., McNiven, S., El'Skaya, A., et al. (2000) *Analytica Chimica Acta*, **405**, 23–9.
222 Rathbone, D.L., Su, D.Q., Wang, Y.F. and Billington, D.C. (2000) *Tetrahedron Letters*, **41**, 123–6.
223 Lai, E.P.C., Fafara, A., VanderNoot, V.A., et al. (1998) *Canadian Journal of Chemistry – Revue Canadienne de Chimie*, **76**, 265–73.
224 Kugimiya, A. and Takeuchi, T. (2001) *Biosensors and Bioelectronics*, **16**, 1059–62.
225 Li, P., Huang, Y., Hu, J.Z., et al. (2002) *Sensors*, **2**, 35–40.
226 Marx, K.A. (2003) *Biomacromolecules*, **4**, 1099–120.
227 Dickert, F.L., Forth, P., Lieberzeit, P. and Tortschanoff, M. (1998) *Fresenius' Journal of Analytical Chemistry*, **360**, 759–62.
228 Cao, L., Li, S.F.Y. and Zhou, X.C. (2001) *Analyst*, **126**, 184–8.
229 Lin, C.-Y., Tai, D.-F. and Wu, T.-Z. (2003) *Chemistry – A European Journal*, **9**, 5107–10.
230 Haupt, K., Noworyta, K. and Kutner, W. (1999) *Analytical Communications*, **36**, 391–3.
231 Hayden, O., Bindeus, R., Haderspöck, C., et al. (2003) *Sensors and Actuators B*, **91**, 316–9.
232 Hedborg, E., Winquist, F., Lundstrom, I., et al. (1993) *Sensors and Actuators A*, **37**, 796–9.
233 Kriz, D. and Mosbach, K. (1995) *Analytica Chimica Acta*, **300**, 71–5.
234 Kriz, D., Kempe, M. and Mosbach, K. (1996) *Sensors and Actuators B*, **33**, 178–81.
235 Panasyuk, T.L., Mirsky, V.M., Piletsky, S.A. and Wolfbeis, O.S. (1999) *Analytical Chemistry*, **71**, 4609–13.
236 Piletsky, S.A., Piletskaya, E.V., Sergeyeva, T.A., et al. (1999) *Sensors and Actuators B*, **60**, 216–20.
237 Sergeyeva, T.A., Piletsky, S.A., Erovko, A.A., et al. (1999) *Analytica Chimica Acta*, **392**, 105–11.
238 Sergeyeva, T.A., Piletsky, S.A., Panasyuk, T.L., et al. (1999) *Analyst*, **124**, 331–4.
239 Chen, G., Guan, Z., Chen, C.T., et al. (1997) *Nature Biotechnology*, **15**, 354–7.
240 Cheng, Z., Wang, E. and Yang, X. (2001) *Biosensors and Bioelectronics*, **16**, 179–85.
241 Seong, H., Lee, H.B. and Park, K. (2002) *Journal of Biomaterials Science – Polymer Edition*, **13**, 637–49.
242 Yamazaki, T., Ohta, S., Yanai, Y. and Sode, K. (2003) *Analytical Letters*, **36**, 75–89.
243 Ramström, O. and Mosbach, K. (1999) *Current Opinion in Chemical Biology*, **3**, 759–64.
244 Severin, K. (2005) *Applications of Molecularly Imprinted Materials as Enzyme Mimics*, in *Molecularly Imprinted Materials: Science and Technology* (eds M. Yan and O. Ramström), Marcel Dekker, New York, pp. 619–40.
245 Robinson, D.K. and Mosbach, K. (1989) *Journal of the Chemical Society D – Chemical Communications*, 969–70.
246 Ohkubo, K., Funakoshi, Y. and Sagawa, T. (1996) *Polymer*, **37**, 3993–5.
247 Liu, X.C. and Mosbach, K. (1998) *Macromolecular Rapid Communications*, **19**, 671–4.

248 Svenson, J., Zheng, N. and Nicholls, I.A. (2004) *Journal of the American Chemical Society*, **126**, 8554–60.

249 Liu, X.C. and Mosbach, K. (1997) *Macromolecular Rapid Communications*, **18**, 609–15.

250 Kalim, R., Schomäcker, R., Yüce, S. and Brüggemann, O. (2005) *Polymer Bulletin*, **55**, 287–97.

251 Zhang, H., Piacham, T., Drew, M., *et al.* (2006) *Journal of the American Chemical Society*, **128**, 4178–9.

3
Nanoparticles Functionalized with Bioactive Molecules: Biomedical Applications

Ivonne Olmedo[1], Ariel R. Guerrero[1], Eyleen Araya[1,2] and Marcelo J. Kogan[1,3]

[1]Universidad de Chile, Departamento de Química Farmacológica y Toxicológica, Facultad de Ciencias Químicas y Farmacéuticas, Olivos 1007, Independencia, Santiago, Chile
[2]University of Barcelona, Department of Physical Chemistry, Marti i Franques 1-13 08028 Barcelona, Spain
[3]Centro para la Investigación Interdisciplinaria Avanzada en Ciencias de Materiales (CIMAT)

3.1
Introduction

Present-day nanomedicine exploits a broad variety of structured nanomaterials as nanoparticles (NPs), which are finding increasing use in various aspects of medicine, ranging from diagnostics, screening, sensing to therapy, drug delivery, imaging and therapy. Some of the biomedical applications that are being actively pursued are localized delivery using drug-loaded biodegradable polymers, magnetic resonance imaging (MRI) using iron oxide NPs, drug elution from nanoporous ceramic coatings and hyperthermia using magnetic NPs [1]. Many different types of NPs are currently being studied for applications in nanomedicine. They can be made from metals, such as gold NPs (GNPs), and other inorganic materials (Figure 3.1a), from carbon as carbon nanotubes (CNTs) [2] (Figure 3.1b) or they can be made from organic substances such as poly(ethylene glycol) (PEG) [3] (Figure 3.1c).

NPs used in therapy and diagnosis must be nontoxic, biocompatible and stable in biological media. Furthermore, they must be selectively addressed to the desired target. NPs can be coated with bioactive molecules, as for example peptides, to make them interact or bind to a biological target, thereby providing a controllable means of "tagging" or targeting it. To ensure that the NPs reach the desired target, it is important to anchor a vector that specifically recognizes the target.

There are a large number of inorganic NPs that can be potentially used in the biomedical field; among them, metallic NPs (MNPs) such as gold, cobalt, iron oxide, core–shell silica/gold and more recently silver have been studied. In addition, during the last few years, biological applications of carbon nanoparticles

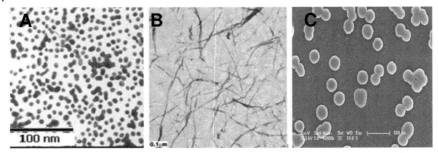

Figure 3.1 Electronic microscopy images of NPs. (a) Transmission electronic microscopy (TEM) micrograph of 5 nm in diameter GNP capped with cyclodextrines (Kogan *et al.*, unpublished results, 2007); (b) TEM images (on carbon-coated grids) of CNTs specifically single-wall CNTs (SWNT) ((Reprinted from Reference [3], Copyright Wiley-VHC Verlag GmbH. Reproduced with permission).); (c) scanning electron microscopy (SEM) of paclitaxel-loaded PLA/MPEG-PLA NPs of blend ratio 100/0. (Reprinted from Reference [3], Copyright Wiley-VCH Verlag GmbH. Reprinted with permission)

(CNPs) such as fullerenes and CNTs have been investigated. NPs are very interesting for biomedical applications, for their size scale is similar to that of biological molecules (e.g. proteins and genes) and structures (e.g. virus and cells), which allows them to arrive close to several biological entities of interest [4]. In this sense, MNPs can be coated with different biomolecules for surface stabilization and, at the same time, to obtain a method for introducing diverse functionality to the surface of NPs, thereby providing a link between biological target and nanostructures for molecular level recognition [5, 6]. On the other hand, organic NPs (ONPs), made of organic substances, also have biological applications involving functionalization, especially for pharmaceutical purposes, as quoted in Horn and Rieger's review [7]. Organic molecules have complex chemical structures formed with covalent bonds, having atoms different from the carbon backbone, which provide functional groups that are not present in bare NPs.

This chapter focuses on the synthesis, properties and functionalization of nanoparticles with bioactive compounds and the biomedical applications.

3.2
MNPs

3.2.1
Gold Nanoparticles

3.2.1.1 Synthesis and Properties
While the chemical properties of GNPs were first studied by Faraday, the main application of gold colloids in the biomedical field has been related to their electronic and optical properties [8, 9]. At present, there are many synthetic routes to

obtain GNPs and most of them start from Au(III) salts, which are reduced to metallic gold using different reduction reagent, such as sodium citrate and $NaBH_4$. The particles can also be modified with either polymers or linkers (e.g. peptides, oligonucleotides, antibodies and other bioactive molecules) through the bond formation with thiols, dithiols and amine groups [1, 10].

The assembly of nanomaterials into specific structures composed of GNPs has become one of the key aims of new materials and molecular devices. Some authors believe that the structural control of the NPs assembly by external conditions should pave the way toward novel nanoscale materials for use in a broad range of applications in materials engineering and science [11, 12]. Normally, a colloidal solution of GNPs, with diameter of 5–30 nm, exhibits a red color due to the optical absorption peak around 520 nm of wavelength; however, when GNPs are aggregated, the absorption peak shifts towards longer wavelengths and the solution turns purple [13]. Light scattering by MNPs is dominated by the collective oscillations of the conduction electrons induced by the incident electric field (light), which is known as the surface plasmon resonance. The specific color (i.e. frequency band) is a function of the size, shape and material properties of the particle [14].

3.2.1.2 Functionalization of GNPs with Bioactive Compounds and Biomedical Applications of Functionalized GNPs

Two main strategies to bind molecules to GNPs have been reported. In strategy A, a bioactive compound is conjugated to GNPs by means of the spontaneous reaction of a thiol (Figure 3.2, Strategy A) with the GNPs surface. Thiols are the most important type of stabilizing molecules for GNPs of any size. It is an accepted assumption that the use of thiols leads to the formation of strong Au–S bonds [15]. By contrast, in strategy B, functionalized GNPs are capped with a linker, which is then activated and functionalized with the biologically active molecule. The linker is a bifunctional molecule containing a thiol, which allows binding to the gold surface, and a functional group (e.g. carboxyl group) that is attached to the molecule.

3.2.1.2.1 Nucleic Acids
Based on the particular optical properties of GNPs, DNA-functionalized GNPs have been used in various forms for the detection of proteins, oligonucleotides, certain metal ions and other small molecules through colorimetric means. Many examples of biosensing using GNPs have been reported. Colorimetric sensors should be simple to design, easy to operate, give a fast color change and have a minimal consumption of materials. In this sense, Liu *et al.*

Figure 3.2 Two strategies to bind a bioactive molecule to GNPs.

designed two sensors, one for adenosine and the other for cocaine detection, based on the disassembly of GNPs aggregates linked by DNA aptamers. Aptamers are nucleic acid based binding molecules that can selectively bind with a broad range of molecules; in addition, DNA aptamers are easy to obtain, are stable to biodegradation and present low vulnerability to denaturation. The authors constructed a cocaine sensor, which was made of GNPs (13 nm) aggregates, containing three components: GNPs functionalized with 3′-thiol modified DNA or GNPs U5′-thiol-modified DNA (5′ Coap Au), a DNA linker molecule that contains two sequences that recognize the attached DNAs (3′ Coap and 5′ Coap) and also a cocaine aptamer molecule (Figure 3.3). The interaction between GNPs conjugated to DNA and DNA linker yielded purple aggregates of GNPs when suspended in buffer solution. These aggregates disassembled within 10s of the soft adding of cocaine to give a red colored solution, that is to say the color change was instantaneous in the presence of cocaine. The interaction between the cocaine with the aptamer produces a change in the aptamer conformation that leads to the separation of GNPs from the DNA linker molecule. The sensitivity of this detection method was not as high as that of the fluorescence-based detection method, but the sensor had a high selectivity [16].

Pavlov *et al.* developed a method to amplify the detection of thrombin in solution and on a surface using GNPs functionalized with an aptamer (the aptamer recognizes thrombin) as a catalytic label for the amplification [17]. Aptamer **1**, (5′-HS-$(CH_2)_9$-TTTTTTTTTTTTTTGGTTGGTGTGGTTGG-3′), was covalently

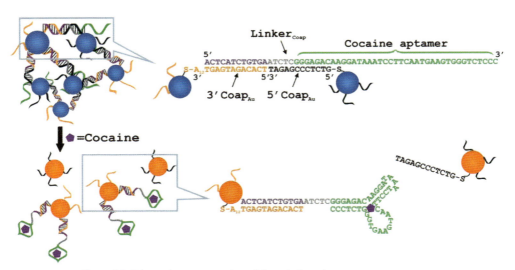

Figure 3.3 Schematic representation of the colorimetric detection of cocaine based on cocaine-induced disassembly of NP aggregates linked by a cocaine aptamer. (Reprinted from Reference [16], Copyright Wiley-VHC Verlag GmbH. Reproduced with permission.)

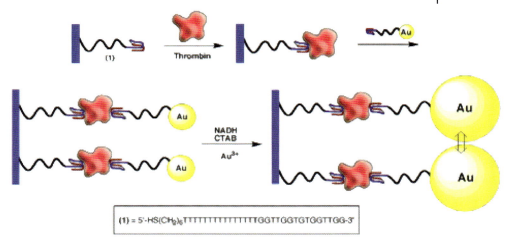

Figure 3.4 Amplified detection of thrombin on surfaces by the catalytic enlargement of thrombin aptamer-functionalized GNPs. (Reprinted with permission from Reference [17]. Copyright 2007 American Chemical Society.)

attached to a maleimide-functionalized siloxane monolayer and then thrombin was bound to the aptamer attached to the surface (Figure 3.4). In contrast, GNPs stabilized by citrate were functionalized with the thiolated aptamer **1**. The **1**-functionalized GNPs were associated to the second thrombin binding site (of the thrombin attached to the siloxane surface) and the resulting GNPs interface was enlarged in a growth solution that included $HAuCl_4$ and reducing reagents. At high concentrations of thrombin, the surface loading of bound thrombin increases, resulting in an increased number of GNPs seed for enlargement, which is reflected by the higher absorbance spectra. The high concentration of GNP–functionalized aptamer–thrombin aggregates allowed the GNPs to mutually interact. Solution phase analysis of thrombin led to the specific optical sensing of thrombin on glass surfaces, which was achieved with a sensitivity limit that corresponded to 2 nM [17].

Xu *et al.* have evaluated the use of DNA-GNP as colorimetric indicators of enzymatic activity [18]. Herein an operationally simple colorimetric endonuclease inhibition assay was realized. This assay enables the real-time monitoring of endonuclease activity and the simultaneous determination of the efficiencies of endonuclease inhibitors. Two separate batches of 13 nm GNPs were prepared and subsequently functionalized with two different thiol modified oligonucleotide strands (complementary to each other), DNA-1 (5′-CTCCCTAATAACAATTTA-TAACTATTCCTA-A10-SH-3′) and DNA-2 (5′-TAGGAATAGTTATAAATTGTT-TTAGGGAG-A10-SH-3′), which were denoted as DNA-GNP-1 and DNA-GNP-2, respectively. These conjugates, DNA-GNP-1 and DNA-GNP-2, can hybridize to form a cross-linked network of GNPs, which is purple in color owing to the redshifted plasmon band of the GNPs—a highly diagnostic feature of aggregate formation. As the endonuclease degrades the DNA-duplex interconnects, particles are released, regenerating a red color due to the dispersed nanoparticles. The color can be observed by UV/Visible spectroscopy at 520 nm. The DNA aggregates were used to evaluate the enzymatic activity of deoxyribonuclease (DNase I). In a typical experiment, DNase I at different concentrations was added to a solution of the aggregates. The solution gradually changed from purple to red. By measuring the absorbance at 520 nm, the nucleic acid hydrolysis catalyzed by DNase I could followed quantitatively [18].

DNA-GNP were used for the detection of mercury, which is a widespread pollutant in various forms (metallic, ionic and as part of organic and inorganic salts complexes) with distinct toxicological profiles. Solvated mercuric ion (Hg^{2+}) is a caustic and carcinogenic material with high cellular toxicity. One study in this field showed a highly selective and sensitive colorimetric detection method for Hg^{2+} that relies on thymidine-Hg^{2+}-thymidine coordination chemistry and complementary DNA-GNP with deliberately designed T-T mismatches. When two complementary DNA-GNPs are combined they form DNA linked aggregates that can dissociate reversibly, with the concomitant purple to red color change (Figure 3.5). For this novel colorimetric Hg^{2+} assay, the authors prepared two types of GNPs, each functionalized with different thiolated-DNA sequences (probe A: 5′HS-C10-A10-T-A103′ and probe B: 5′HS-C10-T10-T-T10 3′), which were complementary except for a single thymidine–thymidine mismatch. First, they added an aliquot of an aqueous solution of Hg^{2+} at a designated concentration to a solution of the DNA-GNP aggregated formed from probe A and probe B at room temperature;

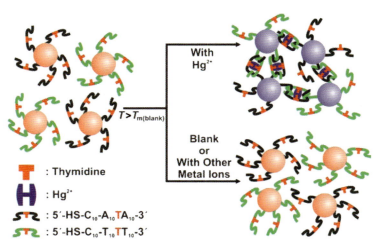

Figure 3.5 Schematic representation of the colorimetric mercury detection method for Hg2+ that relies on thymidine-Hg2+-thymidine coordination chemistry and complementary DNA-GNP with deliberately designed T-T mismatches. (Reprinted from Reference [19]. Copyright Wiley-VHC Verlag GmbH. Reproduced with permission.)

they then heated the solution while monitoring its extinction at 525 nm. Without Hg^{2+} (blank) the aggregates melt with a dramatic purple to red color change. However, in presence of Hg^{2+} the aggregates melt at a temperature greater than that of the blank, because of the strong coordination of Hg^{2+} to the two thymidines that make up the T-T mismatch, thereby stabilizing the double DNA strands containing the T-T single base mismatches present in GNPs' surface. They also evaluated the selectivity of the system for Hg^{2+} by testing the response of the assay to other metal ions. In consequence, a colorimetric method to detect Hg^{2+} with high selectivity and sensitivity has been developed. In addition, it is enzyme-free and does not require specialized equipment other than a temperature control unit. Another advantage is the high water solubility of the oligonucleotide-modified GNP probes, which allow this assay to be performed in aqueous media without the need of organic cosolvents [19] (Figure 3.5).

In the same way, and because of the high toxicity of Pb^{2+}, a sensor for simple Pb^{2+} detection and quantification is highly desirable. Liu et al. employed a Pb^{2+}-dependent DNAzyme to assemble GNPs and also demonstrated that the particles showed different assembly states controlled by the concentration of Pb^{2+} in the system. Many of the DNAzymes have shown metal-dependent activities and can, therefore, be used to design metal sensors. The DNAzyme is composed of a substrate strand and an enzyme strand, the substrate strand is a DNA/RNA chimer with a single RNA linkage that serves as the cleavage site; in the presence of Pb^{2+}, the substrate is cleaved into two pieces by the enzyme. To elaborate a colorimetric biosensor for Pb^{2+}-dependent DNAzyme detection, GNPs functionalized with 5′-thiol-modified 12-mer DNA (DNA_{Au}) were used as color reporting groups. DNA GNPs, DNA substrate and the enzyme mentioned before can assemble to form blue colored GNP aggregates through DNA base pairing interaction. Pb^{2+} detection was performed by heating the sensor to above 50 °C and subsequently cooling it slowly to room temperature in the presence of Pb^{2+}; the DNA substrate strand was cleaved by the enzyme in the cooling process. Because cleaved substrate can no longer assemble GNPs, a red color due to separated GNPs was observed. However, if no Pb^{2+} was present, the system (substrate strand, enzyme and functionalized GNPs) was reassembled and showed a blue color. The sensor worked over a wide pH range, with optimal performance near physiological pH. In addition, the system was highly reproducible [20].

Gene expression detection can provide powerful insights into the chemistry and physiology of biological systems; furthermore, an abnormality-expressed gene can be used as a new drug target or as a genetic marker for disease diagnosis. Baptista et al. reported a novel biosensor for the specific detection of gene expression through the use of GNP probes, that is to say, GNPs functionalized with thiol-DNA-oligonucleotides [21]. The method used noncross-linking aggregation of gold nanoprobes to detect gene expression directly from total messenger RNA (mRNA) extracted. The authors chose *Saccharomyces bayanus* yeast (a growth form of eukaryotic microorganisms classified in the kingdom Fungi), commonly used in the wine industry, as model organism and RNA source. *S. bayanus* posses a gene, FSY1, that encodes a specific protein. FSY1 expression is strongly regulated by the available carbon source in growth medium. FSY1 gene expression detection with gold nanoprobes was carried out with an experimental setup that consisted of several total RNA extracts from yeasts grown under two different conditions (expression or repression of gene) and GNPs functionalized with a complementary sequence to FSY1 mRNA. The colorimetric sensor consisted of visual and spectrophotometric comparison of the solution before and after salt (NaCl) induced gold nanoprobe aggregation (Figure 3.6): (i) blank, the gold nanoprobe alone; (ii) negative control (FSY1neg), in which FSY1 expression was repressed and (iii) positive result sample (FSY1pos) containing total mRNA from cells in which FSY1 was abundantly expressed. The results showed that, after NaCl addition, both blank and FSY1neg turned from the initial pink to a blue-purple color, denoting gold nanoprobe aggregation. However, for FSY1pos, the solution maintained the original pink color after aggregation salt addition, even after standing overnight at room temperature. This increased stability was explained by the fact that all the gold nanoprobes were hybridized with the complementary target mRNA. This method detected FSY1 mRNA expression within 15 minutes by no color change

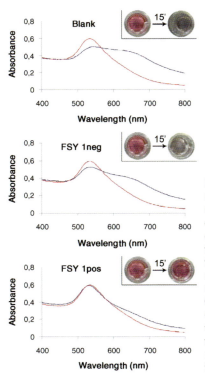

Figure 3.6 FSY1 mRNA colorimetric detection method. The inset in each panel represents sample color change after 15 min of adding a concentrated NaCl solution for a final concentration of 2 M. UV/Visible spectra were taken before (red trace) and 15 min after (blue trace) addition of the NaCl solution. "Blank" refers to nanoprobe alone; "FSY1pos" refers to sample containing total RNA from cells grown in 0.5% fructose, in which FSY1mRNAexpression occurs; "FSY1neg" is a negative control containing total RNA from cells grown in 0.5% glucose, in which FSY1 mRNA expression is repressed. (Reprinted from Reference [21], with permission from Elsevier.)

of a solution containing total RNA extracted from yeasts and a FSY1 specific gold nanoprobe compared with other two samples, where a color change from pink to purple was observed. In addition, the inexpensive experimental setup, short development time (without the need of signal amplification or temperature control) and that the color change could be assessed visually make it an extremely simple and fast method for direct detection of the gene expression [21].

3.2.1.2.2 **Peptides and Proteins** GNPs were functionalized with a Tat protein - derived peptide sequence to translocate them into the cell nucleus of hTERT-BJ1 human fibroblasts for cellular imaging GNPs. First, the GNPs were functionalized with the amino acid tiopronin, which acts as a linker, and the Tat-peptide sequence was linked to the tiopronin (Figure 3.7).

Transmission electron microscopy (TEM) images showed that GNPs conjugated with Tat peptide (nuclear localization sequence) were mainly located in the nucleus of fibroblast and also revealed no appreciable cytotoxic effects [22]. In contrast, Tkachenko *et al.* functionalized GNPs with four different nuclear localization peptides ending with a Cys residue (which contains a free thiol) to evaluate the cellular trajectories of peptide gold complexes. Functionalized GNPs were

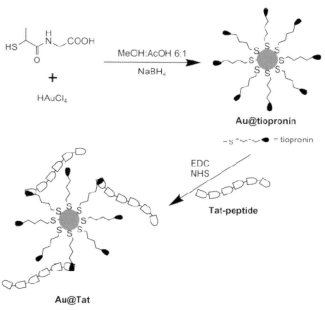

Figure 3.7 Functionalization of GNPs with Tat-peptide. (Reprinted with permission from Reference [22]. Copyright 2007 American Chemical Society.)

modified with four peptides, each one with a different nuclear localization signal. Cell and nuclear targeting in HeLa cells by GNPs conjugated was demonstrated by video-enhanced color differential interference contrast microscopy and TEM images [23].

Liu et al. determined the optimum conditions for assembling stable peptide-poly(ethylene glycol) PEG/GNP conjugates and formulated an empirical understanding of how PEG length and peptide coverage influence cellular internalization in HeLa cells. Two peptides derived from the adenovirus fiber protein were studied: NLS (CGGFSTSLRARKA) a nuclear localization sequence and RME (CKKKKKKSEDEYPYVPN), which can induce receptor-mediated endocytosis. In addition, three different thiol-modified PEGs with different average molecular weight and four particle diameters were explored. The systematic studies reported revealed important differences in the stabilities of GNPs modified with PEG and mixed PEG/peptide monolayer; longer PEGs, smaller NPs and large PEG:GNP ratios favor more stable solutions. An analysis of cell uptake showed that the number of PEG 5000/RME conjugates inside the HeLa cells after 6 hours was greater than with the other conjugates, probably because PEG 5000/RME conjugates appear to contain a net positive charge, which may provide a favorable interaction with the cell membrane [24].

A simple strategy for specific recognition of different motifs in a bioanalytical assay is the introduction of specific recognition group on the GNP surface. Lévy et al. made NPs functionalized with biotin as well as particles modified a peptide analog of biotin, Strep-tag II: WSHPQFEK. They designed a linker peptide with strong affinity for gold, with the ability to self-assemble into a dense layer that excludes water, and remain soluble and stable. This pentapeptide was called CALNN (Cys-Ala-Leu-Asn-Asn). GNPs were conjugated using a peptide mixture containing mainly CALNN and, in lesser proportion, CALNNGKbiotinG or CALNNGG-strep-tag. To detect the presence of biotin and of its peptide analog they used a streptactin peroxidase; a recombinant protein containing Strep-tag was used as a positive control and CALNN-GNP conjugate as a control for non-specific interactions. The results indicated that GNPs functionalized with biotin and the peptide analog of biotin specifically interacted with streptactin peroxidase [25].

In the field of neurodegenerative diseases, which are characterized by spontaneous self-assembly of proteins into an insoluble fibrous deposit, Kogan et al. have described an NPs-mediated heating to dissolve amyloid deposits of $A\beta_{1-42}$, a small protein involved in Alzheimer's disease (AD) pathogenesis. These deposits were remotely and locally dissolved through the combined use of weak microwave fields and gold nanoparticles. GNPs were linked to the peptide H-Cys-Leu-Pro-Phe-Phe-Asp-NH$_2$ forming the conjugate GNP-Cys-PEP, which contains the sequence H-Leu-Pro-Phe-Phe-Asp-NH$_2$ (PEP) that can selectively recognize the Aβ aggregates. The conjugate was incubated with a solution of $A\beta_{1-42}$, where fibrils spontaneously started growing and forming precipitates; at different times and then at different stages of growing, weak microwave field (0.1 W) were applied. The GNP-Cys-PEP conjugate was able to bind to the fibrils, absorbed the radiation and dissipated energy, causing desegregation of the amyloid deposits and aggregates, leading to the formation of smaller species such as dimers, trimers and amorphous aggregates. These findings were supported by TEM images of the aggregates before and after the irradiation, thioflavin T assay and size-exclusion chromatography (SEC) analysis, which showed an increase in low molecular weight species as monomers and oligomers [26].

Huang *et al.* developed a highly specific sensing system for platelet-derived growth factors (PDGFs) that uses GNPs. PDGF is a growth factor protein found in human platelets. There are three known isoforms and all of them bind specifically and with different degrees to two receptors, namely the PDGF α- and β-receptors. Binding of the receptors to PDGF activates intracellular tyrosine kinase signals and this process is directly related to cell proliferation. Two thiol-modified aptamer (Apt) 35-mer-DNA oligonucleotides were attached to GNPs and then the conjugates (Apt-GNP) were equilibrated with bovine serum albumin (BSA) to minimize any nonspecific adsorption of PDGFs. The system was stable in solution containing up to 3.0 M NaCl. Aliquots containing different concentrations of PDGFs in the presence of Apt-GNP were maintained at room temperature for 2 hours. After this time, the Apt-GNP solutions changed from red to purple in the presence of PDGFs, because the PDGF molecules acted as bridges that linked Apt-GNP together (interparticle cross-linking), which caused aggregation of Apt-GNP to a greater extent. The results presented imply that there are practical applications of Apt-GNP in cancer diagnosis and protein analysis [27].

3.2.2
Nanoshells and Metal Heterodimers

Gold nanoshells, which are spherical NPs that consist of a dielectric core (silica) surrounded by a thin metal shell (gold or silver), possess a tunable plasmon resonant response that gives intense optical absorption and scattering. The plasmon resonance frequency of the NPs can be varied by modifying the relative dimensions of core and the shell layers. Hirsch *et al.* used core–shells to develop a rapid, specific and selective whole blood assay, with the capacity to detect various selected analytes. To fabricate the gold/silica nanoshells, silica NPs with surface amine groups were synthesized and then decorated with very small gold colloid via adsorption to the amine groups. Briefly, gold/silica nanoshells were grown by reacting $HAuCl_4$ with the silica colloid particles. Gold adsorbed on the surface of silica particles acted as nucleation sites, forming a complete metal shell. After that, gold nanoshells of 96 nm-diameter core with a 22 nm gold shell and resonant at 720 nm were fabricated. The presence of analytes was detected optically in a solution of nanoshells–conjugated antibodies by monitoring the decrease in single nanoshells plasmon resonance extinction. To obtain these conjugates, the antibodies were tethered to the surface of nanoshells using thiolated PEG linkers (PEG amine with 2-aminothiolane). A stable solution of disperse antibody–conjugated nanoshells possessed a strong extinction peak at 720 nm. Upon addition of analyte, the 720 nm peak began to diminish as a result of spectral redshifting from redistributed particle–particle aggregate resonances. As consequence of this phenomenon and by measuring the reduced extinction resonance of the particles, the investigators were able to efficiently identify and quantify an immunoglobulin (IgG), in saline, serum and whole blood using nanoshells conjugated to antibodies via PEG linker [28].

Recently, researchers have begun to explore heterostructured NPs by integrating multiple NP components into a single nanosystem. Choi *et al.* have developed heterodimer NPs of FePt-Au with multifunctionalities [29]. To achieve uniform FePt-Au heterodimer synthesis, first FePt of 6 nm were synthesized through a chemical reaction between Fe^0 and Pt^{2+}. Subsequently, these NPs were allowed to react with $AuCl(PPh_3)$; the Pt-containing NP seeds served as a catalytic surface that facilitated the successive growth of gold and this way the heterodimers of FePt-Au were obtained (Figure 3.8). For biomedical applications, water solubility and biocompatibility are important requisites. Consequently, the authors adopted two different types of thiolated-terminated PEG ligands with one end composed of dihydrolipoic acid (DHLA) with the other remaining as -OH (DHLA-PEG-OH) or modified to an $-NH_2$ group (DHLA-PEG-NH_2). Both ligands were attached to hydrophobic heterodimer NPs; the ligand became strongly bound to the GNP surface via chelating bonds of thiol groups. The hydrophilicity and biocompatibility of heterodimer NPs can be attributed to the PEG region of the ligands. The amino groups of DHLA-PEG-NH_2 ligands were used for further bioconjugation with neutravidin and HmenB1 antibodies, which specifically recognize polysialic acids (PSAs) in tumor cells, for *in vitro* molecular magnetic resonance imaging (MRI). They first examined the feasibility of heterodimer NPs as probes for avidin–biotin interaction in chip based detection. The results showed a specific biorecognition reaction between neutravidin-coated-FePt-Au NPs and biotin on pre

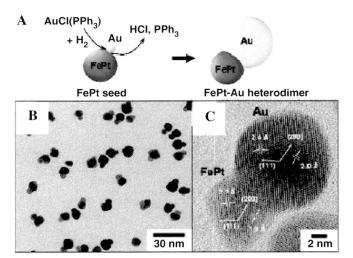

Figure 3.8 (a) Schematic of FePt-Au heterodimer formation; (b) TEM image; and (c) HR-TEM image of FePt-Au heterodimer NPs. (Reprinted with permission from Reference [29]. Copyright 2007 American Chemical Society.)

patterned chip. In contrast, the HmenB1 antibody-FePt-Au NPs conjugates were tested in two different lines with overexpression of PSAs and control cells without PSAs. In this case, the results showed a MRI contrast of polysialic acids' overexpressed on neuroblastoma cells by FePt-Au-HMenB1 antibody conjugates, while no MRI contrast was observed from PSAs negative control cells. This protocol produced excellent multicomponent hybrid NPs in terms of size, shape and crystallinity. These particles were also efficiently conjugated to two different linkers for better solubility and biocompatibility [29].

3.2.3
Iron Oxide NPs

3.2.3.1 Synthesis and Properties

The application of iron oxide NPs (IONPs) for *in vitro* diagnosis has been practiced for nearly 40 years, principally using maghemita and magnetite iron oxides (Fe_2O_3 and Fe_3O_4, respectively) with single domains of about 5–20 nm diameter. These IONPs can form homogeneous suspensions called ferrofluids, which can interact with an external magnetic field and be positioned to a specific area inside the body. A superparamagnetic behavior of the particles, at room temperature, is highly desired because they do not retain any magnetism after removal of a magnetic field. This property facilitates their biomedical application in MRI, cellular therapy, tissue repair, drug delivery and hyperthermia [30, 31].

Many synthetic processes and techniques to obtain IONPs have been developed, which are classified in physical and chemical methods. Initially, methods such as gas phase deposition and electron beam lithography were used. However, their main drawback is the inability to control the nanometric size of particles. In contrast, chemical routes are simpler, more efficient and tractable, with appreciable control over the size and, sometimes, the shape of particles. Most chemical syntheses are based on the coprecipitation of Fe^{2+} and Fe^{3+} aqueous salt solutions by addition of a base [32, 33].

3.2.3.2 Functionalization of IONPs

3.2.3.2.1 Peptides or Proteins
For biomedical applications, magnetic IONPs must be coated with biocompatible molecules during or after the synthesis process to prevent aggregation, a change in structure or biodegradation when exposed to the biological environment. Surface modification of superparamagnetic contrast agents with HIV-1 Tat peptide has emerged as an effective technique for intracellular magnetic labeling because the conjugation of the Tat peptide to the IONPs facilitates their cellular uptake. In addition, cells labeled with these conjugates can be readily detected by MRI [34]. Recently, ultrasmall superparamagnetic iron oxide NPs (USPIO) have been conjugated to HIV. Tat peptide to label CD4+ T cells for MRI, preserving proliferative, regulatory and migratory behavior of these cells *in vitro*. USPIO were prepared for conjugation of the Tat peptide by functionalization of the dextran on the magnetic NPs' surface. The uptake and loading of the

USPIO NPs in CD4+ T cells was examined using inductively coupled plasma optical emission spectrometry (ICP-OES). This technique allows measuring the iron content of treated cells. A concentration-dependent increase in cell labeling was observed when Tat-derivatized nanoparticles were employed in a 5-min reaction, with no uptake of iron when unconjugated USPIO were used. The imaging potential for MRI of this contrast agent was also measured; Tat-USPIO provided effective contrast enhancement *in vitro* while the unlabelled cells did not yield any contrast and in addition cell viability was reduced at high concentration of Tat-USPIO nanoparticles; above 50% cell death was visible by light microscopy 1 hour after incubation with the highest concentration (10 μL Fe/10^6) [35].

Other study in the same area investigated the feasibility of using MRI to monitor T cells *in vivo* after loading T cells with monocrystalline superparamagnetic iron oxide nanoparticles (SPIONs) modified with a peptide sequence from the Tat protein of HIV-1 in mice. The Tat peptide sequence was conjugated to a small iron oxide (5 nm) core coated with cross-linked (CL) aminated dextran, yielding CLIO-Tat nanoparticles. Tat peptide was modified to carry a fluorescein isothiocyanate (FITC) tag to follow the peptide during conjugation and as a marker for fluorescence microscopy. This study showed that the presence of FITC-conjugated CLIO-Tat in T cells was dose-dependent, that is, 97% of cells were FITC positive after loading with 8000 ng mL^{-1} of FITC-CLIO-Tat nanoparticles. Confocal microscopy images of labeled cells demonstrated that there was a strong FITC staining in most loaded T cells (above 95%); FITC label was distributed in clusters throughout the cytoplasm and nuclei of the T cells, moreover there was no alteration in their cytoplasmic or nuclear morphology, as shown by confocal microscopy. Finally, to measure the effects of T cells labeled with Tat-peptide derived NPs on the spleen by MRI, a group of six mice were injected intravenously with a suspension of T cells loaded with 8000 ng mL^{-1} of FITC-CLIO Tat. Changes in image intensity caused by the agent were measured by MRI, proving that these particles can be used to analyze T cell distribution events *in vivo* [36].

Imaging of soft tissues structures of the musculoskeletal system has become the domain of MRI due to its superiority over other imaging techniques. In the MRI field, IONPs have been broadly used as contrast agents, because the inclusion of small particles within certain tissues can increase the signal intensities obtained by this technique. A study in this area carried out by Shie *et al.* showed the preparation and application of amine surface-modified IONPs as an MRI contrast agent [37]. First, they prepared water-soluble, dispersed Fe_3O_4 NPs, which subsequently were functionalized with NH_3^+ groups. To demonstrate the utility of these functionalized IONPs in MRI, the authors evaluated the *in vivo* contrast effect of the magnetite NPs in a rat model using a 1.5 T clinical MRI imager. IONPs were administered to the rat through the tail vein and the images were taken immediately and 4 hours later, followed by every 24 hours until the images were complete. The signal intensity of the tissue in each test was determined by standard region-of-interest measurement of a cross-section of the tissue using the provided image quantification software. The images showed that the new synthesized IONPs may be useful as a hepatic MRI contrast agent with satisfactory fast response and prolonged contrast effect useful for sustained monitoring of disease progression without repetitive dosing. In addition, hemolysis analysis in human whole blood and *in vitro* cytotoxicity using 3-[4,5-dimethylthiazol-2-yl]-2,5-diphenyltetrazolium bromide (MTT) assay were also studied. In hemolysis analysis different aliquots of IONPs were added to a volume of human whole blood, while in the cell viability assay Cos-7 monkey kidney cells were cultivated in plates. The results of MTT assay revealed no effect in cell viability at the concentration studied (0.128–8 nM) and the analysis performed in human whole blood revealed that only a mild hemolysis could be detected in the highest iron concentration (0.1 M), a concentration higher than that required for the MRI contrast enhancement [37].

A cellular magnetic-linked immunosorbent assay (CMALISA) has been developed as an application of MRI for *in vitro* diagnosis. To validate this method, three contrast agents with affinity for the integrin family of adhesion molecules were synthesized by grafting to USPIO one of the following specific ligands: (i) the CS1 fragment of fibronectin (USPIO-g-CS1); (ii) the peptide GRGD (USPIO-g-GRGD); and (iii) a nonpeptide small molecule with RGD mimetic (USPIO-g-mimRGD). The integrins are a ubiquitously expressed class of cell surface receptor involved in the cell–cell and cell–matrix interaction and represent an interesting therapeutic target because they have an important role in diverse pathologies, such as atherosclerosis, acute renal failure and cancer. The tripeptide sequence RGD is a common cell recognition motif, which is part of integrin binding ligands; the CS1 fragment contains the motif LDV, which has high affinity for integrins. In this way, the affinity of these contrast agents was tested by CMALISA on Jurkat cells and on rat mononuclear cells (MNC). The cells were fixed on a normal ELISA (enzyme-linked immunosorbent assay) plate and analyzed by MRI after incubation with the three contrast agents. The concentration of bound contrast agent was estimated and their apparent constants of affinity for integrins were also evaluated. All the results realized by MRI confirmed a specific interaction of integrin-targeted contrast agents with their receptors as compared to the control. CMALISA was proposed to convert the MRI data into concentrations, and offers a second application for the contrast agents. This new magnetic based immunoassay method can be performed in a high-throughput setting and it allows the accurate detection and quantification of the cells receptors. The procedure is also fast and permits the automatic running of the screening protocol [38].

Tissues and cell-specific targeting can be achieved by employing NPs coated with a ligand recognized by a receptor on the target cell. A study realized by Gupta *et al.* evaluated *in vitro* cell response caused by IONPs. In that work SPIONs of specific shape and size with tailored surface chemistry (coated with two different proteins) have been prepared and characterized. The influence of IONPs and their conjugates on human dermal fibroblast *in vitro* have also been assessed. The magnetic NPs were synthesized by coprecipitation of ferrous and ferric salt solution on a base and stabilized by oleic acid. Lactoferrin and ceruloplasmin were coupled covalently at the NPs surface by using carbodiimide methods. The effect of IONPs and their conjugates on cell adhesion was determined with fibroblast cell suspension incubated with or without NPs, while to determine cell cytotoxicity/viability an MTT assay was applied on cells incubated with different IONPs concentrations. At the same time, fibroblast cell morphology using scanning electron microscopy (SEM) and cytoskeleton organization using immunofluorescence were also evaluated. The results showed that lactoferrin or ceruloplasmin-derivatized superparamagnetic IONPs were targeted at the surface of the fibroblast cells with little effect on the cytoskeleton morphology, that is to say, that each NPs type (different surface characteristics) caused a distinctly different cell response. From cell culture studies it was observed that the underivatized IONPs reduced cell adhesion and viability. One possible explanation for this result is that these NPs were taken up by the cells as a result of endocytosis, causing cell death, possibly by some intracellular mechanism. In contrast, IONPs derivatized with lactoferrin or ceruloplasmin showed a low degree of toxicity in the MTT assay, which was attributed to the fact that these ligands (ceruloplasmin and lactoferrin) were targeted at the surface receptors without being internalized. Finally, SEM studies showed that lactoferrin and ceruloplasmin derivatized IONPs were highly adhesive to the cell surface receptors. These experiments suggested that the cell response can be directed via a specifically engineered particle surface [39].

A method to detect protein amyloid plaques (Aβ) in the brains of transgenic mice overexpressing amyloid precursor protein (APP) or both mutant APP and presenilin-1 (APP/PS1) by magnetic resonance micro-imaging has also been reported. This method uses the Aβ_{1-40} peptide magnetically labeled with monocrystalline iron oxide nanoparticles (MIONs). Dextran-coated MION particles were linked to the Aβ_{1-40} peptide to assess their ability to bind Aβ_{1-42} peptide amyloid aggregates. The authors selected this peptide because it is the major constituent of Alzheimer's disease plaque amyloid. In an *in vivo* assay, the magnetically labeled peptides were co-injected with mannitol, an agent used to increase blood–brain barrier (BBB) permeability. This preparation was injected directly into the common carotid artery of AD-transgenic mice and nontransgenic controls. The results showed that systemic injection of Aβ_{1-40} peptide adsorbed onto MIONs with mannitol can be used to detect Aβ plaques by µMRI in the brain of AD transgenic mice. They also proved that the Aβ_{1-40} peptide, in this case, is essential for the targeting to Aβ plaques, because plaques were detected when MION particles were injected alone in transgenic mice [40].

Zhao *et al.* have prepared novel polymer (O-CMC) coated magnetic NPs (O-MNPs) conjugating Tat as drug/gene carrier [41]. They chose OCMC (MW: 40 000) polymer as the coating agent because it is biocompatible, biodegradable, nontoxic, water-soluble and it also has some unique antitumor and antibacterial bioactivities. In addition, its bifunctional groups – carboxyl and amino groups – can be readily covalently coupled with diverse bioactive molecules. The HIV-Tat peptide is an 86-amino acid polypeptide essential for viral replication. It has been shown to freely travel through cellular and nucleic membranes, and its membrane translocational property is dominated by a short signal GRKKRRQRRR (Tat). After O-MNPs synthesis, the particles were derivatized with a peptide sequence from the HIV-tat protein obtaining O-MNP-tat conjugate. To evaluate the O-MNP-tat as a drug carrier, MTX (antitumor drug) was incorporated as model drug and MTX-loaded-O-MNP-tat were prepared. Antitumor tests were performed in U-937 human lymphoma cell line and the cytotoxicity was measured with colorimetric assay based on the use of MTT tetrazolium salt. The results confirmed that the MTX loaded O-MNP-tat showed a better antitumor effect than MTX-O-MNP without tat sequence, because the membrane translational property of tat produced high levels of cell internalization. For that reason, the authors hypothesized that the combination of magnetic target characteristics with the translocational property could be beneficial in cancer therapy [41].

3.2.3.2.2 Polymers At present, cancer diseases are one of the most important causes of death, and combined chemotherapy is now used for most cancer treatments. The introduction of new targeting delivery for cancer treatment has allowed Kumar *et al.* to develop hyaluronic acid-Fe_2O_3 (maghemite) hybrid magnetic NPs for targeted delivery peptides [42]. Hyaluronic acid (HA) plays an important role in biological organisms and has different favorable characteristics, such as lower immunogenicity; moreover, it is found in all mammalian species, which is very important in designing a general and biocompatible delivery carrier to be used in the human body. In addition, various tumors (e.g. epithelial, stomach, colon and acute leukemia) overexpress HA-binding receptors. Maghemite NPs were produced by a typical coprecipitation method in aqueous medium without any surfactant. The NPs had an average diameter of <40 nm and narrow size distributions. To obtain the hybrid HA-Fe_2O_3 NPs, first a set of HA-NPs was prepared by self-assembling microemulsion techniques; subsequently, these HA-NPs were used for hybridization with maghemita NPs to produce the hybrid conjugate. Atrial natriuretic peptide (ANP) labeled with FITC encapsulated within hybrid NPs was employed for peptide transfer studies, using alveolar type II epithelial cells and HEK293 as culture system. The cells were examined under a fluorescence microscope 24, 48 and 72 hours after incubation with labeled NPs. The results demonstrated that hybrid HA-Fe_2O_3 NPs were capable of delivering the peptide in almost all (100%) cells. Interestingly, the results also showed that the particles delivered the peptide to the nucleus of the cells. Normally, alveolar type II epithelial and HEK293 cells do not uptake ANP by itself, providing evidence that, when encapsulated in its hybrid NPs, the peptide is taken up by the same cells mentioned before. The foremost significance of these findings is that the hybrid NPs could be utilized to deliver drugs to specific cells. In this sense, is possible to obtain a hyaluronic–maghemita–antitumor drug conjugates for cancer treatment. Applying an external localized magnetic field gradient it would be possible to attract drug-loaded magnetic HA-Fe_2O_3 hybrid particles from the blood circulation to target the NPs to the particular tumor site, so that the entire drug can be delivered to the desired tissue and to the specific cell effectively [42].

3.2.4
Silver NPs

Silver NPs have attracted considerable interest because of their potential applications in biological sciences. Many methods such as chemical reduction, radiolytic

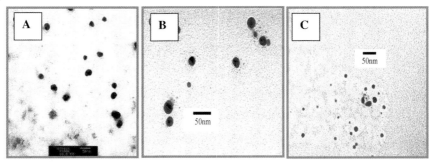

Figure 3.9 TEM image and corresponding particle analysis of PGA-stabilized nanosized silver colloid: (a) 0.5 wt% PGA; (b) 1 wt% PGA; and (c) 2 wt% PGA-stabilized silver colloid. (Reprinted from Reference [44], with permission from Elsevier.)

process and polyol methods have been developed for the synthesis of silver NPs [43]. In this sense, Yu presented a synthesis of silver NPs using a macromolecular and polyanionic Na$^+$-poly (γ- glutamic acid) (PGA) silver nitrate complex (Figure 3.9) [44]. Their antibacterial activity against *Pseudomonas aeruginosa* was also evaluated. AgNO$_3$ was chosen as precursor and the PGA and dextrose served as stabilizers and reducing agent, respectively, in synthesis of silver NPs. The antibacterial activity tests were assessed by the proliferation of Gram positive (MRSA) and Gram negative *(P. aeruginosa)* bacteria. The silver NPs were small, within 15–60 nm, with a uniform shape and very narrow size distribution; PGA limited further aggregation of silver NPs and stabilized the dispersed NPs in solution. The results of antibacterial activity against the Gram positive and Gram negative bacterium showed that there was no bacteria residue on the PGA silver NPs specimens after 24 hours cultivation. These results demonstrated that this new method of fabricating polyanionic PGA-stabilized nanosized silver NPs is promising and has potential for biomedical application [44].

3.2.5
Quantum Dots

Quantum dots (QDs) are nanometer-scale semiconductors crystals and are defined as particles with physical dimensions smaller than the exciton Bohr radius. Generally, they are composed of groups II–VI or III–V elements of the periodic table. QDs have a long fluorescent lifetime after excitation, which may be taken advantage of in time-gated imaging. They can be synthesized with diameters from a few nanometers to a few micrometers. The choices of shell and coating are particularly importance, as the shell stabilizes the nanocrystal, while the coating confers properties to the QDs that allow their incorporation into a desired application [45].

Actually, QDs have emerged as a novel and promising class of fluorophore for cellular imaging. Zhu *et al.* have developed an immunofluorescent detection system for *Cryptosporidium parvum* and *Giarda lamblia* pathogens using QDs [46]. Herein, QDs were successfully demonstrated to be an excellent fluorophore in the detection of microbial cells. The presence of these pathogens in various water sources is commonly determined by using immunofluorescent antibody techniques. However, different water samples sometimes contain inert particles or algal cells with strong autofluorescence, which can impede the specific detection of *C. parvum* and *G. lamblia*. Consequently, two strategies have been used to label *C. parvum* and *G. lamblia* cells with QD antibody conjugates. The results revealed that QD labeling exhibited better photostability and higher brightness than the two most commonly used commercial staining kits (Figure 3.10). This novel detection system could provide quantitative measurement with great sensitivity and photostability and, potentially, could revolutionize microbial detection in environmental microbiology studies [46].

Gerion *et al.* reported the case where four different sequences of DNA were linked to four nanocrystal samples having different colors of emission in the range 530–640 nm. Semiconductor nanocrystals linked to oligonucleotides have been studied before [47]; these biocompounds retain the properties of both nanocrystals and DNA. In the second case, DNA attached to QDs preserve the ability to hybridize to their complements. The authors focused their work on the activity of such DNA–nanocrystal conjugates and the ability to sort them using the attached DNA. To perform selective hybridization studies, or sorting experiments, the authors used a micrometer-size pattern grown on top of a silicon wafer (gold squares were grown on top of a silicon wafer using lithography techniques) derivatized with a thiolated oligonucleotides and four DNA–nanocrystal conjugates. Each emission color (green, orange, yellow, red) of nanocrystal conjugates was prepared with different a sequence of DNA following a literature procedure [48]. The results revealed that fluorescence from the gold pattern shows up when the oligonucleotides on a nanocrystal surface were complementary to the ones on the metal surface. If no DNA was present on the particles or on the gold substrate, the fluo-

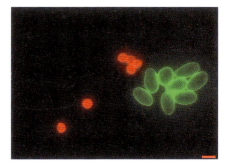

Figure 3.10 Dual-color image of QD 605-labeled C. parvum (red) and QD 565-labeled G. lamblia (green). Scale bar: 10 μm. (Reprinted from Reference [46] with permission from the American Society for Microbiology.)

rescence emission was vanishingly small with only a few patterns having a faint, barely detectable fluorescence. The successful result of the one color hybridation experiment suggested that DNA–nanocrystal conjugates could be of use in multicolor microarray studies [49].

Visualization of the endocytic pathways is very important to evaluate drug and gene delivery. For this purpose, dextran and bovine serum albumin (BSA) labeled with a fluorophore such as rhodamine are often used. Hanaki *et al.* examined albumin from 10 species regarding their efficacy as the adjuvant to disperse CdSe/ZnS-core/shell QDs capped with 11-mercaptoundecanoic acid (MUA) in biological buffers. They prepared two different MUA-QDs: QD560 that emitted yellow and MUA-QD640 that emitted red, which subsequently were mixed with different serum albumins forming MUA-QD/albumin complexes. To examine cellular uptake of QD/albumin complexes, Vero cells, established from the kidney of a normal adult monkey, were incubated with the complexes for 2 hours. To study the intracellular distribution or dynamics of MUA-QDs, the authors selected QD640 because they could be excited with green to yellow light sources so as to minimize the damage to living cells. A time-lapse study of intracellular distribution revealed that QD640 remained in vesicles in the perinuclear region in mitotic phase; after cell division, the vesicles could be observed not only in perinuclear region but also in the neighborhood of the cell membrane. The dispersion ability of MUA-QDs by one of the 10 albumins was then examined in biological buffers. The results showed Sheep Serum Albumine was the most effective for the dispersion. The authors also checked the cell toxicity of MUA-QDs by MTT assay, since cadmium and selenium are toxic compounds. It was shown that at 0.4 mg mL^{-1} the QDs conjugate did not affect the viability of Vero cells [50].

Protein microarray technologies provide a means of investigating the proteomic content of clinical biopsy specimens to determine the relative activity of key nodes within cellular signaling pathways. Geho *et al.* have described the use of inorganic fluorescent NPs (QDs) conjugated to streptavidin in a reverse-phase protein microarray format for signal pathway profiling [51]. Reverse-phase protein microarrays (RPMAs) are an emerging high-throughput technology that offers a means to quantitatively measure both protein levels and posttranslational modifications of signaling proteins in clinical specimens. In the RPMA, proteins extracted from cellular lysates are arrayed onto a nitrocellulose substrate and probed with a primary antibody (Figure 3.11). In turn, a biotinylated secondary antibody recognizes the presence of the primary antibody. The biotinyl groups are then detected by streptavidin linked to reporter molecules such as enzymes system or QDs. Jurkat T-cell lysates were prepared and later arrayed on nitrocellulose-coated FAST glass slides in serial dilutions with a negative control included (sample buffer only) and, once the slides were arrayed, they were immunostained. The primary polyclonal antibody used in that study was specific for intracellular signaling protein. The secondary biotinylated goat anti-rabbit IgG H+L antibody was incubated with the arrays for 15 min. For amplification purposes the slides were incubated with streptavidin–biotin complex. For slides developed using QDs and catalyzed reporter

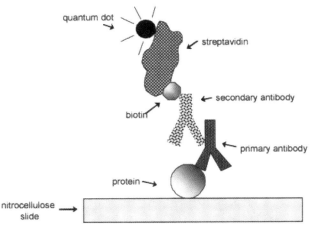

Figure 3.11 Reverse-phase microarray. In the RPMA, proteins extracted from cellular lysates are arrayed onto a nitrocellulose substrate and probed with a primary antibody. In turn, a biotinylated secondary antibody recognizes the presence of the primary antibody. The biotinyl groups are then detected by streptavidin linked to reporter molecules such as QDs or enzymes such as HRP that carry out catalyzed reporter deposition for signal amplification. (Reprinted with permission from Reference [51]. Copyright 2007 American Chemical Society.)

amplification instead of the streptavidin enzyme system, pegylated-QDs-655-PEG (polyethylene)-Sav was incubated for 15 min. The slides probed with QDs were visualized using 254 nm epifluorescent UV. Streptavidin-conjugated QDs were incorporated into a standard reverse-phase microarray assay to determine their utility as reporter molecules for this type of high-throughput proteomic test. It was found that streptavidin-conjugated QDs had superior detection characteristics in RPMA. In addition, the hyperspectral imaging of the QDs microarray enabled unamplified form detection of proteins within Jurkat cellular lysates. The reported system performed well at higher signal intensities, and the dynamic range was significantly broader than the colorimetric methods [51].

3.2.6 Nanowires

Semiconductor nanowires are nanoscale building blocks that could, through bottom-up assembly, enable diverse applications, principally in nanoelectronics, photonics and bio/chemical sensor field [52]. In comparison to spherical NPs, nanowires offer additional degrees of freedom in self-assembly due to their inherent shape anisotropy. Furthermore, the ability to synthesize particles with different segments along the length of a nanowire provides the opportunity to introduce multiple chemical or biological functionalities by exploiting the selective binding of different ligands to the different segments of a nanowire [53].

Yang *et al.* presented a novel method for the preparation of surface molecularly imprinted size-monodisperse nanowires. The molecularly imprinted polymers have already been used successfully for mimicking natural receptors and for the synthesis of polymers carrying binding sites with high affinity toward drug, peptides and proteins. The imprinted molecule, glutamic acid, was immobilized on the pore walls of a silane-treated nanoporous alumina membrane. The nanoporous was filled with a pyrrole mixture and the polymerization was then initiated. The alumina membrane was subsequently removed by chemical dissolution, leaving behind polypyrrole nanowires with glutamic acid binding cavities at the surface. Molecularly imprinted pyrrole has been proven to be a promising material for molecular recognition of amino acids. In this work the glutamic acid recognition ability by imprinted nanowires was investigated by the steady-state binding method. The results revealed that the imprinted nanowires indeed exhibited a higher capacity to bind glutamic acid than the control nanowires formed in the absence of an imprinting glutamic acid. The nanowires also showed a high selectivity for this amino acid in a study realized in the presence of phenylalanine and arginine. This is a unique kind of imprinted material; relatively monodisperse nanowires with a moderately high surface area and imprinted binding sites located at, or close to, the surface (Figure 3.12). Without question, these nanowires would

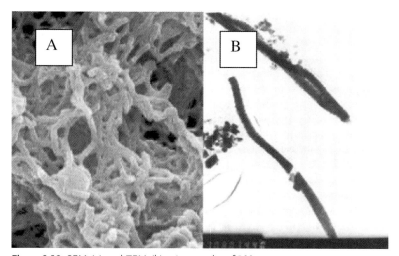

Figure 3.12 SEM (a) and TEM (b) micrographs of 100-nm-diameter polypyrrole nanowires after the removal of alumina template membranes. (Reprinted with permission from Reference [54]. Copyright 2007 American Chemical Society.)

greatly increase the usefulness of molecularly imprinted polymers for immunoassays and related biological applications [54].

3.3
CNTs

Carbon nanotubes (CNTs), as a new class of nanomaterial, were discovered in 1991 by Iijima [55] and have also been employed in biosensors. Such an application is attributed to their unique electrical properties, which make a redox active close to the surface of proteins, and enable direct electron transfer between proteins and electrode [56]. CNTs are highly conductive (rapid electron transfer) nanomaterials with great promise for applications in biochemical sensing [57–59]. Several successful sensors based on CNTs have been reported for the detection of substances, including for NADH [58], glucose [59], cytochrome c [60] and thymine [61].

Chen et al. have developed a biosensor for H_2O_2 based on multi-wall carbon nanotubes (MWNTs) and GNPs. Acid-pretreated, negatively charged MWNTs were first modified on the surface of a glassy carbon (GC) electrode, then positively charged hemoglobin (Hb) was adsorbed onto MWNT films by electrostatic interaction. The {Hb/GNP}n multilayer films were finally assembled onto a Hb/MWNT film through a layer-by-layer assembly technique (Figure 3.13). The assembly of Hb and GNP was characterized with cyclic voltammetry (CV), electrochemical impedance spectroscopy (EIS) and transmission electron microscopy (TEM). The direct electron transfer of Hb is observed on the Hb/GNP/Hb/MWNT/GC electrode, which exhibits excellent electrocatalytic activity for the reduction of H_2O_2 to construct a third-generation mediator-free H_2O_2 biosensor. Compared to H_2O_2 biosensors based only on carbon nanotubes, the proposed biosensor modified with MWNTs and GNPs displays a broader linear range and a lower detection limit for H_2O_2. Rapid and accurate determination of H_2O_2 is of a great importance because it is not just the product of reactions catalyzed by many highly selective oxidases but also an essential compound in food, pharmaceutical and environmental analyses [62].

Du et al. have developed a simple method for immobilization of acetylcholinesterase (AChE) – covalent bonding to a multiwall carbon nanotube (MWNT) – cross-linked chitosan composite (CMC) – and a sensitive amperometric sensor for rapid detection of acetylthiocholine (ATCl) has been based on this [63]. One of the products of hydrolysis of acetylthiocholine chloride (ATCl) with acetylcholinesterase (AChE) is thiocholine. Glutaraldehyde was used as cross-linker to covalently bond the AChE, and efficiently prevented leakage of the enzyme from the film. Detection of thiocholine can be used to assess the activity of AChE, a biomarker of the effect of pesticides [organophosphates (OPs) and carbamates] that inhibit cholinesterases. Analysis of ATCl is, therefore, of great importance, particularly in the development of sensors for detection of environmental pollutants such as OPs and carbamates [64].

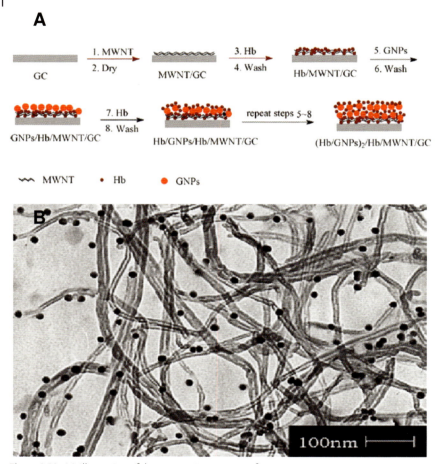

Figure 3.13 (a) Illustration of the preparation process of modified electrode; (b) TEM image GNP/Hb/MWNT film. (Reprinted from Reference [56], with permission from Elsevier.)

3.4
Organic Nanoparticles (ONPs)

The synthesis, functionalization and biomedical applications of ONPs is a broad field. In this section we give one approach to the subject, illustrated with some examples.

3.4.1
Synthesis and Properties of ONPs

The most straightforward approach to synthesizing ONPs is to start from molecules that already have a desired group or property. Another possibility is to begin with different organic molecules and subsequently add functionality through a core–shell strategy. Many ways have been described to synthesize ONPs. Milling processes can be employed to obtain nanoscale materials, but the resulting materials usually have undesirable residues from the milling process itself and, in addition, greater dispersion in sizes and shapes is obtained. ONP synthesis is usually based on precipitation processes, where polymers are often used as aids, acting as a "glue" to bind organic molecules. In such cases, the resulting material is called a pseudolatex. Without using polymers, the resulting material is still in a colloidal state; if it is made in water, it is called hydrosol. Horn and Rieger [7] describe five general strategies for ONP production in water environments, namely, for hydrosols: (i) those formed of an active compound the hydrophobic organic compound is first solubilized in an organic solvent and (ii) by solubilizing the organic compound in a hydrophilic solvent. On the other hand, a pseudolatex can be made (i) by mixing the active compound with a polymer – the organic compound is solubilized in an organic solvent; (ii) by solubilizing the organic compound in an amphiphilic solution; and (iii) by solubilizing the organic compound in a hydrophilic solvent.

Laser ablation can also be employed to obtain ONP, as shown by Li *et al.* [65] who synthesized ONPs made of phthalocyanines, obtaining sizes between 60 and 100 nm. This alternative has not been explored very much, possibly because it has the same problems that typical milling processes have. ONPs stability is based roughly on the same principles as their metallic counterparts, although the models used to describe this stability, like the classic DLVO theory, have their limitations in describing all the factors that are involved, as reported in Destrée and Nagy's review [66] and also in Horn and Rieger's review [7]. Nucleation processes are not as well understood for ONPs as they are for MNPs; more research is required. ONPs are characterized roughly through the same means as MNPs, chiefly TEM, SEM and, to a lesser extent, atomic force microscopy (AFM). Depending on how they interact with light, UV/Visible spectroscopy can be used to detect absorbance maxima that are characteristic for each ONPs type and size [7] (Figure 3.14). If another property is to be measured, for example, fluorescence, photoluminescence, and so on, then it should be included in the characterization protocols.

Good reasons to choose ONPs for biomedical applications are that organic materials are generally less toxic than metallic materials and, if the proper molecules are chosen for their synthesis, they are less likely to be recognized by the immune system. For example, ONPs do not catalyze oxidative stress reactions, whereas several MNPs do. In this sense, we should recall that many MNPs are functionalized with organic molecules to avoid or diminish toxicity.

For ONPs, functional groups from their constituents give different properties to the molecule, and it also means that the functionalization strategy can take a different approach. ONPs can be made of a substance with the required properties, thus avoiding a capping or core–shell strategy; NPs are often made of an already functional organic compound. Core–shell ONPs, however, are also often found, as described below; which strategy should be used is chosen by convenience.

Some ONPs are regular and polished but, usually, they have more irregular sizes and shapes, as well as broader size distributions, and in general their average sizes

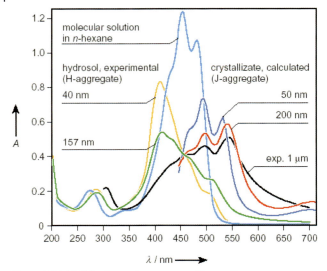

Figure 3.14 UV/Vis absorption spectra of 5 ppm β-carotene. Influence of aggregate structure and particle size compared with the molecular solution in n-hexane. (Reprinted from Reference [7]. Copyright Wiley-VHC Verlag GmbH. Reproduced with permission.)

are often larger than those of MNPs [67] (Figure 3.15). They are also not as electron-dense as their metallic counterparts, which makes them to appear dimmer, for example, in a TEM image, and the possibility of magnetic targeting is, of course, ruled out. In short, we cannot expect ONPs to behave like MNPs, but they can still provide very valuable functionality for biomedical purposes, as the enormous and increasing number of research papers in the field shows.

3.4.2
Functionalization Strategies

Here we discuss how biological molecules are linked to ONPs. To do this, different strategies can be employed depending on what is desired and on the starting molecules. Three defined approaches can be distinguished (Figure 3.16):

1. To build a nanoparticle that is made entirely of the functional organic molecule, like some of the fluorescent ONPs from Wang Lun's group [67, 68].
2. To build a nanoparticle, and include the functional molecule in the bulk, producing ONPs in a one-pot reaction. This method allows most of the functional molecules to be in the bulk, while some will be available on the surface. A very promising work related to this strategy is that of Langer's group, published in a paper by Farokhzad [69] where they incorporate docetaxel (Dtxl), an anticancer drug, in a one-pot reaction to a polymer containing free carboxyl

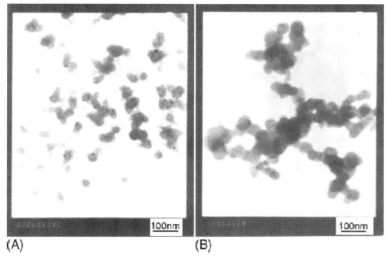

Figure 3.15 FTEM images of unwrapped anthracene NPs (a) and wrapped anthracene/poly-acrylamide nanoparticles (b). (Reprinted from Reference [74], with permission from Elsevier.)

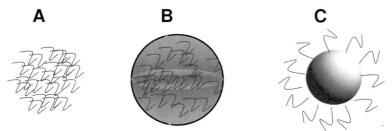

Figure 3.16 Three defined approaches for functionalization of ONPs: (a) to build a nanoparticle that is made entirely of the functional organic molecule (represented in red); (b) to build a nanoparticle, and include the functional molecule (represented in red) in the bulk (represented in gray) of the nanoparticle; (c) to build a nanoparticle in a core (gray)-shell (red) strategy.

groups (Figure 3.17). The result is a dispersion of ONPs containing the drug and carboxyl groups. These carboxyl groups are employed to add, in a second step through chemical reaction, an aptamer, that is a DNA oligonucleotide that provides specificity. Thus, their ONPs are bifunctional: anticancer and specific. The results are low cytotoxicity and little side effects in a mice model. In addition, this method allows nanoparticles to produce *in situ* release of the functional substance when they arrive at their target. This strategy is especially helpful in drug delivery. Another of many examples of this strategy is has been

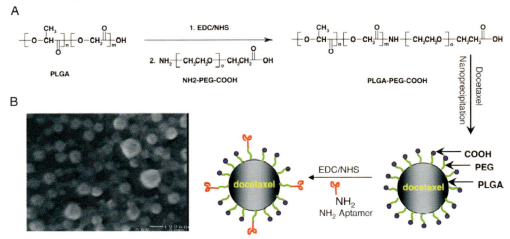

Figure 3.17 Development of Dtxl-encapsulated pegylated PLGA NP-Apt bioconjugates. (a) Schematic representation of the synthesis of PLGA-PEG-COOH copolymer and strategy of encapsulation of Dtxl. The authors developed Dtxl-encapsulated, pegylated NPs by the nanoprecipitation method. These particles have a negative surface charge attributable to the carboxylic acid on the terminal end of the PEG. The NPs were conjugated to amine-functionalized A10 PSMA. Apt by carbodiimide coupling chemistry. (b) Representative scanning electron microscopy image of resulting Dtxl-encapsulated NPs is shown. [EDC, 1-ethyl-3-(3-dimethylaminopropyl)carbodiimide; HS, N-hydroxysuccinimide.] (Reprinted from Reference [70]. Copyright 2007 National Academy of Sciences, USA.)

given by Duan et al. [70]. The authors describe first the synthesis of the polymer in acetone and later the incorporation of the functional molecule, 5-fluorouracil (an anticancer drug). Subsequently, acetone is removed, producing the precipitation of the nanoparticles, for future release of the anticancer drug from the nanoparticles.

3. To build a nanoparticle in a core–shell strategy. The cores are synthesized first and then the functional molecules are added as a shell by simple adhesion (physical adsorption in an insolubilization strategy). Examples are given in the literature [71, 72]. Organic molecules forming the core of an ONP usually include functional groups to which other molecules are added, such as peptides/proteins and nucleic acids.

Which strategy to adopt will depend on the requirements of the user. If the system is to be used *in vivo*, several requirements are added. Some recipes may lead to toxic, non-biocompatible nanoparticles, or other undesirable properties. A careful study should be made to choose a strategy that suits the investigator's needs.

3.4.3
ONPs Types and Applications

Many criteria can be considered when trying to classify ONPs, but in this discussion we consider a few major types of ONPs for biomedical purposes. Physically, viruses comply with all the requirements to be considered as nanoparticles (solid state and nanometric size), but they will not be discussed here. Fischlechner and Donath [73], and Singh, González and Manchester [74] have written comprehensive reviews on the subject.

3.4.3.1 Fluorescent ONPs
Interaction with light is one of the most interesting properties in ONP research. A significant number of the research papers are related to molecules absorbing light, or emitting either fluorescence or photoluminescence. Light interactions with ONPs is an interesting research field for biological applications, as ONPs can be used as tracers or markers of different compounds and, therefore, of processes. In this approach, we discuss next issues of biocompatibility and toxicity for *in vivo* use.

Fluorescence phenomena are enhanced by the conjugation of fluorescent molecules with nanoparticles. This has been shown in several publications [58, 75] and such ONPs are designed to take advantage of this phenomenon. Good examples of ONP usage in combination with their fluorescence properties have been described in a series of papers by Wang Lun's research group in China, for the qualitative and quantitative analysis of several biologically interesting species, including proteins [76, 77], nucleic acids [78], hexavalent chromium [79, 80], glutathione [81] and nitrites [82], achieving good analytical quality parameters, in a very interesting approach.

As with MNPs, the optical properties of ONPs are size dependent. In this sense, it is interesting to mention the work of Fu and Yao's [83], who have studied ONPs made of a fluorescent compound, 1-phenyl-3-[(dimethylamino)styryl]-5-[(dimethylamino)phenyl]-2-pyrazoline (PDDP). They found that the optical properties depended strongly on particle size, showing a remarkable bathochromic displacement in the absorbance maxima of nanoparticles, the displacement increasing with size from 50 to 310 nm. Fluorescence emission and decay were also studied, showing changes similar to those in the absorbance spectra. The shift in absorbance and fluorescence maxima must be taken into account for tracing purposes.

3.4.3.2 Cancer-Aimed ONPs
Drug delivery for cancer therapies is a long-standing research target as the usual side effects of conventional therapies are highly undesirable. ONPs show characteristics that make them quite promising in reducing the side effects of traditional chemotherapy.

Among the major protagonists here are poly(ethylene glycol) (PEG) nanoparticles, and PEG combined with other polymers, aimed at treating different varieties of cancer, seeking delivery of antitumoral drugs. Van Vlerken, Vyas and Amiji have recently reviewed PEG and poly(ethylene oxide) (PEO) nanoparticles for antitumoral drug release [84], reviewing how cancer therapies are enhanced by the use of nanoparticles. Polymers other than PEG and PEO have also been tried. Numerous research papers have been published recently [85–92] and this number will surely increase in the next few years. Interestingly, a literature search yields a huge number of reports, most of them produced between 2005 and 2007 (the present chapter was written in 2007), with the number increasing over time. A general point of question for effective *in vivo* usage of ONPs is their often short lifetimes, which are reduced by opsonization processes activated by the action of leukocytes and their derivative cells. To increase the action time of ONPs, "stealth" properties are being pursued, where "stealth" is understood as a means of hiding from the immune system. Several papers are dedicated to this particular point, like that of Béduneau's group [93].

3.4.3.3 Delivery of ONPs through the Blood–Brain Barrier (BBB)

Another important application of ONPs is the delivery of xenobiotics into the brain, by going through the blood–brain barrier (BBB). Several papers describe this use [94–97], although further innovation in this field is required as systems currently being investigated still have little absorption and short half-lives. One way around this problem has been found by giving nanoparticulate systems for brain delivery by nasal administration, thus avoiding first-pass metabolism.

Mishra and collaborators have described the use of ONPs to deliver azidothymidine (AZT) into the brain [98]. AZT is an antiviral employed classically for therapy against HIV; its high water solubility diminishes its chances of crossing the BBB. ONPs enhance BBB crossing by protecting it and making it more oil-soluble. AIDS, however, remains a little explored use for ONPs.

3.4.3.4 Nucleic Acids/Gene Delivery

Nucleic acids and gene delivery is another recent field of research open for nanoparticulate systems. Gene therapy is another expanding field, and ONPs are good delivery agents that different from viruses, which are commonly used for this purpose. The literature again shows an increasing number of publications in this area. Opsonization here is a problem again, and, in this respect, hydrophilic and non-ionic polymers like PEG and poly-L-lysine (PLL) are useful. For example, Gu *et al.* have described PEGylated nanoparticles for the delivery of plasmid DNA [99], with good transfection rates. Moffatt and Cristiano [100] and other groups have found similar results, foreseeing an optimistic future in this field.

Lipid nanoparticles have also been used for gene delivery. Findings with this type of ONP, as well as their limitations, are explained in Li and Szoka's review [101], emphasizing the size requirements to obtain good penetration.

3.4.3.5 Other Biomedical Uses of ONPs

A nanoparticulate system may be useful in the delivery of drugs that are poorly soluble in water. Recent papers describe this use, making it a promising research area. For example, Chan and collaborators [102] have described a colloidal system useful for insoluble drugs; the system is cleavable in acidic media, leading to selective delivery of drugs. Similar approaches have also been explored by other authors [103], especially for all-*trans* retinoic acid [104] and itraconazole [105].

Destrée, Ghijsen and Nagy [106] have described microemulsion-synthesized ONPs designed for transdermal drug delivery, a little explored possibility. The possibilities are good, but the idea requires further study; so far, *in vitro* and *in vivo* evaluations of the system are still required. Zhao and collaborators [107] have described ONPs with a porous structure, their ONPs loaded with hemoglobin, and intended for oxygen delivery, another possibility that has been poorly explored. A few papers describe the use of ONPs specifically for oral administration. Bala *et al.* have described ONPs for encapsulation of ellagic acid [108], the latter being a known antioxidant. Ellagic acid suffers first pass metabolism, thus reducing its oral bioavailability, while their nanoparticles show improved bioavailability. A similar effort has been published by Sonaje *et al.* [109]. Another interesting paper on the subject is that of Prego *et al.* [110] who use ONPs to improve oral bioavailability of peptides, which are known to have the same metabolism problem mentioned before, as well as having little absorption through the intestinal epithelium.

3.5
Conclusions

This chapter gives an overview of the different strategies used to functionalize nanomaterials with bioactive molecules. The chemical reactions used to link active molecules to NPs are similar to those used in other fields of chemistry, such as polymers, oligosaccharides and peptides. However, various organic reactions that could be used for the functionalization of nanomaterials have not yet been explored. Notably, more techniques are required to characterize functionalized nanomaterials, such as TEM, AFM and so on, than are used in organic and inorganic chemistry. It is also important to mention that one of the main concerns is how the bioactive molecule interacts and packs on a nanoparticle surface. The interaction of the linked bioactive molecules with the NPs could change the interaction with the biological target, thereby reducing the molecular recognition. Applications of NPs in biomedicine are promising; however, several concerns, such as toxicity, stability and tagging, must be addressed. Importantly, there are few toxicity studies. Cytotoxicity tests, to date, are not complete in terms of assessing the effects of nanomaterials on cellular metabolism.

Acknowledgments

This work was supported by FONDECYT 1061142 and FONDAP 11980002 (17 07 0002) grants.

List of Abbreviations

Aβ	amyloid plaques
AChE	acetylcholinesterase
AD	Alzheimer's disease
AFM	atomic force microscopy
APP	amyloid precursor protein
Apt	aptamer
ANP	atrial natriuretic peptide
ATCl	acetylthiocholine chloride
AZT	azidothymidine
BBB	blood–brain barrier
BSA	bovine serum albumin
CALNN	Cys-Ala-Leu-Asn-Asn
CL	cross-linked
CNPs	carbon nanoparticles
CNTs	carbon nanotubes
CMALISA	cellular magnetic-linked immunosorbent assay
CMC	chitosan composite
CV	cyclic voltammetry
DHLA	dihydrolipoic acid
DNase I	deoxyribonuclease
Dtxl	docetaxel
ELISA	enzyme-linked immunosorbent assay
EIS	electrochemical impedance spectroscopy
FITC	fluorescein isothiocyanate
GC	glassy carbon
GNPs	gold NPs
HA	hyaluronic acid
ICP-OES	inductively coupled plasma optical emission spectrometry
IgG	immunoglobulin
IONPs	iron oxide NPs
MIONs	monocrystalline iron oxide nanoparticles
MNC	mononuclear cells
MNPs	metallic NPs
MRI	magnetic resonance imaging
MTT	3-[4,5-dimethylthiazol-2-yl]-2, 5-diphenyltetrazolium bromide
MUA	11-mercaptoundecanoic acid
MWNTs	multiwall carbon nanotubes
NLS	CGGFSTSLRARKA peptide
NPs	nanoparticles
ONPs	organic nanoparticles
OPs	organophosphates
PDGFs	platelet-derived growth factors
PDDP	1-phenyl-3-[(dimethylamino)styryl]-5-[(dimethylamino)phenyl]-2-pyrazoline

PEG	poly(ethylene glycol)
PEO	poly(ethylene oxide)
PEP	H-Leu-Pro-Phe-Phe-Asp-NH2
PGA	polyanionic Na+-poly (γ-glutamic acid)
PLL	poly-L-lysine
PSAs	polysialic acids
QDs	quantum dots
mRNA	messenger RNA
RME	CKKKKKKSEDEYPYVPN peptide
RPMAs	reverse-phase protein microarrays
SEC	size-exclusion chromatography
SEM	scanning electron microscopy
SPIONs	superparamagnetic iron oxide nanoparticles
TEM	transmission electron microscopy
USPIO	ultrasmall superparamagnetic iron oxide

References

1 Kogan, M.J., Olmedo, I., Hosta, L., et al. (2007) *Nanomedicine*, **2**, 287–306.
2 Ballesteros, B., De la Torre, G., Rahman, E.C., et al. (2007) *Journal of the American Chemical Society*, **129** (16), 5061–8.
3 Xie, Z., Lu, T., Chen, X., et al. (2007) *Journal of Applied Polymer Science*, **105**, 2271–9.
4 Liu, W.T. (2006) *Journal of Bioscience and Bioengineering*, **102**, 1–7.
5 Sharron, G.P., He, L. and Natan, M.J. (2003) *Current Opinion in Chemical Biology*, **7**, 609–15.
6 Fukumori, Y. and Ichikawa, H. (2006) *Advanced Powder Technology*, **17**, 1–28.
7 Horn, D. and Rieger, J. (2001) *Angewandte Chemie International Edition*, **40**, 4330–61.
8 Turkevich, J., Stevenson, P.C. and Hillier, J. (1951) *Discussions of the Faraday Society*, **11**, 55–75.
9 Sonvico, F., Dubernet, C., Colombo, P. and Couvreur, P. (2005) *Current Pharmaceutical Design*, **11**, 2091–105.
10 Zhu, T., Vasilev, K., Kreiter, M., et al. (2003) *Langmuir*, **19** (22), 9518–25.
11 Higuchi, M., Ushiba, K. and Kawaguchi, M. (2007) *Journal of Colloid and Interface Science*, **308**, 356–63.
12 Ryadnov, M.G., Ceyhan, B., Niemeyer, C.M. and Woolfson, D.N. (2003) *Journal of the American Chemical Society*, **125**, 9388–94.
13 Wang, H., Brandl, D.W., Nordlander, P. and Halas, N.J. (2007) *Accounts of Chemical Research*, **40**, 53–62.
14 Schultz, D.A. (2003) *Current Opinion in Biotechnology*, **14**, 13–22.
15 Yeganeh, M.S., Dougal, S.M., Polizzotti, R.S. and Rabinowitz, P. (1995) *Physical Review Letters*, **74**, 1811–14.
16 Liu, J. and Lu, Y. (2006) *Angewandte Chemie International Edition*, **45**, 90–4.
17 Pavlov, V., Xiao, Y., Shlyahovsky, B. and Willner, I. (2004) *Journal of the American Chemical Society*, **126**, 11768–9.
18 Xu, X., Han, M.S. and Mirkin, C.A. (2007) *Angewandte Chemie International Edition*, **46**, 3468–70.
19 Lee, J.-S., Han, M.S. and Mirkin, C.A. (2007) *Angewandte Chemie International Edition*, **46**, 4093–6.
20 Liu, J. and Lu, Y. (2004) *Chemistry of Materials*, **16**, 3231–8.
21 Baptista, P., Doria, G., Henriquez, D., et al. (2005) *Journal of Biotechnology*, **119**, 111–17.
22 De la Fuente, J. and Berry, C.C. (2005) *Bioconjugate Chemistry*, **16**, 1176–80.
23 Tkachenko, A.G., Xie, H., Liu, Y., et al. (2004) *Bioconjugate Chemistry*, **15**, 482–90.
24 Liu, Y., Shiptom, M.K., Ryan, J., et al. (2007) *Analytical Chemistry*, **79** (6), 2221–9.

25 Lèvy, R., Thanh, T.K., Doty, R.C., et al. (2004) *Journal of the American Chemical Society*, **126**, 10076–84.
26 Kogan, M.J., Bastus, N.G., Amigo, R., et al. (2006) *Nano Letters*, **6**, 110–15.
27 Huang, C.C., Huang, Y.-F., Cao, Z., et al. (2005) *Analytical Chemistry*, **77**, 5735–41.
28 Hirsch, L.R., Jackson, J.B., Lee, A., et al. (2003) *Analytical Chemistry*, **75**, 2377–81.
29 Choi, J.-S., Jun, Y.-W., Kim, ··, et al. (2006) *Journal of the American Chemical Society*, **128**, 15982–3.
30 Pankhurst, Q.A., Connolly, J., Jones, S.K. and Dobson, J. (2003) *Journal of Physics D: Applied Physics*, **36**, R167–81.
31 Berry, C.C. and Curtis, S.G. (2003) *Journal of Physics D: Applied Physics*, **36**, R198–206.
32 Tartaj, P., Morales, M., Veintemillas-Verdaguer, S., et al. (2003) *Journal of Physics D: Applied Physics*, **36**, R182–97.
33 Gupta, A.K. and Gupta, M. (2005) *Biomaterials*, **26**, 3995–4021.
34 Zhao, M., Kircher, M.F., Josephson, L. and Weissleder, R. (2002) *Bioconjugate Chemistry*, **13**, 840–4.
35 Garden, O.A., Reynolds, P.R., Yates, J., et al. (2006) *Journal of Immunological Methods*, **314**, 123–33.
36 Dodd, C.H., Hsu, H.-C., Chu, W.-J., et al. (2001) *Journal of Immunological Methods*, **256**, 89–105.
37 Shieh, D.B., Cheng, F.G., Su, C.-H., et al. (2005) *Biomaterials*, **26**, 7183–91.
38 Burtea, C., Laurent, S., Roch, A., der Elst, L. and Muller, R.N. (2005) *Journal of Inorganic Biochemistry*, **99**, 1135–44.
39 Gupta, A.K. and Curtis, S.G. (2004) *Biomaterials*, **25**, 3029–40.
40 Wadghiri, Y.Z., Sigurdsson, E.M., Sadowski, M., et al. (2003) *Magnetic Resonance in Medicine*, **50**, 293–302.
41 Zhao, A., Yao, P., Kang, C., Yuan, X., Chang, J. and Peiyu, P. (2005) *Journal of Magnetism and Magnetic Materials*, **295**, 37–43.
42 Kumar, A., Sahoo, B., Montpetit, A., et al. (2007) *Nanomedicine*, **3** (2), 132–7.
43 Shin, H.S., Yang, H.J., Kim, S.B. and Lee, M.S. (2004) *Journal of Colloid and Interface Science*, **274** (1), 89–94.
44 Yu, D.-G. (2007) *Colloids and Surfaces B*, **59**, 171–8.
45 Jamieson, T., Bakhshi, R., Petrova, D., et al. (2007) *Biomaterials*, **28**, 4717–32.
46 Zhu, L., Ang, S. and Liu, W.-T. (2004) *Applied and Environmental Microbiology*, **70** (1), 597–8.
47 Michalet, X., Ekong, R., Fougerousse, F., et al. (1997) *Science*, **277**, 1518–23.
48 Michalet, X., Pinaud, F., Lacoste, T.D., et al. (2001) *Single Molecules*, **2** (4), 261–76.
49 Gerion, D., Parak, W.J., Williams, S.C., et al. (2002) *Journal of the American Chemical Society*, **124**, 7070–4.
50 Hanaki, K., Momo, A., Oku, T., et al. (2003) *Biochemical and Biophysical Research Communications*, **302**, 496–501.
51 Geho, D., Lahar, N., Gurnani, P., et al. (2005) *Bioconjugate Chemistry*, **16**, 559–66.
52 Liebre, C.M. (2002) *Nano Letters*, **2** (2), 1–2.
53 Salem, A.K., Chen, M., Hayden, J., et al. (2004) *Nano Letters*, **4** (6), 61163–5.
54 Yang, H.-H., Zhang, S.-Q., Tan, F., et al. (2005) *Journal of the American Chemical Society*, **127**, 1378–9.
55 Iijima, S. (1991) *Nature*, **354**, 56–8.
56 Chen, S., Yuan, R., Chai, Y., et al. (2007) *Biosensors and Bioelectronics*, **22**, 1268–74.
57 Tasis, D., Tagmatarchis, N., Bianco, A. and Prato, M. (2006) *Chemical Reviews*, **106** (3), 1105–36.
58 Zhang, M.G., Smith, A. and Gorski, W. (2004) *Analytical Chemistry*, **76**, 5045–50.
59 Ye, J.S., Wen, Y., Zhang, W.D., et al. (2004) *Electrochemistry Communications*, **6**, 66–70.
60 Wang, J.X., Li, M.X., Shi, Z.J., et al. (2002) *Analytical Chemistry*, **74**, 1993–7.
61 Wang, Z.H., Wang, Y.M. and Luo, G.A. (2003) *Electroanalysis*, **15**, 1129–33.
62 Wang, J., Lin, Y. and Chen, L. (1993) *Analyst*, **118**, 277–80.
63 Du, D., Huang, X., Cai, J., et al. (2007) *Analytical and Bioanalytical Chemistry*, **387**, 1059–65.
64 Suprun, E., Evtugyn, G., Budnikov, H., et al. (2005) *Analytical and Bioanalytical Chemistry*, **383**, 597–604.
65 Li, B., Kawakami, T. and Hiramatsu, M. (2003) *Applied Surface Science*, **210**, 171–6.

66 Destrée, C. and Nagy, J.B. (2006) *Advances in Colloid and Interface Science*, **123**, 353–67.
67 Wang, L.Y., Wang, L., Xia, T.T., et al. (2005) *Spectrochimica Acta A*, **61**, 2533–8.
68 Wang, L.Y., Wang, L., Xia, T.T., et al. (2004) *Analytical Sciences*, **20**, 1013–17.
69 Farokhzad, O.C., Cheng, J.J., Teply, B.A., et al. (2006) Proceedings of the National Academy of Sciences of the United States of America, 103, 6315–20.
70 Duan, J.F., Du, J. and Zheng, Y.B. (2007) *Journal of Applied Polymer Science*, **103**, 2654–9.
71 Wang, L., Xia, T.T., Wang, L.Y., et al. (2005) *Microchimica Acta*, **149**, 267–72.
72 Wang, L., Xia, T.T., Liu, J.S., et al. (2005) *Spectrochimica Acta A*, **62**, 565–9.
73 Fischlechner, M. and Donath, E. (2007) *Angewandte Chemie International Edition*, **46**, 3184–93.
74 Singh, P., González, M.J. and Manchester, M. (2006) *Drug Development Research*, **67**, 23–41.
75 Sun, Y.Y., Liao, J.H., Fang, J.M., et al. (2006) *Organic Letters*, **8**, 3713–16.
76 Xia, T.T., Wang, L., Bian, G.R., et al. (2006) *Microchimica Acta*, **154**, 309–14.
77 Wang, L.Y., Wang, L., Dong, L., et al. (2005) *Spectrochimica Acta A*, **61**, 129–33.
78 Wang, L., Xia, T.T., Wang, L.Y., et al. (2005) *Microchimica Acta*, **149**, 267–72.
79 Wang, L.Y., Wang, L., Xia, T.T., et al. (2004) *Analytical Sciences*, **20**, 1013–17.
80 Wang, L., Xia, T.T., Liu, J.S., et al. (2005) *Spectrochimica Acta A*, **62**, 565–9.
81 Wang, L.Y., Wang, L., Xia, T.T., et al. (2005) *Spectrochimica Acta A*, **61**, 2533–8.
82 Wang, L., Dong, L., Bian, G.R., et al. (2005) *Analytical and Bioanalytical Chemistry*, **382**, 1300–3.
83 Fu, H.B. and Yao, J.N. (2001) *Journal of the American Chemical Society*, **123**, 1434–9.
84 Van Vlerken, L.E., Vyas, T.K. and Amiji, M.M. (2007) *Pharmacological Research*, **24**, 1405–14.
85 Watanabe, M., Kawano, K., Yokoyama, M., et al. (2006) *International Journal of Pharmaceutics*, **308**, 183–9.
86 Xu, P.S., Van Kirk, E.A., Murdoch, W.J., et al. (2006) *Biomacromolecules*, **7**, 829–35.
87 Layre, A., Couvreur, P., Chacun, H., et al. (2006) *Drug Development and Industrial Pharmacy*, **32**, 839–46.
88 Dong, Y.C. and Feng, S.S. (2006) *Journal of Biomedical Materials Research A*, **78A**, 12–19.
89 Xie, Z.G., Lu, T.C., Chen, X.S., et al. (2007) *Journal of Applied Polymer Science*, **105**, 2271–9.
90 Zhang, L.Y., Yang, M., Wang, Q., et al. (2006) *International Journal of Pharmaceutics*, **308**, 183–9.
91 Duan, J.F., Du, J. and Zheng, Y.B. (2007) *Journal of Applied Polymer Science*, **103**, 2654–9.
92 Renoir, J.M., Stella, B., Ameller, T., et al. (2006) *Journal of Steroid Biochemistry*, **102**, 114–27.
93 Béduneau, A., Saulnier, P., Anton, N., et al. (2006) *Pharmacological Research*, **23**, 2190–9.
94 Gao, X.L., Tao, W.X., Lu, W., et al. (2006) *Biomaterials*, **27**, 3482–90.
95 Kim, H.R., Gil, S., Andrieux, K., et al. (2007) *Biomacromolecules*, **8** (3), 793–9.
96 Zhang, Q.Z., Zha, L.S., Zhang, Y., et al. (2006) *Journal of Drug Targeting*, **14**, 281–90.
97 Lu, W., Wan, J., She, Z.J. and Jiang, X.G. (2007) *Journal of Controlled Release*, **118**, 38–53.
98 Mishra, V., Mahor, S., Rawat, A., et al. (2006) *Journal of Drug Targeting*, **14**, 45–53.
99 Gu, W.W., Xu, Z.H., Gao, Y., et al. (2006) *Nanotechnology*, **17**, 4148–55.
100 Moffatt, S. and Cristiano, R.J. (2006) *International Journal of Pharmaceutics*, **321**, 143–54.
101 Li, W.J. and Szoka, F.C. (2007) *Pharmacological Research*, **24**, 438–49.
102 Chan, Y., Bulmus, V., Zareie, V.H., et al. (2006) *Journal of Controlled Release*, **115**, 197–207.
103 Lee, W.C., Li, Y.C. and Chu, I.M. (2006) *Macromolecular Bioscience*, **6**, 846–54.
104 Jeong, Y.I., Kim, S.H., Jung, T.Y., et al. (2006) *Journal of Pharmaceutical Sciences*, **95**, 2348–60.
105 Vaughn, J.M., McConville, J.T., Burgess, D., et al. (2006) *European Journal of Pharmaceutics and Biopharmaceutics*, **63**, 95–102.

106 Destrée, C., Ghijsen, J. and Nagy, J.B. (2007) *Langmuir*, **23** (4), 1965–73.
107 Zhao, J., Liu, C.S., Yuan, Y., *et al.* (2007) *Biomaterials*, **28**, 1414–22.
108 Bala, I., Bhardwaj, V., Hariharan, S., *et al.* (2006) *Journal of Drug Targeting*, **14**, 27–34.
109 Sonaje, K., Italia, J.L., Sharma, G., *et al.* (2007) *Pharmacological Research*, **24**, 899–908.
110 Prego, C., Torres, D., Fernandez-Megía, E., *et al.* (2006) *Journal of Controlled Release*, **111**, 299–308.

Part Two Solid-Supported Reagents and Scavengers

4
Oxidizing and Reducing Agents

Samuel Beligny and Jörg Rademann

Leibniz Institute for Molecular Pharmacology, Robert-Rössle-Str. 10, 13125 Berlin, Germany

4.1
Introduction

Oxidations and reductions are pivotal transformations in organic synthesis. A considerable amount of work has been devoted towards the development of novel oxidation and reduction techniques, and numerous catalysts have become standard tools for the organic chemist. Polymer-supported reagents have been introduced with the initial idea of simplifying the life of the synthetic chemist, especially limiting the tedious purification procedures inherent to every synthetic work, and supported reagents could become central in the efforts to make chemistry cleaner and "greener." Therefore, unsurprisingly, considerable effort has been devoted to the development of solid-supported version of oxidative and reductive reagents to facilitate the clean and efficient synthesis of new chemical libraries by combining the advantages of both reagents and support techniques.

Earlier work on polymeric redox reagents has been reviewed extensively by Sherrington [1], while more recent developments are covered by the reviews of Ley [2], Kirschning, and Cozzi [3, 4]. Since the turn of the millennium, the field of polymer reagents has witnessed an enormous growth, resulting in an exponential increase in publications. Since 2000, polymer reagents have been used for the first time regularly for the preparation of complex products and in library synthesis. Therefore, we have had to be selective in the choice of the presented work. This chapter is focused on the most representative and striking examples, combining a review of recent innovations in the field with applications of the newly developed and older resins. Functional resins for oxidation and reduction that have found broader practical application or opened powerful options for the synthesis of highly functional molecules have been selected. In some cases these polymer reagents were found as a solution to synthetic problems, enabling the preparation of libraries of biologically interesting compounds.

The Power of Functional Resins in Organic Synthesis. Judit Tulla-Puche and Fernando Albericio
Copyright © 2008 WILEY-VCH Verlag GmbH & Co. KGaA, Weinheim
ISBN: 978-3-527-31936-7

4.2
Considerations Concerning the Nature of the Solid Support Used for Polymer-Supported Redox Reagents

All polymer-supported chemistry, including classical solid phase synthesis, relies heavily upon the nature and the properties of the solid support materials, or polymers used. Most reactions within polymers can be modeled by second-order rate kinetics in combination with diffusion of the soluble components into spherical compartments, the beads [5]. In many cases the adhesion of reagents to the polymer surface can play a significant role [6]. Several other properties of the solid support may greatly influence on polymer-supported reactions, including swelling, solvent compatibility, loading capacity and the flexibility of polymer-attached functional groups [7]. For polymer-supported catalysts and ion exchange resins, which are highly relevant in redox resins, leaching of metal ions is an important factor, and for covalently attached reagents the recyclability of the reactive polymer can be of primary interest. In addition, the cost and the commercial availability of polymers and polymer reagents often have a significant impact on their use.

Most polymeric redox reagents have been developed on microporous polystyrene, typically cross-linked with 1% divinylbenzene. Few examples have been reported for macroporous polystyrene, silica or other supports such as high-loaded cross-linked polyethylene imine (Ultraresins). Some problems specific for redox reactions can also arise from the reactivity of the polymer support itself. In cross-linked polystyrene, for example, benzylic positions can be oxidized at elevated temperatures and thus can account for a competing reaction pathway [8]. Further reactivities are found for other solid supports as well.

In summary, when designing and using solid-supported reagents for redox reactions, the nature and the properties have to be carefully considered to optimize reagent preparation, loading and reactivity of the polymer supports.

4.3
Oxidizing Resins

Section 4.3.1 reports on newly developed supported reagents that have been introduced after the year 2000, as well as their applications. In Section 4.3.2 innovative applications of older polymer reagents are also presented and discussed.

4.3.1
Novel Oxidative Resins

4.3.1.1   Solid-Supported Hypervalent Iodine Reagents
In the last few years, hypervalent iodine reagents, most notably IBX and Dess-Martin periodane, have become reagents of choice for oxidations of alcohols to the corresponding aldehyde or ketone [9]. The first polymer-supported IBX reagents (Figure 4.1) having the advantages of being non-explosive and usable in most

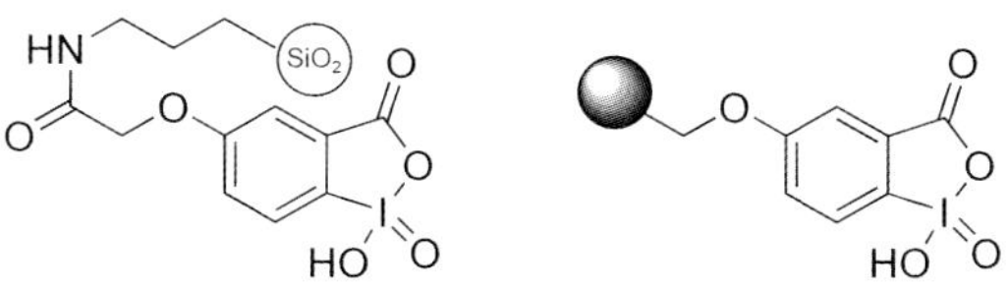

Figure 4.1 Supported IBX reagents.

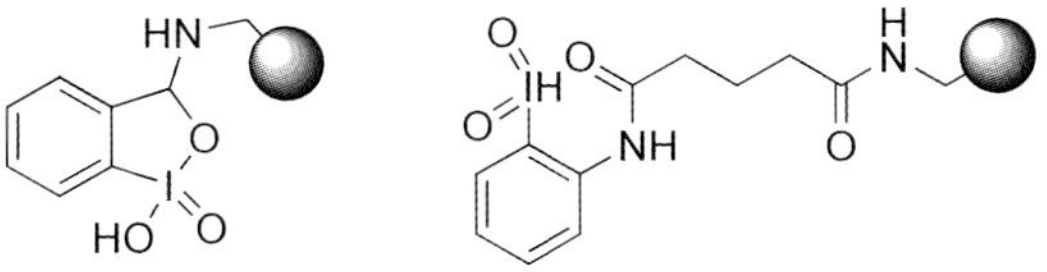

Figure 4.2 Supported IBX analogues.

common organic solvents were reported simultaneously by two groups in 2001 [10, 11]. Both groups attached the resin through the aromatic core. The Giannis group used aminopropylsilica-gel as a solid matrix to give reagent **1**. The Rademann group chose polystyrene and obtained a high-loading resin **2**. Janda and coworkers opted for similar resin types to obtain soluble and insoluble polymer-supported reagents. All catalysts showed very good activity, selectivity and were readily recycled by oxidation.

The practicability of catalyst **2** has been demonstrated in the synthesis of peptide aldehydes libraries targeting caspases 1 and 3 [12]. The peptide aldehydes were prepared using a succession of solid phase, and polymer-assisted solution phase synthesis, and were obtained in excellent purity.

Since this first breakthrough, other solid-supported IBX reagents have been prepared with the same strategy of attachment [13], most notably a high loading polymer prepared by the Lei group, showing improved oxidative activity when used with a small amount of triethylamine [14].

Other attachment strategies were explored based on the IBX amides and esters developed in the Zhdankin group [15]. Their polymer-supported versions **3** were prepared and gave reagents with similar activity (Figure 4.2) [16]. Another recent development was the preparation of *N*-(2-iodylphenyl)-acylamide (NIPA) resin **4** [17], based on the soluble and stable IBX analog developed in the Zhdankin laboratories. This hypervalent iodine(V) reagent is a very efficient oxidative reagent with good loading levels, allowing very good and rapid conversions.

4.3.1.2 Supported TEMPO

The use of TEMPO (2,2,6,6-tetramethylpiperidine-1-oxyl) as an oxidant for the oxidation of primary and secondary alcohols to the corresponding aldehyde or ketone in combination with primary oxidants has also received significant attention in the past few years [18]. It is, therefore, not surprising to see the amount of effort that has been devoted to the synthesis of immobilized versions of this family of reagents (Figure 4.3) [19], on a wide variety of supports (e.g. silica-supported

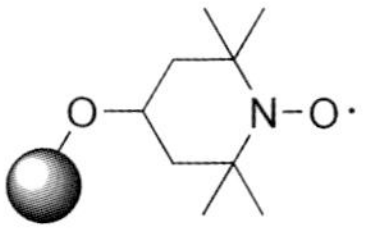

Figure 4.3 Supported TEMPO.

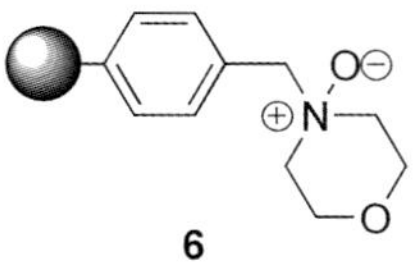

5

6

Figure 4.4 Polymer-supported NMO.

TEMPO, polystyrene-supported TEMPO, sol-gel TEMPO, PEG-TEMPO, supported oxammoium salts, etc.). These catalysts have shown good activity and selectivity, and were also readily recyclable. The aerobic oxidation of alcohols with supported TEMPO employing oxygen as co-oxidant has been explored [19k, 19l, 20], either in the presence of $Co(NO_3)_2$ and $Mn(NO_3)_2$ (Minisci's conditions) or in the presence of Cu(II), giving interesting results.

4.3.1.3 Supported Co-oxidants

4.3.1.3.1 Polymer-Supported NMO

A major problem in using solid-supported oxidants catalytically, concomitantly with a stoichiometric amount of soluble organic co-oxidant, is the necessity to remove and discard the by-product arising from the co-oxidant. This means extra purification steps, which preclude the benefit of using solid-supported reagents. This also goes against the development of cleaner technologies.

An interesting option to get around this problem is, instead of having the catalyst attached to the solid support, to have the co-oxidant attached and to use the catalyst in its unattached form. This idea was exploited in the group of William Kerr in Glasgow [21] with the development of an easily accessible polymer-supported version of N-methylmorpholine N-oxide (NMO) **6** (Figure 4.4). It was used as the co-oxidant for the TPAP oxidation and presents the advantage of eliminating the amine by-products from the reaction, therefore simplifying the purification process. The resin could be re-oxidized and recycled without loss of activity.

Supported NMO was used in the Ley group in their impressive total synthesis of epothilone C [22]. This stereoselective convergent synthesis incorporates polymer reagents, catalysts, scavengers and catch-and-release techniques, being a magnificent demonstration of the power of functional resins. In the initial route, the synthesis of the thiazole fragment required the dihydroxylation of alkene **7** followed by oxidative cleavage to give aldehyde **9** (Scheme 4.1). The dihydroxylation was performed using a catalytic amount of osmium tetroxide and polymer-supported NMO as co-oxidant. The osmium tetroxide was scavenged using polymer-supported pyridine, giving the diol in almost quantitative yield and clean enough to be used in the second step without further purification. The oxidative

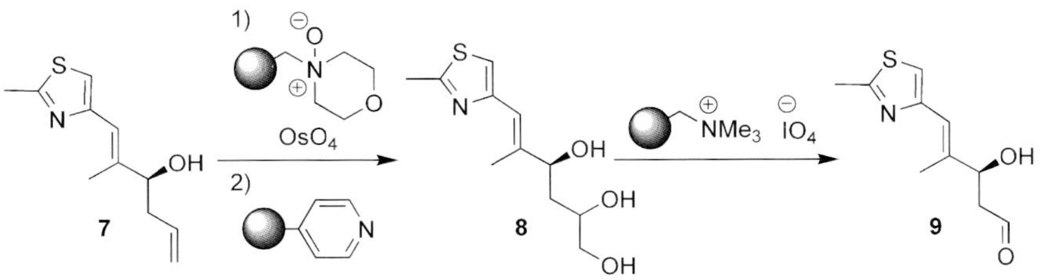

Scheme 4.1 Application of supported NMO as co-oxidant in dihydroxylation.

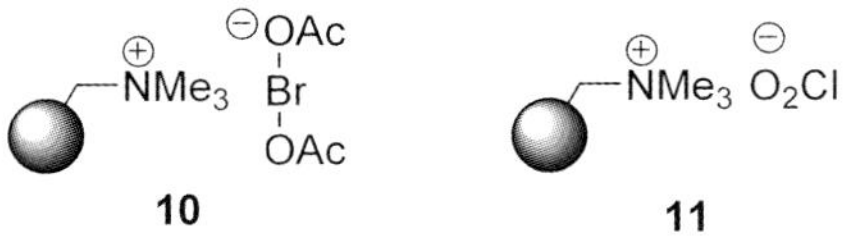

Figure 4.5 Immobilized bisacetoxybromate(I) (**10**) and chlorite (**11**).

cleavage was completed using polymer-supported periodate, which afforded thiazole **9** cleanly and almost quantitatively.

4.3.1.3.2 **Supported Co-oxidants for TEMPO Oxidation** The use of polymer-supported co-oxidants for the TEMPO-catalyzed oxidation of alcohols has given very interesting results. The Kirschning group has developed an immobilized version of bisacetoxybromate(I) **10** (Figure 4.5) anions, which proved to be an excellent as co-oxidant for TEMPO-mediated oxidation of alcohols to the corresponding aldehydes [23]. This method is especially interesting since the purification step is kept to a minimum; only a trace of TEMPO is sufficient to have the reaction running, and it is not only efficient with simple alcohols but also with more challenging compounds.

Use of polymer-bound chlorite **11** was shown to be more efficient for the oxidation of secondary alcohols to the corresponding ketone [23c]. The method was applied to a series of complex synthetic intermediates and gave excellent results. Immobilized chlorite was shown to be also a very efficient co-oxidant in the conversion of primary alcohols into the corresponding carboxylic acid [24]. This method is particularly attractive due to the ease of purification, the excellent yields and purity obtained also on more complex structures. What makes these techniques particularly interesting is that they have found applications in the synthesis of complex molecules. It was the method of choice for the synthesis of intermediate **13**, the core of azadirachtin, in studies towards the synthesis of this natural product by the Nicolaou group [25]. This bicyclic aldehyde was obtained very cleanly using this method (Scheme 4.2).

The same system also proven to be very efficient for the synthesis of **16**, the glycosidic moiety of Altromycin B [26]. The method described by Kirschning proved to be the optimal oxidative method, giving quantitative yields without the requirement of any purification, whereas other oxidation techniques gave only mediocre results (Scheme 4.3). It was also the method of choice for the chiral-pool

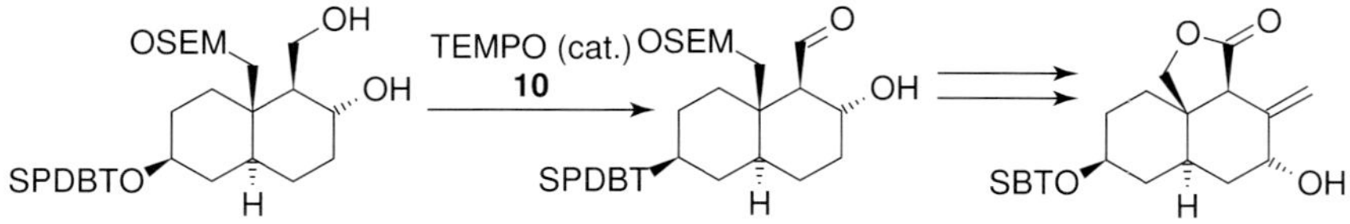

Scheme 4.2 Use of immobilized bisacetoxybromate(I) **10** in synthesis.

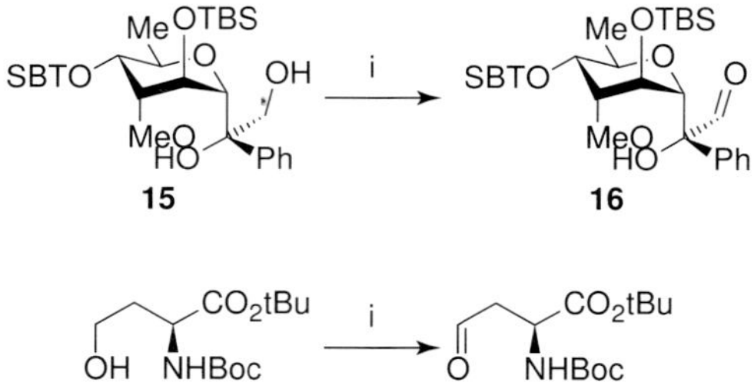

i: TEMPO (cat.), **10**

Scheme 4.3 Further use of immobilized bisacetoxybromate(I) **10** in synthesis.

synthesis of (2S,4R)-4-hydroxyornithine [27], a non-proteinogenic amino acid widely found in nature, giving aldehyde **18**.

4.3.1.4 Asymmetric Oxidation

The development of solid-supported chiral oxidants is a challenging area that has yielded interesting results in the development of a chiral supported dioxiran precursor. The preparation of non-racemic epoxides has been extensively studied in recent years since they are important building blocks in stereoselective synthesis. A supported dioxirane precursor based on α-fluorotropinones was shown to promote the epoxidation of alkenes [28, 29]. The reactant was anchored on mesoporous MCM-41 and amorphous silicas. It has shown comparable activity to its homogenous counterpart and good stability on recycling. The enantiomerically enriched version efficiently promotes the enantioselective epoxidation of alkenes, with ee values up to 80% (Scheme 4.4).

4.3.1.5 Oxidation with Multi Supported Reagents

Another option is the use of both solid-supported versions of the catalyst and the co-oxidant. The Toy group in Hong-Kong has developed a multipolymer system for the TEMPO-catalyzed alcohol oxidation in which both the pre-catalyst and the co-oxidant are attached to a polymer. As it was clearly impossible to use two insoluble supported reagents, the idea was to use an insoluble polymer in conjunction with a soluble one. An insoluble polymer-supported diacetoxyiodosobenzene (PSDIB), an analog of **10**, and a soluble polymer-supported TEMPO were used,

Scheme 4.4 Multipolymer Pinnick oxidation.

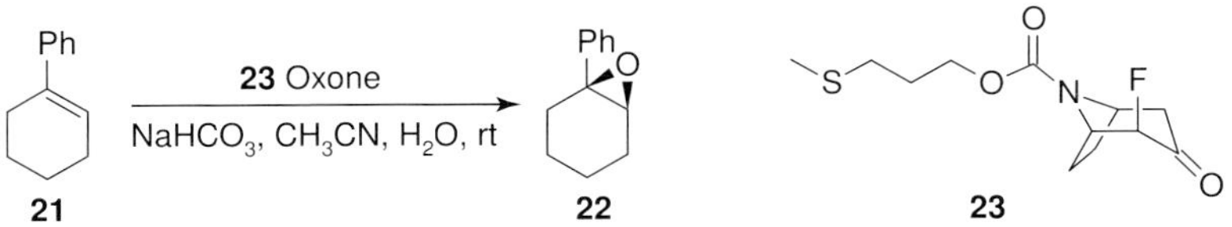

Scheme 4.5 Asymmetric epoxidation with supported chiral dioxiran precursor.

enabling the clean oxidation of alcohols to the corresponding aldehydes and ketones [30].

A multipolymer version of the aerobic TEMPO oxidation of alcohols in the presence of Cu(II) was also developed in the Toy group using immobilized TEMPO and polymer-supported 2,2′-bipyridine, which form the oxidative organometallic complex with copper [31].

The combination of two or more functional resins in a single vessel to promote two successive reactions has been shown to be possible. The Ley group has developed a method for the oxidation of primary alcohols to the corresponding acid simultaneously using three supported species in a single vessel [32]: a catalytic amount of polymer-supported TEMPO for the oxidation of the alcohol to the corresponding aldehyde; polymer chlorite and polymer-supported hydrogen phosphate for oxidation of the resulting aldehyde to the acid. This method gave excellent results, and only requires filtration to afford the clean product. The Achilefu group took advantage of this technique, using the solid-supported Pinnick type oxidation [33], in the synthesis of norzooanemonin (**20**, Scheme 4.5) [34].

4.3.1.6 Nanoparticles and Polymer Incarcerated Oxidants

A problem inherent to polymer-supported metal reagents is leaching, which leads to catalyst deactivation and product contamination by the metal. In recent years, nanoparticles have gained a lot of attention. Metal nanoparticles possess the advantage of having high catalytic activity due to their large surface area. Highly recyclable nanopalladium catalysts showing very good activity for the aerobic oxidation under atmospheric pressure of molecular oxygen, converting alcohols into the corresponding aldehydes, ketones and also carboxylic acids, have been developed recently [35]. Another very interesting solution to this problem was found by Kobayashi with the formation of polymer-micelle incarcerated metals such as ruthenium, showing activity in the oxidation of sulfides and alcohols, with minimal leaching [36].

4.3.1.7 High-Loading Resins

An important limitation of functional resins is their relatively low loading (0.5–2 mmol g^{-1} for most commercial reagents), which bar the use of such reagents from large-scale chemistry. To circumvent this limitation, the Rademann group has developed a high loading resin based on polycationic ultraresin [37]. This resin was used as a support to carry periodate [38]. The functional resin **24** (Figure 4.6) was obtained with a loading of up to 5.4 mmol g^{-1}, and was shown to be a polyvalent reagent, enabling the oxidation of sulfides, hydroquinones, hydrazines and the oxidative cleavage of diols.

4.3.2
Applications of Functional Resins for Oxidation

4.3.2.1 Polymer-Supported Chromic Acid

As mentioned in the introduction, we intend to show how functional resins can find practical use in the synthesis of complex molecules. Polymer-supported chromic acid (PS-CrO$_3$), a commercially available reagent, has long been known [39]. However, it is still seldom used in synthesis. The usefulness of this reagent was demonstrated in the synthesis of rosiglitazone (**27**, Scheme 4.6), an antihyperglycemic agent [40]. One of the key steps is the oxidation of alcohol **25** to the corresponding aldehyde **26**. Polymer-supported chromic acid gave very clean conversion without over-oxidation.

4.3.2.2 Applications of Supported TPAP

The TPAP oxidation developed by Griffith and Ley [41] has proven to be a very important tool in organic synthesis. The polymer-supported perruthenate (PSP) version, first developed in the Ley group [42], enables the use of the reagent either stoichiometrically or catalytically in tandem with a co-oxidant. The power of PSP

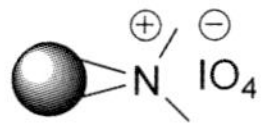

Figure 4.6 Periodate ultraresin.

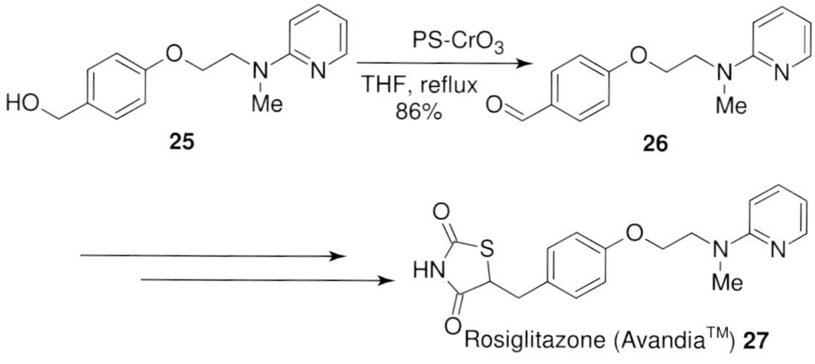

Scheme 4.6 Application of polymer-supported chromic acid in the synthesis of rosiglitazone (**27**).

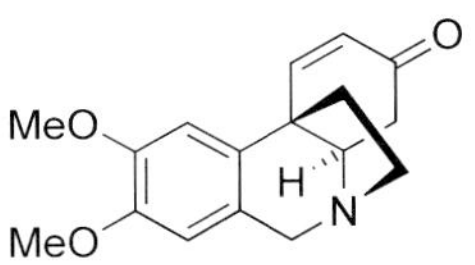

Figure 4.7 The natural product oxomaritidine.

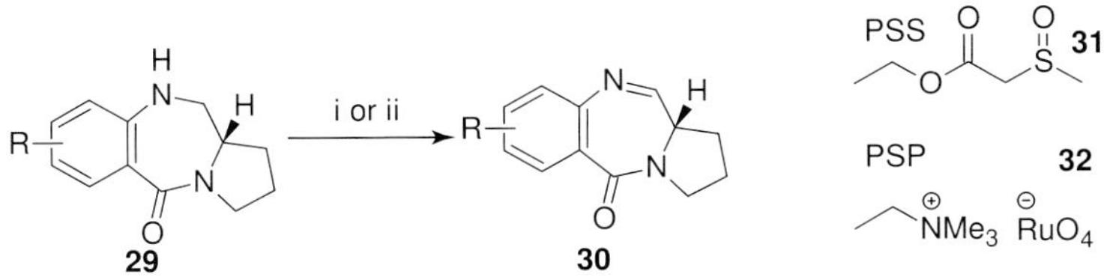

Scheme 4.7 Application of PSP for the oxidation of amines to the corresponding imine in the synthesis of PBDs.

i) PSS, (COCl)$_2$/Et$_3$N, CH$_2$Cl$_2$, -50°C to rt, 3-3.5 h, 50-64%.
ii) PSP, NMO, CH$_2$Cl$_2$, rt, 8-9h, 62-75%

has been exploited in the total synthesis of natural products such as oxomaritidine (**28**, Figure 4.7) by the Ley group. This alkaloid was prepared using a flow process involving seven synthetic steps, including a TPAP oxidation using an excess of PSP, linked into one sequence [43].

New oxidatives applications of PSP have been disclosed recently. PSP [as well as polymer-supported sulfide (PSS) [44] mediated Swern oxidation] can efficiently oxidize secondary amines to their corresponding imines [45]. Various pyrrolo[2,1-c][1,4]benzodiazepines **30** (PBDs), a family of compounds showing anticancer and antibiotic activity, have been cleanly prepared following this method (Scheme 4.7).

4.3.2.3 Aerobic Oxidation with Polymer-Supported Catalysts

Another way of circumventing the problems linked with organic co-oxidants is the use of molecular oxygen as the co-oxidant [46]. Molecular oxygen presents numerous advantages. It is environmentally friendly, technically attractive as it does not require any removal step, and it enables the use of solid-supported catalyst.

An aerobic version of the TPAP oxidation has been developed [47] and adapted by the Ley group to PSP [48]. This method was applied by this group in the aforementioned total synthesis of epothilone C [22]. The synthesis of the thiazole fragment (Scheme 4.8) requires the oxidation of primary alcohol **33** to the corresponding aldehyde **34**. This was achieved using catalytic PSP under an oxygen atmosphere.

Notably, the co-oxidant problem also occurs for TEMPO-catalyzed oxidations. For the aerobic oxidation of alcohols with supported TEMPO see the end of Section 4.3.1.2.

4.3.2.4 Osmium Tetroxide

Osmium tetroxide is one of the most efficient reagents for double bond dihydroxylation to give the corresponding vicinal diol. This reagent is, however, not perfect

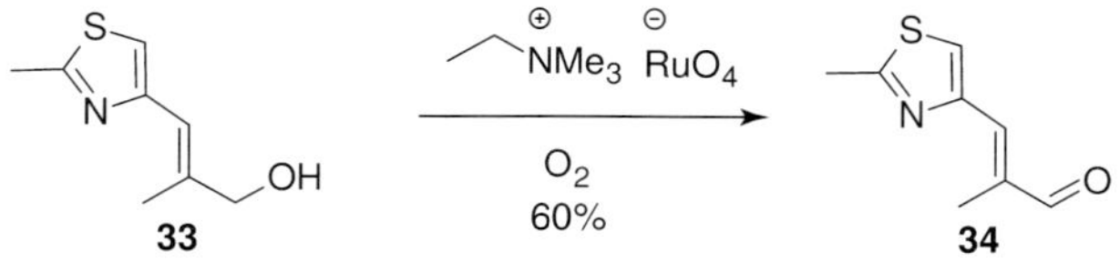

Scheme 4.8 Aerobic oxidation using PSP.

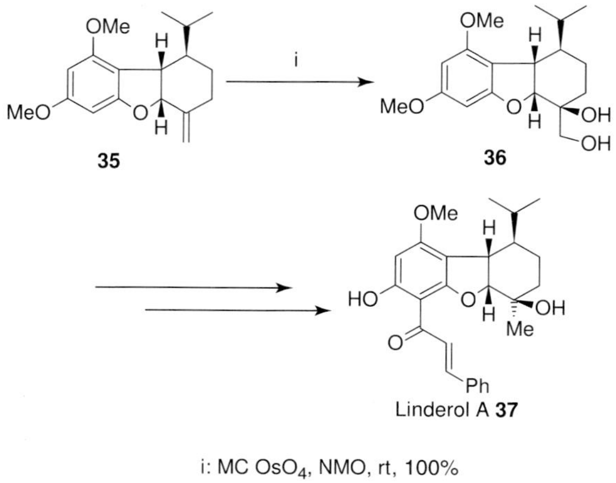

Scheme 4.9 Use of MC OsO₄ in the synthesis of linderol A (**37**).

i: MC OsO₄, NMO, rt, 100%

as it is highly toxic, volatile, expensive, it has a limited shelf life and cannot be recovered, which generally prevents it from being used in industry. Kobayashi has developed a microencapsulated osmium tetroxide (MC OsO₄), which circumvents all these problems [49]. This version of the catalyst is non-toxic, stable in air and recoverable. It has also found application in the synthesis of several natural products, and could be used with chiral inducers for asymmetric dihydroxylation [50].

One of the key steps in the first total synthesis of linderol A (**37**, Scheme 4.9), a compound with potent inhibitory activity on melanin biosynthesis, is a stereoselective dihydroxylation. MC OsO₄ gave excellent results as the cis-diol was obtained quantitatively as a single isomer [51].

4.4
Polymer-Supported Reducing Reagents

4.4.1
Novel Reagents

4.4.1.1 Chiral Boranes

The development of chiral functional resins for enantioselective reductions has given rise in the last few years to very powerful reagents with, sometimes, activity and selectivity comparable to the unbound analog.

The enantioselective reduction of prochiral ketones by chiral catalysts and reagents has received tremendous attention. The recovery and purification of those catalysts is, however, problematic. Immobilization on a solid support offers an interesting solution. Chiral sulfonamides have proven to be powerful reagents for the asymmetric borane reduction of ketones. The Zhao group in Shangai has developed a polymer-supported version of such reagents, showing comparable reactivity to the unbound analog [52]. This reagent was used by the Kamal group to prepare optically pure vasicinones [53], which are biologically interesting compounds that show good bronchodilatory activity. Enantioselective reduction of dione **38** with polymer-supported chiral sulfonamide **39** gave (−)-vasicinone **40** in excellent yield and with over 90% ee (Scheme 4.10).

Polymer-supported chiral sulfonamides also proved to be efficient catalysts for the enantioselective reduction of β-keto-nitriles [54]. This was demonstrated in the synthesis of anti-depressant drugs ®-fluoxetine **43** and ®-duloxetine (Scheme 4.11).

The CBS catalyst, devised by Corey, Bakshi and Shibata, has proven to be very useful in synthesis. The Schore group has prepared two polymer-supported

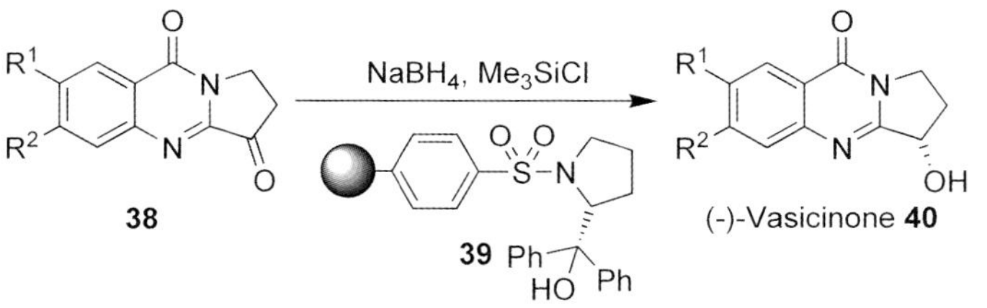

Scheme 4.10 Asymmetric reduction using polymer-supported chiral sulfonamide **39**.

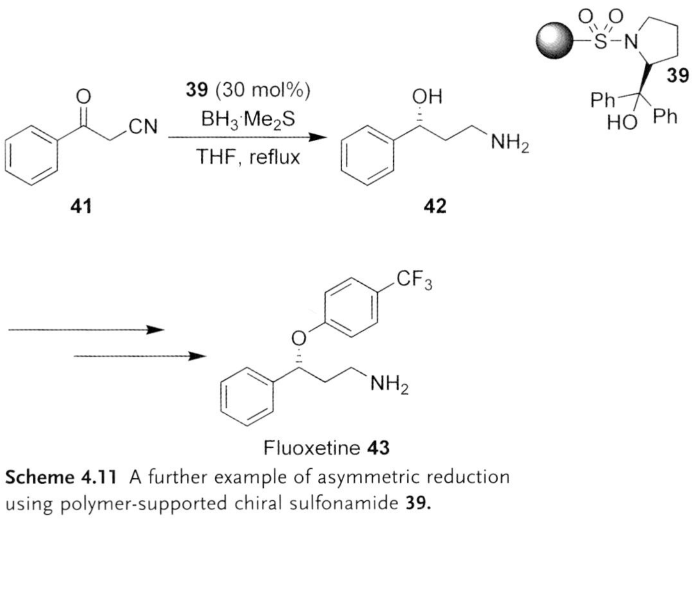

Scheme 4.11 A further example of asymmetric reduction using polymer-supported chiral sulfonamide **39**.

versions of this catalyst [55], a pendant-bound version **44** and a cross-linked version **45** (Figure 4.8). The cross-linked version proved to be the best as it affords enantioselectivities almost identical to those of the solution-phase model.

This polymer-supported reagent was used in an improved synthesis of *d*-biotin, an important compound in the treatment of type 2 diabetes; following the Hoffmann–Roche approach [56], the key synthetic step introducing stereogenic centers into the molecule was performed on a 20-g scale using the polymer-supported version of CBS to afford the desired intermediate **47** in high yield and excellent enantiomeric excess (Scheme 4.12).

4.4.1.2 Polymer-Supported Ru-TsDPEN

Another very powerful reducing agent for the enantioselective reduction of ketones is Noyori's (1*S*,2*S*)- or (1*R*,2*R*)-*N*-(*p*-tolylsulfonyl)-1,2-diphenylethylenediamine (TsDPEN) in combination with [RuCl$_2$(*p*-cymene)$_2$]. The Wang group has prepared solid-analog **49** of this reagent (Figure 4.9), which exhibit high activities and enantioselectivities for heterogeneous asymmetric transfer hydrogenation of aromatic

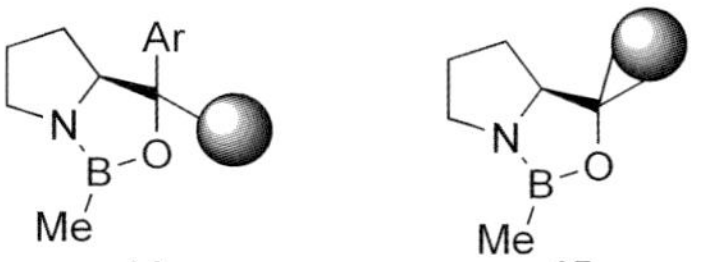

Figure 4.8 Solid-supported CBS.

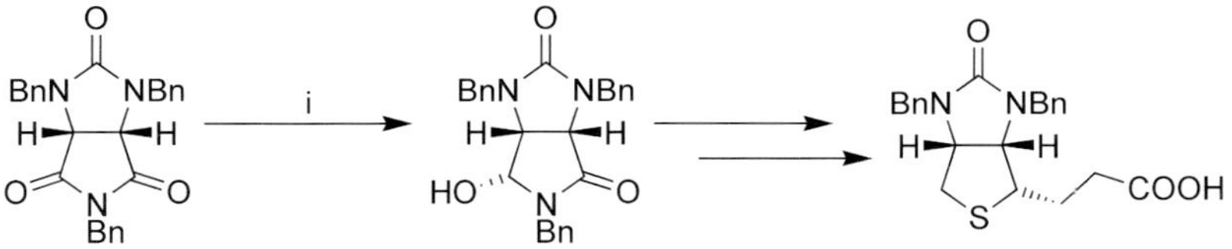

i: **45**, THF, reflux, 82%
Scheme 4.12 Use of solid-supported CBS (**45**) in the synthesis of *d*-biotin.

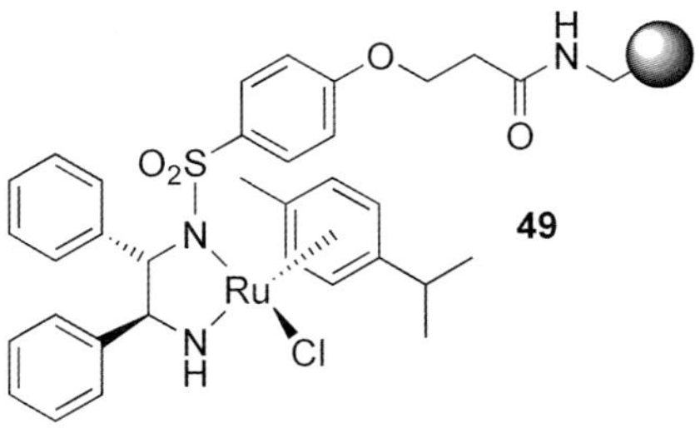

Figure 4.9 Polymer-supported Ru-TsDPEN.

ketones [57]. It has been used in the synthesis of ®-fluoxetine **43**, giving the product with excellent optical purity.

4.4.1.3 Polymer-Supported Hydrazine Hydrate

One of the methods of choice for reduction of aromatic nitro compounds to the corresponding, synthetically important, aromatic amines is the use of hydrazine hydrate. Lu's group have developed a recoverable, recyclable polymer version of hydrazine hydrate that shows excellent activity, remaining very active even after eight cycles of use (Scheme 4.13), and avoiding the release of toxic hydrazine hydrate [58].

4.4.1.4 Nanoparticles and Polymer Incarcerated Reagents

Nanoparticles reagents have also been developed for reductions. Nickel nanoparticles were prepared by the Yus group as an efficient reagent for the reduction of α,β-unsaturated carbonyl compounds, giving the corresponding saturated compound in excellent yield with high purity [59]. The system was composed of nickel(II) chloride, lithium metal, a catalytic polymer-supported arene and ethanol (Scheme 4.14). Platinum and palladium nanoparticles [35a, 60], as well as polymer incarcerated palladium [61], have also been successfully used in hydrogenating both alkenes and alkynes in the presence of hydrogen gas.

4.4.1.5 Polycationic Ultra-borohydride

A drawback of supported borohydride, often, is the low loading. The Rademann group has developed a high loading polycationic ultra-borohydride resin based on their polycationic ultraresin (Section 4.3.1.7), with a reducing activity of up to 12 mmol g^{-1} [38]. This resin showed high activity in the reduction of aldehydes, ketones and nitro-olephines, giving the products in very good yields and excellent purity. This resin was also successfully used in reductive amination.

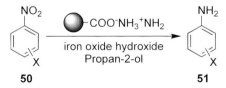

Scheme 4.13 Use of polymer-supported hydrazine hydrate.

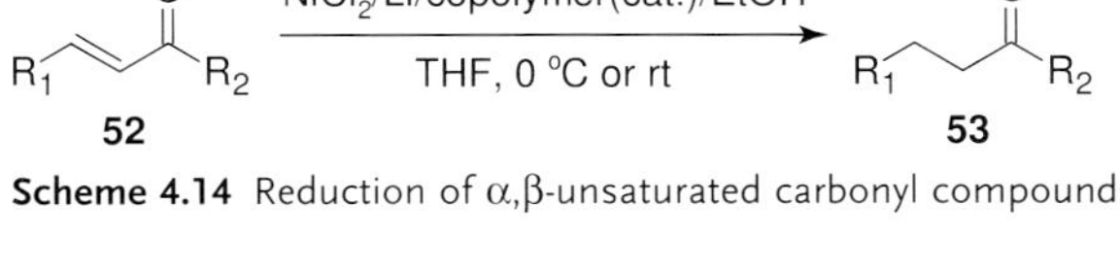

Scheme 4.14 Reduction of α,β-unsaturated carbonyl compounds using nickel nanoparticles.

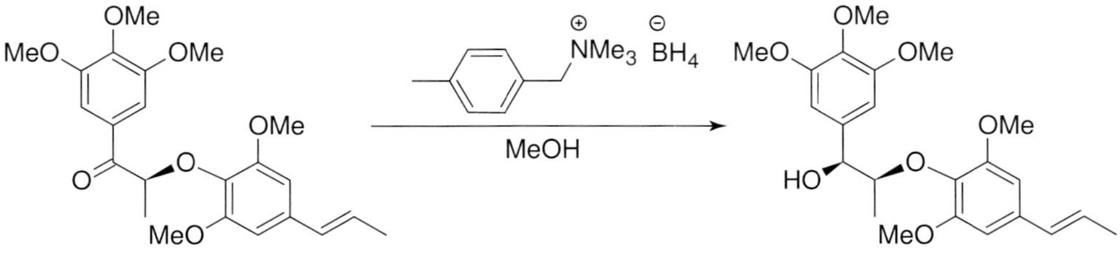

Scheme 4.15 Use of polymer-supported borohydride in synthesis.

4.4.2
Solid–Supported Reducing Agents and Their Applications

One of the most widely used supported reagents for reduction is polymer-supported borohydride. An interesting example is the final step of the total synthesis of polysphorin **55**, an anti-malarial natural product (Scheme 4.15) [62]. The stereoselective reduction of ketone **54** occurred with excellent diastereomeric excess when PS-BEMP was used, affording the **55** in 90% yield.

4.5
Conclusion

This chapter has demonstrated the progress and many improvements made in the development of novel oxidative and reducing resins over the past few years, leading to more and more applications in this area. Perhaps the most remarkable achievement is the production of efficient chiral functional resins with activities and selectivity comparable to the unbound reagents. Some of these reagents were used in the synthesis of complex natural products and of libraries of compounds. Nevertheless, published applications are generally confined to small volume syntheses, for example for applications in medicinal chemistry. What remains to be exploited is the use of polymer reagents on a large scale and in industry.

References

1 Akelah, A. and Sherrington, D.C. (1981) *Chemical Reviews*, **81**, 557–87.
2 Ley, S.V., Baxendale, I.R., Bream, R.N., et al. (2000) *Journal of the Chemical Society – Perkin Transactions 1*, 3815–4195.
3 Kirschning, A., Monenschein, H. and Wittenberg, R. (2001) *Angewandte Chemie – International Edition*, **40**, 650–79.
4 For recent reviews, see also: (a) Cozzi, F. (2006) *Advanced Synthesis Catalysis*, **348**, 1367–90. (b) Benaglia, M., Puglisi, A. and Cozzi, F. (2003) *Chemical Reviews*, **103**, 3401–29.
5 Egelhaaf, H.J. and Rademann, J. (2005) *Journal of Combinatorial Chemistry*, **7**, 929–41.

6 Rademann, J., Barth, M., Brock, R., *et al.* (2001) *Chemistry – A European Journal*, **7**, 3884–9.

7 (a) Sherrington, D.C. (1998) *Chemical Communications*, 2275–86. (b) Sherrington, D.C. (2001) *Journal of Polymer Science Part A – Polymer Chemistry*, **39**, 2364–77.

8 Nicolaou, K.C., Baran, P.S. and Zhong, Y.L. (2001) *Journal of the American Chemical Society*, **123**, 3183–5.

9 Wirth, T. (2005) *Angewandte Chemie – International Edition*, **44**, 3656–65.

10 (a) Mulbaier, M. and Giannis, A. (2001) *Angewandte Chemie – International Edition*, **40**, 4393–4. (b) Mulbaier, M. and Giannis, A. (2003) *Arkivoc*, 228–36. (c) Sorg, G., Mengel, A., Jung, G. and Rademann, J. (2001) *Angewandte Chemie – International Edition*, **40**, 4395–7. (d) Reed, N.N., Delgado, M., Hereford, K., *et al.* (2002) *Bioorganic and Medicinal Chemistry Letters*, **12**, 2047–9.

11 For other polymer-supported hypervalent iodine reagents, see: Togo, H. and Sakuratani, K. (2002) *Synlett*, 1966–75.

12 Sorg, G., Thern, B., Mader, O., *et al.* (2005) *Journal of Peptide Science*, **11**, 142–52.

13 Lei, Z., Denecker, C., Jegasothy, S., *et al.* (2003) *Tetrahedron Letters*, **44**, 1635–7.

14 Lei, Z.Q., Ma, H.C., Zhang, Z. and Yang, Y.X. (2006) *Reactive and Functional Polymers*, **66**, 840–4.

15 Zhdankin, V.V., Koposov, A.Y., Netzel, B.C., *et al.* (2003) *Angewandte Chemie – International Edition*, **42**, 2194–6.

16 (a) Chung, W.J., Kim, D.K. and Lee, Y.S. (2003) *Tetrahedron Letters*, **44**, 9251–4. (b) Chung, W.J., Kim, D.K. and Lee, Y.S. (2005) *Synlett*, 2175–8. (c) Lecarpentier, P., Crosignani, S. and Linclau, B. (2005) *Molecular Diversity*, **9**, 341–51. (d) Jang, H.S., Chung, W.J. and Lee, Y.S. (2007) *Tetrahedron Letters*, **48**, 3731–4.

17 (a) Ladziata, U., Koposov, A.Y., Lo, K.Y., *et al.* (2005) *Angewandte Chemie – International Edition*, **44**, 7127–31. (b) Ladziata, U. and Zhdankin, V.V. (2007) *Synlett*, 527–37.

18 For a review on TEMPO, see: Sheldon, R.A. and Arends, I.W.C.E. (2004) *Advanced Synthesis Catalysis*, **346**, 1051–71.

19 (a) Ciriminna, R., Bolm, C., Fey, T. and Pagliaro, M. (2002) *Advanced Synthesis Catalysis*, **344**, 159–63. (b) Weik, S., Nicholson, G., Jung, G. and Rademann, J. (2001) *Angewandte Chemie – International Edition*, **40**, 1436–9. (c) Brunel, D., Fajula, F., Nagy, J.B., *et al.* (2001) *Applied Catalysis A: General*, **213**, 73–82. (d) Dijksman, A., Arends, I.W.C.E. and Sheldon, R.A. (2001) *Synlett*, 102–4. (e) Fey, T., Fischer, H., Bachmann, S., *et al.* (2001) *Journal of Organic Chemistry*, **66**, 8154–9. (f) Ciriminna, R., Bolm, C., Fey, T. and Pagliaro, M. (2002) *Advanced Synthesis Catalysis*, **344**, 159–63. (g) Tanyeli, C. and Gumus, A. (2003) *Tetrahedron Letters*, **44**, 1639–42. (h) Pozzi, G., Cavazzini, M., Quici, S., *et al.* (2004) *Organic Letters*, **6**, 441–3. (i) Ferreira, P., Hayes, W., Phillips, E., *et al.* (2004) *Green Chemistry*, **6**, 310–12. (j) Ferreira, P., Phillips, E., Rippon, D., *et al.* (2004) *Journal of Organic Chemistry*, **69**, 6851–9. (k) Ferreira, P., Phillips, E., Rippon, D. and Tsang, S.C. (2005) *Applied Physics B – Lasers and Optics*, **61**, 206–11. (l) Gheorghe, A., Matsuno, A. and Reiser, O. (2006) *Advanced Synthesis Catalysis*, **348**, 1016–20. (m) Gheorghe, A., Cuevas-Yanez, E., Horn, J., *et al.* (2006) *Synlett*, 2767–70. (n) Luo, J.T., Pardin, C., Lubell, W.D. and Zhu, X.X. (2007) *Chemical Communications*, 2136–8.

20 (a) Sheldon, R.A., Arends, I.W.C.E., Brink, G.J. and Dijksman, A. (2002) *Accounts of Chemical Research*, **35**, 774–81. (b) Minisci, F., Recupero, F., Rodino, M., *et al.* (2003) *Organic Process Research and Development*, **7**, 794–8. (c) Minisci, F., Recupero, F., Cecchetto, A., *et al.* (2004) *European Journal of Organic Chemistry*, 109–19. (d) Gilhespy, M., Lok, M. and Baucherel, X. (2005) *Chemical Communications*, 1085–6. (e) Benaglia, M., Puglisi, A., Holczknecht, O., *et al.* (2005) *Tetrahedron*, **61**, 12058–64. (f) Jiang, N. and Ragauskas, A.J. (2006) *Journal of Organic Chemistry*, **71**, 7087–90. (g) Gilhespy, M., Lok, M. and Baucherel, X. (2006) *Catalysis Today*, **117**, 114–19.

21 (a) Brown, D.S., Campbell, E., Kerr, W.J., *et al.* (2000) *Synlett*, 1573–6. (b) Brown, D.S., Kerr, W.J., Lindsay, D.M., *et al.* (2001) *Synlett*, 1257–9.

22 Storer, R.I., Takemoto, T., Jackson, P.S., et al. (2004) *Chemistry – A European Journal*, **10**, 2529–47.

23 (a) Sourkouni-Argirusi, G. and Kirschning, A. (2000) *Organic Letters*, **2**, 3781–4. (b) Brunjes, M., Sourkouni-Argirusi, G. and Kirschning, A. (2003) *Advanced Synthesis Catalysis*, **345**, 635–42. (c) Kloth, K., Brunjes, M., Kunst, E., et al. (2005) *Advanced Synthesis Catalysis*, **347**, 1423–34.

24 For similar reactions using poly[4-(diacetoxyiodo)styrene] see: (a) Sakuratani, K. and Togo, H. (2003) *Synthesis*, 21–3. (b) Tashino, Y. and Togo, H. (2004) *Synlett*, 2010–12.

25 Nicolaou, K.C., Roecker, A.J., Monenschein, H., et al. (2003) *Angewandte Chemie – International Edition*, **42**, 3637–42.

26 Pasetto, P. and Franck, R.W. (2003) *Journal of Organic Chemistry*, **68**, 8042–60.

27 Rudolph, J., Hannig, F., Theis, H. and Wischnat, R. (2001) *Organic Letters*, **3**, 3153–5.

28 Sartori, G., Armstrong, A., Maggi, R., Mazzacani, et al. (2003) *Journal of Organic Chemistry*, **68**, 3232–7.

29 Or other supported dioxiranes precursor, see: Kan, J.T.W. and Toy, P.H. (2004) *Tetrahedron Letters*, **45**, 6357–9.

30 But, T.Y.S., Tashino, Y., Togo, H. and Toy, P.H. (2005) *Organic and Biomolecular Chemistry*, **3**, 970–1.

31 Chung, C.W.Y. and Toy, P.H. (2007) *Journal of Combinatorial Chemistry*, **9**, 115–20.

32 Yasuda, K. and Ley, S.V. (2002) *Journal of the Chemical Society – Perkin Transactions 1*, 1024–5.

33 Bal, B.S., Childers, W.E. and Pinnick, H.W. (1981) *Tetrahedron*, **37**, 2091–6.

34 Berezin, M. and Achilefu, S. (2007) *Tetrahedron Letters*, **48**, 1195–9.

35 (a) Park, C.M., Kwon, M.S. and Park, J. (2006) *Synthesis*, 3790–4. (b) Yasuhiro, Y., Nakao, R. and Rhee, H. (2007) *Journal of Organometallic Chemistry*, **692**, 420–7.

36 (a) Miyamura, H., Akiyama, R., Ishida, T., et al. (2005) *Tetrahedron*, **61**, 12177–85. (b) Matsumoto, T., Ueno, M., Kobayashi, J., et al. (2007) *Advanced Synthesis Catalysis*, **349**, 531–4.

37 Barth, M. and Rademann, J. (2004) *Journal of Combinatorial Chemistry*, **6**, 340–9.

38 Barth, M., Shah, S.T.A. and Rademann, J. (2004) *Tetrahedron*, **60**, 8703–9.

39 Cainelli, G., Cardillo, G., Orena, M. and Sandri, S. (1976) *Journal of the American Chemical Society*, **98**, 6737–8.

40 Li, X., Abell, C., Warrington, B.H. and Ladlow, M. (2003) *Organic and Biomolecular Chemistry*, **1**, 4392–5.

41 (a) Griffith, W.P., Ley, S.V., Whitcombe, G.P. and White, A.D. (1987) *Journal of the Chemical Society D – Chemical Communications*, 1625–7. (b) Ley, S.V., Norman, J., Griffith, W.P. and Marsden, S.P. (1994) *Synthesis*, 639–66.

42 Hinzen, B. and Ley, S.V. (1997) *Journal of the Chemical Society – Perkin Transactions 1*, 1907–8.

43 Baxendale, I.R., Deeley, J., Griffiths-Jones, C.M., et al. (2006) *Chemical Communications*, 2566–8.

44 For articles on PSS, see: (a) Liu, Y.Q. and Vederas, J.C. (1996) *Journal of Organic Chemistry*, **61**, 7856–9. (b) Harris, J.M., Liu, Y.Q., Chai, S.Y., et al. (1998) *Journal of Organic Chemistry*, **63**, 2407–9.

45 Kamal, A., Devaiah, V., Reddy, K.L. and Shankaraiah, N. (2006) *Advanced Synthesis Catalysis*, **348**, 249–54.

46 For reviews on the use of molecular oxygen in the oxidation of alcohols, see: (a) Mallat, T. and Baiker, A. (2004) *Chemical Reviews*, **104**, 3037–58. (b) Zhan, B.Z. and Thompson, A. (2004) *Tetrahedron*, **60**, 2917–35.

47 (a) Lenz, R. and Ley, S.V. (1997) *Journal of the Chemical Society – Perkin Transactions 1*, 3291–2. (b) Marko, I.E. Giles, P.R., Tsukazaki, M., et al. (1997) *Journal of the American Chemical Society*, **119**, 12661–2.

48 Hinzen, B., Lenz, R. and Ley, S.V. (1998) *Synthesis*, 977–9.

49 (a) Nagayama, S., Endo, M. and Kobayashi, S. (1998) *Journal of Organic Chemistry*, **63**, 6094–5. (b) Kobayashi, S., Ishida, T. and Akiyama, R. (2001) *Organic Letters*, **3**, 2649–52. (c) Kobayashi, S. and Akiyama, R. (2003) *Chemical Communications*, 449–60. (d) Ishida, T., Akiyama, R. and Kobayashi, S. (2003) *Advanced Synthesis Catalysis*, **345**, 576–9.

50 (a) Kobayashi, S., Ishida, T. and Akiyama, R. (2001) *Organic Letters*, **3**, 2649–52. (b)

Kobayashi, S. and Sugiura, M. (2006) *Advanced Synthesis Catalysis*, **348**, 1496–504.

51 (a) Yamashita, M., Ohta, N., Kawasaki, I. and Ohta, S. (2001) *Organic Letters*, **3**, 1359–62. (b) Yamashita, M., Ohta, N., Shimizu, T., *et al.* (2002) *Journal of Organic Chemistry*, **68**, 1216–24.

52 (a) Hu, J.B., Zhao, G., Yang, G.S. and Ding, Z.D. (2001) *Journal of Organic Chemistry*, **66**, 303–4. (b) Hu, J.B., Zhao, G. and Ding, Z.D. (2001) *Angewandte Chemie – International Edition*, **40**, 1109–11.

53 Kamal, A., Devaiah, V., Shankaraiah, N. and Reddy, K.L. (2006) *Synlett*, 2609–12.

54 Wang, G.Y., Liu, X.S. and Zhao, G. (2005) *Tetrahedron-Asymmetry*, **16**, 1873–9.

55 Price, M.D., Sui, J.K., Kurth, M.J. and Schore, N.E. (2002) *Journal of Organic Chemistry*, **67**, 8086–9.

56 Chen, F.E., Jia, H.Q., Chen, X.X., *et al.* (2005) *Chemical and Pharmaceutical Bulletin*, **53**, 743–6.

57 Li, Y.Z., Li, Z.M., Li, F., *et al.* (2005) *Organic and Biomolecular Chemistry*, **3**, 2513–18.

58 Shi, Q.X., Lu, R.W., Jin, K., *et al.* (2006) *Green Chemistry*, **8**, 868–70.

59 Alonso, F., Osante, I. and Yus, M. (2006) *Synlett*, 3017–20.

60 Oyamada, H., Akiyama, R., Hagio, H., *et al.* (2006) *Chemical Communications*, 4297–9.

61 Akiyama, R. and Kobayashi, S. (2003) *Journal of the American Chemical Society*, **125**, 3412–13.

62 Lee, A.L. and Ley, S.V. (2003) *Organic and Biomolecular Chemistry*, **1**, 3957–66.

5
Base and Acid Reagents

Tadashi Aoyama

Tadashi AOYAMA, Department of Materials and Applied Chemistry, College of Science and Technology, Nihon university, 1-8-14 Kanda Surugadai, Chiyoda-ku, Tokyo 101-8308, Japan

5.1
Base Reagents

For environmental reasons, it is highly desirable to replace homogeneous acid and base catalysts for organic synthesis with the greener chemistry of heterogeneous acid and base catalysts. Numerous studies have been devoted to develop heterogeneous acid catalysts, but fewer studies have been made using base supported reagents than with acid supported reagents. This is because homogeneous chemical processes that employ acid reagents and catalysts are environmentally more destructive than base reagents and catalysts. In addition, base reagents such as an alkaline hydroxide are inexpensive and their waste products are relatively easy to dispose after neutralization and, consequently, they are highly competitive as catalysts.

Several base supported reagents using various inorganic solid supports, such as alumina, silica, zeolite and MgO, have been studied. Their properties and applications are covered in several detailed reviews and papers. The present section describes the recent application of inorganic solid supported base reagents for organic synthesis. Alumina supported potassium fluoride (KF/Al_2O_3), which was introduced by Ando *et al.*, has been widely used as a heterogeneous super base for many reactions, and many applications have been reported. Recently, silica surface bound organic bases, a hybrid organic–inorganic material, were developed as heterogeneous base catalysts. Much effort has been devoted to the development of similar catalysts.

5.1.1
KF/Al_2O_3

KF/Al_2O_3 was first reported in 1979 by Ando *et al.*, who showed its use in the *O*-alkylation of aliphatic alcohols [1]. Since then, KF/Al_2O_3 has been widely used in

The Power of Functional Resins in Organic Synthesis. Judit Tulla-Puche and Fernando Albericio
Copyright © 2008 WILEY-VCH Verlag GmbH & Co. KGaA, Weinheim
ISBN: 978-3-527-31936-7

various reactions, for instance carbon–oxygen, carbon–nitrogen and carbon–carbon bond formation reactions, and it is the most popular base among inorganic solid supported base reagents. Recently, it has been reported to act as an effective promoter for Suzuki couplings and Sonogashira couplings.

5.1.1.1 Generation of Dichlorocarbene

Generation of dihalocarbene by using both KF/Al_2O_3 and phase-transfer catalysts was first reported by Ting et al. [2]. This method, however, was less effective than that using only a phase-transfer catalyst. Therefore, the reactions of dihalocarbenes using this method have received little attention. However, Ishino, Komatsu et al. have shown the N-formylation of secondary amines using KF/Al_2O_3 under mild conditions [3]. A proposed pathway is the reaction of secondary amines with dichlorocarbene generated in situ from chloroform and KF/Al_2O_3, followed by hydrolysis by water contained in the latter to afford the N-formylated products (Scheme 5.1).

The reaction of imines with chloroform in the presence of KF or Al_2O_3 did not proceed at all, whereas the use of commercially available KF/Al_2O_3 led to the gem-dichloroaziridines. Although conventional methods for generation of dichlorocarbene require a large excess of chloroform, the present method is able to reduce the amount of chloroform considerably [4] (Scheme 5.2).

5.1.1.2 Acceleration of the Metal-Catalyzed Reaction

Basu et al. have reported palladium-catalyzed selective cross coupling of bromopyridines and amines on the surface of potassium fluoride supported basic aluminium oxide [5]. For example, amination of 2,6-dibromopyridine by this method afforded the monoaminated products predominantly, even after prolonged the reaction time and using excess amine, whereas previous method [6] gave only bis-aminated products (Scheme 5.3).

Heteroatom–carbon bond formation reactions were also reported using the reagents system, 1,10-phenanthroline and KF/Al_2O_3 in the presence of CuI [7].

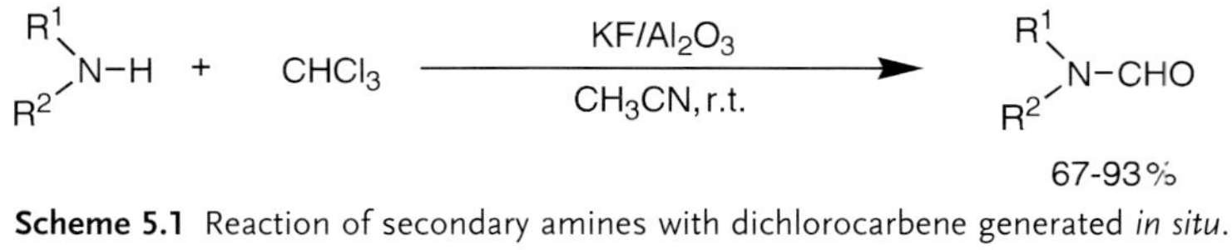

Scheme 5.1 Reaction of secondary amines with dichlorocarbene generated in situ.

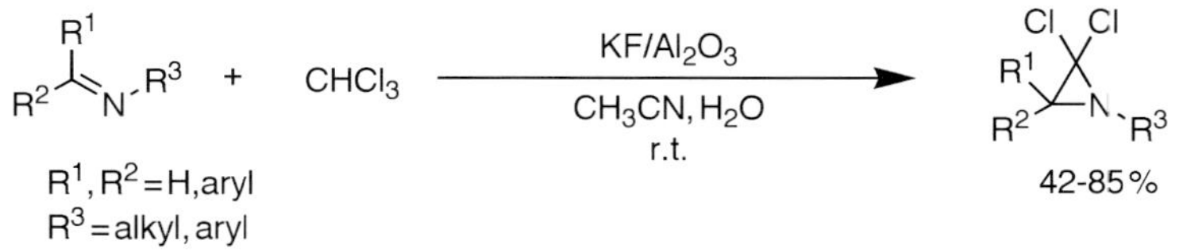

Scheme 5.2 Reaction of imines and chloroform in the presence of commercially available KF/Al_2O_3.

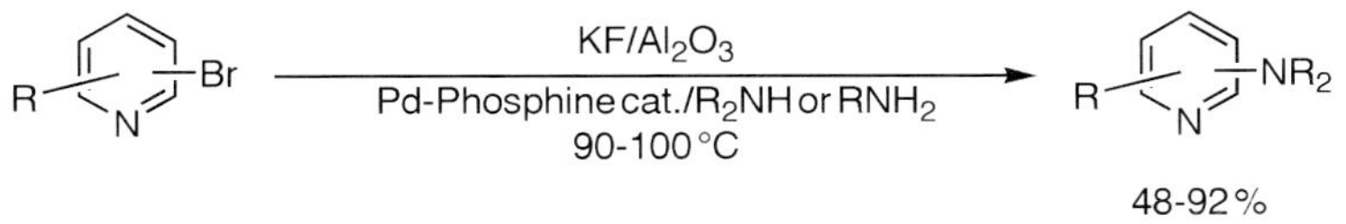

Scheme 5.3 Palladium-catalyzed selective cross coupling of bromopyridines and amines on potassium fluoride supported basic aluminium oxide.

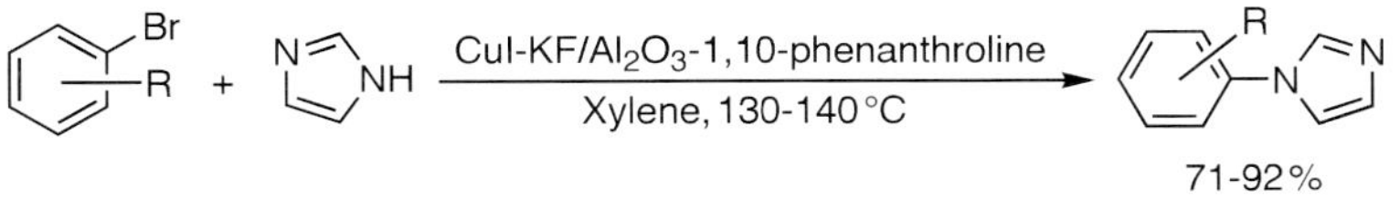

Scheme 5.4 Heteroatom–carbon bond formation using CuI-KF/Al_2O_3-1,10-phenanthroline.

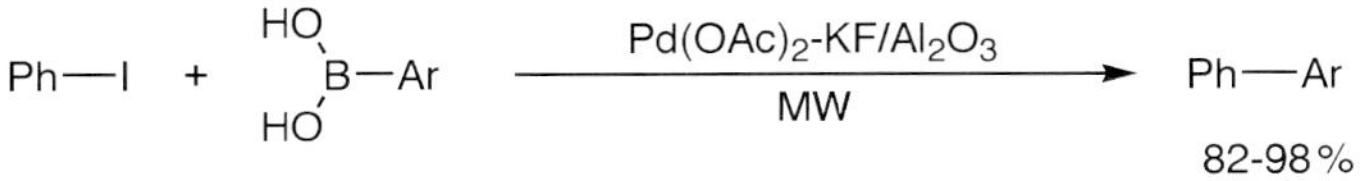

Scheme 5.5 Microwave-assisted reaction of iodobenzene with arylboronic acids.

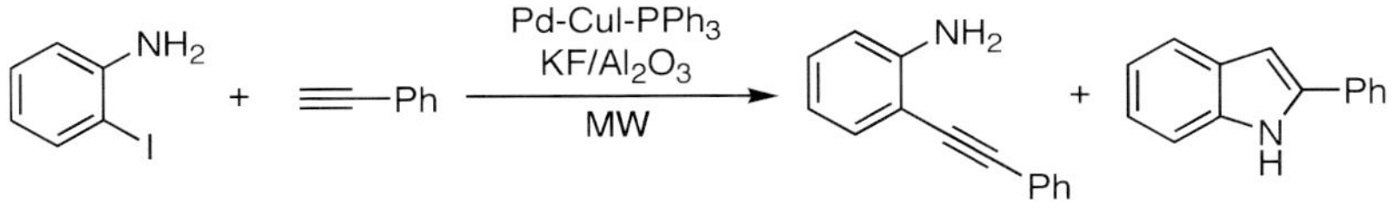

	Overall yield (%)	Ratio
amine : alkyne = 1:1	86	100 : 0
amine : alkyne = 2:1	84	54 : 46

Scheme 5.6 Reaction of *o*-iodoaniline and alkyne under microwave irradiation.

This system catalyzed *N*-arylation of imidazole and pyrazole and benzimidazole with aryl bromides. *N*-Amidation and *O*-arylation were also catalyzed using this reagents system [8, 9] (Scheme 5.4).

5.1.1.3 With Microwave

Palladium-catalyzed reactions such as Suzuki, Heck, Trost-Tsuji and Stille have been modified in the presence of KF/Al_2O_3 without solvent under microwave irradiation conditions by Villemin *et al.* [10]. These reactions rapidly afforded the product in moderate to excellent yields. For instance, a reaction of iodobenzene with arylboronic acids was completed in 2–15 min. Recently, improved methods under microwave irradiation have been reported [11–16] (Scheme 5.5).

Kabalka *et al.* also developed the Glaser coupling [17] and Sonogashira [18] reactions under microwave irradiation without solvent. A molar ratio of *o*-iodoaniline to alkyne of 2 : 1 gave a mixture of coupling and cyclization products (Scheme 5.6).

The reaction of benzaldehyde with triethylsilane in the presence of KF/Al$_2$O$_3$ in DMF gave only benzyloxytriethylsilane [19]. However, when hexane or THF were used instead of DMF, the Tishchenko reaction occurred to give, mainly, benzyl benzoate (Scheme 5.7).

5.1.2
Silica-Gel Bound Organic Bases

Silica-gel bound organic bases, such as diamines, ammonium hydroxide, guanidines and 4-methylpyridine, have become of interest in recent years. The best advantage of these silica-gel bound organic bases is that they can be reused for subsequent reactions (Figure 5.1).

Nitroalkenes are important precursors for the synthesis of insecticides, fungicides and pharmaceuticals. The reaction of benzaldehyde and nitromethane in the presence of diamino-functionalized mesoporous catalyst gave nitroalkenes in excellent yields; the catalyst could be reused six times without loss of activity [20]. Knoevenagel and aldol reactions were also catalyzed by the same catalyst, which could be reused five times [21].

MCM-41 containing amines have been used as catalysts for glycidol ring-opening reactions with fatty acids under mild conditions [22].

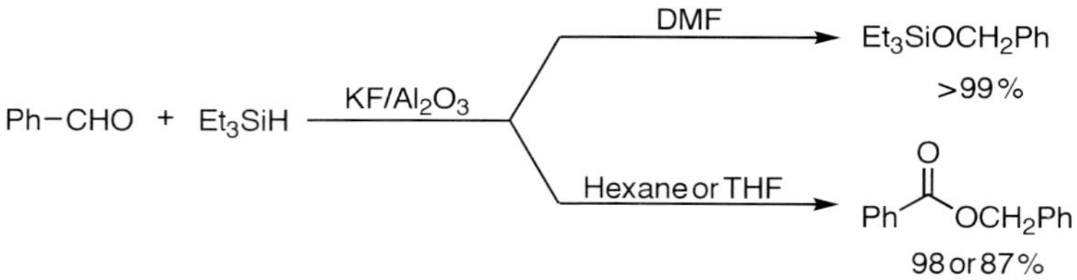

Scheme 5.7 Effect of solvent on the reaction of benzaldehyde with triethylsilane in the presence of KF/Al$_2$O$_3$.

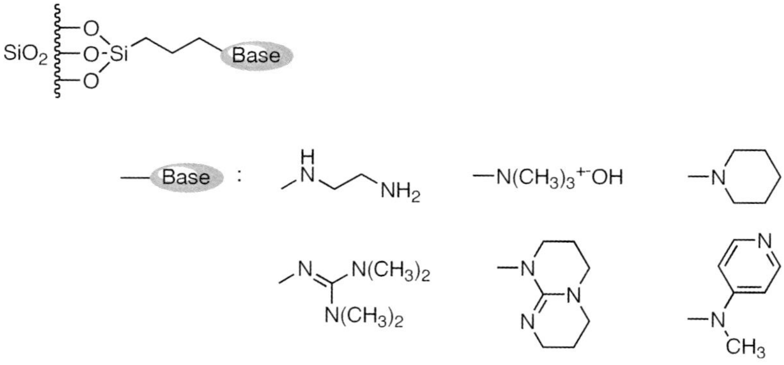

Figure 5.1 Some silica-gel bound organic bases.

5.1 Base Reagents

Quaternary ammonium hydroxides immobilized on MCM-41 (MCM-41OH1) [23] act a stronger base catalyst than amine analogs. The condensation of salicylaldehyde with diethyl-2-pentenedicarboxylate in the presence of this catalyst led to a mixture of coumarin-3-acrylates and chromene derivatives [24]. The ratio of the formation of coumarin-3-acrylates and chromene derivatives depended on the base strength of the catalyst (Scheme 5.8).

Knoevenagel reactions and epoxidation of enones were achieved by using silica-gel immobilized guanidines. 3-Nonenoic acid, which is precursor of γ-nonanoic lactone, was produced from malonic acid and heptanal in the presence of silica-gel bound 1,1,3,3-tetramethylguanidine [25]. This catalyst and MCM-TBD catalyzed the epoxidation of enones in the presence of H_2O_2. MCM-TBD has been prepared from MCM-41 covered with 3-trimethoxylsilylpropoxymethyloxirane and 1,5,7-triazabicyclo[4.4.0]dec-5-ene (TBD) by Jacobs et al. [26] (Scheme 5.9).

MCM-TBD was used for a consecutive alkylation-Knoevenagel and aza-Michael–Knoevenagel reaction [27], a selective preparation of carbamates and alkyl carbonate [28] and a cycloaddition of CO_2 to epoxides [29]. For instance, a mixture of 2-phenyloxirane and MCM-TBD in acetonitrile was charged in an autoclave, CO_2 was then introduced (50 bar) and the mixture was heated at 140 °C for 70 h, to give the corresponding product in 90% yield. The reaction using MCM-TBD is slower than that with homogeneous 7-methyl-1,5,7-triazabicyclo[4.4.0]dec-5-ene, but has the great advantage that MCM-TBD can easily be recovered from the reaction mixture and reused for three further cycles (Scheme 5.10).

Neat SiO_2-TBD also acted as a catalyst for Nef reactions [30]. A treatment of sec-nitroalkanes with neat SiO_2-TBD allowed direct conversion into the corresponding ketones (Scheme 5.11).

4-(Dimethylamino)pyridine (DMAP) has been widely used as an efficient catalyst for various reactions, and several approaches for immobilization of DMAP on

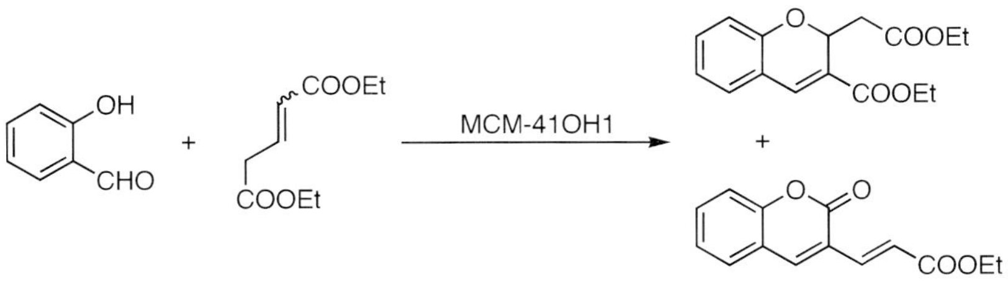

Scheme 5.8 Condensation of salicylaldehyde with diethyl-2-pentenedicarboxylate in the presence of MCM-41OH1.

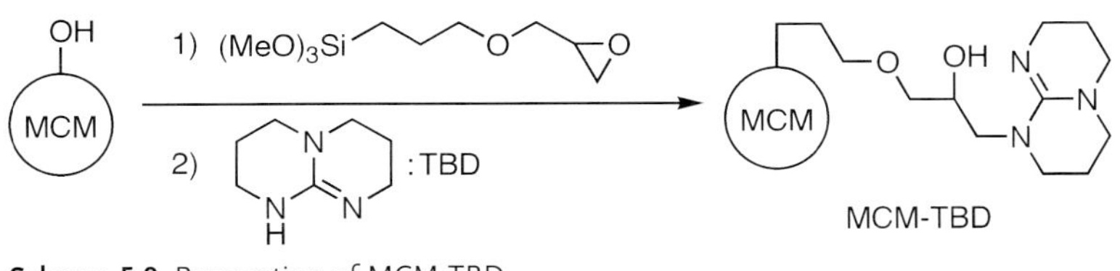

Scheme 5.9 Preparation of MCM-TBD.

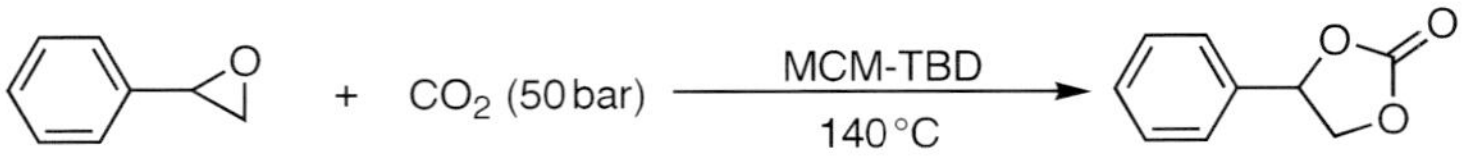

Scheme 5.10 Cycloaddition of CO_2 to 2-phenyloxirane using MCM-TBD.

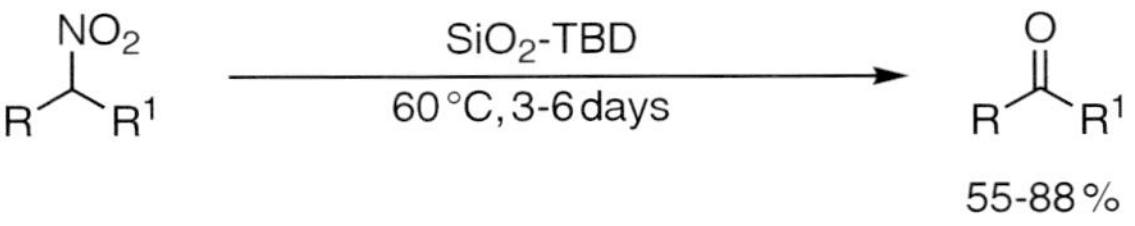

Scheme 5.11 Conversion of nitroalkanes into the corresponding ketones using SiO_2-TBD.

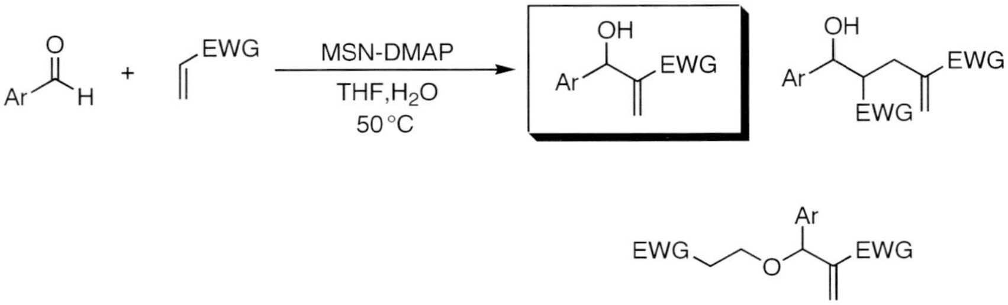

Scheme 5.12 Baylis–Hillman reaction employing MSN-DMAP.

various organic and inorganic solid supports have been reported [31–41]. The catalytic activity of these immobilized DMAP, however, is lower than that of DMAP only. Lin *et al.* have developed an effective immobilized DMAP (MSN-DMAP) using mesoporous silica nanosphere [42]. MSN-DMAP was an effective catalyst for Baylis–Hillman, acylation and silylation reactions. Usually, the Baylis–Hillman reaction requires stoichiometric or excess amounts of DMAP and many by-products are formed. In contrast, no by-products are observed when MSN-DMAP was used for the reaction (Scheme 5.12).

5.1.2.1 For Electroorganic Synthesis

Fuchigami *et al.* have developed an electrolytic system using solid-supported bases for *in situ* generation of a supporting electrolyte from methanol or acetic acid [43–46]. This system can achieve methoxylations of carbamates, furans, phenols and substituted benzenes, acetoxylations, Kolbe-reactions and non-Kolbe reactions. Among various solid-supported organic bases tested, silica-gel immobilized piperidine was the most effective for methoxylations, Kolbe-reactions and non-Kolbe reactions. Silica-gel immobilized morpholine was effective for acetoxylations. These solid-supported organic bases are easily recovered and can be reused at least 10 times without loss of activity (Figure 5.2).

5.1.3
Others

KF/Al_2O_3 is widely used as a solid base in a wide variety of reactions, whereas application of CsF/Al_2O_3 to organic reactions is not so well reported. Figueras

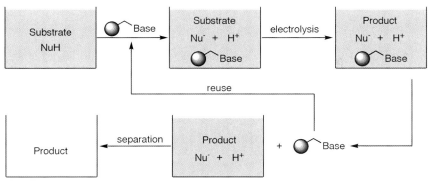

Figure 5.2 Electrolytic system using solid-supported bases for *in situ* generation of a supporting electrolyte.

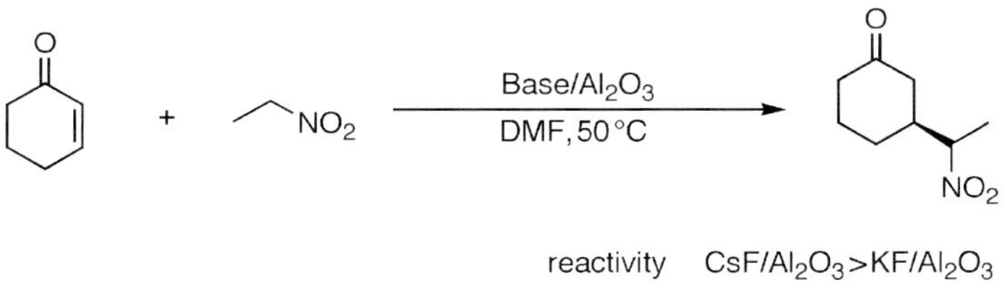

reactivity   CsF/Al$_2$O$_3$ > KF/Al$_2$O$_3$

Scheme 5.13 Relative reactivity of CsF/Al$_2$O$_3$ and KF/Al$_2$O$_3$ for the Michael reaction of EtNO$_2$.

et al. have reported that CsF/Al$_2$O$_3$ is more effective than KF/Al$_2$O$_3$ for Michael reaction of EtNO$_2$ [47] (Scheme 5.13).

CsF/Al$_2$O$_3$ was also used for deacylation reactions [48]. Among several inorganic supported CsF systems, Celite-supported CsF has been used for many reactions, such as *N*-alkylation, oxidation of thiols, addition reaction and Williamson synthesis [49–54].

KOH/kaolin combined with a chiral catalyst was used for asymmetric alkylation of glycine imine esters, and recycled for subsequent reactions without loss of activity over three times [55, 56].

Asymmetric synthesis of β-phosphono-malonates via KOH/Fe$_2$O$_3$ catalyzed phospha-Michael reaction to Knoevenagel acceptors has been achieved by Enders and Tedeschi [57]. The reaction did not proceed when only KOH was used as a catalyst. Fe$_2$O$_3$ was the best diastereoselective support among the other supports tested, Al$_2$O$_3$, ZnO, Cu$_2$O, MnO$_2$ and MgO.

Na$_2$CO$_3$/Al$_2$O$_3$ has been used as a deacylating reagent of α-bromo-β-ketoesters and α-bromo-β-diketones [58]. When NaOH/Al$_2$O$_3$ was used for the reaction instead of Na$_2$CO$_3$/Al$_2$O$_3$, the expected α-bromoesters and α-bromoketones were formed in low yield. Na$_2$CO$_3$/SiO$_2$ was also effective, as well as Na$_2$CO$_3$/Al$_2$O$_3$.

5.2
Acid Reagents

Acid catalysts have been used in such important industrial process as reforming, paraffin isomerization, catalytic cracking and so on. Solid acid catalysts have many advantages, such as easy handling, decrease of plant corrosion and environmentally safe disposal. Many solid acids, therefore, have been developed. From the point of green chemistry, the development of reusable supported acid catalysts has attracted much attention. For instance, acid-supported reagents prepared at the end of twentieth century have been developed to replace traditional homogeneous reaction conditions with easier, milder reaction conditions. In the syntheses of thioacetals and dithioacetals from aldehydes, acidic supported reagents prepared in the twenty-first century have not only these advantages but are also reusable. Silica gel-supported heteropoly acid ($H_3PW_{12}O_{40}/SiO_2$) [59] and silica gel-supported polyphosphoric acid (PPA/SiO_2) [60] can be reused several times. $H_3PW_{12}O_{40}/SiO_2$ was recycled five times in the synthesis of oxathioacetals without loss of catalytic activity, and PPA/SiO_2 could be reused nine times to afford the product in at least in 80% yield (Table 5.1).

5.2.1
SSA

An acidic catalyst, prepared from silica gel and chlorosulfuric acid, has been developed by Zolfigol *et al.* and named "silica sulfuric acid (SSA)" [67] (Scheme 5.14).

SSA is a better proton source than polymer-supported sulfonic acids such as polystyrene sulfonic acid and Nafion-H. Many reactions using SSA have been reported, such as multicomponent reactions [68–76], carbon–oxygen [77–86], carbon–nitrogen [87–93] and carbon–sulfur [94, 95] bond formation or cleavage.

Wu *et al.* achieved stereoselective synthesis of β-amino ketones by using SSA [96]. This three-component reaction of arylaldehydes, anilines and cyclohexanone in the presence of SSA gave, preferentially, anti β-amino ketones in ethanol; the SSA could be reused for subsequent reactions (Scheme 5.15).

SSA is an efficient catalyst for thia-Michael addition to electron-deficient and sterically unhindered conjugated enones under solvent-free and mild conditions [95]. The reactions of thiols with enones such as cyclohexenone, cyclopentenone and ethyl vinyl ketone afforded the corresponding adducts in excellent yields. However, when chalcones were used as an enone, the corresponding adducts were obtained in very poor yields (Scheme 5.16).

A new and convenient method was found for the one-pot synthesis of symmetrical and unsymmetrical linear imides in the presence of SSA. The amidation of alcohols with nitriles in the presence of an acidic catalyst is known as the Ritter reaction. Salehi *et al.* have modified this method and developed the one-pot synthesis of a wide variety of linear imides instead of alcohols [97] (Scheme 5.17).

Similar catalysts, silica-gel bound aliphatic and aromatic sulfonic acids, have also been developed and used for several reactions. For instance, propanesulfonic

Table 5.1 Silica-gel supported reagents.

Property	Reagent (reported year)							
	SOCl$_2$[61] (1986)	CoBr$_2$[62] (1994)	TaCl$_5$[63] (1997)	Cu(OTf)$_2$[64] (1999)	AlCl$_3$[65] (2003)	PPA [60] (2004)	HClO$_4$[66] (2006)	H$_3$PW$_{12}$O$_{40}$[59] (2006)
Stability of supported reagent	Sensitivity (water)	Stable	Sensitivity (water)	Stable	Sensitivity (water)	Stable	Stable	Stable
Preparative method	Ad.[a]	Mixing	Mixing	Mixing	Ad.[a]	Ev.[b]	Ev.[b]	Ev.[a]
Reusable					Three times	Nine times	Twice	Five times
Reaction condition	r.t./5 h	r.t./10 min	r.t./2 min	r.t./0.5 h (neat)	reflux/1.2 h	r.t./0.5 h	r.t./10 min	r.t./20 min (neat)
Molar ratio (Aldehyde: Thiols)	1:3	1:1.2	1:1.2	1:2	1:1.2	1:1.2	1:1.2	1:1.2

a Adsorption.
b Evaporation.

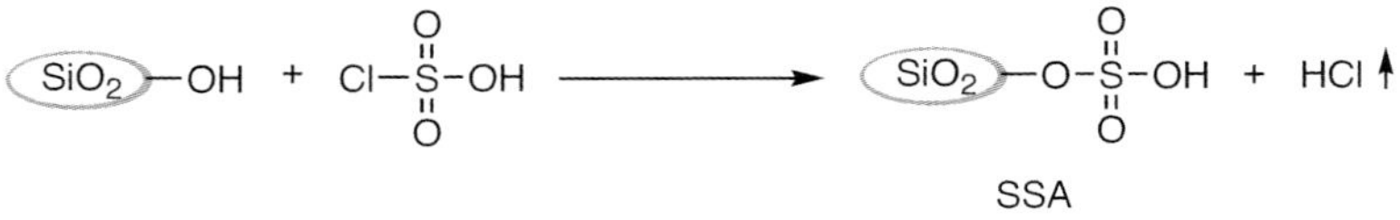

Scheme 5.14 Formation of silica sulfuric acid.

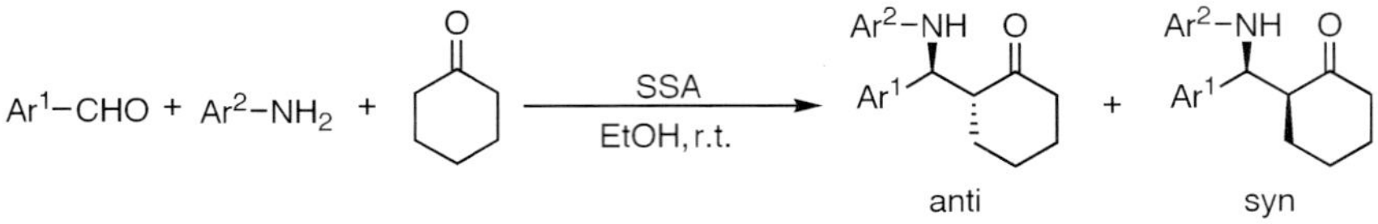

Scheme 5.15 Stereoselective synthesis of β-amino ketones.

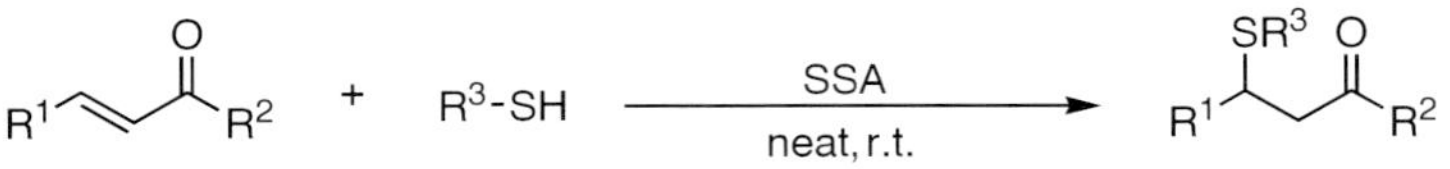

Scheme 5.16 thia-Michael addition to conjugated enones under solvent-free, mild conditions.

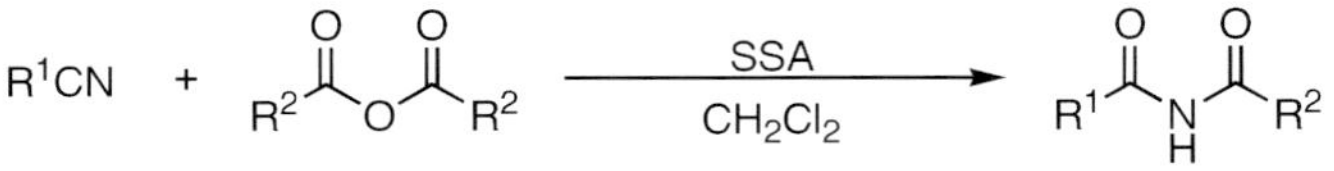

Scheme 5.17 One-pot synthesis of linear imides.

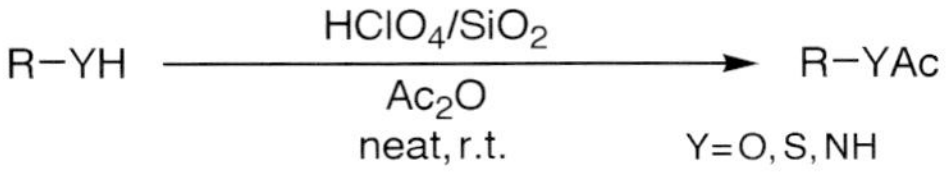

$$R-YH \xrightarrow[\text{neat, r.t.}]{\text{HClO}_4/\text{SiO}_2, \text{Ac}_2\text{O}} R-YAc \quad Y=O, S, NH$$

Scheme 5.18 Acetylation using $HClO_4/SiO_2$.

acid modified silica gel was developed in 1989 [98] and used recently as a recyclable catalyst for chemoselective deprotection of *tert*-butyldimethylsilyl ethers [99], bromination of arene [100], tetrahydropyranylation [101], acetalization and acetylation [102].

5.2.2
HClO$_4$/SiO$_2$

Chakraborti *et al.* have reported that silica-gel supported perchloric acid is an effective catalyst for the acetylation of phenols, thiols, alcohols and amines [103]. Generally, acetylation of these compounds requires large amount of Ac$_2$O. The acetylation using HClO$_4$/SiO$_2$, however, was completed with an equivalent amount of Ac$_2$O. HClO$_4$/SiO$_2$ was easily recovered and could be reused in five consecutive reactions without loss of activity (Scheme 5.18).

HClO$_4$/SiO$_2$ has been used for many acid-catalyzed reactions, such as protection of carbonyl compounds [104–108], selective deprotections [109, 110], Ferrier rearrangement [111, 112], glycosylation [113, 114] and stereoselective synthesis of enaminones and enamino esters [115]. For instance, it catalyzed not only a cleavage reaction but direct acetylation of benzylidene acetals [116]: the cleavage reaction of 2,3-di-O-acetyl-4,6-benzylidene-α-D-glucopyranoside proceeded in the presence of $HClO_4/SiO_2$ in MeCN. When Ac_2O was used instead of MeCN, direct conversion of benzylidene acetals into the acetylated compound occurred. The recovered catalyst could be reused three times in the direct deprotection and acetylation reactions and the yields were over 90% in each case (Scheme 5.19).

5.2.3
HBF$_4$/SiO$_2$

Silica-gel supported fluoroboric acid (HBF$_4$/SiO$_2$) has been used as a reusable catalyst for the protection of aldehydes [117] and alcohols [118], and for the synthesis of 1,5-benzodiazepines [119]. It has also been used for regioselective ring-opening reaction of epoxides with nitrogen heterocycles [120]. When the reaction of styrene oxide with pyrrole was carried out in the presence of HBF$_4$/SiO$_2$, pyrrole attacked the benzylic position at styrene oxide. In contrast, similar reactions using imidazole resulted in attack at another carbon atom of the styrene oxide (Scheme 5.20).

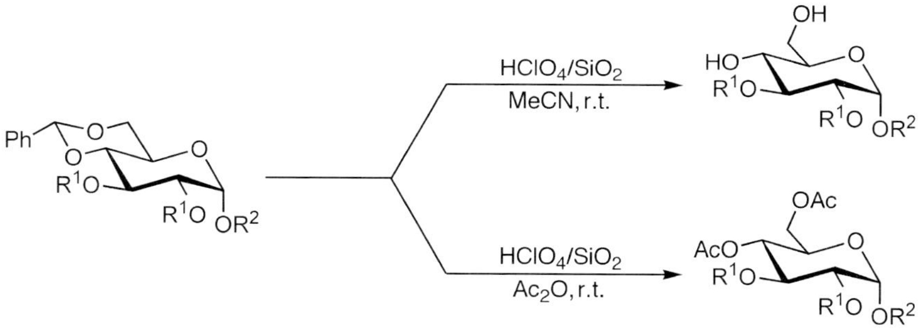

Scheme 5.19 Cleavage and acetylation of 2,3-di-O-acetyl-4,6-benzylidene-α-D-glucopyranoside with HClO$_4$/SiO$_2$.

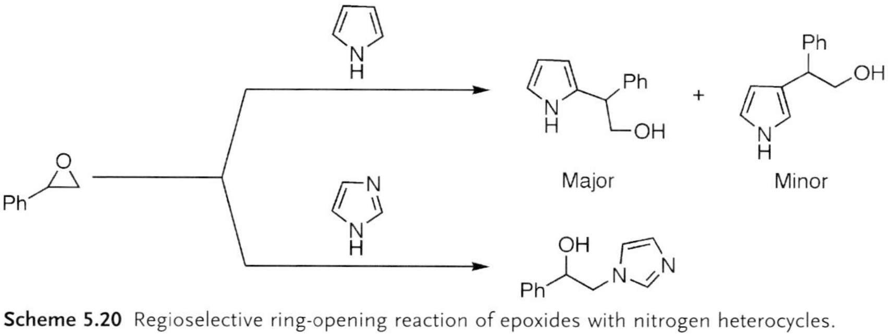

Scheme 5.20 Regioselective ring-opening reaction of epoxides with nitrogen heterocycles.

5.2.4
Silica-Gel Supported Heteropoly Acid

Since the end of twentieth century, supported acid catalysts, which were prepared from heteropoly acid and organic or inorganic solid, have been developed. To date, a great variety of organic or inorganic solids have been used as supports for heteropoly acids, such as clay, carbon, silica gel and zirconia and so on. Silica gel is most frequently employed as a support. Silica-gel supported 12-tungstophosphoric acid ($H_3PW_{12}O_{40}/SiO_2$) is a useful catalyst for many acid-catalyzed reactions [121–123]. For instance, it has been used as an effective catalyst for a four-component coupling process for the synthesis of β-acetamido ketones (Scheme 5.21) [122].

This catalyst also promoted a three-component coupling reaction under improved protocol conditions for the Biginelli reaction. The recovered catalyst, which is washed with acetonitrile and dried at 150 °C/0.5 Torr for 1.5 h, can be reused four successive times (Scheme 5.22) [123].

5.2.5
Others

Silica-supported polytrifluoromethanesulfosiloxane (SiO_2-Si-SCF_3) has been prepared by Jiang et al., and has been used to catalyze the Friedel–Crafts benzylation of arene with benzyl alcohol [124]. The selectivity of the reaction was influenced by catalyst concentration, reaction time and arene/benzyl alcohol molar ratios. In the range 30:1 to 10:1 of toluene/benzyl alcohol molar ratios, the yield of alkylated product was quantitative, but with a molar ratio of 5:1, 37% the intramolecular condensed product, dibenzyl ether, was formed (Scheme 5.23).

P_2O_5/SiO_2 was used for the solvent-free nitration of aromatic compounds. Activated aromatic compounds were converted into the corresponding nitro-aromatic compounds at room temperature under solvent-free conditions in excellent yields in 2–7 min, and deactivated aromatic compounds were converted into nitrobenzenes in moderate to good yields in 15–20 min [125].

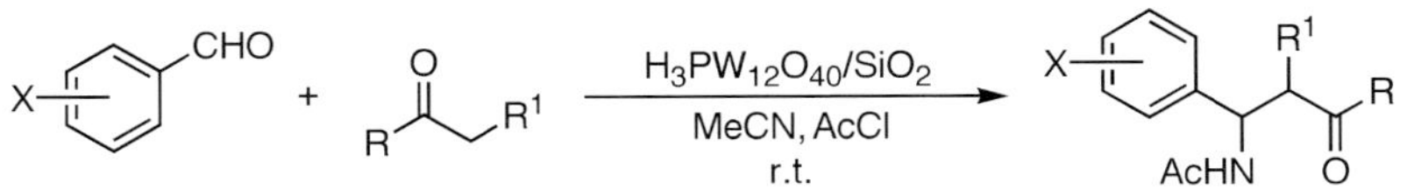

Scheme 5.21 Synthesis of β-acetamido ketones.

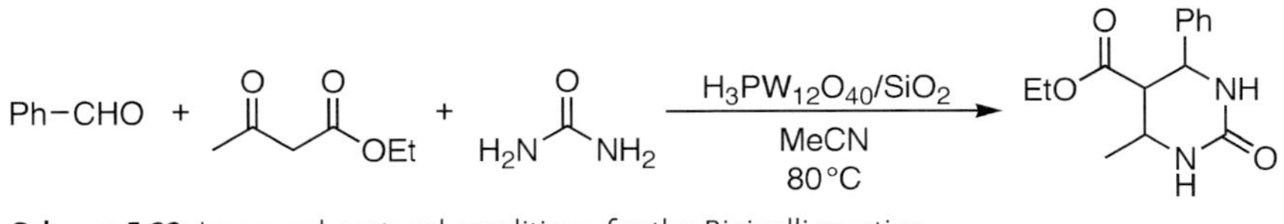

Scheme 5.22 Improved protocol conditions for the Biginelli reaction.

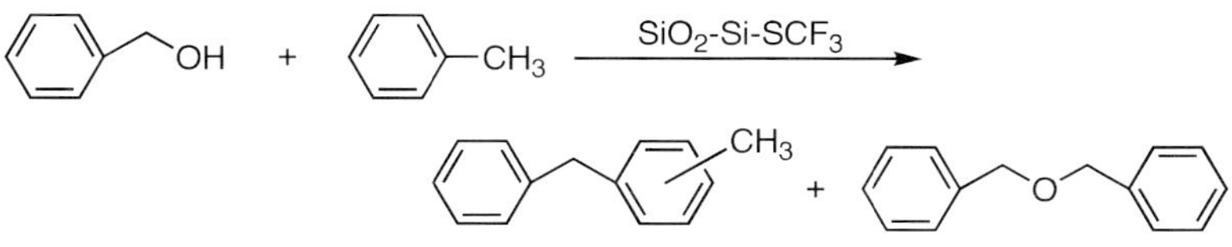

Scheme 5.23 Effect of the molar ratio of reactants on selectivity in the Friedel–Crafts benzylation of toluene with benzyl alcohol.

	Yield (%)	
BnOH:PhMe = 10:1	100	0
BnOH:PhMe = 5:1	63	37

$NaHSO_4/SiO_2$, which has been used as a catalyst for the selective acetylation of diols [126], has been used more recently for the synthesis of heterocyclic compounds. For example, a three-component reaction for the synthesis of quinazolinones was catalyzed with $NaHSO_4/SiO_2$ [127].

Phosphoric acid immobilized on silica (H_3PO_4/SiO_2) catalyst has been widely exploited over many years [128–130]. It has also been used as a reusable catalyst for solvent-free imino-aldol three-component coupling reaction in recent years [131].

Silica-gel supported polyphosphoric acid (PPA/SiO_2) has been developed as a reusable catalyst – oxathio- and dithio- acetals were synthesized by using PPA/SiO_2. The recovered catalyst can be reused nine successive times, after drying each time at 100 °C for 2 h under reduced pressure [60].

5.3
Applications

Many reactions in organic synthesis are reactions of nucleophiles and electrophiles, in other words they are acid–base reactions. Usually, the nucleophilicity or the electrophilicity of a substrate is enhanced by strong acid or base to promote a reaction. However, a reaction can be also promoted when weak acid and base activate both a nucleophile and an electrophile at the same time. In a homogeneous medium, acid and base could not coexist because of neutralization. Therefore, it is not expected to promote a reaction by acid and base cooperation. However, acid and base supported on an inorganic solid make it possible to promote a reaction by acid and base cooperation.

Recently, Lin et al. have demonstrated a cooperative catalytic system, using bifunctionalized mesoporous silica nanosphere materials, that consists of an acid, the ureidopropyl group and a base, the 3-[2-(2-aminoethylamino)ethylamino]propyl group [132]. This catalyst promoted aldol, Henry and cyanosilylation reactions. In these reactions, the secondary-amine as a base site activated the nucleophiles. On the other hand, the acid site could activate the carbonyl group to nucleophilic attack through double hydrogen bonding. The catalysts were reused three times without purification (Figure 5.3).

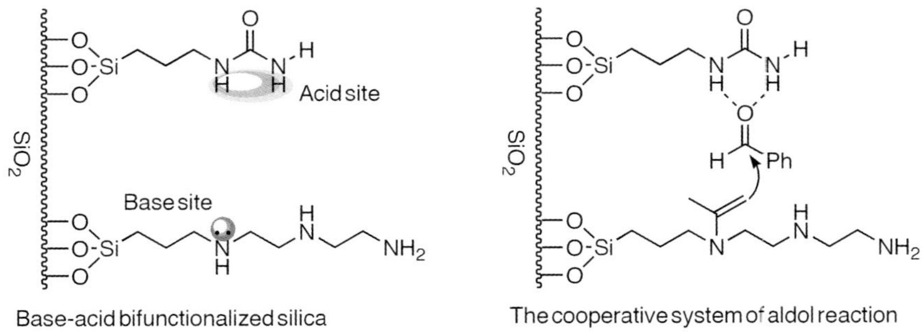

Figure 5.3 Example of an acid–base system and its application to the aldol reaction.

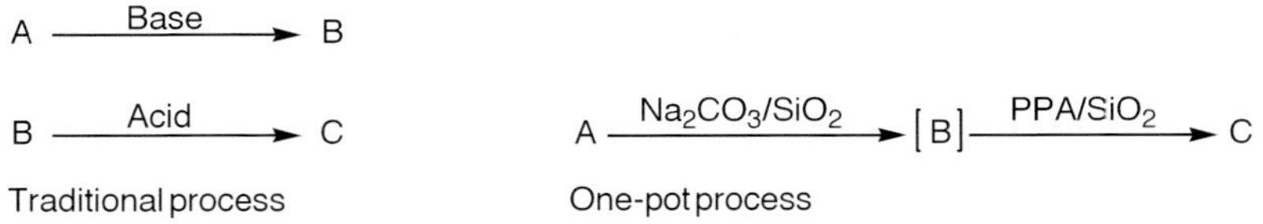

Figure 5.4 Outline of traditional and one-pot processes.

Acid- and base-catalyzed reactions in one-pot have also been developed by using a combination of acid- and base-supported reagents. In traditional process, the acid-catalyzed and base-catalyzed reactions have to be carried out separately. Therefore, the reaction and isolation of products have to be carried out more than once to synthesize target compounds, because an acid and a base can not coexist in a homogeneous solution. Recently, we have shown a base- and acid-catalyzed multistep reaction process in one-pot by using a couple of PPA/SiO$_2$ and Na$_2$CO$_3$/SiO$_2$ [133, 134]. Large amounts of acid and base as supported reagents exist on the internal surface of inorganic supports; the acid and base could not contact each other. Therefore, acid- and base-supported reagents are available for reaction in the same vessel (Figure 5.4).

A one-pot synthesis of benzo[b]thiophenes has been achieved using the Na$_2$CO$_3$/SiO$_2$ and PPA/SiO$_2$ couple [133]. When Na$_2$CO$_3$ and PPA were used instead of Na$_2$CO$_3$/SiO$_2$ and PPA/SiO$_2$ for this reaction, the reaction did not proceed. Several benzo[b]thiophenes and naphtho[2,1-b]thiophenes were synthesized in one-pot in the presence of Na$_2$CO$_3$/SiO$_2$ and PPA/SiO$_2$ in good to excellent yields (Table 5.2).

2,4-Disubstituted thiochromans were also synthesized by using the Na$_2$CO$_3$/SiO$_2$-PPA/SiO$_2$ system [134]. In general, thiochromans were synthesized by thia-Claisen rearrangement of allyl phenyl sulfide and an acid-catalyzed intramolecular cyclocondensation of thiol and β-arylthio aldehyde, which was prepared from arylthiol and α,β-unsaturated aldehyde under basic conditions. Jafarzadeh et al. and Ishino et al. synthesized these thiochromans directly from arylthiols and α,β-unsaturated aldehydes under acidic conditions. PPA/SiO$_2$ also catalyzed this direct

Table 5.2 One-pot synthesis of benzo[b]thiophenes.

Reagents system		Yield (%)
Na$_2$CO$_3$	PPA	0
Na$_2$CO$_3$	PPA/SiO$_2$	27
Na$_2$CO$_3$/SiO$_2$	PPA	46 (41)[a]
Na$_2$CO$_3$/SiO$_2$	PPA/SiO$_2$	80

a A figure in parentheses indicates the yield of the intermediate.

Table 5.3 Synthesis of thiochromans.

Entry	Ar	R^1	R^2	R^3	Selectivity (%) (*trans*-isomer)	Yield (%)
1	p-MeC$_6$H$_4$	Me	H	Me	90.9	83
2	p-MeC$_6$H$_4$	Pr	H	Me	96.2	76
3	p-MeC$_6$H$_4$	Ph	H	Me	>99.9	63
4	Ph	Me	H	H	80.0	64
5	Ph	Pr	H	H	84.7	73
6	Ph	Ph	H	H	>99.9	88

reaction, but a small amount of by-product, which was formed under acidic conditions, was observed. In the reaction using Na$_2$CO$_3$/SiO$_2$-PPA/SiO$_2$, Na$_2$CO$_3$/SiO$_2$ preferentially promoted Michael addition and by-product was not detected; the *trans*-isomer of thiochromans was formed selectively, due to intramolecular cyclocondensation occurring on the surface of PPA/SiO$_2$. This system is a first example; it is not only an effective method for the one-pot synthesis of thiochromans, but is also a highly stereoselective synthesis of thiochromans that takes advantage of a supported reagent (Table 5.3).

These developments, such as bifunctionalized catalyst and multi-step reaction process using a combination of supported reagents, are very interesting as a new strategy of organic synthesis. They will contribute, as powerful tools, to the development of combinatorial and green chemistry.

5.4
Conclusions

Base- and acid-supported reagents, which were developed in the second half of the twentieth century as easily handled and environmentally safe reagents, are still being used in many organic reactions. Even now, they are being developed in response to recent reaction technology. These developed reagents should be able to promote more effectively reactions that have attracted recent attention, such as a solvent-free and microwave-assisted reactions. In many cases, base- and acid-supported reagents offer advantages over the corresponding base and acid in solution reactions, and the easy removal of the supported-reagents by simple filtration facilitates the isolation of products.

In addition, a technical use of these base- and acid-supported reagents has also allowed cooperative base–acid reactions and the coexistence of base and acid reactions. Furthermore, the use in one-pot of many kinds of supported reagents is possible and, therefore, it is expected that novel one-pot tandem reaction process will be constructed.

There is little doubt that the use of base- and acid-supported reagents will see further development.

List of Abbreviations

DMAP	4-(dimethylamino)pyridine
DMF	dimethylformamide
MCM-41	Mobil's Composition of Matter-41
MSN	mesoporous silica nanosphere
PPA	polyphosphoric acid
SSA	silica sulfuric acid
TBD	1,5,7-triazabicyclo[4.4.0]dec-5-ene
THF	tetrahydrofuran

References

1 Yamawaki, J. and Ando, T. (1979) *Chemistry Letters*, 755–8.
2 Ting, X. and Liu, C. (1988) *Youji Huaxue*, **8**, 511–13.
3 Mihara, M., Ishino, Y., Minakata, S. and Komatsu, M. (2003) *Synthesis*, 2317–20.
4 Mihara, M., Ishino, Y., Minakata, S. and Komatsu, M. (2005) *Journal of Organic Chemistry*, **70**, 5320–2.
5 Basu, B., Jha, S., Mridha, N.K. and Bhuiyan, Md.M.H. (2002) *Tetrahedron Letters*, **43**, 7967–9.
6 Wagaw, S. and Buchwald, S.L. (1996) *Journal of Organic Chemistry*, **61**, 7240–1.
7 Hosseinzadeh, R., Tajbakhsh, M. and Alikarami, M. (2006) *Tetrahedron Letters*, **47**, 5203–5.
8 Hosseinzadeh, R., Tajbakhsh, M., Mohadjerani, M. and Mehdinejad, H. (2004) *Synlett*, 1517–20.
9 Hosseinzadeh, R., Tajbakhsh, M., Mehdinejad, H. and Alikarami, M. (2005) *Synlett*, 1101–4.
10 Villemin, D. and Caillot, F. (2001) *Tetrahedron Letters*, **42**, 639–42.
11 Marquez, H., Loupy, A., Calderon, O. and Perez, E.R. (2006) *Tetrahedron*, **62**, 2616–21.
12 Wang, M., Li, P. and Wang, L. (2004) *Synthetic Communications*, **34**, 2803–12.
13 Wang, J.-X., Yang, Y. and Wei, B. (2004) *Synthetic Communications*, **34**, 2063–9.
14 Yan, J., Wang, Z. and Wang, L. (2004) *Journal of Chemical Research*, 71–3.

15. Reddy, B.V.S., Srinivas, R., Yadav, J.S. and Ramalingam, T. (2002) *Synthetic Communications*, **32**, 219–23.
16. Sabitha, G., Abraham, S., Reddy, B.V.S. and Yadav, J.S. (1999) *Synlett*, 1745–6.
17. Kabalka, G.W., Wang, L. and Pagni, R.M. (2001) *Synlett*, 108–10.
18. Kabalka, G.W., Wang, L., Namboodiri, V. and Pagni, R.M. (2000) *Tetrahedron Letters*, **41**, 5151–4.
19. Baba, T., Kawanami, Y., Yuasa, H. and Yoshida, S. (2003) *Catalysis Letters*, **91**, 31–4.
20. Kantam, M.L. and Sreekanth, P. (1999) *Catalysis Letters*, **57**, 227–31.
21. Choudary, B.M., Kantam, M.L., Sreekanth, P., et al. (1999) *Journal of Molecular Catalysis A – Chemical*, **142**, 361–5.
22. Cauvel, A., Renard, G. and Brunel, D. (1977) *Journal of Organic Chemistry*, **62**, 749–51.
23. Rodriguez, I., Iborra, S., Corma, A., et al. (1999) *Chemical Communications*, 593–4.
24. Rodriguez, I., Iborra, S., Rey, F. and Corma, A. (2000) *Applied Catalysis A: General* **194–195**, 241–52.
25. Blanc, A.C., Macquarrie, D.J., Valle, S., et al. (2000) *Green Chemistry*, **2**, 283–8.
26. Subba Rao, Y.V., De Vos, D.E. and Jacobs, P.A. (1997) *Angewandte Chemie – International Edition in English*, **36**, 2661–3.
27. Coelho, A., El-Maatougui, A., Raviña, E., et al. (2006) *Synlett*, 3324–8.
28. Carloni, S., De Vos, D.E., Jacobs, P.A., et al. (2002) *Journal of Catalysis*, **205**, 199–204.
29. Barbarini, A., Maggi, R., Mazzacani, A., et al. and (2003) *Tetrahedron Letters*, **44**, 2931–4.
30. Ballini, R., Fiorini, D., Maggi, R., et al. (2006) *Synlett*, 1849–50.
31. Hierl, M.A., Gamson, E.P. and Klotz, I.M. (1979) *Journal of the American Chemical Society*, **101**, 6020–2.
32. Delaney, E.J., Wood, L.E. and Klotz, I.M. (1982) *Journal of the American Chemical Society*, **104**, 799–807.
33. Shinkai, S., Tsuji, H., Hara, Y. and Manabe, O. (1981) *Bulletin of the Chemical Society of Japan*, **54**, 631–2.
34. Tomoi, M., Akada, Y. and Kakiuchi, H. (1982) *Makromolekulare Chemie – Rapid Communications*, **3**, 537–42.
35. Menger, F.M. and McCann, D.J. (1985) *Journal of Organic Chemistry*, **50**, 3928–30.
36. Deratani, A., Darling, G.D. and Frechet, J.M.J. (1987) *Polymer*, **28**, 825–30.
37. Deratani, A., Darling, G.D., Horak, D. and Frechet, J.M.J. (1987) *Macromolecules*, **20**, 767–72.
38. Guendouz, F., Jacquier, R. and Verducci, J. (1988) *Tetrahedron*, **44**, 7095–108.
39. Bergbreiter, D.E. and Li, C. (2003) *Organic Letters*, **5**, 2445–7.
40. Bergbreiter, D.E., Osburn, P.L. and Li, C. (2002) *Organic Letters*, **4**, 737–40.
41. Corma, A., Garcia, H. and Leyva, A. (2003) *Chemical Communications*, 2806–7.
42. Chen, H.-T., Huh, S., Wiench, J.W., et al. (2005) *Journal of the American Chemical Society*, **127**, 13305–11.
43. Tajima, T. and Fuchigami, T. (2005) *Journal of the American Chemical Society*, **127**, 2848–9.
44. Tajima, T. and Fuchigami, T. (2005) *Angewandte Chemie – International Edition*, **44**, 4760–3.
45. Tajima, T. and Fuchigami, T. (2005) *Chemistry – A European Journal*, **11**, 6192–6.
46. Tajima, T. and Fuchigami, T. (2006) *Yuki Gosei kagaku kyokaishi*, **64**, 406–14.
47. Clacens, J.-M., Genuit, D., Veldurthy, B., et al. (2004) *Applied Catalysis B – Environmental*, **53**, 95–100.
48. Hein, M., Miethchen, R. and Schwäbisch, D. (1999) *Journal of Fluorine Chemistry*, **98**, 55–60.
49. Hayat, S., -ur-Rahman, A., Choudhary, M.I., et al. (2001) *Tetrahedron*, **57**, 9951–7.
50. Shah, S.T.A., Khan, K.M., Fecker, M. and Voelter, W. (2003) *Tetrahedron Letters*, **44**, 6789–91.
51. Shah, S.T.A., Khan, K.M., Heinrich, A.M., et al. (2002) *Tetrahedron Letters*, **43**, 8603–6.
52. Shah, S.T.A., Khan, K.M., Heinrich, A.M. and Voelter, W. (2002) *Tetrahedron Letters*, **43**, 8281–3.
53. Lee, J.C. and Choi, Y. (1998) *Synthetic Communications*, **28**, 2021–6.
54. Polshettiwar, V. and Kaushik, M.P. (2004) *Catalysis Communications*, **5**, 515–18.

55 Yu, H., Takigawa, S. and Koshima, H. (2004) *Tetrahedron*, **60**, 8405–10.
56 Yu, H. and Koshima, H. (2003) *Tetrahedron Letters*, **44**, 9209–11.
57 Tedeschi, L. and Enders, D. (2001) *Organic Letters*, **3**, 3515–17.
58 Aoyama, T., Takido, T. and Kodomari, M. (2004) *Tetrahedron Letters*, **45**, 1873–6.
59 Firouzabadi, H., Iranpoor, N., Jafari, A.A. and Jafari, M.R. (2006) *Journal of Molecular Catalysis A–Chemical*, **247**, 14–18.
60 Aoyama, T., Takido, T. and Kodomari, M. (2004) *Synlett*, 2307–10.
61 Kamitori, Y., Hojo, M., Masuda, R., et al. (1986) *Journal of Organic Chemistry*, **51**, 1427–31.
62 Patney, H.K. (1994) *Tetrahedron Letters*, **35**, 5717–18.
63 Chandrasekhar, S., Takhi, M., Ravindra, Y.R., et al. (1997) *Tetrahedron*, **53**, 14997–5004.
64 Anand, R.V., Saravanan, P. and Singh, V.K. (1999) *Synlett*, 415–16.
65 Tamami, B. and Parvanak, K.B. (2003) *Synthetic Communications*, **33**, 4253–8.
66 Agnihotri, G. and Misra, A.K. (2006) *Tetrahedron Letters*, **47**, 3653–8.
67 Zolfigol, M.A. (2001) *Tetrahedron*, **57**, 9509–11.
68 Yakaiah, T., Reddy, G.V., Lingaiah, B.P.V., et al. (2005) *Synthetic Communications*, **35**, 1307–12.
69 Chen, W.Y. and Lu, J. (2005) *Synlett*, 2293–6.
70 Khodaei, M.M., Khosropour, A.R. and Fattahpour, P. (2005) *Tetrahedron Letters*, **46**, 2105–8.
71 Salehi, P., Dabiri, M., Zolfigol, M.A. and Baghbanzadeh, M. (2005) *Synlett*, 1155–7.
72 Khodaei, M.M., Khosropour, A.R. and Fattahpour, P. (2006) *Journal of Chemical Research*, 682–4.
73 Azizian, J., Mohammadi, A.A., Soleimani, E., et al. (2006) *Journal of Heterocyclic Chemistry*, **43**, 187–90.
74 Salehi, P., Dabiri, M., Zolfigol, M.A. and Baghbanzadeh, M. (2005) *Tetrahedron Letters*, **46**, 7051–3.
75 Salehi, P., Dabiri, M., Zolfigol, M.A. and Fard, M.A. (2003) *Tetrahedron Letters*, **44**, 2889–91.
76 Salehi, P., Dabiri, M., Zolfigol, M.A. and Fard, M.A. (2003) *Heterocycles*, **60**, 2435–40.
77 Jin, T.S., Wang, H.X., Wang, K.F. and Li, T.S. (2004) *Synthetic Communications*, **34**, 2993–9.
78 Jin, T.S., Zhao, Y., Liu, L.B. and Li, T.S. (2005) *Journal of Chemical Research*, 438–9.
79 Pore, D.M., Desai, U.V., Mane, R.B. and Wadgaonkar, P.P. (2004) *Synthetic Communications*, **34**, 2135–42.
80 Hajipour, A.R., Zarei, A., Khazdooz, L. and Pourmousavi, S.A. (2006) *Indian Journal of Chemistry*, **45B**, 305–8.
81 Shirini, F., Zolfigol, M.A. and Mohammadi, K. (2003) *Phosphorus Sulfur and Silicon and the Related Elements*, **178**, 1617–21.
82 Shirini, F., Zolfigol, M.A. and Mohammadi, K. (2003) *Phosphorus Sulfur and Silicon and the Related Elements*, **178**, 2357–61.
83 Salehi, P., Dabiri, M., Zolfigol, M.A. and Fard, M.A. (2004) *Phosphorus Sulfur and Silicon and the Related Elements*, **179**, 1113–21.
84 Hajipour, A.R., Zarei, A., Khazdooz, L., et al. (2005) *Synthesis*, 3644–8.
85 Desai, U.V., Thopate, T.S., Pore, D.M. and Wadgaonkar, P.P. (2006) *Catalysis Communications*, **7**, 508–11.
86 Niknam, K., Zolfigol, M.A., Khorramabadi-Zad, A., et al. (2006) *Catalysis Communications*, **7**, 494–8.
87 Zolfigol, M.A., Mirjalili, B.F., Bamoniri, A., et al. (2004) *Bulletin of the Korean Chemical Society*, **25**, 1414–16.
88 Zolfigol, M.A., Madrakian, E. and Ghaemi, E. (2002) *Molecules*, **7**, 734–42.
89 Zolfigol, M.A. and Choghamarani, A.G. (2003) *Phosphorus Sulfur and Silicon and the Related Elements*, **178**, 1623–9.
90 Azizian, J., Karimi, A.R., Kazemizadeh, Z., et al. (2005) *Synthesis*, 1095–8.
91 Shaabani, A. and Rahmati, A. (2006) *Journal of Molecular Catalysis A–Chemical*, **249**, 246–8.
92 Kiasat, A.R., Kazemi, F. and Mehrjardi, M. (2006) *Asian Journal of Chemistry*, **18**, 969–72.
93 Salehi, P., Dabiri, M., Zolfigol, M.A., et al. (2006) *Tetrahedron Letters*, **47**, 2557–60.

94 Hajipour, A.R., Zarei, A., Khazdooz, L. and Ruoho, A.E. (2006) *Synthesis*, 1480–4.
95 Pore, D.M., Soudagar, M.S., Desai, U.V., et al. (2006) *Tetrahedron Letters*, **47**, 9325–8.
96 Wu, H., Shen, Y., Fan, L., et al. (2007) *Tetrahedron*, **63**, 2404–8.
97 Habibi, Z., Salehi, P., Zolfigol, M.A. and Yousefi, M. (2007) *Synlett*, 812–14.
98 Badley, R.D. and Ford, W.T. (1989) *Journal of Organic Chemistry*, **54**, 5437–43.
99 Karimi, B. and Zareyee, D. (2005) *Tetrahedron Letters*, **46**, 4661–5.
100 Das, B., Venkateswarlu, K., Krishnaiah, M. and Holla, H. (2006) *Tetrahedron Letters*, **47**, 8693–7.
101 Shimizu, K., Hayashi, E., Hatamachi, T., et al. (2004) *Tetrahedron Letters*, **45**, 5135–8.
102 Shylesh, S., Sharma, S., Mirajkar, S.P. and Singh, A.P. (2004) *Journal of Molecular Catalysis A–Chemical*, **212**, 219–28.
103 Chakraborti, A.K. and Gulhane, R. (2003) *Chemical Communications*, 1896–7.
104 Khan, A.T., Choudhury, L.H. and Ghosh, S. (2006) *Journal of Molecular Catalysis A–Chemical*, **255**, 230–5.
105 Kamble, V.T., Jamode, V.S., Joshi, N.S., et al. (2006) *Tetrahedron Letters*, **47**, 5573–6.
106 Kumar, R., Tiwari, P., Maulik, P.R. and Misra, A.K. (2006) *Journal of Molecular Catalysis A–Chemical*, **247**, 27–30.
107 Rudrawar, S., Besra, R.C. and Chakraborti, A.K. (2006) *Synthesis*, 2767–71.
108 Khan, A.T., Parvin, T. and Choudhury, L.H. (2006) *Synthesis*, 2497–502.
109 Tiwari, P. and Misra, A.K. (2006) *Tetrahedron Letters*, **47**, 3573–6.
110 Agarwal, A. and Vankar, Y.D. (2005) *Carbohydrate Research*, **340**, 1661–7.
111 Tiwari, P., Agnihotri, G. and Misra, A.K. (2005) *Carbohydrate Research*, **340**, 749–52.
112 Agarwal, A., Rani, S. and Vankar, Y.D. (2004) *Journal of Organic Chemistry*, **69**, 6137–40.
113 Du, Y., Wei, G., Cheng, S., et al. (2006) *Tetrahedron Letters*, **47**, 307–10.
114 Mukhopadhyay, B., Collet, B. and Field, R.A. (2005) *Tetrahedron Letters*, **46**, 5923–5.
115 Das, B., Venkateswarlu, K., Majhi, A., et al. (2006) *Journal of Molecular Catalysis A–Chemical*, **246**, 276–81.
116 Agnihotri, G. and Misra, A.K. (2006) *Tetrahedron Letters*, **47**, 3653–8.
117 Kamble, V.T., Bandgar, B.P., Joshi, N.S. and Jamode, V.S. (2006) *Synlett*, 2719–22.
118 Chakraborti, A.K. and Gulhane, R. (2003) *Tetrahedron Letters*, **44**, 3521–5.
119 Bandgar, B.P., Patil, A.V. and Chavan, O.S. (2006) *Journal of Molecular Catalysis A–Chemical*, **256**, 99–105.
120 Bandgar, B.P. and Patil, A.V. (2007) *Tetrahedron Letters*, **48**, 173–6.
121 Kozhevnikova, E.F., Rafiee, E. and Kozhevnikov, I.V. (2004) *Applied Catalysis A: General A*, **260**, 25–34.
122 Rafiee, E., Shahbazi, F., Joshaghani, M. and Tork, F. (2005) *Journal of Molecular Catalysis A–Chemical*, **242**, 129–34.
123 Rafiee, E. and Shahbazi, F. (2006) *Journal of Molecular Catalysis A–Chemical*, **250**, 57–61.
124 Zhou, D.-Q., Yang, J.-H., Dong, G.-M., et al. (2000) *Journal of Molecular Catalysis A–Chemical*, **159**, 85–7.
125 Hajipour, A.R. and Ruoho, A.E. (2005) *Tetrahedron Letters*, **46**, 8307–10.
126 Nishiguchi, T., Kawamine, K. and Ohtsuka, T. (1992) *Journal of Organic Chemistry*, **57**, 312–16.
127 Das, B. and Banerjee, J. (2004) *Chemistry Letters*, **33**, 960–1.
128 Krawietz, T.R., Lin, P., Lotterhos, K.E., et al. (1998) *Journal of the American Chemical Society*, **120**, 8502–11.
129 Babu, G.P., Murthy, R.S. and Krishnan, V. (1997) *Journal of Catalysis*, **166**, 111–14.
130 Fougret, C.M. and Hölderich, W.F. (2001) *Applied Catalysis*, **207**, 295–301.
131 Lock, S., Miyoshi, N. and Wada, M. (2004) *Chemistry Letters*, **33**, 1308–9.
132 Huh, S., Chen, H.-T., Wiench, J.W., et al. (2005) *Angewandte Chemie–International Edition*, **44**, 1826–30.
133 Aoyama, T., Takido, T. and Kodomari, M. (2005) *Synlett*, 2739–42.
134 Aoyama, T., Okada, K., Nakajima, H., et al. (2007) *Synlett*, 387–90.

6
Nucleophilic, Electrophilic and Radical Reactions

Magnus Johansson[1] and Nina Kann[2]

[1]Lead Generation, AstraZeneca R&D Mölndal, 43183 Mölndal, Sweden
[2]Chalmers University of Technology, Organic Chemistry, Department of Chemical and Biological Engineering, 41296 Göteborg, Sweden

6.1
Introduction

This chapter deals with both initial developments as well as more recent advances in the area of polymer-supported nucleophilic and electrophilic reagents, and reagents intended for radical-mediated reactions. Reviews by Kirschning [1] and Ley [2] cover this area up to 2001, and we also refer to the Ley group for reports on the application of polymer-supported reagents in total synthesis [3–5]. More details can also be found in reviews by Bradley [6] and Drewry [7].

6.2
Polymer-Supported Nucleophilic Reagents

6.2.1
Polymer-Supported Secondary Amines

Several polymer-supported secondary amines have been reported as linkers, scavengers and nucleophilic catalysts. Surprisingly, few examples of their use as polymer-bound reagents have been described. However, Carpino and Williams reported already in 1978 the use of a polymer-bound piperazine for the deprotection of Fmoc-protected amines [8]. The reagent is prepared by alkylation of Boc-protected piperazine with chloromethyl-polystyrene, followed by acidic removal of the Boc-group (Scheme 6.1). The dibenzofulvene produced after deprotection is then scavenged by the piperazine. The reagent is easily regenerated with the help of *n*-butyllithium.

Polymer-bound piperazine can also be used for the deprotection of phosphane-, phosphate- and phosphinite-borane complexes to give the corresponding free

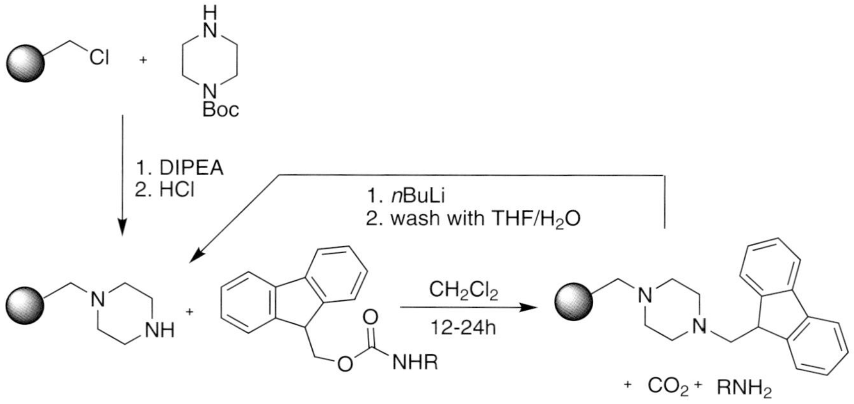

Scheme 6.1 A polymer-bound piperazine for Fmoc-deprotection.

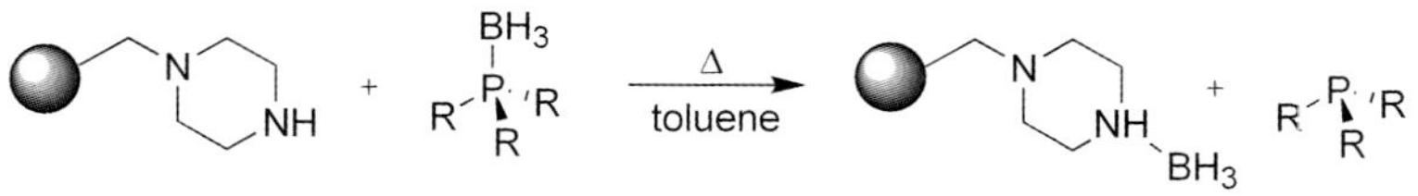

Scheme 6.2 Deprotection of borane-protected phosphorus compounds.

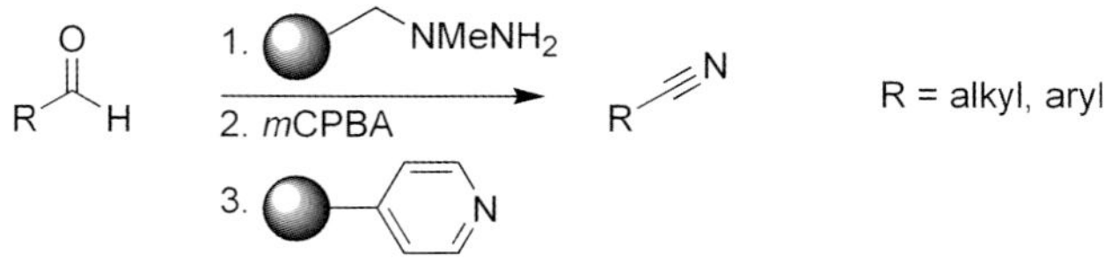

Scheme 6.3 Conversion of aldehydes into nitriles, employing a polymer-bound hydrazine.

phosphorus compound and a polymer-supported amine-borane (Scheme 6.2) [9]. This procedure shows great promise because of the ease of isolation of the corresponding free phosphane without any further elaboration of this air-sensitive product.

In the dehydration of aldehydes to form nitriles, Ley and colleagues used a polymer-supported hydrazine (Scheme 6.3). The immobilized hydrazine was reacted with aldehydes to give a hydrazone. Subsequent oxidation with *m*CPBA (*meta*-chloroperoxybenzoic acid) rendered the corresponding N-oxide, which spontaneously eliminated to form the desired nitrile [10]. Polyvinylpyridine was used as a scavenger for excess *m*CPBA.

The application of secondary amines as organocatalysts in different processes makes use of the inherent nucleophilicity of these compounds, but in this case they function as catalysts rather than reagents, and are thus covered in Chapter 4 of this book. Polymer-bound tertiary amines have so far only found use as deprotecting agents for phosphorus-borane complexes (see above) [9]. They are widely used as scavengers and linkers, however.

6.2.2
Reactions with Polymer-Bound Phosphanes

Another nucleophilic entity widely used both in solution and on polymer support is phosphorus. The handling of phosphanes both *pre-* and *post-*synthesis is considered problematic, *pre* due to their facile oxidation (in the case of electron-rich phosphanes), *post* due to the formation of the corresponding oxide that makes purification of the desired product troublesome. In this case, the use of a polymer-bound phosphane shows many advantages over the solution phase reaction: the phosphane is easier to handle, and purification of the desired product along with removal of the undesired phosphane oxide is simplified.

Phosphanes are widely used as reagents in organic synthesis, triphenylphosphane being the most commonly applied due to its stability towards oxidation. The polymer-supported analog has so far found use in the transformation of alcohols into alkyl halides and acids into acid halides, using either carbon tetrachloride or carbon tetrabromide as the halogen source (Scheme 6.4) [7, 11–14]. This system can also successfully transform primary amides and oximes into nitriles, whereas secondary amides are transformed into imidoyl chlorides (Scheme 6.5) [15].

Ley and coworkers used this method to carry out the cyclization of 1,4-disubstituted-(thio)semicarbazides, forming 2-amino-1,3,4-oxadiazoles (Scheme 6.6). A polymer-bound tertiary base was needed to facilitate the ring closure [16].

In summary, the most obvious advantages of the CCl_4 or CBr_4/PS-TPP (polystyrene-supported triphenylphosphane) system are the ease of purification

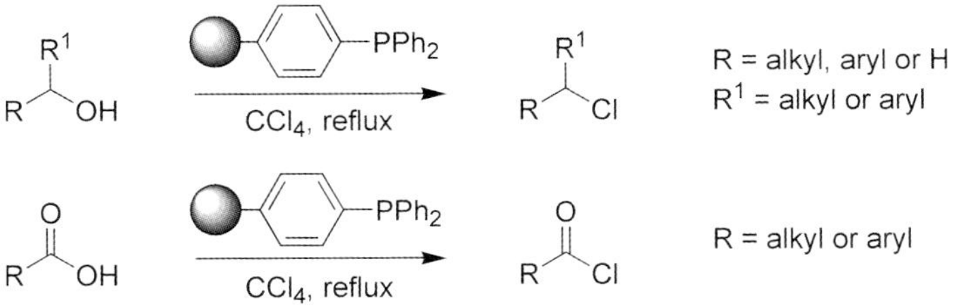

Scheme 6.4 Phosphane-assisted halogenation of alcohols and carboxylic acids.

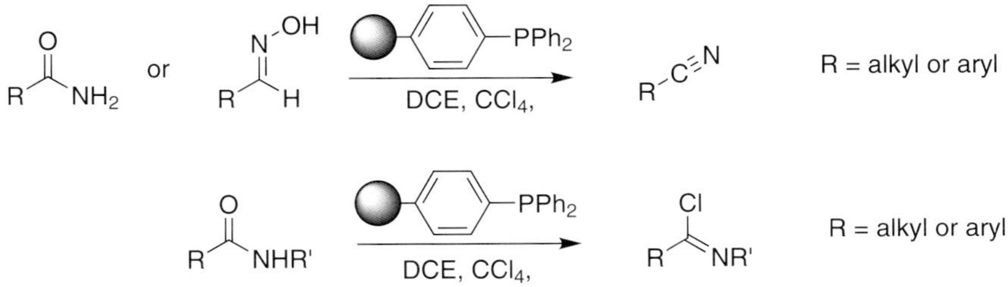

Scheme 6.5 Transformations of amides and oximes involving a polymer-supported phosphane.

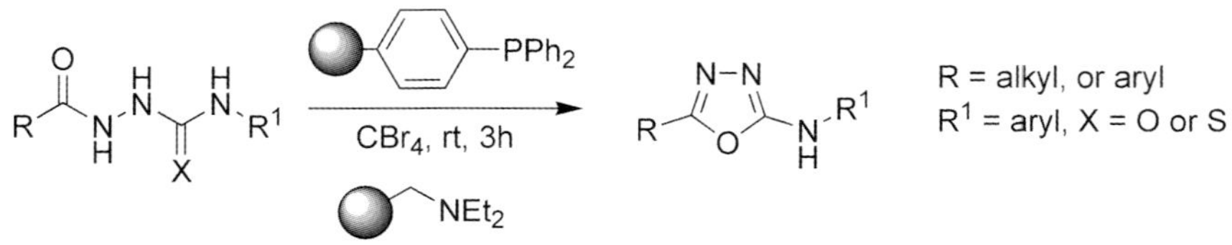

Scheme 6.6 Polymer-assisted ring closure to form 1,3,4-oxadiazoles.

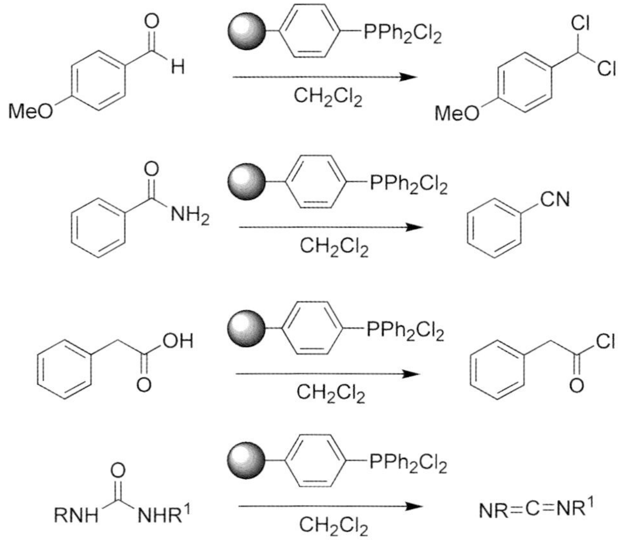

Scheme 6.7 Various transformations involving polymer-bound triphenylphosphane dichloride.

and the neutral conditions achieved, as compared to alternative reagents such as thionyl chloride, phosphorus oxytrichloride or a strong mineral acid.

By oxidation of polymer-supported triphenylphosphane, the corresponding oxide is formed. This intermediate will, upon treatment with phosgene or triphosgene, form the pentavalent dichlorotriphenylphosphane derivative, a strong dehydrating agent [17, 18]. Phosphane dihalogenide complexes can be used to prepare acid halogenides, carbodiimides, nitriles, halohydrins, alkyl halides, formic acid esters and amides (Scheme 6.7) [19–22]. The analogous iodine and bromine derivatives show a similar reactivity pattern [23–25]. A more recent publication reports the use of polymer-supported triphenylphosphane ditriflate as a potent dehydrating agent, forming amides, esters, anhydrides, azides, nitriles, ethers, epoxides and thioacetates. The phosphane oxide obtained after the reaction can be treated with triflic anhydride to regenerate the parent reagent [26, 27].

Another well established reaction in solution-phase chemistry using a nucleophilic phosphane is the Staudinger reaction. A polystyrene-supported alternative proved to be less efficient, however. The hydrolysis of the intermediate iminophosphorane on solid support was impractically slow, whereas a fluorous-tagged triphenylphosphane gave the corresponding amine in good yield and purity [28].

In the ozonolysis of double bonds, an intermediate ozonide is formed. This intermediate can readily be cleaved using polymer-supported triphenylphosphane (Scheme 6.8). The Janda group compared polystyrene-bound triphenylphosphane to a poly(ethylene glycol) (PEG)-bound and found that the degeneration of ozonides is more efficient using the PEG-based phosphane. In addition, the obtained phosphane oxide could be regenerated via reduction using freshly prepared alane in THF [29]. Ley and coworkers applied this methodology to the synthesis of epothilones [25].

Nucleophilic phosphanes can also be used for olefin isomerization. Nitroolefins having a substituent on the nitro group, readily available via the nitro aldol reaction followed by dehydration, often give a mixture of *(E)*- and *(Z)*-isomers. By adding catalytic amounts of polymer-supported triphenylphosphane, the *(Z)*-isomer could be isomerized to the corresponding *(E)*-isomer, in most cases with total *(E)*-selectivity [30].

Another well-known process that utilizes a nucleophilic phosphane is the Mitsunobu reaction, that is, the reaction between an acidic partner and an alcohol, typically facilitated by an azodicarboxylate and a phosphane. Two options are possible, anchoring of the electrophilic part to the solid support, dealt with in the next section, or anchoring of the nucleophilic phosphane. Georg *et al.* used polystyrene-bound triphenylphosphane and DEAD (diethyl azodicarboxylate) in their synthesis of aryl ethers [31]. Alcohols were reacted successfully with electron-rich and electron-deficient phenols, giving the desired products in good yield and purity. More recently, Wilhite and coworkers disclosed an efficient protocol for the synthesis of pyridine ethers using ADDP [1,1'-(azodicarbonyl)dipiperidine] and polymer-supported triphenylphosphane (Scheme 6.9) [32]. Both methods eliminate purification problems caused by triphenylphosphane oxide, but chromatography is still needed.

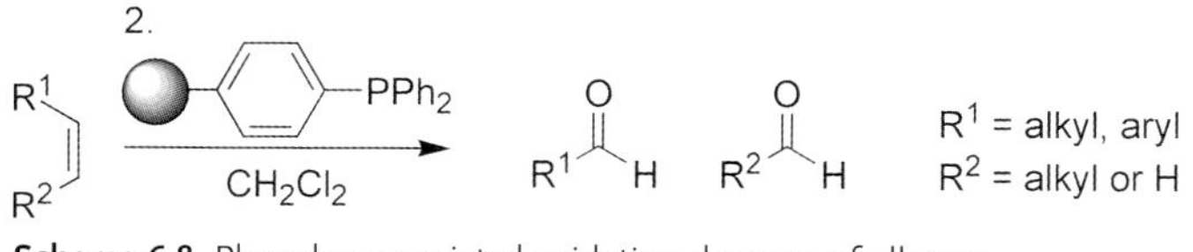

Scheme 6.8 Phosphane-assisted oxidative cleavage of alkenes.

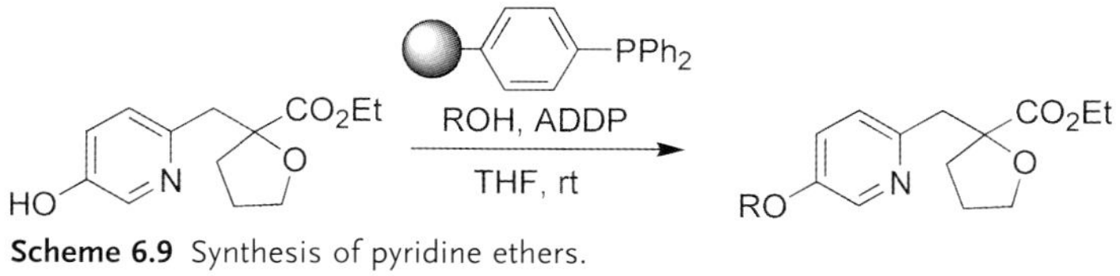

Scheme 6.9 Synthesis of pyridine ethers.

6.2.3
The Wittig Reaction and Related Olefination Reactions

Polymer-supported Wittig reagents are commonly encountered in the literature, and are easily prepared by reacting polymer-bound triphenyl/alkyl phosphanes with alkyl/benzyl halides (Scheme 6.10) [33–35]. The use of a supported phosphane in Wittig-type olefinations removes at least two purification steps, the phosphonium salt can be purified by extensive washing of the polymer and the obtained phosphane oxide is readily removed after the reaction by simple filtration. However, the problems associated with controlling the *(E/Z)*-olefin geometry of the olefin are not solved by using a polymer-assisted reaction, and thus still remain a potential problem.

In contrast, the related carbon–carbon bond formation using polymer-supported Horner–Wadsworth–Emmons (HWE) reagents gives rise to fairly good *(E)*-selectivity [36]. Two attachment points to the solid support are possible, either to the ester or to the phosphonate. Whereas the phosphonate-linkage method will effect a direct cleavage of the product, the ester-linked material has to be hydrolyzed off the resin. The Nicolaou group demonstrated the versatility of the HWE-reaction using a solid supported phosphonate, performing olefinations at different stages in their synthesis of a library of muscone-derivatives. The *(E/Z)*-selectivity was typically around 9:1 [37]. Using ring-opening metathesis polymerization (ROMP), Barrett and coworkers prepared a phosphonate-containing polymer. Screening different bases revealed that tetramethylguanidine (TMG), *tert*-butyltetramethylguanidine (Barton's base) and lithium hexamethyldisilazide (LiHMDS) produce the desired olefin in good yields. However, Barton's base proved to have the widest scope. Reacting different aldehydes under these conditions gave an *(E/Z)*-ratio of >99:1 in the case of αβ-unsaturated ethyl esters, while αβ-unsaturated nitriles were produced in an *(E/Z)*-ratio of >70:30, in all cases with excellent yields and purities (Scheme 6.11) [38].

In the preparation of a combinatorial library of α,β-unsaturated carboxylic acids, a group from Rhone Poulenc Rorer utilized the HWE-olefination reaction. Starting from 1-substituted phosphonoacetic acid anchored to polystyrene, a 48-membered library containing a diverse set of aldehydes (aliphatic, aromatic and basic) was produced, with an *(E/Z)*-ratio that was highly dependant on the α-substituent of the phosphonate, where hydrogen gave a 9:1 ratio in favor of the *(E)*-product

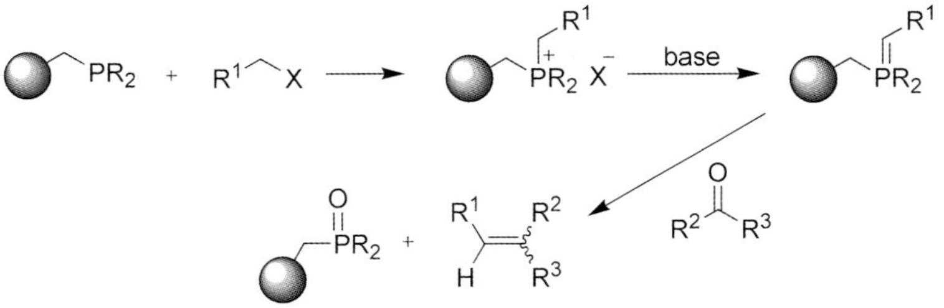

Scheme 6.10 Polymer-assisted Wittig olefination.

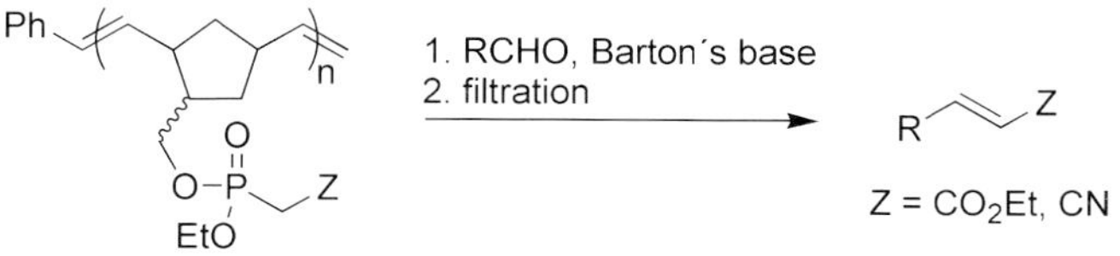

Scheme 6.11 Horner–Wadsworth–Emmons reactions using a phosphonate attached to a ROMP-polymer.

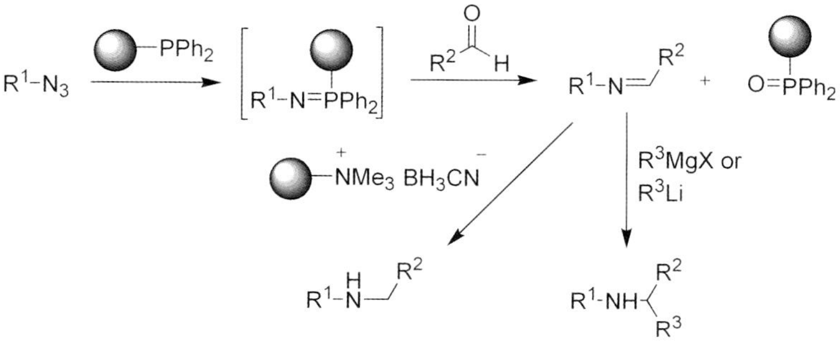

Scheme 6.12 Polymer-bound phosphane-mediated aza-Wittig reaction.

whereas a bulky substituent gave more (Z)-product. Yields and purities were good to excellent [39].

The aza-Wittig process, related in terms of the preparation of the requisite iminophosphorane to the Staudinger reaction mentioned earlier, is another example of a reaction where a polymer-supported phosphane can be applied. Instead of hydrolysis, which is a slow process in the solid phase, an aldehyde is reacted to form an imine via an aza-Wittig process. The *in situ* aza-Wittig reaction has been reported by Hemming and coworkers, who used polystyrene-supported triphenylphosphane in the Staudinger ligation, followed by an aza-Wittig reaction between the supported iminophosphorane and different aldehydes, thus facilitating cleavage from the solid-support (Scheme 6.12). The obtained imines were directly reduced by solid-supported borohydride or reacted with a Grignard reagent or an organolithium reagent [40]. This process has also found wide application in annulations and ring-closure reactions to form various heterocyclic ring systems [41–45].

6.2.4
Nucleophiles Anchored to Anion Exchange Resins

A great number of nucleophilic reagents are, for obvious reasons, anchored to an anion exchange resin containing a quaternary amine functionality. Different organic and inorganic ions can be attached by a simple ion exchange process. After use, they are easily regenerated using a second ion exchange reaction. The Ley group among others has utilized these reagents in several total syntheses of complex natural products [14, 16, 25]. Table 6.1 lists a selection of reagents of this type.

Table 6.1 Nucleophiles anchored to a quaternary amine.

Reagent	Substrate	Product	Reference
⬤–NMe$_3^+$ RCOCHCOR^1 $^-$	MeI	RCOCHMeCOR^1	[58]
⬤–NMe$_3^+$ CN$^-$	R-X, X = Br	R-CN	[59]
⬤–NMe$_3^+$ NaCHO$_3^-$	RCH$_2$-X, X = Br or I	RCH$_2$-OH	[16, 60–62]
⬤–NMe$_3^+$ SCOMe$^-$	R-X, X = Cl, Br or OTs	R-SC(=O)Me	[63]
⬤–NMe$_3^+$ N$_3^-$	R-X, X = Cl, Br or OTs β-substituted styrene oxides	R-N$_3$ α-azidophenethyl-alcohols	[14, 64]
⬤–NMe$_3^+$ NO$_2^-$	R-X, X = Cl, Br	R-NO$_2$	[58]
⬤–NMe$_3^+$ NCS$^-$	R-X, X = Cl, Br	R-SCN	[59]
⬤–NMe$_3^+$ NCO$^-$	R-X, X = Cl, Br	R-N(CO)-OR^1	[65]
⬤–NMe$_3^+$ OH$^-$	R$_2$NCOCF$_3$	R$_2$NH	[16]
⬤–NMe$_3^+$ ArO$^-$	R-X, X = Cl, Br or OMs	R-OAr	[58]
⬤–NMe$_3^+$ ArCOO$^-$	R-X, X = Cl,	R-OOCAr	[66]
⬤–NMe$_3^+$ ArS$^-$	BrCH$_2$Cl	ArS-CH$_2$Cl	[67]
⬤–NMe$_3^+$ PhSe$^-$	R-X, X = Cl, Br or I	R-PhSe	[14, 68]
⬤–NMe$_3^+$ X$^-$ X = F, Cl Br or I	R-X	R-X (exchange)	[69]

6.3
Reagents for Electrophilic Reactions

6.3.1
Reagents for Alkylation and Silylation

The ability of the aryl sulfonate moiety to act as a good leaving group has been exploited in a method for alkylating nucleophiles reported by Reitz and coworkers [58]. An arylsulfonyl resin, developed by Maryanoff and colleagues [59], was treated with different alcohols to afford polymer-bound sulfonate esters (Scheme 6.13). Modification of the side-chain via various reactions (Grignard addition, Wittig reaction, Suzuki coupling, reductive amination, hydride reduction) is possible at this stage and was demonstrated in a later article [60]. Subsequent treatment with amines, thiolates or imidazole afforded the corresponding alkylated nucleophiles in varying yields and purities. Best results were obtained with secondary amines, thiolate and imidazole, while primary amines gave somewhat lower yields but high selectivity for the monoalkylated product. Anilines, being less nucleophilic, were not alkylated to any great extent.

Although triazenes have been more extensively used as linkers [61–64], polymer-bound alkyl triazenes have been employed by several groups as a means of transferring an alkyl moiety to a heteroatom, thus avoiding the use of toxic and explosive diazoalkanes. Rademann *et al.* have prepared polymer-bound triazenes via reduction of a *p*-nitroarene resin with $SnCl_2$, followed by formation of the corresponding diazonium ion and treatment with different amines (Scheme 6.14) [65]. Esterification of a range of carboxylic acids of varying properties, in terms of steric hindrance, pK_a, size, functionality and stability, was then carried out using these reagents, affording the corresponding esters in excellent yields (>90% in nearly all cases). In a later report, the same group demonstrated the utility of the triazene reagents in the esterification of amino acids and peptide derivatives, and verified that the mild reactions conditions did not cause any racemization during the reaction [66].

Bräse and coworkers used a similar concept to prepare sulfonic esters via polymer-bound triazenes [67]. The methodology was subsequently expanded to phosphoric and phosphinic acids with good results [68]. For the alkylation of sodium sulfonates, this method has the advantage that sulfene formation is

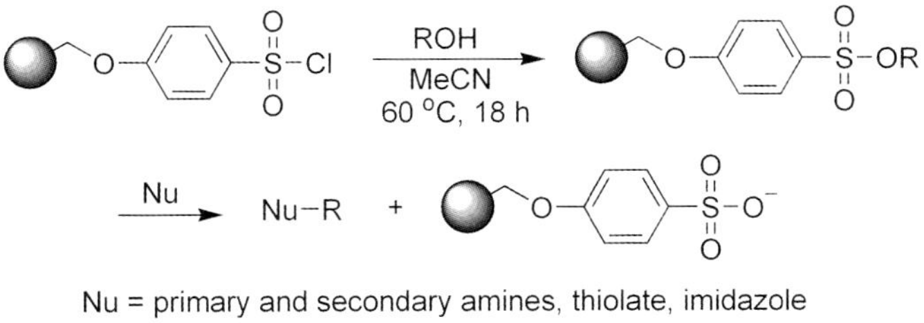

Nu = primary and secondary amines, thiolate, imidazole

Scheme 6.13 Alkylation of amines and thiolate via polymer-supported arylsulfonate esters.

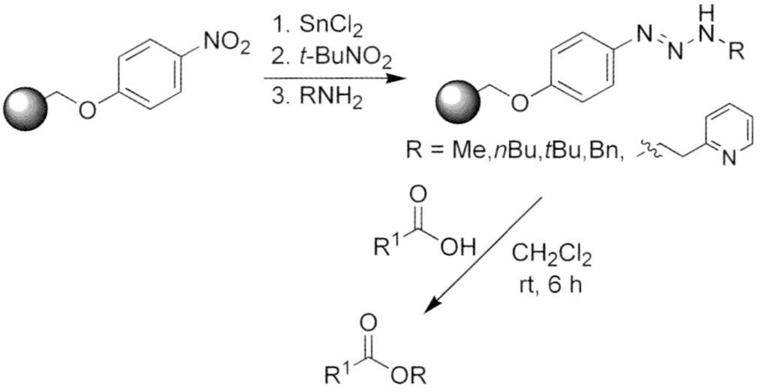

Scheme 6.14 Alkylation of carboxylic acids using polymer-bound alkyl triazenes.

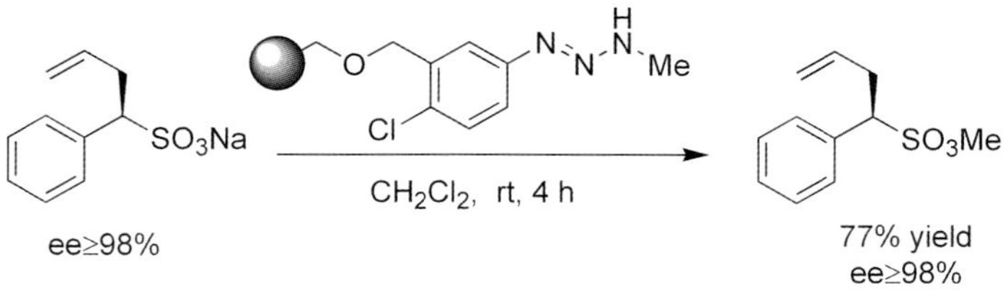

Scheme 6.15 Racemization-free alkylation of a chiral sodium sulfonate.

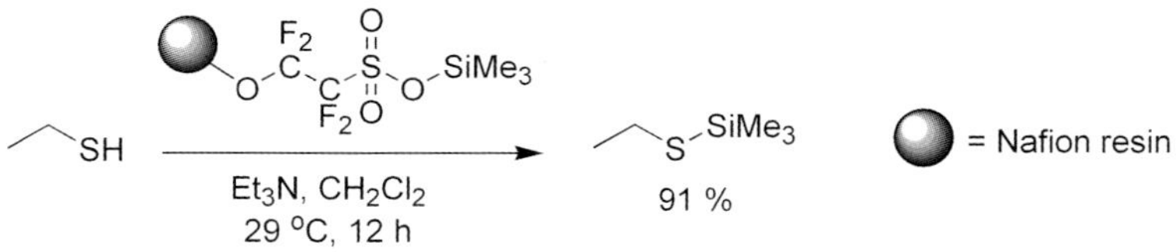

Scheme 6.16 Silylation of heteroatoms using a polymer-bound TMS-triflate equivalent.

avoided, thus allowing the alkylation of chiral sulfonates without concomitant racemization. A chiral sulfonate was methylated in 77% yield and high purity using a reagent of this type (Scheme 6.15); several more examples were reported. Alkyl halides can also be prepared using a similar approach, by treating the solid-supported triazene with the corresponding trimethylsilyl halide [69].

Polymer-bound reagents for silylation have been reported. Murata and Noyori treated Nafion, a perfluorinated resin-sulfonic acid, with trimethylsilyl chloride to afford a polymer-bound trimethylsilyl ester that functions as a polymer-bound equivalent of trimethylsilyl trifluoromethanesulfonate, converting alcohols, ethanethiol and diethylamine into the corresponding silylated compounds in high yields (Scheme 6.16) [70]. Acetic acid could also be silylated, albeit with moderate conversion. The corresponding polystyrene-bound reagent is commercially available.

6.3.2
Reagents for Acylation and Sulfonylation

Polymer-supported sulfonyl chlorides can be used for the selective monosulfonylation of anilines, as demonstrated by the Ley group [71, 72]. Commercially available polystyrene-DMAP (4-dimethylaminopyridine) was treated with different aryl sulfonyl chlorides to form polymer-bound sulfonylation reagents. These were reacted with substituted anilines to form an array of hydroxamic acids (Scheme 6.17), using a synthetic strategy involving polymer-bound reagents in all steps. The products were subsequently evaluated as histone deacetylase inhibitors.

As mentioned earlier, the use of polymer-supported reagents can facilitate the tedious work-up generally involved in the Mitsunobu reaction due to difficulties in separating the product from the phosphane oxide as well as the reduced form of the azodicarboxylate reagent. The anchoring of phosphanes to a solid support has been covered in Section 6.2.2; however, the attachment of the alkyl azodicarboxylate reagent to a polymer is an alternative strategy, having the additional advantage of rendering these potentially explosive reagents easier to handle. Vederas and coworkers reported the first polymer-bond reagent of this type [73]. Commercially available hydroxymethylpolystyrene resin was treated with phosgene in the presence of pyridine to form a chloroformate (Scheme 6.18). Reaction

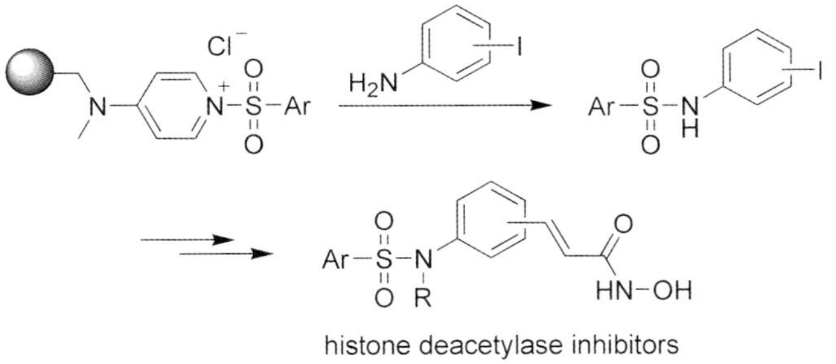

Scheme 6.17 Selective monosulfonylation of anilines using a polymer-supported reagent.

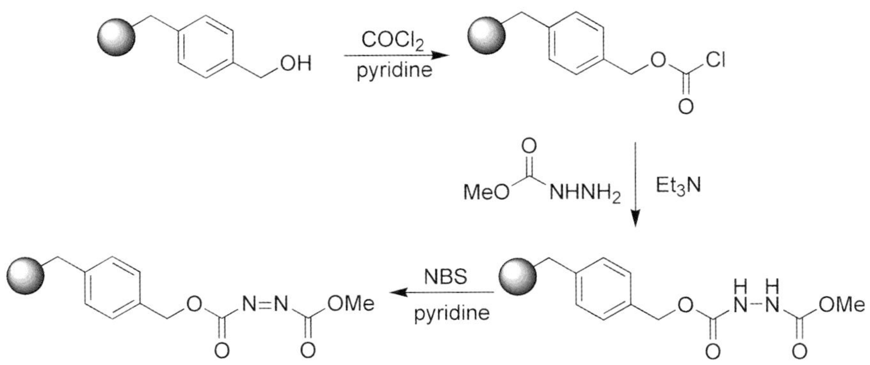

Scheme 6.18 Preparation of methyl azodicarboxylate resin for the Mitsunobu reaction.

with methyl carbazate and triethylamine was followed by oxidation with N-bromosuccinimide (NBS) to form a polymer-bound methyl azodicarboxylate. Mitsunobu reactions involving several different types of acidic components (carboxylic acids, succinimide, ethyl cyanoacetate) with primary and secondary alcohols were then performed to evaluate the utility of the polymer-bound reagent. Yields for the reactions were found to be approximately 20% lower than the corresponding reaction using a soluble reagent, with the exception of a cyclolactonization, which gave a markedly higher yield with the polymer-bound reagent. However, purification was greatly simplified, involving only precipitation of the phosphane oxide with diethyl ether and subsequent filtration. The active form of the reagent could be regenerated by treatment with NBS/pyridine, showing no decrease in activity over five redox cycles.

6.3.3
Nitrogen Electrophiles on Solid Phase

Azides are versatile molecules that have both nucleophilic and electrophilic properties, acting as 1,3-dipoles in cycloaddition reactions for example [74]. Unfortunately, due to their toxic and explosive nature, they are unpleasant to handle. Rebek and coworkers have prepared a tosyl azide attached to Amberlite and found it to be remarkably more stable than the parent solution phase tosyl azide [75]. The reagent was applied in diazo transfer reactions, affording the desired α-diazodicarbonyl products in excellent yields in most cases. A more recent report by Green *et al.* describes a polystyrene-supported benzenesulfonyl azide and its application in the preparation of α-diazoketones [76]. The authors found that the reagent gave good yields in the diazo transfer reactions (Scheme 6.19), except when the substrate contained bulky substituents like *tert*-butyl. In most cases, no purification other than filtration was necessary after the reaction was completed. The α-diazoketones can subsequently be used to generate carbenes.

Diazotation of primary aromatic amines can also be performed using a polymer-supported reagent. Ley and coworkers used a polymer-bound nitrite in the synthesis of azo dyes [77]. The nitrate reagent can be prepared via ion exchange, starting from tetraalkylammonium chloride resin, but is also commercially available. Upon treatment with acid, the reagent generates an electrophilic species (in general N_2O_3) that converts anilines into diazonium ions (Scheme 6.20). Further transfor-

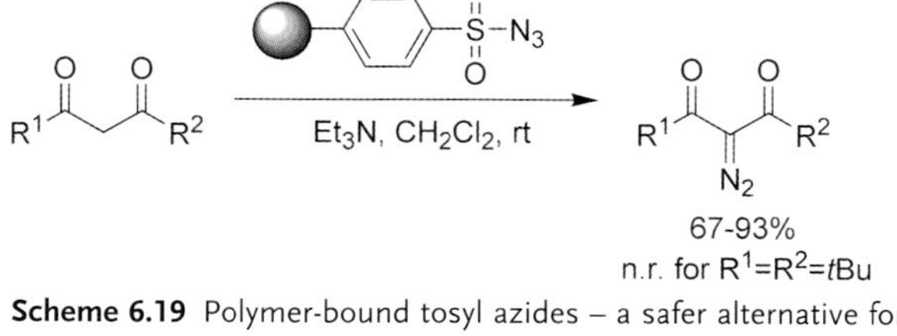

Scheme 6.19 Polymer-bound tosyl azides – a safer alternative for diazo transfer reactions.

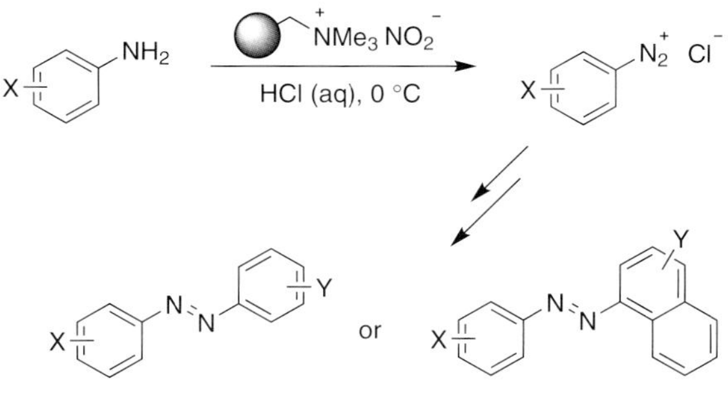

$X = H, OMe$ or NO_2 $Y = OH, NMe_2$

Scheme 6.20 Diazotation using a polymer-bound nitrite.

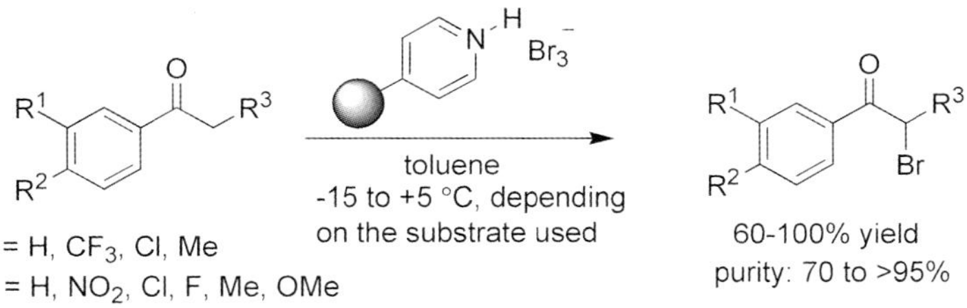

$R^1 = H, CF_3, Cl, Me$
$R^2 = H, NO_2, Cl, F, Me, OMe$
$R^3 = H, Me$

Scheme 6.21 Bromination of activated methylene compounds.

mations then afforded the desired azo dyes. This strategy minimizes the amounts of aqueous waste produced in the diazotation reaction, thus providing a greener approach to the target compounds.

6.3.4 Electrophilic Bromination Reagents

Electrophilic bromination of activated methylene compounds can be effected using polymer-supported pyridinium bromide perbromide (PSPBP) as shown by Fréchet *et al.* [78] and also by other groups [79, 80]. To focus on some more recent examples, Ley and coworkers treated acetophenones with PSPBP, affording the corresponding monobrominated compounds (Scheme 6.21) [81]. The reaction temperature needed to be carefully controlled to avoid dibromination. Further transformations, also involving polymer-supported reagents, then generated various different benzophenones. PSPBP was also applied towards the preparation of substituted 1,4-benzodioxanes as well as 2-amino-4-aryl-1,3-thiazoles [82]. More examples of polymer-supported bromination reactions can be found in an earlier review [1].

6.4
Radical Reactions Using Supported Reagents

One distinct advantage of working with polymer-bound reagents is the possibility to anchor toxic, explosive or volatile materials that are difficult or unpleasant to handle. Stannanes constitute one example of reagents that are gratifying to apply in the solid supported form, not only owing to their toxicity but also because they are in many cases difficult to separate from the product after completed reaction, resulting in loss of material. One of the first successful examples of immobilized stannanes was reported by the Neumann group, who used two different strategies for preparing the reagent (Scheme 6.22) [83]. The first method incorporated the stannyl chloride functionality already into the monomer before polymerization. The authors emphasize the importance of using a dimethylene spacer between the tin atom and the aromatic ring to increase the stability of the reagent, thus avoiding the more labile aryl or benzyl stannanes. Also, a macroporous structure involving 9–10% cross-linking was found to be more suitable for enabling reagents to penetrate the polymer. Several reagents were evaluated for the conversion of the stannyl chloride into the corresponding hydride, with dibutylaluminium hydride giving the best results. The second method involved chloromethylation of unfunctionalized macroporous polystyrene (12% cross-linking), followed by conversion into a phosphonium salt and a subsequent Wittig reaction to render a vinylated polymer. Hydrostannylation, followed by reduction with lithium aluminum hydride then afforded the polymer-bound stannane. Ensuing publications describe the application of such polymers in different radical processes such as dehalogenation, dehydroxylation and deamination reactions [84], and also Giese coupling [85] of aryl and alkyl halides with acrylonitrile [86].

Several other types of polymer-bound stannanes have since been reported, including a polymer-bound distannane used in radical-mediated cyclizations and radical addition to acetylenes [87], and stannanes with longer tethers to the polymer [88, 89] applied in the Barton–McCombie deoxygenation of alcohols [90, 91] and

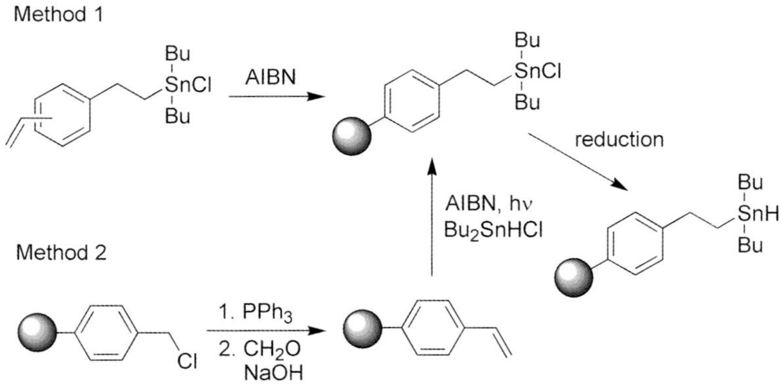

Scheme 6.22 Preparation of a polymer-bound stannane.

in reductive dehalogenation. Enholm *et al.* have described a polymer-supported allyl stannane for radical-mediated allylation of alkyl halides, although a soluble non-cross-linked polystyrene was used in this case [92]. An interesting alternative to traditional polymer-supported stannanes is the use of inorganic solid networks containing Sn-H, which have the advantages that a higher tin loading can be obtained and no leaching of the reagent is observed [93]. The supports in many cases compared well to the corresponding solution phase tin hydrides in reduction reactions.

Triorganogermanium hydrides can also be used to mediate radical reactions, and Bowman and coworkers have developed polystyrene- as well as Quadragel-supported germanium reagents for this purpose [94]. The reagents were applied in Barton–McCombie deoxygenation as well as in radical-mediated cyclization reactions (Scheme 6.23), where the Quadragel-anchored reagent, incorporating a short poly(ethylene glycol) chain between the polystyrene and the germanium hydride, was found to give better results.

A somewhat different approach to a polymer-supported radical source was described by Giacomelli and coworkers, who used an *N*-hydroxythiazole 2(3)-thione anchored to a Wang resin [95]. The reagent was prepared in solution and equipped with a pendant COOH group to facilitate attachment to the resin. The applicability of the reagent in radical-mediated reactions was then investigated via a Hunsdiecker reaction (Scheme 6.24). Treatment of the reagent with an acyl

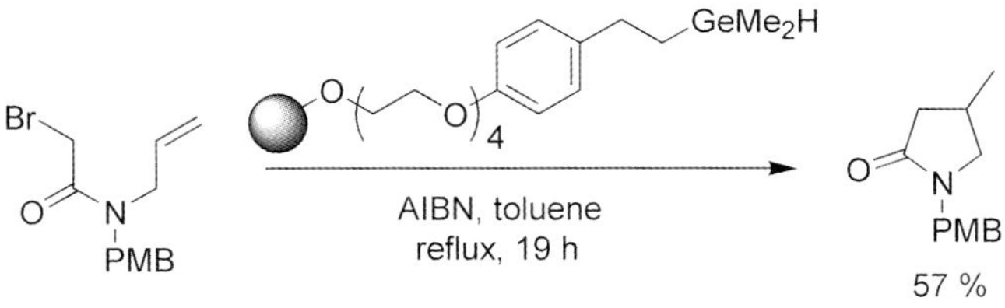

Scheme 6.23 A polymer-supported germanium reagent for radical-mediated cyclization.

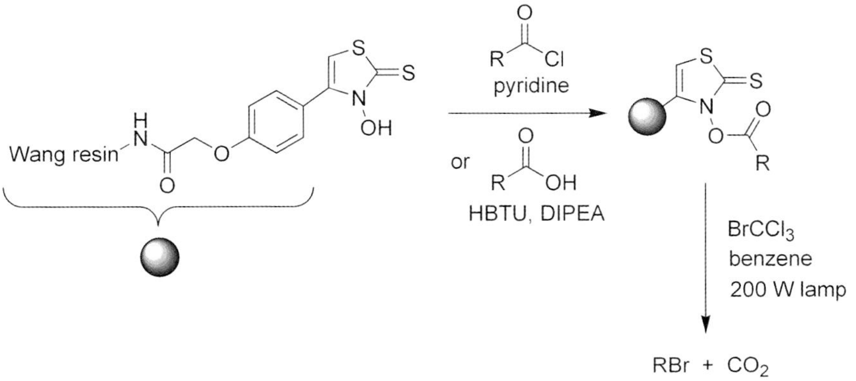

Scheme 6.24 A radical-mediated Hunsdiecker reaction.

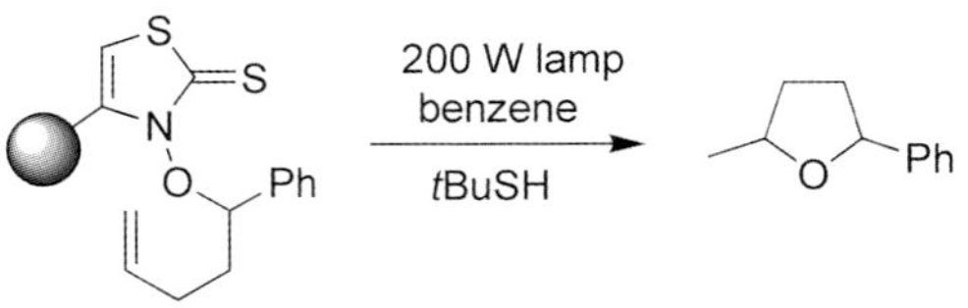

Scheme 6.25 Cyclization using a polymer-supported radical source.

chloride in pyridine [or, alternatively, reaction with a carboxylic acid under standard coupling conditions (HBTU/DIPEA)] generated an O-acylthiohydroxamate. Irradiation with a 200 W discharge lamp in the presence of $CBrCl_3$ in benzene liberated CO_2 gas, simultaneously creating the radical of the alkyl group, which was subsequently brominated to form the corresponding alkyl bromide. Radical cyclization reactions were also successful, albeit somewhat slow, affording pure derivatized tetrahydrofurans after simple filtration of the resin followed by concentration of the resulting solution (Scheme 6.25).

6.5
Conclusions

In summary, polymer-bound reagents for nucleophilic, electrophilic and radical reactions can in many instances be seen as more practical alternatives to the corresponding reaction using soluble reagents, facilitating removal of by-products in the Mitsunobu reaction for instance and enabling long reaction sequences to be carried out without extensive purification of the intermediates. The anchoring of toxic and/or highly reactive reagents such as stannanes and azides to a polymer also provides a safer option for the handling of such compounds. Many reagents have not been adapted for polymer-assisted solution phase chemistry, however, and so there still remains ample room for development in this area.

List of Abbreviations

ADDP	1,1'-(azodicarbonyl)dipiperidine
AIBN	azoisobutyronitrile
Ar	aryl
Bn	benzyl
Boc	tert-butoxycarbonyl
Bu	butyl
DCE	dichloroethane
DEAD	diethyl azodicarboxylate
DIPEA	N,N-diisopropylethylamine
DMAP	4-dimethylaminopyridine
Et	ethyl

Fmoc	9-fluorenylmethoxycarbonyl
HBTU	O-(benzotriazol-1-yl)-N,N,N',N'-tetramethyluronium hexafluorophosphate
HWE	Horner–Wadsworth–Emmons
LiHMDS	lithium hexamethyldisilazide
*m*CPBA	*meta*-chloroperoxybenzoic acid
Me	methyl
Ms	methanesulfonyl
NBS	*N*-bromosuccinimide
Nu	nucleophile
PEG	poly(ethylene glycol)
Ph	phenyl
PMB	4-methoxybenzyl
PSPBP	polystyrene-supported pyridinium bromide perbromide
PS-TPP	polystyrene-supported triphenylphosphane
ROMP	ring-opening metathesis polymerization
THF	tetrahydrofuran
TMG	tetramethylguanidine
TMS	trimethylsilyl
Ts	4-toluenesulfonyl

References

1. Kirschning, A., Monenschein, H. and Wittenberg, R. (2001) *Angewandte Chemie – International Edition in English*, **40**, 650–79.
2. Ley, S.V., Baxendale, I.R., Bream, R.N., et al. (2000) *Journal of the Chemical Society – Perkin Transactions 1*, 3815–4195.
3. Ley, S.V. and Baxendale, I.R. (2002) *The Chemical Record*, **2**, 377–88.
4. Ley, S.V., Baxendale, I.R., Brusotti, G., et al. (2002) *Fármaco*, **57**, 321–30.
5. Ley, S.V. and Baxendale, I.R. (2002) *Nature Reviews Drug Discovery*, **1**, 573–86.
6. McNamara, C.A., Dixon, M.J. and Bradley, M. (2002) *Chemical Reviews*, **102**, 3275–99.
7. Drewry, D.H., Coe, D.M. and Poon, S. (1999) *Medicinal Research Reviews*, **19**, 97–148.
8. Carpino, L.A., Williams, J.R. and Lopusinski, A. (1978) *Journal of the Chemical Society D – Chemical Communications*, 450–1.
9. Sayalero, S. and Pericas, M.A. (2006) *Synlett*, 2585–8.
10. Baxendale, I.R., Ley, S.V. and Sneddon, H.F. (2002) *Synlett*, 775–7.
11. Landi, J.J. Jr. and Brinkman H.R. (1992) *Synthesis*, 1093–5.
12. Hodge, P. and Richardson, G. (1975) *Journal of the Chemical Society D – Chemical Communications*, 622–3.
13. Regen, S.L. and Lee, D.P. (1975) *Journal of Organic Chemistry*, **40**, 1669–70.
14. Baxendale, I.R., Brusotti, G., Matsuoka, M. and Ley, S.V. (2002) *Journal of the Chemical Society – Perkin Transactions 1*, 143–54.
15. Harrison, C.R., Hodge, P. and Rogers, W.J. (1977) *Synthesis*, 41–3.
16. Baxendale, I.R., Ley, S.V. and Martinelli, M. (2005) *Tetrahedron*, **61**, 5323–49.
17. Relles, H.M. and Schluenz, R.W. (1974) *Journal of the American Chemical Society*, **96**, 6469–75.
18. Wells, A. (1994) *Synthetic Communications*, **24**, 1715–19.
19. Caputo, R., Ferrer, C., Noviello, S. and Palumbo, G. (1986) *Synthesis*, 499–501.

20 Caputo, R., Cassano, E., Longobardo, L. and Palumbo, G. (1995) *Tetrahedron*, **51**, 12337–50.
21 Caputo, R., Ferreri, C. and Palumbo, G. (1987) *Synthetic Communications*, **17**, 1629–36.
22 Caputo, R., Corrado, E., Ferreri, C. and Palumbo, G. (1986) *Synthetic Communications*, **16**, 1081–7.
23 Caputo, R., Cassano, E., Longobardo, L., et al. (1995) *Synthesis*, 141–3.
24 Akelah, A. and El-Borai, M. (1980) *Polymer*, **21**, 255–7.
25 Storer, R.I., Takemoto, T., Jackson, P.S. and Ley, S.V. (2003) *Angewandte Chemie – International Edition*, **42**, 2521–5.
26 Elson, K.E., Jenkins, I.D. and Loughlin, W.A. (2004) *Tetrahedron Letters*, **45**, 2491–3.
27 Fairfull-Smith, K.E., Jenkins, I.D. and Loughlin, W.A. (2004) *Organic and Biomolecular Chemistry*, **2**, 1979–86.
28 Lindsley, C.W., Zhao, Z., Newton, R.C., et al. (2002) *Tetrahedron Letters*, **43**, 4467–70.
29 Sieber, F., Wentworth, P. Jr., Toker, J.D., et al. (1999) *Journal of Organic Chemistry*, **64**, 5188–92.
30 Stanetty, P. and Kremslehner, M. (1998) *Tetrahedron Letters*, **39**, 811–12.
31 Tunoori, A.R., Dutta, D. and Georg, G.I. (1998) *Tetrahedron Letters*, **39**, 8751–4.
32 Humphries, P.S., Do, Q.-Q.T. and Wilhite, D.M. (2006) *Beilstein Journal of Organic Chemistry*, **2**, 21.
33 Bernard, M. and Ford, W.T. (1983) *Journal of Organic Chemistry*, **48**, 326–32.
34 Hughes, I. (1996) *Tetrahedron Letters*, **37**, 7595–8.
35 Bolli, M.H. and Ley, S.V. (1998) *Journal of the Chemical Society – Perkin Transactions 1*, 2243–6.
36 Martina, S.L.X. and Taylor, R.J.K. (2004) *Tetrahedron Letters*, **45**, 3279–82.
37 Nicolaou, K.C., Pastor, J., Winssinger, N. and Murphy, F. (1998) *Journal of the American Chemical Society*, **120**, 5132–3.
38 Barrett, A.G.M., Cramp, S.M., Roberts, R.S. and Zecri, F.J. (1999) *Organic Letters*, **1**, 579–82.
39 Salvino, J.M., Kiesow, T.J., Darnbrough, S. and Labaudiniere, R. (1999) *Journal of Combinatorial Chemistry*, **1**, 134–9.
40 Hemming, K., Bevan, M.J., Loukou, C., et al. (2000) *Synlett*, 1565–8.
41 Charette, A.B., Boezio, A.A. and Janes, M.K. (2000) *Organic Letters*, **2**, 3777–9.
42 Gil, C. and Bräse, S. (2005) *Chemistry – A European Journal*, **11**, 2680–8.
43 Grieder, A. and Thomas, A.W. (2003) *Synthesis*, 1707–11.
44 Lopez-Cremades, P., Molina, P., Aller, E. and Lorenzo, A. (2000) *Synlett*, 1411–14.
45 Snider, B.B. and Zhou, J. (2005) *Journal of Organic Chemistry*, **70**, 1087–8.
46 Gelbard, G. and Colonna, S. (1997) *Synthesis*, 113–6.
47 Harrison, C.R. and Hodge, P. (1980) *Synthesis*, 299–301.
48 Cardillo, G., Orena, M., Porzi, G. and Sandri, S. (1981) *Synthesis*, 793–4.
49 Cardillo, G., Orena, M. and Sandri, S. (1986) *Journal of Organic Chemistry*, **51**, 713–7.
50 Cardillo, G., Orena, M., Sandri, S. and Tomasini, C. (1985) *Tetrahedron*, **41**, 163–7.
51 Cainelli, G., Contento, M., Manescalchi, F. and Mussatto, M.C. (1981) *Synthesis*, 302–3.
52 Hassner, A. and Stern, M. (1986) *Angewandte Chemie International Edition in English*, **25**, 478–9.
53 Cainelli, G., Manescalchi, F. and Panunzio, M. (1979) *Synthesis*, 141–4.
54 Salunkhe, M.M., Sande, A.R., Kanade, A.S. and Wadgaonkar, P.P. (1997) *Synthetic Communications*, **27**, 2885–91.
55 Ramadas, K. and Janarthanan, N. (1999) *Synthetic Communications*, **29**, 1003–7.
56 Weber, J.V., Faller, P., Kirsch, G. and Schneider, M. (1984) *Synthesis*, 1044–5.
57 Bongini, A., Cainelli, G., Contento, M. and Manescalchi, F. (1980) *Journal of the Chemical Society, Chemical Communications*, 1278–9.
58 Rueter, J.K., Nortey, S.O., Baxter, E.W., Leo, G.C. and Reitz, A.B. (1998) *Tetrahedron Letters*, **39**, 975–8.
59 Zhong, H.M., Greco, M.N. and Maryanoff, B.E. (1997) *Journal of Organic Chemistry*, **62**, 9326–30.
60 Baxter, E.W., Rueter, J.K., Nortey, S.O. and Reitz, A.B. (1998) *Tetrahedron Letters*, **39**, 979–82.
61 Young, J.K., Nelson, J.C. and Moore, J.S. (1994) *Journal of the American Chemical Society*, **116**, 10841–2.

62. Nelson, J.C., Young, J.K. and Moore, J.S. (1996) *Journal of Organic Chemistry*, **61**, 8160–8.
63. Bräse, S., Enders, D., Köbberling, J. and Avemaria, F. (1998) *Angewandte Chemie – International Edition*, **37**, 3413–15.
64. Bräse, S., Köbberling, J., Enders, D., et al. (1999) *Tetrahedron Letters*, **40**, 2105–8.
65. Rademann, J., Smerdka, J., Jung, G., Grosche, P. and Schmid, D. (2001) *Angewandte Chemie – International Edition*, **40**, 381–5.
66. Smerdka, J., Rademann, J. and Jung, G. (2004) *Journal of Peptide Science*, **10**, 603–11.
67. Vignola, N., Dahmen, S., Enders, D. and Bräse, S. (2001) *Tetrahedron Letters*, **42**, 7833–6.
68. Vignola, N., Dahmen, S., Enders, D. and Bräse, S. (2003) *Journal of Combinatorial Chemistry*, **5**, 138–44.
69. Pilot, C., Dahmen, S., Lauterwasser, F. and Bräse, S. (2001) *Tetrahedron Letters*, **42**, 9179–81.
70. Murata, S. and Noyori, R. (1980) *Tetrahedron Letters*, **21**, 767–8.
71. Vickerstaffe, E., Warrington, B.H., Ladlow, M. and Ley, S.V. (2003) *Organic and Biomolecular Chemistry*, **1**, 2419–22.
72. Bapna, A., Vickerstaffe, E., Warrington, B.H., et al. (2004) *Organic and Biomolecular Chemistry*, **2**, 611–20.
73. Arnold, L.D., Assil, H.I. and Vederas, J.C. (1989) *Journal of the American Chemical Society*, **111**, 3973–6.
74. Kolb, H.C., Finn, M.G. and Sharpless, K.B. (2001) *Angewandte Chemie – International Edition*, **40**, 2004–21.
75. Roush, W.R., Feitler, D. and Rebek, J. (1974) *Tetrahedron Lett.*, **19**, 1392–2.
76. Green, G.M., Peet, N.P. and Metz, W.A. (2001) *Journal of Organic Chemistry*, **66**, 2509–11.
77. Caldarelli, M., Baxendale, I.R. and Ley, S.V. (2000) *Green Chemistry*, **2**, 43–5.
78. Fréchet, J.M.J., Farrall, M.J. and Nuyens, L.J. (1977) *Journal of Macromolecular Science – Chemistry*, **A11**, 507–14.
79. Cacchi, S., Caglioti, L. and Cernia, E. (1979) *Synthesis*, 64–6.
80. Bongini, A., Cainelli, G., Contento, M. and Manescalchi, F. (1980) *Synthesis*, 143–6.
81. Habermann, J., Ley, S.V. and Smits, R. (1999) *Journal of the Chemical Society – Perkin Transactions 1*, 2421–3.
82. Habermann, J., Ley, S.V., Scicinski, J.J., et al. (1999) *Journal of the Chemical Society – Perkin Transactions 1*, 2425–7.
83. Gerigk, U., Gerlach, M., Neumann, W.P., et al. (1990) *Synthesis*, 448–52.
84. Gerlach, M., Jordens, F., Kuhn, H., et al. (1991) *Journal of Organic Chemistry*, **56**, 5971–2.
85. Giese, B. (1983) *Angewandte Chemie – International Edition in English*, **22**, 753–64.
86. Bokelmann, C., Neumann, W.P. and Peterseim, M. (1992) *Journal of the Chemical Society – Perkin Transactions 1*, 3165–7.
87. Harendza, M., Lessmann, K. and Neumann, W.P. (1993) *Synlett*, 283–5.
88. Ruel, G., The, N.K., Dumartin, G., et al. (1993) *Journal of Organometallic Chemistry*, **444**, C18–C20.
89. Dumartin, G., Ruel, G., Kharboutli, J., et al. (1994) *Synlett*, 952–4.
90. Boussaguet, P., Delmond, B., Dumartin, G. and Pereyre, M. (2000) *Tetrahedron Letters*, **41**, 3377–80.
91. Barton, D.H.R. and McCombie, S.W. (1975) *Journal of the Chemical Society – Perkin Transactions 1*, 1574–85.
92. Enholm, E.J., Gallagher, M.E., Moran, K.M., et al. (1999) *Organic Letters*, **1**, 689–91.
93. Jenkins, P.M. and Tsang, S.C. (2004) *Green Chemistry*, **6**, 69–71.
94. Bowman, W.R., Krintel, S.L. and Schilling, M.B. (2004) *Synlett*, 1215–18.
95. De Luca, L., Giacomelli, G., Porcu, G. and Taddei, M. (2001) *Organic Letters*, **3**, 855–7.

7
Coupling and Introducing Building Block Reagents

Rafael Chinchilla and Carmen Nájera

Universidad de Alicante, Departamento de Química Orgánica and Instituto de Síntesis Orgánica (ISO), Facultad de Ciencias, ctra, San Vicente s/n, 03690 San Vicente del Raspeig, Alicante, Spain

7.1
Introduction

Probably the most important advantage in using a functionalized polymer as a reagent is the simplification of product work-up, separation and isolation, their use in organic synthesis being a continuously expanding topic. With quite insoluble cross-linked polymer resins, a simple filtration can be used for the isolation and washing, thus avoiding time-consuming and expensive chromatographic separations, something that is, nowadays, crucial when the preparation of a large array of compounds is required, as in combinatorial chemistry. Therefore, supported reagents can be used also in excess without giving rise to additional problems in the work-up procedure. In addition, resins provide the possibility of automation in the case of repetitive stepwise syntheses and the facility of carrying out reactions in flow reactors on a commercial scale. Moreover, the supported reagent can be reused after the facile separation, or will lead to anchored residual products that, if the appropriate chemistry is available, could be recycled.

There are, though, also several disadvantages. Probably the most important is the additional time and cost in synthesizing a supported reagent. However, more and more supported reagents are becoming commercially available, and the regeneration and recovery of the supported reagent is now always taken into account. Another frequent serious problem when dealing with polymeric reagents is the occurrence of slower reactions and lower yields than their soluble non-polymeric counterparts. Appropriate choice of support and reaction conditions can overcome this. In addition, the ultimate capacity of a functionalized polymer is restricted and this may be important in preparative organic chemistry when stoichiometric (or more) quantities of the supported reagent are required. Difficulties in the characterization of reactions on insoluble polymers can arise, where traditional analytical techniques can be useless.

The Power of Functional Resins in Organic Synthesis. Judit Tulla-Puche and Fernando Albericio
Copyright © 2008 WILEY-VCH Verlag GmbH & Co. KGaA, Weinheim
ISBN: 978-3-527-31936-7

Despite these potential drawbacks, numerous polymeric reagents applicable to all kind of synthetic transformations have been developed and considerable scope now exists for their exploitation in routine synthetic chemistry [1]. This chapter deals mainly with polymer-supported reagents that have been employed for the activation of a carboxyl functionality, thus creating an acylation species ready to couple with an electrophile and liberating the polymeric support. The chapter includes sections devoted to different types of anchored coupling reagents that can activate the carboxyl group by formation of an active ester without the intervention of any other coupling reagent. The last section deals with polymer-supported species that create an anchored, isolable active ester after reaction with a soluble coupling reagent.

7.2
Supported Carbodiimide and Isourea Reagents

Carbodiimides are traditionally one of the most frequently employed dehydrating reagents for the creation of the ester or amide bond, despite known problems related to difficulties in removing the corresponding generated urea or unreactive acylurea by-products [2]. Therefore, the use of polymer-supported carbodiimides is quite appropriate for the easy elimination of these by-products from the reaction medium by simple filtration, as they will remain anchored to the solid phase. In addition, if an ester or amide formation is intended an excess of acid can be used, as the remainder will also stay anchored as the corresponding isourea.

Although polycarbodiimides have long been known as polymers with film- and fiber-forming capabilities [3], probably the first example of an insoluble carbodiimide employed as a coupling agent is the use of polyhexamethylenecarbodiimide (**1**) (obtained by catalytic decarboxylation of 1,6-diisocyanatohexane) in the peptide coupling reaction of N-protected amino acids with amino acid ester hydrochlorides in the presence of triethylamine as base and dichloromethane as solvent [4]. However, the Merrifield chloromethylated cross-linked polystyrene has been the most popular and useful starting resin for the preparation of these type of coupling reagents. Polystyrene-bound carbodiimide **2** (R = iPr) was the first of these type of resins [5] and was employed for the preparation of some palladium(II) and platinum(II) dipeptide complexes [6]. However, this polymeric carbodiimide **2** (R = iPr) failed in other amide-forming coupling process [7], which is attributed to a reagent low quality after being obtained following the reported procedure [8]. In addition, the related supported carbodiimide **2** (R = Et) was prepared and used successfully as peptide coupling reagent in the formation of several dipeptides [9].

$$\mathrm{\{N=C=N-(\ )_6\}_n}$$

1

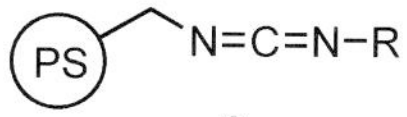

7.2 Supported Carbodiimide and Isourea Reagents

PS—CH₂—N=C=N—R
2
[R = Et, iPr, Cy]

PS = Crosslinked Polystyrene

The resin-supported carbodiimide **2** (R = Cy), related to the popular solution phase reagent dicyclohexylcarbodiimide (DCC), has been the most successfully employed polymeric carbodiimide of this series, especially in the presence of additives to accelerate the coupling reaction and avoid the acylisourea-unreactive acylurea rearrangement [2]. This carbodiimide has been used for esterification reactions, as exemplified in the reaction of dithiane-containing alcohol **3** with Fmoc-protected valine in the presence of a catalytic amount of N,N-dimethylaminopyridine (DMAP) to give ester **4** [10] (Scheme 7.1). This polymer-supported reagent **2** (R = Cy) has also been used without any additive in the amidation reaction of 3,4-diaminocyclopentanol scaffolds with 2-(methylsulfanyl)acetic acid [11].

There are examples of the use of supported carbodiimide **2** (R = Cy) for the formation of N-hydroxybenzotriazole (HOBt)-derived active esters from N-Boc-protected amino acids (see below), which have been subsequently employed in coupling reactions for the synthesis of dipeptide p-nitroanilides and dipeptide diphenyl phosphonates [12, 13]. Such HOBt-active esters have also been generated from pyrazinone-derived acids using **2** (R = Cy) [14], as well as from difluorophenyl acetic acids [15] for further amidation reactions in the parallel synthesis of tissue factor VIIa inhibitors.

However, more frequently employed has been the commercial polystyrene-supported carbodiimide **5** (0.9–1.4 mmol g^{-1}), another bound variant of DCC, although showing similar reactivity. In this supported reagent the tether to the polystyrene backbone has greater activity than the N-methylene linkage present in reagent **2** (R = Cy).

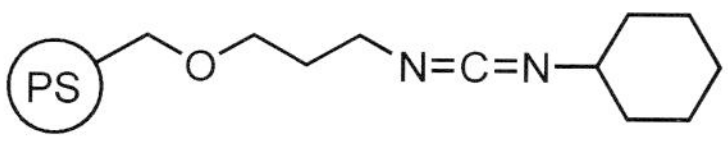

PS—O—(CH₂)₃—N=C=N—Cy
5

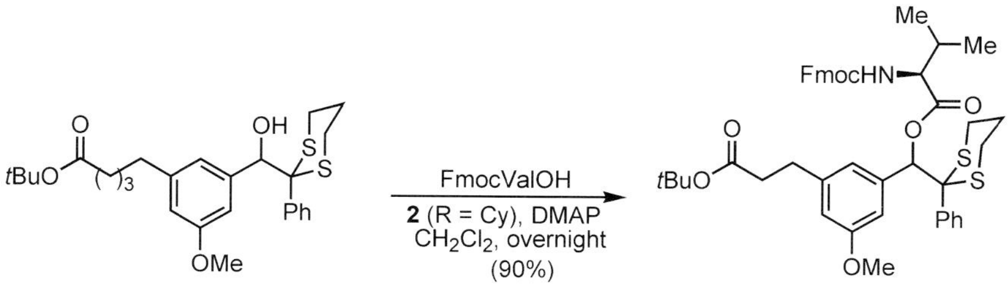

Scheme 7.1 Esterification reaction using supported carbodiimide **2** (R = Cy).

The high reactivity of this polymeric-related DCC **5** is particularly noticeable in the macrolactonization reaction of 12, 13, 15 and 16-carbon ω-hydroxy acids **6** (Scheme 7.2). Here, the combination of carbodiimide **5** and DMAP gave rise to the corresponding macrolactones **7** in identical yields to when DCC was used – the yield being even higher for the 13-membered lactone **7a** (only a 32% yield was obtained using DCC) [16].

Resin **5** has also proved efficient in the clean preparation of amides when HOBt is present as additive [17], with reaction times being shortened when using microwave heating [18]. Other examples of the use of this resin are in the synthesis of primary amides from 3,4-dihydropyrimidin-2-(1H)-one 5-carboxylic acids, in the presence of HOBt as additive [19], and the synthesis of active esters from 1-hydroxybenzotriazole (HOBt) (Section 7.6.3) for peptide couplings for the synthesis of α-ketothiazoles such as tissue factor VIIa inhibitors [20]. A mono-amidation reaction using resin **5** performed on o-phenylenediamines to give o-amidophenylamines, using 1-hydroxy-7-azabenzotriazole (HOAt) as additive, gave final products that have been used to prepare libraries of benzimidazoles [21]. An example of the use of this reagent in an amidation reaction for the preparation of a pharmacologically interesting compound is the synthesis of UK-427857 (Maraviroc) (**10**), a potent antagonist of the CCR5 receptor in the treatment of HIV. In this preparation, the last step involves additive-absent carbodiimide **5**-promoted coupling of triazole-containing amine template **8** with 4,4-difluorocyclohexanecarboxylic acid (**9**) (Scheme 7.3) [22]. Other applications of supported carbodiimide **5** are as activating agent for the coupling of amines and N,N'-bis(tert-butoxycarbonyl)thiourea in the synthesis of guanidines [23] and as additive in the preparation of acylsulfonamides using 4-(N,N-dimethylamino)pyridine (DMAP) [24].

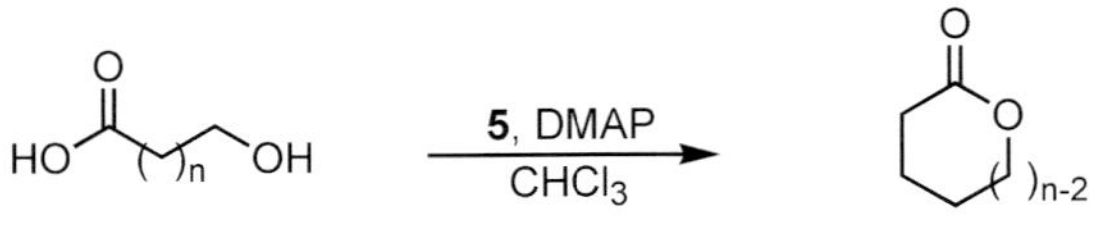

6a (n = 10)
6b (n = 11)
6c (n = 13)
6d (n = 14)

7a (52%)
7b (77%)
7c (95%)
7d (96%)

Scheme 7.2 Macrolactonization reaction using polymeric carbodiimide **5**.

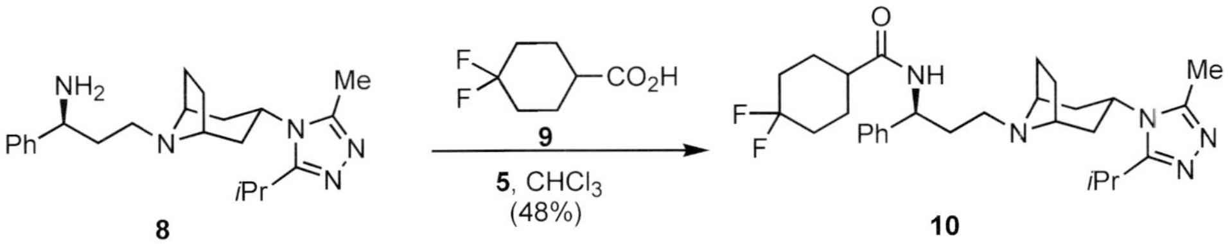

Scheme 7.3 Synthesis of Maraviroc (**10**), an antagonist of the CCR5 receptor.

Polystyrene-supported tethered 1-ethyl-3-(3-dimethylamino-propyl)carbodiimide hydrochloride (PS-EDC or PS-EDCI, **11**) is also a commercially available, frequently used resin-supported carbodiimide [25], which is highly effective in the coupling of carboxylic acids and amines, in the absence of any additive; the use of chloroform as solvent is essential.

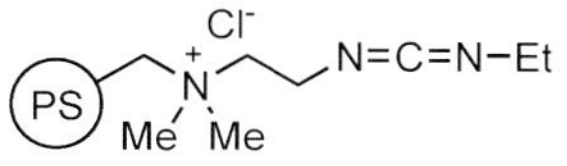

The usefulness of such supported carbodiimides is clearly illustrated in the application of resin **11** for the formation of Mosher amides **13** in deuteriochloroform as solvent, as shown in the general procedure involving the *(S)*-Mosher acid (**12**) depicted in Scheme 7.4. This procedure allowed a clean, fast determination of the optical purity of amines by NMR analysis after direct filtration of the residual polymer-supported urea **14** [26].

Due to its effectiveness in forming the peptide bond, this reagent has been used for the production of high-throughput lead discovery libraries [27]. Other examples of operational simplicity achieved using this supported coupling reagent are the formation of peptide bonds in the synthesis of disulfide peptides as substrates for trypanothione reductase, purified after laborious chromatographies when using solution phase chemistry [28]. In addition, tyrosine-derived diphenol monomers such as **17**, which can be used for the preparation of pseudo-polypeptides in biomaterial sciences, have been obtained by solid phase amide coupling of desaminotyrosine (**15**) and L-tyrosine hexyl ester (**16**) promoted by PS-EDC (**11**) (Scheme 7.5) [29]. In this case, the purity of the final product was higher than when using an analogous solution phase process. Moreover, PS-EDC (**11**) has also been useful for promoting the condensation of carboxylic acids with *N*-hydroxysuccinimide

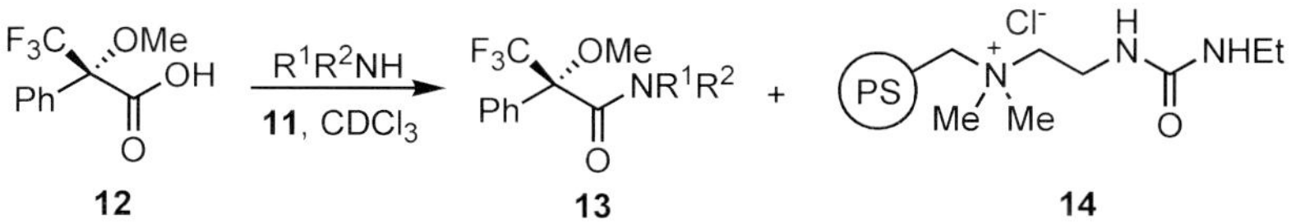

Scheme 7.4 Synthesis of Mosher amides using supported carbodiimide **11**.

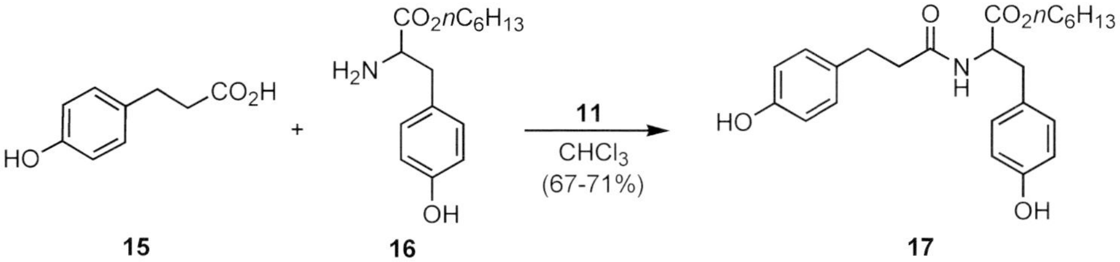

Scheme 7.5 Amidation reactions using the polymeric carbodiimide PS-EDC (**11**).

(HOSu or NHS) or pentafluorophenol to give water-soluble activated esters, which can not be purified by extraction [30]. It has also been employed with sulfonamides to afford acylsulfonamide libraries, this last case requiring the addition of DMAP as additive [31]. The formation of thiol esters [32] and also the synthesis of benzoxazinone and benzoxatine scaffolds, via cyclodehydration, has been performed using PS-EDC (**11**) [33].

There are also examples of the preparation of soluble supported carbodiimides, which after the coupling reaction give an anchored urea suitable for removal by precipitation after a solvent change. However, they are not commercially available and their use has been limited. Thus, carbodiimide **18** has been prepared by ring-opening metathesis polymerization (ROMP) of the corresponding monomer, showing a very high loading (2.5 mmol g^{-1}, 77% of its theoretical capacity), and has been used in esterification, amidation and anhydride-forming reactions [34]. A further use is in the preparation of a dendritic polyglycerol-supported *N*-cyclohexyl carbodiimide and its utility in esterification and amidation reactions [35].

18

Carbodiimide **2** (R = Cy) has been employed for the preparation of isolable supported *O*-alkylisourea reagents **19**, which are suitable for the clean high yielding generation of methyl, allyl, benzyl and 4-nitrobenzyl esters after reaction with carboxylic acids in THF as solvent at 60 °C and filtration of the supported urea [36–38]. Reagent **19** (R = Me) is synthesized simply by treatment of carbodiimide **2** (R = Cy) with methanol under microwave heating [36, 37] or, more generally, by reaction of **2** (R = Cy) with the corresponding alcohol in the presence of a catalytic amount of copper(II) triflate under thermal or microwave heating [38]. However, the corresponding solid-supported *tert*-butyl isourea (**19**, R = *t*Bu) could not be prepared.

R = Me, Allyl, Bn, 4-O$_2$NC$_6$H$_4$CH$_2$

19

7.3
Supported Aminium and Uronium Reagents

In peptide coupling chemistry, tetramethylurea-derived aminium salts from 1-hydroxybenzotriazole (HOBt), such as the *N*-[(1*H*-benzotriazol-1-yl)(dimethylamino)methylene]-*N*-methylmethanaminium hexafluorophosphate and tetrafluoroborate

N-oxide salts (HBTU and TBTU, respectively) [39], or from 1-hydroxy-7-azabenzotriazole (HOAt) such as *N*-[(dimethylamino)-1*H*-1,2,3-triazolo[4,5-*b*]pyridino-1-ylmethylene]-*N*-methylmethanaminium tetrafluoroborate *N*-oxide (HATU) [40], are well established reagents. They are especially devoted to peptide coupling reactions due to their efficiency and the low degree of undesirable racemization produced in the final peptide compared to the use of classical carbodiimide-coupling methods. Therefore, as the polystyrene-supported HOBt is an often used polymeric reagent (Section 7.6.3) [41], its transformation in a supported HOBt and tetramethylurea-derived aminium salt analog to HBTU and TBTU resulted directly. Thus, the reaction of polystyrene-2% divinylbenzene copolymer resin P-HOBt (**20**) with tetramethylchloroformamidinium tetrafluoroborate (**21**) (4 equivalents) in the presence of triethylamine gave polymeric *N*-[(1*H*-benzotriazol-1-yl)(dimethylamino)methylene]-*N*-methylmethanaminium tetrafluoroborate *N*-oxide (P-TBTU, **22**) (Scheme 7.6) [42].

This resin **22** was used in the coupling reaction between Boc and Cbz-protected amino acids and amino acid esters hydrochlorides using acetonitrile as solvent and pyridine as base at room temperature, with P-HOBt (**20**) being recovered after the reaction completion by filtration, as shown in the preparation of the dipeptide BocGly-PheOEt (**23**) (Scheme 7.7) [42]. The reaction times were longer and the yields were slightly lower when using P-TBTU (**22**) compared with using non-supported TBTU. The extent of racemization was measured by the epimerization degree on the tripeptide generated by coupling CbzGlyPheOH and ValOMe (Anteunis' test) [43] and by the racemization obtained in the dipeptide formed by coupling L-BzLeuOG and GlyOEt (Young's test) [44]. No epimerization was found in the first case (similarly to when using non-supported TBTU) and a 15% yield of the racemized D-isomer in the second (lower than when using non-supported TBTU). In addition, reagent **22** was considerably more hydrolytically stable than non-supported TBTU as coupling agent, such that even wet acetonitrile was an appropriate solvent. However, sterically $C^{\alpha\alpha}$ di-substituted hindered amino acid

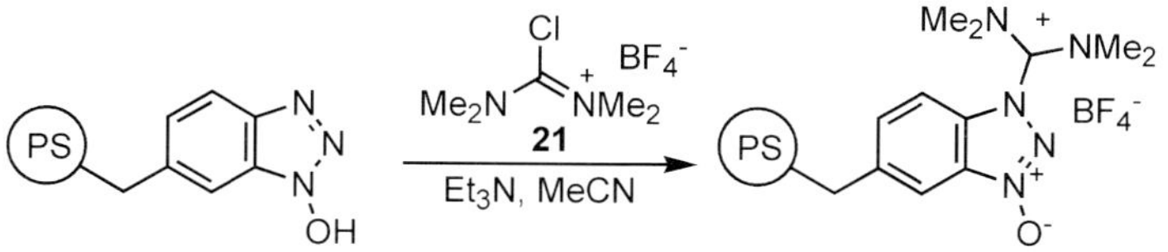

Scheme 7.6 Preparation of polymer-supported TBTU (**22**).

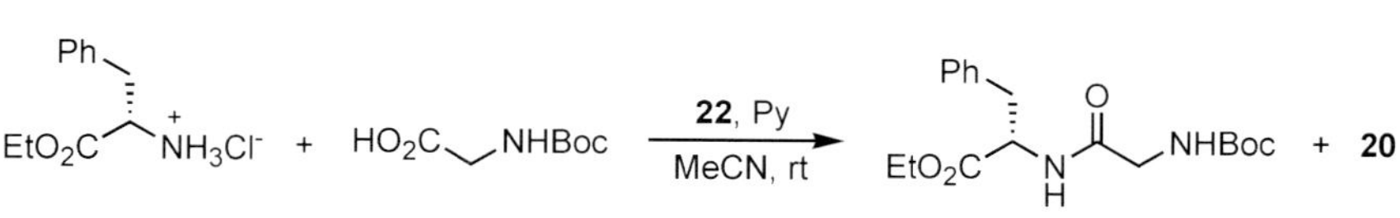

Scheme 7.7 Synthesis of a protected dipeptide using P-TBTU (**22**) as coupling agent.

derivatives, such as those from 2-aminoisobutyric acid (Aib), gave low coupling yields.

The use of a hexachloroantimonate related reagent P-HBTU (**25**) for the formation of hydroxamic acid derivatives, such as **26**, has also been reported (Scheme 7.8) [45]. Thus, treatment of sulfanonamido-substituted cinnamic acid **24** with supported iminium coupling reagent **25** gave directly the corresponding HOBt-derived anchored active ester. After washing, it was treated with *O*-(tetrahydro-2*H*-pyran-2-yl)hydroxylamine (THPONH$_2$) in the presence of diisopropylethylamine to release the hydroxamate in solution, which was then hydrolyzed to the corresponding hydroxamic acid **26**. Such sulfanonamido-substituted cinnamic acids are precursors of histone deacetylase inhibitors.

A copolymer of polystyrene and maleic anhydride containing the *N*-hydroxysuccinimide moiety (P-HOSu, **27**) (Section 7.6.2) has been employed similarly to P-HOBt (**20**) for the reaction with tetramethylchloroformamidinium tetrafluoroborate and hexafluorophosphate (**28**), in the presence of pyridine as base, to generate the corresponding supported uronium salts P-TSTU and P-HSTU (**29**), respectively (Scheme 7.9) [46]. These polymeric reagents have been used as supported analogues of the peptide coupling reagents 2-succinimido-1,1,3,3-tetramethyluronium tetrafluoroborate and hexafluorophosphate (TSTU and HSTU, respectively), affording moderate yields when performing the coupling of Boc and Cbz-protected amino acids and amino acid ester hydrochlorides using pyridine as base and acetonitrile as solvent at 50 °C. The yields of the corresponding obtained peptides were lower than when using their non-supported counterparts, although with lower racemization levels according to Anteunis' and Young's tests. The residual P-HOSu (**27**) was recovered quantitatively once the coupling reaction was finished by precipitation with hexane, filtration and washing with dilute HCl, and reused for the preparation of fresh reagents.

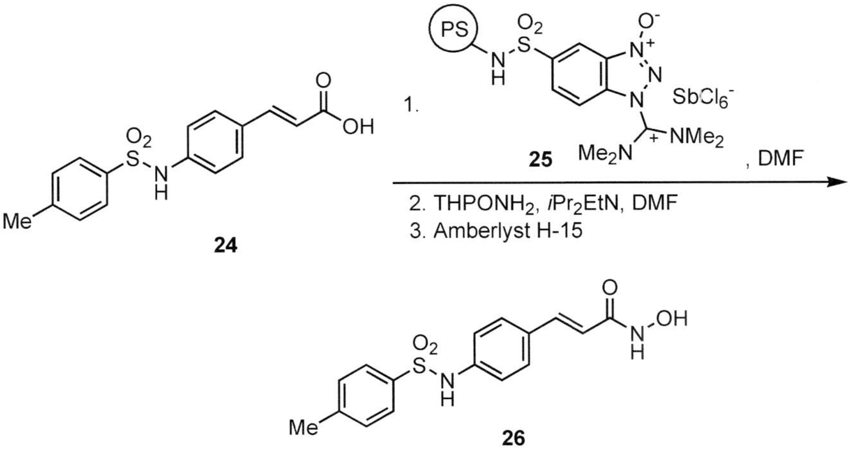

Scheme 7.8 Synthesis of a cinnamic acid-derived hydroxamate using active esters generated with P-HBTU (**25**).

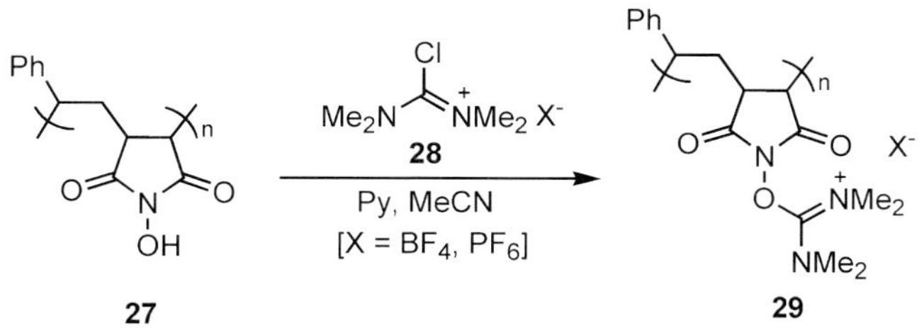

Scheme 7.9 Preparation of the polymeric uronium salts P-TSTU and P-HSTU (**29**).

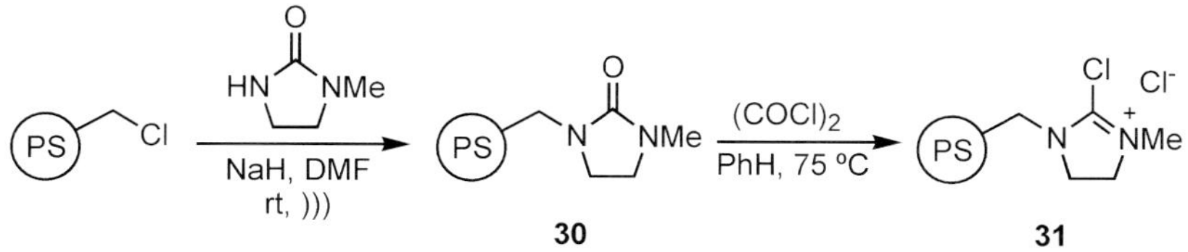

Scheme 7.10 Preparation of the supported coupling reagent P-DMI (**31**).

A resin-supported version of the coupling reagent 2-chloro-1,3-dimethylimidazolinium chloride (DMC) has been obtained by anchoring 1-methyl-2-imidazolidinone (DMI) to the Merrifield resin using sodium hydride in DMF as solvent under sonication, giving the corresponding P-DMI (**30**) (Scheme 7.10) [47]. Chlorination of resin **30** with oxalyl chloride in benzene gave the desired P-DMC (**31**), which could not be fully characterized because of its partial regeneration to P-DMI (**30**) with moisture during purification, conversion being estimated as 80–90% by FT-IR. This supported reagent has proved effective in the esterification of hindered components such as pivalic acid or *tert*-butyl alcohol using triethylamine as base in dichloromethane as solvent. The resulting supported urea by-product P-DMI (**30**) was separated by filtration and reused up to three times. In addition, this P-DMC (**31**) was employed in the coupling of N-protected amino acids and amino acid esters to afford the corresponding dipeptides with low levels of undesired epimerization [47].

Fluoroformamidinium salts such as tetramethylfluorochloroformamidinium hexafluorophosphate (TFFH) [48] afford highly reactive acid fluorides when reacting with carboxylic acids; these acid fluorides are suitable for the synthesis of amides derived from mono- and sterically hindered disubstituted amino acids in acceptable yields. Related to these formamidinium salts are the ROMP-derived fluoroformamidinium hexafluorophosphate **32** and **33**, prepared from formamidinium norbornene monomers by ROMP or graft copolymerized onto polystyrene cores, respectively [49]. These supported formamidinium salts have been used to prepare hindered amides in the case of reagent **32**, and dipeptides and tripeptides in the case of reagent **33**, in the presence of the Hünig base, always achieving high yields of the coupled product. However, reagent **33** gave rise to complete epimerization and racemization when performing Anteunis' and Young's tests, respectively, therefore limiting its use to hindered α,α-disubstituted amino acid residues when dealing with peptide synthesis.

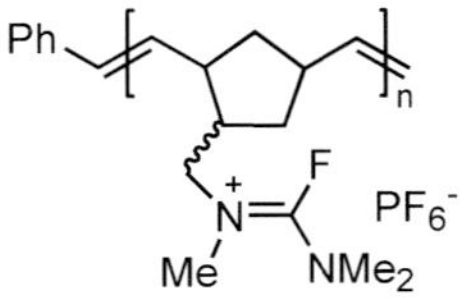

32

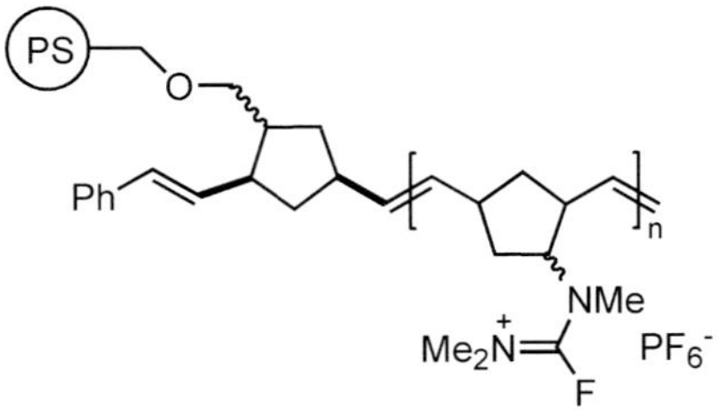

33

7.4
Supported Phosphorous-Containing Reagents

Benzotriazol-1-yl-N-oxy-tris(dimethylamino)phosphonium hexafluorophosphate (BOP) has been one of the most used phosphonium salts, with applications in the activation of the carboxylic acid function for peptide synthesis [50, 51], being specially efficient for the coupling of α,α-dialkyl amino acids, such as Aib derivatives [52]. The supported version of this BOP reagent has been prepared by reaction of a commercially available P-HOBt (**34**) (1 mmol g^{-1} resin loading) (Section 7.6.3) with 5 equivalents of bromotris(dimethylamino)phosphonium hexafluorophosphate (BrOP) in the presence of triethylamine as base, affording the corresponding P-BOP (**35**) (Scheme 7.11) [53]. The effective resin loading, determined by the yield of the coupling reaction of Fmoc-βAlaOH with isopropylamine, was found to be around 0.6–0.7 mmol g^{-1}, depending on the nature (and the dryness) of the employed solvent.

This P-BOP (**35**) has been used as a solid-supported reagent for peptide coupling reactions, the racemization tests affording lower values than when using the non-supported analog BOP. This reagent is also a suitable activating reagent for

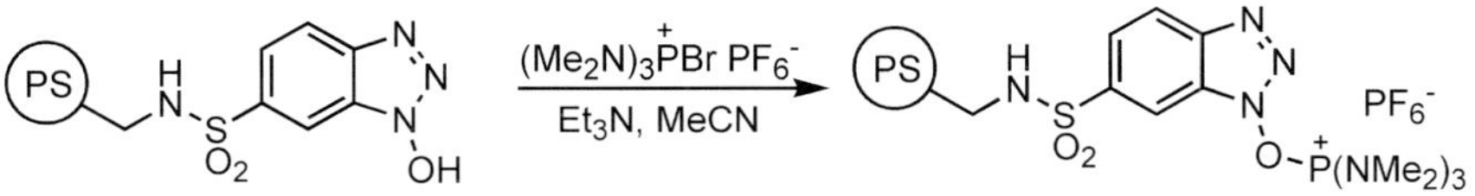

Scheme 7.11 Preparation of the supported phosphonium salt P-BOP (**35**).

difficult peptide coupling reactions involving α,α-dialkyl amino acids, such as the coupling of BocAibOH and ValOMe, which was obtained in 76% yield even working at room temperature.

Polymer-supported triphenylphosphine ditriflate (**37**) has been prepared by treatment of polymer bound (polystyrene-2% divinylbenzene copolymer resin) triphenylphosphine oxide (**36**) with triflic anhydride in dichloromethane, the structure being confirmed by gel-phase ^{31}P NMR [54, 55] (Scheme 7.12). This reagent is effective in various dehydration reactions such as ester (from primary and secondary alcohols) and amide formation in the presence of diisopropylethylamine as base, the polymer-supported triphenylphosphine oxide being recovered after the coupling reaction and reused. Interestingly, with amide formation, the reactive acyloxyphosphonium salt was preformed by addition of the carboxylic acid to **37** prior to addition of the corresponding amine. This order of addition ensured that the amine did not react competitively with **37** to form the unreactive polymer-supported aminophosphonium triflate.

Related anchored 1,1,3,3-tetraphenyl-2-oxa-1,3-diphospholanium bis-triflate (**39**) has been prepared by reaction of brominated poly(styrene-co-divinylbenzene) resin **38** with the phosphorous anion generated from 1,2-bis(diphenylphosphino)ethane and sodium naphthalenide followed by further oxidation and reaction with triflic anhydride (Scheme 7.13) [55]. This supported reagent has also been employed, to a lesser extent than **37**, for the formation of esters and amides by reaction of carboxylic acids with primary alcohols and amines, respectively.

The combination of polystyrene-bound triphenylphosphine and carbon tetrachloride has been used for the condensation of N-alkoxycarbonyl α-amino acids and primary amines, including amino acid esters, in the presence of N-methylmorpholine as base and refluxing dichloromethane as solvent [56]. In this case, supported triphenylphosphane oxide was isolated by filtration after the coupling reaction. The nature of the intermediate involved in the condensation was supposed to be the acid chloride, as these derivatives are found when heating polystyrene-supported dichlorotriphenylphosphorane with carboxylic acids [57, 58]. However, evidence supported by infrared spectroscopy suggests the formation of

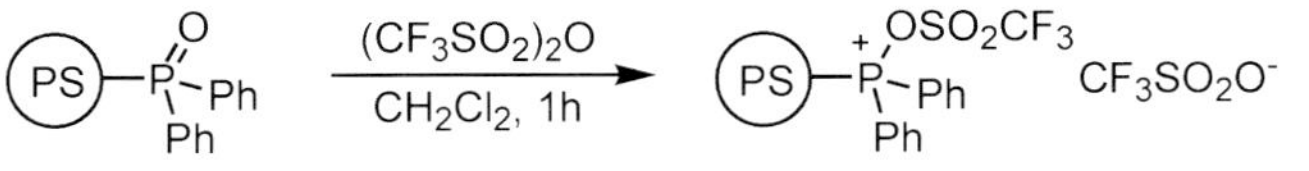

Scheme 7.12 Preparation of polymer-supported triphenylphosphine ditriflate (**37**).

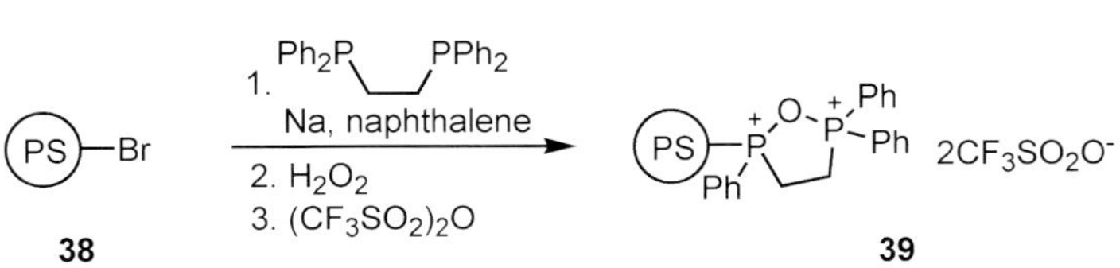

Scheme 7.13 Preparation of polymeric 2-oxa-1,3-phospholanium bis-triflate **39**.

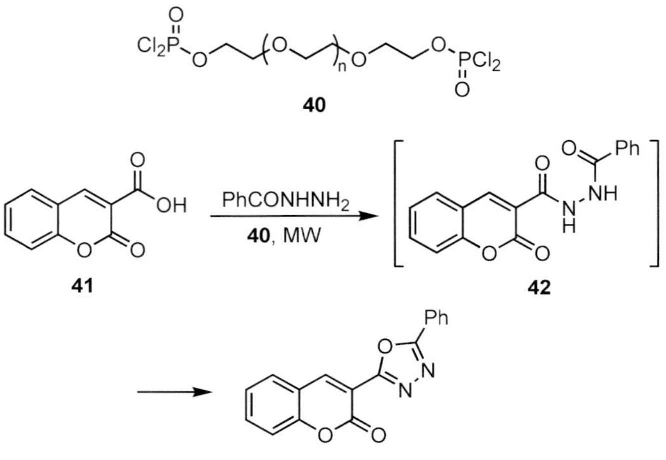

Scheme 7.14 Synthesis of a coumarin 1,3,4-oxadiazole derivative using the PEG-supported dichlorophosphate **40**.

a resin-bound mixed phosphinic anhydride. A variation of this procedure is the use of polystyrene-bound triphenylphosphine-iodine complexes, which has enabled the readily coupling of N-protected amino acids and α-aminoacyl esters in high yield [59]. The reaction is carried out by mixing the supported phosphane and iodine in dichloromethane at room temperature and adding the coupling partners; no detectable racemization is observed when performing the coupling of FmocPheOH and GlytBu. In addition, the PEG-supported dichlorophosphate **40** has been obtained by reaction of PEG-6000 with phosphorus oxychloride, and has been employed in solid solvent-free coupling of coumarin-3-carboxylic acid (**41**) and benzoic acid hydrazines under microware irradiation, giving rise to coumarin-derived 1,3,4-oxadiazoles such as **43** in 7 min after in situ dehydration of the coupled intermediate **42** (Scheme 7.14) [60]. In this case, the product **43** was isolated by washing away the polymer reagent with water, the PEG being recovered by extraction from the aqueous solution.

7.5
Other Supported Coupling Reagents

Among coupling reagents that generate mixed anhydrides as activated acyl species for coupling reactions, 2-ethoxy-1-ethoxycarbonyl-1,2-dihydroquinoline (EEDQ) has long been used in solution phase synthesis [61]. The supported version of this reagent has been obtained by copolymerization of 6-(prop-1-en-2-yl)quinoline, polystyrene and 1,4-divinylbenzene as cross-linker and further reaction of the resulting 6-anchored quinoline with ethyl chloroformate and ethanol to give resin **44** [62]. When this resin reacts with a carboxylic acid, an intermediate **45** is formed,

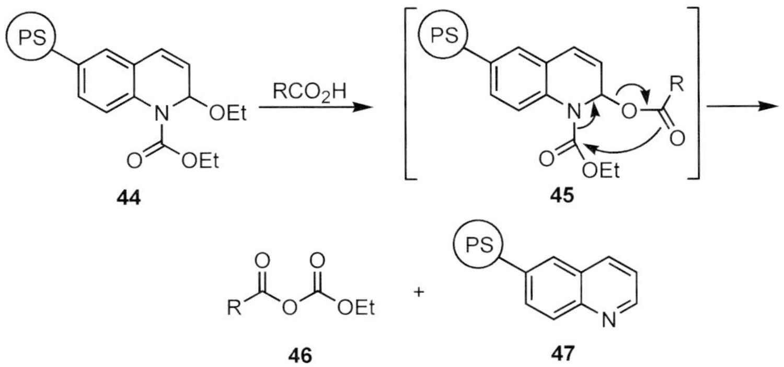

Scheme 7.15 Formation of mixed anhydrides using supported dihydroquinoline **44**.

which suffers a rapid breakdown by way of a six-membered transition state to give the mixed anhydride **46** and the recoverable initial supported quinoline **47** (Scheme 7.15). This has been applied to peptide synthesis when an N-protected amino acid is used in the presence of the free amino group of an amino acid ester [62].

The EEDQ analog 2-isobutoxy-1-isobutoxycarbonyl-1,2-dihydroquinoline (IIDQ) has been prepared [63], taking into account that mixed anhydrides from isobutyl chloroformate are preferable to those from to ethyl chloroformate for peptide synthesis [64]. The supported version of IIDQ (**48**) has been obtained recently after reaction of the corresponding polystyrene-attached quinoline (obtained by reaction of 6-hydroxyquinoline with the Merrifield resin using potassium carbonate as base) with isobutyl chloroformate in the presence of Hünig's base, and quenching the reactive intermediate with isobutanol [65]. This resin-bound reagent (**48**) (1.6 mmol per g loading) proved stable under laboratory storage for 2 months and was used in amidation reactions performed in acetonitrile as solvent. Using this reagent, no influence of the order of addition of reagent and substrates was observed, something frequently important when dealing with this type of mixed anhydride-forming agents. Even hindered amino acids acid derivatives such as BocAibOH gave high yields, and no racemization was observed when the coupling of CbzGlyPheOH and ValOMe (Anteuni's test) was carried out. Filtration after completion of the reaction allowed recovery of the initial isoquinoline, which was reused for the preparation of new **48**. This resin-anchored reagent has been employed in a flow reactor coupling process, creating anhydrides of Fmoc-amino acids that subsequently react with P-HOBt (**34**), forming activated amino esters ready for a peptide coupling step [66].

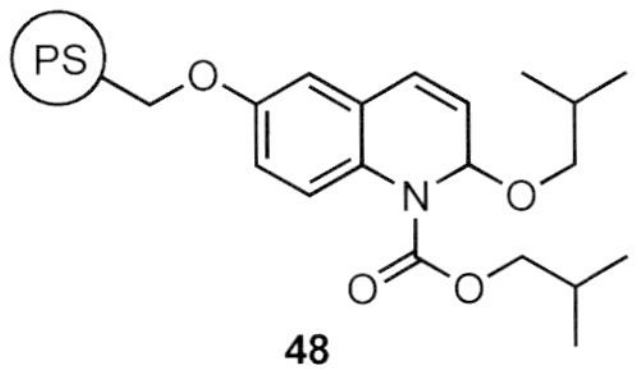

N-Alkyl-2-halopyridinium salts have been extensively used as activated agents for carboxylic acids [67], N-methyl-2-chloropyridinium iodide (Mukaiyama reagent) being particularly employed to convert carboxylic acids into esters and amides [68]. However, chromatographic separation is generally needed to completely remove the by-products. Although sulfonic acid columns can be used to scavenge the residual Mukaiyama reagent and the by-product (N-methyl-2-pyridone), this strategy is not compatible with molecules possessing basic sites. Therefore, a supported version of this reagent, which could simplify the purification operations, has been developed. Thus, polystyrene-supported Mukaiyama reagent **49** has been obtained by reaction of Merrifield's resin with 2-chloropyridine in the presence of potassium iodide (Scheme 7.16) and has been used in the dehydration of thioureas, forming carbodiimides, and also in the guanylation of primary amines [69].

However, some difficulties found when reproducing the preparation of reagent **49**, with acceptable loadings, prompted the necessity of employing a better leaving group than a halide. Therefore, hydroxymethylphenoxymethyl polystyrene crosslinked with 1% divinylbenzene (Wang resin, **50**) was activated with trifluoromethanesulfonic anhydride, forming a triflate ester, which was immediately substituted by the present 2-chloropyridine to generate the supported Mukaiyama reagent **51** with complete conversion (Scheme 7.17) [70].

This supported reagent (**51**) proved to be effective for the high-yielding synthesis of esters or amides from carboxylic acids and alcohols or amines (primary and secondary), in the presence of triethylamine or diisopropylethylamine as base and dichloromethane as solvent. An example is the amidation reaction of 2-(biphenyl-4-yl)acetic acid (**52**) with a poorly nucleophilic amine such as methylphenylamine, giving the corresponding amide **53** and the easily separated residual anchored 2-pyridone **54** (Scheme 7.18). However, a limitation of reagent **51** is its observed low effect when used in the synthesis of oligopeptides containing more than two amino acid residues [70]. Thus, the coupling of a C-terminal dipeptide (CbzGlyPheOH) with ValOMe afforded a poor isolated yield (<15%).

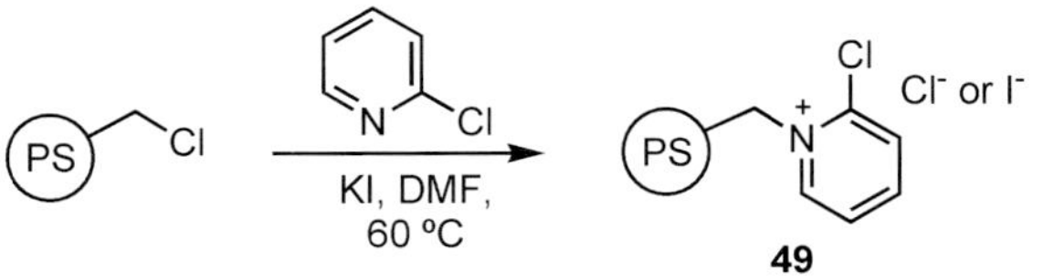

Scheme 7.16 Preparation of polystyrene resin-supported Mukaiyama reagent **49**.

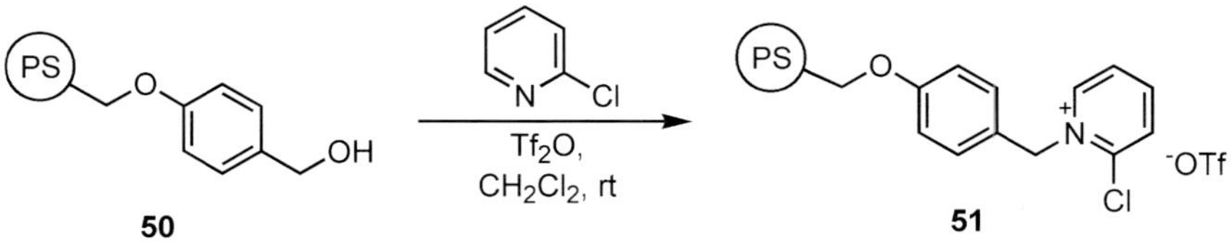

Scheme 7.17 Preparation of Wang resin-supported Mukaiyama reagent **51**.

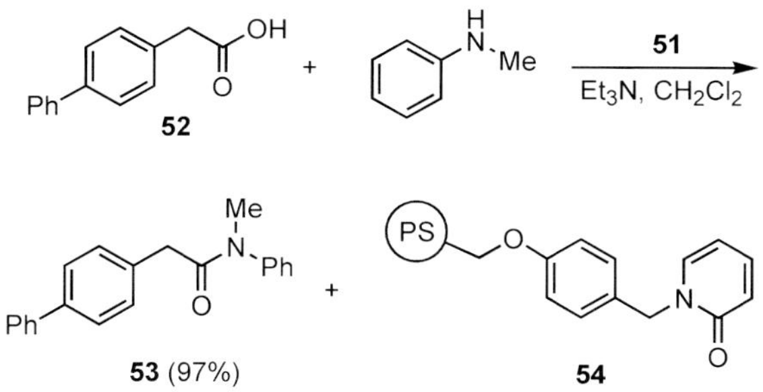

Scheme 7.18 Amidation reaction using polymer-anchored Mukaiyama-related reagent **51**.

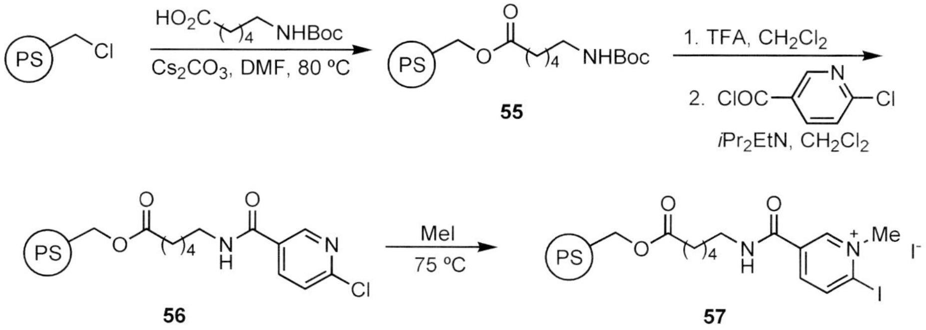

Scheme 7.19 Preparation of tethered Mukaiyama reagent **57**.

A differently anchored Mukaiyama reagent is the N-methylpyridinium iodide salt **57** [71], which has been obtained by reaction of the Merrifield resin with N-Boc-aminocaproic acid in the presence of cesium carbonate to give the supported ester **55** (Scheme 7.19). Further Boc-deprotection and reaction with 6-chloronicotinoyl chloride in the presence of Hünig's base furnished the anchored 2-chloropyridine **56**, which was transformed into the final N-methylpyridinium salt **57** after N-methylation in neat methyl iodide. This supported reagent has been used in the rapid microwave-assisted esterification of carboxylic acids and alcohols in the presence of triethylamine as base, with dichloromethane as solvent at 80 °C, the products being obtained in high purity after simple resin filtration [72].

Triazines, such as 2,4,6-trichloro[1,3,5]triazine and 2-chloro-4,6-dimethoxy[1,3,5]triazine are known selective coupling reagents for amide synthesis [73]. These chlorotriazines are generally activated with N-methylmorpholine (NMM) to give a morpholinium salt that reacts with a carboxylic acid to give an activated ester, which is relatively stable to oxygen nucleophiles but reacts easily with amines to give amides. The supported triazine coupling reagent **59** has been prepared by reacting 2,4,6-trichloro[1,3,5]triazine to a Wang resin loaded with Gly (**58**) in the presence of diisopropylethylamine as base (Scheme 7.20) [74]. Reaction of resin

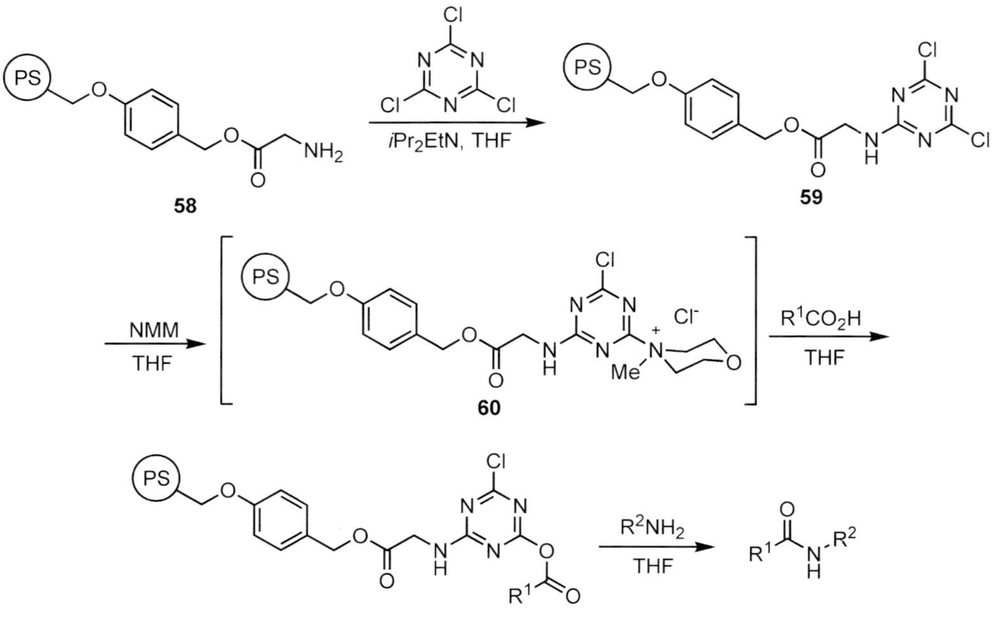

Scheme 7.20 Preparation of supported triazine coupling reagent **59** and its use in amidation reactions.

59 with NMM generated the supported morpholinium salt **60**, which reacted with aliphatic and aromatic carboxylic acids to give the resin loaded with the carboxylate **61**, which can even be dried and stored. Reaction of anchored activated ester **61** with amines in the presence of NMM gave the corresponding amides in good yields (60–87%).

However, when the supported triazine **59** was employed for the preparation of the dipeptide formed by coupling BocValOH and PheOMe, the yield was rather low and impurities were observed. Therefore, the influence of the nature of the polymer was investigated by preparing different resin-anchored triazines and performing the coupling between activated ester **62** and PheOMe (Scheme 7.21) [74]. The cross-linked polystyrene-poly(ethylene glycol) gave a more flexible support, being able to accommodate the reagent far from the beads, thereby improving the reactivity and giving a higher yield of the final dipeptide **63** (95%), with less than 5% of racemization.

Poly(ethylene glycol) (PEG) 4000 has been employed for the preparation of a PEG-anchored version of the previous dichlorotriazine reagents after its reaction with 2,4,6-trichloro[1,3,5]triazine in the presence of NMM [75]. This polymeric reagent (**64**) has been used for the preparation of anchored benzoic acid derivatives in the presence of NMM and THF as solvent, the supported activated esters being precipitated by addition of ether. Their further reaction with aliphatic and aromatic amines as bases afforded the corresponding amides in good yields, although sec-

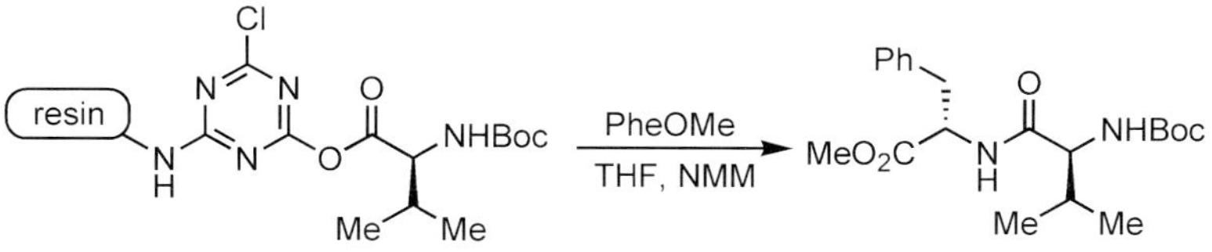

Scheme 7.21 Comparison of the efficiency of a dipeptide formation using an active ester from a dichlorotriazine anchored to different resins.

Starting resin	Yield (%)
Polystyrene-CH$_2$NH$_2$	25
Wang-O$_2$CCH$_2$NH$_2$	64
Polystyrene-O$_2$CCH$_2$NH$_2$	51
Polystyrene-PEG-NH$_2$	95

ondary amines gave lower yields. These PEG-supported dichlorotriazines have also been used as electrophilic scavengers, removing nucleophilic reactants such as alcohols, thiols, triphenylphosphane and triphenylphosphane oxide [76]. In addition, tetra(ethylene glycol) has been condensed with two molecules of 2,4,6-trichloro[1,3,5]triazine to give a tetra(ethylene glycol) bis(dichlorotriazinyl) ether, which has been copolymerized with tris(2-aminoethyl)amine. The resulting monochlorinated copolymer has a very high loading of the dehydrocondensing activity (ca. 3 meq g^{-1}) and has been used in the preparation of amides by a procedure consisting of, first, addition of NMM to the copolymer and further mixing with the carboxylic acid and the amine. Protic solvents can be used, although attempts to recycle the copolymer failed [77].

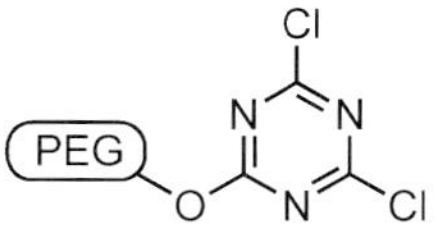

2,4,6-Trichloro[1,3,5]triazine has also been attached to the Wang resin. Reaction of the resulting supported dichlorinated triazine with 2-(2-aminoethoxy)ethanol and further treatment again with an excess of 2,4,6-trichloro[1,3,5]triazine afforded resin **65**, which has been shown to be effective in the conversion of carboxylic acids into acid chlorides, using triethylamine as base in dichloromethane or acetone as solvent [78]. The obtained acid chlorides can be easily transformed into ester or amides after removal of the resin by filtration and addition of an alcohol or an amine. From the isolated yields of the esters or amides it could be deduced that the yields of the obtained acid chlorides are around 70–90%. However, a chiral amino acid was completely racemized under this resin treatment.

A special case of acylation of amines occurs when using the chiral supported 1,2-disulfonamide **66** (0.58 mmol g^{-1} loading), which can be considered the first polymeric reagent employed in the kinetic resolution of primary amines through enantioselective acetylation of one of the enantiomers in the racemic mixture [79]. The Merrifield-derived chiral scaffold could be quantitatively recovered and recycled.

7.6
Active Ester-Forming Polymeric Reagents

The active ester methodology has been much applied mainly to form peptide bonds under mild conditions in both liquid- and solid-phase synthesis. In liquid-phase synthesis, most frequently the activation of a carboxyl group has employed HOSu, followed by esters of electron-withdrawing-containing phenols (nitrophenols of pentafluorophenols), the most reactive HOBt esters usually being unstable and used in solid-phase synthesis. The preparation of these active esters requires the presence of a coupling reagent such as DCC or DIC.

When dealing with polymer-supported analogues of these active ester-forming XOH species, a lower reactivity is expected, which is attributed to a relatively slow mass transport within the polymer pores that can not keep up with the rate of the acylation reaction. Therefore, the obtained polymeric active esters usually have a long shelf-life and are rather insensitive to moisture. To overcome this lower reactivity, the supported active esters from highly reactive HOBt derivatives have been the most frequently used. This section deals with these supported active ester-forming species.

7.6.1
Phenol-Derived Polymers

The older solid supported reagents employed in the preparation of anchored active esters useful in peptide synthesis have been the derivatives of o-nitrophenol. Thus, divinylbenzene cross-linked poly-4-hydroxy-3-nitrostyrene (**67**) [80] has been used for anchoring N-protected amino acids to resin **67**, using the DCC method in DMF as solvent [81]. The insoluble polymeric active esters **68** (1–1.5 mmol of amino acid per g of resin), once separated by filtration, could be stored at room temperature without decomposition and were used for the coupling with amino acid esters in DMF, giving the corresponding peptide **69**, with the liberated resin **67** being separated by centrifugation (Scheme 7.22). Removal of the N-blocking group from the formed peptide enabled repetition of the coupling reaction with an insoluble active ester of another N-blocked amino acid, thus elongating the peptide chain, an example being found in the preparation of the nonapeptide hormone bradykinin [82]. Resin **67** was also employed for the synthesis of cyclic peptides, performed by DCC-promoted coupling of N-benzyloxycarbonyl derivatives with a terminal carboxylic acid functionality to polymeric nitrophenol **67** [83]. Subsequent N-deprotection of the active ester, followed by cyclization and de-anchoring once the terminal amino group was liberated gave rise to the final products. Moreover, the supported benzoate ester from resin **67** has been used to introduce a benzoyl group after reaction with the lithium carbanions from acetophenone or 2-phenylacetonitrile [84].

Other o-nitrophenol-containing resins have been prepared with the aim of increasing the distance between the reactive center and the macromolecular backbone, which should accelerate the active ester formation by achieving an easier approach of the reagents. Thus, the Friedel–Crafts alkylation of styrene-divinylbenzene copolymer with 4-hydroxy-3-nitrobenzyl chloride promoted by aluminium trichloride gave 4-hydroxy-3-nitrobenzylated polystyrene (**70**) (approximately 30% of the aromatic rings of the polymer were substituted according to elemental

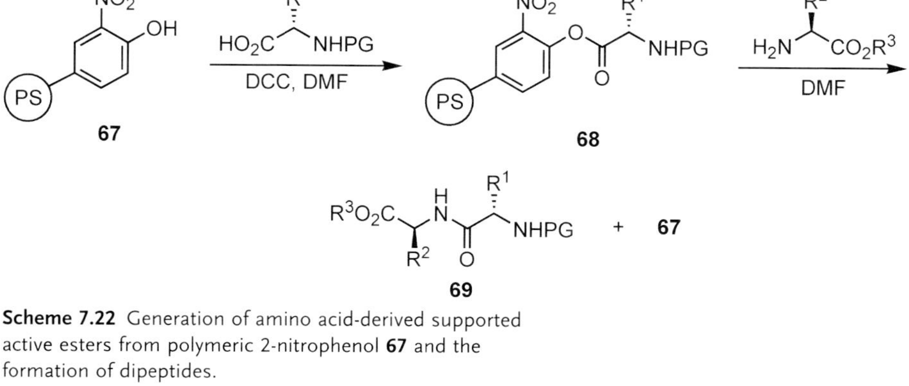

Scheme 7.22 Generation of amino acid-derived supported active esters from polymeric 2-nitrophenol **67** and the formation of dipeptides.

analysis). This resin has been used in the preparation of supported active esters of N-blocked amino acids using DCC as coupling agent for the synthesis of di-, tri- and tetrapeptides, although the coupling reaction still required hours for completion [85]. In addition, the luteinizing hormone-releasing hormone pyroGlu-His-Trp-Ser-Tyr-Gly-Leu-Arg-Pro-Gly-NH$_2$ has been synthesized by the insoluble stepwise couplings of protected amino acid active esters of resin **70**, or by the soluble stepwise couplings of protected amino acids using DCC. In both methods, the same intermediate peptides are formed, but the overall yield using the polymeric reagent was 40%, whereas that for the DCC method was only 7% [86]. Moreover, resin **70** has been used to prepare the corresponding active ester from 9H-fluorene-9-carboxylic acid for the acylation of the amino group of amino acid derivatives prior to HPLC analysis [87].

70

Other supported active ester-forming reagents not only lengthened the distance between the nitrophenol moiety and the polymer backbone but also increased the reactivity by introducing another electron-withdrawing functionality at the phenol ring, as it is known that the electrophilicity of an active ester is related to the acidity of the parent alcohol [88]. This is the case of the supported nitrophenol **71**, which has been obtained by treating the Merrifield resin with 3-hydroxy-4-nitrobenzoic acid in the presence of triethylamine in ethanol as solvent [89] (ca. 0.6 mmol o-nitrophenol per g of resin). This resin has been employed for the preparation of active esters of Boc-protected amino acids, using DCC as coupling agent, for the synthesis of a tetrapeptide. Moreover, 4-hydroxy-3-nitrobenzophenone-including resin **72** was obtained by aluminium trichloride promoted Friedel–Crafts acylation of cross-linked polystyrene with 4-fluoro- or 4-chloro-3-nitrobenzoyl chloride, followed by generation of the corresponding phenol after halogen substitution by a hydroxy group [90]. In this resin **72**, a carbonyl function at the o-nitrophenol ring provides an almost 40-fold rate increase of the active ester from benzoic acid in the reaction with *tert*-butylamine compared to an o-nitrophenol ester such as resin **70** [90]. Polymeric reagent **72** has also been employed for the formation of active esters for the acylation of 2-aminopyridines and 2-aminothiazoles [91], as well as for the preparation of active esters of 3,5-dinitrobenzoic acid for the derivatization of aliphatic amines, amino-alcohols and amino acids prior to HPLC analysis [92].

71

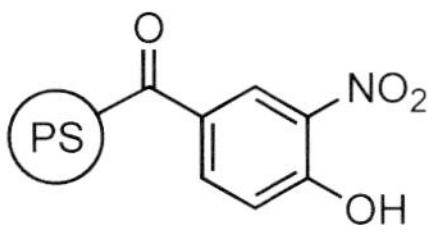

72

A more powerful related supported acylating agent has been obtained by treating polystyrene beads with 3-nitro-4-methoxybenzenesulfonyl chloride in the presence of aluminium trichloride (Scheme 7.23) [93]. The methoxy group of the resulting resin **73** was hydrolyzed using a 40% solution of benzyltrimethylammonium hydroxide in water (Triton B), affording the phenolic resin **74** (1.6–1.8 mmol g^{-1} loading). This resin was employed for the generation of active esters of N-protected amino acids for acylation reactions. Remarkably, the presence of a sulfonyl group increases the relative reaction rate of the active ester from benzoic acid 230-fold in the reaction with *tert*-butylamine compared to just 40-fold when a carbonyl group is present, such as in resin **72**.

When the Merrifield resin was suspended in DMF and refluxed with a sodium hydroxide methanolic solution of the *O*-carbonate ethyl ester of 4-mercaptophenol, the corresponding supported carbonate resin was obtained, which was hydrolyzed to give the phenol-sulfide resin (**75**, Scheme 7.24) (0.91 mmol g^{-1} loading) [94]. Polymer **75** was employed for the formation of the corresponding active esters of N-protected amino acids using a mixed anhydride or a DCC-induced esterification. Once the supported amino acid ester was formed, and before the peptide coupling with the amino group of another amino acid, the sulfide group was oxidized to a sulfone using hydrogen peroxide, thus creating an electron-withdrawing functionality on the aromatic ring and, therefore, increasing its ability as a leaving group. After the dipeptide formation, resin **76** was liberated.

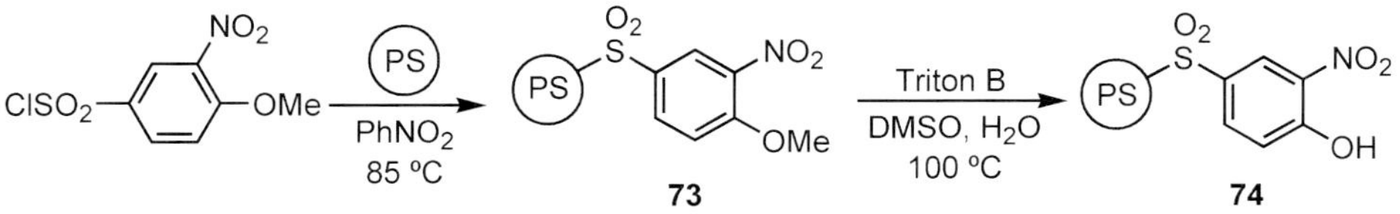

Scheme 7.23 Preparation of sulfonylated 2-nitrophenolic resin **74**.

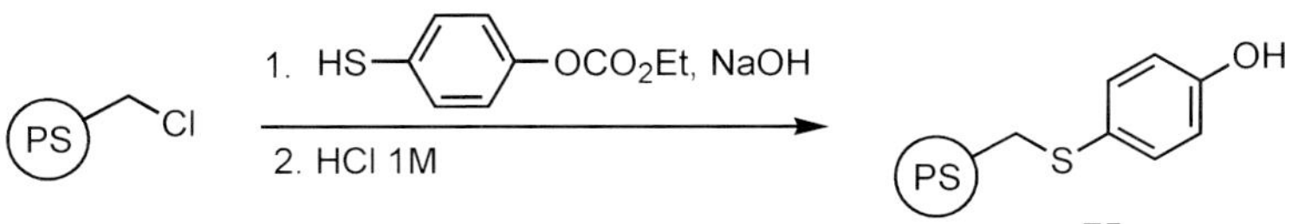

Scheme 7.24 Synthesis of phenol-sulfide resin **75**.

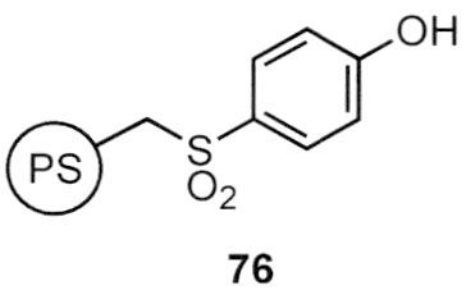

76

Merrifield resin was reacted with tetrachlorohydroquinone to give a polymeric monobenzyl ether of tetrachlorohydroquinone (**77**) (0.7 mmol g^{-1} loading). The esters of this functionalized polymer act as polymeric active esters, which have been used for the acylation of amines and the synthesis of some peptides when using the corresponding active esters of N-protected amino acids [95].

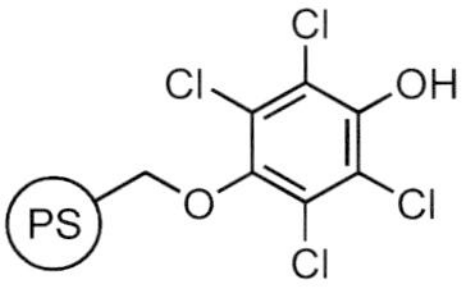

77

The tetrafluorophenol-containing resin **79** has been prepared by reaction of cross-linked aminomethylated polystyrene with 2,3,5,6-tetrafluoro-4-hydroxybenzoic acid (**78**) in the presence of 1,3-diisopropylcarbodiimide (DIC) as coupling reagent and HOBt as additive; the loading was determined using ^{19}F NMR (84%) (Scheme 7.25) [96]. Resin **79** has proved effective for DIC/DMAP-promoted acylation, generating highly reactive supported active esters **80**, and also for reaction with sulfonyl chlorides in the presence of diisopropylethylamine, affording activated polymeric sulfonate esters. This activated resin reacted with a wide scope of N-nucleophiles, including primary and secondary amines and anilines, in the preparation of amide and sulfonamide libraries. In addition, a resin identical to **79**, although changing polystyrene by TentaGel (a grafted copolymer of low cross-linked polystyrene matrix on which PEG is grafted), has proved useful for the

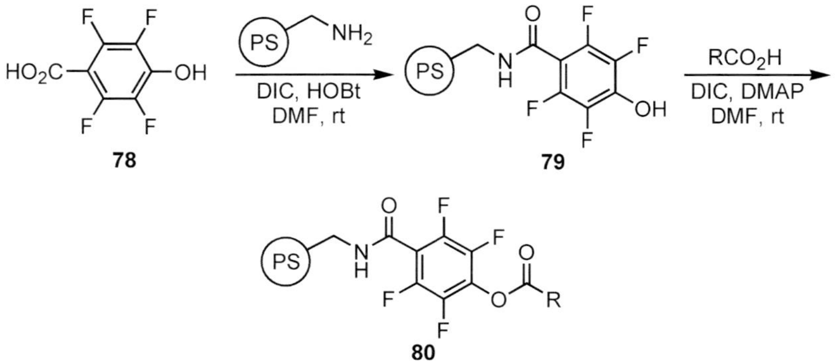

Scheme 7.25 Preparation of tetrafluorophenol-containing resin **79** and formation of its active esters.

preparation of tetrafluorophenol active esters employed for dipeptide couplings in aqueous solutions with a minimum of active ester hydrolysis [97].

7.6.2
N-Hydroxysuccinimide-Derived Polymers

HOSu-derived esters tend to be more stable and easier to store than other active ester reagents. Therefore, supporting the HOSu moiety on solid supports is an attractive idea for achieving easy purifications [98]. Thus, poly(ethylene-co-maleic anhydride) (**81**) has been used as starting material for the preparation of poly(ethylene-co-N-hydroxymaleimide) (**82**) after reaction with hydroxylamine hydrochloride in the presence of pyridine as base (Scheme 7.26) [99]. This polymeric HOSu (**82**) was found to be soluble in DMF and DMSO but insoluble in water or alcohols or dimethoxyethane (DME). The polymer was used for the preparation of the corresponding active esters **83** after reaction with N-Boc-amino acids, employing DCC as coupling agent or using a mixed anhydride methodology with preliminary reaction of **82** with isobutyl chloroformate. These supported active esters (**83**) were used in the synthesis of peptides after suspension of the polymer in DME and reaction with the free amino group of esterified amino acids or peptides.

To increase the insolubility of polymeric active esters **83** when preparing hexa- and higher peptides (which required the use of solvents such as DMF), the polymeric HOSu **82** was cross-linked by exposure to high-energy electrons [99]. The HOSu-containing polymer **82** has also being cross-linked by treating the starting poly(ethylene-co-maleic anhydride) (**81**) (average molecular wt 20 000–30 000) with polyamines such as hydrazine, spermine or spermidine before the reaction with hydroxylamine [100]. Thus, the spermidine cross-linked poly(ethylene-co-N-hydroxymaleimide) (**82**) gave an average particle size of 200–300 mesh and was insoluble in water, alcohols, MeCN, DMF or acetic acid. In addition, the mechanical stability increased, with the particulate structure of the polymer being retained under normal stirring. This cross-linking has also been performed with aromatic diamines such as benzidine, giving macroreticular granules that are subsequently treated with hydroxylamine; insoluble active esters of N-Boc-amino acids have also been prepared [101].

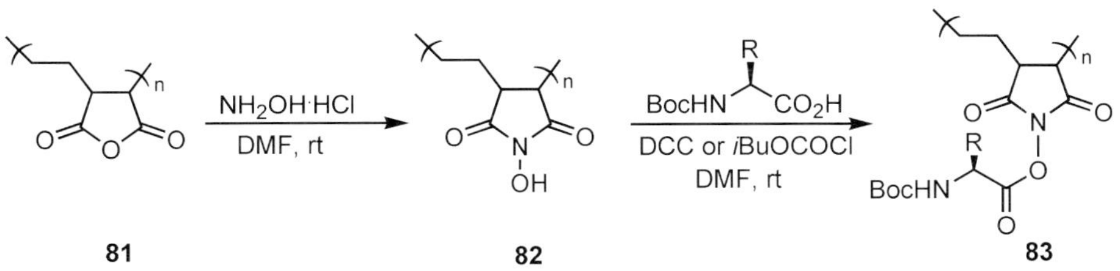

Scheme 7.26 Synthesis of polymeric N-hydroxysuccinimide **82** and preparation of anchored active esters of N-Boc-protected amino acids.

The cross-linked polymer **27** (cross-linker not represented) containing the HOSu residue for anchored DCC-promoted active ester formation was also obtained by copolymerization of *N*-acetoxymaleimide with styrene and divinylbenzene, followed by hydrolysis [102]. In addition, supported HOSu-active esters have been prepared by a method consisting of the synthesis of *N*-hydroxymaleimide esters of N-protected amino acids and subsequent copolymerization of these esters with styrene, divinylbenzene and 4-chlorophenylmaleimide [103].

Noncross-linked copolymers of styrene and *N*-hydroxymaleimide **27** (Section 7.3) have been prepared more recently after reaction of poly(styrene-*alt*-maleic anhydride) (average molecular wt 550 000) with 50% aqueous hydroxylamine at 90 °C [104]. This polymeric reagent **27** (1.5 mmol g^{-1} loading) was insoluble in many solvents and presented good mechanical stability, with no cross-linking being necessary. This polymeric HOSu (**27**) was employed as an additive for the *in situ* generation of active esters in the DCC-mediated peptide coupling reaction between Boc, Cbz and Fmoc-protected amino acids and amino acid esters in MeCN as solvent, the supported additive **27** being recovered after complete precipitation by adding a certain amount of hexane. The extent of racemization employing this additive was examined using Anteunis' test; the degree of epimerization in the final tripeptide being reduced from 19:1 in the absence of additive to 9:1. As commented above (Section 7.3), this polymeric HOSu (**27**) has been used for the synthesis of uronium-type salts, which are useful as peptide coupling reagents [46], as well as for the preparation of polymeric reagents for the protection of amino functionality with the 9-fluorenylmethoxycarbonyl (Fmoc) [105], 2,7-di-*tert*-butyl-9-fluorenylmethoxycarbonyl (Dtb-Fmoc) [106], allyloxycarbonyl (Alloc) and propargyloxycarbonyl (Proc) [107] groups.

The HOSu-containing polymer **85** has been obtained by ROMP of *O*-silylated oxanorbornene monomer **84** using Grubbs' catalyst and terminating the reaction with ethyl vinyl ether, followed by further desilylation, as the non-silylated monomer did not undergo the ROMP process (Scheme 7.27) [108]. The final polymer **85** was

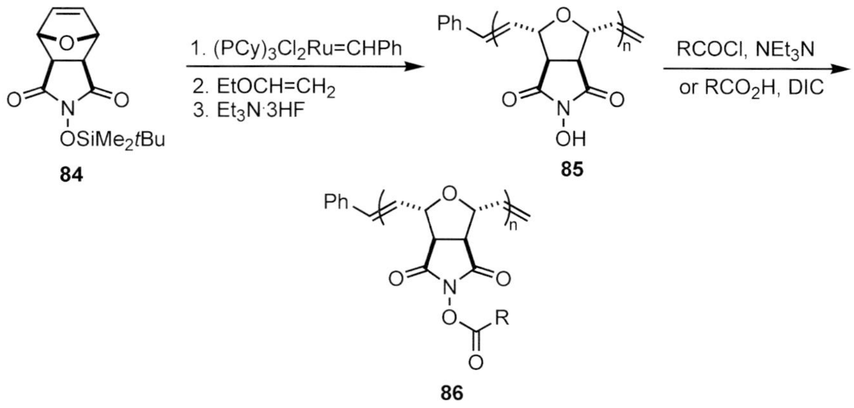

Scheme 7.27 ROMP-prepared polymeric *N*-hydroxysuccinimide **85** and synthesis of their active esters.

obtained quantitatively (5.5 mmol g^{-1} loading) and was acylated with acid chlorides or with carboxylic acids in the presence of DIC to give the supported active esters **86**, which reacted with amines to give the corresponding amides and the recovery of polymer **85**, which could be reused. An alternative methodology for generating the supported active esters **86** has been the preparation of a monomer consisting of an active ester of the non-silylated equivalent to **84** followed by the ROMP process [108]. This procedure has been used for the facile, purification-free synthesis of Mosher amides [109].

The ROMPgel resin **87** (3.3 mmol g^{-1} HOSu loading) was obtained using the Wang resin (1.1 mmol g^{-1} loading) as solid support and using a ROMP process based on an acetylated oxanorbornene monomer related to compound **84**, followed by liberation of the HOSu resin after a subsequent acetylation reaction. Resin **87** has been employed for the parallel formation of active esters, using HBTU as coupling agent in DMF as solvent. The reaction takes place after keeping the resin inside a microreactor made of high-grade polypropylene mesh sidewalls and cap and a radio tag identifier known as IRORI. MiniKans [110]. The resin could be reused several times with no apparent loss of activity or purity, although a drawback was the gradual loss of resin from the microreactor.

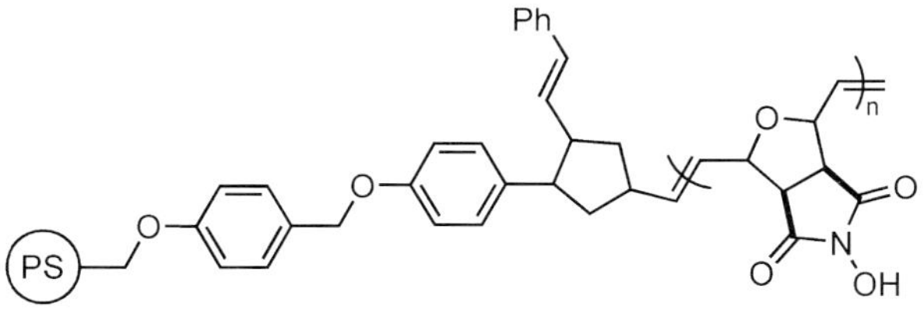

87

Two available chloromethylpolystyrene resins, the traditional Merrifield resin, and the highly cross-linked polystyrene grafted with poly(ethylene glycol) Argopore-Cl have been transformed into their respective thiol resins **88** by reaction with thiourea followed by basic hydrolysis of the *S*-alkyl isothiourea intermediate. The conversion into **88** is quantitative based on solid-phase NMR and elemental analysis (Scheme 7.28). An excess of *N*-hydroxymaleimide in the presence of pyridine in DMF as solvent was used to cap the solid phase thiol, affording the HOSu-containing resin **89** [111]. Polymeric reagent **89** was used to generate resin-bound

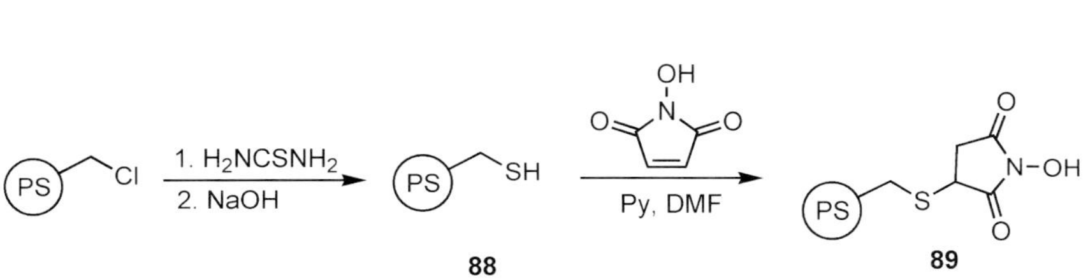

Scheme 7.28 Preparation of *N*-hydroxysuccinimide-containing polystyrene resin **89**.

active esters by acylation with carboxylic acids (3 equivalents) in the presence of ethyl 3-(dimethylamino)propylcarbodiimide (EDAC). After filtration and washing the supported active esters, these resins were coupled with primary, branched primary and secondary amines to give the corresponding amides in excellent yields. The choice of solvent was important to ensure sufficient resin swelling ($CHCl_3$/DMF 1:1 v/v for the Merrifield-derived resin and dimethylformamide for the Argopore-derived resin).

An example of the use of resin **89** (Merrifield-derived) is the formation of the corresponding anchored active ester **91** after reaction with progesterone-11-hemisuccinate (**90**) (Scheme 7.29). Subsequent reaction of the isolated active ester **91** with benzylamine gave the progesterone amide **92** and the supported HOSu **89** [111].

The commercially available polystyrene-thiophenol resin **93** has been used to anchor the 1-pyrenebutyric acid HOSu ester by means of its N-hydroxymaleimide ester derivative **94** as a way of introducing a fluorescent label (Scheme 7.30). Thus, reaction of resin **93** with 1-pyrenebutyric-derived N-hydroxymaleimide ester **94** in the presence of diisopropylethylamine gave the supported active ester **95**. Further reaction of isolated resin **95** with an amine such as tryptamine **96** gave the fluorescent amide **97** and the HOSu resin **98** [112]. Other fluorescent labels such as 6-carboxyfluorescein diacetate, as well as biotin, have been introduced to different primary amines using the corresponding anchored active esters. The hydroxysuccinimide resin **98** can also be prepared by treating supported thiophenol **93** with N-hydroxymaleimide in the presence of diisopropylethylamine [112]. This resin was used as an alternative to the former method for creating anchored active esters from the corresponding acids using DCC; however, much lower yields of the final amides were obtained [112].

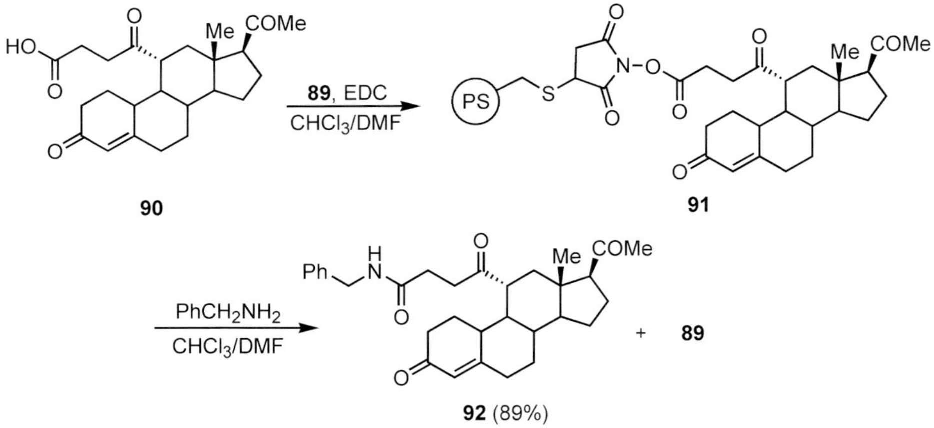

Scheme 7.29 Synthesis of a progesterone-derived amide via a solid-supported active ester formation using N-hydroxysuccinimide resin **89**.

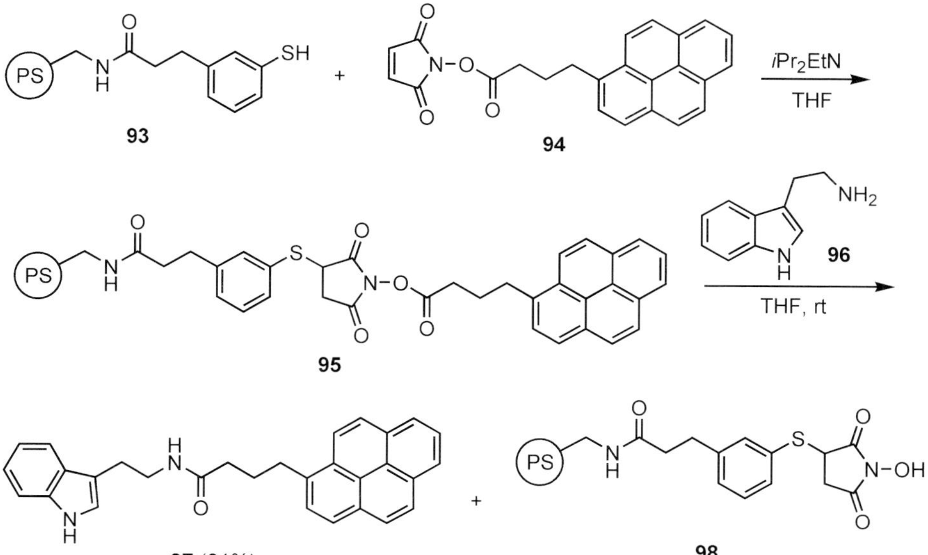

Scheme 7.30 Introduction of a fluorescent label into 2-(1H-indol-3-yl)ethanamine (**96**) using an anchored N-hydroxysuccinimide active ester formation methodology.

The polymer-supported N-hydroxysuccinimide **98** has been employed for the solution phase column chromatography-free parallel synthesis of carbamates [113], an example of the methodology employed being shown in Scheme 7.31. Thus, treatment of resin **98** with bischlorotrimethyl carbonate (BTC), followed by filtration and washing, afforded the polymer-supported HOSu-chloroformate **99**. Further treatment of resin **99** with an alcohol, such as benzyl alcohol, gave the polymer-supported HOSu-carbonate **100** after filtration, which then reacted with a primary amine such as 4-phenylbutylamine to furnish the corresponding carbamate **101** and the recovery of resin **98** after filtration. When tertiary alcohols were employed, the reaction of the corresponding supported carbonate with the amine was severely hindered and no final carbamate was observed.

Other examples of N-hydroxysuccinimide-including polymers dedicated to the formation of active esters for acylation reactions include the double succinimide-containing resin **104**. This reagent has been prepared by reaction of cross-linked aminomethylated polystyrene with 1,2,3,4-cyclopentanetetracarboxylic anhydride (**102**), in the presence of triethylamine as base, and further treatment of supported anhydride **103** with hydroxylamine (Scheme 7.32) [114]. Using resin **104**, a series of active esters that are highly stable upon storage were prepared using carboxylic acids and coupling reagents such as EDC or DIC; both primary and secondary amines were acylated in just 20 min in almost quantitative yield.

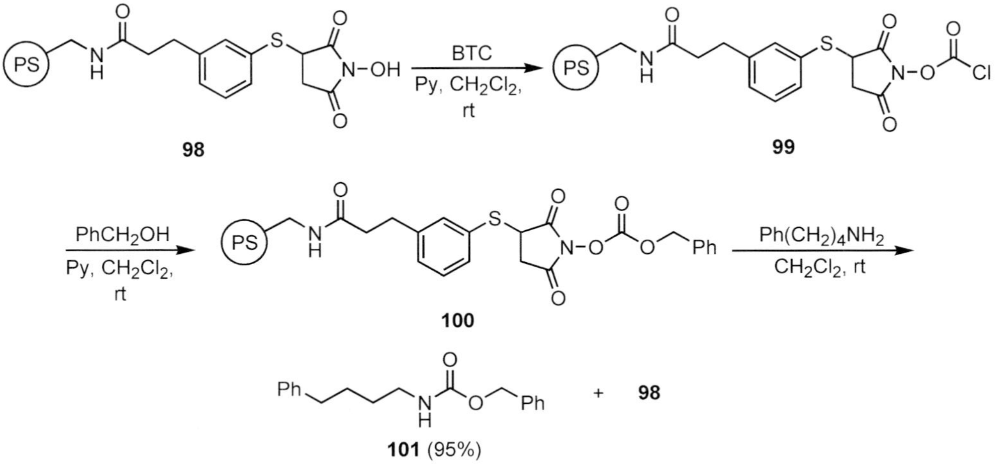

Scheme 7.31 Preparation of a carbamate-protected amine using the polymeric N-hydroxysuccinimide-derived benzyloxycarbonylation reagent **100**.

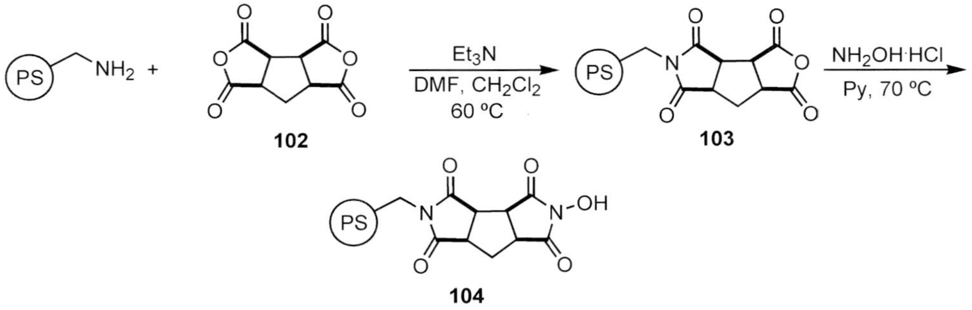

Scheme 7.32 Preparation of N-hydroxysuccinimide-containing resin **104**.

7.6.3
N-Hydroxybenzotriazole-Derived Polymers

Active esters obtained from 1-hydroxybenzotriazole (HOBt) are considered more reactive than those from other common active ester-forming agents. Thus, polymeric HOBt active esters appear to be about 100 times more reactive than the corresponding polymeric nitrophenol esters. However, this value rises to 8000 for their soluble analogues [90], thus showing the lowering of reactivity when active esters are solid-supported.

Polystyrene-bound 1-hydroxybenzotriazole (**20**) [41] (Section 7.3) was originally prepared by Friedel–Crafts alkylation of macroporous polystyrene or copolystyrene-2% divinylbenzene using 3-nitro-4-chlorobenzyl bromide or 3-nitro-4-chlorobenzyl alcohols in the presence of aluminium trichloride. The so-obtained

7.6 Active Ester-Forming Polymeric Reagents

supported 4-chloro-3-nitrobenzylated polystyrene **105** was treated with hydrazine at 100 °C to give resin **106**, which after cyclization with hydrochloric acid gave rise to polymer-supported 1-hydroxybenzotriazole (**20**) (P-HOBt) (1.25–1.70 mmol OH per g loading) [115]. Resin **20** was coupled, using DCC as coupling agent, to many N-blocked amino acids, yielding polystyrene-bound active esters **107** (Scheme 7.33), which were stable for at least two months when stored in a desiccator. These derivatives proved useful in peptide coupling reactions, as demonstrated in the synthesis (among others) of the tetrapeptide Boc-Leu-Leu-Val-Tyr(OBz)-OBz, which has also been obtained using active esters from (4-hydroxy-3-nitro)benzylated polystyrene (**70**) [85] with high efficiency but with longer reaction times (36–40 h) – in the present case all the couplings being completed within 20 min. The P-HOBt (**20**) can be recovered after each acylation reaction and reused without loss of activity.

The loading of the prepared P-HOBt (**20**) should not be higher than 1.5 mmol g^{-1} as its efficiency drops significantly at higher values [116]. This apparently surprising result arises from the possible reaction of the hydrazine and nitro groups of resin **106** situated on different polymer chains under acidic conditions, which is most probable if their presence is frequent, thus forming hydroxytriazene moiety-containing structures of type **108** (Scheme 7.34). The hydroxy group on these systems (**108**) showed much lower reactivity than when belonging to the hydroxy-benzotriazole system.

Other examples of the use of the polymeric esters from P-HOBt (**20**) for the preparation of peptides after reaction with amino group-free amino acids or pep-

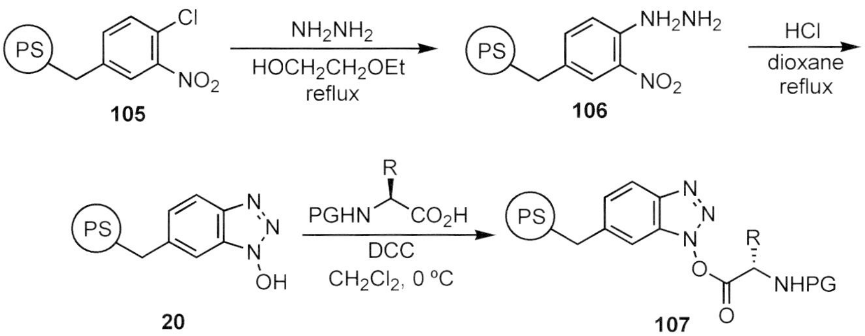

Scheme 7.33 Synthesis of P-HOBt (**20**) and formation of supported active esters from N-protected amino acids.

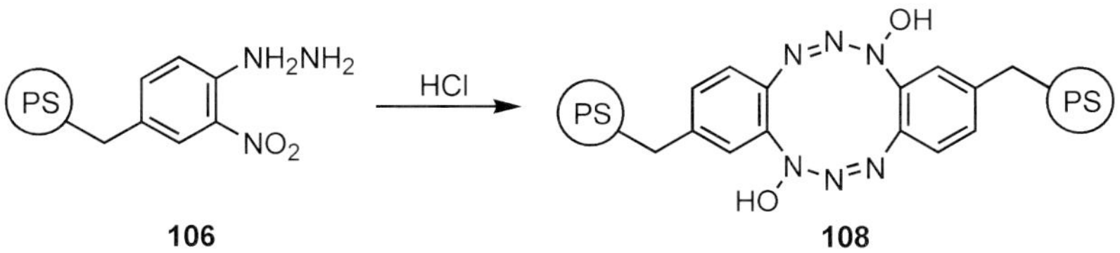

Scheme 7.34 Undesirable dimerization of polymeric nitrohydrazine **106**.

tides [117] or even with polyoxyethylene esters of amino acids can be found [118], as in the synthesis of the C-terminal half of immunologically active thymic polypeptide thymosin α_1 [119]. In this last case, two suitably blocked segments, Boc-Asp(OtBu)-Leu-Lys(2Cbz)-Glu(OBz)-Lys(2Cbz)-Lys(2Cbz)-OH and Boc-Glu(OBz)-Val-Val-Glu(OBz)-Glu(OBz)-Ala-Glu(OBz)-Asn-OBz, were prepared racemization-free entirely by using the P-HOBt activation in each coupling step. Some other thymosin-like peptides have also been obtained via these resin supported active esters [120], as well as some peptides designed as potential antisickling agents [121].

Racemization-free peptide coupling of P-HOBt (**20**) esters can be carried out with an amino acid or peptide temporarily C-terminal-protected by a phase-transfer reagent [122]. This avoids a deprotection step, as the C-deprotection is achieved when the peptide is extracted. In addition, macroporous P-HOBt (**20**), which could be stored undesiccated for two years and still retain 90% of its activity, has been used in the high-yielding synthesis of primary and secondary amides from acetic acid, benzoic acid and pyrene-1-carboxylic acid through their corresponding active esters, with various solvents, even water, being suitable [123]. Scheme 7.35 illustrates and example, with the DCC-promoted formation of the supported active ester **110** from pyrene-1-carboxylic acid (**109**) and its subsequent reaction with the most reactive amino group of a dinucleophile such as 3-aminopropan-1-ol to give the amidation product **111** and recovery of P-HOBt (**20**). Moreover, Boc-Ala and di-Boc-His P-HOBt active esters have been coupled with N-cetyl N',N''-dimethylethylenediamine for the synthesis of amino acid-functionalized surfactants [124].

Medium-ring lactams (**113**) of 7-, 9-, 11-, and 13-membered rings have been prepared in moderate yields by an intramolecular cyclization of P-HOBt esters **112** derived from the corresponding Boc-protected amino acids, after the corresponding N-deprotection (Scheme 7.36) [125]. In addition, HOSu esters **115** have been obtained from P-HOBt-derived active esters **114**, this method affording betters results than when using the polymeric carbodiimide PS-EDC (Scheme 7.36) [126]. P-HOBt (**20**) has also been used for the generation of immobilized carbonates from the Boc, Cbz or Fmoc protecting groups, thus allowing the

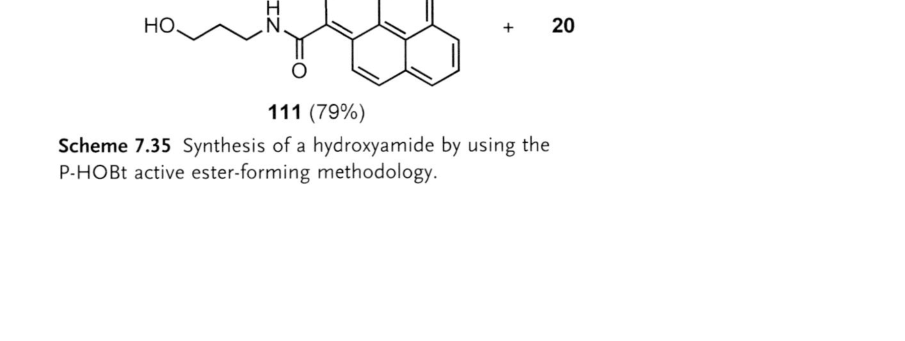

Scheme 7.35 Synthesis of a hydroxyamide by using the P-HOBt active ester-forming methodology.

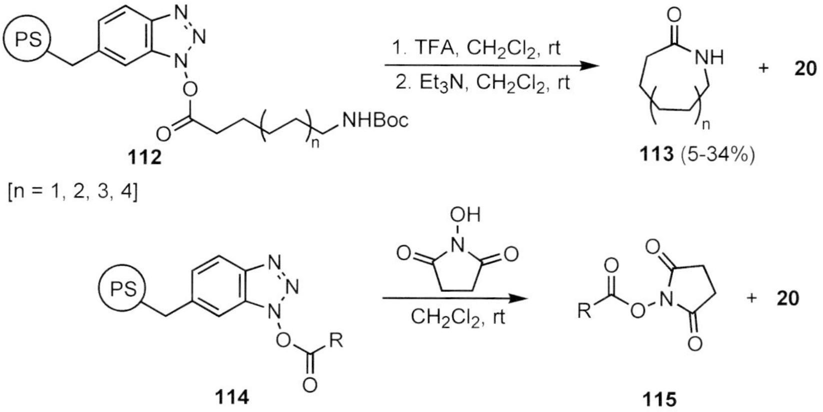

Scheme 7.36 Synthesis of medium-ring lactams and N-hydroxysuccinimide esters employing P-HOBt-derived active esters.

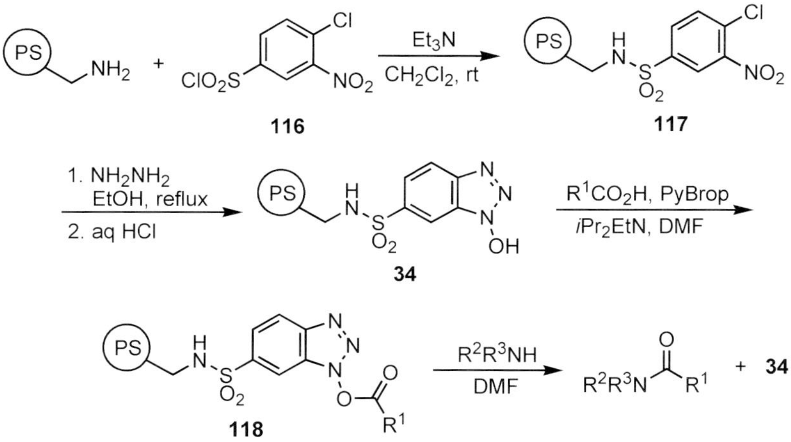

Scheme 7.37 Preparation of polystyrene-resin sulfonamide-containing P-HOBt (**34**) and the synthesis of amides using their active esters.

corresponding N-protection of primary and secondary amines and recovery of the supported HOBt [127].

Obviously, the incorporation of an electron-withdrawing substituent in the polymer-supported benzotriazole ring should enhance its reactivity. Thus, the sulfonamide-substituted HOBt resin **34** (Section 7.3) has been obtained by reaction of aminomethylated divinylbenzene cross-linked polystyrene with the sulfonyl chloride **116** to give supported sulfonamide **117**, followed by typical hydroxybenzotriazole formation (Scheme 7.37) [128]. The effect induced by the sulfonamide group was determined by measuring the acidity of non-polymeric analogues, showing a pK_a of 3.59 for the sulfonamide-substituted HOBt, and a pK_a of 4.64

for HOBt, thus anticipating a higher electrophilicity of an active ester when promoting acylation reactions [128]. This HOBt derivative-tethered polymer (**34**) is commercially available (0.8–1 mmol g^{-1} loading) and has been used in the rapid, clean parallel synthesis of amide libraries through active esters **118**, prepared from the corresponding carboxylic acids using bromotris(pyrrolidino)phosphonium hexafluorophosphate (PyBrop) as coupling agent [128]. This amide library-preparation procedure employing active esters from resin **34** has been optimized with an automated synthesizer and using the cheaper N,N'-diisopropylcarbodiimide (DIC) as coupling agent [129]. In addition, the efficiency of the active esters from P-HOBt **34** in amidation reactions has also been shown in the synthesis of complex natural product-like molecules containing a spiroketal scaffold [130].

The former HOBt-including resin **34** has also been used in a combined solid phase and solution synthesis of a library of α,α-disubstituted-α-acylaminoketones, which are interesting compounds as acdysone agonists [131]. An example is shown in Scheme 7.38, where the polymeric active ester **119**, obtained by coupling resin **34** with the corresponding carboxylic acid using the DIC/DMAP combination, reacts with the α-aminoacetone trifluoroacetate salt **120** (obtained by cleavage of a carbamate linked resin-bound aminoketone) in the presence of Hünig's base to give the α,α-dimethyl-α-benzoylaminoketone **121** after removing the supported HOBt **34** by filtration. In addition, active esters from this P-HOBt (**34**) have been employed for the clean and high-yielding preparation of low molecular weight hydroxamic acids after reaction with O-protected or free hydroxylamine, with subsequent recycling of the spent resin **34** [132].

The increased reactivity of HOBt-derived esters achieved by adding an electron-withdrawing group at the aromatic nucleus is also shown when using the commercially available HOBt-6-carboxamidomethyl polystyrene resin (**122**) (1.45 mmol g^{-1} loading). Thus, this polymeric HOBt has been used, after coupling with the corresponding acid **123**, for the preparation of supported active esters **124** incorporating a fluorescent 4-acetamido-1,8-naphthalimide moiety, which can be employed in the labeling of amines such as benzylamine by formation of the corresponding amide **125** (Scheme 7.39) [133]. The reaction rate constant for this

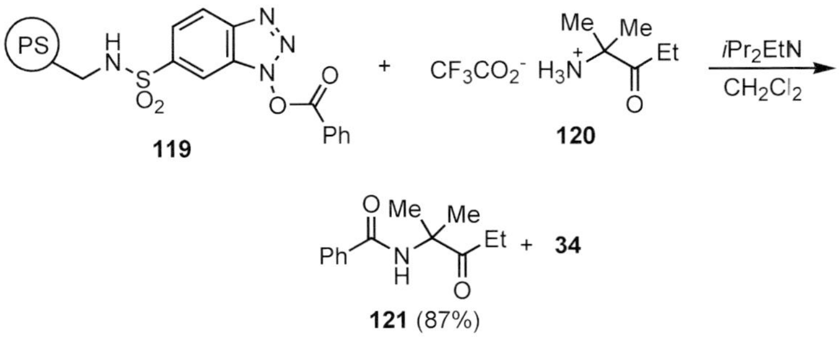

Scheme 7.38 Synthesis of a α,α-dimethyl-α-benzoylaminoketone by benzoylation using a P-HOBt-derived active ester.

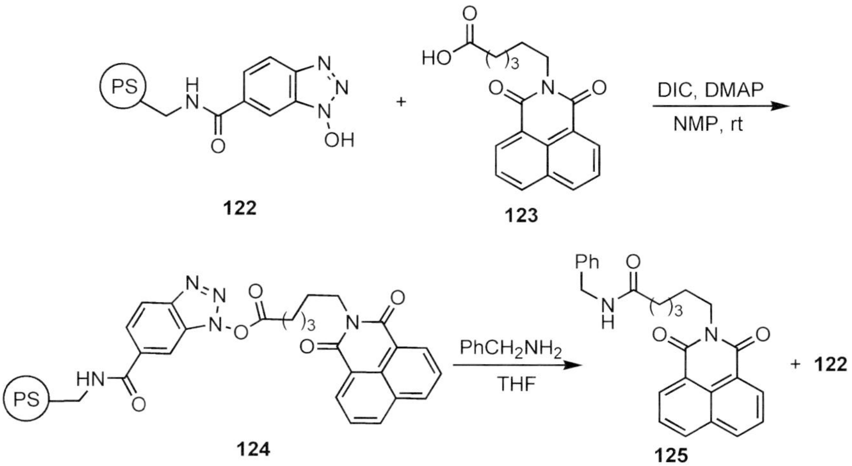

Scheme 7.39 Fluorescence labeling of benzylamine using P-HOBt-derived supported active ester methodology.

amidation reaction was much higher for the active esters from HOBt resin **122** than for the same esters obtained using HOSu-derived resin **89** or nitrophenol resin **72**. In addition, for the active esters from those different resins, a very high selectivity was found for aminolysis over hydrolysis or alcoholysis, which suggest that this labeling strategy might be applied to aqueous conditions.

A convergent synthesis of the potent selective inhibitor of the enzyme phosphodiesterase sildenafil (Viagra) has been based on polymer supported reagents [134]. In this synthesis, the HOBt-supported resin **126** has been used for the isolation and preparation of the resin-bound active ester **128**, performed by coupling polymer **126** with the benzoic acid sulfonamide derivative **127** by means of PyBrop (Scheme 7.40). Subsequent reaction of active ester **128** with aminopyrazole **129** gave rise to the clean synthesis of amide **130**, which has been transformed into sildenafil (**131**) after a base-promoted pyrimidinone formation.

A polymeric HOBt (**132**), which is soluble in dichloromethane, DMF or water and bears a poly(ethylene glycol) (PEG) chain (average molecular wt 4000), has been prepared from PEG-NH$_2$ and 1-hydroxybenzotriazole-5-carboxylic acid using DCC as coupling agent [135]. The corresponding active esters were prepared by reaction with an excess of carboxylic acid anhydrides and used for the synthesis of the model tetrapeptide Cbz-Leu-Ala-Gly-Val-O*t*Bu [136].

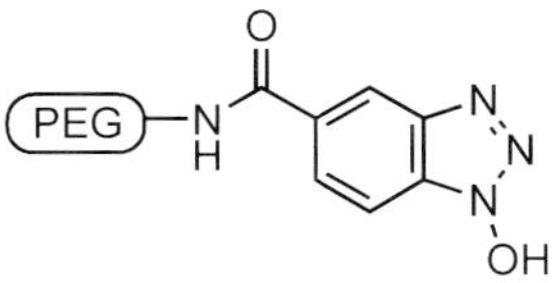

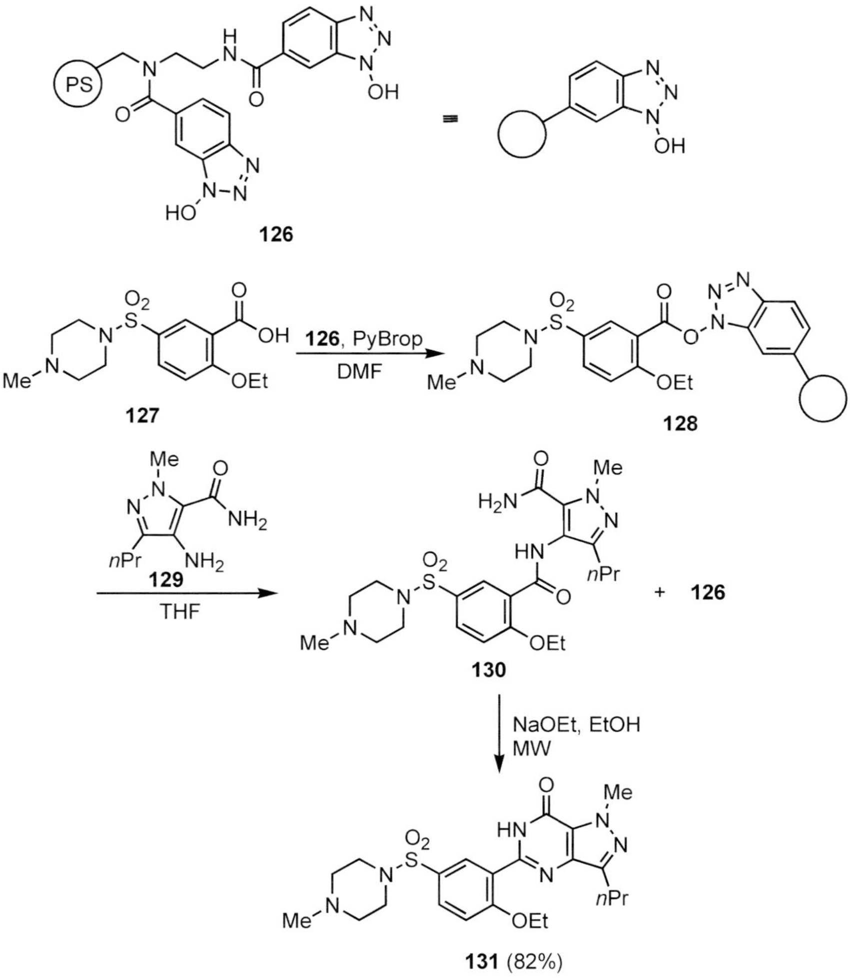

Scheme 7.40 Synthesis of sildenafil (**131**) using a solid supported active ester from P-HOBt **126**.

7.6.4
Other Active Ester-Forming Polymers

The 4-nitrobenzophenone oxime polymer **133**, Kaiser resin, has been obtained by Friedel–Crafts acylation of cross-linked polystyrene using 4-nitrobenzoyl chloride in the presence of aluminium trichloride, followed by reaction with hydroxylamine [137]. Resin **133** has been used for carbodiimide-promoted coupling with N-protected amino acids. However, its role as active ester-forming reagent is limited, its cleaving reaction by nucleophilic displacement using amino acid esters being rather slow and requiring catalysis by carboxylic acids. Its use in coupling reactions has been mainly limited to acting as a solid support in the solid-phase synthesis of peptides [137–139], although its supported esters can be used in the acylation of more nucleophilic amines such as *tert*-butyldimethylsilyl-*O*-hydroxylamine, thus driving to hydroxamic acids after desilylation [140]. The Kaiser resin (**133**) has also been used for the preparation of the supported phosgenated derivative **134**, after reaction with the phosgene equivalent BTC [141]. Phosgenated resin **134** was stable on prolonged exposure to air, unlike typical chloroformate behavior, and reacted with primary amines to afford the corresponding supported carbamates. These compounds gave ureas upon heating in the presence of primary or secondary amines, with recovery of the oxime Kaiser resin (**133**).

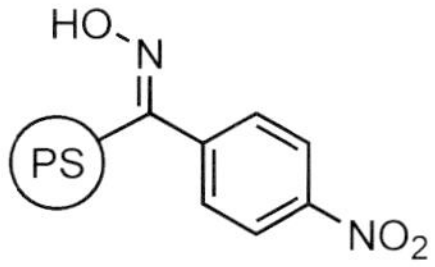

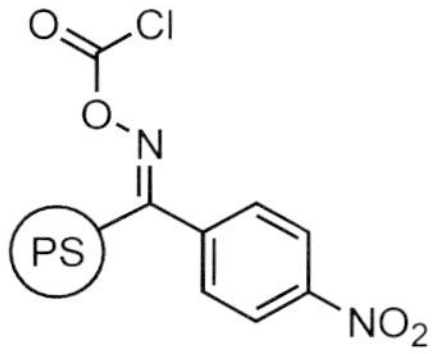

Polymer-bound pyrimidinones such as **135** (0.5 mmol g^{-1} loading) have been obtained using Merrifield resin as the starting anchoring polymer, and have been used for the synthesis of the acylating agents **136** after reaction with the corresponding acid chlorides in dichloromethane as good swelling solvent, in only 5 min at 80 °C under microwave irradiation (Scheme 7.41) [142]. These active esters (**136**) have been used for the rapid acylation of primary and secondary amines,

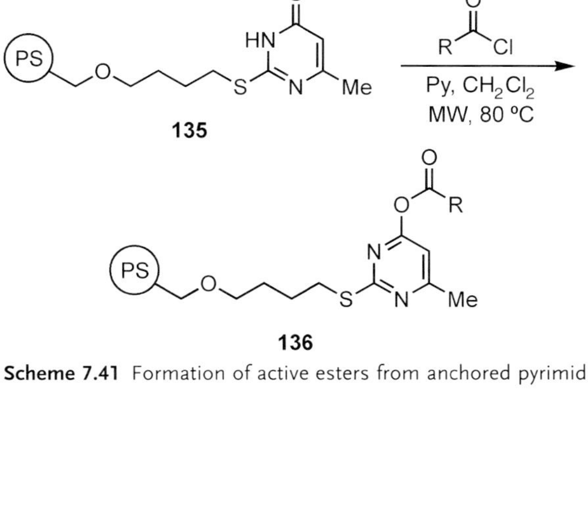

Scheme 7.41 Formation of active esters from anchored pyrimidinone **135**.

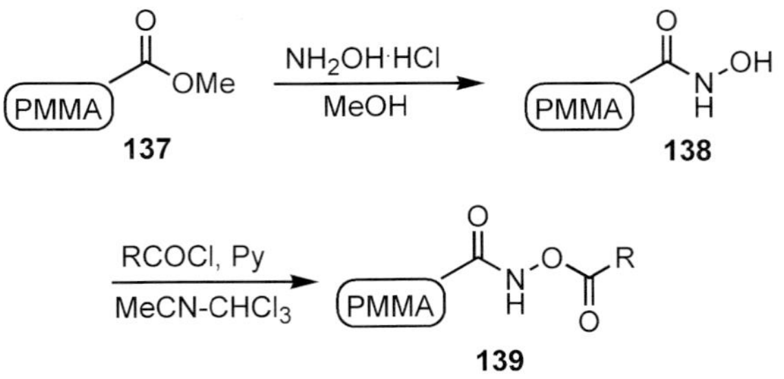

Scheme 7.42 Preparation of PMMA-supported hydroxamic acid-derived active esters.

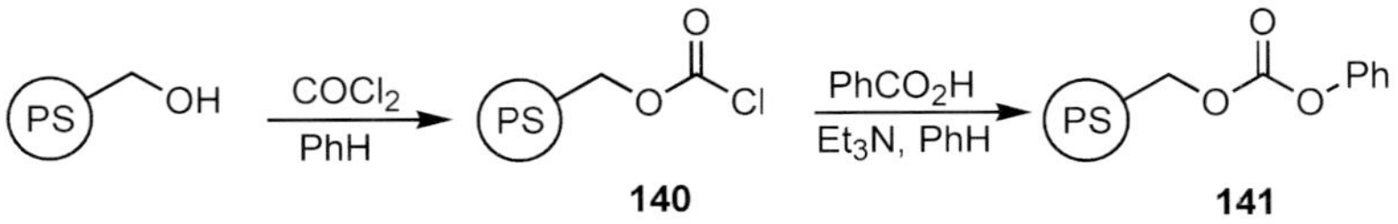

Scheme 7.43 Preparation of the supported mixed anhydride benzoylation reagent **141**.

as well as alcohols, phenols and thiophenols, working also under microwave irradiation.

Differently cross-linked poly(methyl methacrylate) (PMMA) (**137**) has been employed for the synthesis of supported hydroxamic acids **138** after treatment with hydroxylamine, and reacted with acyl chlorides in the presence of pyridine to give the supported hydroxamic acid esters **139**, which are suitable as solid acyl transfer reagents when reacting with alkyl and aryl amines (Scheme 7.42) [143].

A resin with mixed carbonic-carboxylic anhydride function has been prepared and used as acylating reagent [144]. Thus, the supported chloroformate **140** was obtained by reaction of hydroxymethylated polystyrene with phosgene in benzene (Scheme 7.43). Further treatment with benzoic acid in the presence of triethylamine in benzene gave rise to the supported mixed anhydride **141**. This polymer has been used in the benzoylation of several amines, although variable amounts of benzoic acid from attack to the internal carbonyl were obtained when using aromatic amines.

7.7
Conclusions

Although solid-supported reagents have been in use for decades, in recent years they have experienced a surge in popularity with the increased emphasis on parallel and combinatorial synthesis as a means to increase productivity in medicinal chemistry. Polymer-bound reagents allow rapid and easy removal of formed supported by-products or reagent excesses from the reaction medium, generally just by simple filtration. Among the present arsenal of polymeric reagents, those employed for performing coupling reactions after the creation of an active ester are especially important, as these transformations are key in synthetic organic chemistry. Thus, cross-linked polystyrene has been the most frequently used resin for anchoring analogues of coupling reagents used in solution, creating polymeric coupling reagents that can, in many cases, overcome the traditional drawbacks of supported reagents, which are mainly related to slow reaction rates. As the demand for rapid and clean synthetic procedures for the preparation of arrays of compounds for pharmaceutical research increases day by day, there is no doubt that new anchored coupling reagents with higher reactivity will be developed, and that many more synthetic procedures involving all these polymeric reagents will show up in the near future.

References

1 (a) Leznoff, C.C. (1978) *Accounts of Chemical Research*, **11**, 327–33. (b) McKillop, A. and Young, D.W. (1979) *Synthesis*, 401–22. (c) McKillop, A. and Young, D.W. (1979) *Synthesis*, 481–500. (d) Akelah, A. and Sherrington, D.S. (1981) *Chemical Reviews*, **81**, 557–87. (e) Akelah, A. (1981) *Synthesis*, 413–38. (f) Sherrington, D.C. and Hodge, P. (1988) *Synthesis and Separations Using Functional Polymers*, John Wiley & Sons Ltd, Chichester. (g) Suttleworth, S.J., Allin, S.M. and Sharma, P.K. (1997) *Synthesis*, 1217–39. (h) Kaldor, S.W. and Siegel, M.G. (1997) *Current Opinion in Chemical Biology*, **1**, 101–6. (i) Drewry, D.H., Coe, D.M. and Poon, S. (1999) *Medicinal Research Reviews*, **19**, 97–148. (j) James, I.W. (1999) *Tetrahedron*, **55**, 4855–946. (k) *Medicinal Research Reviews*, **19**, 97–148. (l) Bhattacharyya, S. (2000) *Combinatorial Chemistry and High Throughput Screening*, **3**, 65–92. (m) Ley, S.V., Baxendale, I.R., Bream, R.N., et al. (2000) *Journal of the Chemical Society – Perkin Transactions 1*, 3815–4195. (n) Albericio, F. and Kates, S.A. (2000) *Solid-Phase Synthesis*, 275–330. (o) Albericio, F., Chinchilla, R., Dodsworth, D.J. and Nájera, C. (2001) *Organic Preparations and Procedures International*, **33**, 203–303. (p) Katritzky, A.R., Suzuki, K. and Singh, S.K. (2004) *Arkivoc*, **i**, 12–35. (q) Bhattacharyya, S. (2005) *Molecular Diversity*, **9**, 253–7.

2 (a) Kurzer, F. and Douraghi-Zadeh, K. (1967) *Chemical Reviews*, **67**, 107–52. (b) Williams, A. and Ibrahim, I.T. (1981) *Chemical Reviews*, **81**, 589–636.

3 Campbell, T.W. and Monagle, J.J. (1962) *Journal of the American Chemical Society*, **84**, 1493.

4 Wolman, Y., Kivity, S. and Frankel, M. (1967) *Journal of the Chemical Society D – Chemical Communications*, 629–30.

5 Weinshenker, N.M. and Shen, C.M. (1972) *Tetrahedron Letters*, 3281–4.

6 Castillo, M., Romero, A. and Ramirez, E. (1983) *Transition Metal Chemistry (London)*, **8**, 262–6.

7 Desai, M.C. and Stramiello, L.M. (1993) *Tetrahedron Letters*, **34**, 7685–8.

8 Weinshenker, N.M., Shen, C.M. and Wong, J.Y. (1988) *Organic Syntheses*, Coll, **VI**, 951–4.

9 Ito, H., Takamatsu, N. and Ichikizaki, I. (1975) *Chemistry Letters*, 577–8.

10 Cano, M., Ladlow, M. and Balasubramanian, S. (2002) *Journal of Organic Chemistry*, **67**, 129–35.

11 Guan, Y., Green, M.A. and Bergstrom, D.E. (2000) *Journal of Combinatorial Chemistry*, **2**, 297–300.

12 Senten, K., Van der Veken, P., Bal, G., et al. (2001) *Tetrahedron Letters*, **42**, 9135–8.

13 Senten, K., Daniëls, L., Van der Veken, P., et al. (2003) *Journal of Combinatorial Chemistry*, **5**, 336–44.

14 Parlow, J.J., Case, B.L., Dice, T.A., et al. (2003) *Journal of Medicinal Chemistry*, **46**, 4050–62.

15 Parlow, J.J., Stevens, A.M., Stegeman, R.A., et al. (2003) *Journal of Medicinal Chemistry*, **46**, 4297–312.
16 Keck, G.E., Sanchez, C. and Walker, C.A. (2000) *Tetrahedron Letters*, **41**, 8673–6.
17 Lannuzel, M., Lamothe, M. and Perez, M. (2001) *Tetrahedron Letters*, **42**, 6703–5.
18 Sauer, D.R., Kalvin, D. and Phelan, K.M. (2003) *Organic Letters*, **5**, 4721–4.
19 Desai, B., Dallinger, D. and Kappe, C.O. (2006) *Tetrahedron*, **62**, 4651–64.
20 Parlow, J.J., Dice, T.A., Lachance, R.M., et al. (2003) *Journal of Medicinal Chemistry*, **46**, 4043–9.
21 Yun, Y.K., Porco, J.A. Jr. and Labadie, J. (2002) *Synlett*, 739–42.
22 Price, D.A., Gayton, S., Selby, M.D., et al. (2005) *Tetrahedron Letters*, **46**, 5005–7.
23 Guisado, O., Martinez, S. and Pastor, J. (2002) *Tetrahedron Letters*, **43**, 7105–9.
24 Wang, Y., Sarris, K., Sauer, D.R. and Djuric, S.W. (2007) *Tetrahedron Letters*, **48**, 5181–4.
25 Miller, M.M. and Boger, D.S. (2005) Polystyrene-1-ethyl-3-(3′-dimethylaminopropane)-carbodiimide Hydrochloride (PS-EDCI), in *Handbook of Reagents for Organic Synthesis, Reagents for Glycoside, Nucleotide and Peptide Synthesis* (ed. D. Crich), John Wiley & Sons Ltd, Chichester, pp. 529–33.
26 Adamczyk, M. and Fishpaugh, J.R. (1996) *Tetrahedron Letters*, **37**, 7171–2.
27 (a) Boger, D.L., Goldberg, J., Jiang, W., et al. (1998) *Bioorganic and Medicinal Chemistry*, **6**, 1347–78. (b) McReynolds, K.D., Bhat, A., Conboy, J.C., et al. (2002) *Bioorganic and Medicinal Chemistry*, **10**, 625–37. (c) Montezaei, R., Ida, S. and Campbell, D.A. (1999) *Molecular Diversity*, **4**, 143–8.
28 Chibale, K., Chipeleme, A. and Warren, S. (2002) *Tetrahedron Letters*, **43**, 1587–9.
29 Gupta, S., Lopina, A. and S.T. (2004) *Journal of Polymer Science Part A – Polymer Chemistry*, **42**, 4906–15.
30 Adamcyk, M., Fishpaugh, J.R. and Mattingly, P.G. (1995) *Tetrahedron Letters*, **36**, 8345–6.
31 Sturino, C.F. and Labelle, M. (1998) *Tetrahedron Letters*, **39**, 5891.
32 Adamczyk, M. and Fishpaugh, J.R. (1996) *Tetrahedron Letters*, **37**, 4305–8.
33 (a) Buckman, B.O., Morrissey, M.M. and Mohan, R. (1998) *Tetrahedron Letters*, **39**, 1487–8. (b) Parlow, J.J. and Flynn, D.L. (1998) *Tetrahedron*, **54**, 4013–16. (c) Flynn, D.L., Devraj, R.V., Naing, W., et al. (1998) *Medicinal Chemistry Research*, **8**, 219–43.
34 Zhang, M., Vedantham, P., Flynn, D.L. and Hanson, P.R. (2004) *Journal of Organic Chemistry*, **69**, 8340–4.
35 Roller, S., Zhou, H. and Hagg, R. (2005) *Molecular Diversity*, **9**, 305–16.
36 Crosignani, S., White, P.D. and Linclau, B. (2002) *Organic Letters*, **4**, 1035–7.
37 Crosignani, S., White, P.D. and Linclau, B. (2002) *Organic Letters*, **4**, 2961–3.
38 (a) Crosignani, S., White, P.D., Steinauer, R. and Linclau, B. (2003) *Organic Letters*, **5**, 835–56. (b) Crosignani, S., Launay, D., Linclau, B. and Bradley, M. (2003) *Molecular Diversity*, **7**, 203–10. (c) Crosignani, S., White, P.D. and Linclau, B. (2004) *Journal of Organic Chemistry*, **69**, 5897–905.
39 König, W. and Geiger, R. (1970) *Chemische Berichte*, **103**, 788–98.
40 Carpino, L.A. (1993) *Journal of the American Chemical Society*, **115**, 4397–8.
41 Nájera, C. (2005) Polystyrene-bound 1-Hydroxybenzotriazole (PS-HOBT), in *Handbook of Reagents for Organic Synthesis, Reagents for Glycoside, Nucleotide and Peptide Synthesis* (ed. D. Crich), John Wiley & Sons Ltd, Chichester, pp. 523–5.
42 Chinchilla, R., Dodsworth, D.J., Nájera, C. and Soriano, J.M. (2000) *Tetrahedron Letters*, **41**, 2463–6.
43 Van der Auwere, C., Van Damme, S. and Anteunis, M.J.O. (1987) *International Journal of Peptide and Protein Research*, **29**, 464–71.
44 Williams, M.W. and Young, G.T. (1963) *Journal of the Chemical Society*, 881–9.
45 Vickerstaffe, E., Warrington, B.H., Ladlow, M. and Ley, S.V. (2003) *Organic and Biomolecular Chemistry*, **1**, 2419–22.
46 Chinchilla, R., Dodsworth, D.J., Nájera, C. and Soriano, J.M. (2003) *Arkivoc*, **x**, 41–7.
47 Disadee, W., Watanabe, T. and Ishikawa, T. (2003) *Synlett*, 115–17.

48 Carpino, L.A. and El-Faham, A. (1995) *Journal of the American Chemical Society*, **117**, 5401–2.
49 Barrett, A.G.M., Bribal, B., Hopkins, B.T., et al. (2005) *Tetrahedron*, **61**, 12033–41.
50 Castro, B., Dormoy, J.-R., Evin, G. and Selve, C. (1975) *Tetrahedron Letters*, 1219–22.
51 Fournier, A., Danho, W. and Felix, A.M. (1989) *International Journal of Peptide and Protein Research*, **33**, 133–9.
52 Frérot, E., Coste, J., Pantaloni, A., Dufour, M.-N. and Jouin, P. (1991) *Tetrahedron*, **47**, 259–70.
53 Filip, S.V., Lejeune, V., Vors, J.-P., et al. (2004) *European Journal of Organic Chemistry*, 1936–9.
54 Elson, K.E., Jenkins, I.D. and Loughlin, W.A. (2004) *Tetrahedron Letters*, **45**, 2491–3.
55 Fairfull-Smith, K.E., Jenkins, I.D. and Loughlin, W.A. (2004) *Organic and Biomolecular Chemistry*, **2**, 1979–86.
56 Landi, J.J. Jr. and Brinkman, H.R. (1992) *Synthesis*, 1093–5.
57 Relles, H.M. and Shluenz, R.W. (1974) *Journal of the American Chemical Society*, **96**, 6469–75.
58 Hodge, P. and Richardson, G. (1975) *Journal of the Chemical Society D – Chemical Communications*, 622–3.
59 Caputo, R., Cassano, E., Longobardo, L., et al. (1995) *Synthesis*, 141–3.
60 Li, Z., Yu, J., Ding, R., et al. (2004) *Synthetic Communications*, **34**, 2981–6.
61 Belleau, B. and Malek, G. (1968) *Journal of the American Chemical Society*, **90**, 1652.
62 Brown, J. and Williams, R.E. (1971) *Canadian Journal of Chemistry – Revue Canadienne de Chimie*, **49**, 3765–6.
63 Kiso, Y. and Yajima, H. (1972) *Journal of the Chemical Society D – Chemical Communications*, 942–3.
64 Vaughan, J.R. Jr. and Osato, R.L. (1952) *Journal of the American Chemical Society*, **74**, 676–8.
65 Valeur, E. and Bradley, M. (2005) *Chemical Communications*, 1164–6.
66 Baxendale, I.R., Ley, S.V., Smith, C.D. and Tranmer, G.K. (2006) *Chemical Communications*, 4835–7.
67 Mukaiyama, T. (1979) *Angewandte Chemie – International Edition in English*, **18**, 707–8.
68 (a) Mukaiyama, T., Usui, M., Shimada, E. and Saigo, K. (1975) *Chemistry Letters*, 1045–8. (b) Bald, E., Saigo, K. and Mukaiyama, T. (1975) *Chemistry Letters*, 1163–6.
69 Convers, E., Tye, H. and Whittaker, M. (2004) *Tetrahedron Letters*, **45**, 3401–4.
70 Crosignani, S., Gonzalez, J. and Swinnen, D. (2004) *Organic Letters*, **6**, 4579–82.
71 Donati, D., Morelli, C., Porcheddu, A. and Taddei, M. (2004) *Journal of Organic Chemistry*, **69**, 9316–18.
72 Donati, D., Morelli, C. and Taddei, M. (2005) *Tetrahedron Letters*, **46**, 2817–19.
73 Kaminski, Z.J. (2000) *Biopolymers (Peptide Science)*, **55**, 140–64.
74 Masala, S. and Taddei, M. (1999) *Organic Letters*, **1**, 1355–7.
75 Xia, M. and Wang, Y. (2003) *Synthetic Communications*, **33**, 403–8.
76 Falchi, A. and Taddei, M. (2000) *Organic Letters*, **2**, 3429–31.
77 (a) Kunishima, M., Yamamoto, K., Watanabe, Y., et al. (2005) *Chemical Communications*, 2698–700. (b) Kunishima, M., Yamamoto, K., Hioki, H., et al. (2007) *Tetrahedron*, **63**, 2604–12.
78 Luo, G., Xu, L. and Poindexter, G.S. (2002) *Tetrahedron Letters*, **43**, 8909–12.
79 Arseniyadis, S., Subhash, P.V., Valleix, A., et al. (2005) *Chemical Communications*, 3310–12.
80 Packham, D.I. (1964) *Journal of the Chemical Society*, 2617–24.
81 Fridkin, M., Patchornik, A. and Katchalski, E. (1966) *Journal of the American Chemical Society*, **88**, 3164–5.
82 Fridkin, M., Patchornik, A. and Katchalski, E. (1968) *Journal of the American Chemical Society*, **90**, 2953–7.
83 Fridkin, M., Patchornik, A. and Katchalski, E. (1965) *Journal of the American Chemical Society*, **87**, 4646–8.
84 Cohen, B.J., Kraus, M.A. and Patchornik, A. (1977) *Journal of the American Chemical Society*, **99**, 4165–7.
85 Kalir, R., Fridkin, M. and Patchornik, A. (1974) *European Journal of Biochemistry*, **42**, 151–6.
86 Fridkin, M., Hazum, R., Rotman, R. and Kock, Y. (1977) *Journal of Solid-Phase Biochemistry*, **2**, 175–82.
87 Zhou, F.X., Krull, I.S. and Feisbuch, B. (1993) *Journal of Chromatography*, **648**, 357–65.

88 Koppel, I., Koppel, J., Leito, I., et al. (1994) *Journal of Chemical Research (S)*, 212–13.
89 Panse, G.T. and Laufer, D.A. (1970) *Tetrahedron Letters*, 4181–4.
90 Cohen, B.J., Karoly-Hafeli, H. and Patchornik, A. (1984) *Journal of Organic Chemistry*, **49**, 922–4.
91 Kim, K. and Le, K. (1999) *Synlett*, 1957–9.
92 Bourque, A.J. and Krull, I.S. (1991) *Journal of Chromatography*, **537**, 123–52.
93 Carpino, L.A., Cohen, B.J., Lin, Y.-Z., et al. (1990) *Journal of Organic Chemistry*, **55**, 251–9.
94 Marshall, D.L. and Liener, I.E. (1970) *Journal of Organic Chemistry*, **35**, 867–8.
95 Narang, C.K., Kachhawaha, V. and Mathur, N.K. (1988) *Reactive Polymers*, **8**, 189–92.
96 Salvino, J.M., Kumar, N.V., Orton, E., et al. (2000) *Journal of Combinatorial Chemistry*, **2**, 691–7.
97 Corbett, A.D. and Gleason, J.L. (2002) *Tetrahedron Letters*, **43**, 1369–72.
98 Goodnow, R.A.Jr. and Shao, H. (2005) Polymer-bound N-hydroxysuccinimide, in *Handbook of Reagents for Organic Synthesis, Reagents for Glycoside, Nucleotide and Peptide Synthesis* (ed. D. Crich), John Wiley & Sons Ltd, Chichester, pp. 525–8.
99 Laufer, D.A., Chapman, T.M., Marlborough, D.I., et al. (1968) *Journal of the American Chemical Society*, **90**, 2696–8.
100 Fridkin, M., Patchornik, A. and Katchalski, E. (1972) *Biochemistry*, **11**, 466–71.
101 Andreev, S.M., Tsiryapkin, V.A., Samoilova, N.A., et al. (1977) *Synthesis*, 303–4.
102 Narita, M., Teramoto, T. and Okawara, M. (1972) *Bulletin of the Chemical Society of Japan*, **45**, 3149–55.
103 Akiyama, M., Shimizu, K. and Narita, M. (1976) *Tetrahedron Letters*, 1015–18.
104 Chinchilla, R., Dodsworth, D., Nájera, C. and Soriano, J.M. (2001) *Tetrahedron Letters*, **42**, 4487–9.
105 Chinchilla, R., Dodsworth, D.J., Nájera, C. and Soriano, J.M. (2001) *Tetrahedron Letters*, **42**, 7579–81.
106 Chinchilla, R., Dodsworth, D.J., Nájera, C. and Soriano, J.M. (2002) *Bioorganic and Medicinal Chemistry Letters*, **12**, 1817–20.
107 Chinchilla, R., Dodsworth, D.J., Nájera, C. and Soriano, J.M. (2003) *Synlett*, 809–12.
108 Barrett, A.G.M., Cramp, S.M., Roberts, R.S. and Zécri, F.J. (2000) *Organic Letters*, **2**, 261–4.
109 Arnaud, T., Barrett, A.G.M., Hopkins, B.T. and Zécri, F.J. (2001) *Tetrahedron Letters*, **42**, 8215–17.
110 Roberts, R.S. (2005) *Journal of Combinatorial Chemistry*, **7**, 21–32.
111 Adamczyk, M., Fishpaugh, J.R. and Mattingly, P.G. (1999) *Tetrahedron Letters*, **40**, 463–6.
112 Katoh, M. and Sodeoka, M. (1999) *Bioorganic and Medicinal Chemistry Letters*, **9**, 881–4.
113 Suminoshi, H., Shimizu, T., Katoh, M., et al. (2002) *Organic Letters*, **4**, 3923–6.
114 Shao, H., Zhang, Q., Goodnow, R., et al. (2000) *Tetrahedron Letters*, **41**, 4257–60.
115 Kalir, R., Warshawsky, A., Fridkin, M. and Patchornik, A. (1975) *European Journal of Biochemistry*, **59**, 66–1.
116 Barrada, A., Cavelier, F., Jacquier, R. and Verducci, J. (1989) *Bulletin de la Société chimique de France*, 511–14.
117 Stern, M., Kalir, R., Patchornik, A., et al. (1977) *Journal of Solid-Phase Biochemistry*, **2**, 131–9.
118 Heusel, G., Bovermann, G., Göhring, W. and Jung, G. (1977) *Angewandte Chemie – International Edition in English*, **16**, 642–3.
119 Mokotoff, M. and Patchornik, A. (1983) *International Journal of Peptide and Protein Research*, **21**, 145–54.
120 Mokotoff, M., Zhao, M., Roth, S.M., et al. (1990) *Journal of Medicinal Chemistry*, **33**, 354–60.
121 Sheh, L., Mokotoff, M. and Abraham, D.J. (1987) *International Journal of Peptide and Protein Research*, **29**, 509–20.
122 Chen, S.-T., Chang, C.-H. and Wang, K.-T. (1991) *Journal of Chemical Research (S)*, 206–7.
123 Dendrinos, K., Jeong, J., Huang, W. and Kalivretenos, A.G. (1998) *Chemical Communications*, 499–500.

124 Moss, R.A., Lukas, T.J. and Nahas, R.C. (1977) *Tetrahedron Letters*, **18**, 3851–4.
125 Huang, W. and Kalivretenos, A.G. (1995) *Tetrahedron Letters*, **36**, 9113–16.
126 Dendrinos, K.G. and Kalivretenos, A.G. (1998) *Tetrahedron Letters*, **39**, 1321–4.
127 Dendrinos, K.G. and Kalivretenos, A.G. (1998) *Journal of the Chemical Society – Perkin Transactions 1*, 1463–5.
128 Pop, I.E., Déprez, B.P. and Tartar, A.L. (1997) *Journal of Organic Chemistry*, **62**, 2594–603.
129 Gooding, O.W., Vo, L., Bhattacharyya, S. and Labadie, J.W. (2002) *Journal of Combinatorial Chemistry*, **4**, 576–83.
130 Kulkarni, B.A., Roth, G.P., Lobkovsky, A. and Porco, J.A. Jr. (2002) *Journal of Combinatorial Chemistry*, **4**, 56–72.
131 Garcia, J., Nicolás, E., Albericio, F., et al. (2002) *Tetrahedron Letters*, **43**, 7495–8.
132 Devocelle, M., McLoughlin, B.M., Sharkey, C.T., et al. (2003) *Organic and Biomolecular Chemistry*, **1**, 850–3.
133 Chang, Y.-T. and Schultz, P.G. (1999) *Bioorganic and Medicinal Chemistry Letters*, **9**, 2479–82.
134 Baxendale, I.R. and Ley, S.V. (2000) *Bioorganic and Medicinal Chemistry Letters*, **10**, 1983–6.
135 Mutter, M. (1978) *Tetrahedron Letters*, **19**, 2839–42.
136 Mutter, M. (1978) *Tetrahedron Letters*, **19**, 2843–6.
137 DeGrado, W.F. and Kaiser, E.T. (1980) *Journal of Organic Chemistry*, **45**, 1295–300.
138 DeGrado, W.F. and Kaiser, E.T. (1982) *Journal of Organic Chemistry*, **47**, 3258–61.
139 (a) Smith, R.A., Bobko, M.A. and Lee, W. (1998) *Bioorganic and Medicinal Chemistry Letters*, **8**, 2369–74. (b) Lumma, W.C. Jr., Witherup, K.M., Tucker, T.J., et al. (1998) *Journal of Medicinal Chemistry*, **41**, 1011–13.
140 Golebiowski, A. and Klopfenstein, S. (1998) *Tetrahedron Letters*, **39**, 3397–400.
141 Scialdone, M.A., Shuey, S.W., Soper, P., Hamuro, Y. and Burns, D.M. (1998) *Journal of Organic Chemistry*, **63**, 4802–7.
142 Petricci, E., Mugnaini, C., Radi, M., et al. (2004) *Journal of Organic Chemistry*, **69**, 7880–7.
143 Sophiamma, P.N. and Sreekumar, K. (1997) *European Polymer Journal*, **33**, 863–7.
144 Van Boom, J.H., Burgess, P.M.J. and Van Deursen, P. (1974) *Journal of the Chemical Society D – Chemical Communications*, 619–20.

8
Supported/Tagged Scavengers for Facilitating High-Throughput Chemistry

Paul R. Hanson[1,2], Thiwanka Samarakoon[1,2] and Alan Rolfe[1,2]

[1]University of Kansas, Department of Chemistry, 1251 Wescoe Hall Drive, Lawrence, KS 66045
[2]The Center for Chemical Methodologies and Library Development at the University of Kansas, 1501 Wakarusa Drive, Lawrence, KS 66047, USA.

8.1
Introduction

The growing demand for facile production of libraries of compounds in desirable amounts and sufficient purity for high-throughput screening (HTS) has presented challenging opportunities in the development of facilitated synthetic protocols that ultimately serve to uncover molecular leads as potential therapeutic agents aimed at improving human health. Arguably, the most effective advances realized in combinatorial chemistry during this past decade were those techniques that successfully integrated the sciences of organic synthesis and purification. Traditional solid phase organic synthesis (SPOS), which encompasses several virtues and is based on a resin-platform, undoubtedly has been the primary driving force utilized to address this demand. While SPOS has its merits for offering a direct purification technique, and is highly amenable to automation, limitations have spurred the advancement of alternate platforms for use in the arena of facilitated synthesis.

To address these limitations, the last decade has seen a paradigm shift in the field whereby "the scaffold is returned" to solution and reagents/scavengers are immobilized [1]. Recent successes in solution phase, multistep total syntheses championed by Ley and coworkers, utilizing exclusively immobilized reagents and scavengers [2], whereby filtration was the sole purification protocol, are a testament to the power of this approach. Despite huge advances in this area, limitations in nonlinear reaction kinetics, low resin-load capacities, means of distributing reagents, and the mechanics and technologies behind multistep parallel solution phase sequences continue to warrant the development of new platforms and improved strategies toward the ultimate goal of facilitating drug discovery.

This chapter highlights the latest developments and key advances in the application of scavengers in modern day synthesis. In particular, emphasis is placed on

The Power of Functional Resins in Organic Synthesis. Judit Tulla-Puche and Fernando Albericio
Copyright © 2008 WILEY-VCH Verlag GmbH & Co. KGaA, Weinheim
ISBN: 978-3-527-31936-7

the removal of excess reagents, intermediates and by-products to yield the desired products in high purity without the use of classical work-up procedures. While this chapter will focus on the application of scavengers, the broader application of scavengers in library production and total synthesis has been highlighted. The overall aim is to showcase the application of both commercial and other types of available supported scavengers on various tags across a range of methodologies.

Over recent years the aforementioned paradigm shift has driven the use of excess reagents to drive reactions to completion to expedite purification in the synthesis of large libraries of compounds. This has generated the need for an efficient process for the removal excess reagent to yield the desired product in high purity. Classical methods of purification such as aqueous extraction, chromatography, crystallization and distillation are time consuming and can be tedious procedures if run in a parallel fashion. To circumvent these problems several scavenging methods have emerged for the removal of excess nucleophiles, electrophiles, transition-metal catalysts and by-products. Scavengers meet this goal since they are highly effective at improving crude reaction purity and can be readily removed at the end of the reaction by simple filtration. A tagged scavenger has a functional group that is complementary to that of an excess reagent, thereby allowing the reaction between the two components. The scavengers can take advantage of both ionic and covalent interactions, and bind to the excess reagent/byproduct. The tag then allows for purification via filtration, eliminating the need for chromatography (Figure 8.1). Many of these scavengers are now available commercially for application across a wide range of synthetic methodologies in both academia and industry.

In the mid-1990s, the importance of using scavengers in organic synthesis was addressed by several independent industrial research groups [1]. Pioneering work by Letsinger [3] and Merrifield [4] on solid phase organic synthesis provided the impetus for the rapid development and application of solid phase methodology in the construction of large libraries. Reagents can be immobilized by tethering to either an insoluble or semi-soluble support material. Conventional supports are based on a divinylbenzene (DVB), cross-linked polystyrene resin, which can be micro- or macroporous depending on the cross-linking. Polystyrene-based resins have been the standard platform for immobilization of reagents and scavengers.

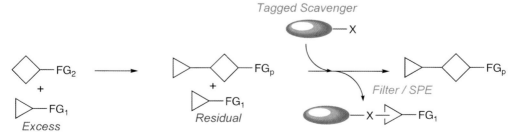

Figure 8.1 Use of scavengers to bind excess reagent/byproduct, eliminating the need for chromatography.

However, in recent years, several different methods have emerged to address the issue of reaction kinetics, functional group load (mmol g^{-1}) and physical properties of the immobilization platform. These include, silica, fluorous, oligomeric, ionic, gel, dendrimer and magnetically-tagged reagents. This chapter will start with a discussion of scavengers aimed at the sequestration of nucleophiles and will be organized along the lines of the immobilization platform.

8.2 Supported/Tagged Reactive Functionalities for the Sequestration of Nucleophiles (Nucleophile/Electrophilic Scavengers)

Electrophilic scavengers contain a reactive electrophilic functional group and, hence, can react with a compatible nucleophile, allowing for its sequestration and subsequent removal. Electrophilic scavengers are also referred to as nucleophile scavengers. Several electrophilic/nucleophile scavengers have been developed to date and are summarized in Table 8.1.

8.2.1 Polystyrene Resin-Supported Scavengers

Since the discovery of solid phase peptide synthesis in 1963 [4], the venerable polystyrene resin has remained the cornerstone of combinatorial chemistry over the years and continues to be utilized as the primary mode of support for immobilizing reagents and scavengers. This section briefly outlines the latest developments of resin-based scavengers.

8.2.1.1 Resin-Supported Isocyanates

Isocyanates have traditionally been a highly utilized amine scavenger. Kaldor and coworkers first reported the polystyrene-based isocyanates [1c]; since this

Table 8.1 Summary of electrophilic/nucleophile scavengers developed to date.

Scavenger	Nucleophile scavenged
Isocyanate	Primary and secondary amines, alcohols, thiols
Acid chlorides/anhydrides	Primary and secondary amines, alcohols, thiols
Aldehydes/diketones	Primary amines, hydrazines
Sulfonyl chlorides	Primary and secondary amines, alcohols, alkoxides
Phosphonyl chlorides	Primary and secondary amines
Epoxides	Thiols and alkoxides
Triazene	Primary and secondary amines, alcohols, thiols
Diazonium salts	Primary and secondary amines

publication, several supported isocyanates have appeared in the literature. Among these, Bradley and coworkers have developed several resin bound isocyanates [5]. In their work, a macroporous resin-bound isocyanate was produced via the reaction between para-diisocyanate (PDI) and a macroporous amino resin. The resin was optimized for maximum swelling (prepared with 40% cross-linking and 300% porogen level in toluene) and was reported to be more reactive than commercially available polystyrene amino methyl isocyanate resins.

Bradley and coworkers next reported the synthesis of an array of supported isocyanates using three different commercially available diisocyanates and the optimally prepared macroporous resin mentioned above. In addition, they employed a gel resin obtained from commercial sources (Scheme 8.1). The Bradley scavengers have higher reactivity and display better reaction kinetics than the commercially available resins. They have been utilized to scavenge amines in a pilot library of DCC-mediated amides, whereby excellent results were achieved in scavenging the residual amines (Figure 8.2).

Conventional polystyrene isocyanates have been successfully utilized for nucleophile scavenging in the synthesis of pharmaceutically attractive molecules. Player and coworkers have reported the use of conventional resin-bound isocyanates in the scavenging of amines in the synthesis of an 80-member library of triazolopyrimidines with purities greater than 90% (Scheme 8.2) [6].

Ellman and coworkers have also reported the use of these resin-bound isocyanates for the scavenging of thiols in the parallel synthesis of a small library of cysteine protease inhibitors (Scheme 8.3) [7]. Simple filtration afforded the hydroxy sulfide products in moderate to good yields. Excess Dess-Martin periodinane, as well as the by-product in the final step, were scavenged using a resin-bound thiosulfate to yield target compounds.

In a recent report, Gregg and coworkers outline the use of conventional resin-bound isocyanates in scavenging amines for the synthesis of a large library of aryl pyrazolo-pyrimidine carboxamides (Scheme 8.4). Various primary as well as secondary amines were utilized in the amidation of activated nitro phenyl ester. Subsequent scavenging of amines with polystyrene isocyanate afforded the target

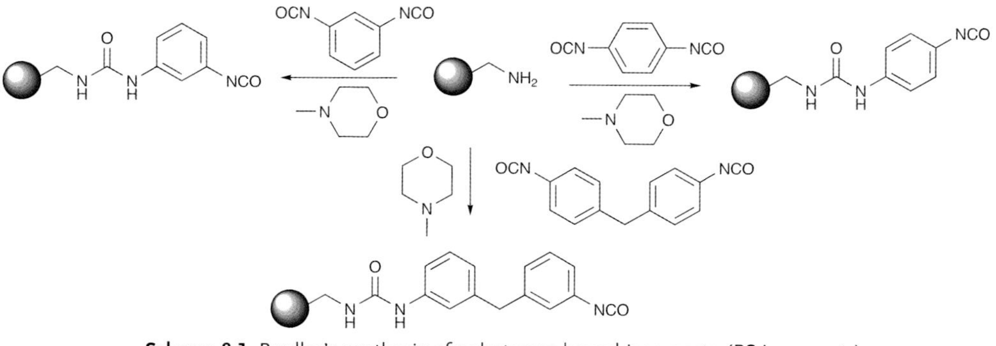

Scheme 8.1 Bradley's synthesis of polystyrene-bound isocyanate (PS-isocyanate).

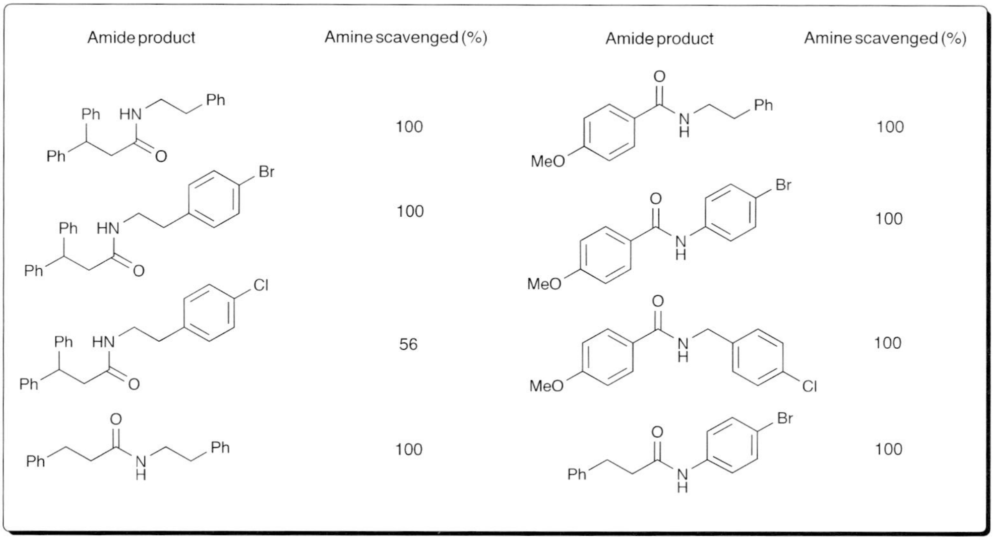

Figure 8.2 Scavenging of excess amines utilizing the Bradley polystyrene-bound isocyanate.

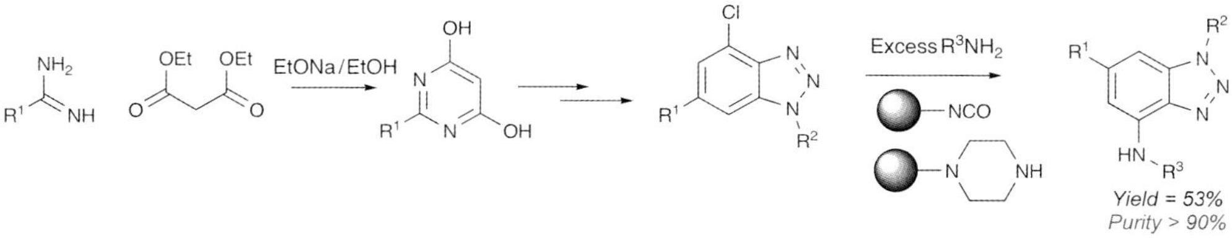

Scheme 8.2 Utilization of PS-isocyanate for amine scavenging in Triazolo pyrimidine library synthesis.

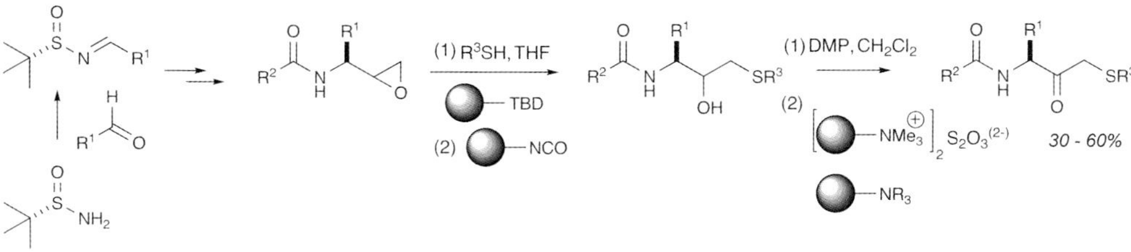

Scheme 8.3 Utilization of PS-isocyanate for amine scavenging in synthesis of lystein protease inhibitor library.

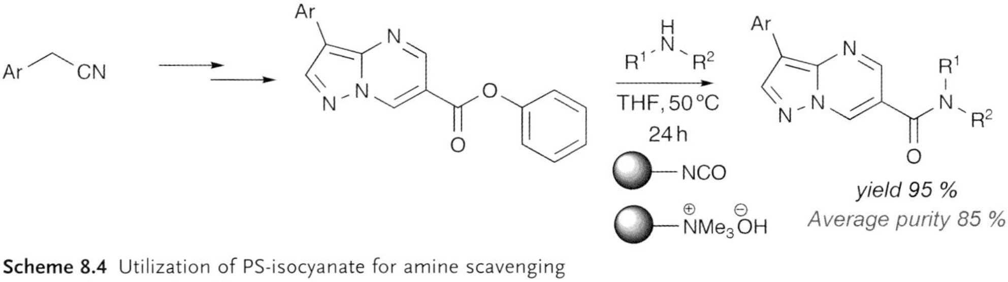

Scheme 8.4 Utilization of PS-isocyanate for amine scavenging in pyrazolo pyrimidine library synthesis.

compounds with purities greater than 85% over a range of 426 compounds (more than 90% of library population) [8].

8.2.1.2 Resin-Supported Anhydrides and Acid Chlorides

Fontaine and coworkers have recently reported the synthesis of a supported aza-lactone via atom radical transfer polymerization (ATRP) [9]. This method involved the preparation of a Wang resin-supported initiator, followed by subsequent ATR polymerization between 2-vinyl-4,4-dimethyl-5-oxazolone (VAZ) and styrene to generate several macroporous, aza-lactone functionalized resins with different architectures. These were shown to scavenge benzyl amines in a highly efficient fashion (Scheme 8.5).

Fréchet and coworkers have reported the development of a functionalized polymer monolith for use in parallel solution phase synthesis in continuous flow applications [10]. In this report, the authors outline the preparation of an azalactone-functionalized monolith for scavenging nucleophiles. This method involves the preparation of a macroporous poly(chloromethylstyrene co-divinylbenzene) monolith via the polymerization of the relevant mixture of monomer, initiator and porogen. These are allowed to react with a free radical initiator (4-cyanovaleric acid), followed by reaction with the monomer of choice, to synthesize the functionalized monolith. The authors have thus prepared monoliths functionalized with VAZ to provide an azalactone-functionalized monolith. These monoliths were then demonstrated to completely remove amines after flowing a solution of amine in THF through the monolith for 30 min. They have also reported the reaction of these monoliths with alcohols as well. A small demonstration library of ureas was prepared and after 8 min of residence time up to 76% of the alkyl amines were found to be scavenged (Scheme 8.6).

8.2.1.3 Resin-Supported Aldehydes

Kaldor and others first reported resin-bound aldehydes [11] for the scavenging of amines. Recently, Miguel and coworker investigated the use and recycling of

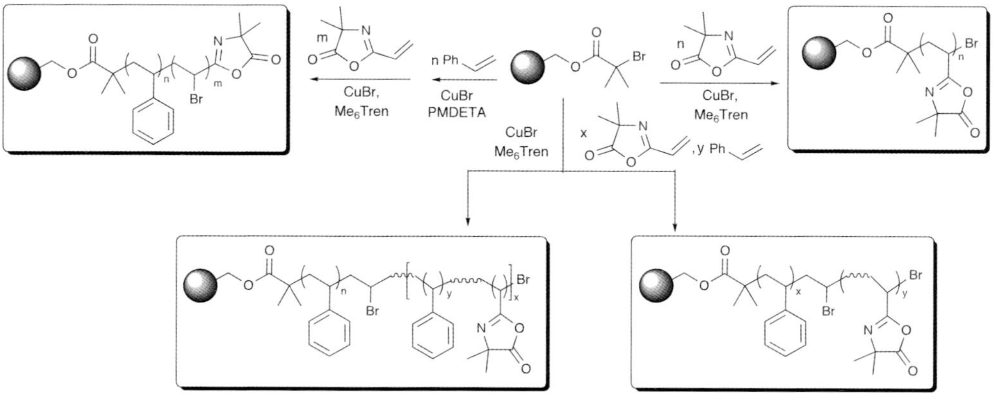

Scheme 8.5 Fontain's synthesis of supported aza-lactone.

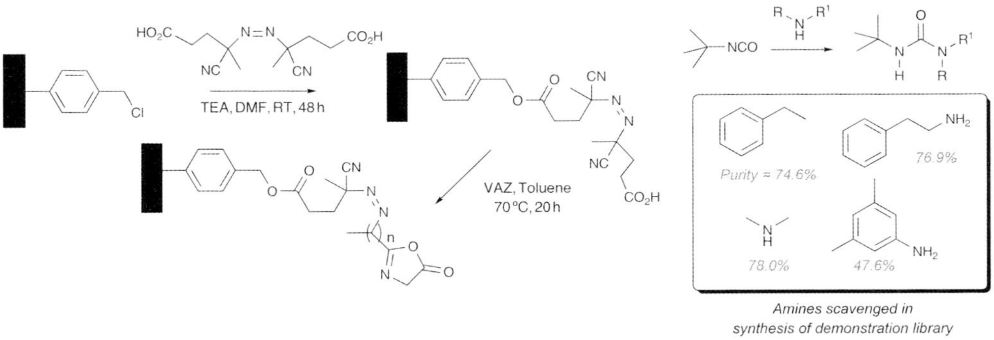

Scheme 8.6 Utilization of supported aza-lactone for amine scavenging.

polystyrene-supported benzaldehyde in amine scavenging [12]. The PS-benzaldehyde has been found to be amenable to three reaction cycles with the corresponding yields being maintained throughout. In addition, these scavengers sequestered primary amines over secondary with remarkable selectivity (Scheme 8.7).

Zhu and coworkers have reported the synthesis of functionalized poly(vinyl alcohol) resins for use as scavengers [13]. This was achieved via inverse suspension polymerization along side epichlorohydrin as a cross-linker. These resins were found to have excellent swelling characteristics in DMF, CH_3OH, dioxane, THF, CH_2Cl_2 and H_2O. These were then functionalized with glutaric aldehyde to provide a polymer-supported aldehyde (Scheme 8.8).

This polymer-supported aldehyde was found to scavenge both alkyl as well as aryl primary amines over secondary amines and exhibit remarkable facility in the synthesis of a demonstration library of amides, ureas and secondary amines (Scheme 8.9).

8.2.1.4 Resin-Supported Diketones

Kirschning and coworkers have recently reported the synthesis of a resin-bound diketone for the sequestration of amines and hydrazines [14]. The resin-bound diketone was synthesized via oxidation of a chloromethyl resin, followed by reductive coupling with 2,4-pentadione, to provide the requisite scavenger resin. This resin was found to be more efficient than supported-aldehydes for amine and hydrazine scavenging and, furthermore, found to be selective for the scavenging of primary amines over secondary amines (Scheme 8.10).

8.2.1.5 Resin-Supported Maleimide

Maleimide is a well-known Michael acceptor, dienophile and dipolarophile and hence is another versatile functional moiety that has found multiple uses in combinatorial chemistry as both a scavenger as well as a template in library synthesis [15]. Barrett [16] and Porco [17] have both reported the synthesis of a polystyrene resin-supported maleimide but did not report its use in the scavenging of nucleophiles. Hall and coworkers have described the synthesis of a supported-maleimide

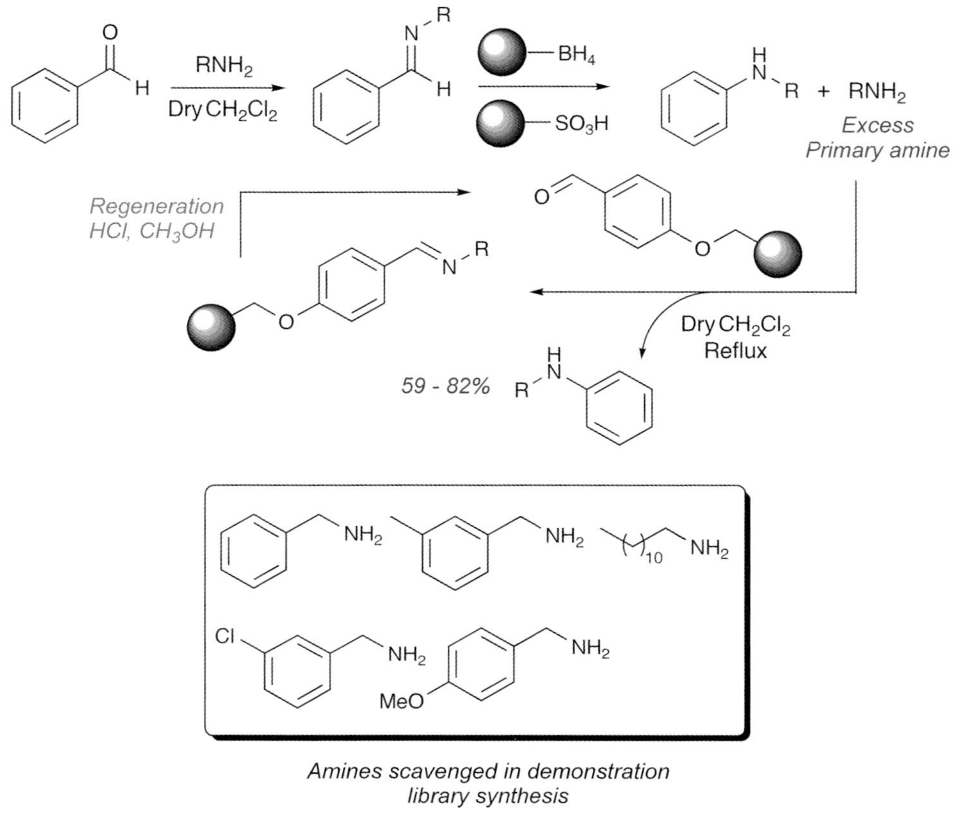

Scheme 8.7 Miguel's recyclable polystyrene-supported aldehyde.

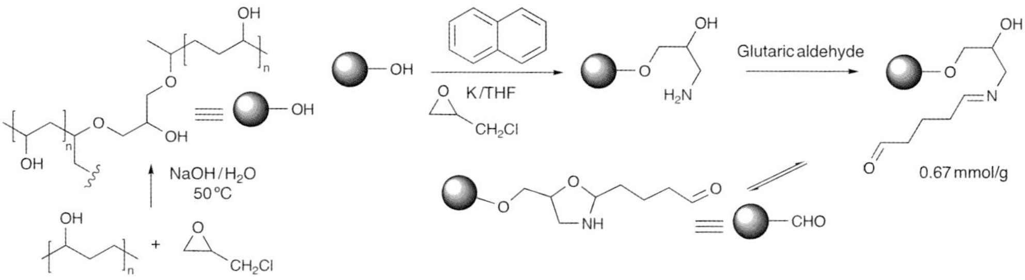

Scheme 8.8 Zhu's utilization of polyvinyl alcohols for the synthesis of supported aldehydes.

reagent for the library synthesis of functionalized pyrrolidines via a [3 + 2] cycloaddition. The reagent is synthesized via the esterification of *p*-nitrobenzoic acid with both Rink and Sasrin resins, followed by reduction of the nitro group and subsequent condensation with maleic anhydride. This resin was found to have a remarkable ability to scavenge alkyl as well as aromatic thiols (Scheme 8.11) [15].

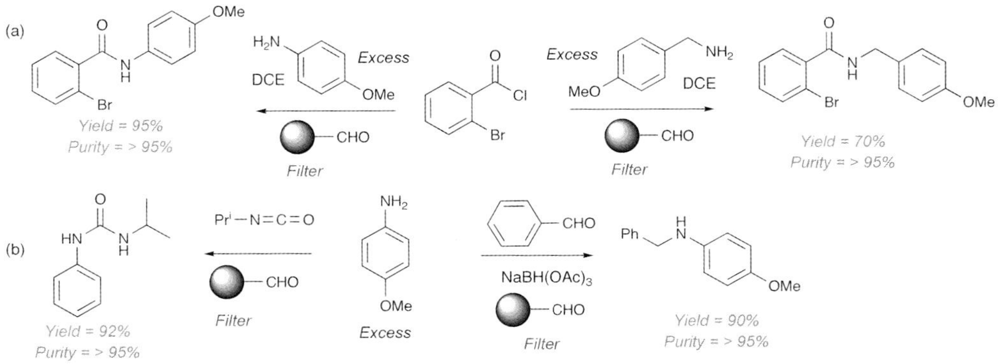

Scheme 8.9 Utilization of the Zhu aldehyde for amine scavenging.

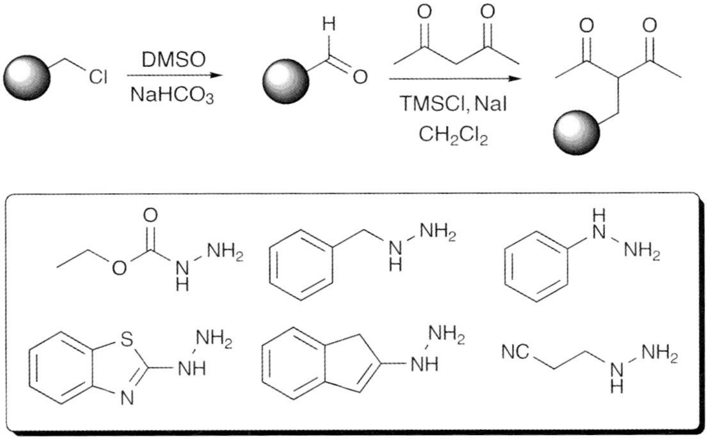

Scheme 8.10 Resin-bound diketones in hydrazine scavenging.

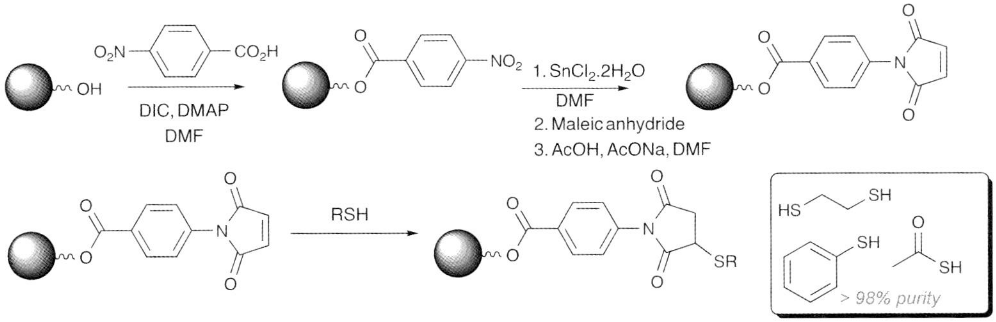

Scheme 8.11 Synthesis of supported maleimides for thiol scavenging.

8.2.1.6 Resin-Supported Diazonium Salts

Bräse and coworkers were the first to introduce triazene linkers for use in SPOC in a seminal report in 2000 [18]. They found immobilized diazonium ions have high stability, which enabled their application as linkers and scavengers. Their synthesis involves the reduction of 3-amino-6-chlorobenzoic acid, which was subsequently coupled onto a Merrifield resin via standard etherification. Subsequent conversion of the amino group into a diazonium salt generated the supported diazonium tetrafluoroborate, which was found to sequester amines (Scheme 8.12).

In related work, Lazny and coworkers report a new economical synthesis of four new polymeric supports with three- and six-carbon atom spacers and triazene linkers derived from *meta*- and *para*-aminophenol. In addition to their use as supports for immobilization of secondary amines, they can scavenge amines (Scheme 8.13) [19].

8.2.1.7 Resin-Supported Halide

Lipschutz and coworkers have recently reported the use of modified, conventional Merrifield resin for scavenging both PPh_3 and $O=PPh_3$ in transition metal-catalyzed cross coupling reactions (Scheme 8.14) [20]. The modified resin was produced by reacting commercially available, high-load, chloromethylated polystyrene, *in situ* with NaI, to produce a more reactive iodo-Merrifield resin, which participated in facile removal of both PPh_3 and $O=PPh_3$.

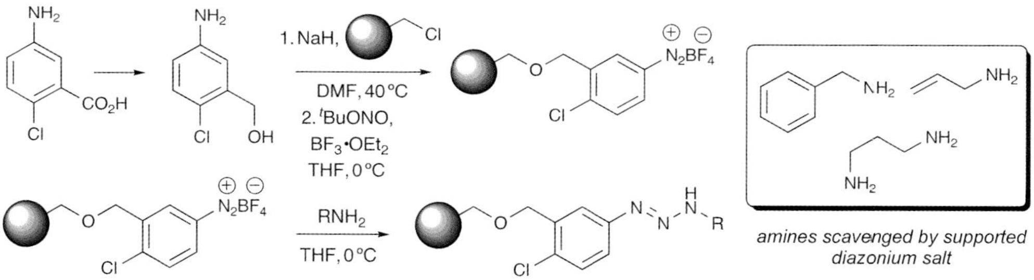

Scheme 8.12 Bräse's synthesis of supported diazonium salts for amine scavenging.

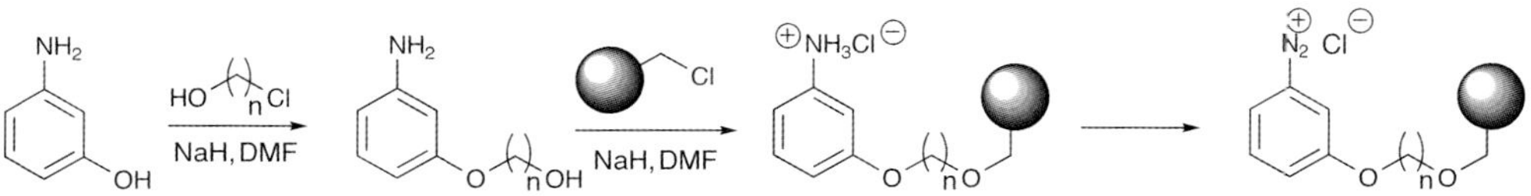

Scheme 8.13 Lazny's synthesis of supported diazonium salts for amine scavenging.

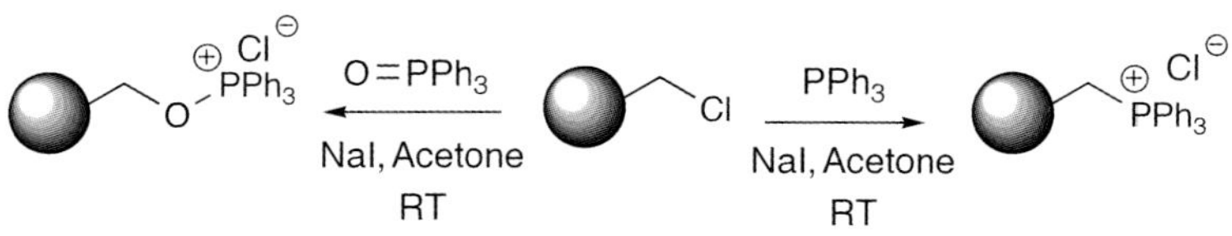

Scheme 8.14 Utilization of Merrifield resin in triphenyl phosphine and triphenyl phosphine oxide scavenging.

8.2.2
PEG-Supported Scavengers

Despite the aforementioned successes, resin-bound reagents and scavengers still suffer from slow reaction kinetics due to heterogeneity, as well as low load levels of traditional resins. These deficiencies have prompted the continued development of alternative platforms for facilitating synthesis. One such platform is the use of poly(ethylene glycol) (PEG)-supports, which was pioneered by Janda and coworkers. PEG-supports have emerged as an attractive alternative since they address a key weakness of their traditional polystyrene-bound counterparts in that they are soluble polymers [21], and hence react under homogenous conditions, thereby improving reaction kinetics. Once the reaction is completed, they are precipitated and filtered off from the desired products. Traditionally, Et_2O and hexanes are the solvents of choice for precipitating PEG-based supports.

8.2.2.1 PEG-Supported Dichlorotriazine (DCT)

Trichlorotriazine is a well-known versatile reagent in organic synthesis and has been utilized to facilitate several different reactions [22]. Due to its highly electrophilic character it reacts readily with nucleophiles and is therefore eminently suitable for scavenging nucleophiles. Taddei and coworkers first reported a polystyrene resin-bound dichlorotrazene (DCT) as a coupling agent for amide synthesis [23]. They also published a subsequent report on the development of a PEG-supported DCT reagent [24]. The reagent was readily prepared by simple condensation of trichlorotriazine with MeO-PEG-OH to provide the soluble PEG-DCT reagent (Scheme 8.15).

The soluble PEG-DCT reagent was utilized to scavenge alcohols in the synthesis of esters and silyl-protected alcohols as well as acetals, and thiols in the synthesis

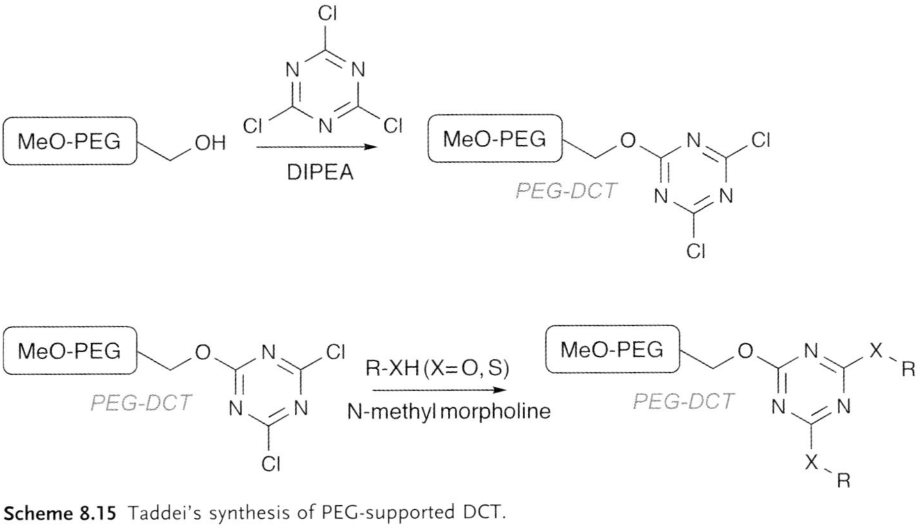

Scheme 8.15 Taddei's synthesis of PEG-supported DCT.

trans-glycosylated carbohydrates, with high efficiency. The authors also reported the sequestration of triphenylphosphine and triphenylphosphine oxide by this PEG-supported DCT reagent (Scheme 8.16). Since this report by Taddei, there have been several developments pertaining to supported DCT for the scavenging of nucleophiles.

8.2.3
Silica-Supported Scavengers

Silica has recently emerged as an alternative support for the immobilization of reagents and scavengers due to several attributes. A key feature is that reaction kinetics are higher than in conventional resins, because functional groups lie on the surface and hence are not embedded and thus limited by diffusion rates. In addition, silica is compatible with a range of solvents (polar and non-polar), while maintaining its insolubility, as a consequence of its non-swelling nature. Moreover, the free flowing nature of silica allows for easy handling while its mechanical stability allows for conventional methods of stirring Si-supported reagents [25]. Woodward and coworker recently reported the synthesis of Si-bound reagents and Si-grafted pellets; among these was a Si-supported isocyanate. The Si-supported isocyanates were not employed in scavenging, but are now commercially sold for that use (Scheme 8.17) [26].

8.2.3.1 Silica-Supported DCT
Recently, Pattarawarapan and coworker reported the synthesis of a Si-supported DCT for scavenging nucleophiles [27]. 3-Aminopropyltrimethoxysilane was allowed to react with commercial silica gel in toluene. TCT (trichloroazine) was then reacted with the Si-immobilized amine to provide the supported DCT reagent. This was found to sequester both amines (primary and secondary) as well as alcohols with remarkable facility and two libraries of sulfonamides and amides were prepared in parallel format. Filtration was the sole purification step (Scheme 8.18).

8.2.4
Fluorous-Tagged Scavengers

Fluorous-tagged reagents and scavengers, championed by Curran, Zhang and others, have gained much prominence over the last few years [28]. Fluorous tagging does not involve the use of conventional resin beads, but entails the attachment of a perfluoroalkyl tag to a reagent. Both heavy fluorous tags (60% or more fluorine by weight) as well as light fluorous tags (40% or less fluorine by weight) are utilized in this process. Workup involves the use of liquid–liquid extraction for the separation of fluorous and non-fluorous substances (heavy fluorous tags) or fluorous silica gel-based solid phase extraction (light fluorous synthesis). Since fluorous silica gel has a bonded phase of C_8F_7 perfluorohydro carbon chains, it retains fluorine-containing molecules via the strong fluorine–fluorine interactions. Since resins are not involved, all chemistry takes place in solution phase, allowing

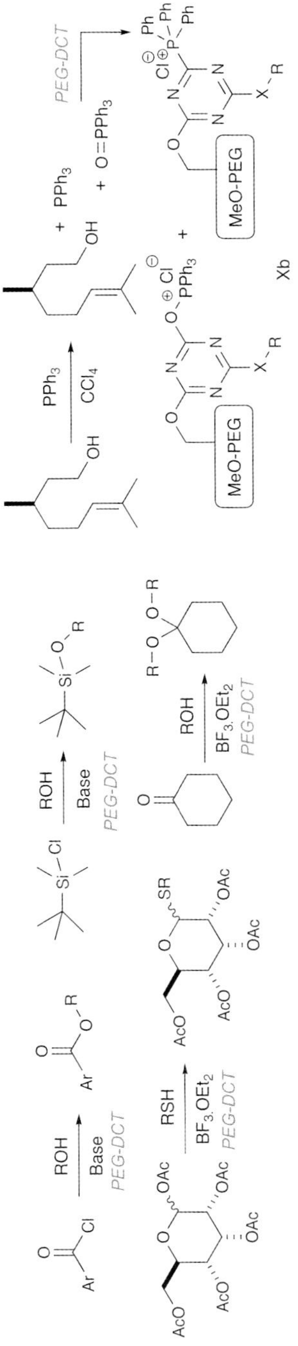

Scheme 8.16 Utilization of PEG-supported DCT in nucleophile scavenging.

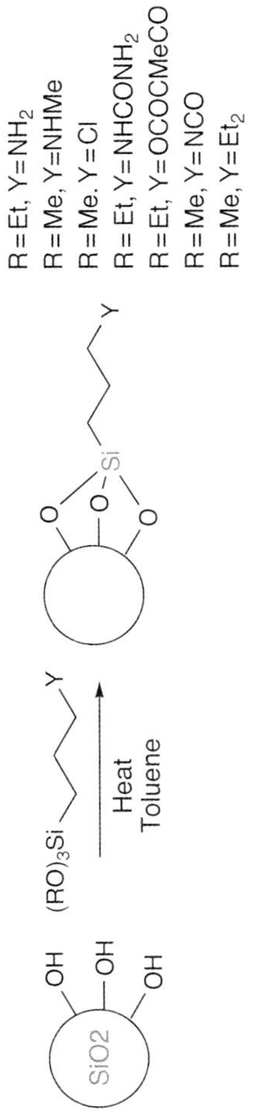

Scheme 8.17 Wood's synthesis of silica-supported DCT.

R=Et, Y=NH$_2$
R=Me, Y=NHMe
R=Me, Y=Cl
R=Et, Y=NHCONH$_2$
R=Et, Y=OCOCMeCO
R=Me, Y=NCO
R=Me, Y=Et$_2$

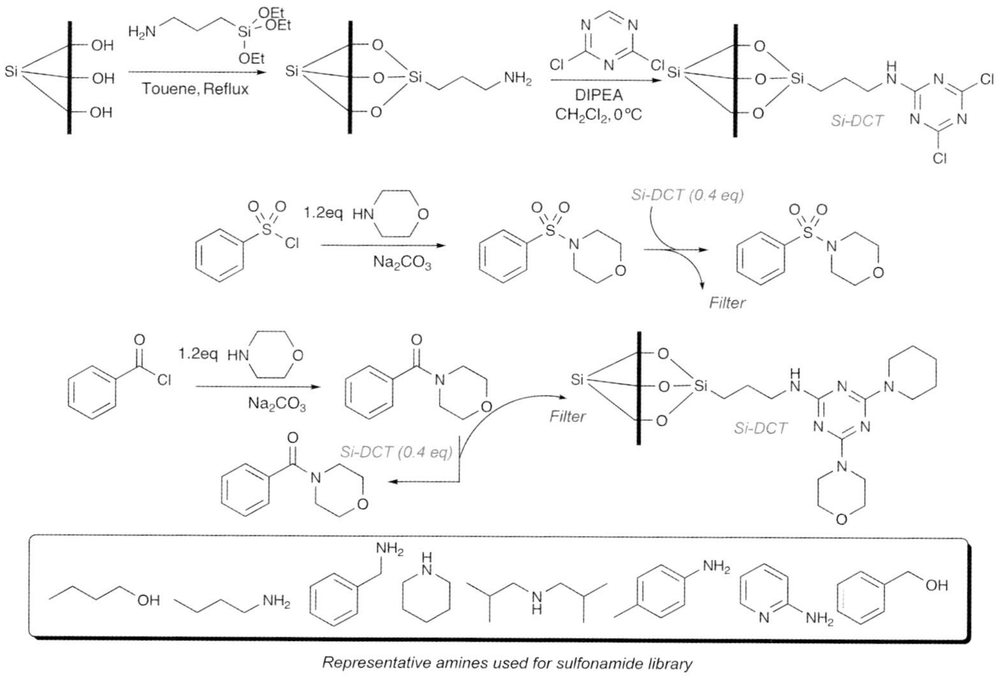

Scheme 8.18 Pattarawarapan's design and use of silica-supported DCT.

for the utilization of normal reaction conditions with minimal of optimization time, and increased reaction kinetics over conventional resin-bound chemistry [29]. Fluorous tags are chemically inert and hence do not partake in any side reactions with exogenous reactants. These features, coupled with their simple preparation, scale-up, and adaptability to automation, have elevated fluorous-tagged scavengers and reagents as highly attractive tools for both traditional and high-throughput chemistry.

8.2.4.1 Fluorous-Tagged Anhydrides and Isocyanates

Zhang and coworkers have developed the use of fluorous-tagged isocyanate as well as isatoic anhydride and acid chloride for the scavenging of amines for use in combinatorial as well as conventional chemistry (Figure 8.3) [30]. All reagents were readily prepared from commercially available sources, and the F-isatoic anhydride was prepared via a simple NaH-mediated alkylation of isatoic anhydride with perfluoroalkyl halide.

Zhang and coworkers elegantly demonstrated the use of F-isocyanates as well as F-isatoic anhydrides in the synthesis of a library of amino alcohols (via amine ring opening of glycidyl epoxide ethers) and a library of ureas and thioureas. Purities were reported to be in excess of 95%, while yields were slightly better when the anhydride was used (Scheme 8.19).

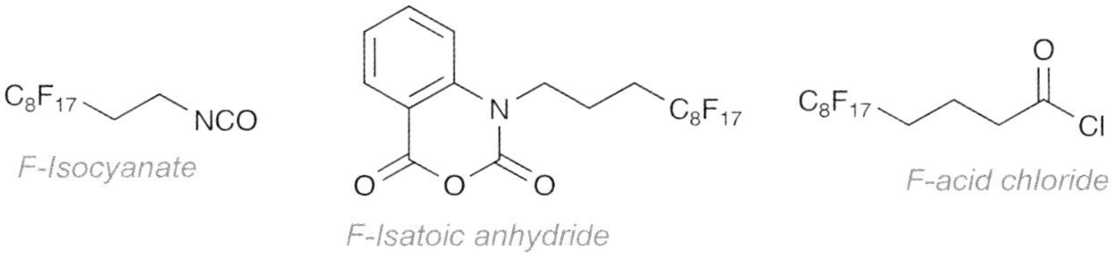

8.2 Supported/Tagged Reactive Functionalities for the Sequestration of Nucleophiles

Figure 8.3 Fluorous-tagged isocyanate, isatoic anhydride and acid chloride for the scavenging of amines.

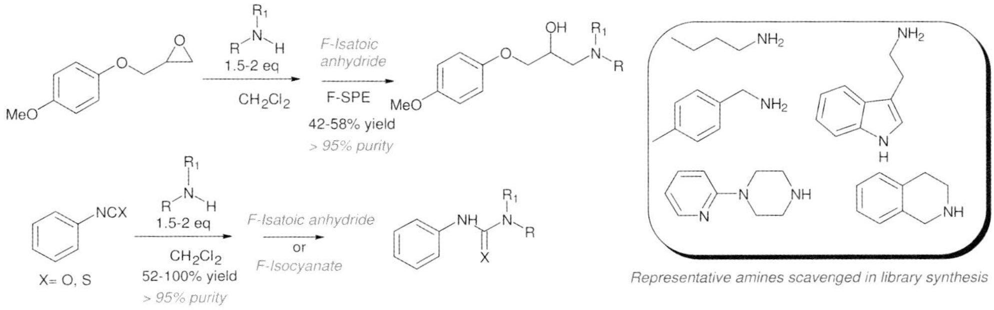

Scheme 8.19 Utilization of fluorous electrophilic scavengers for amine sequestration in library synthesis.

Zhang and coworkers workers have also demonstrated the use of F-acid chlorides alongside the concurrent use of the corresponding conventional resin-bound scavengers in the synthesis of a small library of sulfonamides with both primary as well as secondary amines. Scavenging was almost complete (<2% amine remained) with the use of F-acid chlorides, while a substantial amount remained in the use of resins. After 24 h, amines were scavenged by the fluorous reagents, while 3–12% amines remained unquenched when resins were used (Scheme 8.20) [31].

8.2.4.2 Fluorous-Tagged Sulfonyl Chlorides and Acid Chlorides and Epoxides

Lindsley and coworkers have reported the synthesis of a suite of fluorous-tagged scavengers for parallel solution phase synthesis (Figure 8.4) [32]. These scavengers have been applied towards the synthesis of various libraries in high efficiency and exceptional purity (>96%). Of particular note is the use of a tagged-epoxide for removal of alkoxide and thiols (Scheme 8.21).

8.2.4.3 Fluorous-Tagged Aldehyde

Ladlow and coworkers have reported the use of fluorous-tagged aldehydes as a protecting group in the synthesis of a library of sulfonamides. The F-aldehyde was prepared via a simple alkylation of 4-hydroxy-2-meoxybanzaldehyde with a perfluoroalkyl halide. The authors have protected various primary amines with the F-aldehyde followed by reduction, sulfonylation and Suzuki coupling and acid-mediated deprotection. Filtration via a fluorous SPE (solid phase extraction) was

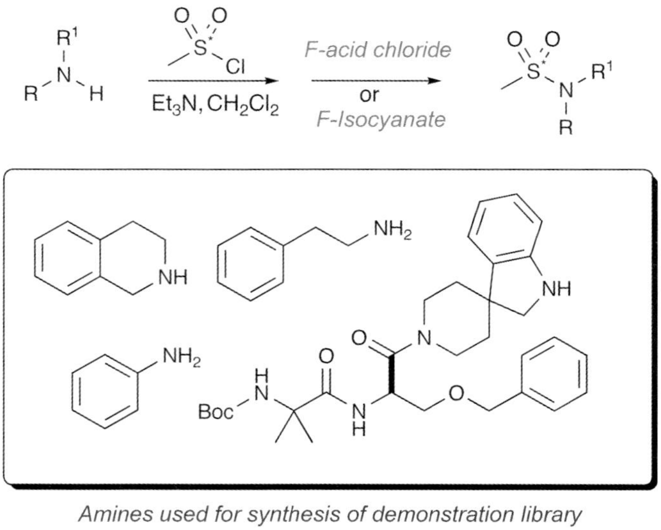

Scheme 8.20 Zhang's utilization of F-acid chlorides and F-isocynates for amine scavenging in comparison studies.

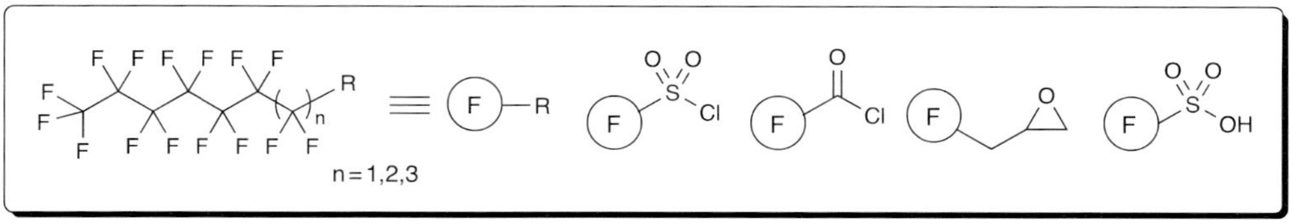

Figure 8.4 Fluorous-tagged scavengers used for parallel solution phase synthesis.

the sole mode of purification (Scheme 8.22) [33]. Although the authors have not reported the use of the F-aldehyde for amine scavenging, the aforementioned results point toward the promise of this reagent for scavenging amines in parallel synthetic protocols.

8.2.4.4 Fluorous-Tagged DCT

Zhang and coworkers have reported the synthesis of two fluorous DCT reagents for parallel synthesis. TCT was allowed to react with a fluorous-tagged alcohol as well as a fluorous-tagged thiol to produce the two reagents shown in Scheme 8.23 [34].

These reagents were utilized in the usual fashion to scavenge thiol as well as amines in the synthesis of a library of sulfides and amides with a high degree of purities and yields (Scheme 8.24). A significant feature of this approach was the purification utilizing plate-to-plate fluorous SPE as well as an automated solid phase extraction on a RapidTrace system, thus demonstrating the capacity of fluorous platforms for automation.

8.2 Supported/Tagged Reactive Functionalities for the Sequestration of Nucleophiles

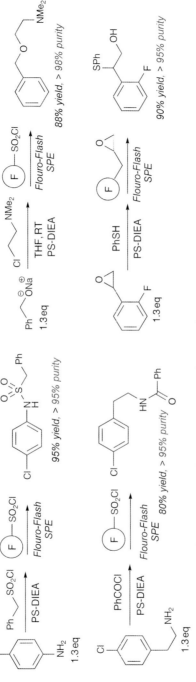

Scheme 8.21 The Lindsley fluorous electrophilic scavenger suite in library synthesis.

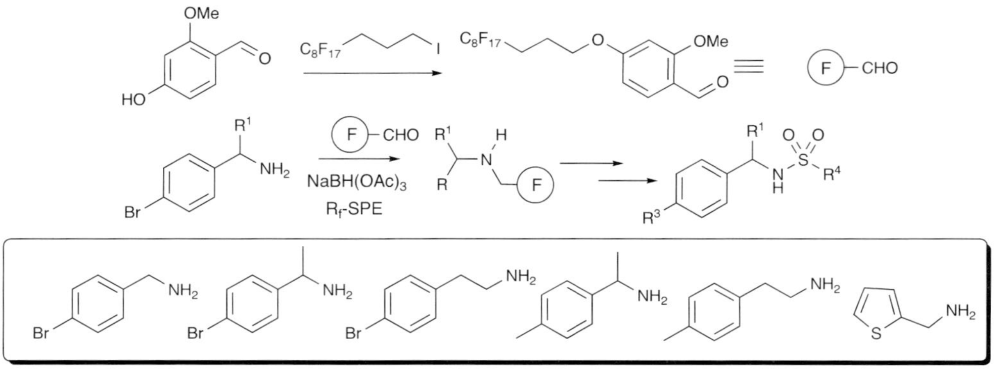

Scheme 8.22 F-aldehydes in amine scavenging.

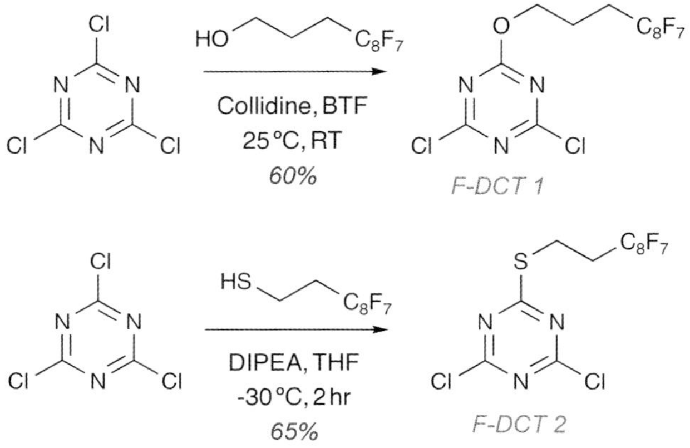

Scheme 8.23 The Zhang F-DCT reagents.

8.2.5
ROMP-Derived Electrophilic Scavengers

Soluble polymers and scavenger resins have emerged as a means of utilizing solution phase reaction kinetics with all the advantages of their solid phase counterparts. During these developments, pioneering work by Barrett and coworkers [35, 36] led to the general use of ring-opening metathesis polymerization (ROMP) for generating high-load, immobilized reagents. Reagent generation involves the polymerization of strained functionalized norbornenyl-tagged functional groups catalyzed by Grubbs ruthenium alkylidene metathesis catalysts (Figure 8.5) [37]. In most cases, norbornenyl-derived cross-linkers were employed to enhance swelling properties and insure insolubility. Occasionally, the polymeric backbone was hydrogenated, using Wilkinson catalyst under high pressure. Barrett's seminal 2002 review [35] discusses a wide variety of ROMP-gel supported reagents.

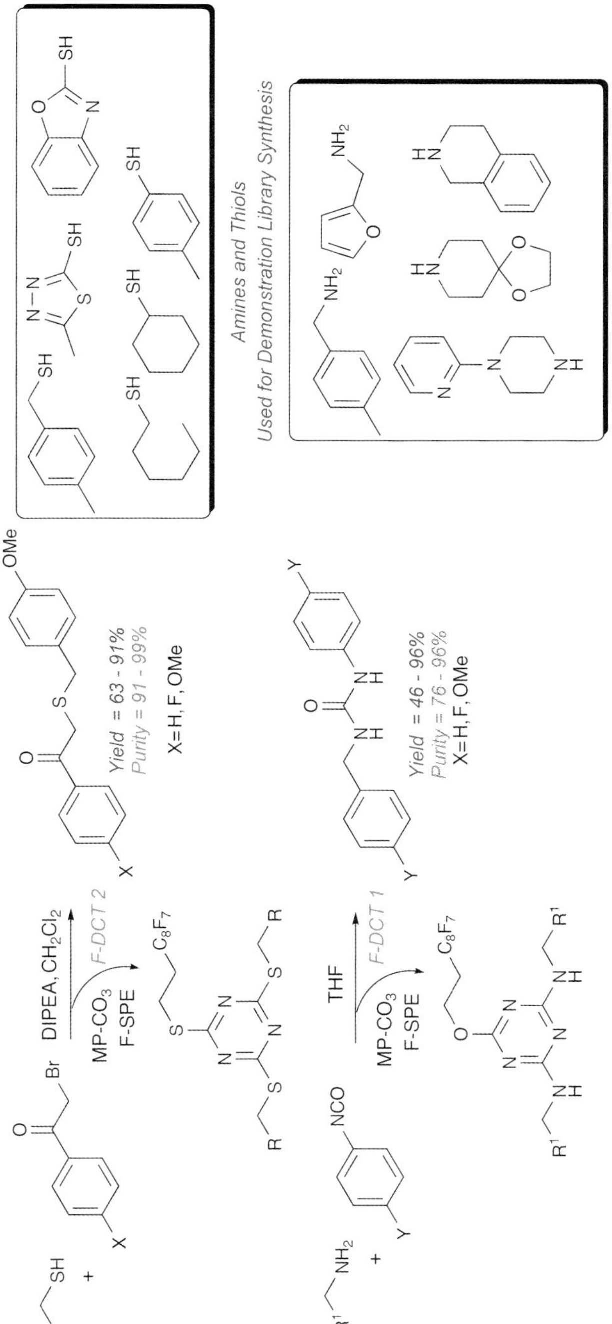

Scheme 8.24 Utilization of F-DCT in amine and thiol scavenging.

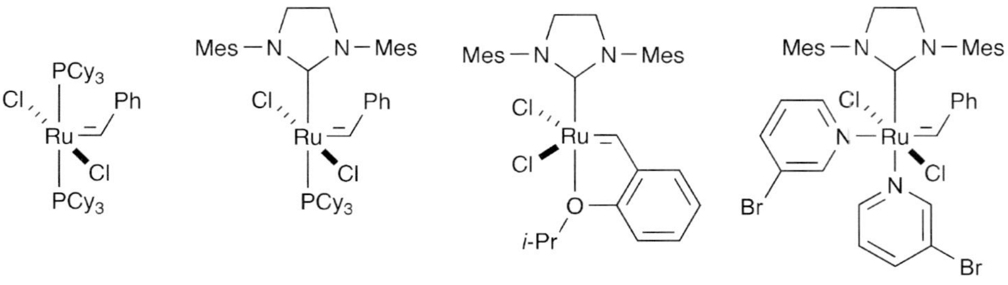

Figure 8.5 Examples of Grubbs ruthenium alkylidene metathesis catalysts.

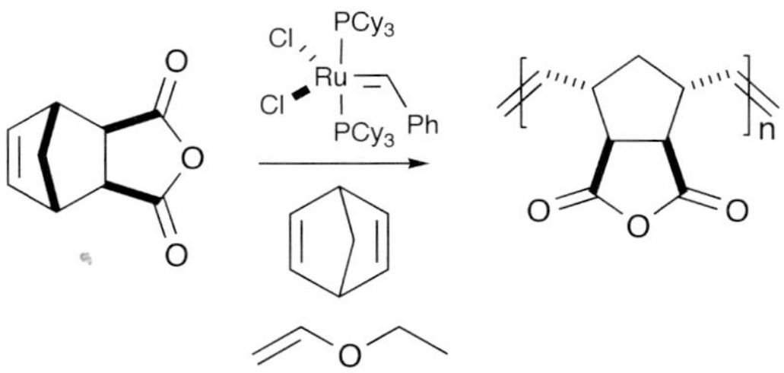

Scheme 8.25 Utilization of oligomeric anhydride for nucleophile scavenging.

8.2.6
ROMP-Gel Anhydride

Barrett and coworkers reported the use of ROM polymerization in the synthesis of a supported anhydride for nucleophile scavenging [38]. The norbornenyl-tagged monomer can either be bought commercially or prepared via a simple Diels–Alder reaction. Polymerization is then carried out in the presence of the Grubbs first-generation ruthenium alkylidene catalyst (Figure 8.5) in the presence of a cross-linking agent (norbornadiene) to produce a ROMP-gel possessing exceptionally high load capacity (Scheme 8.25).

Barrett and coworkers found that the ROMP-gel anhydrides efficiently scavenge various amines as well as alcohols with high efficiency and purity (Scheme 8.26).

Hanson and coworkers recently reported the synthesis of several high load reagents via ROMP techniques to provide supported reagents possessing solubility profiles that are tunable and which can be used in parallel solution phase synthesis. These high load reagents are generated by the ROM polymerization of norbornene- or 7-oxonorbornene-based monomers employing different Grubbs catalysts (Figure 8.5) to produce oligomers of varying lengths and solubility profiles. The diverse solubility profile of these reagents allows reactions to be conducted in common reaction solvents such as DCM, THF, DMF and $CHCl_3$. Following reaction/transformation, precipitation with an appropriate solvent

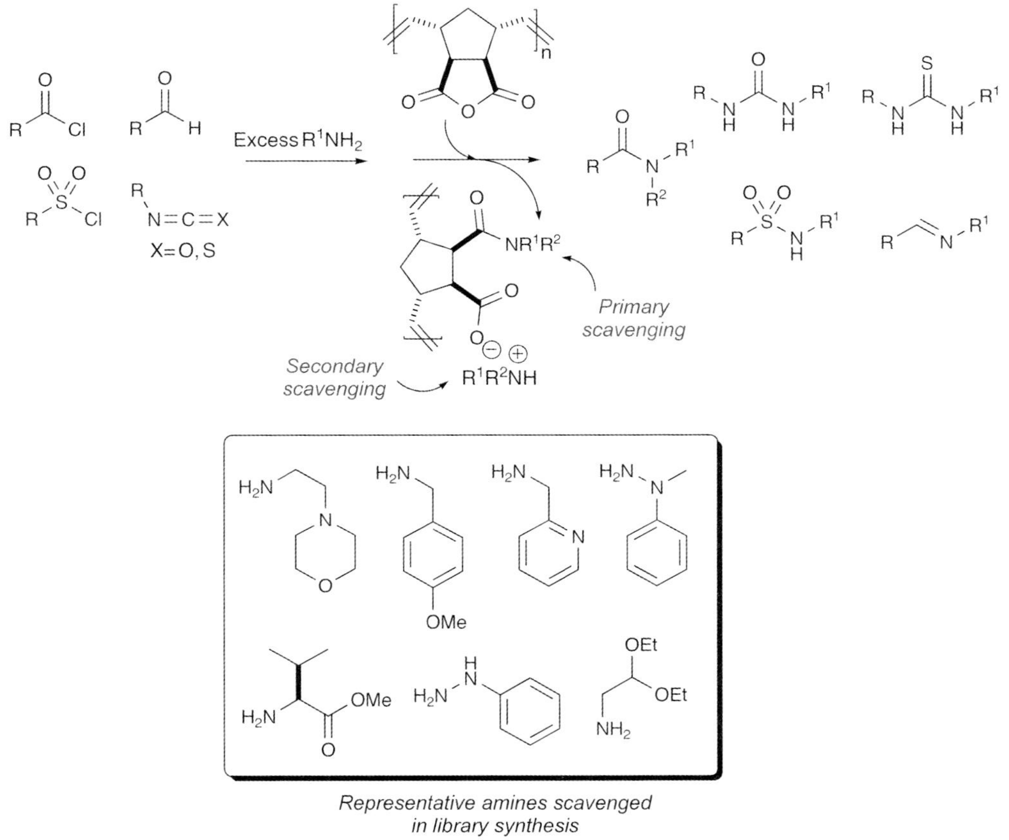

Scheme 8.26 Utilization of oligomeric anhydride in library synthesis.

(Et$_2$O, MeOH, or EtOAc), followed by filtration, provides products in good to excellent yields and purities. Notably, precipitation with EtOAc provides a wider solubility profile – a crucial feature in producing libraries of polar compounds.

8.2.6.1 Oligomeric Bis-acid Chloride

Hanson and coworkers have reported a high-load, oligomeric bis-acid chloride (OBAC) as a general nucleophile scavenger that was capable of removing alcohols and thiols in addition to amines [39]. The requisite monomer, *trans*-bicyclo[2.2.1]hept-5-ene-2,3-dicarbonyl dichloride, was conveniently prepared in a two-step sequence beginning with a Diels–Alder reaction between fumaric acid and cyclopentadiene followed by chlorination using oxaloyl chloride and catalytic DMF. Subsequent ROM polymerization with either first- or second-generation Grubbs catalysts [37] yields OBAC reagents of varying chain lengths (Scheme 8.27).

The OBAC scavenger efficiently removed excess amines (1°, 2°), alcohols (1°, 2°, allylic, propargylic and benzylic) and thiols after a common benzoylation event,

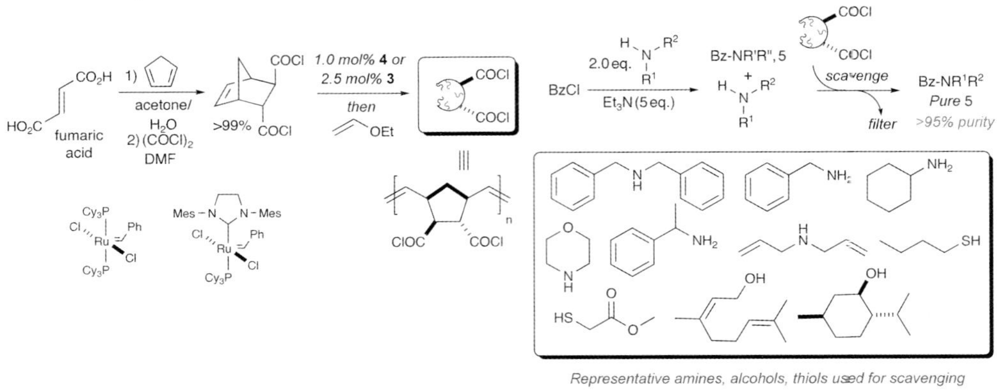

Scheme 8.27 Soluble oligomeric bis acid chloride (OBAC).

yielding products with greater than 95% purity. The synthetic utility of the OBAC scavenger was assessed by comparing its efficiency against a commercially available polystyrene-based isocyanate (PS-NCO) resin. The "loading" of OBAC scavengers is higher than the corresponding isocyanate polystyrene resin (9.1 vs ~1.3 mmol g^{-1}), thus the amount of OBAC reagent required for each scavenging reaction was determined to be seven times lower than the amount of isocyanate resin for a given experimental procedure. In a typical comparison experiment, 35 mg of OBAC versus 250 mg of the polystyrene-based isocyanate was found to be optimal for scavenging. While both OBAC and PS-NCO performed similarly for scavenging amines and thiols, OBAC was deemed more efficient for scavenging alcohol.

8.2.6.2 Oligomeric Sulfonyl Chloride (OSC)

Hanson and coworkers also reported the development of a high load, soluble oligomeric sulfonyl chloride (OSC) for scavenging of amines (Scheme 8.28) [40]. The OSC was synthesized by Diels–Alder reaction of vinylsulfonyl chloride and cyclopentadiene to generate monomer in 75–90% yield, followed by ROM polymerization with the second-generation Grubbs catalyst [37] to produce a free flowing solid. This scavenger is soluble in CH$_2$Cl$_2$, THF and DMF, and is insoluble in Et$_2$O, EtOAc and MeOH. The scavenging ability of OSC was investigated in the benzoylation and tosylation of various amines. Following the benzoylation/tosylation event, the oligomer was added as a solution to remove the excess amine (1°, 2° and benzylic), followed by precipitation of the scavenger to furnish the amides and sulfonamides in excellent yields and purities. Moreover, the solubility profile of this oligomer lends itself to facile dispensing via an automated liquid handling system, offering a distinct advantage over traditional solid-phase scavenging agents (Scheme 8.28).

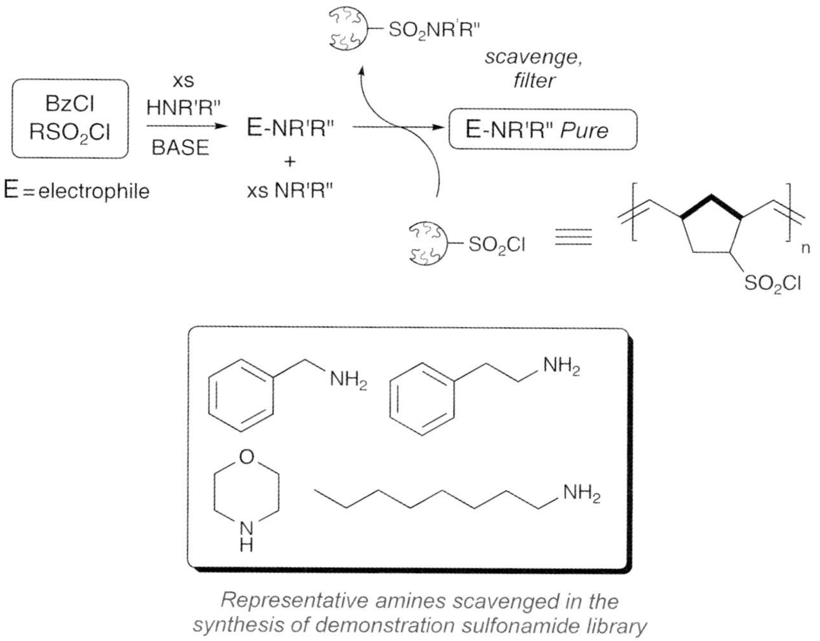

Scheme 8.28 Soluble oligomeric sulfonyl chloride (OSC).

8.2.6.3 Oligomeric Phosphonyl Chloride

Hanson and coworkers have recently reported the development of a high-load, ROMP-derived phosphonyl chloride (OPC) scavenging agent for scavenging amines (Scheme 8.29) [41]. This reagent can be readily generated via the ROM polymerization of bicyclo[2.2.1]hept-5-en-2-ylphosphonic dichloride, which is conveniently assembled from the Diels–Alder reaction of cyclopentadiene and vinylphosphonic dichloride. The OPC reagent was exploited in the rapid, efficient scavenging of primary and secondary amines that are present in excess following a common benzoylation event at room temperature (30–60 min) or under microwave conditions for shorter duration (<5 min). In addition, the scavenger was found to efficiently scavenge amines in the presence of alcohols with remarkable selectivity.

8.3 Supported/Tagged Reactive Functionalities for the Sequestration of Electrophiles (Electrophile/Nucleophilic Scavengers)

Nucleophilic scavengers contain a reactive nucleophilic functional group and, hence, can react with a compatible electrophile, allowing for its sequestration. Therefore, these are also referred to as electrophile scavengers.

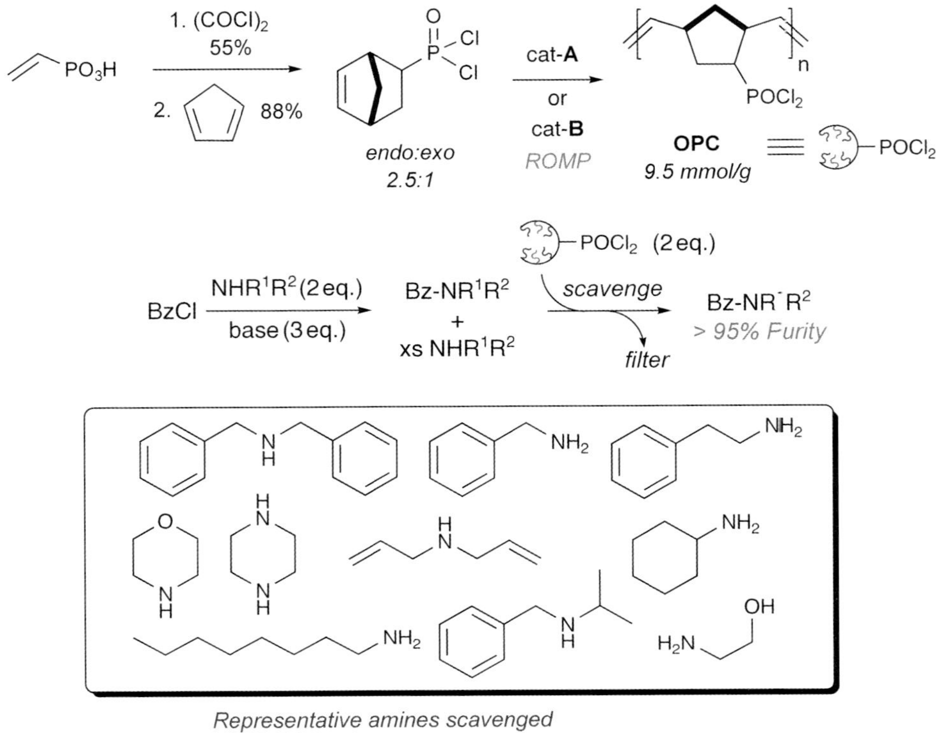

Scheme 8.29 Soluble oligomeric phosphonyl chloride (OPC).

Several electrophile/nucleophilic scavengers have been developed to date and are summarized in Table 8.2.

8.3.1
Polystyrene Resin-Based Scavengers

8.3.1.1 Supported Amine

Amine scavengers have been one of the most developed and widely used to remove a range of electrophiles on several different platforms. Traditionally developed on a polystyrene bead, amine scavengers have been utilized in natural products or library synthesis to yield high purity via a facilitated protocol. In 2005, Liu reported the use of a polystyrene-bound amine scavenger for the removal of excess aldehydes to provide imidazole derivatives in high purity without the need for classical purification (Scheme 8.30) [42].

Palladium-catalyzed organic synthesis has been utilized throughout the pharmaceutical industry, taking advantage of its C–C bond formation capability. However, the issue of residual levels of palladium has been a major concern due to its ability to attenuate biological screens. To address this, several scavengers

8.3 Supported/Tagged Reactive Functionalities for the Sequestration of Electrophiles

Table 8.2 Summary of electrophile/nucleophilic scavengers developed to date.

Scavenger	Electrophile scavenged
Amine	Acids, acid chlorides, sulfonyl chlorides
Thiol	Acid chlorides, sulfonyl chlorides
Alcohol	Acid chlorides, sulfonyl chlorides
Hydrazines	Aldehydes, ketones
Diene	Dienophiles
Dienophile	Diene

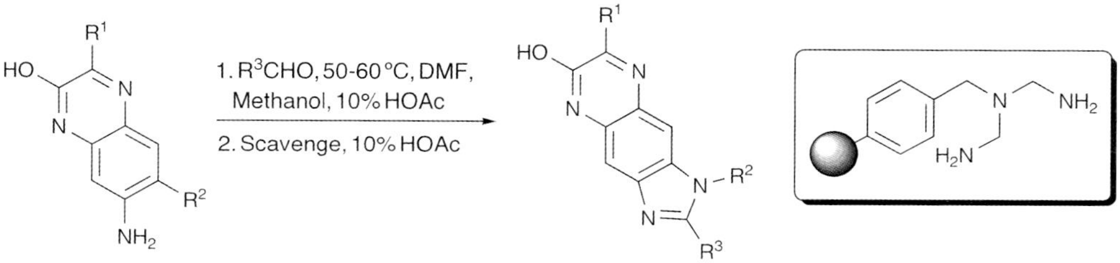

Scheme 8.30 Removal of excess aldehydes utilizing a polystyrene-bound amine.

were developed and applied for the removal of residual palladium. One such example was reported by Ogura, who demonstrated a range of polymer-bound ethylene diamines as scavengers of both palladium(0) and (II) species [43]. The levels of residual palladium from Suzuki–Miyaura coupling were reduced from 2000–3000 to 100–300 ppm by application of these scavengers. Additionally, a commercially available chelating resin, DIANON CR20, was evaluated and gave comparable results as the polymer-bound scavengers (Scheme 8.31). However, a longer reaction time was needed to obtain comparable results using equivalent loadings of DIANON CR20.

8.3.1.2 Supported Thiols

Hong *et al.* have reported a simple protocol for the removal of 9-methylene-9*H*-fluorene, a by-product in the deprotection of Fmoc-protected amines [44]. Several thiol-containing polystyrene resin scavengers were utilized to remove this by-product in a mixture with DBU and product (Scheme 8.32). This methodology provided an efficient procedure for the removal of Fmoc groups, yielding amines in high purity without the need for chromatography.

8.3.1.3 Supported Phosphines

In a similar fashion, efforts for the development and application of supported phosphines as scavengers to remove various electrophiles have been reported. Reports demonstrate the diversity of resin-tags used, and their applications as both

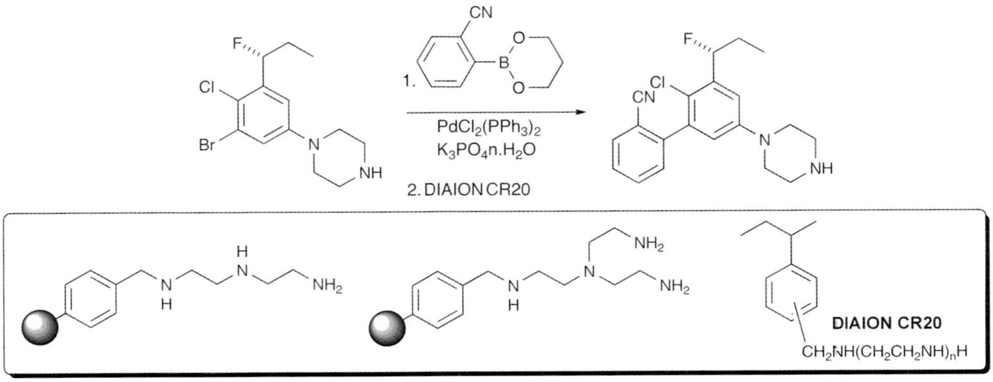

Scheme 8.31 Scavenging of residual palladium from the Suzuki-Miyaura coupling.

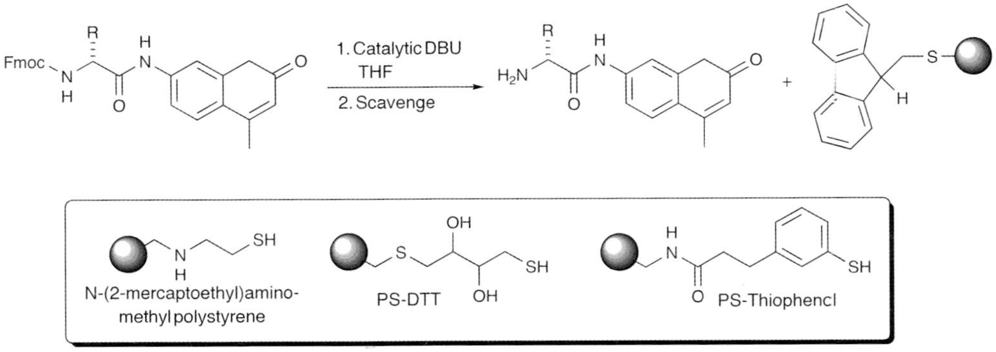

Scheme 8.32 Removal of 9-methylene-9H-fluorene by-product using thiol.

supported reagents and scavengers [45]. Hii reported the application of three phosphine-functionalized polymers as effective scavengers of palladium [46]. Notably, the most effective was an inexpensive and commercially available PS-PPh$_2$ which gave the desired products in purity of >98.5% by ICP-AES analysis (Scheme 8.33).

As previously described in the residual removal of palladium, the residual presence of other metals from catalysis has also been an area of concern. The emergence of the ruthenium-based Grubbs catalyst has led to its application in the synthesis of complex natural products and libraries. The removal of residual catalyst from the product, however, has remained a problem. One approach to circumvent this problem was reported by Breinbauer, who scavenged residual ruthenium using an inexpensive polymer-bound chelate phosphine [47]. Scavenging of crude mixtures removed ruthenium to give the desired product in >95% purity (Scheme 8.34).

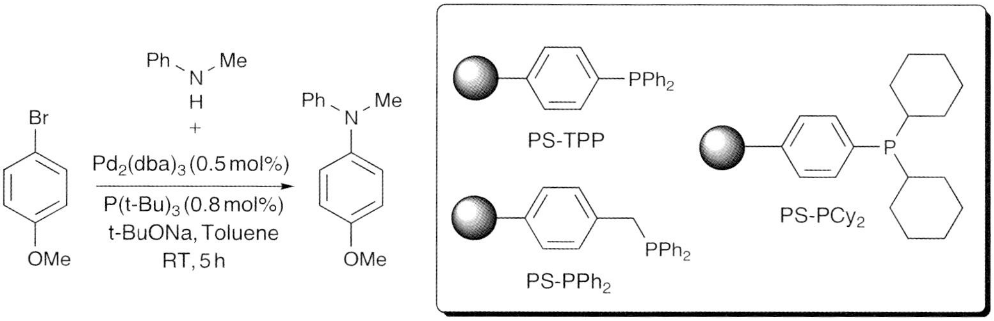

8.3 Supported/Tagged Reactive Functionalities for the Sequestration of Electrophiles | 209

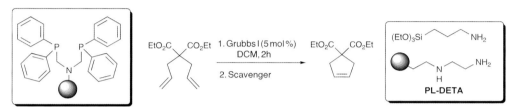

Scheme 8.33 Phosphine-functionalized polymers as effective scavengers of palladium.

Scheme 8.34 Efficient scavenging of residual ruthenium from ring-closing metathesis.

8.3.1.4 Supported Anthracene for Dienophile Scavenging

In 2004, Porco described the application of a polymer-supported anthracene for the scavenging of excess dienophile from Diels–Alder reactions [48]. This scavenger was not only demonstrated with a range of dienophiles, but also its application was related in the synthesis of flavonoid dienes and other classes of cyclic molecules (Scheme 8.35).

8.3.1.5 Supported Hydrazines

Wessjohann *et al.* reported the application of polymer-supported benzylhydrazines as a reversible scavenger resin for aromatic aldehydes (Scheme 8.36) [49]. These "protected aldehydes" can undergo additional transformations and eventually be released to either reveal the original aldehyde functionality or be released in a diversification step.

8.3.1.6 Supported Acids/Bases for Scavenging

Supported resins bearing acidic and basic functionality have been used to scavenge immobilize excess reagents and by-products utilizing ionic interactions. Such resins are available commercially as either free flowing powers or as a preformed cartridge (SPE). Massi described the efficient synthesis of 3,4-dihydropyrimidin-2-(1*H*)-ones utilizing a multicomponent Biginelli reaction in combination with polymer-assisted solution phase chemistry (Scheme 8.37) [50]. In this example, a solid bound ytterbium(III) reagent is used to catalyze the multicomponent reaction of aldehydes, 1,3-dicarbonyl-containing compounds and ureas. The key step in this

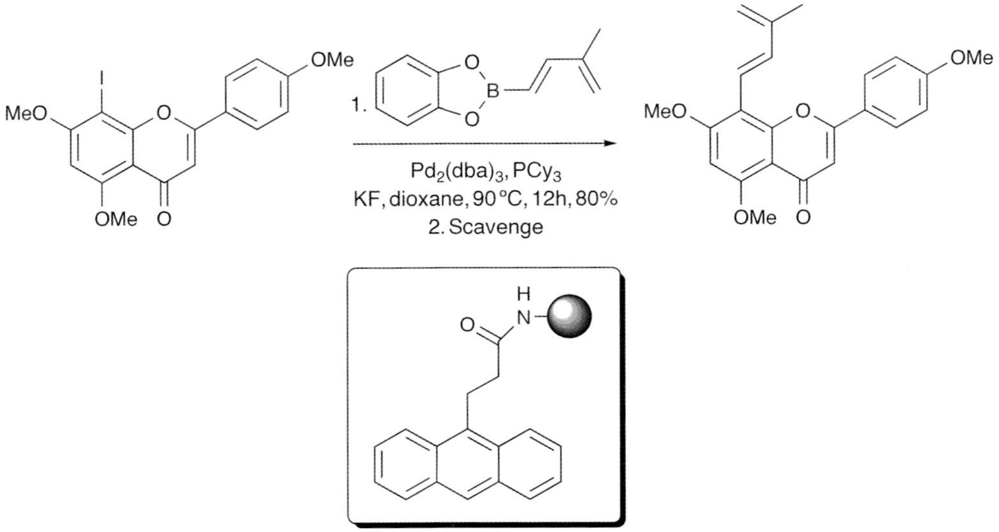

Scheme 8.35 Scavenging of excess dienophile from Diel-Alder reactions using a polymer-supported anthracene.

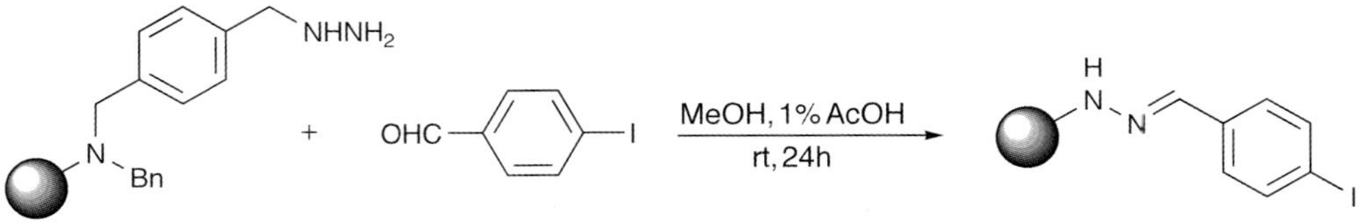

Scheme 8.36 Polymer-supported benzylhydrazines as a reversible scavenger resin for aromatic aldehydes.

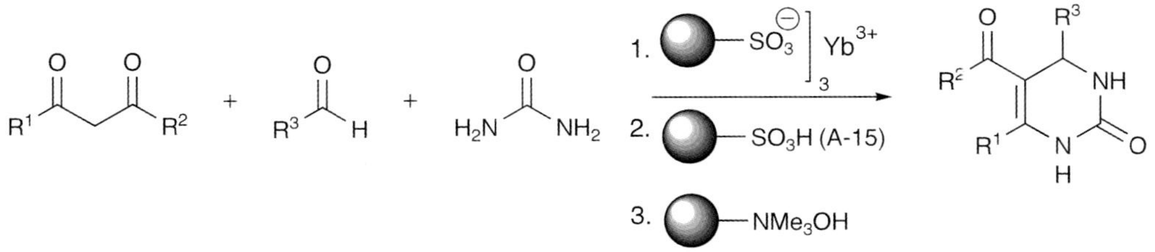

Scheme 8.37 Immobilized Amberlyst resins as the key purification step avoiding classical purification.

parallel synthesis was avoiding classical purification by utilizing a mixed-resin bed containing strongly acidic resin Amberlyst 15 (A-15) and the strongly basic resin Amberlyst 900 OH to scavenge excess urea and by-products.

Recently, Djuric, Sauer and coworkers described the application of a silica-bound *p*-toluene sulfonic acid (Si-SCX) cartridge for scavenging of DMAP used in the synthesis of acylsulfonamides [51]. Reactions were delivered to a packed SPE, allowing gravity elution to effectively remove the DMAP, providing the desired products in high purity by LC/MS (Scheme 8.38).

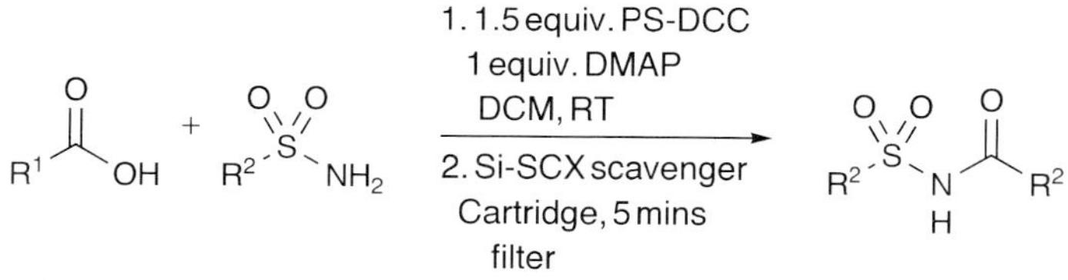

8.3 Supported/Tagged Reactive Functionalities for the Sequestration of Electrophiles | 211

Scheme 8.38 Scavenging of DMAP using a silica-bound p-toluene sulfonic acid (Si-SCX) cartridge.

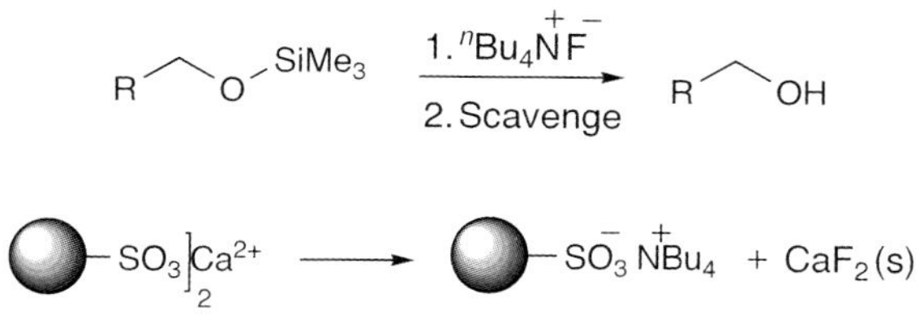

Scheme 8.39 Non-aqueous approach to the removal of tetrabutylammonium salts.

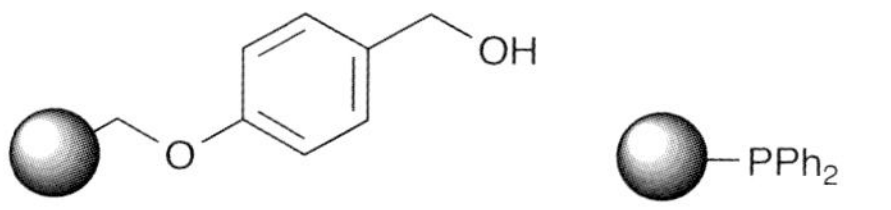

Figure 8.6 Resin-bound PPh$_3$ and benzyl alcohol scavengers.

Tetrabutylammonium salts can be difficult to remove from crude reactions without the use of an aqueous work-up. One non-aqueous approach to remove them was reported by Parlow, who applied a calcium sulfonate scavenger [52]. The resin reacts with excess tetrabutylammonium fluoride to afford insoluble calcium fluoride, which the acidic polymer protonates to afford the corresponding salt, thereby eliminating the need for standard aqueous work-up (Scheme 8.39).

8.3.1.7 Scavengers in "Tablets"

Anderson reported another novel methodology for application of resin-bound reagents and scavengers [53], which were prepared as tablets mixed with polystyrene beads. This methodology is applicable in both solution phase and solid phase protocols. The tablets can be formed in any shape that is desired. Results tentatively showed that the tablets could act as "sealed packages", embedding sensitive functional groups and protecting them from moisture and oxygen. A range of reagents and scavengers, such as resin-bound PPh$_3$ and benzyl alcohols, were effectively applied as scavengers (Figure 8.6). This mode of packaging also has a safety feature in that it reduces worker exposure to toxic, hazardous and dusty materials. Additionally, since the tablets do not have to be weighed, it significantly speeds up parallel solution and solid phase synthesis.

8.3.1.8 Magnetically-Tagged Nanoparticle for Impurity Removal

As previously discussed, the effective removal of palladium is a drawback in the use of this effective and selective reagent. More recently, Matos has reported a magnetically-tagged scavenger for the removal of palladium [54]. Here, the thiol component of the scavenger is coupled to magnetite (Fe_3O_4) nanoparticles that can be used as an effective scavenger (>99%) in either organic or aqueous media. The particles have approximate turnover numbers (TON) of 800 and have been demonstrated in at least five successive reactions.

8.3.1.9 Phase Switching–Bipyridyl Tagging Approaches to Scavengers

Ley reported the application of phase-switch purification for removal of scavengers in a catch and release strategy [55]. In this approach, a scavenger bearing a bipyridyl moiety is used to sequester excess reagent from the reaction. The scavenger is not immobilized on a resin, silica or on a fluorous tag. Instead it is a homogenous scavenger that is captured by the irreversible chelation of the bipyridyl tag using a resin-bound copper(II) species (Figure 8.7). This approach utilizes the benefits of both homogenous reaction kinetics and the ability to remove the scavenger in a heterogeneous method (Scheme 8.40). Furthermore, this approach has been expanded for scavenging of reagents tagged with a bipyridyl motif.

8.3.1.10 PEG-Supported Amine Scavenger Microgels

Building on the use of mesoporous silicate tags, Janda reported the application of a microgel bound scavengers, which are composed of cross-linked poly(styrene) prepared by radical polymerization [56]. Microgels are readily soluble in many organic solvents and eventually precipitated after scavenging of excess reagent. A tris(2-aminoethyl)amino microgel was applied as a scavenger of excess isocyanate, which was used in the formation of the urea (Scheme 8.41).

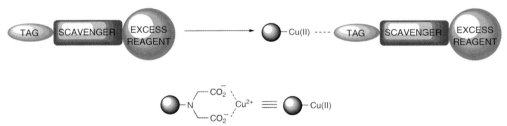

Figure 8.7 Phase-switch purification for removal of scavengers in a catch and release strategy.

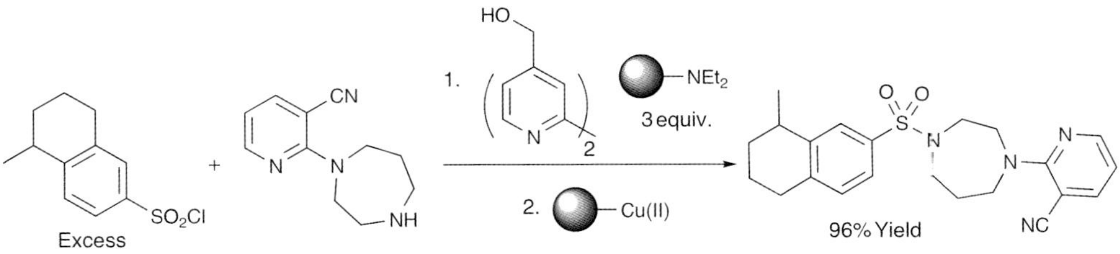

Scheme 8.40 Removal of scavengers in a catch-and-release strategy.

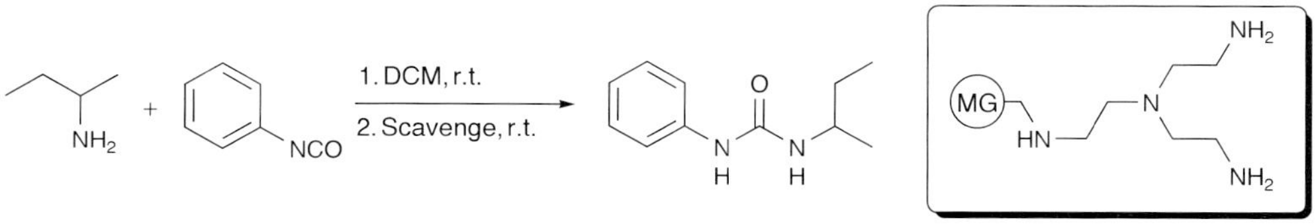

Scheme 8.41 Tris(2-aminoethyl)amino microgel as an efficient scavenger of excess isocyanate.

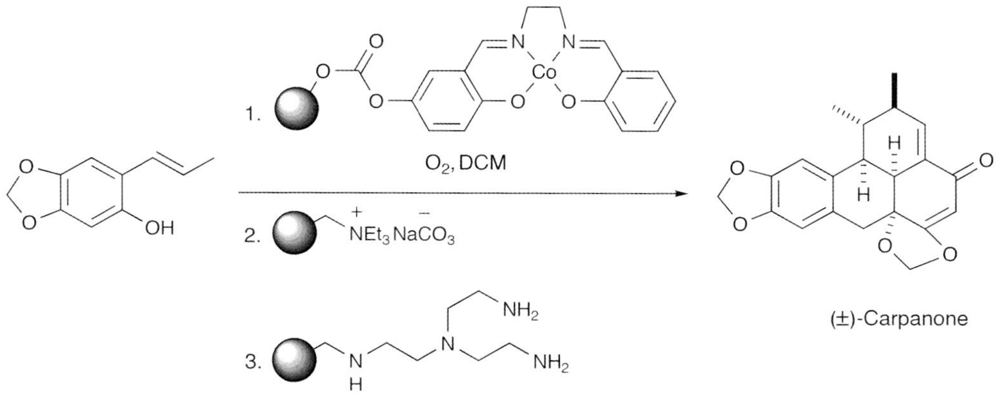

Scheme 8.42 Total synthesis of Carpanon utilizing resin-bound reagents and polymers.

8.3.1.11 Application of Polystyrene Resin-Supported Scavengers towards the Synthesis of Natural Products

Natural products and their derivatives have interested the pharmaceutical industry over the years, giving an abundant supply of drug leads and scaffolds. Classical approaches to the synthesis of natural products combine numerous steps with standard isolation, work-up and purification, which is inefficient and time consuming. One of the key validations of scavengers has been their application in total synthesis. Recent progress in combinatorial chemistry has led to an increase in reports of natural product synthesis using resin-bound reagents and scavengers.

In 2002, Ley reported the application of resin-bound reagents and polymers towards the synthesis of carpanone [57]. In the final steps towards carpanone, a resin-bound Co(salen) catalyst was used to give the desired intermediate along with the formation of a small amount of aldehyde by-product. To remove this by-product, a resin-bound tris-amine scavenger was used, yielding the desired product in high purity (Scheme 8.42).

A further example of the application of resin-bound reagents and scavengers was reported by Ley for the total synthesis of (–)-obliquine [58]. This general and straightforward synthesis led to the preparation of a range of naturally occurring amaryllidaceae alkaloids (Scheme 8.43). Furthermore, this approach allowed for the rapid synthesis of analogues in high yield and purity without the need to resort to extensive chromatography.

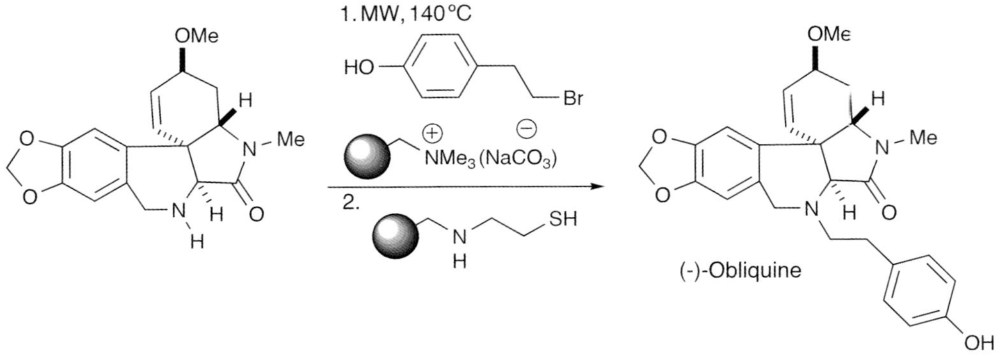

Scheme 8.43 Total synthesis of (−)-Obliquine utilizing resin-bound reagents and polymers.

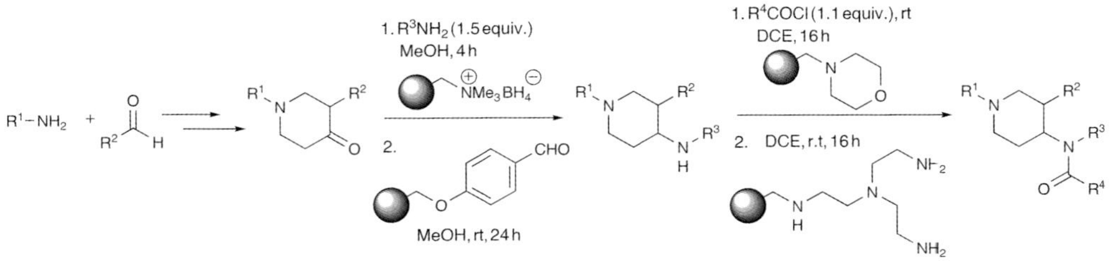

Scheme 8.44 Synthesis of an acylaminopiperdine library utilizing immobilized reagents and scavengers.

8.3.1.12 Application of Polystyrene Resin-Supported Scavengers towards Library Synthesis

Recently, there has been an increase in the amount of reports that demonstrate the application of both resin-bound reagents and scavengers towards libraries of drug-like molecules. The first full paper utilizing mainly polymer-supported reagents and scavengers was by Meppen for the synthesis of an acylaminopiperidine library [59]. An example of this process can be highlighted in the final stages, where a ketone undergoes reductive amination, which is scavenged with a polymer-bound aldehyde to remove excess amine (Scheme 8.44). In the final step, acylation is undertaken with excess reagent to push the reaction to completion, then removed using a resin-bound tris-amine. This sequence yields products with high levels of diversification and good purity over a six-step synthesis utilizing resin-supported chemistry.

In 2007, Aubé reported the application of the synthesis of oxazolines and dihydrooxazines libraries using resin-bound scavengers via two methods [60]. In the first method, a polymer-bound PPh$_3$ scavenger was used to scavenge excess hydroxyalkyl azide. Alternatively, a polymer-bound tosyl hydrazine was used to scavenge excess aldehyde (Scheme 8.45).

In a final example, Ladlow [61] reported an automated, polymer-assisted strategy for the synthesis of 2-alkylthiobenzimidaoles and N,N'-dialkylbenzimidazolin-2-ones (Scheme 8.46). This approach incorporates in-line purification utilizing resin-

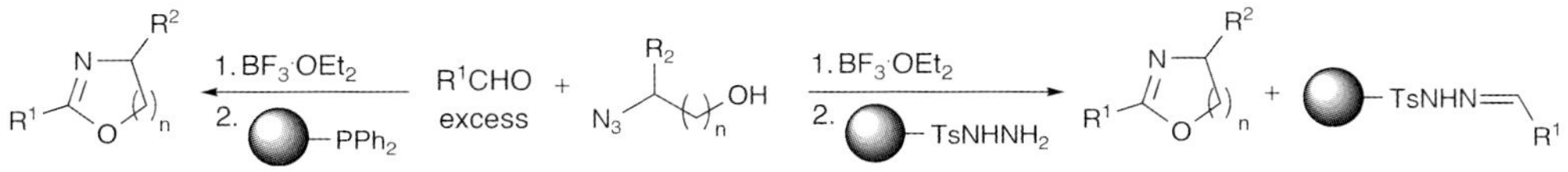

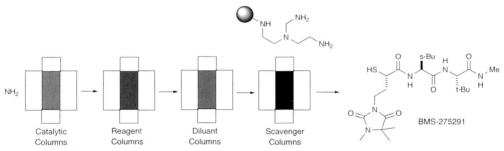

Scheme 8.45 Synthesis of dihydrooxazines libraries using resin-bound scavengers.

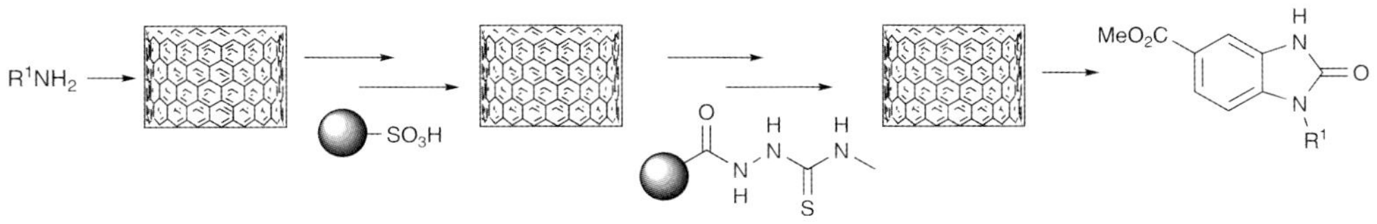

Scheme 8.46 In-line purification utilizing resin-bound scavengers towards library synthesis.

Figure 8.8 A solid phase "synthesis machine."

bound scavengers to synthesize a 96-member library of 2-alkylthiobenzimidaoles and a 72-member library N,N'-dialkylbenzimidazolin-2-ones.

8.3.1.13 Application of Polystyrene Resin-Supported Scavengers in a Flow-through Platform towards Combinatorial Libraries and Natural Product Synthesis

Continuous flow synthesis and its application to both library production and natural product synthesis is a rapidly growing area. Flow-through has many virtues, including (i) in-line monitoring for rapid reaction optimization, (ii) full automation potential, (iii) safety, (iv) continuous scale up and (v) integration with custom cartridges that allows for the use of solid-phase reagents and scavenging of reagents and by-products. Utilizing this technology, Lectka described the application of both resin-bound reagents and scavengers in a flow-through motif towards the synthesis of metalloproteinase inhibitor BMS-275291 [61]. In this solid phase "synthesis machine", a combination of cartridges is used, containing resin-bound catalyst, reagents, reaction beds and finally scavenger columns (Figure 8.8). In their method, a resin-bound trisamine was utilized to scavenge the Fmoc by-product that is removed during the process. This concept allows for the introduction of substrates, resulting in enhanced complexity via simple percolation through linked columns, ultimately affording desired product without the need for classical purification.

Ley reported a continuous flow platform for the preparation of a demonstration library of 5-amino-4-cyanopyrazoles and their conversion into 4-aminopyrazolopyrimidines [62]. In this method, they applied a novel flow microwave device connected to a down stream cartridge loaded with resin-supported amine to sequester excess 2-(ethoxymethylene)propanedinitrile (Scheme 8.47). The products were converted by a batch mode into structurally more complex pyrazolopyrimidine structures.

In 2006, Ley reported the application of a continuous flow-through platform for the preparation of the neolignan natural product Grossamide (Scheme 8.48) [63]. This was the first report of the enantioselective total synthesis of Grossamide utilizing a fully automated and scalable flow reactor with supported-reagents and scavengers. This principle has a wider impact on multiple step assembly of libraries and natural products.

Further examples have been reported by Ley in recent years, utilizing both designer cartridges of reagents and scavengers for both the synthesis of small libraries and natural products [2a, 64].

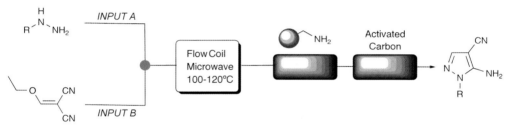

Scheme 8.47 Resin-supported amine in a continuous flow platform.

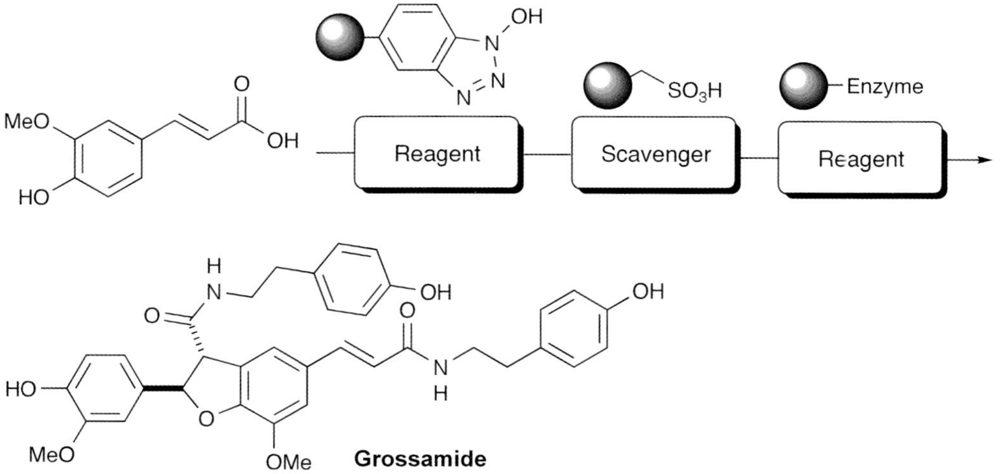

Scheme 8.48 Total synthesis of Grossamide utilizing a fully automated reactor with supported reagents and scavengers.

8.3.2
Silica-Supported Scavengers

8.3.2.1 Silica-Supported Amines

As stated above, silica tags have seen increased use in solid phase synthetic chemistry, with amine resins specifically being a major contributor to the development and advancement of this immobilized tag. Rousseau reported the application of high-load amino and diamino functionalized mesoporous silica as a highly effective scavenger of sulfonyl chlorides, aromatic isocyanides and acid chlorides [65]. Synthesis of this scavenger is based on the condensation of propyltrimethoxysilanes with tetraethoxysilane, leading to highly functionalized scavengers (Scheme 8.49). Unlike polystyrene-tagged amine scavengers, these Si-tagged scavengers have demonstrated a range of solvents while maintaining scavenging load.

Acylation and benzylation reactions are two of the most common reactions for diversification of core scaffolds. To counteract the use of excess acylating reagents, several scavengers have been developed and used in the facilitated synthesis of acylated compound libraries. In 2003, Khmelnitsky reported the efficient removal of excess acyl donors following enzymatic acylations [66]. Scavenging was accomplished with various commercially available amino-functionalized scavengers tagged on both silica and polystyrene resins (Figure 8.9). These scavengers can be

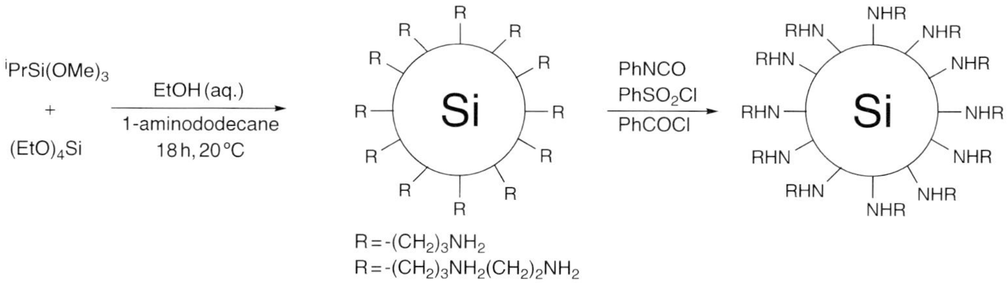

Scheme 8.49 Functionalized amino mesoporous silica as an effective scavenger of sulfonyl chlorides, aromatic isocyanides and acid chlorides.

Electrophiles Scavenged

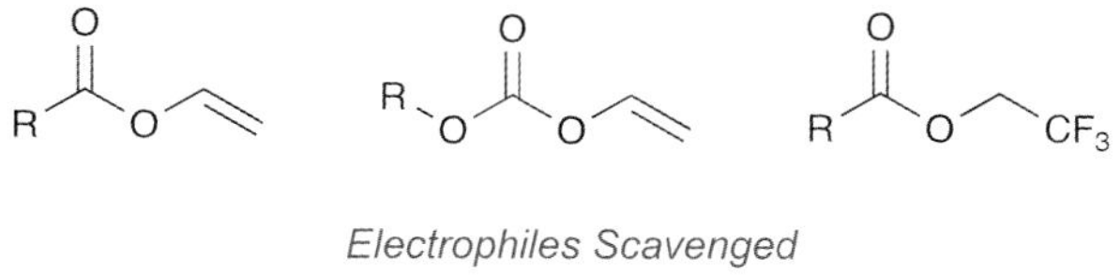

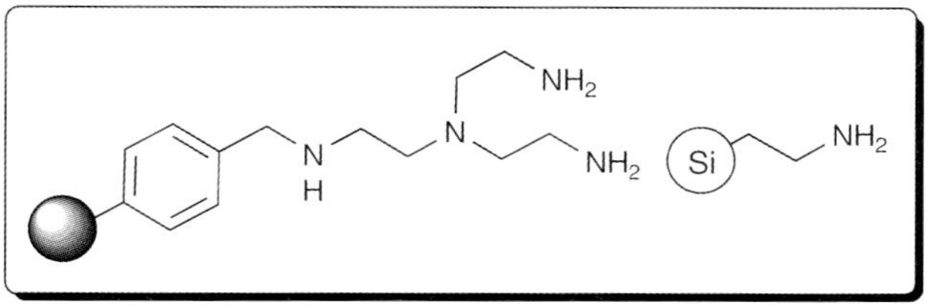

Figure 8.9 Scavengers used for removal of excess acyl donors.

applied to a wide variety of electrophilic donors, revealing a highly efficient method of acyl scavenging, affording the acylated product in high yield.

In 2006, Crudden demonstrated the application of a mesoporous silicate resin-bound amine scavenger for the removal of residual ruthenium catalyst [67]. Mesoporous silicates modified with aminopropyltriethoxy silane-derivatized silicates removed 99.99% of ruthenium in a single treatment. Evaluation of this scavenger against a more traditional polystyrene-bound and silica amine resin was reported. Results showed that overall the tagged mesoporous silicate scavenger removed more of ruthenium than did the corresponding silica and resin immobilized scavengers.

As discussed previously, a range of different supported scavengers have been utilized as an effective protocol to give highly pure compounds. In 2003, Kappe evaluated a range of amine-functionalized scavengers in the application of microwave-assisted scavenging towards acylated dihydropyrimidine libraries [68]. Although they all commonly have the same amine functionality, the mode of immobilization was investigated to evaluate the effect of the tag on overall scavenging ability. Scavenging of both excess Boc_2O and unwanted by-products was undertaken using polystyrene and silica supported diamines, aminomethylated-functionalized SynPhase Lanterns, and diethylenetriamine StratoSphere plugs (Scheme 8.50). During the synthesis of N3-functionalized dihydropyrimidines libraries, each scavenger was assessed for the removal of excess anhydride after microwave-assisted benzoylation. Kinetic evaluation showed that classical reagents, such as polymer supported and Si-bound scavengers, required less equivalents and shorter reaction times. More recent platforms such as the aforementioned StratoSphere plugs and SynPhase Lanterns required either more time and/or equivalents of scavenger.

8.3.2.2 Silica-Supported Thiols

In 2007, Wilson reported the application of highly functionalized, multidentate thiol-based Si-supported scavenger for the removal of palladium species from active pharmaceutical ingredients [69]. The removal of palladium from numerous types of reactions was demonstrated, including protective group removal, as well as removal from Suzuki, Heck and Buchwald–Hartwig coupling reactions (Scheme 8.51). Scavenging was achieved at room temperature to reduce levels of palladium to below 5 ppm (unoptimized). Furthermore, these Si-based scavengers had broad

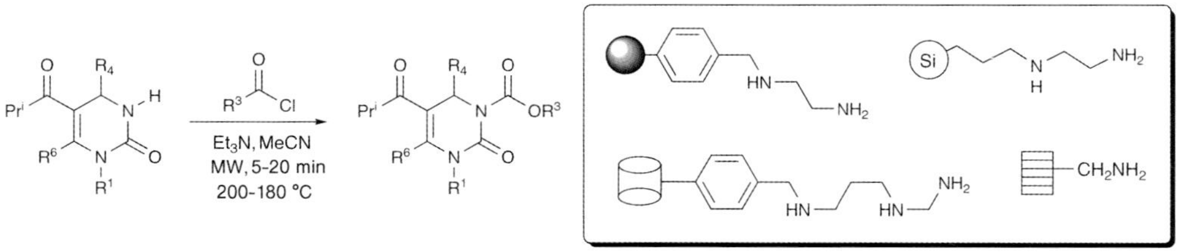

Scheme 8.50 Kinetic evaluation of polystyrene and silica supported diamines with SynPhase Lanterns and StratoSphere plugs.

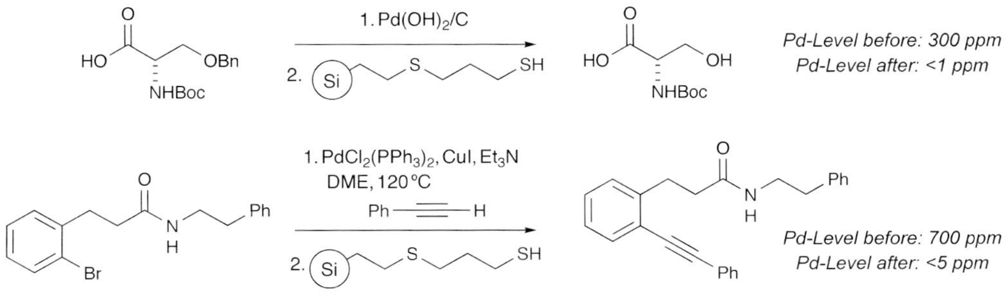

8.3 Supported/Tagged Reactive Functionalities for the Sequestration of Electrophiles | 219

Scheme 8.51 Highly functionalized, multidentate thiol-based Si-supported scavenger.

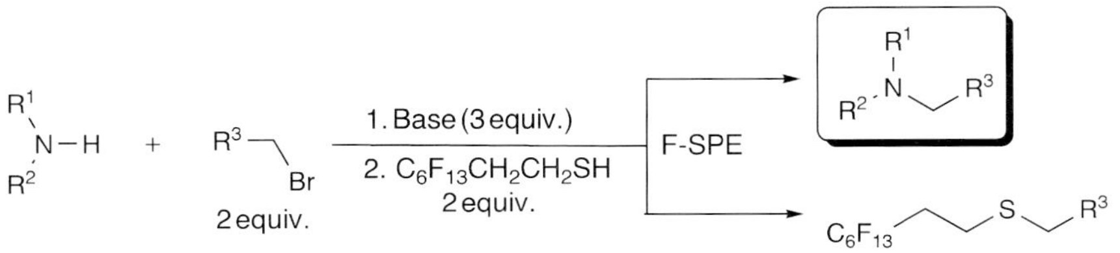

Scheme 8.52 Removal of alkylation reagents utilizing a fluorous-tagged thiol scavenger.

solvent compatibility, stability, ease of handling and application to large-scale synthesis.

8.3.3
Fluorous-Tagged Scavengers

8.3.3.1 Fluorous-Tagged Thiols

Like amine scavengers, Chen described the application of a commercially available fluorous-tagged thiol scavenger for the removal of alkylation reagents [70]. A range of tertiary amines was synthesized via alkylation using excess α-bromoketones/benzyl bromides (Scheme 8.52). The desired compounds were isolated by scavenging excess reagent, and product and scavenger were separated by solid phase extraction on fluorous silica gel.

Like Kappe in 2003 [68], Zhang in 2005 reported a kinetic study for the evaluation of fluorous-tagged scavengers with their polystyrene resin counterpart [71]. Kinetic studies were undertaken for both the fluorous-tagged thiol as a scavenger for α-bromoketones and the fluorous-tagged triphenylphosphine used in the bromination of alcohols (Scheme 8.53). When the reactions were carried out involving fluorous-tagged scavengers, the reactions were homogenous, providing solution-phase reaction kinetics. However, scavengers tagged with a solid-support were heterogeneous, and the reaction kinetics was greatly affected by the nature of the solid-support and the reaction environment. Overall, significantly greater amounts of scavenger and more time were required when using a solid-supported scavenger.

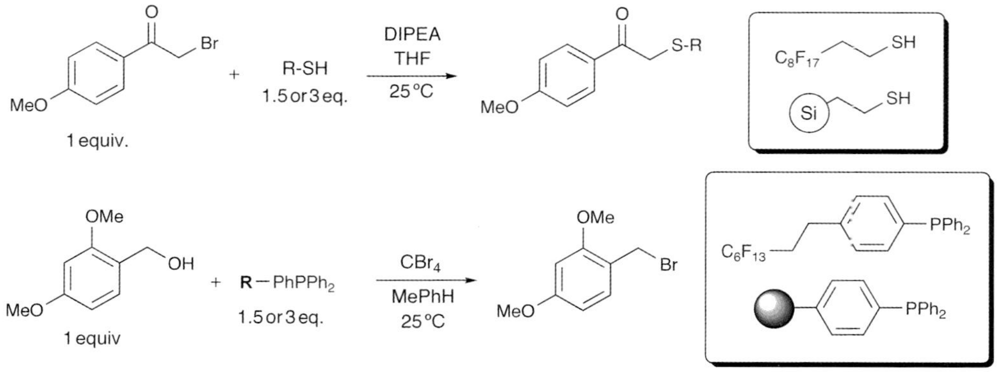

Scheme 8.53 Evaluation of fluorous-tagged scavengers with their polystyrene resin counterpart.

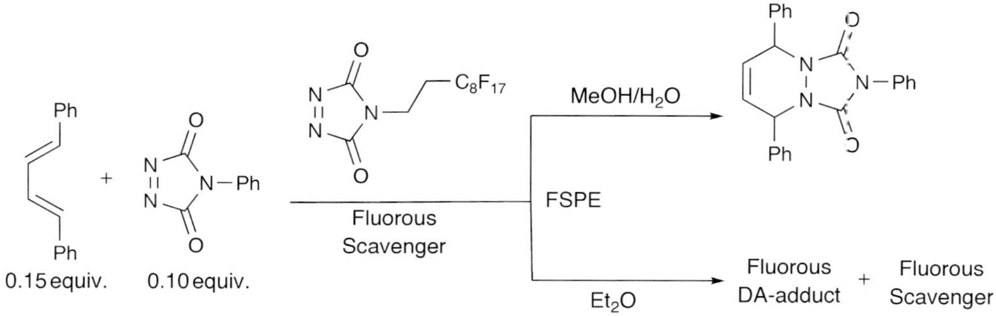

Scheme 8.54 Application of a fluorous-supported dienophile as a scavenger for dienes.

8.3.3.2 Fluorous-Tagged Dienophile

In recent years there has been a greater focus on the application of Diels–Alder reactions as a core protocol in synthesizing highly functional and stereocomplex molecules. However, to push a reaction to completion, the use of excess reagents is usually desirable, thus presenting a challenge in purification. Curran reported the first synthesis and application of a fluorous-supported dienophile, which were utilized as scavengers for dienes (Scheme 8.54) [72]. Scavenging was achieved in minutes under microwave conditions in which removal of the scavenger was achieved using a fluorous solid phase extraction (FSPE) workup.

8.3.4
Supported Ionic Platforms for Scavenging Electrophiles

8.3.4.1 Task Specific Ionic Liquid Amine

An alternative approach to scavenging that does not utilize poly(ethylene glycol) (PEG), silica or fluorous tags is ionic liquids. One such account was the application of quaternary ammonium salts bearing amino functionality reported by Stien in 2002 [73]. These high-load scavengers were utilized in scavenging excess electrophiles such as Boc$_2$O, tosyl chloride and phenyl isocyanate (Scheme 8.55). In these cases, an aqueous solution of the scavenger was added to the crude reaction, which sequestered excess Boc$_2$O and trafficked it into the aqueous phase to provide the desired product in high purity.

Stien also reported the synthesis and application of quaternary ammonium-supported reagents (TAMA-Cl and BAX-sulfate) for scavenging acids and excess electrophiles [74]. BAX-sulfate is a highly crystalline scavenger reagent, which reacts with electrophiles (Boc$_2$O, TsCl, PhNCO), quenches acids and precipitates quantitatively with the addition of diethyl ether. After scavenging with BAX-sulfate, the workup stage requires simple filtration. For example, protection of benzylamine gave a quantitative yield of product but the crude mixture retained 0.48 equivalent of Boc$_2$O. After scavenging with BAX-sulfate no excess reagent was detected, affording the desired product in high purity (Scheme 8.56).

Another approach to ionic scavengers was reported by Song, who demonstrated the application of an imidazole-based ionic liquid for the removal of excess reagents such as benzyl chloride, methanesulfonyl chloride, aniline and p-toluidine (Scheme 8.57) [75]. The advantages of this method compared to that of resin-bound scavengers include lower amounts of scavenger and rapid scavenging (30–120 min), thus demonstrating the potential of ionic liquid scavengers in combinatorial chemistry.

Wang reported an example of a water soluble, ionic scavenger for the sequestration of excess acid chlorides in the synthesis of benzamides and sulfonamides [76].

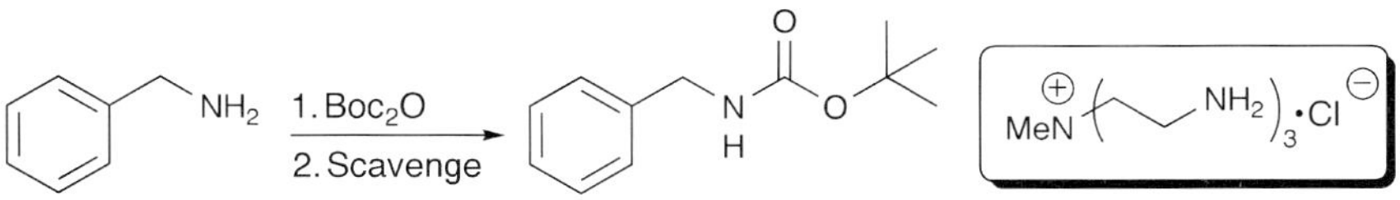

Scheme 8.55 Scavenging excess electrophiles with quaternary ammonium salts bearing amino functionality.

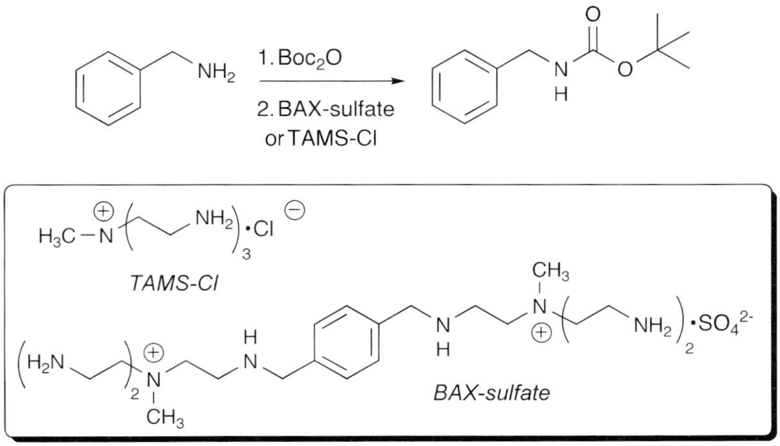

Scheme 8.56 Application of TAMA-Cl and BAX-sulfate as scavengers of excess boc anhydride.

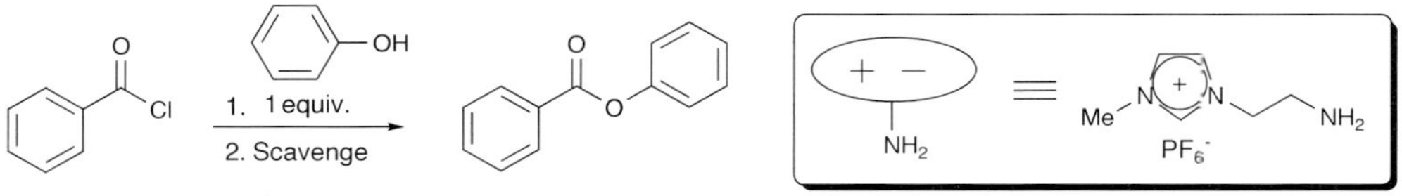

Scheme 8.57 Imidazole-based ionic liquid for the removal of excess benzyl chloride.

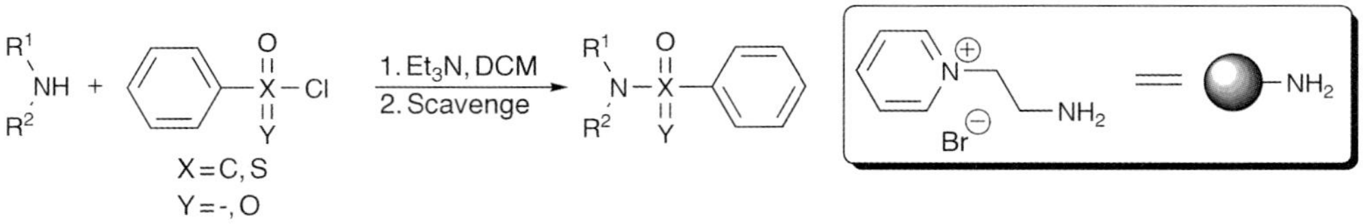

Scheme 8.58 Sequestration of excess acid chlorides using a water soluble, ionic scavenger.

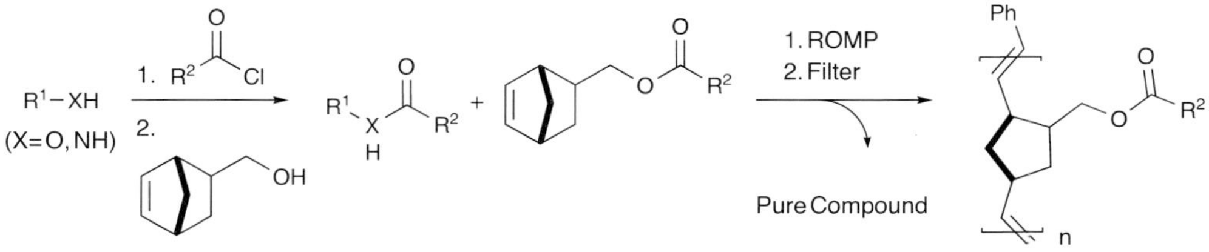

Scheme 8.59 Scavenging of excess acid chlorides using a norbornenyl-tagged alcohol in a Capture-ROMP-Filter protocol.

This inexpensive scavenger can be prepared on a large scale and effectively eliminates excess acyl chlorides by simple aqueous extraction (Scheme 8.58). The desired compounds were isolated in high yield and purity, making this a versatile reagent for combinatorial solution-phase synthesis of amide libraries.

8.3.5
ROMP-Derived Nucleophilic Scavengers

In 2002, Hanson demonstrated that the timing of the metathesis event can be exploited in scavenging. In this work, a norbornenyl-tagged alcohol was utilized as a scavenger to remove excess acid chlorides from the crude reaction [77]. Upon completion of the scavenging event, the captured electrophile was subjected to ROM polymerization. Subsequent removal of the polymer via filtration afforded the desired product in high purity (Scheme 8.59).

More recently, three different ROMP-derived PPh_3 reagents have been reported, including (i) a high load ROMP-gel PPh_3 by Barrett [78], (ii) a ROMP-derived soluble oligomeric triphenylphosphine (OTPP) by Hanson [79] and (iii) a ROMP-derived bis-phosphine oligomer PPh_3 by Luh [80]. Although none of these reports indicate their scavenging ability, all undoubtedly have potential as high-load, scavengers of electrophiles.

8.4
Conclusion

In this chapter we have discussed and summarized recent developments and applications of immobilized scavengers in organic synthesis. Building on earlier work using polystyrene resins, a new array of tagged and immobilized reagents/scavengers has emerged. Their application in facilitated protocols aimed at drug discovery, as well as their use in natural product synthesis, points to a very bright future. Despite these successes, continued effort is needed to integrate and improve this technology into current and future platforms. In particular, progress in flow-through technology, coupled with recent advances made in the development of immobilized scavengers and reagents, allow for new and exciting possibilities for accelerating drug discovery.

References

1 (a) Regen, S.L. and Lee, D.P. (1975) *Journal of Organic Chemistry*, **40**, 1669–70. (b) Kaldor, S.W., Fritz, J.E., Tang, J. and McKinney, E.R. (1996) *Bioorganic and Medicinal Chemistry Letters*, **6**, 3041–4. (c) Kaldor, S.W., Siegel, M.G. and Fritz, J.E. (1996) *Tetrahedron Letters*, **37**, 7193–6. (d) Flynn, D.L., Crich, J.Z., Devraj, R.V., *et al.* (1997) *Journal of the American Chemical Society*, **119**, 4874–81. (e) Booth, R.J. and Hodges, J.C. (1997) *Journal of the American Chemical Society*, **119**, 4882–6. (f) Shuttleworth, S.J., Allin, S.M. and Sharma, P.K. (1997) *Synthesis*, 1217–39. (g) Booth, R.J. and Hodges, J.C. (1999) *Accounts of Chemical Research*, **32**, 18–26. (h) Ley, S.V., Baxendale, I.R., Bream, R.N., *et al.* (2000) *Journal of the Chemical Society–Perkin Transactions 1*, 3815–4195. (i) Monenschein, A., Kirschning, H. and Wittenberg, R. (2001) *Angewandte Chemie–International Edition*, **40**, 650–79. (j) Eames, J. and Watkinson, M. (2001) *European Journal of Organic Chemistry*, 1213–24. (k) Yoshida, J.-I. and Itami, K. (2002) Tag strategy for separation and recovery. *Chemical Reviews*, **102**, 3693–716. (l) van Heerbeek, R., Kamer, P.C.J. and van Leeuwen, P.W.N.M. and Reek, J.N.H. (2002) *Chemical Reviews*, **102**, 3717–56.

2 (a) Baxendale, I.R., Brusotti, G., Matsuoka, M. and Ley, S.V. (2002) *Journal of the Chemical Society–Perkin Transactions 1*, 143–54. (b) Storer, R.I., Takemoto, T., Jackson, P.S., *et al.* (2004) *Chemistry–A European Journal*, **10**, 2529–47. (c) Baxendale, I.R., Ley, S.V. and Piutti, C. (2002) *Angewandte Chemie–International Edition*, **41**, 2194–7. (d) Parlow, J.J. and Flynn, D.L. (1998) *Tetrahedron*, **54**, 4013–31.

3 Letsinger, R.L. and Kornet, M.J. (1963) *Journal of the American Chemical Society*, **85**, 3045–6.

4 Merrifield, R.B. (1963) *Journal of the American Chemical Society*, **85**, 2149–54.

5 (a) Galaffu, N., Sechi, G. and Bradley, M. (2005) *Molecular Diversity*, **9**, 263–75. (b) Galaffu, N. and Bradley, M. (2005) *Tetrahedron Letters*, **46**, 859–61.

6 Baindur, N., Chadha, N. and Player, M.R. (2003) *Journal of Combinatorial Chemistry*, **5**, 653–9.

7 Lee, A. and Ellman, J.A. (2003) *Organic Letters*, **3**, 3707–9.

8 Gregg, B.T., Tymoshenko, D.O., Razzano, D.A. and Johnson, M.R. (2007) *Journal of Combinatorial Chemistry*, **9**, 507–12.

9 Fournier, D., Pascual, S., Montembault, V., Haddleton, D.M. and Fontaine, L. (2006) *Journal of Combinatorial Chemistry*, **8**, 522–30.

10 Tripp, J.A., Stein, J.A., Svec, F. and Fréchet, J.M.J. (1999) *Organic Letters*, **2**, 195–8.

11 (a) Kalder, S.W., Siegel, M.G., Fritz, J.E., Dressman, B.A., Hahn, P. *Tetrahedron Lett.* **37**, 1996, 7193–7196. (b) Kobayashi, S. and Aoki, Y. (1998) *Tetrahedron Letters*, **39**, 7345.

12 Guinó, M., Brulé, E. and de Miguel, Y.R. (2003) *Journal of Combinatorial Chemistry*, **5**, 161–5.

13 Wang, Z., Luo, J., Zhu, X.X., Jin, S and Tomaszewski, M.J. (2004) *Journal of Combinatorial Chemistry*, **6**, 961–6.

14 Schön, U., Messinger, J., Merayo, N., et al. (2003) *Synlett*, **7**, 983–6.

15 Hoveyda, H.R. and Hall, D.G. (2001) *Organic Letters*, **3**, 3491–4.

16 Barrett, A.G.M., Boffey, R.J., Frederiksen, M.U., et al. (2001) *Tetrahedron Letters*, **42**, 5579.

17 Wang, X., Parlow, J.J. and Porco, J.A. (2000) *Organic Letters*, **2**, 3509.

18 Dahmen, S. and Bräse, S. (2000) *Angewandte Chemie – International Edition*, **39**, 3681–3.

19 Lazny, R., Nodzewska, A. and Klosowski, P. (2004) *Tetrahedron*, **60**, 121–30.

20 Lipshutz, B.H. and Blomgren, P.A. (2001) *Organic Letters*, **3**, 1869–71.

21 For reviews on soluble polymers, see: (a) Gravert, D.J. and Janda, K.D. (1997) *Chemical Reviews*, **97**, 489–509. (b) Toy, P.H. and Janda, K.D. (2000) *Accounts of Chemical Research*, **33**, 546–54. (c) Dickerson, T.J., Reed, N.N. and Janda, K.D. (2002) *Chemical Reviews*, **102**, 3325–44. (d) Haag, R. (2001) *Chemistry – A European Journal*, **7**, 327–35. (e) Haag, R., Sunder, A., Hebel, A. and Roller, S. (2002) *Journal of Combinatorial Chemistry*, **4**, 112–19. (f) Bergbreiter, D.E. (2002) *Chemical Reviews*, **102**, 3345–84.

22 (a) Selected references: Sandler, S.R. (1970) *Journal of Organic Chemistry*, **55**, 3967–8. (b) Venkataraman, K. and Wagle, D.R. (1979) *Tetrahedron Letters*, **20**, 3037–40. (c) Bandgar, B.P. and Pandit, S.S. (2002) *Tetrahedron Letters*, **43**, 3413–14. (d) Venkataraman, K. and Wagle, D.R. (1980) *Tetrahedron Letters*, **21**, 1893–6. (e) Gold, H. (1960) *Angewandte Chemie*, **72**, 956–9. (f) Chakrabarti, J.K. and Hotten, T.M. (1972) *Journal of the Chemical Society D – Chemical Communications*, 1226–7.

23 Masala, S. and Taddei, M. (1999) *Organic Letters*, **1**, 1355–7.

24 Falchi, A. and Taddei, M. (2000) *Organic Letters*, **2**, 2663–6.

25 http://www.silicycle.com (accessed 10/2007).

26 Tomefte, R.S. and Woodward, S. (2004) *Tetrahedron Letters*, **45**, 39–42.

27 Pattarawarapan, M. and Singhatana, S. (2006) *Chiang Mai Journal of Science*, **33**, 203–9.

28 (a) Studer, A., Hadida, S., Ferritto, R., Kim, S.-Y., Jeger, P., Wipf, P. and Curran, D.P. (1997) *Science*, **275**, 823–6. (b) Curran, D.P. (1999) *Medicinal Research Reviews*, **19**, 432–8. (c) Curran, D.P. (2000) *Pure and Applied Chemistry*, **72**, 1649–53.(d) Gladysz, J.A., Curran, D.P. and Horváth, I.T. (2004) *Handbook of Fluorous Chemistry*, Wiley-VCH Verlag GmbH, Weinheim.

29 Zhang, W. (2003) *Tetrahedron*, **59**, 4475–89.

30 Zhang, W., Chen, C.H.-T. and Nagashima, D. (2003) *Tetrahedron Letters*, **44**, 2065–8.

31 Zhang, A.S., Elmore, C.S., Egan, M.A., Melillo, D.G. and Dean, D.C. (2005) *Journal of Labelled Compounds and Radiopharmaceuticals*, **48**, 203–8.

32 Lindsley, C.W., Zhao, Z. and Leister, W.H. (2002) *Tetrahedron Letters*, **43**, 4225–8.

33 Villard, A.-L., Warrington, B.H. and Ladlow, M. (2004) *Journal of Combinatorial Chemistry*, **6**, 611.

34 Lu, Y. and Zhang, W. (2006) *QSAR and Combinatorial Science*, **25** (*8*), 728–31.

35 Barrett, A.G.M., Hopkins, B.T. and Köbberling, J. (2002) *Chemical Reviews*, **102**, 3301–24.

36 For the use of both RCM and ROMP, in Combinatorial Chemistry, see: (a) Harned, A.M., Probst, D.A. and Hanson, P.R. (2003) *The use of Olefin Metathesis in Combinational Chemistry: Supported and Chromatography – Tree Synthesis*, in *Handbook of Metathesis*, Vol. 2 (ed. R.H. Grubbs), Wiley-VCH Verlag GmbH, Weinheim, pp. 361–402.(b) Flynn, D.L., Hanson, P.R., Berk, S.C. and Makara, G.M. (2002) *Current Opinion in Drug Design and Development*, **5**, 571–9.

37 (a) Nguyen, S.T. and Trnka, T.M. (2003) The discovery and development of well-defined, ruthenium-based olefin metathesis catalysts, in *Handbook of Metathesis*, Vol. **1** (ed. R.H. Grubbs),

Wiley-VCH Verlag GmbH, New York, pp. 61–85.(b) Schwab, P., France, M.B., Ziller, J.W. and Grubbs, R.H. (1995) *Angewandte Chemie – International Edition in English*, **34**, 2039–41. (c) Wu, Z., Nguyen, S.T., Grubbs, R.H. and Ziller, J.W. (1995) *Journal of the American Chemical Society*, **117**, 5503–11. (d) Schwab, P., Grubbs, R.H. and Ziller, J.W. (1996) *Journal of the American Chemical Society*, **118**, 100–10. (e) Scholl, M., Ding, S., Lee, C.W. and Grubbs, R.H. (1999) *Organic Letters*, **1**, 953–6.

38 Arnauld, T., Barrett, A.G.M., Cramp, S.M., et al. (2000) *Organic Letters*, **2**, 2663–6.

39 Moore, J.D., Byrne, R.J., Vedantham, P., et al. (2003) *Organic Letters*, **5**, 4241–4.

40 Zhang, M., Moore, J.D., Flynn, D.L. and Hanson, P.R. (2004) *Organic Letters*, **6**, 2657–60.

41 Herpel, R.H., Vedantham, P., Flynn, D.L. and Hanson, P.R. (2006) *Tetrahedron Letters*, **47**, 6429–32.

42 Zhang, J., Zhang, L., Zhang, S., et al. (2005) *Journal of Combinatorial Chemistry*, **7**, 657–64.

43 Urawa, Y., Miyazawa, M., Ozeki, N. and Ogura, K. (2003) *Organic Process Research and Development*, **7**, 191–5.

44 Sheppeck, J.E.II , Kar, H. and Hong, H. (2000) *Tetrahedron Letters*, **41**, 5329–33.

45 (a) Caldarelli, M., Habermann, J. and Ley, S.V. (1999) *Bioorganic and Medicinal Chemistry Letters*, **9**, 2049–52. (b) Storer, R.I., Takemoto, T., Jackson, P.S. and Ley, S.V. (2003) *Angewandte Chemie – International Edition*, **42**, 2521–5.

46 Guino, M. and Hii, K.K. (2005) *Tetrahedron Letters*, **46**, 6911–13.

47 Westhus, M., Gonthier, E., Brohm, D. and Breinbauer, R. (2004) *Tetrahedron Letters*, **45**, 3141–2.

48 Lei, X. and Porco, J.A. Jr. (2004) *Organic Letters*, **6**, 795–8.

49 Zhu, M., Ruijter, E. and Wessjohann, L.A. (2004) *Organic Letters*, **6**, 3921–4.

50 Dondoni, A. and Massi, A. (2001) *Tetrahedron Letters*, **43**, 7975–8.

51 Wang, Y., Sarris, K., Sauer, D.R. and Djuric, S.W. (2007) *Tetrahedron Letters*, **48**, 5181–4.

52 Parlow, J.J., Vazquez, M.L. and Flynn, D.L. (1998) *Bioorganic and Medicinal Chemistry Letters*, **8**, 2385.

53 Ruhland, T., Holm, P. and Anderson, K. (2003) *Journal of Combinatorial Chemistry*, **5**, 842–50.

54 Rossi, L.M., Vono, L.R.L., Silva, P.F., et al. (2007) *Applied Catalysis A: General*, **330**, 139–44.

55 Siu, J., Baxendale, I.R., Lewthwaite, R.A. and Ley, S.V. (2005) *Organic and Biomolecular Chemistry*, **3**, 3140–60.

56 Shimomura, O., Clapham, B., Spanka, C., et al. (2002) *Journal of Combinatorial Chemistry*, **4**, 436–41.

57 Baxendale, I.R., Lee, A.-L. and Ley, S.V. (2002) *Journal of the Chemical Society – Perkin Transactions 1*, 1850–7.

58 Baxendale, I.R. and Ley, S. (2005) *Industrial and Engineering Chemistry Research*, **44**, 8588–92.

59 Creswell, M.W., Bolton, G.L., Hodges, J.C. and Meppen, M. (1998) *Tetrahedron*, **16**, 3983–98.

60 Chaudhy, P., Schoenen, F., Neuenswander, B., et al. (2007) *Journal of Combinatorial Chemistry*, **9**, 473–6.

61 (a)France, S., Bernstein, D., Weatherwax, A. and Lectka, T. (2005) *Organic Letters*, **7**, 3009–12. (b) Vickerstaffe, E., Warrington, B.H., Ladlow, M. (2005) *J. Comb. Chem.*, **7**, 385–397.

62 Smith, C.J., Iglesias-Sigüenza, F.J., Baxendale, I.R. and Ley, S.V. (2007) *Organic and Biomolecular Chemistry*, **5**, 2758–61.

63 Baxendale, I.R., Griffiths-Jones, C.M., Ley, S.V. and Tranmer, G.K. (2006) *Synlett*, **3**, 427–30.

64 (a) Storer, R.I., Takemoto, T., Jackson, P.S., et al. (2004) *Chemistry – A European Journal*, **10**, 2529–47. (b) Storer, R.I., Takemoto, T., Jackson, P.S. and Ley, S.V. (2003) *Angewandte Chemie – International Edition*, **42**, 2521. (c) Vickerstaffe, E., Warrington, B.H., Ladlow, M. and Ley, S.V. (2004) *Journal of Combinatorial Chemistry*, **6**, 332–9.

65 Macquarrie, D.J. and Rousseau, H. (2003) *Synlett*, **2**, 244–6.

66 Usyatinsky, A.Y., Astakhova, N.N. and Khmelnirsky, Y.L. (2003) *Biotechnology and Bioengineering*, **82**, 379–85.

67 McEleney, K., Allen, D.P., Holiday, A.E. and Crudden, C.M. (2006) *Organic Letters*, **8**, 2663–6.

68 Dallinger, D., Yu, N. and Kappe, C.O. (2003) *Molecular Diversity*, **7**, 229–45.

69 Galaffu, N., Man, S.P., Wilkes, R.D. and Wilson, J.R.H. (2007) *Organic Process Research and Development*, **11**, 406–13.
70 Zhang, W., Curran, D.P. and Chen, C.H.-T. (2002) *Tetrahedron*, **58**, 3871.
71 Chen, H.-T.C. and Zhang, W. (2005) *Molecular Diversity*, **9**, 353–9.
72 Werner, S. and Curran, D.P. (2003) *Organic Letters*, **5**, 3293–6.
73 Ghanem, N., Martinez, J. and Stien, D. (2002) *Tetrahedron Letters*, **43**, 1692–5.
74 Ghanem, N., Martinez, J. and Stien, D. (2004) *European Journal of Organic Chemistry*, 84–9.
75 Song, G., Cai, Y. and Peng, Y. (2005) *Journal of Combinatorial Chemistry*, **7**, 561–6.
76 Lei, M., Tao, X.-L. and Wang, Y.-G. (2006) *Helvetica Chimica Acta*, **89**, 532–6.
77 Moore, J.D., Harned, A.M., Henle, J., *et al.* (2002) *Organic Letters*, **4**, 1847–9.
78 Barrett, E., Årstad, A.G.M., Hopkins, B.T. and Kübberling, J. (2002) *Organic Letters*, **4**, 1975–7.
79 Harned, A.M., He, H.S., Toy, P.H., Flynn, D.L. and Hanson, P.R. (2004) *Journal of the American Chemical Society*, **127**, 52–3.
80 Yang, Y.-C. and Luh, T.-Y. (2003) *Journal of Organic Chemistry*, **68**, 9870–3.

9
Metal Scavengers

Aubrey Mendonca

ChemRoutes Corporation, 9719-42 Avenue, Edmonton, Alberta, Canada T6E 5P8

9.1
Background

Metal ions play a key role in pharmaceutical drug discoveries. In pharmaceutical research, asymmetric reactions are quite popular for the synthesis of medicinal compounds. Metals are being used as catalysts for most asymmetric reactions to obtain high selectivity, productivity and stability in achieving a desired product effectively and efficiently. When using metal ions as catalysts in a reaction mixture, we often end up with these metals as impurities in the final product. Sometimes it is difficult to remove these metal impurities from the final product. This is a critically important task in pharmaceutical process research, where final products must meet stringent purity requirements. With the growing use of organometallic reagents and catalysts in pharmaceutical synthesis, the removal of trace metal impurities from reaction mixture becomes important. In recent years, several academic and industrial laboratories have become actively involved in developing silica and polymer based polyfunctional resins to scavenge different metal ions from the reaction mixture entirely or to bring them to an acceptable ppm range.

In contrast, metal ions are present in many essential nutrients. They are being increasingly used as prevalent components of diagnostic or therapeutic agents to treat or study various diseases and metabolic disorders [1, 2]. Recently, metal ions have played an important role in diagnosis through photodynamic therapy (PDT) and magnetic resonance imaging (MRI) and in human pathology therapies. The list of such useful metal ions is continuing to grow; it includes commonly known metals such as zinc, copper and manganese and also some metals that were thought to be poisonous, such as selenium and molybdenum [3].

Heavy metals are also present in the environment, having been introduced through several industrial processes [4]. Their presence in excess can be harmful to humans and can have a negative impact on the environment. Heavy metals are non-degradable, and so they are accumulated in the environment, [5, 6] mostly through industrial wastes. Since the accumulation of many toxic chemicals

The Power of Functional Resins in Organic Synthesis. Judit Tulla-Puche and Fernando Albericio
Copyright © 2008 WILEY-VCH Verlag GmbH & Co. KGaA, Weinheim
ISBN: 978-3-527-31936-7

threatens public health, research and development has increased to remedy this environmental situation. In the United States alone, fuel and power industries produce 2.4 million tons of heavy metal waste, while agriculture and waste disposal produce another 2 million tons per year. These metals can have a large environmental, health, and economic impact [7]. The US National Priorities List (NPL) lists more than 60% of sites of heavy metals as containing contaminants of concern. Many of these metals, such as arsenic, lead, mercury and cadmium, rank high on the EPA (Environmental Protection Agency) priority list.

9.2
Medicinal Chemistry Uses of Metals

At the interface between medicine and inorganic chemistry is emerging a branch of chemistry called medicinal inorganic chemistry. This is a fairly new branch of bioinorganic chemistry, which is growing exponentially and needs much attention. Medicinal inorganic chemistry covers metal-based drugs, metal-containing diagnostic aids, and the medicinal recruitment of endogenous metal ions.

Traditional medical practitioners have been using metals in their medical practice for centuries. The benefits of precious metals, such as gold and silver, attracted ancient Chinese, Egyptian, Greek and Indian healers to use them in cures of various sorts. Copper and iron have also been widely used in metal-based therapies. There has always been a curious connection between the discovery of a new precious element and its quick movement into medicinal practice [8]. Recent uses of transition metal ions in photodynamic therapy (PDT) and magnetic resonance imaging (MRI) are quite successful. Most notably are platinum and gadolinium complexes, which are used in the treatment of testicular cancer and in magnetic resonance imaging (MRI) techniques.

9.2.1
New Chemistries

Asymmetric metal catalysis is a great tool in pharmaceutical research and development, which is generating tremendous academic and industrial interest. During a metal catalytic step in a synthetic route, several critical considerations should be kept in mind, such as material availability, speed of development, ease of implementation, access to the technology and cost. In short, asymmetric metal catalysis must be competitive against alternative methods for producing a chiral target. All this is balanced against performance factors, such as activity, selectivity, productivity and stability in achieving a desired transformation effectively and efficiently. Although asymmetric metal catalysis has great advantages it carries with it the problem of having to deal with residual metals. As the use of organometallic catalysts in synthesizing chiral and non-chiral compounds has been increasing, there is a greater need for chemists to find ways to remove metal-related impurities. Adding to the problem is even more severe regulatory restrictions on the level of metals allowed in pharmaceutical products. Routine testing with a validated method is required for specific metal residues from catalysts. This testing is in addition to regulatory requirements that may specify general heavy-metal tests relating to overall production quality, including metal from all contamination sources. Catalysis is now such a major feature in pharmaceutical production that trace-metal reduction to low parts-per-million levels is an increasingly important challenge for the industry. Specifically, transition-metal catalysts are being explored in various reactions. On the positive side, the catalyst use is developed and optimized; the performance improves and requires decreasing loadings. Nevertheless, chemists still may find dealing with residual metals to be a stumbling block in developing new synthetic routes.

9.2.2
Examples of Drugs Involving Use of Metals as Catalysts

As discussed earlier, catalysis is an important and versatile tool for chemical synthesis and is gaining importance in drug development. Heterogeneous or homogenous catalysts that contain transition metals are often used in chemical reactions. A few years ago, Merck established its catalysis laboratory with a simple aim, namely, the early and rapid implementation of efficient catalytic processes in API synthesis. The company replaced *ad hoc* efforts with a centralized catalysis laboratory that works closely with the chemists who design overall synthetic routes. The laboratory takes advantage of automated and high-throughput experimentation.

Asymmetric hydrogenation was chosen as an initial emphasis. Despite being the most powerful form of asymmetric catalysis, there are actually a very limited number of asymmetric hydrogenations being used in drug manufacturing. Recently, rhodium-catalyzed hydrogenation of an unprotected enamine using a ferrocenyl phosphine Josiphos-type ligand, a key step, has been employed for the process development for new drugs used for the treatment of diabetes. Similarly, a ruthenium-based catalysis has been used in asymmetric hydrogenation of N-sulfonylated-α-dehydroamino acids [9]. This is a highly enantioselective reaction in the synthesis of an anthrax lethal factor inhibitor, streamlining the synthetic process by avoiding lengthy protecting and deprotecting steps.

Another example is from Pfizer, who has also established an in-house metal catalyst development effort for chiral drugs, which include the world's biggest selling drug, Lipitor. The group has designed the C1-symmetrical bisphosphine ligand trichickenfootphos (TCFP), which is used in synthesizing new g-amino acid-based drug, Pregabalin. Asymmetric hydrogenation with TCFP is good for various substrates, offering high enantioselectivities and high activity under mild conditions. TCFP is proving to be versatile and has been used it in the rhodium-catalyzed asymmetric hydrogenation of β-acetamido dehydroamino acids to produce chiral β-amino acids [10].

Developments in asymmetric metal catalysis often involve efforts among pharmaceutical researchers, catalyst developers and academic scientists. Close collaboration between academic laboratories and industry has helped to ensure that discoveries are quickly scaled.

9.3
Process Chemistry Uses of Metals

In the pharmaceutical industry, process research is based on the design of efficient, environmentally friendly and economically viable chemical syntheses for drug substances. During chemical syntheses one often ends up having some metal impurities. Effective removal of these impurities is often a bottleneck in process research. Currently, precious-metal catalysts have a wide range of applications in fine chemical and pharmaceutical processes, allowing for clean and efficient routes to products in high yields. This enhances the efficiency and productivity of a chemist in drug discovery. However, requirements of very low limits on residual metal impurities in drug substances mean that metal catalyst removal from the product is vital. Similarly, the high value of precious metals means that their efficient recovery is also critical to the economic performance of a process.

9.3.1
Large-Scale Removal of Metals in Chemistry

Almost all transition metals are being used in chemical synthesis. Of these, Pd, Ru, Pt, Rh, Fe, Zn, Cu and Mg are most commonly used in metal based catalysts. Hence they are the most common metal impurities observed in pharmaceutical research, and an increasingly important problem in organic synthesis. Crystallization, extraction, precipitation and adsorption are the preferred large-scale methods for removing these impurities. These methods should be monitored closely during process streams. Additionally, the isolated products should then be examined for its ability to work in the next step. Chromatography is another commonly used method in pharmaceutical research for removal of metal impurities in large-scale processes. Chromatography is an adsorption based technique that offers a fast and versatile solution to many impurity problems. There have been some advances in chromatography techniques, such as introduction of chiral column chromatography, which selectively adsorbs one substance over the other. Recently, colorimetric analyses and Inductive Coupled Plasma Mass Spectrometry (ICP-MS) have been used to detect and remove the metal ions from reaction mixtures. In the case of batch processing, the metal scavenger bound to a solid support is added to the reaction mixture after completion. The process relies on chemically driven reactions where the excess reagents and reaction by-products react with, and bind to, the metal scavenger. The solution, which now only contains product, is separated from the resin bound impurities by filtration.

9.3.2
Alternatives to Functional Resins

Selective removal of unwanted impurities from pharmaceutical process streams using adsorbents is quite a common approach. It is improving over time as new materials and experiments become available. Recently, an approach for microplate evaluation of process adsorbents has been introduced. This approach allows for rapid selection of the most appropriate adsorbent and conditions for carrying out a given purification task. Adsorbent screening is carried out by the placing of polypropylene microcentrifuge tubes, which contain different candidate adsorbents, into the rack, along with an empty tube containing no adsorbent. A solution of a sample process stream is added to each tube, and the rack is agitated for few minutes. The tubes are then centrifuged and a sample of the supernatant from each tube is taken for analysis, to determine the concentration of the product and impurity. Ecosorb C-941, a carbon based product developed by Graver Technologies, is one of the best known examples for selective absorbance of rhodium metal; similarly, DarcoKB-B is famous for removing palladium metal, in particular from Suzuki–Miyaura biaryl coupling reactions [11].

9.4 Players in the Field

There are several players in this field who have developed silica and polymer based functionalized metal scavengers – SiliCycle, Polymer Labs, Reaxa and Aldrich are key players in the medicinal and process chemistry fields. ChemRoutes Corporation has carried out significant research effort in this field, commercialized by the above-mentioned players, and is now developing novel technologies in the metal scavenging area for the nanotechnology field. Examples of the chemistries commercialized by the various players are given below.

9.4.1 ChemRoutes

ChemRoutes is now focusing on the nanotechnology area, working on developing irregular silica as well as on a uniform spherical NanoSilicaTM bead with a broad range of functional groups.

9.4.2 Silicycle

Removing residual transition metals after reaction completion is a major task for many chemists in the pharmaceutical industry. The toxic nature of transition metals means their concentration has to be reduced to the single digit ppm level for the material to be used *in vivo* and in clinical work. Silicycle Inc. is a silica gel company who has commercialized several functionalized silica based metal scavengers for the purification of active pharmaceutical ingredients. Scheme 9.1 shows the process for using silica based metal scavengers.

Silicycle offers a wide range of silica-based metal scavenger products (Figure 9.1) to simplify the purification step and remove various transition metals. These metal scavengers are based on functionalized silica designed to react and bind excess

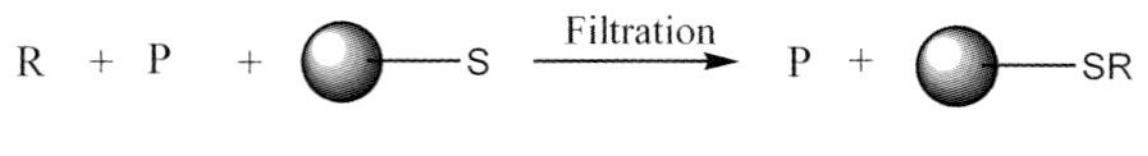

232 | *9 Metal Scavengers*

R + P + ⬤—S $\xrightarrow{\text{Filtration}}$ P + ⬤—SR

R = Residual transition metal
P = Product
S = Silica metal scavenger

Scheme 9.1 Metal scavenger process.

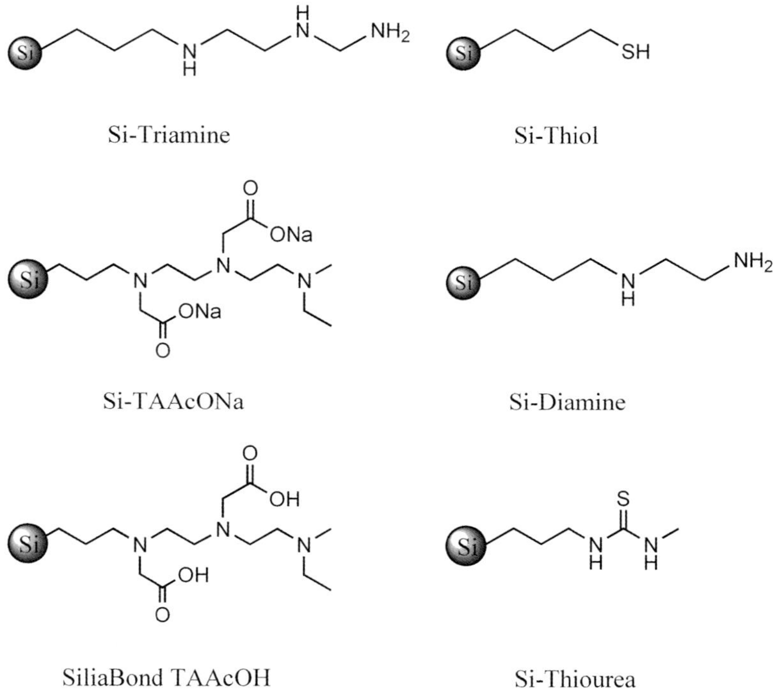

Si-Triamine

Si-Thiol

Si-TAAcONa

Si-Diamine

SiliaBond TAAcOH

Si-Thiourea

Figure 9.1 Silicycle metal scavenger products.

reagents and/or by-products. Most of Silicycle metal scavengers are used in medicinal and process chemistry.

When selecting a metal scavenger for the reaction, every component in the reaction mixture must be considered since solvent, ligand, product, residual reagents and by-products are some of the key factors that can affect scavenger performance.

9.4.3
Aldrich

Aldrich, another player in this field, has developed a few functionalized metal scavenger resins. They are used in a wide range of solution-based synthetic reac-

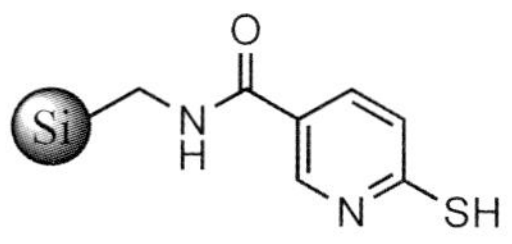

6-Thionicotinamide

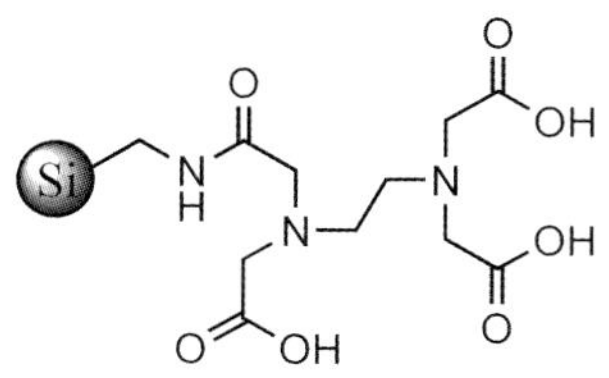

Ethylenediaminetriacetic acid

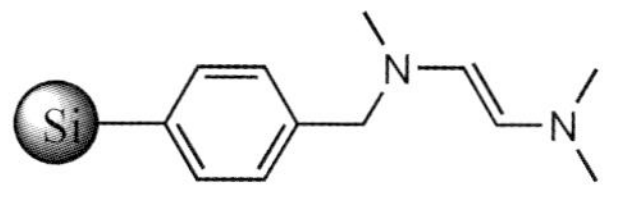

N,N,N'-Trimethylethylenediamine

Figure 9.2 Metal scavengers products introduced by Aldrich.

tions. Some of these metal scavengers bound to silica are shown in Figure 9.2. These metal scavengers efficiently bind to the metal ions at room temperature in organic or aqueous solution. The scavenged metals bound to the silica are then easily removed upon filtration.

9.4.4
Reaxa

Reaxa is also working in this field and has developed two quite efficient metal scavenger product lines, QuadraSil and QuadraPure. These two products are based on functionalized beads and silicas, which extract the metal contaminants in batch or flow process streams. In process chemistry, QuadraPure is use in pre-packed cartridges which are compatible with the flash system. Both QuadraPure and QuadraSil are GMP (Good Manufacturing Practice) compliant with high purity, chemical stability and extensive washing to eliminate potential metal substance. QuadraSil and QuadraPure metal scavengers are shown in Figures 9.3 and 9.4, respectively.

9.4.5
Polymer Laboratories/Varian

Polymer Laboratories (PL), which is now a part of Varian, Inc., has mainly focused its expertise in copolymerization technology to develop a high-performance, monodispersed base material. This specially functionalized macroporous polystyrene product has a much higher loading capacity than functionalized silicas and is compatible with all polar and non-polar, protic and aprotic solvents.

StratoSpheres are their main product line, which has a range of specially functionalized polymeric beads that are highly effective at removing charged and uncharged metal species from various organic solvents and analytical mixtures.

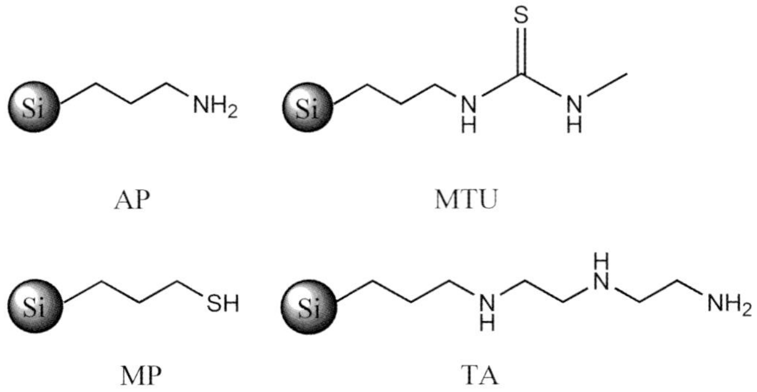

Figure 9.3 Quadrasil scavenger products.

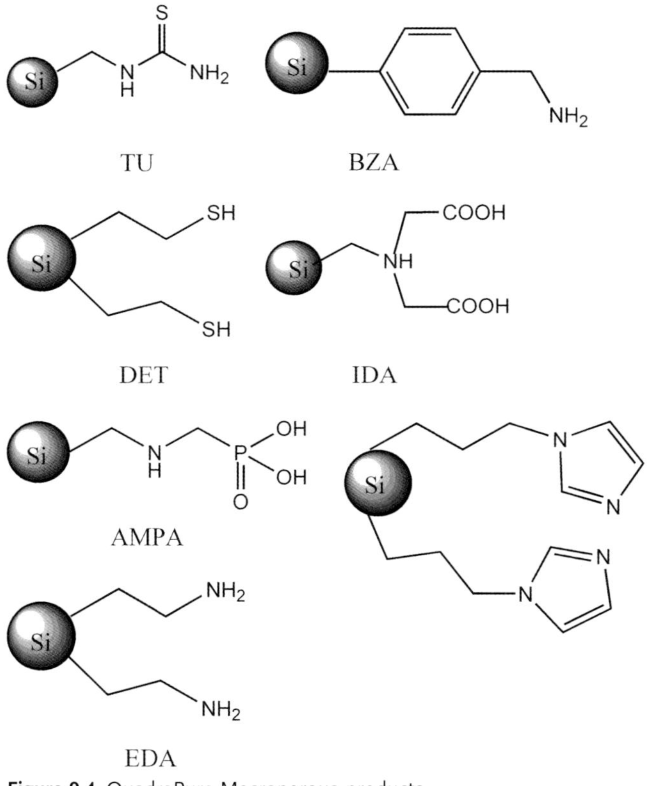

Figure 9.4 QuadraPure Macroporous products.

With just one pass of the reaction mixture through a tube containing the functionalized polymer, under gravity drain, metals such as platinum, palladium, tin, lead and ruthenium can be reduced down to trace, or even zero, levels. If the metal species being sequestered is highly colored, then in most cases the device will be self-indicating, showing the user when the device is reaching its capacity. This metal removal product is based around a 45 μm monodisperse macroporous polymeric material that offers a higher loading than conventional functionalized silica media. Figure 9.5 summarizes metal scavenger products introduced by Polymer Laboratories.

9.4.6
Degussa

Degussa has developed two silica based metal scavenger products, Deloxan(R)THP II and Deloxan(R) MP (Figure 9.6). Both products are based on organofunctional polysiloxane beads with properties that make them ideal metal scavengers. The Deloxan(R)THP II has thiourea, while Deloxan(R) MP has a thiol functionality. The surface density of organofunctional polysiloxanes groups is responsible for metal scavengers. Both Deloxan(R)THP II and Deloxan(R) MP metal scavengers are available in fine and irregular size and used in product development in batch mode application. Deloxan(R)THP II and Deloxan(R) MP show high affinity towards transition metals, especially Pd, Pt, Rh, Ag, Au, Zn, Cu(I), Cd, Hg, Ru, Ir, Cu(II) and Fe(III).

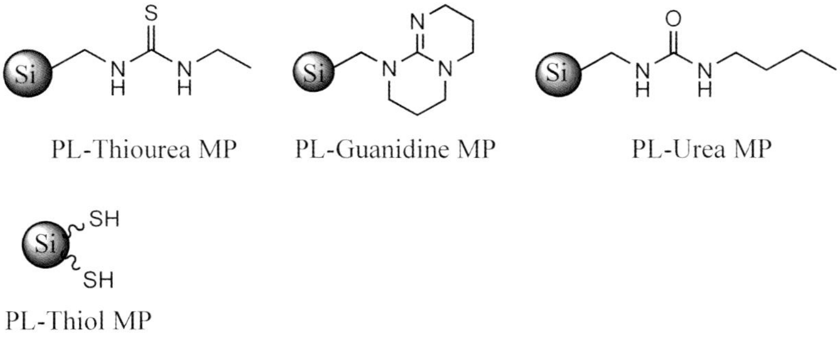

PL-Thiourea MP   PL-Guanidine MP   PL-Urea MP

PL-Thiol MP

Figure 9.5 StratoSpheresTM metal scavenger by Polymer Laboratories.

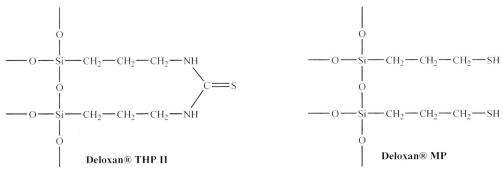

Figure 9.6 Structures of Deloxan(R) THP II and Deloxan(R) MP.

9.4.7
Johnson Matthey Chemicals

Johnson Matthey Chemicals, a polymer manufacture company, has developed a fiber technology called Smopex(R), which removes metal contamination from process streams. Smopex(R) scavengers have a unique advantage in that the active scavenger groups are almost exclusively located on the surface of the fiber, allowing the fast and efficient removal of metals from a wide range of process solutions. Smopex(R) is made of graft copolymerization of polyolefin fibers using functionalized monomers with metal binding properties bound to the exterior of the fiber. Scheme 9.2 shows the structure of Smopex(R).

The open fibrous structure of the Smopex(R) polymer enables greater metal recoveries in faster times. It selectively removes ionic and non-ionic species from aqueous and organic solutions. Smopex(R) is used in processes that have liquors containing low levels of precious or non-precious metals, even down to ppb levels. Some Smopex(R) benefits include it being mechanically and chemically stable, easy to separate from solution, stable across broad pH ranges and easy to handle. There are several Smopex(R) products based on the functional groups attached to the fiber. Smopex(R) is successfully used in pharmaceutical, agrochemical, fine chemical and plating for metal removal.

9.5
Silica Manufacture of the Metal Scavenger Resins

Silica has been used for fast and easy separation and purification of mixtures, and is made from the linkage of repetitive units of silicon. Silica is not affected in any way by any organic solvent since the pore structure is rigid and permanent. Silica has mainly been used in different types of chromatography. Silicycle has developed techniques that allow reactive functional groups to graft onto silica and then end-cap the residual silanol groups for an inert non-swelling support. There are two main types of silica metal scavenger products, irregular and uniform.

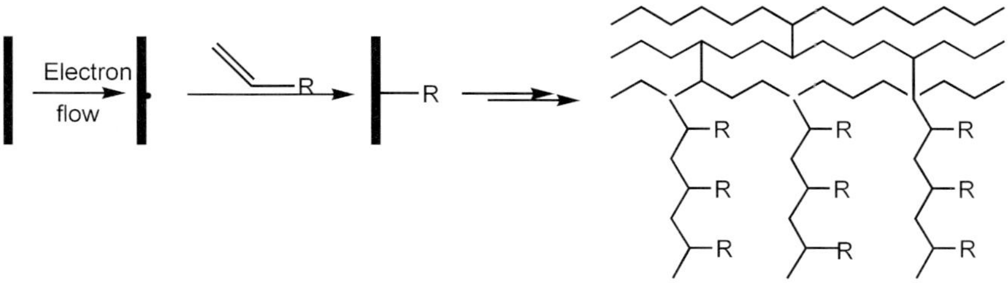

Scheme 9.2 Smopex(R) structure.

9.5.1
Irregular Silica

Irregular silica has broad particle and pore distribution with a low metal content. This form of silica is mainly used in flash chromatography, thin-layer chromatographic plates, and gravity chromatography. Based on particle size and grade we can separate irregular silica from regular/uniform silica. Degussa Deloxan(R)THP II and Deloxan(R) MP silica based scavenger supports are available on irregular silica that has particle sizes of 0.3–1.4 mm, and are usually used in fixed-bed mode applications.

9.5.2
Uniform Silica

Uniform silica is used in obtaining high product purity, since it has tight pore and narrow particle size. NanoSilica and Quadrasil are examples of functionalized, uniform silica metal scavengers. Both scavengers offer extremely rapid removal of metal catalyst residues. The products are based on high-quality spherical silica beads with well-defined porosity that can be used in aqueous or organic solution and in batch or flow processes. Their average particle size is 54 mm, which is much smaller than irregular silica, and has an average surface area of $715\,m^2\,g^{-1}$.

9.6
Polymer Manufacture of the Metal Scavenger Resins

9.6.1
Polystyrene Based Resins

Polystyrene (PS) based scavengers are the most widely available forms of solid supported scavengers due to their low cost of manufacture. However, the polystyrene backbone greatly influences the behavior of the scavenger in terms of solvent compatibility and reaction rate. The metal binding capacity of the resins depends on the inner microstructure of the matrix, which affects the swelling of the beads, and the length of spacer. The adsorption of metal ions by the resins can be explained by the coordination between ligand and metal ions rather than an anion exchange mechanism.

Precious metals, in normal oxidation states, form stable complexes with ligands containing "soft" donor atoms. Resins that contain nitrogen and, especially, sulfur atoms in their functional groups attached to the polymer matrix make good scavenger resins. Several coordinating and chelating resins based on sulfur-bonded tetrazolium and dithizone, grafted on a polystyrene matrix, have been synthesized during the past few years. These resins have shown to be of high selectivity and exhibit faster kinetics for the extraction of precious metals. For example, incorporating thiourea into a polystyrene divinylbenzene polymer matrix affords an improvement in selectivity for binding precious metals such as gold and silver over copper and iron in both batch and column operations [12].

9.6.2
Macroporous Based Resin

Macroporous resins are compatible with organic, aqueous, protic and aprotic media. They have low swelling characteristics, making the resin ideal for use in fixed bed applications. StratoSpheres and QuadraPure are widely used macroporous based resins. The QuadraPure consists of functionalized macroporous polystyrene beads (Figure 9.7).

Being low-swell and robust beads, both brands of resins can be used in either batch or flow manufacturing processes. They have several advantages such as being able to be used in purification, easy filtration, GMP compliance, capability to absorb target metals only, and efficient metal recovery from the resin.

9.6.3
Gel-Type Resin

Ionic liquids have attracted great attention as an alternative reaction media in organic synthesis. These highly polar solvents often lead to enhanced reaction kinetics and product selectivity [13]. Despite promising and spectacular results, practical drawbacks, such as product isolation and recovery of the catalyst, are still important barriers to their wide spread use in medicinal and process chemistry.

Polyionic gel beads consist of highly polar microenvironments, suitable for both efficient metal scavenging and active heterogeneous catalyst preparation. They display an efficient metal-soaking ability, which depends on strong non-covalent interactions between metal and ions within the polyionic gel.

Scheme 9.3 shows an example of an ionic gel adsorbing metal ion. A suspension of ionic gel (**1**) was gently stirred at 30 °C in a solution (orange color) of Pd-(OAc)$_2$ in dimethylformamide. The orange supernatant became colorless after a few

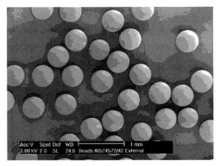

Figure 9.7 QuadraPure beads. Source: We acknowledge Reaxa for the use of this figure, which is taken from a Reaxa QuadraPure presentation.

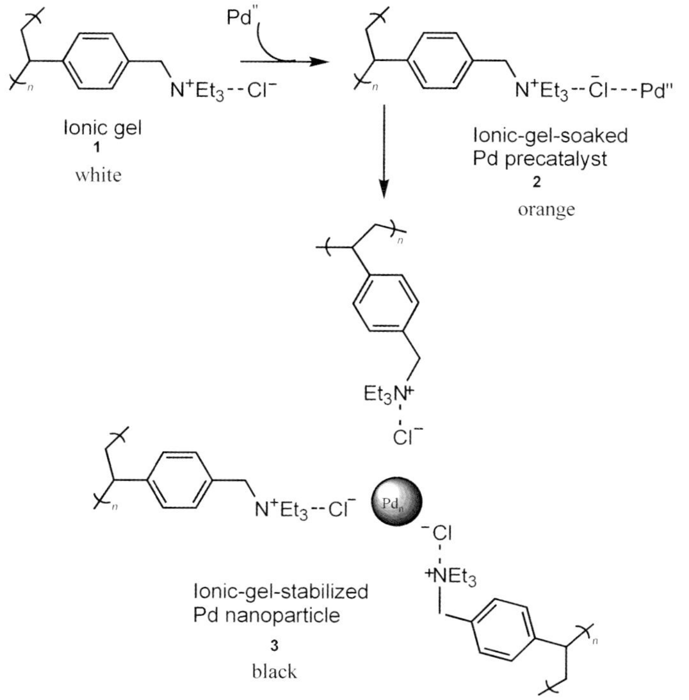

Scheme 9.3 Ionic gel preparation.

minutes, whereas the polyionic resin became dark orange. This observation shows the high affinity of ionic gels towards metals. In the next step, the efficiency of three polyionic gels were seen in a Suzuki coupling reaction, where the color change of polyionic gels from orange to black indicates the presence of Pd particles [14].

9.7
Palladium Removal in Organic Chemistry

In medicinal chemistry, transition metal catalysis is one of the most interesting and under-utilized classes of organic transformations due to the toxicity of the metal itself and its usual presence in the final product. Certain transition metals are starting to see regular use, though they cause numerous problems for the pharmaceutical process chemist during scale-up. Palladium is probably the most commonly used transition metal catalyst and can be used in various synthetic transformations. Heck [15] and Suzuki cross-coupling reactions [16, 17] are two of the better known reactions. The residual Pd is not an issue for the *in vitro* screens; however, if the target chemical compound proves to be a potential drug-like candidate, then a method of metal removal must be found before the *in vivo* work can begin. Silica bound and polymer bound metal scavengers like thiol, thiourea, triamine acetic acid (TAA) and triamine are widely used to remove Pd. Some important parameters, such as temperature, solvent, nature of complex, number of equivalents, pore size and reaction time, should be considered in optimizing the processes when using these supports in Pd scavenging. Each of the above parameters shows a direct or indirect effectiveness on the scavenging of palladium complexes in organic or aqueous solutions. The scavenging process can be optimized, by adjusting several combinations of the above-mentioned variables. Another important factor to consider is the oxidation state of the metal and the ligand.

In addition to the silica or polymer based resins mentioned above, polymer based resins with specific chemistries at the functional group have also been widely used as palladium scavengers for synthetic organic reactions. These include macroporous polystyrene-bound trimercaptotriazine (TMT). It, along with the thiol resin, has been found to be highly effective in reducing the concentration of palladium in both aqueous and non-aqueous solutions [18].

9.8
Platinum Removal in Organic Chemistry

Along with palladium metal, platinum has also been widely used as a catalyst in the pharmaceutical industries for asymmetric reactions in product development. The recovery of noble metals from waste solutions is a very important economical and ecological problem that has attracted recent attention. Recently, increasing effort has been devoted to the preparation of ion-exchange resins containing selective functional groups, chelating and coordinating resins that are useful for platinum extraction.

Poly(vinylbenzyl chloride) resins bearing diamine and guanidine ligands have an excellent sorption ability towards platinum ions from hydrochloric acid solutions that is proportional to anion-exchange capacity [19]. These resins are characterized by their surface area, which includes pore radius, porosity in swollen state, water uptake, content of chloride and nitrogen and amino group capacity.

Several specialized silica-based functionalized platinum scavengers such as N-acetyl-L-cysteine, 2-aminoethyl sulfide, 2-mercaptoethyl ethyl sulfide, 3-mercaptopropyl ethyl sulfide, pentaerythritol 2-mercaptoacetate ethyl sulfide and triamine ethyl sulfide amide, developed by Strem Chemicals, are also available. These scavengers are highly stable and available in pure forms, which successfully scavenge platinum metal ions in batch processes.

9.9
Ruthenium and Tin Removal in Organic Chemistry

With the diversity and reliability of organic synthetic reactions using metal as catalysts increasing greatly over the last few decades[20], the removal of active metal species, such as ruthenium and tin, is essential to provide clean and safe compounds for screening. Ruthenium-based catalysts are mainly used in olefin metathesis, one of the recent landmark discoveries in modern synthetic chemistry. There are several products on the market that are effective in scavenging ruthenium and tin metal species, reducing these metal concentrations to an acceptable level.

Thiol and thiourea containing polymeric media (Figure 9.5) are highly effective in removing ruthenium, as well as tin, metal ions. These can reduce the concentration of ruthenium and tin metal species to less than 1 ppm after just one pass through the resin bed.

Smopex(R) is also successfully used to remove ruthenium metal in reaction mixtures. It has a high affinity towards Ru ions in aqueous solution, and one can recover 100% of the scavenged Ru metal ion from the reaction mixture.

9.10
Other Metals Removal

Rhodium, iron, zinc, copper and magnesium are some of the other commonly used metals in organic synthesis to develop drugs.

Polystyrene–divinylbenzene polymer matrices, with a thiourea functional group, show good selectivity towards iron and copper, and effectively bind and remove them, both in batch and column operations. The adsorption mechanism can be explained by the coordination between ligand and metal ions [12].

Silica-based metal scavengers, such as pentaerythritol 2-mercaptoacetate ethyl sulfide and 3-mercaptopropyl ethyl sulfide, which have been developed by Phosphonics are used to remove rhodium and iron across all oxidation states. These scavengers work well in virtually all solvents, including water, and are available in various particle sizes.

There have also been some examples where a single unit of an amino acid, such as cysteine, can be used to bind and eliminate metal ions. These have been shown to have a greater affinity toward zinc and magnesium. Recently, amino acids used in short peptides have been immobilized on solid supports and used for metal binding. Supports for such immobilization include silica, polymer resins and membranes. These amino acids are ideal building blocks for metal chelation system [21]. They provide a wide range of binding functionality and show high affinity towards copper, zinc, magnesium and calcium.

9.11
Conclusion

As medicinal and process chemistry adopt their routes to the use of metals, with or without the use of microwaves, to hasten the reaction, increase yields, or give cleaner products, we hope this chapter summarizes the metal scavenger technologies and products available on the market from a broad range of suppliers. Technologies based on silica or polymer based supports are constantly advancing and the one chosen depends on various factors. Medicinal chemists focus on purity of the final compound after scavenging, whereas in process chemistry, in addition to the above requirement, the cost of the scavenger and metal recovery play an important role.

List of Abbreviations

GMP	Good Manufacture Practice
ICP-MS	inductive coupled plasma mass spectrometry
MRI	magnetic resonance imaging
NPL	National Priorities List
PDT	photodynamic therapy
PL	Polymer Laboratories
PS	polystyrene
TAA	triamine acetic acid
TCFP	trichickenfootphos
TMT	trimercaptotriazine

References

1 Orvig, C. and Abrams, M.J. (eds) (1999) Special issue on Medicinal Inorganic Chemistry. *Chemical Reviews*, **99**, 2201–4.
2 Farrell, N.P. (ed.) (1999) *Uses of Inorganic Chemistry in Medicine*, Royal Society of Chemistry, Cambridge.
3 Subcommittee on the Tenth Edition of the Recommended Dietary Allowances, Food and Nutrition Board, Commission on Life Sciences, National Research Council3. . (1989) *Recommended Dietary Allowances: 10th Edition (Dietary Reference Intakes)*, National Academy Press, Washington, DC.
4 Forstner, U. and Wittmann, G.T.W. (1981) *Metal Pollution in the Aquatic Environment*, Springer-Verlag, New York.
5 Ireland, M.P. (1991) *Biological Monitoring of Heavy Metals*, John Wiley & Sons, Inc., New York.
6 Vernet, J.P. (1992) *Impact of Heavy Metals on the Environment*, Elsevier, New York.
7 Gadd, G.M. and White, C. (1993) *Trends in Biotechnology*, **11** (8), 353–9.
8 Howard-Lockand, H.E. and Lock, C.J.L. (1987) *Uses in therapy*, in *Comprehensive Coordination Chemistry*, Vol. **6** (eds G. Wilkinson, R.D. Gillardand and J.A. McCleverty), Pergamon, New York, p. 755.
9 Shultz, C.S., Dreher, S.D., Ikemoto, N., et al. (2005) *Organic Letters*, **7** (16), 3405–8.
10 Wu, H.-P. and Hoge, G. (2004) *Organic Letters*, **6**, 3645–7.
11 Welch, C.J., Albaneze-Walker, J., Leonard, et al. (2005) *Organic Process Research and Development*, **9**, 198–205.
12 Zuo, G. and Muhammad, M. (1995) *Reactive Polymers*, **24**, 165–81.
13 (a) Sheldon, R. (2001) *Chemical Communications*, 2399–407. (b) Dupont, J., de Souza, R.F. and Suarez, P.A.Z. (2002) *Chemical Reviews*, **102**, 3567–92.
14 Thiot, C., Schmutz, M., Wagner, A. and Mioskowski, C. (2006) *Angewandte Chemie*, **118**, 2934–7.
15 De Meijere, A. and Meyer, F.E. (1994) *Angewandte Chemie – International Edition in English*, **33**, 2379–411.

16 Molander, G.A. and Bernardi, C.R. (2002) *Journal of Organic Chemistry*, **67**, 8424–9.
17 Wallow, T.I. and Novak, B.M. (1994) *Journal of Organic Chemistry*, **59**, 5034–7.
18 Rosso, V.W., Lust, D.A., Bernot, P.J., et al. (1997) *Organic Process Research and Development*, **1**, 311–14.
19 Jermakowicz-Bartkowiak, D. (2005) *Reactive and Functional Polymers*, **62**, 115–28.
20 Archibald, B., Brummer, O., Devenney, M., Gorer, S., Jandeleit, B., Uno, T., Weinberg, W.H., Weskamp, T. (2004) *Combinatorial Methods in Catalysis*, in *Handbook of Combinatorial Chemistry*, Vol. 2 (eds K.C. Nicolaou, R. Hanko and W. Hartwig), John Wiley & Sons, Inc., New York, p. 885.
21 Malachowski, L., Stair, J.L. and Holcombe, J.A. (2004) *Pure and Applied Chemistry*, **76**, 777–87.

Part Three Resin-Bound Catalysts

10
Polymer-Supported Organocatalysts

Belén Altava, M. Isabel Burguete and Santiago V. Luis

University Jaume I-CSIC, Department of Inorganic and Organic Chemistry, A. U. of Advanced Organic Materials, Avda Sos Baynat s/n, E-12071 Castellón, Spain

10.1
Introduction

From the different catalytic approaches chemists have devised to achieve the efficient transformation of organic substrates, organocatalysis is, probably, the most difficult one to define since the term was originally coined, as soon as in 1928, by Langenbeck [1, 2]. Very roughly, catalytic methodologies can be classified into two large areas: biocatalysis and chemocatalysis. Catalytic biotransformations take place with the participation of catalytic entities of complex biomolecules or aggregated structures of biomolecules, with enzymes, whole cells or microorganisms being classical examples [3]. In contrast, chemocatalytic processes can occur with the contribution of abiotic entities that belong to two general classes: the surfaces of some materials and molecular catalysts. In the first case, the active catalytic sites are located on the surfaces of bulk inorganic materials such as noble metals (Pd, Pt, etc.) [4, 5], clays [6] or zeolites [7, 8]. The most important industrial applications of chemocatalysis are currently being carried out, probably, with such catalysts. Within this area we can consider both processes for the production of bulk chemicals, such as those taking place in the petrochemical industry [9], and processes for the manufacture of fine chemicals or intermediates for the pharmaceutical industry [10].

With molecular catalysis, a discrete molecular entity is involved in the catalytic reaction. Nevertheless, we need to be aware that, in many instances, the true catalytic species is not a single molecular entity but aggregates of larger complexity [11, 12]. This is important as, in some cases, the differential behavior of polymer-supported catalysts originate in changes in the aggregation state of the catalytic species [13]. The twentieth century saw a spectacular development of molecular catalysis procedures to achieve a vast variety of chemical transformations. These include highly selective processes, including some for which an essentially

The Power of Functional Resins in Organic Synthesis. Judit Tulla-Puche and Fernando Albericio
Copyright © 2008 WILEY-VCH Verlag GmbH & Co. KGaA, Weinheim
ISBN: 978-3-527-31936-7

complete enantioselectivity can be obtained [14]. Most examples of molecular catalysis that have been developed involve the use of metal centers at the catalytic sites, their properties being modulated through coordination to the appropriate ligands. In general, however, the impressive performance of biocatalysts in terms of selectivity and, particularly, activity cannot be reached. Detailed study of the mechanism of action of enzymes and related systems led to the realization that many biocatalytic sites did not contain metals participating in the catalytic cycle but, on the contrary, only organic fragments are responsible for the catalytic function. This understanding was soon transformed by chemists in the search for organic molecules that could act by themselves as efficient catalysts for different organic transformations. Although there is a clear bioinspired origin for the concept of organocatalysis [15], the search for molecules with this function clearly extends beyond those resembling the active fragments found in biocatalysts.

Accordingly, organocatalysis has very often been defined in a "negative" way as that taking place in the absence of metals participating in the catalytic process and excluding the involvement of biocatalysts. Nowadays, the best accepted definition for organocatalysis is most likely that of "the acceleration of chemical reactions with a substoichiometric amount of an organic compound which does not contain a metal atom" [16]. This definition has been often constrained by taking into account that the organic catalyst must be "...of relatively low molecular weight and simple structure..." [17]. The inclusion of such constrains allows us to clearly define the essential elements of a metal-free catalytic process that is not being carried out by an enzyme or related biocatalysts. We need to bear in mind, however, that the concepts of low molecular weight and simple structure can be rather subjective. Perhaps two of the best known families of organocatalysts, as we will see later on, are those of cinchona alkaloids and small peptides, having structural features that sometimes can fit only with difficulty within the former constraints.

The conceptual introduction of organocatalysis as a tool for organic synthesis took place relatively late in the twentieth century, but since the beginning this area of work has attracted much interest, in particular when enantioselective transformations are involved [16, 18–25]. This is not surprising as the possibility of the direct transfer of the chiral information from a simple organic molecule to an achiral substrate in a catalytic process has always been very attractive.

Since the beginning, the development of different kinds of organocatalysts was accompanied by different attempts to prepare the appropriate heterogeneized counterparts using both inorganic and organic substrates [17, 26, 27]. In general, the advantages considered to approach the preparation of this kind of functional materials have been similar to those for other supported reagents and catalysts. The main objective for immobilization of organic catalysts has been the simplification of the work up with the facilitation of product separation. The ability for recycling and reusing a supported organocatalyst has been exploited less often than in the case of supported organometallic catalysts. This can be associated with the consideration of organocatalysts being more simple and accessible species. This is not always true as organic catalysts with increasingly complex structures are currently being developed. We also need to pay attention to the fact that in many cases the catalytic species are used with very high loadings (30–40% loadings can be often found), which justifies the need for a complete recovery of the catalyst. As with other catalytic processes, this can be achieved much more easily with supported species [28].

The main drawback for the immobilization of this kind of catalysts usually lies in the fact that covalent attachment to the support requires the involvement of one of the functional groups of the organic compound. In many cases the organocatalyst acts in a multifunctional way. The transformation of one of its functional groups can hamper the appropriate participation of this group or modify the conformational preferences and, accordingly, the proper disposition of the active functional groups [29, 30]. Different approaches have been developed, as we will see in the following sections, to overcome this limitation.

Organocatalysts have been heterogenized on a large variety of supports, including inorganic materials such as silica, clays or zeolites and organic polymers. The recent use of fluorous solvents and ionic liquids has allowed the development of very interesting approaches to the preparation of catalysts supported on those phases [31]. A detailed analysis of the use of inorganic supports or the use of fluorous or ILs phases is beyond the scope of this chapter. Here, we simply mention some of the achievements in those fields when necessary for comparison with the results obtained using polymeric supports. In the case of organic polymers, many different polymeric backbones have been considered. But, as in other areas of work, polystyrene-divinylbenzene (PS-DVB) matrices are the most usual when insoluble polymers are required, followed by the use of polyacrylic materials (PA) (Figure 10.1).

For the preparation of soluble supported organocatalysts, noncross-linked polystyrene (PS) and poly(ethylene glycol) derivatives (PEG) are the most usual

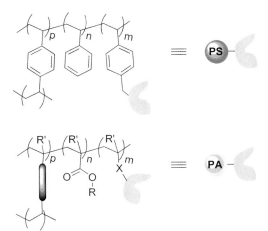

Figure 10.1 The most usual insoluble functional polymers considered in this work.

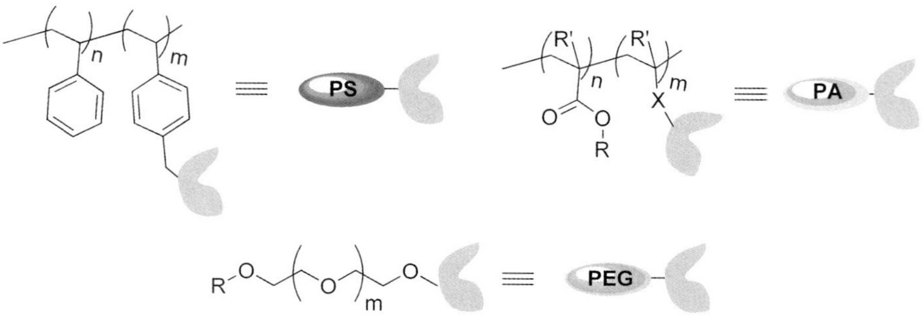

Figure 10.2 The most usual soluble functional polymers considered in this work.

supports [31]. However, the use of poly(ethylene glycol) matrices is becoming more frequent than the use of linear polystyrene (Figure 10.2).

10.2
Polymer-Supported Acidic Catalysts

10.2.1
Catalysis by Strongly Acidic Ion Exchange Resins

Acid catalysis is one of the most usual catalytic processes employed to provide low energy routes for organic reactions. Thus, the use of strongly acidic organic molecules to catalyze organic transformations can be considered as the simplest organocatalytic transformation. The easy accessibility of acidic ion exchange resins allowed, very early on, the use of such materials as acid catalysts in organic chemistry [32, 33]. They represent, most likely, the first example of polymer-supported catalysts and the first instance in which a functional polymer was used for facilitated chemical synthesis. At the same time, such catalysts are, by far, the most important polymer-bound catalysts at an industrial level [34]. This is based on the simplicity and reasonable stability of the resins being used and on the great industrial importance of acid-catalyzed reactions such as hydrolysis, dehydration, hydration, epoxidation, esterification, etherification, ether cleavage, ketal formation, alkylation, isomerization and so on. This explains why initial research efforts in this area were carried out in industry.

10.2.1.1 Conventional Polystyrene Sulfonic Resins
In this context the term "strongly acidic resins" refers, essentially, to sulfonated resins (**1**), the functional groups of which can be considered as polymer-bound analogues of a well-known organic Brønsted acid such as *p*-toluenesulfonic acid (**2**) (Figure 10.3).

The use of low-molecular-weight acids as catalysts for organic reactions continues being very important in the chemical industry, but the use of sulfonated polystyrene has clear advantages: (i) simplification of the work-up and easier

Figure 10.3 A sulfonated resin (**1**) and the corresponding functional group p-toluenesulfonic acid (**2**).

Scheme 10.1 Desulfonation of resin **1**.

storage, (ii) prevention of the corrosion that usually accompanies the use of strong acids and (iii) simplification of associated waste removal procedures. When reasonable long-term stability is achieved for the corresponding resins, for the process under consideration, the higher cost of the materials is not a limiting factor and the use of ion exchange resins as acid catalysts is favored.

Industrial processes use both microrreticular gel-type sulfonated resins and macroporous resins [35]. In the first case the polymeric matrix contains low levels of the cross-linking agent (DVB <8%), and polymer beads or particles do not possess any permanent porosity in the dry state. Macroporous resins are prepared, in contrast, with a higher level of cross-linking agent (DVB 8–20% or even higher) and in the presence of a porogenic agent. Removal of the porogenic agents after polymerization is completed leaves permanent pores in the resulting resins. Industrial ion exchange sulfonated resins possess high functionalization degrees. Essentially, all of the benzene rings of the polystyrene backbone are sulfonated, giving loadings of ca. 4.5 mmol-acid g^{-1} (in the dry state). According to the very polar nature of the functional groups, the resulting resins are very compatible with polar media, unlike other polystyrene resins, and usually contain variable amounts of water. For some applications the presence of water needs to be avoided and this can be achieved through azeotropic distillation.

One important factor that can limit the application of such resins is their thermal stability. In general, they are stable up to about 130 °C. At this temperature desulfonation starts and is accelerated at higher temperatures (Scheme 10.1). The initial process takes places mainly at the sulfonated DVB subunits according to the higher electron density of those polysubstituted aromatic rings. In this regard, macroporous resins, having a higher content in the cross-linking agent, are slightly less thermally stable. Sometimes, however, this drawback is outweighed by the fact that they possess better mechanical properties than gel-type resins.

Industrial processes for which this kind of catalysts were originally used include the hydration of propene and the conversion of isopropanol into diisopropyl ether [36], the cleavage of ethers such as MTBE to produce pure isobutene [37] and the alkylation of phenol with isobutene and isobutene oligomers to afford p-t-butylphenol (as stabilizer for plastics) and isoalkylphenols (for detergents) (Scheme

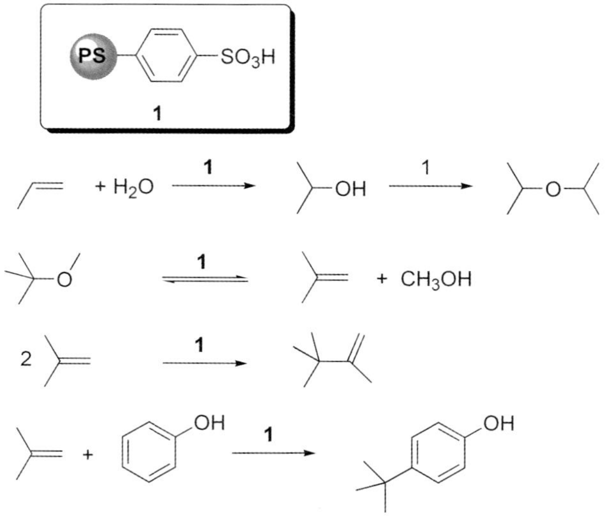

Scheme 10.2 Industrial processes that employ the catalyst **1**.

10.2) [38–40]. The use of powdered strongly acid microporous resins for the oligomerization of isobutene requires strictly anhydrous conditions, but the corresponding technology was known as soon as in 1966 [41]. Nevertheless, the most important application of sulfonated resins has been the preparation of mixed ethers from low molecular alcohols (methanol) and tertiary olefins or alcohols (e.g. isobutene) to produce high-octane gasoline additive ethers [42, 43]. The high potential shown by this kind of catalysts for etherification, esterification and transesterification reactions [44] has also facilitated their use for the preparation of biofuels [45].

In general synthetic organic chemistry, the use of this kind of heterogeneous acid catalysts has been, very often, displaced by perfluorinated sulfonic resins (see below), but applications of conventional acidic ion exchange resins for organic transformations requiring acid catalysis continue to appear. In general, the most common use in the laboratory for polystyrene acidic resins is as catalysts for protection–deprotection protocols, in particular for carbonyl compounds and alcohols. The processes (Scheme 10.3) show that their use can be very efficient and selective for complex molecules incorporating several protective groups. In the first example, the synthesis of compound **2**, the terminal acetonide group can be selectively cleaved with Dowex 50WX8 in the presence of other protecting groups such as internal acetonide, OBn or methyl ester [46]. Selective cleavage of the MOM protecting group is illustrated in the second example for complex gibberellinic acid derivatives (**3**). This can be carried out by Dowex 50WX2 with complete elimination of C/D ring system rearrangements occurring when *p*-toluenesulfonic acid is used as the catalyst [47]. The same resin can selectively transform symmetrical diols into the corresponding monoprotected derivatives (preparation of **4**).

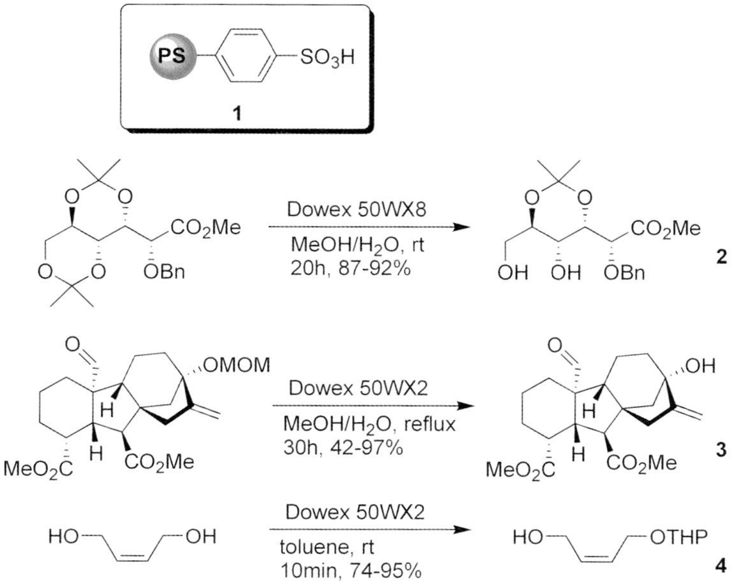

Scheme 10.3 Use of ion exchange resins for organic transformations requiring acid catalysis.

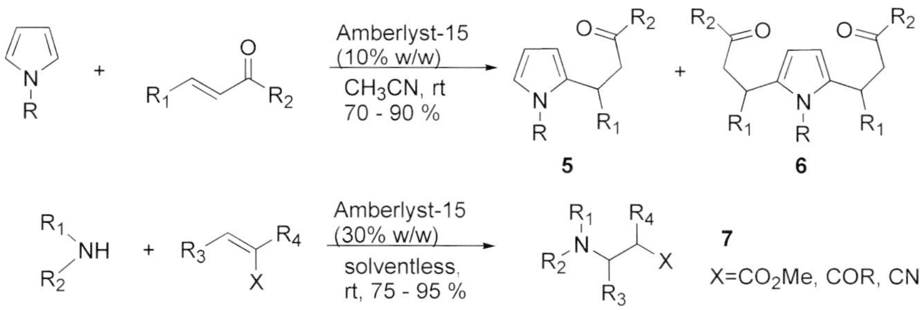

Scheme 10.4 Catalytic application of the commercially available macroporous Amberlyst-15.

Recent examples, for instance, of the catalytic application of the commercially available macroporous Amberlyst-15 include the Michael addition of pyrroles to α,β-unsaturated ketones (Scheme 10.4) [48]. In this process, the acid ion exchange resin (dry, 10% w/w) allows on to obtain mono and dialkylated pyrroles **5** and **6** in reasonable yields. Similarly, this catalyst (dry, 30% w/w) can catalyze the aza-Michael reaction of amines with α,β-unsaturated ketones, esters and nitriles to afford **7** in 75–95% yields under solvent-free conditions. Interestingly, yields were significantly lower using typical solvents such as DCM (dichloromethane), CH_3CN, THF, DMF or EtOH [49]. Recycling the catalyst is possible in both cases, but a smooth decrease in the yield is observed for each new run.

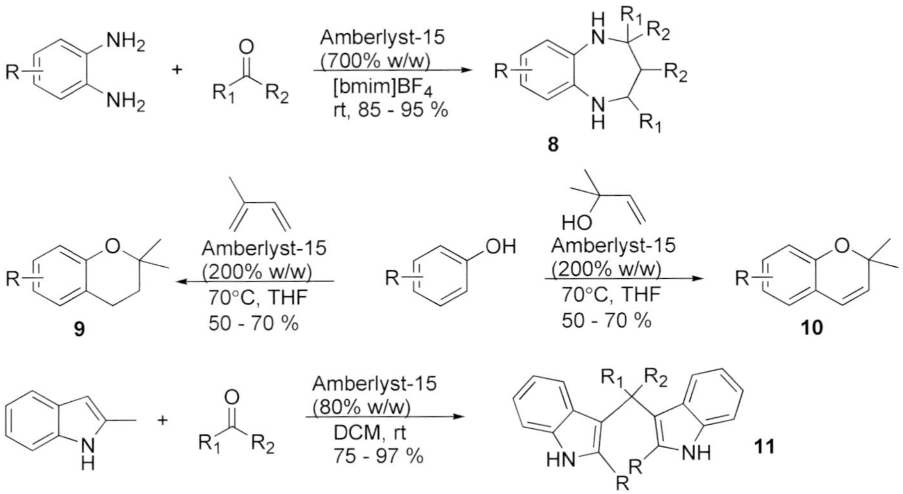

Scheme 10.5 Synthesis of heterocyclic structures employing Amberlyst-15 resin.

The same catalyst has been used for the synthesis of heterocyclic structures such as 1,5-benzodiazepines (Scheme 10.5) [50], benzopyrans [51], 1,8-dioxo-octahydroxanthenes and 1,8-dioxo-decahydroacridines [52], for the preparation of bis and tris(1H-indol-3-yl)methanes [53], and for the one-pot multicomponent synthesis of β-acetamido ketones [54]. 1,5-Benzodiazepines (**8**) have been synthesized efficiently using the same solid acidic catalyst in ionic liquids. Best results were obtained for 1-butyl-3-methylimidazolium tetrafluoroborate ([bmim]BF$_4$). Here, the resin is used in excess (7×) and recycling is again accompanied by reduced yields, which is ascribed to the presence of water formed in the reaction [50]. The rate-inhibiting effect of water seems to be a common feature for reactions catalyzed by cation exchange resins [55]. The methodology used for the one-step synthesis of benzopyrans (**9** and **10**) has allowed the straightforward preparation of some naturally occurring bioactive chromenes (**10** R = 4-OH, 4-MeCO, 6-OH) [51]. For the condensation of indoles and carbonyl compounds to produce **11** it was shown that Amberlyst 15 is a more efficient catalyst than silica supported sodium hydrogen sulfate, but reuse of the resin is reported to require reactivation [53]. Notably, all examples use large amounts of resin (w/w) to achieve efficient transformations.

10.2.1.2 Modification of Conventional Sulfonated Polystyrene Resins

Attempts to modify the structure of commercial strongly acidic ion-exchange resins to obtain improved catalysts were soon developed using different approaches. As in any other catalytic system the properties to be optimized include the activity, selectivity and stability of the catalyst. The acid character of sulfonic groups can be increased by the introduction of electron-withdrawing groups in the aromatic rings of the polystyrene chains. In this regard, nitrated and halogenated sulfonated

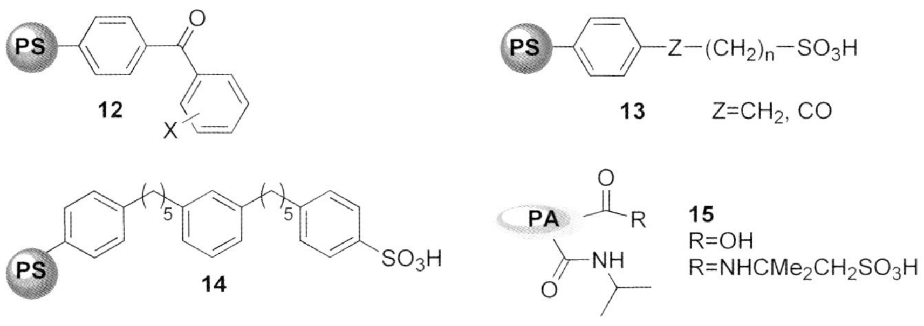

10.2 Polymer-Supported Acidic Catalysts

Figure 10.4 Modified conventional sulfonated polystyrene resins.

resins have been described, as well as the sulfonation of acylated polystyrene resins (**12**, Figure 10.4) [32, 56]. In general, the strength of the acid sites, and accordingly catalytic activities, increases with the level of sulfonation. Persulfonated sulfonic resins containing more than one sulfonic group per aromatic ring are the most active [57]. The solvent significantly influences the acidity. Thus, sulfonated resins are both stronger acids and more active catalysts than sulfonated silicas in water, but the reverse is found in acetonitrile or cyclohexane [58]. A second approach for developing increased acidities has been the combination with strong Lewis acids such as $AlCl_3$, BF_3 or SbF_5, but such materials (polymer-supported super acids) are clearly beyond the scope of this chapter [59]. The thermal stability was improved for resins containing nitro or halogen groups as well as for persulfonated polymers. The increase of acidity of those resins is accompanied by a rise in the temperature of desulfonation up to 180 °C [32]. An alternative methodology to obtain resins with higher desulfonation temperatures is the substitution of arylsulfonic groups by alkylsulfonic groups as in **13**, even at the expenses of a slight decrease in acidity.

With resins **13** containing long aliphatic spacers [60], separation of the active site from the polymeric PS backbone also favors the accessibility of reagents to the catalytic site, compensating for the decrease in activity caused by the lower acidity [61]. PS resins can be sulfonylated directly with the use of 1,3-propane sultone [62]. Alternatively, halide groups in haloalkyl polymers can be converted into sulfonic acid groups by reaction with a thiol and further oxidation of the mercaptoalkylated polymer [63], or polymers functionalized with groups containing enolizable hydrogens can be alkylated with vinyl sulfonic acid esters which, after saponification, provide the corresponding alkyl sulfonic resins [64]. More recently, resins with long aromatic-aliphatic spacers such as **14** have been prepared, but the resulting resins have been only studied as their lanthanide salts [65].

Partial sulfonation under mild conditions has been reported to improve the activity and selectivity for some processes [66–69]. Thus, for the esterification of dodecanoic acid with 3-phenylpropan-1-ol in water, sulfonated polystyrene resins with low substitution degrees gave better results than those with higher loadings [69]. It was suggested that conventional acidic ion exchange resins are too hydrophilic and have a low affinity for apolar substrates under those conditions. Sulfonic resins with low functionalization degrees, being less hydrophilic are more active for the reactions involving such apolar substrates. This hypothesis was confirmed for the hydrolysis of thioesters in refluxing water. Neither sulfonic ion exchange resins like Dowex 50W-X2 nor perfluorinated sulfonic polymers (Nafion) are active in promoting this reaction. In contrast, the reaction was efficiently catalyzed by a resin containing only 0.46 mmol g^{-1} of sulfonic acid groups [70]. Based on the former results, Kobayashi developed a new family of sulfonic resins containing long aliphatic chains (octadecyl) on the sulfonated aromatic rings to increase their hydrophobicity. Such functional polymers have shown, in particular the ones with lower loadings (0.2–0.4 mmol g^{-1}), a high efficiency in water for hydrolytic and protection–deprotection processes [69, 71, 72].

Other approaches include the partial metal exchange of sulfonic groups with transition metal cations [73]. Nanostructured strong acid sulfonic resins based on lyotropic liquid crystals have been reported to exhibit selectivities one-order of magnitude higher than Amberlyst or Nafion polymers for the esterification of benzyl alcohol with 1-hexanoic acid [74]. In many cases, however, the benefits obtained with those modified sulfonic acid polystyrene resins are overcompensated by the higher costs of their preparation and this explains why commercial conventional ion exchange resins continue being the most used acid catalysts in this field (along with Nafion and related resins). As an alternative, the continuous dehydration of alcohols in supercritical fluids using conventional acid resins has been shown to proceed with very high selectivity [75]. The need to develop efficient proton exchange membranes for fuel cells is currently becoming a new driving force for the preparation of new, modified sulfonic acid polymers [76–78].

Weakly acidic ion exchange resins containing carboxylic groups have been much less studied as catalysts. An interesting example is that of polymer **15**, a linear poly(N-isopropylacrylamide), which is soluble in cold water but insoluble in water heated above the critical solution temperature, facilitating its recovery and reuse. Such materials have been tested as catalysts for the hydrolysis of ketals [79].

10.2.2
Catalysis by Perfluorinated Sulfonic Acid Polymers

A significant step towards the use of strongly acidic resins as acid catalysts in organic chemistry was given by Olah and others with the use of Nafion and related polymers for different processes in organic chemistry [80, 81]. Nafion (**16**, Figure 10.5) is a copolymer of tetrafluoroethene and perfluoro[2-(fluorosulfonylethoxy) propyl] vinyl ether developed by DuPont [82].

In comparison, for instance, with Amberlyst 15, Nafion resins have lower loadings (ca. 1 versus 4.7 mmol g^{-1} in the dry state) and are non-porous, providing surface areas below 1 m^2 g^{-1} (ca. 50 m^2 g^{-1} for Amberlyst 15). Conversely, Nafion polymers are clearly more acidic than conventional acidic ion exchange resins [83, 84]. It has been estimated that Nafion can be compared in acidity with 85% sulfuric acid, whereas Amberlyst 15 is a weaker acid than 60% sulfuric acid [85]. They are

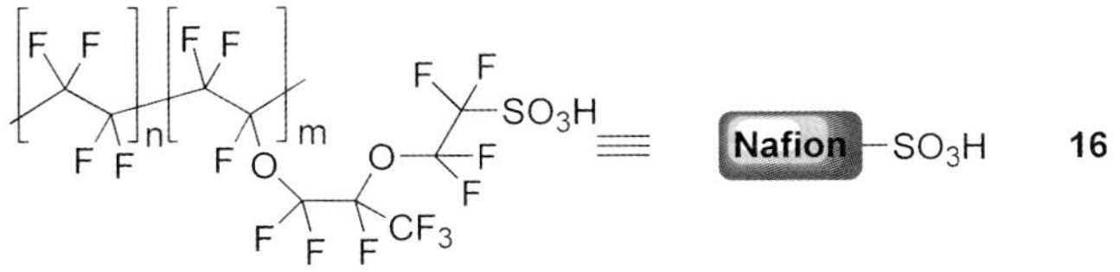

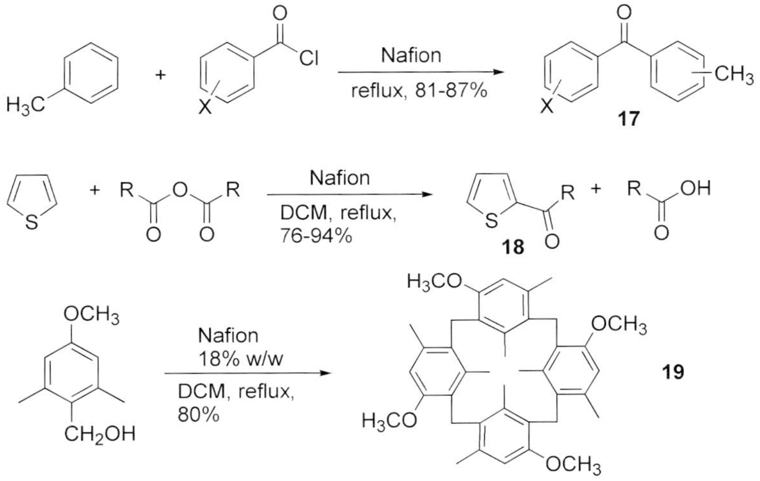

Scheme 10.6 Use of Nafion as catalyst for organic transformations.

also significantly more chemically and thermally stable. In this regard they are similar to Teflon resins. From a practical point of view, Nafion resins can be used as acid catalysts at temperatures up to 180 °C; desulfonation only occurs above 210 °C [86]. To improve their dispersion and, accordingly, their activity, Nafion polymers have been supported over different materials of high surfaces areas such as silica, alumina, pore glass or Chromosorb T [87]. In particular, Nafion-SiO$_2$ composites (for instance SAC-13) are becoming one of the most used materials when application of solid acid catalysts are required [88, 89].

This kind of strongly acidic ion exchange material has been used as catalysts for various organic transformations. Numerous Friedel–Crafts type reactions have been studied. Thus, acylation of toluene with different benzoyl chlorides produced the corresponding benzophenones in 81–87% (Scheme 10.6) [90]. Average values found for the ortho:meta:para isomer distribution were 75:3:22. The experimental procedure requires refluxing the reagent mixture in the presence of the solid. The catalyst showed a decreased activity at lower temperatures and could be reused. The exact nature of the resin (m and n values in **16**, Figure 10.5) was shown to affect its activity [91]. Acylation with aliphatic acid anhydrides is also possible. Starting from mesitylene or thiophene, the corresponding alkyl aryl ketones were obtained in 72–94% yields. The best result corresponds to the reaction of thiophene with butyric anhydride (**18**, R = Pr), whilst only a 75% yield was observed with the use of Amberlyst 15 [92]. Carboxylic acids can be used as acylating agents at higher temperatures and catalyst loadings [93]. This allows a simple strategy exploiting intramolecular Friedel–Crafts processes to afford polycyclic aromatic hydrocarbons [94, 95]. Starting from aromatic and aliphatic sulfonic acids, sulfones can be prepared in 30–82% yields, even if this process requires relatively drastic conditions, in particular for non-activated arenes (160–165 °C, pressure, 50% w/w) and continuous removal of the water formed [96]. Alkylation of aromatics is also possible using Nafion resins as the catalysts. A large variety of alkylating agents has been studied, including alkyl halides, chloroformates and oxalates, alkenes and alcohols [97, 98]. One interesting example is the benzylation of arenes with benzyl alcohols [99]. Here, the trimerization and tetramerization of methoxybenzyl alcohols is accomplished very efficiently to obtain some interesting cyclic structures such as **19**. Other complex structures such as [9,9′]spirobixanthene have been prepared by related procedures [100]. As with other catalysts, the Friedel–Crafts alkylation is a complex process, involving different competing side reactions, and the efficiency of the process varies very much for each individual reaction. Thus, methylation of benzene at 185 °C with methanol only affords a 4.1% yield of the expected product while formation of ethyl benzene from benzene and ethylene at 175 °C takes place in 88% yield [101, 102]. The alkylation of benzene with long-chain olefins to form linear alkyl benzenes has been achieved very efficiently with the use of Nafion-SiO$_2$ composites. At 80 °C conversions of 99% can be obtained, with selectivities towards the linear alkylbenzenes of >95% [103]. Adamantylation of substituted benzenes is also possible with Nafion resins, directly or supported on zeolites or silica, but, in this case, the use of an Amberlyst resin is advantageous as the reaction proceeds with almost exclusive formation of *p*-adamantylatedbenzenes [104]. The continuous fixed-bed catalytic alkylation of *m*-cresol with different alkylating agents has been achieved in supercritical CO$_2$ with Nafion-SiO$_2$ composites (SAC-13) [105]. The reverse dealkylation of alkylaromatics, in particular de-terbutylation, has also been accomplished through transalkylation reactions [106, 107]. Deacylation (deacetylation and decarboxylation) is also possible for compounds having strong steric interactions of the carboxyl or acetyl group with ortho substituents [108].

As with polystyrene sulfonic resins, Nafion-based acid catalysts are highly efficient for hydration and dehydration processes and, in general, for condensation reactions that occur with the formation of water or similar secondary products. Formation of ethers has been studied for various alcohols [109–111]. Dehydration of 1,4- and 1,5-diols at 135 °C affords the corresponding cyclic ethers such as **20** in excellent yields (Scheme 10.7), while 1,3-diols experience different transformations depending on their structure [112]. The dehydration of 1,2-diols mainly proceeds via the pinacol rearrangement. Further condensation of the initially formed carbonyl compound and unreacted diol affords 1,3-dioxolanes [113]. The catalyst could be efficiently reused following a reactivation protocol. Formation of aryl ethers is also possible, and the synthesis of dibenzofurans **21** (X = O) from 2,2′-dihydroxybiphenyls has been reported (Scheme 10.7) [114]. The related reaction

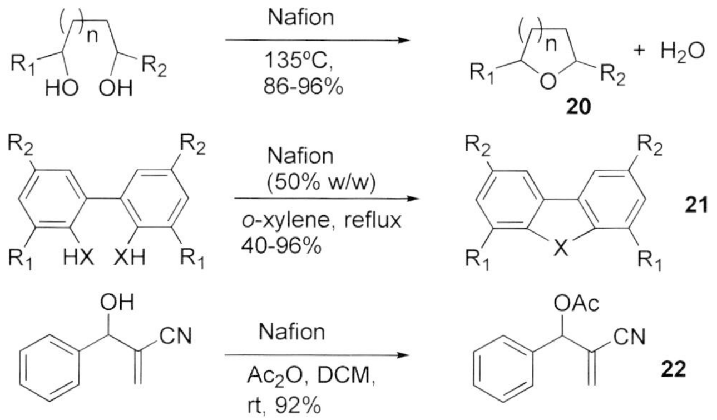

Scheme 10.7 Use of Nafion as catalyst in condensation reactions.

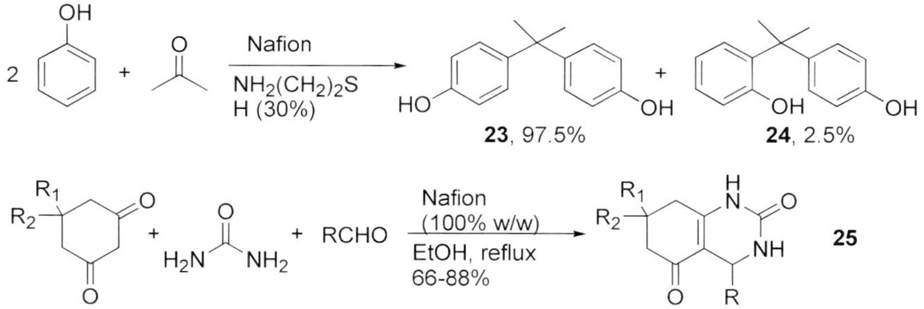

Scheme 10.8 Aldol condensation (top) and Biginelli reaction (bottom) catalyzed by Nafion.

from 2,2′-diaminobiphenyls affords the corresponding carbazoles **21** (X = NH) [115]. Polymeric ethers can be formed by polymerization of THF at room temperature in the presence of Nafion [116]. Acetals and ketals, in particular those prepared from trimethyl orthoformate, and ethylene dithioketals have been prepared in 83–100% yields. As the reverse hydrolytic reaction is also feasible, this has allowed the development of different protocols for the protection and deprotection of carbonyl compounds [117–121]. Reductive cleavage of acetals and ketals to ethers has been performed with triethylsilane and Nafion [122]. This concept has been extended for the formation–cleavage of aldehyde diacetates [123], methoxymethyl, tetrahydropyranyl, and benzyl ethers [124, 125], ester, thioester and amide bond formation [89, 126–129] and so on [118, 119] As an example, the use of Nafion allows the preparation of acetate **22** with excellent yields in the presence of acid sensible functional groups (Scheme 10.7) [128].

Aldol condensation of acetone catalyzed by perfluorinated sulfonic resins has been studied, but conversions are in general low as the reaction is under thermodynamic control [130]. A very efficient preparation of bisphenol A, however, with 99.7% selectivity towards the *p,p*′ isomer (**23**), could be developed using a polymer partly neutralized (30%) with 2-mercaptoethylamine (Scheme 10.8) [131]. Other condensations of aromatic compounds with ketones have been reported [132, 133]. For the Biginelli reaction, a complex multicomponent condensation providing access to interesting heterocyclic structures such as dihydropyrimidinones and octahydroquinazolinones (**25**, Scheme 10.8), Nafion was found to be the most efficient solid acid catalyst [134, 135]. In contrast, the reaction of resorcinol with ethyl acetoacetate or with acrylic acid over Nafion-SiO$_2$ nanocomposites afforded 7-hydroxycoumarines in high yields under mild reaction conditions [136].

Nafion resins have been used not only for the opening of epoxides but also for their isomerization to aldehydes or ketones [137]. Various other rearrangements and isomerizations are catalyzed by this solid acid, in some cases with selectivities higher than those obtained with other solid catalysts [138–140]. Other reactions that have been studied include the Peterson methylenation of carbonyl compounds [141], hetero-Michael additions to unsaturated ketones [142], the Koch-type carbonylation of alcohols to form carboxylic acids [143], dimerization of α-methylstyrene [144], addition of carboxylic acids to olefins [145] and Diels–Alder reactions [146]. Notably, in most cases, reutilization of the catalyst is considered but only after an appropriate washing protocol to regenerate its acidity/activity.

10.2.3
Catalysis by Functional Polymers Containing Specific Acidic Sites

Besides the above-mentioned acidic ion exchange resins, several different functional groups have been introduced on polymeric backbones to develop materials with weak acidic sites that can afford acid catalysis of specific processes selectively and under mild conditions. Perhaps the simplest, conceptually, of those functional groups are N-protonated supported amines, in particular pyridinium salts **26** and **27** (Figure 10.6). The commercial availability of poly(vinyl pyridine) resins has, undoubtedly, facilitated their use. Different modifications have been prepared, including lightly [147] and highly cross-linked [148, 149] resins **26**, along with linear polymers **27** [150]. These acidic catalysts have been successfully used for the THP protection of alcohols [150], for their deprotection [147], for the ketalization

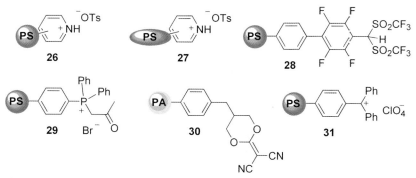

Figure 10.6 Functional polymers containing specific acidic sites.

of aldehydes and ketones and for esterification reactions [148]. The catalytic activity was lost by changing the anion or using commercial anion exchange resins. Cross-linked polymers could be recycled without affecting their activity.

The tetrafluorophenyl bis(trifluoromethanesulfonyl)methane moiety in polymer **28** (Figure 10.6) also has a strong Brønsted acidity. This catalyst can act at very low loadings (0.1–3% molar) showing, at room temperature, an activity higher than that of Nafion composites for several benchmark reactions. Essentially quantitative transformations were obtained for esterification and acetalization reactions [151]; **28** was also able to catalyze Friedel–Crafts acylations, the aldol condensation between benzaldehyde and acetone trimethylsilyl enol ether and other related processes. In some cases the catalyst was reused for up to ten successive runs. Polymers containing the acetonylphosphonium moiety, like **29**, also display catalytic activity for the protection of alcohols and phenols and for their regeneration from O-protected derivatives. Upon methanolysis of the corresponding THP, 1-ethoxyethyl and tetrahydrofuranyl ethers can be obtained in 70–98% yields after prolonged treatment for two days in MeOH at room temperature [152]. The same polymer has been used to catalyze the acetalization and thioacetylation of aldehydes in DCM (10% catalyst, rt, 48 h) [153].

The nature of acidic sites in resins **30** and **31** is very different (Figure 10.6). Polymer **30** contains π-acidic sites and was prepared by copolymerization of a styrene monomer containing the dicyanoketene acetal fragment (67%) and an acrylic cross-linking monomer (33% of ethylene glycol dimethylacrylate: EGDMA). The resulting resin was an excellent and selective catalyst for different deprotection processes for acetals and silyl ethers [154–156]. The same catalyst was active for the addition of silylated nucleophiles to aldehydes and ketones and N-arylaldimines [157, 158]. Notably, **30** was significantly more active than the analogous soluble counterpart. This highlights that effects of heterogenization are more complex than those just related to diffusion of substrates on the polymeric matrix [29, 159]. In contrast, supported trityl perchlorate **31** contains Lewis acid sites that can promote, at low catalyst loadings (3 mol%), the addition of silyl enol ethers to benzaldehyde and different acetals at low temperatures (−78 °C) [160].

10.3
Polymer-Supported Basic Catalysts

10.3.1
Catalysis by Anion Exchange Resins

Many more chemical processes have been developed using solid acid catalysts than with solid base catalysis, in particular at an industrial scale. However, most of those examples correspond to processes involving inorganic materials [161]. The first reports on the use of polymer-supported basic catalysts can be traced back to the use of anion exchange resins containing basic anions. The simplest, most general structure for this kind of resins is given by **32** ($X^- = OH^-$, Scheme 10.9)

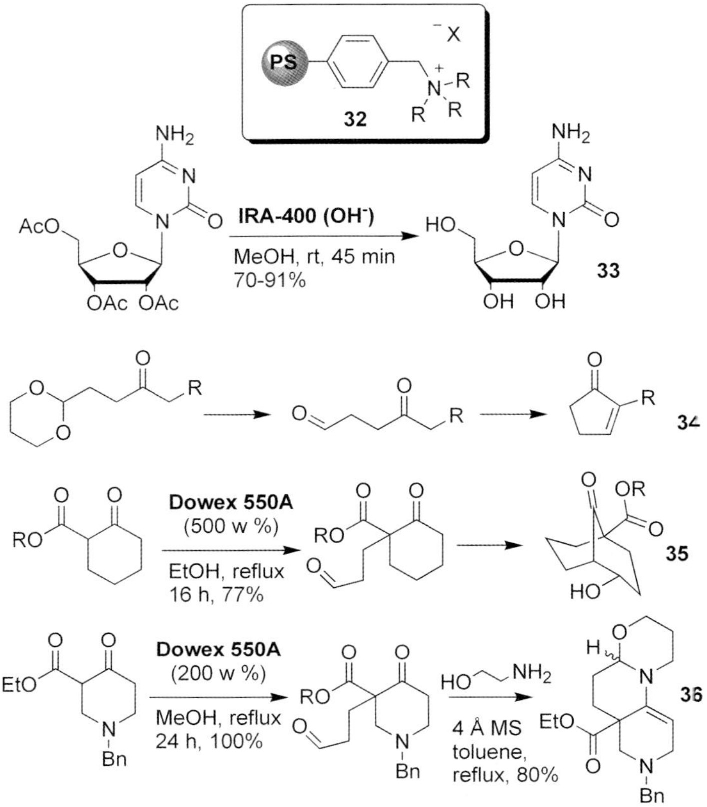

Scheme 10.9 Examples of reactions catalyzed by anion exchange resins.

whose functional group is a quaternary ammonium salt in its hydroxide form. Other basic anions, such as fluoride can be present instead of hydroxide. Many applications of those basic resins involve deprotection reactions and related processes [33, 119]. An example is the formation of nucleosides (**33**, Scheme 10.9) from the corresponding O-peracetylated derivatives that takes place at room temperature in methanol with high yields (45 min) using the resin IRA-400 in its hydroxide form [162]. A second family of reactions that can be catalyzed by strongly basic ion exchange resins is condensation processes. The aldol condensation of acetone has been studied using the hydroxide form of Amberlite IRA-900. The rate-determining step of the reaction was diffusion inside the polymeric matrix. As with acidic resins, the amount of water present is very important. The presence of water increases the selectivity and stability of the catalyst, but decreases the rate of the reaction [163]. Other aldol condensations, including cross-aldolic processes have been also reported [164]. An interesting example is the preparation of 2-alkyl-2-cyclopenten-1-ones **34** (Scheme 10.9). In this case, advantage is taken of the fact that support on insoluble polymeric matrices allows the simultaneous use of reagents or catalysts otherwise incompatible (acids and bases in this case). Thus, the use of mixed ion exchange resins allows the initial acid-catalyzed deprotection of the aldehyde group being followed by the base-catalyzed aldol condensation to afford the final product **34** [165]. Different groups have described other condensation reactions based on the deprotonation of active methylene groups, including Michael processes [166], and Knoevenagel-type condensations of malononitrile or ethylcyanoacetate with benzaldehydes or their Schiff bases [167, 168]. In the synthesis of **35** Dowex basic ion exchange resins can accomplish a one-pot sequence in which the initial Michael adduct experiences a further intramolecular aldol reaction to afford the bicyclic product [166]. Alternatively, in particular when heterocyclic dicarbonyl compounds are the starting materials, the Michael adduct can react with amines and the intermediate aldimines experience an intramolecular Mannich reaction to provide highly functionalized heterocyclic structures such as **36** (Scheme 10.9). Notably, in many of those condensations the ion exchange resin is used in amounts that exceed the stoichiometry ratio. Substitution of the hydroxide by other anions is sometimes important. Thus, carbonylation of methanol to form methyl formate is possible with resins like Amberlite (**26**), but in this case methoxide is a better counter-anion than hydroxide, as it improves the thermal stability of the polymer and prevents hydrolysis of the product [169].

10.3.2
Catalysis by Non-chiral Polymer-Supported Amines and Phosphines

Amines and phosphines are the most traditional examples of organic bases. A large variety of resins of polymers containing amine and phosphine groups have been reported, but originally many of them were prepared as potential ligands for metals or, later on, as scavengers for compounds having aldehyde groups or other functionalities having complementary reactivity [170–172]. Commercially available weakly basic ion exchange resins are based on polymers containing tertiary amine functionalities that upon protonation are converted into the corresponding anion exchangers, but very few applications of those materials as basic catalysts have been reported [33]. The use of polymeric equivalents of imidazole or pyridine as basic catalysts, in particular for esterolytic processes, was developed as one of the first applications in this field [173, 174].

Polymer-supported equivalents of the widely used organic base 4-(dimethylamino)pyridine (DMAP) were soon developed, but many of their reported applications are as a stoichiometric base. Resin **37** (Scheme 10.10), containing a poly(ethylene imine) matrix was the first supported system to be prepared. Those materials were more efficient catalysts than DMAP itself, under the same conditions, for the hydrolysis of p-nitrophenyl esters in aqueous solution [175, 176].

The same reaction was studied with soluble polymer **38**, prepared by polymerization of 4-(N,N-diallylamino)pyridine. This catalyst was even more efficient than **37** or 4-(pyrrolidino)pyridine, a base with much higher activity than DMAP [177].

For synthetic applications, DMAP analogues supported on PS-DVB resins have been the most widely used. Polymer **39** was prepared from a 4% cross-linked

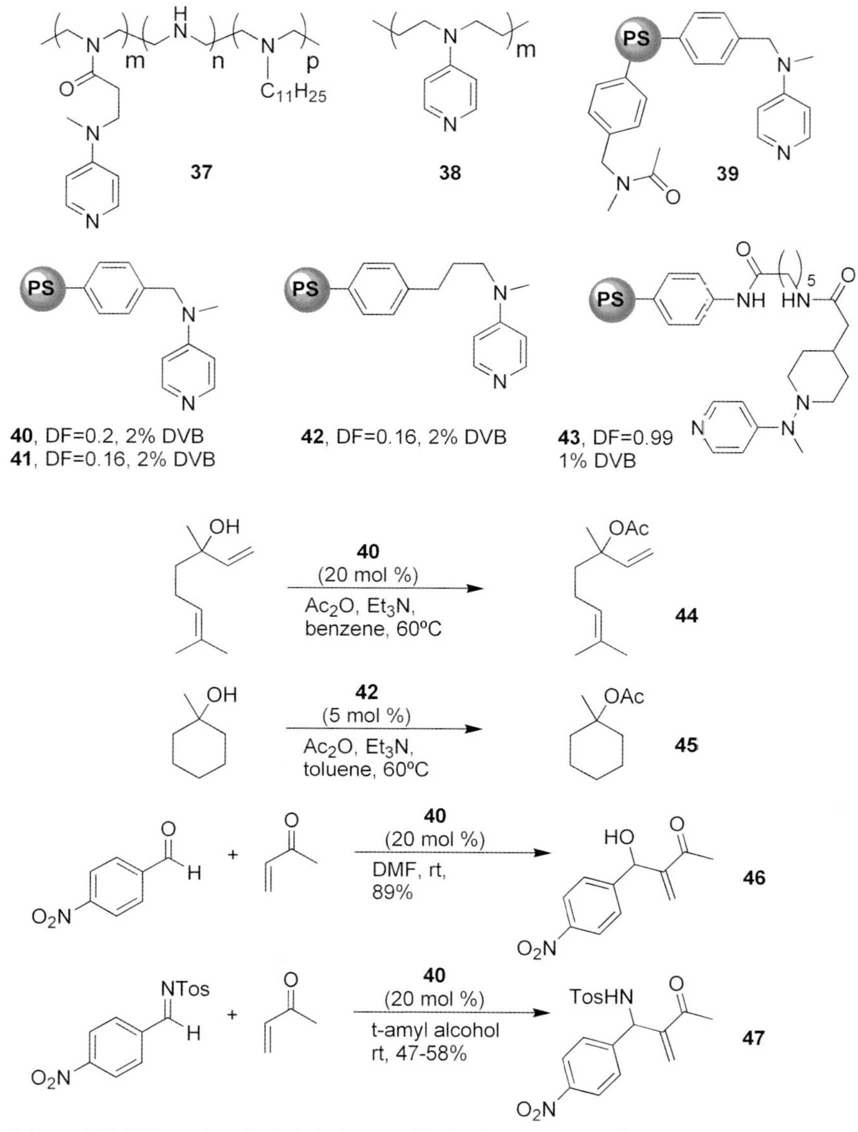

Scheme 10.10 Examples of catalysis by non-chiral polymer-supported amines.

chloromethylated polystyrene by initial treatment with an excess of methylamine and then reaction with 4-chloropyridine. Under those conditions less than 30% of the amino groups reacted to form the DMAP-like functionality and the unreacted groups were acetylated to avoid their interference in the catalytic processes. Compound **39** catalyzed the esterification of carboxylic groups with methanol but at a slower rate than DMAP. An alternative resin containing a long aliphatic spacer was also prepared, but did not produce any appreciable improvement [178]. To avoid the problems associated to unreacted groups, Tomoi prepared a similar catalyst (**40**) by polymerization of 4-(*N*-methyl-*N*-*p*-vinylbenzylamino)pyridine with styrene and DVB. This material was tested for the acetylation of linalool to form **44** and showed an activity only slightly lower than DMAP (Scheme 10.10) [179]. Fréchet and coworkers developed a different strategy for the preparation of a related polymer (**41**) through the amination of chloromethylated polystyrene with 4-methylaminopyridine. They also prepared, in a similar way, resin **42** in which the functional group is separated from the polymer matrix by a short propylenic spacer. Both catalysts were assayed for the acetylation of 1-methylcyclohexanol to afford **45** (Scheme 10.10). Compound **42** displayed a catalytic activity similar to that of DMAP and slightly higher than that of **41** [180]. The former results illustrate the influence that the method for the preparation of a given supported catalyst can have on its overall performance [181, 182]. More recently, such supported DMAP equivalents have been applied as catalysts for both the Baylis–Hillman [183] and aza-Baylis–Hillman reactions to produce compounds with structures such as **46** and **47**, respectively (Scheme 10.10) [184]. In both cases catalyst recycling required either the reactivation with NaOH or the very careful adjustment of the reaction conditions.

The preparation of **45** was used by Verducci as the benchmark reaction to check polymer **43** containing 4-piperidine pyridine fragments. The activity of this catalyst used in 20 mol.% was lower than that of DMAP. Although the catalyst was recycled ten times a 30% decrease in its efficiency was observed. The use of different spacers did not modify significantly the results, but the catalytic activity was clearly decreased by raising the catalyst loading [185].

In addition to **37** and **38**, other soluble noncross-linked polystyrene resins containing DMAP equivalent functions have been described. Linear functional polystyrene **48** (Scheme 10.11) was prepared from a commercial chloromethylated resin and used for esterification, silylation and tritylation of alcohols with a catalytic activity slightly lower than that of molecular DMAP [186]. As activities are not increased in those linear functional polymers, a comparison of results between similar soluble and cross-linked supported DMAP seems to suggest that diffusional factors are not the main reason for reduced activity in those supported systems. Very recently, Toi has contributed to this field by preparing bifunctional resins **49** by polymerization of the corresponding functional monomers. Besides OH, different functional groups (OAc, CH_2OH, OMe) have been introduced as the secondary function. These resins have been used as catalysts for the Morita–Baylis–Hillman reaction, the best results always being observed for compound **49** containing phenolic subunits. Thus, for instance, in the reaction of 2-cyclopenten-1-one with 4-nitrobenzaldehyde, compound **52** (Scheme 10.11) was obtained in 81% yield with **49**, whereas the resin with hydroxymethyl groups afforded a 69% yield, and much lower yields were obtained for polymers not containing hydroxyl groups (30–56%). This has been rationalized in terms of the cooperative participation of the hydroxyl group either by hydrogen-bond activation of the Michael acceptor or by stabilization of the enolate intermediate [187]. A similar increase in

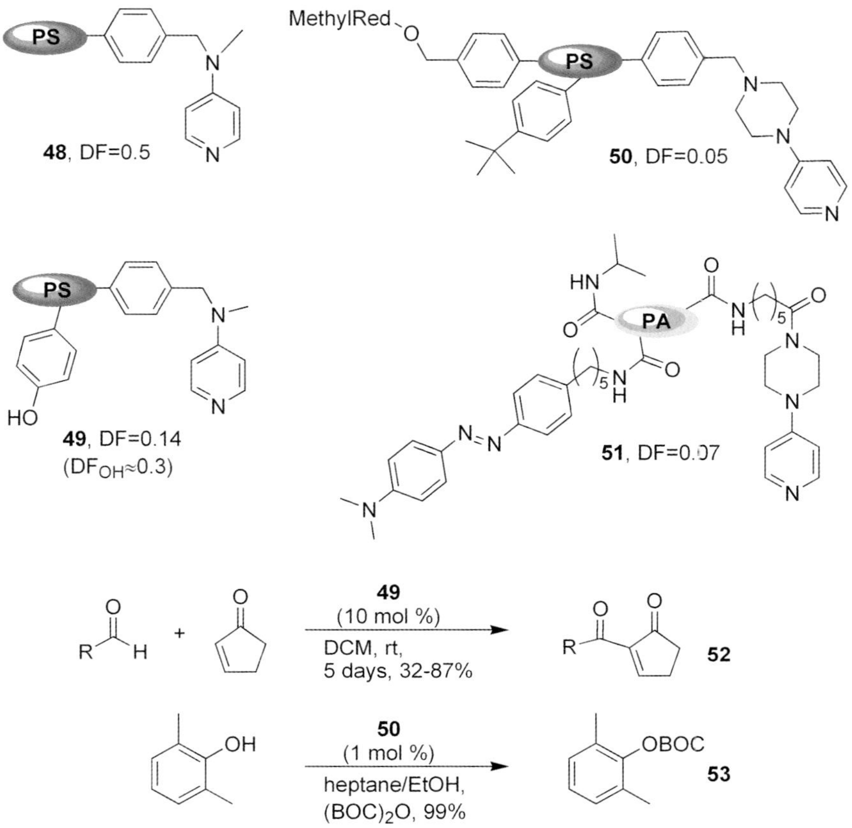

Scheme 10.11 Other soluble noncross-linked polystyrene resins containing DMAP equivalent functions.

catalytic activity has been found, for instance, with MCM-41 postfunctionalized with aminopropyl groups but maintaining silanol groups properly located [188]. The introduction of *tert*-butyl groups in the polymeric network, as with **50** ($DF_{t\text{-}Bu}$ = 0.95), produces linear polymers with solubility properties that greatly facilitate separation and recovery without requiring precipitation and filtering. Resins **50** are soluble in heptane and the reaction of BOC protection of 2,6-dimethylphenol is carried out in a heptane/ethanol mixture (1:1) phase. Addition of water produces phase separation and the catalyst in the heptane phase can be directly reused. The yields increase continuously for successive runs up, reaching 99% at the fifth cycle. Subsequently, these optimum results are maintained. Decreased yields for the first runs are attributed to the product being retained in the heptane layer. Once saturation is attained the optimum yields are observed. The presence of a methyl red dye attached to the polymer (DF_{dye} = 0.002) confirmed that polymer **50** is confined to the apolar phase [189].

Soluble resin **51**, based on an acrylamide backbone, was also prepared by Bergbreiter, using an approach similar to that considered for **50**. In this case the polymer is soluble in polar solvents. The presence of the azo dye (DF_{dye} = 0.0002) is used to check that less than 0.1% of the resin remains in the solution after precipitation with hexanes. This catalyst was able to both catalyze efficiently the acetylation of 1-methylcyclohexanol to give **45** and afford BOC protection of phenols to prepare structures similar to **53** in DCM at room temperature (Scheme 10.11). These reactions required very low catalyst loadings (0.2–5 mol.%). No problems were found for the recycling of **51** [190].

DMAP functionalities have also been attached to the core of highly branched but soluble, multiarm star polymers [191]. This system can catalyze efficiently the Baylis–Hillman reaction for the preparation of **46**. More interestingly, confirmation of the catalytic sites at the core of those star polymers allowed the simultaneous use of a similar polymer containing sulfonic groups without observing any mutual deactivation. Thus, **46** can be prepared starting from the dimethylacetal precursor of the aldehyde, through a one-pot cascade reaction involving sequential acid-catalyzed acetal hydrolysis followed by the DMAP catalyzed Baylis–Hillman reaction.

Simple polystyrene supported triphenylphosphine (**54**, Scheme 10.12) is commercially available and has been used for many synthetic transformations, in particular as ligands for immobilization of metal complexes [172, 192, 193]. Its basicity can also be exploited as a mild basic catalyst for different organic reactions. Thus, polymer **54** catalyzed efficiently the Trost γ addition of several 1,3-dicarbonylic compounds to methyl 2-butynoate to afford tricarbonylic derivatives with the general structure **58** (Scheme 10.12). The reaction takes place with good yields in a mixed water/toluene (5:1) medium at 90 °C. Complete conversion of the starting materials requires 20 h, but the reaction can be greatly accelerated by microwave heating (45 min) [194]. A high loading of the catalyst (35%) is used under both experimental conditions. A polystyrene resin using 1-bis(4-vinylphenoxy)butane as the cross-linking agent instead of DVB (JandaJel resin **55**) was applied to promote the aza-Baylis–Hillman reaction in a process similar to that described above for the synthesis of **47**. For this process the catalyst loading is 10 mol.% and the desired products are obtained in 63–99% yield after 10–48 h in THF at room temperature [195]. Those results seem to be significantly affected by the presence of the polar cross-linking agent and by the degree of functionalization of the polymer. Both factors need to be considered when designing a polymeric catalyst as, in some instances, they play a critical role in determining the efficiency of the corresponding species [182]. Toi has also studied the preparation of bifunctional basic catalysts similar to **49** in which the DMAP-like moiety was substituted by a triphenylphosphine groups. The trends observed were similar to those described in the case of **50** [187].

The same strategy described for polymer **50** was used with soluble supported phosphine **56** in which the DMAP moiety has been substituted by a triphenylphosphine group. This catalyst was assayed for the addition of 2-nitroproprane to methyl acrylate, affording **59** (Scheme 10.12). As with **50** the reaction was carried

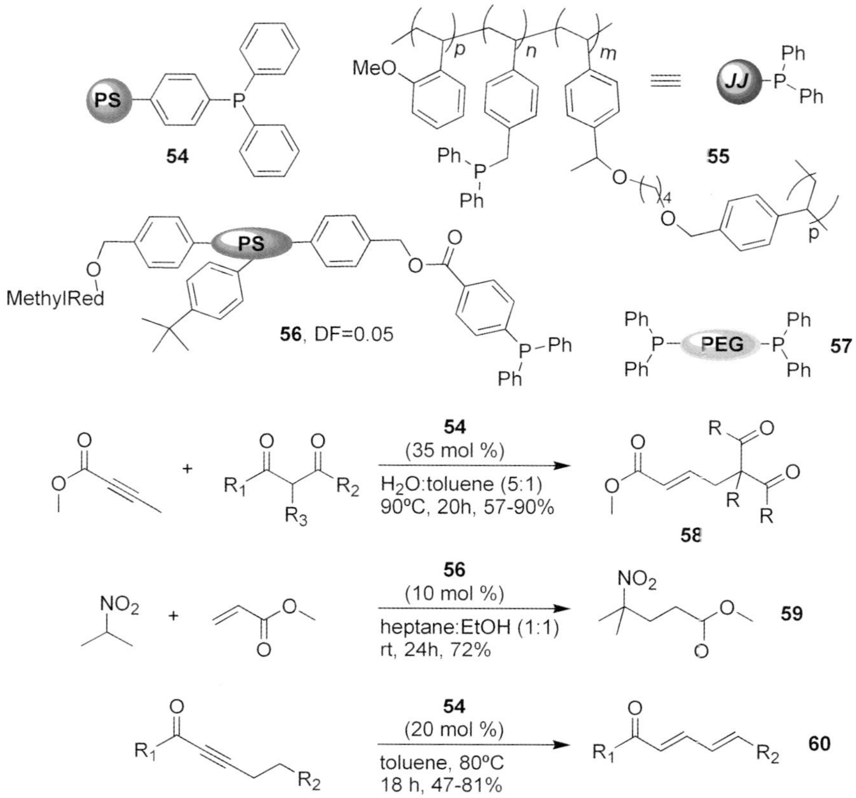

Scheme 10.12 Examples of catalysis by non-chiral polymer-supported phosphines.

out in a heptane/ethanol mixture and phase separation takes place upon addition of water. Once product saturation of the heptane phase occurs after the fifth cycle a maximum yield of 72% was reached [189]. A second soluble phosphine (**57**) was prepared by Shi from PEG bis-mesylate by reaction with lithiated diphenylphosphine. The aza-Baylis–Hillman reaction was used again to assess the catalytic behavior of this supported phosphine. The results obtained were only moderate, similar to those obtained with catalysts **40** and **41**. Betters results were found for this process using phenyl acrylate instead of methyl vinyl ketone. This allowed the synthesis α-methylene-β-amino esters in 50–87% yields [184].

Supported phosphine **54** also acts as an efficient catalyst for isomerization processes. The isomerization of nitroolefins to obtain the (*E*)-isomer has been reported to occur in quantitative yields and without the activation of other side reactions [196]. Similarly, the isomerization of α,β-ynones to (*E*,*E*)-α,β-γ,δ-dienones (**60**) is catalyzed by **54** with high efficiency and stereoselectivity (Scheme 10.12) [197]. The recovery and reuse of these catalysts is generally reported, but this is strongly limited, as in many other applications involving phosphines, by their oxidative degradation.

Other PS-DVB functional polymers that have been studied as basic catalysts include those containing alkylguanidine groups (**61**, **62**), piperazine (**63**), formamides (**64**, **65**), phosphoramide (**66**) and iminophosphorane (**67**) (Scheme 10.13). Resins **61** and **62** catalyze transesterification processes of industrial interest [198]. Compound **62** has been reported to also act as a good catalyst for different C–C forming reactions, such as the addition of nitromethane to saturated and unsaturated ketones and for the addition of diethylphosphite to carbonyl compounds [199]. A strong deactivation is observed, however, when recycling is attempted because of the extensive loss of the functional moieties with use. Polymer **63** catalyzes efficiently various Knoevenagel condensations with a strong simplification of the work up associated with the use of piperidine for those reactions [200].

Polystyrene supported formamide **64** was more reactive than **65** for the allylation of aldehydes with allyltrichlorosilane to afford **69** (Scheme 10.13) [201]. Good yields after 24 h could be obtained only when an excess of catalyst was used. Recycling of the catalyst was possible without appreciable loss of activity for three runs, but only when agitation was carried out with an automatic shaker. The yield decreased dramatically for the third cycle when a magnetic stirrer was used. A similar phenomenon was detected when iminophosphorane **67** was studied for different

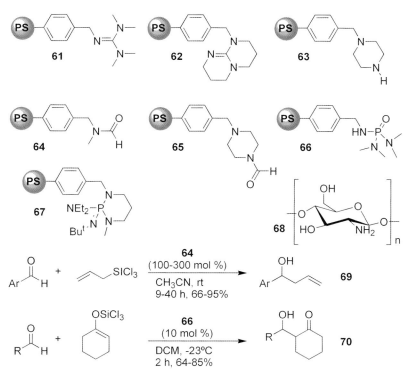

Scheme 10.13 Other PS-DVB functional polymers and their use as basic catalysts.

Michael additions [202]. Phosporamide **66** was prepared as an insoluble equivalent for HMPA and was studied for the aldol condensation between trichlorosilyl enol ethers and aldehydes to give keto alcohols **70** (Scheme 10.13). Moderate to good yields and selectivities were obtained. In the presence of 10% of the catalyst at −23 °C the rate was significantly accelerated, albeit the syn/anti stereoselectivity was decreased by a factor of 5 [203].

In a very different approach, a chitosan biopolymer hydrogel containing amino and hydroxy groups (**68**) has been efficiently employed for aldol and Knoevengal reactions under biphasic conditions. The gel beads can be filtered and recycled for three consecutive runs, maintaining the same overall efficiency.

10.3.3
Polymer-Supported Chiral Basic Catalysts

Numerous chiral amines, many of them of natural origin are known. Thus, an obvious step forward in the development of polymer-supported basic catalysts is the substitution of the achiral amino fragments considered above by chiral amino moieties to prepare the corresponding heterogeneous chiral basic catalysts. The first studies in this field were carried out using polymers containing chiral fragments derived from cinchona alkaloids **71** (Figure 10.7). To anchor the chiral fragment two main approaches have been used. The first exploits the presence of the vinyl group at C-3, whilst the second takes advantage of the hydroxyl group in C-9. Attachment through the nitrogen of the amino functionalities is also possible, but in this case ammonium moieties are obtained that have been mainly studied for phase transfer catalysis (PTC), as discussed below.

Initial achievements in this field were carried out by Kobayashi as early as 1978. Copolymerization of **71** with acrylonitrile produced linear polymers **72** (Scheme 10.14), which are soluble in polar aprotic solvents, that were assayed for different Michael additions. In general, good conversions were obtained, but enantioselectivities were always below 60% ee [204–206]. In a related approach by Oda, different spacers were introduced between the polyacrylonitrile backbone and the chiral fragment, which resulted in an increase of selectivity up to 65% ee [207]. Addition of thiol groups in PS-DVB resins to the double bond allowed the preparation of the corresponding insoluble polymers **73**, which were assayed by Hodge for the addition of thiols to unsaturated ketones and nitrostyrene. Again, selectivities were

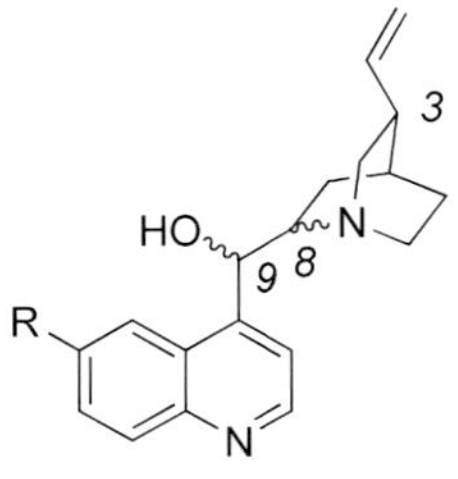

a) Quinine: R=OMe, 8S,9R
b) Quinidine: R=OMe, 8R,9S
c) Cinchonine: R=H, 8R,9S
d) Cinchonidine: R=H, 8S,9R

Figure 10.7 Chiral fragments derived from cinchona alkaloids.

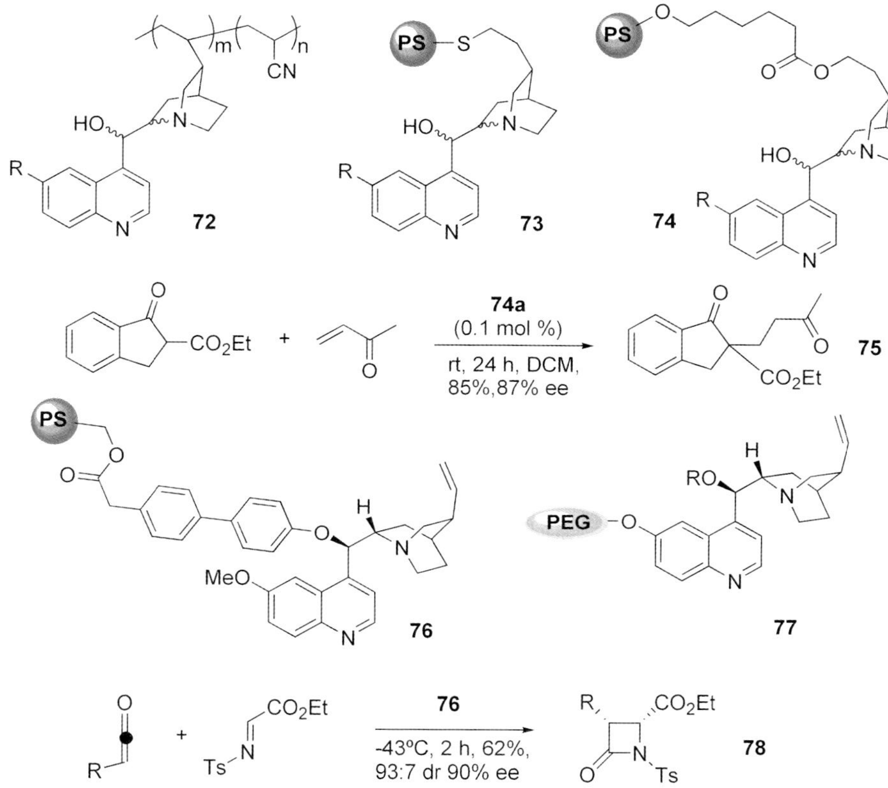

Scheme 10.14 Polymer-supported chiral basic catalysts and their use.

moderate (up to 45% ee for the resin derived from **71c**) but the catalyst could be recycled and reused three times without any decrease in yield or selectivity [208]. Much better results were reported (up to 85% yield and 87% ee) when the double bond of **71a** was hydroxylated at the terminal position and the resulting alcohol esterified with a PS-DVB resin containing a carboxyl functionality separated from the backbone by a long spacer (**74**) [209].

The second approach, using the hydroxyl at C-9, for the attachment has been assayed with the introduction of polymerizable acryloyl derivatives to obtain the corresponding homo and copolymers [210, 211]. Alternatively, esterification with appropriately functionalized PS-DVB resins has been also employed [212]. Enantioselectivities found for different reactions were moderate or low (<40% ee). In the first case homopolymers were significantly more efficient than copolymers. Perhaps the most relevant example of this class is the report of Lectka on the synthesis of β-lactams **78** (Scheme 10.14) through the sequential use of different columns containing the corresponding insoluble catalysts or scavengers. In this case, the polymer-containing quinine fragments **76** gave the best results, acting as a catalyst for the [2 + 2] Staudinger reaction of a ketene and an imine, affording a

93 : 7 cis : trans ratio, with a 90% ee for the major diastereomer. As has been observed in other cases, the system requires a few initial runs before reaching its optimum performance, but after this period the catalyst maintained its efficiency for up to 60 cycles [213]. Application of this methodology has been extended to new reactions and to the preparation of different molecules of pharmacological interest [214, 215].

A very different approach was followed by Cozzi for the preparation of soluble polymer **77** based on a PEG backbone. Preparation of this polymer requires the initial demethylation of quinine and further attachment to PEG through the phenolic oxygen. The interference of the oxygen atoms of the PEG by competitive hydrogen bonding to the substrate was used to rationalize the low selectivities found (22% ee) for the addition of thiophenol to cyclohexenone [216]. The important changes in performance observed upon anchoring through the different strategies, obviously involving a modification of some of the functional groups of the starting alkaloid, suggest that the catalytic mechanism is far more complex than with a simple general base.

The use of chiral equivalents of other supported bases considered in Section 10.3.2 is also attractive, but was not achieved until very recently. Substitution of the dimethylamino subunit in DMAP by a α-methyl proline fragment provides a chiral structure that can be easily attached to different supports by standard methodologies using the carboxyl function. This has allowed the preparation of polymers **79** (Scheme 10.15) from standard PS resins of different functionalization degrees and from a Wang resin [217]. According to previous studies in solution, the benchmark reaction selected for those polymers was the kinetic resolution of *cis*-1,2-cyclohexanol mono-4-dimethylaminobenzoate [218]. This was carried out in DCM at room temperature with a defect of the acylating agent (isobutyric anhydride). Best results were observed at 67% conversion, achieving a 93% ee for the unreacted alcohol **83** (Scheme 10.15). No significant effects were found for the changes in loading or in the nature of the support. Recycling of the catalyst was accompanied by a decrease in its efficiency and, in general, the supported systems were less efficient than their soluble counterparts. Different chiral phosphoramides, such as **80** and **81**, have been prepared by homopolymerization of the corresponding vinylic monomer or by copolymerization in the presence of variable amounts of styrene. In this case the benchmark reaction considered was the allylation of benzaldehyde with allyl trichlorosilane in the presence of an excess of diisopropylethylamine. Resins **81** were more efficient than **80** both in terms of activity (82–84% vs 62–63%) and enantioselectivity (62–63% ee vs 49–51% ee). No attempt was made to study the reuse of those systems. Rather interestingly, in this case the performance (conversion and enantioselectivity) of those supported phosphoramide was better than that of the corresponding soluble analogues. This has been attributed to the possibility of site–site intrapolymeric interactions that facilitates their actuation as bis-phosphoramides.

The latter benchmark reaction for the preparation of **84** was also used recently by Reggelin to study a completely different kind of chiral polymeric bases (Scheme 10.15). In their work, the authors used pyridyl *N*-oxide substituted helically chiral poly(methacrylate)s **82** prepared by helix sense selective anionic polymerization of

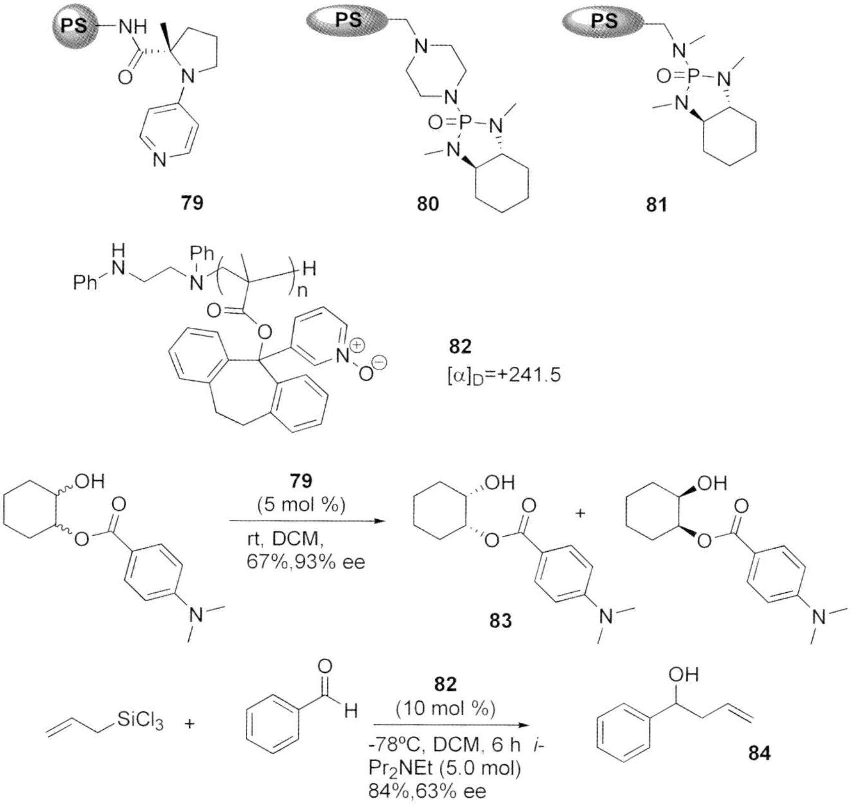

Scheme 10.15 Use of chiral equivalents of other supported bases.

the corresponding acrylic monomers in the presence of a chiral non-racemic base [219]. Only the pyridine-containing monomers could be polymerized efficiently, and this fragment was oxidized to the N-oxide with m-CPBA. Those polymers maintaining their helicity on the required time scale were able to act as organic basic catalysts for the addition of allyltrichlorosilane to benzaldehyde and to induce a low enantioselectivity (19% ee) on the final product **84**. Although the selectivity achieved is rather low, those results clearly open new ways to the development of chiral polymeric catalysts.

10.4
Polymer-Supported Phase Transfer Catalysts

10.4.1
Non-chiral PTC Using Insoluble Supports

Many organic reactions involve the combination of apolar substrates with polar reagents that are only sparingly soluble in the organic solvents required to carry

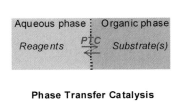 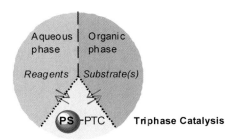

Figure 10.8 Phase transfer catalysis.

out selectively the reaction. One simple catalytic methodology to facilitate such reactions is the addition of a phase transfer catalyst (Figure 10.8). This can transfer the polar reagent from an aqueous solution (or from the solid state) to the organic phase. In this way the effective concentration of the reagent in the organic phase is increased, leading to an enhancement of reaction rate. The opposite protocol, in which the apolar substrate is transferred to the polar phase, is also possible, but it is less usual. In some instances, the use of the appropriate phase transfer catalyst can also modify the selectivity of the corresponding reaction. Many processes have been studied under phase transfer conditions and this methodology has become a standard one in organic synthesis [220, 221].

Nevertheless, the separation of the catalyst at the end of the reaction and, if possible, its recycling is often a limiting factor for the application of PTC. The chemical nature of the catalyst makes it at least partially soluble both in polar and apolar solvents and higher catalyst loadings are often used to maximize the effects on the reaction rates. This led very soon to the development of polymer-supported phase transfer catalysts [222]. When using insoluble supports, an additional phase is added to the former biphasic system and, accordingly, the term "triphase catalysis" was coined (Figure 10.8) [223–225].

In most cases gel-type microporous resins have been used for this purpose and the reaction takes place within the polymer gel phase. This is clearly associated to the main limitations found for the use of those polymer-supported catalysts, which are related to the effective diffusion of reagents and substrates into the polymeric matrix. For a polymer-supported PTC, the following steps need to take place for the reaction to occur: (i) the substrate(s) and the reagents need to be transported to the surface of the polymeric particles; (ii) the substrate(s) and the reagent must diffuse from the particle surface to reach the active site; (iii) the transformation takes place at the active site; (iv) the products have to diffuse back to the surface of the polymeric particle and (v) the products are transported either to the bulk organic or the aqueous phase according to their solubility characteristics. According to the heterogeneous nature of the process, with three phases being involved, any of the mass transfer or diffusion steps can be rate limiting [225–228]. Thus, unsurprisingly, initial results revealed an important decrease in the activity of the supported system when compared with that of the homogeneous analog.

The chemical structures of the active sites in PS-DVB supported PTC can be roughly classified in three main groups: (i) ammonium salts closely related to traditional anion exchange resins (**85**); (ii) phosphonium salts (**86**) [229–231]; (iii) crown ethers [232, 233], and their open chain equivalents (podands), cryptands and other synthetic receptors (**87–89**) (Scheme 10.16). The onium salts act by transferring the active anionic species to the gel phase by anion exchange with the original counterion (X⁻ in **85** and **86**). Neutral crown ethers and related species act by selective complexation of the corresponding cation (usually alkaline or alkaline-earth cations). The counter anion is transferred to the gel phase as the counter anion is required to maintain the electroneutrality. In general, both cryptands and polynitrogenated receptors have been less used. The presence of several nitrogen atoms usually results in polyalkylation of the receptor with the chloromethyl groups of the Merrifield resin, giving polymers with complex functional structures and with a significantly increased cross-linking and a reduced mobility of the active sites [234, 235].

Taking into account the former discussion on the difficulties associated with the efficient use of polymer-supported PTC, it is easy to understand that an appropriate design of the overall process is essential for success. Several factors have been found to be critical for the performance of the catalyst. For onium salts, the use of a bulky R groups is preferred, in particular for nucleophilic displacement reactions. For the substitution of bromide by iodide in 1-bromooctane, reaction rates

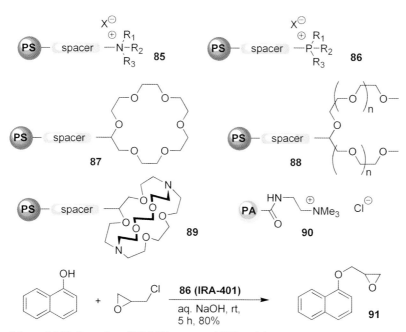

Scheme 10.16 Examples of PS-DVB supported PTC and the use of one of them in the synthesis of epoxide **91**.

were much higher for the tributyl derivatives of **85** and **86** than for the trimethyl derivatives [236]. Crown ethers show similar activity to **85** and **86** when soft anions are involved, but they are less active for hard anions [237]. The use of long aliphatic spacers always produces more active catalysts [238]. This can be linked to the improved accessibility of the catalytic sites and to a modification of the hydrophobic/hydrophilic balance of the reaction site. Similarly, lightly cross-linked polymers are more appropriate than those with a high level of cross-linking [239]. Higher functionalization degrees of the polymers are detrimental for the catalytic activity for onium-functionalized resins [240, 241]. For crown ethers, however, optimum results can be obtained with highly loaded resins [242]. With onium resins, high loadings make the polymer more compatible with water than with organic solvents. On the other hand, the reaction works better with resins with small particle sizes, when the organic components, in particular the solvent, swell efficiently the polymeric matrix and when concentrated aqueous solutions of the inorganic reagent are used. Moreover, as the reactions involve two immiscible phases, vigorous stirring is required. This factor is, however, limited by the need to maintain the integrity of the polymeric particles, in particular if recycling is intended [225, 243].

Polymer-supported PTC has been applied to a large variety of organic reactions [244–247], including classical substitution reactions such as the formation of nitriles from halogenides, halogen exchange and formation of ethers and thioethers, in particular when phenols are involved. Other important reactions include the nucleophilic opening of epoxides, the synthesis of dichlorocyclopropanes, the oxidation of alcohols to carbonyl groups using water-soluble anionic oxidants and different elimination reactions. An example is the synthesis of the epoxide **91**, an intermediate for the synthesis of propanolol, an anti-oxidant and β-blocker, by alkylation of 1-naphthol (Scheme 10.16) [248]. In this case a commercial ion exchange resin (Amberlite IRA-401 in its OH$^-$ form) is used as the PTC, and the final product is obtained by epoxide ring opening of **91** with isopropylamine.

An interesting approach to minimize diffusional and mass transfer limitations has been developed by Ford with the use of insoluble polymer colloids. Those are submicroscopic anion exchange resins with very high surface/mass ratios that can be recovered at the end of the reaction by coagulation followed by filtration or by direct ultrafiltration [249–251]. Insoluble polymers derived from acrylamide have been also studied for PTC, especially those containing quaternized ammonium groups. In particular, resin **90** containing a 2% of DVB as cross-linking agent has shown catalytic activity similar to that of resins **86** for different processes [252].

10.4.2
Chiral PTC Using Insoluble Supports

Cinchona alkaloids are by far the chiral fragments most used for the challenge of achieving chiral PTC using insoluble supports. For this purpose, one of the nitrogen atoms of **71**, usually the aliphatic one, is quaternized, either by using this atom to anchor the polymeric backbone (**92**, Scheme 10.17) or by its appropriate

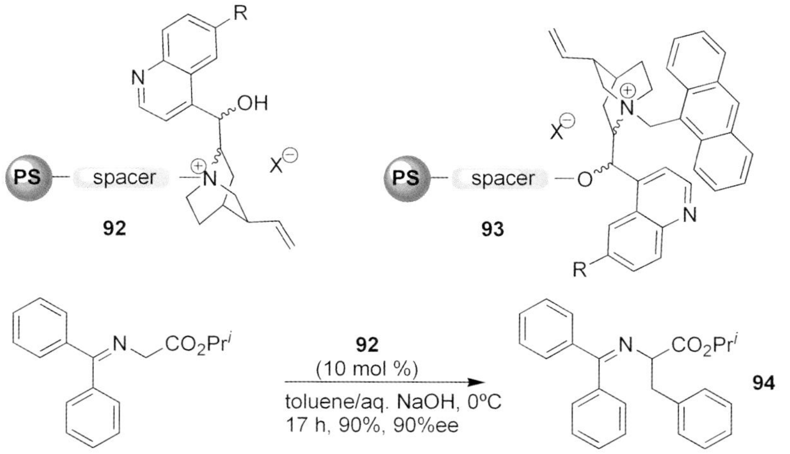

Scheme 10.17 Insoluble supports for chiral PTC and their use in the alkylation of amino acids imines.

alkylation when the attachment is made by one of the strategies above considered (Section 10.3.3). Initial studies were carried out using resins **92** for different synthetic procedures with disappointing results. Thus, the epoxidation of chalcone using NaOH and 30% H_2O_2 in a two-phase liquid system was carried out with very low selectivities (<4%) independently of the use of short (–CH_2–) or long spacers (–$(CH_2)_{12}$–) [253]. Similarly, only a 27% ee was achieved for the Michael addition of methyl (1-oxoindan-2-yl)-carboxylate to 3-buten-2-one catalyzed by resin **92** derived from alkaloid **71d** [254].

A much better performance was observed for the alkylation of amino acids imines following the O'Donnell–Corey–Lygo methodology [255]. Benzylation of the benzophenone imine of glycine *i*-propyl ester to afford compound **94** was catalyzed by resin **92** (1% cross-linked, spacer = CH_2) derived from **71d** (R = H, 8*S*,9*R*) with 90% yield and 90% ee (Scheme 10.17) [256, 257]. The reaction was carried out at 0 °C in toluene using 25% aqueous NaOH. Changes in the nature of the ester alkyl residue, the alkylating agent or the reaction conditions always led to a decrease in selectivity [256, 258]. The introduction of longer aliphatic spacers between the quinuclidinium nitrogen and the polymeric backbone produced, in this case, some interesting results. Using **92** (1% cross-linked, from **71d**, spacer = –$(CH_2)_n$- with n = 4, 6, 8) the major enantiomer obtained for **94** (*tert*-butyl ester) had the (*R*) instead of the (*S*) configuration obtained in the former case [259]. Moreover, the same configuration was observed when the pseudoenantiomeric alkaloid **71c** was used. Under optimized conditions the selectivity (up to 81%) was slightly lower than that for the resin not containing the spacer.

The alternative method of anchoring through the hydroxyl group was used for the preparation of polymers **93** [260]. In these polymers, the quaternized nitrogen atom bears a 9-anthracenylmethyl group, a structural feature that had been observed to improve the selectivity in solution. By working in toluene at −50 °C and using solid CsOH in excess the authors were able to obtain for the *tert*-butyl ester with a selectivity (94% ee) almost equivalent to that reported for the soluble catalyst. Notably, the enantioselective catalytic alkylation of polymer-supported glycine imine *tert*-butyl ester has been reported recently [261]. In this approach, the substrate is anchored to a solid phase while a soluble chiral phase transfer catalyst is used.

Other chiral fragments such as *N*-methylephedrine, *N,N*-dimethyl α-methylbenzylamine or strychnine have been immobilized, as quaternary ammonium salts, on PS-DVB and tested for different reactions under PTC conditions, but only low levels of enantioselectivity were obtained [262–264].

10.4.3
PTC Using Soluble Supports

The use of soluble supports for the attachment of PTC catalysts is a logical step to minimizing mass transfer and diffusional problems while keeping the advantages for recovery, purification and work-up of supported systems. From the different potential supports, PEG continues being the one most often selected. In fact, PEG itself is a PTC, according to its hydrophilic/hydrophobic balance [265]. At the same time, PEG monomethyl ethers grafted to PS-DVB have been reported as very effective phase transfer catalysts [266–268]. Nowadays, most efforts in this area concentrate on the anchoring of chiral fragments with potential PTC properties to develop efficient procedures for chiral PTC.

The use of PEG-supported ammonium or phosphonium salts for PTC was introduced relatively late in this field. Preparation of compounds **95** and **96** and the corresponding monofunctional derivatives (i.e. **97**) were reported in 1991 (Scheme 10.18) [269]. These catalysts were shown to be more active than PEG or tetrabutylammonium bromide for several elimination reactions. More recently,

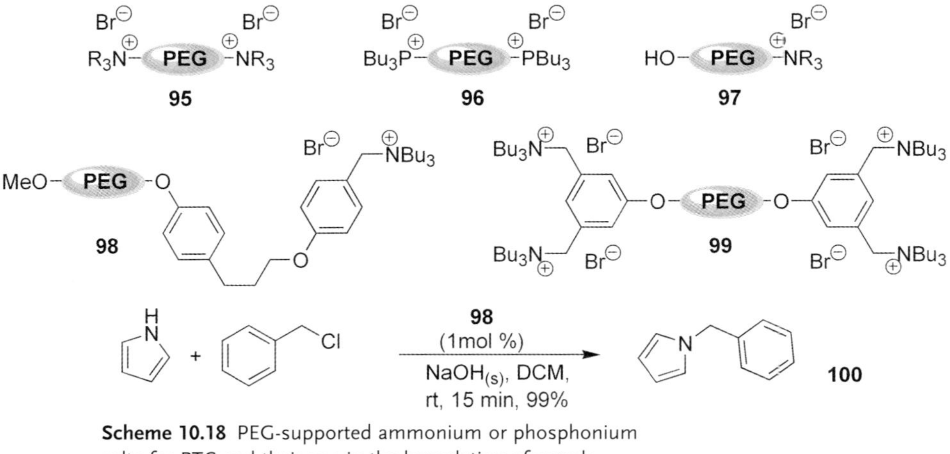

Scheme 10.18 PEG-supported ammonium or phosphonium salts for PTC and their use in the benzylation of pyrrole.

Benaglia and Cozzi described the preparation and use of compound **98**, in which the ammonium group is separated from the PEG backbone by a spacer [270, 271]. The activity observed for **98** was higher than for related PS-DVB systems or for PEG-derivatives containing shorter linkers. Thus, for instance, benzylation of pyrrole can be completed (99% yield) after less than 15 min at room temperature using solid NaOH and DCM containing 1% of **98**. After three recycles, a 93% yield of **100** was obtained under the same conditions (Scheme 10.18). An important problem for those PEG-based PTCs is their very low degrees of functionalization (0.182 mmol g^{-1} for **98**) [26]. To overcome this drawback, which necessitates the use of large weight amounts of the functional PEG even for low catalyst molar ratios, the tetrafunctionalized catalyst **99** was prepared [272]. This catalyst presents a much higher loading (0.679 mmol g^{-1}) and requires weight amounts almost four times lower than for **98**. At the same time, its catalytic activity was significantly increased. Thus, for the same benchmark reaction leading to the preparation of **100**, an 85% yield was obtained in 5 min, under the same conditions but with only 0.1% molar catalyst loading [273].

Chiral versions of PEG-bound PTCs have been prepared by the attachment of cinchona alkaloids. Catalysts **101** and **102** (Figure 10.9), related to non-cationic **77**, were prepared starting from monomethylated PEG5000 and using spacers similar to that of **98** [216]. Both materials were tested as catalysts for the synthesis of the *tert*-butyl ester related to **94** and similar compounds, using the standard O'Donnell–Corey–Lygo procedure. Under optimized conditions (DCM, solid CsOH, −78 °C, 60 h, 10% mol catalyst loading) both systems were less efficient than the insoluble

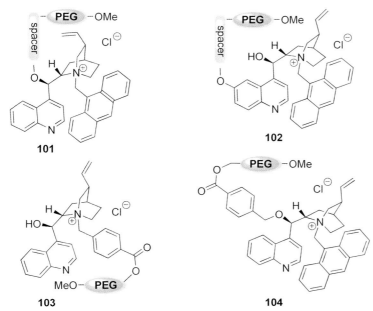

Figure 10.9 Chiral versions of PEG-bound PTCs.

polymer-supported analog **93**. Only with **101** was a reasonable enantioselectivity observed (64% ee, 75% yield). The decrease in performance of these PEG-bound catalysts has been ascribed, as in other cases, to the capacity of the PEG chain to interact with inorganic cations, modifying the microenvironment of the catalytic site and the exact mechanism of the reaction [216, 270, 274, 275]. Almost simultaneously, Cahard described the synthesis and study, for the same reaction, of PEG-bound catalysts **103** and **104** (Figure 10.9) using a different spacer [259]. Again, both catalysts showed a lower performance than analogous insoluble PS-bound systems **92** and **93**: 80% yield and 81% ee (*S*-isomer) for **103** (starting from **71d**; 10% of catalyst, *tert*-butyl ester, 50% aq. KOH/toluene, 0 °C, 15 h); 67% yield and 71% ee for **104d** (solid CsOH/toluene, -60 °C, 72 h). Strong variations were observed in the outcome of the reaction by changing the reaction conditions, the solvents, the alkylating agents, the nature of the PEG or when the pseudoenantiomeric alkaloid **71c** was used instead of **71d**. Recycling could not be efficiently accomplished for **103** and **104**, apparently due to the lability of the ester linkage.

Some examples of the use of functional soluble polyethylene supports have also been reported. Bergbreiter and coworkers used ethylene oligomers (M_W = 1000–3000) containing terminal active groups such as ammonium, phosphonium or crown ethers [276]. These materials have thermoresponsive solubility. Being only soluble in organic solvents at high temperatures (ca. 100 °C), they easily separate from the reaction mixture by cooling. This allowed their recovery and reuse for up to four cycles without modification of their catalytic activity. The phosphonium and crown ether derivatives displayed an activity similar to or higher than that of the related PS-bound PTCs. The utility of some of those systems is limited, however, for their thermal instability at the working temperatures required. Polyethylene chains have also been introduced on the hydroxyl end of PEG-bound ammonium salts **97** to take advantage of their thermoresponsiveness [277]. The resulting catalysts presented a reduced activity relative to that of the parent **97** for nucleophilic substitution reactions.

10.5
Polymer-Supported Oxidation Catalysts

10.5.1
Non-chiral Polymer-Supported Oxidation Catalysts

The potential for the use of supported species for oxidation processes was very early realized. Many traditional oxidizing agents have important drawbacks such as are low solubilities, presence of highly toxic or hazardous materials and problematic separation of byproducts at the end of the reaction. Many of those limitations can be overcome through the use of supported reagents or catalysts. Much effort has been dedicated to the preparation of inorganic materials such as mixed oxides, functionalized silicas, clays or zeolites. The delay in the development of the corresponding oxidizing systems supported on organic polymers has been often associated to the presence of structural fragments (i.e. benzylic positions in PS-DVB resins) with a limited oxidative stability. Despite this, efficient polymer-supported oxidative reagents have been prepared and some of them have become commercially available [278]. These include the use of supported peracids, the ionic immobilization of inorganic oxidants, in particular supported equivalents of pyridinium chromates, or hypervalent iodine compounds. In many instances it is relatively easy to consider a catalytic version for such oxidations involving functional polymers as reagents. For oxidants supported as anions on ion exchange resins or related materials, the use of polymer-supported PTC conditions is the obvious methodology and this has been briefly mentioned above. For other situations, the main approach is the use of resins containing functional groups that are selective oxidants and that can be easily regenerated in the presence of a terminal oxidizing agent like Oxone, H_2O_2 or, ideally, O_2. Many of them involve metal complexes in which the metal center switches between different oxidation states in the catalytic cycle, but examples in which an organic functional group is the one involved in the process are also known. The best studied of such functional polymers include nitroxyl radicals, acids/peracids, ketones and hypervalent iodine derivatives.

Polymer-supported 2,2,6,6-tetramethyl-1-piperidinyloxy radical (TEMPO) has been studied for the oxidation of primary and secondary alcohols using various terminal oxidants. This corresponds with the well-known applications of this and related molecules in solution organic synthesis [279]. Immobilization of TEMPO was initially carried out in 1985 on both cross-linked and linear PS to give polymers **105** and **106** (Figure 10.10)[280, 281]. Those polymers presented high loadings, ranging from DF 39 to 100 [282]. Those resins were used to efficiently catalyze the oxidation of benzyl alcohol to benzaldehyde at room temperature using potassium ferricianide or cupric chloride as the terminal oxidant with 4% catalyst loadings. The resins with lower DFs were the ones with a better performance for this reaction. The introduction on the polymeric backbone of additional functional functionalities to increase its hydrophilicity had a clearly positive effect on the catalytic activity of the resulting resins **107** and **108** (Figure 10.10). This can be associated with the fact that the reaction was carried out in water/organic solvent mixtures.

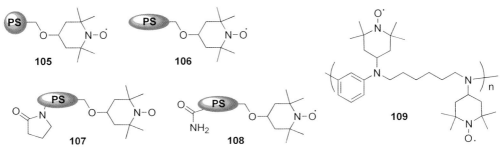

Figure 10.10 Examples of polymer-supported 2,2,6,6-tetramethyl-1-piperidinyloxy radical (TEMPO).

The soluble polynitrogenated commercial polymer Chimassorb 944 was used more recently to prepare the corresponding resin **109** [283, 284]. This catalyst was effective, at low loadings (1 mol.%), for the selective oxidation of alcohols using bleach or cuprous chloride/air as terminal oxidants, with short reaction times and 80–99% yields. Resin **109** can be recovered by precipitation with *tert*-butyl methyl ether, which allowed its reuse at least twice. This catalyst was found to be more active than other supported TEMPO systems, including those developed from silica and other inorganic supports [285, 286]. Recently, Toy has shown how a JandaJel supported TEMPO can also act as an efficient catalyst (1 mol.%) for the oxidation of alcohols to aldehydes and ketones in a multipolymer system in which PS-supported diacetoxyiodosobenzene is the terminal oxidant [287]. An oxoammonium resin, the expected oxidation product from polymer-supported TEMPO, has been used as a stoichiometric reagent for the efficient oxidation of alcohols [288].

Polymer-supported arsonic acids (**110**, Scheme 10.19) can be prepared with different functionalization degrees from brominated PS-DVB by lithiation and reaction with triethylarsenite followed by oxidation with hydrogen peroxide [289]. The Baeyer–Villiger oxidation of cyclic ketones could be accomplished with high yields and selectivities at 60–90 °C using catalytic amounts of those polymers (ca. 1–3 mol.%) in dioxane using 90% H_2O_2. The activity of **110** was, however, clearly lower than that for benzenearsonic acids in solution. The use of 30% H_2O_2 is also possible working under triphasic conditions with $CHCl_3$ as the organic solvent. Under those conditions the kinetic of the process was much slower, but at the same time retards hydrolysis of the ester formed. The same polymers can be effi-

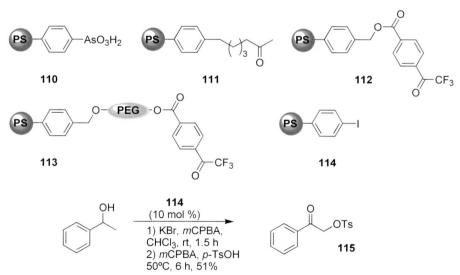

Scheme 10.19 Non-chiral polymer-supported oxidation catalysts and the use of one them in the direct oxidation of alcohols to α-tosyloxyketones (**115**).

ciently applied to the epoxidation of alkenes having different substitution patterns [290]. Polymer-bound phenylselelinic acids (PS-SeO$_2$H) have been prepared by reaction of mercuriated PS-DVB (2% DVB) with SeO$_2$ [291]. In the presence of 30% H$_2$O$_2$ those resins are converted into the corresponding peroxyselelinic acids. Thus, using H$_2$O$_2$ as terminal oxidant, these polymers catalyze (1.5 mol.%) the Baeyer–Villiger oxidation of ketones and the epoxidation of alkenes. In the later case, however, the products open to the corresponding trans diols. Using tert-butyl hydroperoxide, polymer-bound phenylseleninic acids can also oxidize benzylic and cinnamyl alcohols to the corresponding aldehydes or ketones, as well as some aromatic systems (hydroquinone/benzoquinone).

Dioxiranes are useful oxidants that can be generated in situ from ketones and Oxone [292]. Formally, the ketone acts catalytically in this process, but under the reaction conditions used it is degraded through a Baeyer–Villiger reaction. This usually requires larger amounts of ketone than expected for a catalytic process. Different PS-DVB polymers containing ketone functionalities (**111–113**, Scheme 10.19) have been prepared and applied to the Oxone oxidation of anilines to nitro derivatives, pyridines to N-oxides and alkenes to epoxides [293, 294]. The supported ketone is always employed in more than stoichiometric amounts but its degradation is hampered upon immobilization, allowing its reuse for successive cycles without loss of performance. The use of supported ketones for those reactions in catalytic, or almost catalytic, amounts has only been described for systems immobilized on inorganic supports [295, 296].

The conversion of ketones into the corresponding α-tosyloxyketones can be accomplished using *m*CPBA and *p*-toluenesulfonic acid in the presence of a catalytic amount of poly(4-iodostyrene) (**114**) [297]. The linear version of this polymer afforded better yields than the related macroporous cross-linked material, but the latter was recovered and recycled more efficiently. The use of an excess of the terminal oxidant allows the direct oxidation of alcohols to α-tosyloxyketones (**115**, Scheme 10.19). A maximum yield of ca. 50% was obtained for the catalytic process (10% catalyst loading).

10.5.2
Chiral Polymer-Supported Oxidation Catalysts

Several examples are known of the enantioselective conversion of alkenes into epoxides with the use of polymer-supported oxidation catalysts. This can be traced to the pioneering work by Julià and Colonna in 1980. They demonstrated that highly enantioselective epoxidations of chalcones and related α,β-unsaturated ketones can be achieved with the use of insoluble poly(α-amino acids) (**116**, Scheme 10.20) as catalysts [298–301]. The so-called Julià–Colonna epoxidation has been the object of several excellent reviews [302–306]. The terminal oxidant is H$_2$O$_2$ in aq. NaOH. With lipophilic amino acids as the components, such as *(S)*-valine or *(S)*-leucine, enantioselectivities as high as 96–97% ee were obtained. The enantioselectivity depends of several factors, including the side-chain of the amino acid, the nature of the end groups and the degree of polymerization. Thus, for instance,

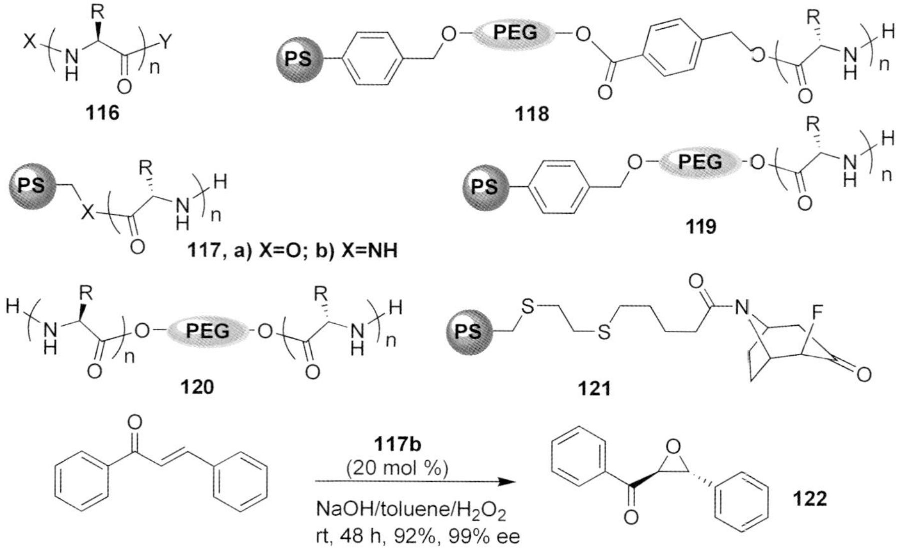

Scheme 10.20 Chiral polymer-supported oxidation catalysts and the use of one of them in the enantioselective epoxidations of chalcones.

the free acids and the *n*-butyl esters performed better than the methyl esters. At least ten amino acid residues are required to obtain high enantioselectivities. This has been associated with the minimum chain length to provide helical properties to the polymer. The recovery and reuse of the catalyst is not simple. This is simpler with polymers having more than 50 residues, but in this case the activity and the selectivity clearly decrease with the reuse [298].

To improve the separation processes and the potential for reuse, different attempts have been made to immobilize short chains of amino acids (>10 residues) onto different supports. The first experiments were carried out by the same authors, who esterified poly[(S)-alanine] to hydroxymethyl PS-DVB (**117a**, $n = 10$, R = CH$_3$) [301]. This catalyst afforded 82% yield with 84% ee, values slightly lower than those of the non-supported analog, and could be easily recovered; however, the enantioselectivity significantly diminished for the second and third run. substitution of the ester linkage by an amide bond (**117b**) was reported by Itsuno to provide a better polymer-supported catalyst [307]. Using the derivative of leucine with 32 residues [**117b**, 2% cross-linked, $n = 32$, R = CH$_2$CH(CH$_3$)$_2$], epoxychalcone **122** is formed in 92% yield and 99% ee (Scheme 10.20). Macroporous and linear PS resins performed worse than **117b** for this reaction. The reaction conditions were further optimized by Roberts to obtain faster reactions (less than 1 h) with high yields and selectivities (85–100%, >95% ee) [308, 309]. This allowed work to be carried out under anhydrous conditions, using urea/H$_2$O$_2$ complex with THF as the solvent and DBU as the base. The use of solid sodium percarbonate substitutes efficiently both the urea/H$_2$O$_2$ complex and DBU. Under such conditions the number of unsaturated ketones to which the reaction can be applied is very much broadened, permitting the application of this methodology to different synthetic procedures, including the total synthesis of some pharmaceuticals [310, 311]. The preparation of catalysts **118** and **119** from polymers containing a PEG spacer (TentaGel resins) has been reported [312, 313]. Catalyst **118** was prepared on a peptide synthesizer. This is important to obtain appropriate poly(amino acid) sequences of controlled length and to study in detail the factors affecting this reaction [314, 315]. The preparation of poly(amino acids) adsorbed on silica has overcome some of the limitations found for the recycling of resins **117** and **118** [316].

Soluble PEG-supported poly(amino acids) **120** containing different PEG moieties and variable lengths for the peptide chains have also been prepared by Roberts [317, 318]. Enantioselectivities of up to 98% ee have been achieved with them. Using a high molecular weight PEG the epoxidation of chalcone with **120** [PEG M_W = 20000, n = 8, R = $CH_2CH(CH_3)_2$] was carried out efficiently in a continuous membrane reactor, allowing up to 25 cycles without any detectable loss in its performance [319]. Similar results were obtained using linear PS as the support [319]. Notably, the number of amino acid residues required to obtain good enantioselectivities is lower when the poly(amino acid) chain is conjugated with PEG than in **116**. Only six residues seem to be required in **118–120**, This has been rationalized in terms of the role played by the PEG moiety in stabilizing the helical arrangement of the amino acid chain, which determines the efficiency of the stereocontrol for this reaction [313, 318]. Such catalysts have been used in a continuous membrane reactor, the activity being kept unaltered after 28 residence times [320].

Poly(β-amino acids) have been also studied for this reaction, but with less success [321].

The first example of the immobilization of a chiral ketone to promote the enantioselective epoxidation of alkenes with Oxone has been reported by Sartori and coworkers [322]. They anchored α-fluorotropinone on KG-60 silica, MCM-41 and a Merrifield resin. The catalysts were tested for the epoxidation of 1-phenylcyclohexene but the polymer-supported fluorotropinone **121** showed a low activity and selectivity. The catalyst immobilized on inorganic supports promoted the stereoselective epoxidation of alkenes with ee values up to 80% and could be reused with the same performance for three runs.

10.6
Polymer-Supported Organocatalysts Based on Amino Acids

Up to now we have grouped organocatalysts according to their reported essential mode of action. In many instances, however, this is an oversimplification and the mechanisms involved are more complex. In many instances, in particular when enantioselective transformations are implicated, organocatalysts behave as multifunctional molecules. This is clearly suggested, for instance, by the significant changes in activity and selectivity obtained with cinchona alkaloids when their "secondary" functional groups are modified. Besides the oxidation protocols described in the former section, amino acids, peptides and peptidomimetics or pseudopeptides have been used for various organocatalytic transformations, in which they participate through complex interactions. A mechanistic description of the corresponding catalytic cycles if often controversial and we will not enter in details on this subject.

10.6.1
Polymer-Supported Proline and Related Catalysts

Undoubtedly, the spectacular results obtained with proline as organocatalyst for different reactions are very much responsible for the arrival of the "golden age of organocatalysis" [16, 323]. Accordingly, many efforts have were made to develop efficient polymer-supported proline organocatalysts. Initial efforts, using gel-type PS supports (1% DVB), were carried out in 1985 [324]. Hydroxyproline was attached to the polymeric backbone through a spacer and the resulting resin (**123**, Scheme 10.21) tested as organocatalyst for an aldol reaction. The results obtained in this first attempt were, however, unsatisfactory (29% yield, 39% ee). More successful was the immobilization of a proline derivative on a JandaJel polymer [325], providing catalyst **124** [326, 327]. This resin was applied to the kinetic resolution of cyclic secondary alcohols by benzoylation. With *trans*-2-phenylcyclohexanol the benzoylated compound was obtained in 44% yield and 96% ee. The catalyst was recovered and reused several times without any significant loss in its performance.

4-Hydroxyproline has been the starting material for the preparation of polymers containing proline moieties via click chemistry (copper catalyzed [3 + 2] cycloaddition of azides and alkynes) by two independent groups [328–330]. Polymer **125** (10 mol.%) promoted the aldol condensation of cyclohexanone and benzaldehyde in water with high diastereoselectivity (95 : 5 anti : syn) and high ee for the major anti diastereoisomer (96%), but with low reactivity (26% after 18 h at rt). The use of good solvents for the resin such as DMF or DMSO increased the activity but at the expenses of a severe decrease in stereo- and enantioselectivity. The reactivity was improved by addition of DiMePEG (10 mol% in water) [328]. The same resin has been also applied (10–20 mol.%) to the enantioselective α-aminoxylation of aldehydes and ketones using nitrosobenzene to afford compounds related to **134** (Scheme 10.21) [329]. Yields were higher for aldehydes than for ketones, but enantioselectivities were excellent in both cases (>95% ee). In the second approach, the proline fragment was attached on dendronized Wang resins. Structure **126** represents the supported second-generation (G2) polyether dendron containing four proline subunits [330]. The selected benchmark reaction was the aldol reaction between benzaldehydes and acetone in DMSO at room temperature (4 days). The results revealed a remarkable influence of the dendronization on the conversion, yield and enantioselectivity. For 4-nitrobenzaldehyde, quantitative conversions, 90–95% yields and 84–85% ee, were obtained for G1 and G2 resins, while the G0 resin afforded less satisfactory results (88% yield, 47% ee). The use of the

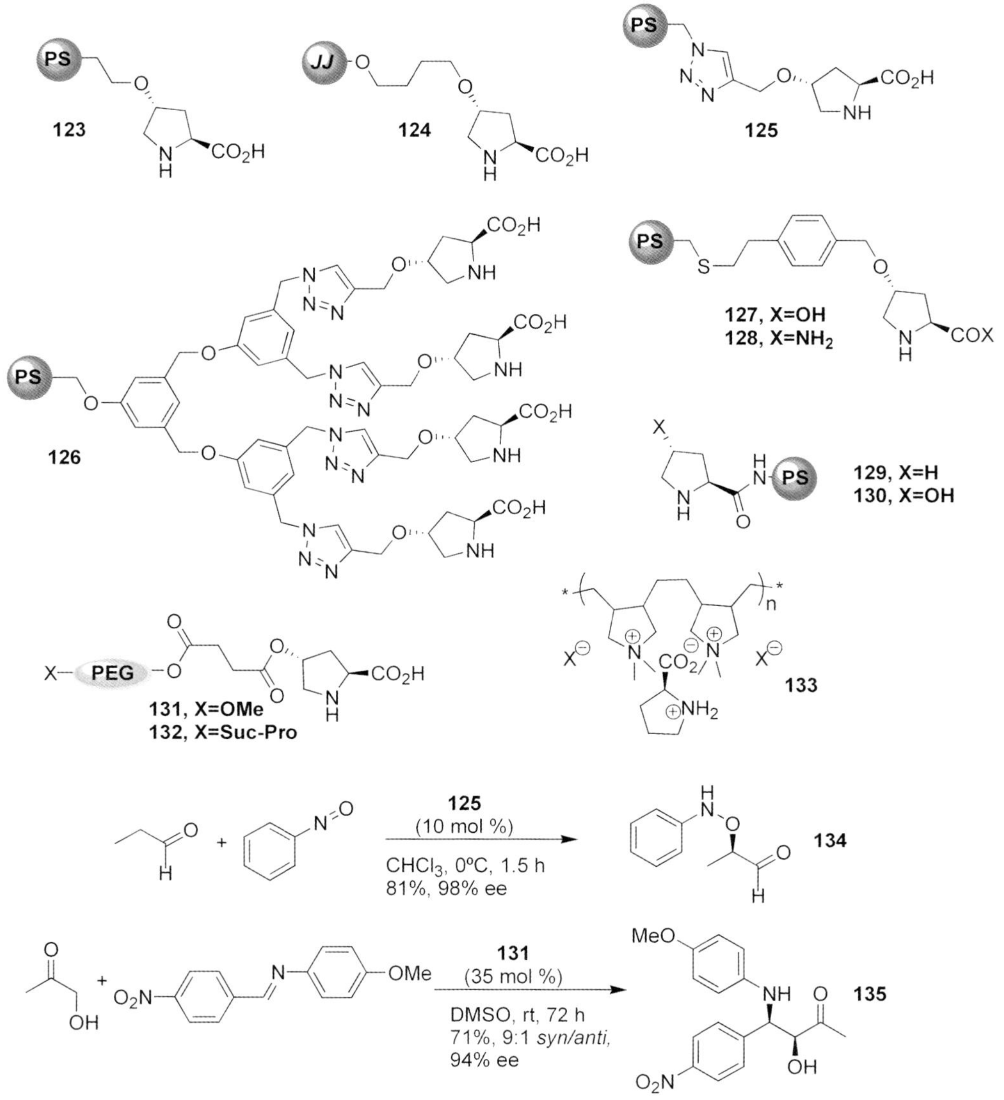

Scheme 10.21 Polymer-supported proline and related catalysts and their use.

G3 polymer did not produce any additional advantage. Unfortunately, recycling of the catalysts is negatively affected by the dendronization. Upon recycling G1–G3 resins, conversion and yields decrease significantly, even if the enantioselectivity remain essentially unaffected.

The aldol reaction in water is also efficiently catalyzed by polymers **127** and **128**, in which the proline moieties are anchored using a different strategy [331]. In this case the N-BOC protected hydroxyproline is reacted with chloromethylstyrene and the vinylic derivative is reacted with a mercaptomethyl PS-DVB resin (1% crosslinked). Addition of the thiol group to the double bond and further N-deprotection affords **127**. Supported prolinamide **128** was prepared in a similar way. Using **127** (10 mol.%) the aldol reaction between cyclohexanone and benzaldehyde required 60 h at room temperature to achieve a 71% conversion with >98% yield (based on conversion), an anti/syn selectivity of 95/5 and an enantioselectivity of 93% ee for the major diastereoisomer. The condensation of this ketone with other aromatic aldehydes also proceeded successfully with high yields, selectivities and enantioselectivities, but failed for aliphatic aldehydes [332]. In general, the use of other ketones works well only for the cyclic ones. Recycling was assayed for five cycles without a significant decrease in performance. Catalysts **127** and **128** (30 mol.%) also catalyzed the α-selenenylation of aldehydes with good yields but with low enantioselectivities, in the range of those observed under homogeneous conditions.

Proline has been attached by its carboxylic group to insoluble PS-DVB, using a 4-methylbenzhydrylamine resin [333]. Resins **129** and **130** were prepared and used (20 mol.%) as catalysts for the enantioselective aldol reaction of acetone and aliphatic aldehydes. Best enantioselectivities with **129** (up to 86% ee) were observed for sterically hindered aldehydes, but at the expense of lower conversions. Chemoselectivities were only moderate (40–75%). In general, **129** performed less efficiently than the soluble analogues. Recovery and reuse was possible but with a slight decrease in performance. Interestingly, the 4-hydroxyproline derivative **130** led to a significant increase in activity but to a substantial decrease in selectivity and ee.

The immobilization of proline of PEG with excellent results was reported by Benaglia and Cozzi in 2001 [334, 335]. Monofunctional derivative **131** was prepared from MeOPEG monosuccinate and hydroxyproline. This catalyst (25–35 mol.%) promoted the aldol reaction of acetone and aldehydes at room temperature at slower rates (up to 80% yield after 40–60 h) but with comparable enantioselectivities (up to 98% ee) to non-supported proline derivatives. The same catalyst has been used to promote related reactions such as the iminoaldol condensation between acetone and imines or the aldol reaction of hydroxyacetone with cyclohexanecarboxaldehyde and imines to afford β-aminoketones, the corresponding *anti*-α,β-dihydroxyketone and *syn*-β-amino-α-hydroxyketones (**135**, Scheme 10.21). In both cases, good enantioselectivities were observed, but yields ranged from low to moderate. In most of the examples considered, the stereoselectivity and enantioselectivity values were comparable with those obtained using proline as the catalyst. Addition of cyclic ketones to nitrostyrene and of 2-nitropropane to cyclohexenone were also assayed with **131**, but only moderate enantioselectivities were obtained [336]. The results were clearly inferior to those obtained with the related non-supported catalysts. Similar results are obtained with difunctional PEG **132**, which has having the advantage of requiring half the weight amount of catalyst. Recycling of **131** and **132** is feasible, but reuse is accompanied by a slight decrease in activity.

Interest in proline as an organocatalyst is reflected in the variety of inorganic and organic supports on which the immobilization has been assayed very recently.

10.6 Polymer-Supported Organocatalysts Based on Amino Acids

Of the inorganic supports, best results were reported for a mesoporous MCM-41 [337]. Support on ionic-liquid phases has been studied by different groups with variable results [338, 339]. Of the non-conventional organic polymers, non-covalent immobilization on poly(diallyldimethylammonium) is notable [340]. Catalysts **133** (15 mol.%) promoted the aldol reaction of acetone and benzaldehydes to afford the corresponding β-hydroxyketones in 50–98% yields and 62–72% ee, which are clearly lower than those reported for other polymer-supported systems. Recycling of the catalysts was possible at least six times without loss of efficiency. More recently, proline has been attached to one DNA strand while an aldehyde was tethered to a complementary DNA sequence and made to react with a non-tethered ketone [341]. To date, the work has focused more on conceptual development than on the analysis of its practical applications in organic synthesis.

No other polymer-supported amino acids other than proline have been reported as organocatalysts, but immobilized imidazolidinones derived from phenylalanine and tyrosine have been reported as efficient catalysts by different groups. The initial report by Benaglia and Cozzi described the preparation of PEG-supported derivative **136** (Scheme 10.22) [342]. This was examined as catalyst (10 mol.%) for the Diels–Alder reaction between acrolein and 1.3-cyclohexadiene. Under optimized conditions, the adduct was obtained with enantioselectivities (92% ee for the major endo isomer) differing little from those reported for the non-supported system. The same catalyst was employed for 1,3-dipolar cycloadditions [343]. Results were reported to be highly dependent on the nature of X in **136**, with the best results being obtained for X = BF_4. As for the Diels–Alder reaction, the related non-supported catalyst is only superior in terms of chemical yields (ca. 10–30% higher) and not in terms of stereo- and enantioselectivity (85:15 trans:cis and up to 87% ee). Efficient recycling could not be achieved, however, in both cases, apparently due to the instability of the catalyst under the reaction conditions.

A similar imidazolidinone was attached by Pihko to a JandaJel resin and studied as catalyst (**137**, 20 mol.%) for the Diels–Alder reaction of different dienes and dienophiles with excellent enantioselectivities (70–99% ee for the major endo adduct) [344]. Except in terms of activity the performance of this catalyst was

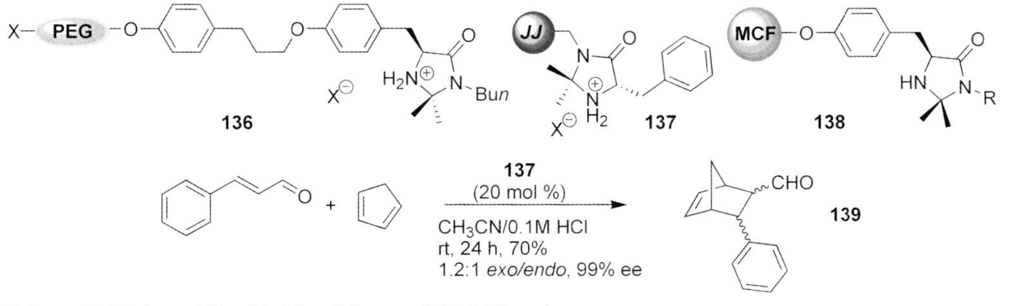

Scheme 10.22 Immobilized imidazolidinones (**136**–**138**) and the use of **137** in the Diels–Alder reaction between the 3-phenylpropenal and cyclopentadiene.

similar or even higher than that of the soluble analogues. In this case, no problems associated with reuse for a second run were reported. The activity of the catalyst was significantly increased when supported onto silica, so that loadings as low as 3.3 mol.% could be used. This silica supported system showed selectivity patterns slightly different from those of **137**. The endo/exo selectivities were improved, but the enantioselectivities decreased for some reactions. More recently, the imidazolidinone derived from tyrosine has been immobilized over polymer-coated siliceous mesocellular foams (MCF) [345]. Catalysts **138** (Scheme 10.22) were examined as promoters for the asymmetric Friedel–Crafts reaction between *N*-methylpyrrole and 3-phenylpropenal and for the Diels–Alder reaction between the same aldehyde and cyclopentadiene (Scheme 10.22). The system showed good reactivity, but enantioselectivities were in general slightly lower than those reported for the former catalysts. The organocatalyst was also anchored directly to the silanol groups of unmodified MCF. In this case good results were obtained when the surface had been partially pre-capped, but the recyclability of **138** was always superior.

10.6.2
Polymer-Supported Peptides and Related Catalysts

In Section 10.5.2 we considered in detail the use of poly(amino acids) as oxidation catalysts for the epoxidation of chalcones. Rather surprisingly, such materials, as homopolymers or anchored to different supports, have been less studied as organocatalysts for other processes. Perhaps the poor results obtained for the enantioselective addition of thiols to unsaturated ketones in the pioneering work by Inoue are responsible for this situation [346].

Cyclic dipeptides have been more extensively studied in this regard. The first report on the immobilization of this class of organocatalysts appeared in 1992. The authors substituted the phenylalanine residue in the original catalyst (cyclo[*(S)*-phenylalanyl-*(S)*-histidyl]) developed by Inoue [347] for a tyrosine residue [348]. The presence of the additional hydroxy group was used for attachment to PS-DVB (1% DVB) to obtain insoluble catalysts **140** (Figure 10.11). Resins **140** were tested for the enantioselective addition of hydrogen cyanide to aldehydes but with poor results (<30% ee). It was shown that the presence of an alkoxy substituent at the 4-position of the aromatic ring has a detrimental effect on the selectivity of the reaction even in solution. To circumvent this problem, a new series of supported cyclic dipeptides were prepared using the non-basic nitrogen atom of the imidazole for anchoring to 2% cross-linked PS-DVB [349]. The attachment was made using spacers of different lengths to favor the accessibility of the catalytic sites. Nevertheless, catalysts **141** gave even poorer results for the same benchmark reaction (<20% ee). Reasonable results (80% yield, 98% ee) were only obtained for a system in which the original cyclic dipeptide was entrapped on a silicon-based matrix, but no efficient recycling of the catalyst could be achieved [350].

A very interesting solid phase combinatorial approach was developed by Jacobsen in 1998 for the identification and optimization of catalysts for the hydrocyana-

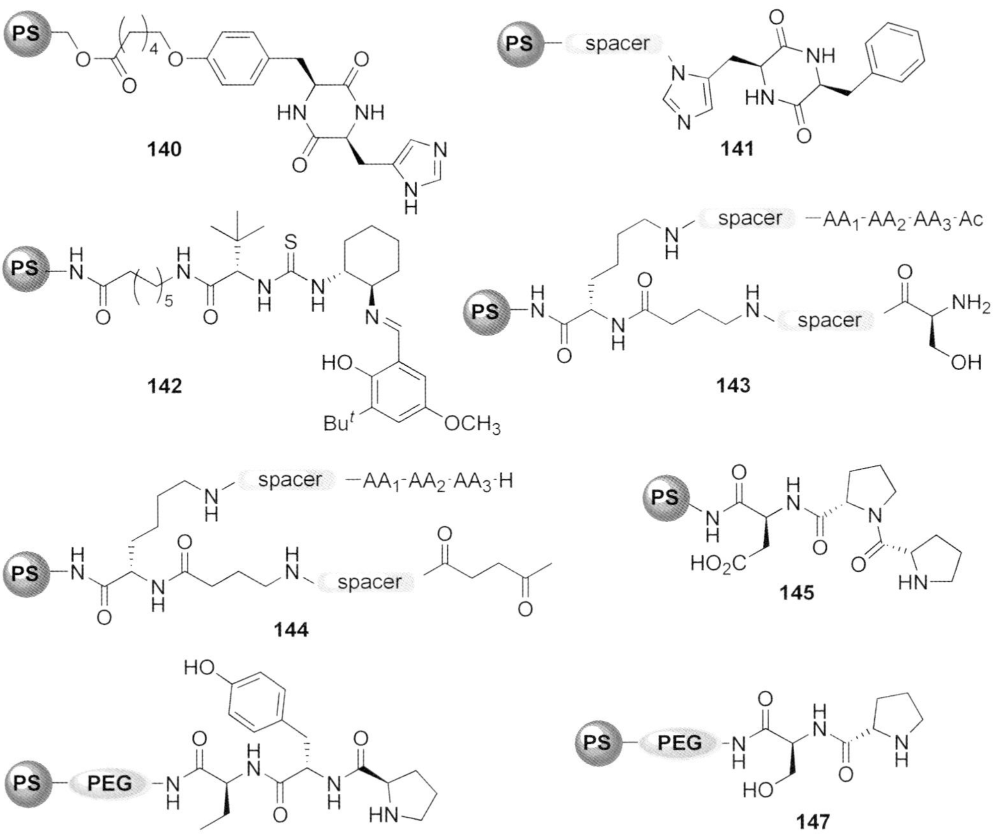

Figure 10.11 Polymer-supported peptides and related catalysts.

tion of imines (the Strecker reaction) [351]. The original goal was to develop a tridentate Schiff base complex based on three structural components: a salicylaldehyde derivative, a chiral diamine and an amino acid, as seen in compound **142** (Figure 10.11). The study of the first library (12 members) of catalysts based on a single ligand and using the metal ion as the element of diversity revealed that the ligand in the absence of any added metal ion was the most enantioselective (19%). This led to the further optimization of this structure as an organocatalyst through the preparation of two successive libraries of 48 and 132 members. The best supported catalyst identified (**142**) afforded an enantioselectivity of 80% ee. The authors then prepared the corresponding homogeneous catalyst, which was slightly more efficient than **142**. The soluble compound similar to **142** with the terminal nitrogen atom of the amino acid N-benzylated afforded a 91% ee after 24h at -78°C in toluene (2 mol.% catalyst loading).

A solid phase combinatorial approach has also been used by Wennemers for the discovery and optimization of tripeptide organocatalysts. Encoded split-and-mix libraries of larger size (up to 3375 members for 15 potential amino acids in each position) were used, which required the development of a novel strategy for a high-throughput analysis of the results [352]. For this purpose the initial screening is made through catalyst–substrate co-immobilization. The second partner of the reaction is labeled with a dye and the reaction between the two components leads to the covalent attachment of the marker. Thus, the presence of a sequence with catalytic activity can be easily followed through the development of color in the corresponding bead. The concept was checked with a simple reaction, as was the acylation of the hydroxyl group of serine bound to a TentaGel-supported bifunctional fragment containing a tripeptide sequence (**143**, Figure 10.11). The pentafluorophenol ester of hydroxyacetic acid with disperse red 1 dye attached to the OH group was used as the acylating agent. This allowed detection of up to 17 tripeptide sequences with catalytic properties for this reaction. Not surprisingly, all of them contained at least one His residue.

This approach was further implemented for the discovery of tripeptide sequences active as catalysts for the aldol reaction. Here, a library of polymers **144** (Figure 10.11) containing an acetone equivalent was prepared [353, 354]. The second component for the aldol reaction was a benzaldehyde having the same dye attached to the 4-position of the aromatic ring via an amide bond. From this study, two sequences were selected: H-L-Pro-D-Ala-D-Asp-NH$_2$ and H-L-Pro-L-Pro-L-Asp-NH$_2$. The second sequence was able to promote the aldol condensation (1 mol.%) between *p*-nitrobenzaldehyde and acetone, in the absence of other solvents, at room temperature, affording a 98% yield after 4 h with 80% ee. The enantioselectivity was raised to 90% by working at −20 °C. Conformational analysis of both tripeptides revealed the adoption of turn-like conformations in which the secondary amine of proline is in close proximity to the carboxylic acid of aspartic acid. The sign of the turn is, however, opposite in the two sequences, which could explain the formation of products with opposite conformations when using each of those tripeptides.

Finally, the H-L-Pro-L-Pro-L-Asp-NH$_2$ sequence was immobilized in different supports: TentaGel, PS-DVB, polyacrylamide (SPAR) and poly(ethylene glycol)-polyacrylamide (PEGA) [355]. TentaGel derived catalyst **145** (Figure 10.11) with low functionalization degrees (0.1–0.2 mmol g^{-1}) proved to be superior to the others. Using 1 mol.% of catalyst, along with 1 mol.% of *N*-methylmorpholine, the aldol product is obtained at room temperature in 93% yield after 2 h with 80% ee. The supported catalyst can be recycled at least three times without a significant drop in catalytic activity or selectivity.

The direct asymmetric aldol reaction has been studied with different di- and tri-peptides sequences supported in PEG-polystyrene resins by two different groups. In both cases the peptides contained proline at the N-terminus. In one case the other components of the sequence were aromatic amino acids. Thus, polymer **146** (20 mol.%) displaying the sequence H-D-Pro-Tyr-Phe-PS produced the aldol product in 90% yield and 33% ee. The enantioselectivity was improved to 73% ee with the addition of 20 mol.% ZnCl$_2$, but required longer reaction times. In the second case, a higher variation of amino acids was considered. Resin **147** with the sequence H-D-Pro-Ser achieved in acetone, at room temperature (24 h), a 94% yield with 63% ee. The use of the tripeptides considered increased the enantioselectivity to 75–77% ee but yields were significantly lowered (13–29%). Using **147** the enantioselectivity was raised to 82% ee by working at −20 °C.

10.7
Polymer-Supported Imidazolium, Thiazolium and Related Structures

The recent development of polymers containing imidazolium, thiazolium and related structures as supported equivalents of ionic liquids (i.e. **148** and **149**, Scheme 10.23) is an interesting development of this field [356–361]. The corresponding salts containing basic counter-anions can act as basic catalysts but can also be transformed, depending on the exact structure, into N-heterocyclic carbenes that can act as nucleophilic organocatalysts [362]. As an example, resins **148**

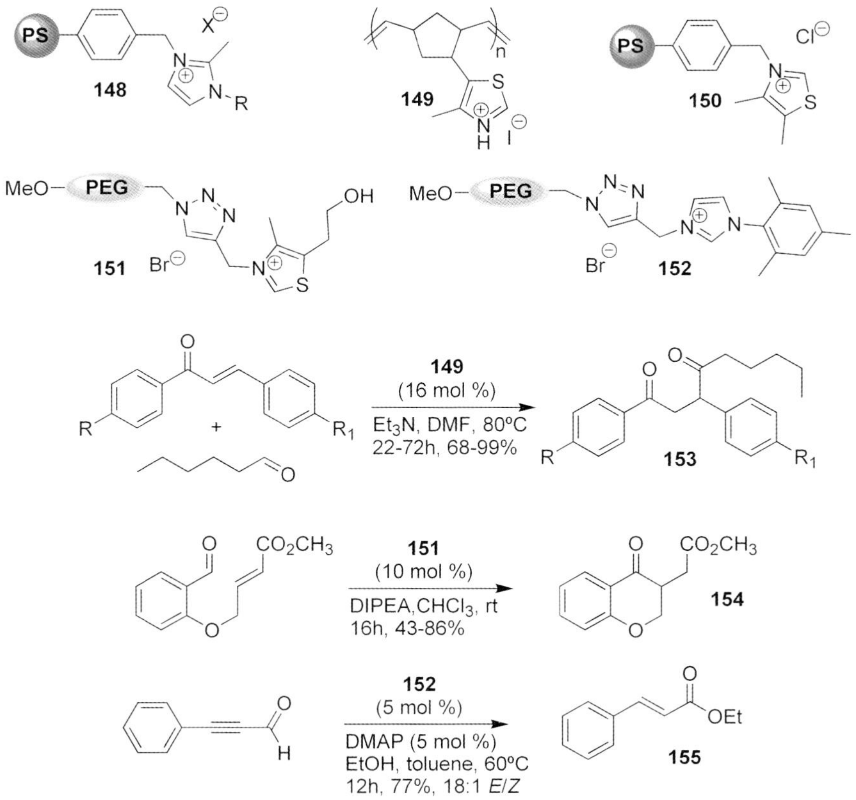

Scheme 10.23 Polymers containing imidazolium, thiazolium and related structures as supported equivalents of ionic liquids a and their uses in synthesis.

(X = OH) act as very efficient bases for the Heck reaction and catalyze efficiently the Henry reaction with excellent yields and selectivities. The catalyst can be used both in batch and under flow conditions and can be directly reused several times with only a minor decrease in activity. The spent polymer can be easily regenerated simply by its treatment with a basic solution [363].

Polymer **149**, prepared by ring-opening polymerization (ROMP) of the corresponding functionalized norbornene, provides a clear illustration of the potential of this kind of materials. This resin catalyzes (12% molar) the reaction of linear aliphatic aldehydes with unsaturated ketones to provide 1,4-dicarbonyl compounds (**153**, Scheme 10.23) and was used for four consecutive cycles without any detectable decrease in performance [358]. A similar dimethylthiazolium structure supported on 2% cross-linked PS-DVB (**150**) was studied as catalyst for the acyloin condensation of a large variety of aldehydes [364]. Catalyst **150** is used in 10 mol.% and the reaction takes place in ethanol at room temperature, with triethylamine as the base, to afford the corresponding α-hydroxyketones in excellent yields. Remarkably, the catalyst can be reused 20 times without losing its activity.

Soluble supported catalysts of this class are also accessible. Thus, various thiazolium, imidazolium and triazolium structures have been recently immobilized on PEG via "click" chemistry [365]. Thiazolium resins (**151**) were the most efficient catalysts (10 mol.%) for the intramolecular Stetter reaction. Compound **154** (Scheme 10.23) was obtained in 81% with the supported catalyst whilst an 84% yield was obtained when the non-supported analog was employed. Imidazolium systems like **152** were also able to promote the carbene-catalyzed diastereoselective redox esterification. In this case, **155** was obtained with higher yields (77%) and stereoselectivities (18:1 $E:Z$ ratio) than when using the non-supported system (63% and 11:1). This can be associated with the positive influence of steric hindrance on the results of this reaction.

10.8
Miscellaneous Organocatalysts

The lack of a really precise definition of the field of organocatalysis makes it difficult to properly categorize all the catalysts and processes that can be included under this umbrella. This leads to some areas of work being included in some cases but not in others. Three such areas are photocatalysis, imprinted polymers and dendrimers. All of them have very specific characteristics and enough entities to be most often considered independently, even if, formally, many of the processes included there could be classified as organocatalytic (metal free). We will not deal in detail with those areas but, instead, include here a brief outline in the context of organocatalysis.

The need to place a photocatalyst onto an appropriate support is very often a requirement for the development of the potential applications of such systems. A large variety of photocatalysts derived from organic compounds have been reported, and many of them are complex structures containing coordinated metal cations.

Nevertheless, examples of metal-free photocatalysts are well known and their immobilization onto polymeric supports has been studied in many instances [366]. Rose Bengal is one such example and its anchoring onto different polymeric matrices can be considered as one of the seminal works in the field of the application of polymers in organic chemistry [367–371]. Covalent and non-covalent immobilization on cross-linked and noncross-linked organic polymers have been assayed with this dye (**156**, Figure 10.12). The resulting supported sensitizer is efficient for different photooxidations with singlet oxygen. These reactions involve mainly the preparation of endoperoxides and products derived from them [366]. Other photocatalysts anchored to polymeric matrices include tetraaryl porphyrins (**157**) [372], methylene blue and triarylpyrilium salts [373]. Besides photoxidative synthetic transformations, such supported photocatalysts have found a useful application in the treatment and decontamination of residual waters, in particular when solar energy can be employed as the light source [374–377].

The polymerization of functional monomers in the presence of an appropriate template is, most likely, one of the most straightforward ways to develop design materials with cavities having enzyme-like properties (Figure 10.13) [378]. If the template is accurately selected the cavity will contain, after removal of the template, the required functional groups located at the right positions and with the proper relative orientations for achieving the desired catalytic activity [379–383]. As much as the catalytic process is maintained metal-free, such imprinted polymers should be considered as organocatalysts. Nevertheless, to maintain adequate topology of the cavity after removal of the template the preparation of highly cross-linked, high molecular weight polymers is necessary. Thus, those catalytic sites, composed if a complex set of high molecular weight moieties, are not easily compatible with the concept of organocatalysts as small organic molecules.

Many efforts have been carried out to translate this straightforward concept into catalytic imprinted materials of practical application. Achieving this goal is,

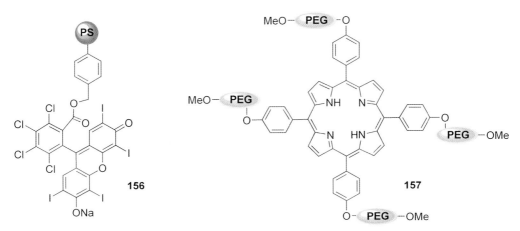

Figure 10.12 Examples of photocatalysts anchored to polymeric matrices.

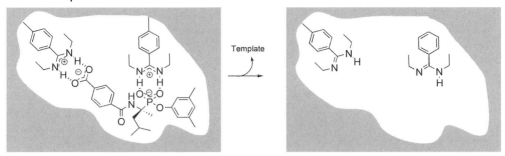

Figure 10.13 Polymerization of functional monomers in the presence of an appropriate template.

however, not so simple. The correct design, selection and preparation of the proper template are, in many cases, the key steps for the whole process. Probably the most useful current paradigm in this regard is the search of templates that are, structurally and electronically, analogues of the transition state for the reaction intended to be activated [384]. The reasoning behind this concept is that in such cases the cavity will be highly efficient in the binding and stabilization of the transition state, thereby decreasing the energy required to reach it. Of course, the application of this concept requires a deep knowledge of the mechanism of the reaction considered – knowledge that is not always available. The activation of different reactions by imprinted polymers has been achieved successfully. Moreover, chemoselective reactions have also been completed with these materials. Current targets in this field focus on the development of catalytic systems fully efficient in terms of stereo- and enantioselectivity and on broadening the scope of the reactions that can be promoted by imprinted polymers [385–387].

Similar considerations are of application to the study of dendritic organocatalytic systems [388]. Different applications have been studied with dendrimers containing organocatalytic functional groups [389–393], and most effort is currently being directed towards the application of those systems to novel reactions and to the achievement of asymmetric dendritic organocatalysts [330, 394, 395].

10.9
Conclusions

In summary, development of functional polymers that can act as efficient organocatalysts is at the origin of the application of modified polymers as reagents and catalysts in chemistry. To date, the most successful accomplishments have been carried out in the area of basic and acid catalysts. Not surprisingly, this area includes some of the most important industrial applications for supported reagents and catalysts. In general, the development of non-chiral polymer-supported organocatalysts is a field of research that involves both industrial and academic research. In many cases, it has been possible to demonstrate how the use of the supported species has some distinct advantages over the analogous homogeneous catalysts. The development of successful chiral polymer-supported organocatalysts is, however, more recent, and has been mainly concentrated on the anchoring of proline derivatives to different kinds of polymeric matrices. In this regard we can envisage that, in future years, much effort will be devoted to the preparation and study of novel functional polymers containing enantioselective organocatalytic moieties in which the nature of the polymeric matrix can provide new mechanisms to enhance the overall efficiency of those systems in terms of activity, selectivity and enantioselectivity. Also in this case, the main goal has to be the development of supported catalysts displaying distinct advantages over the analogous non-supported species. Given the chemical nature of most chiral organocatalysts studied up to now, the impressive potential of the application of combinatorial methodologies will clearly contribute significantly, in the near future, to this field, as has been clearly illustrated by the initial examples reported in the literature.

List of Abbreviations

[bmim]	1-butyl-3-methylimidazolium
Ala	alanine
Asp	aspartic acid
BOC	di-*tert*-butyl dicarbonate
DBU	1,8-diazabicyclo[5.4.0]undec-7-ene
DCM	dichloromethane
DF	functional degree
DiMePEG	dimethylpoly(ethylene glycol)
DIPEA	*N,N*-diisopropylethylamine
DMAP	4-(dimethylamino)pyridine
DMF	dimethylformamide
DMSO	dimethyl sulfoxide
DNA	deoxyribonucleic acid
DVB	divinylbenzene
EGDMA	ethylene glycol dimethylacrylate
MCF	mesocellular foams
m-CPBA	*meta*-chloroperoxybenzoic acid
MeOPEG	methoxypoly(ethylene glycol)
MOM	methoxymethyl
MTBE	methyl *tert*-butyl ether
PA	polyacrylic acid
PEG	poly(ethylene glycol)
PEGA	poly(ethylene glycol)-polyacrylamide
Phe	phenylalanine
Pro	proline
PS	polystyrene
PS-DVB	polystyrene–divinylbenzene

PTC	phase transfer catalysis
ROMP	ring-opening polymerization
Ser	serine
TEMPO	2,2,6,6-tetramethyl-1-piperidinyloxy radical
THF	tetrahydrofuran
THP	tris(hydroxypropyl)phosphine
Tyr	tyrosine

References

1 Langenbeck, W. (1928) *Angewandte Chemie*, **41**, 740–5.
2 Langenbeck, W. (1931) *Angewandte Chemie*, **451**, 97–9.
3 Drauz, K. and Waldmann, H. (2002) *Enzyme Catalysis in Organic Synthesis: A Comprehensive Handbook*, 2nd edn, Wiley-VCH Verlag GmbH, Weinheim.
4 Ertl, G., Knözinger, H. and Weitkamp, J. (1997) *Handbook of Heterogeneous Catalysis*, Wiley-VCH Verlag GmbH, Weinheim.
5 Nishimura, S. (2001) *Handbook of Heterogeneous Catalytic Hydrogenation for Organic Synthesis*, John Wiley & Sons, Inc., New York.
6 Balogh, M. and Laszlo, P. (1993) *Organic Chemistry Using Clays*, Springer-Verlag, New York.
7 Van Bekkum, H., Flanigen, E.M. and Jansen, J.C. (1991) *Introduction to Zeolite Science and Practice*, Elsevier, Amsterdam.
8 Corma, A. and García, H. (1997) *Catalysis Today*, **38**, 257–308.
9 Speight, J.G. (1999) *The Chemistry and Technology of Petroleum*, Marcel Dekker, New York.
10 Sheldon, R.A. and Van Bekkum, H. (2001) *Fine Chemicals through Heterogeneous Catalysis*, Wiley-VCH Verlag GmbH, Weinheim.
11 Kato, N., Mita, T., Kanai, M., et al. (2006) *Journal of the American Chemical Society*, **128**, 6768–9.
12 Burguete, M.I., Collado, M., Escorihuela, J. and Luis, S.V. (2007) *Angewandte Chemie – International Edition*, **46**, 9002–5.
13 Altava, B., Burguete, M.I., Fraile, J.M., et al. (1996) *Tetrahedron*, **52**, 9853–62.
14 Foote, C.S. and List, B. (2000) Special issue on catalytic asymmetric synthesis. *Accounts of Chemical Research*, **33**, 323–40.
15 Breslow, R. (1982) *Science*, **218**, 532–7.
16 Dalko, P.I. and Moisan, L. (2004) *Angewandte Chemie – International Edition*, **43**, 5138–75.
17 Cozzi, F. (2006) *Advanced Synthesis & Catalysis*, **348**, 1367–90.
18 Dalko, P.I. and Moisan, L. (2001) *Angewandte Chemie – International Edition*, **40**, 3726–48.
19 Houk, K.N. and List, B. (2004) Special issue on asymmetric organocatalysis. *Accounts of Chemical Research*, **37**, 487–631.
20 List, B. and Bolm, C. (2004) Special issue on organocatalysis. *Advanced Synthesis Catalysis*, **346**, 1007–249.
21 Berkessel, A., Groger, H. and McMillan, D. (2005) *Asymmetric Organocatalysis: from Biomimetic Concepts to Applications in Asymmetric Synthesis*, Wiley-VCH Verlag GmbH, Weinheim.
22 Guillena, G. and Ramón, D.J. (2006) *Tetrahedron Asymmetry*, **17**, 1465–92.
23 Dalko, P.I. (2007) *Enantioselective Organocatalysis: Reactions and Experimental Procedures*, Wiley-VCH Verlag GmbH, Weinheim.
24 Seayad, J. and List, B. (2005) *Organic and Biomolecular Chemistry*, **3**, 719–24.
25 Gaunt, M.J., Johansson, C.C.C., McNally, A. and Vo, N.T. (2007) *Drug Discovery Today*, **12**, 8–27.
26 Benaglia, M., Puglisi, A. and Cozzi, F. (2003) *Chemical Reviews*, **103**, 3401–29.
27 Benaglia, M. (2006) *New Journal of Chemistry*, **30**, 1525–33.

References

28 Gladysz, A. (2002) Special issue on recoverable reagents and catalysts. *Chemical Reviews*, **102**, 3215–892.

29 Altava, B., Burguete, M.I., García-Verdugo, E., et al. (2001) *Reactive and Functional Polymers*, **48**, 25–35.

30 Fan, Q.H., Wang, R. and Chan, A.S.C. (2002) *Bioorganic and Medicinal Chemistry Letters*, **12**, 1867–71.

31 Bergbreiter, D.E. and Sung, S.D. (2006) *Advanced Synthesis & Catalysis*, **348**, 1352–66.

32 Widdecke, H. (1988) *Design and Industrial Applications of Polymeric Acid Catalysts*, in *Synthesis and Separations Using Functional Polymers* (eds D.C. Sherrington and P. Hodge), John Wiley & Sons, Inc., New York, pp. 149–79.

33 Gelbard, G. (2005) *Industrial and Engineering Chemistry Research*, **44**, 8468–98.

34 Harmer, M.A. and Sun, Q. (2001) *Applied Catalysis A: General*, **221**, 45–62.

35 Guyot, A. (1988) *Synthesis and Structure of Polymer Supports*, in *Synthesis and Separations Using Functional Polymers* (eds D.C. Sherrington and P. Hodge), John Wiley & Sons, Inc., New York, pp. 1–42.

36 Nippon Oil, (1974) German Patent, 2,403,196.

37 Widdecke, H., Klein, J. and Haupt, U. (1986) *Makromolekulare Chemie-Macromolecular Symposia*, **4**, 145–55.

38 Klein, J. and Widdecke, H. (1979) *Chemie Ingenieur Technik*, **51**, 560–8.

39 Widdecke, H. and Klein, J. (1981) *Chemie Ingenieur Technik*, **53**, 954–7.

40 For commercial applications based on acid/base catalyzed chemical processes see, for instance: Mitsutani, A. (2002) *Catalysis Today*, **73**, 57–63.

41 Kroenig, W. and Scharfe, G. (1966) *Erdoel Kohle*, **19**, 497–8.

42 Carlyle, R.M. (1982) *Chemistry and Industry*, 561–4.

43 Tejero, J., Fité, C., Iborra, M., et al. (2006) *Applied Catalysis A: General*, **308**, 223–30.

44 Bozek-Winkler, E. and Gmehling, J. (2006) *Industrial and Engineering Chemistry Research*, **45**, 6648–54.

45 Kiss, A.A., Dimian, A.C. and Rothenberg, G. (2006) *Advanced Synthesis & Catalysis*, **348**, 75–81.

46 Park, K.H., Yoon, Y.J. and Lee, S.G. (1994) *Tetrahedron Letters*, **35**, 9737–40.

47 Seto, H. and Mander, L.N. (1992) *Synthetic Communications*, **22**, 2823–8.

48 Das, B., Damodar, K. and Chowdhury, N. (2007) *Journal of Molecular Catalysis A–Chemical*, **269**, 81–4.

49 Das, B. and Chowdhury, N. (2007) *Journal of Molecular Catalysis A–Chemical*, **263**, 212–15.

50 Yadav, J.S., Reddy, B.V.S., Eshwaraiah, B. and Anuradha, K. (2002) *Green Chemistry*, **4**, 592–4.

51 Kalena, G.P., Jain, A. and Banerji, A. (1997) *Molecules*, **2**, 100–5.

52 Das, B., Thirupathi, P., Mahender, I., Reddy, V.S. and Rao, Y.K. (2006) *Journal of Molecular Catalysis A–Chemical*, **247**, 233–9.

53 Ramesh, C., Banerjee, J., Pal, R. and Das, B. (2003) *Advanced Synthesis & Catalysis*, **345**, 557–9.

54 Das, B. and Reddy, K.R. (2006) *Helvetica Chimica Acta*, **89**, 3109–11.

55 du Toit, E., Nicol, E. and Nicol, W. (2004) *Applied Catalysis A: General*, **277**, 219–25.

56 Bringué, R., Iborra, M., Tejero, J., et al. (2006) *Journal of Catalysis*, **244**, 33–42.

57 Hart, M., Fuller, G., Brown, D.R., et al. (2001) *Catalysis Letters*, **72**, 135–9.

58 Koujout, S. and Brown, D.R. (2004) *Catalysis Letters*, **98**, 195–202.

59 Magnotta, V.L., Gates, B.C. and Schuit, G.C.A. (1976) *Journal of the Chemical Society, Chemical Communication*, 342–3.

60 Luis, S.V., Burguete, M.I. and Altava, B. (1995) *Reactive and Functional Polymers*, **26**, 75–83.

61 Ekerdt, J.G. (1986) *Polymeric Reagents and Catalysts*, in Symposium Series **308** (ed. W.T. Ford), American Chemical Society, Washington, pp. 68–83.

62 Döscher, F., Klein, J., Pohl, F. and Widdecke, H. (1983) *Makromolekulare Chemie–Macromolecular Chemistry and Physics*, **184**, 1585–96.

63 Döscher, F., Klein, J. and Widdecke, H. (1982) *Makromolekulare Chemie–*

64 Brouwer, D.M. and Van de Vondervoort, E.M. (1986) *Makromolekulare Chemie – Macromolecular Chemistry and Physics*, **187**, 2103–10.
65 Nagayama, S. and Kobayashi, S. (2000) *Angewandte Chemie – International Edition*, **39**, 567–9.
66 Altava, B., Burguete, M.I., Collado, M., et al. (2001) *Tetrahedron Letters*, **42**, 1673–5.
67 Hart, M., Fuller, G., Brown, D.R., et al. (2002) *Journal of Molecular Catalysis A – Chemical*, **182**, 439–45.
68 Koujout, S., Kiernan, B.M., Brown, D.R., et al. (2003) *Catalysis Letters*, **85**, 33–40.
69 Manabe, K. and Kobayashi, S. (2002) *Advanced Synthesis & Catalysis*, **344**, 270–3.
70 Iimura, S., Manabe, K. and Kobayashi, S. (2003) *Organic Letters*, **5**, 101–3.
71 Iimura, S., Manabe, K. and Kobayashi, S. (2003) *Organic and Biomolecular Chemistry*, **1**, 2416–18.
72 Aoyama, T., Takido, T. and Kodomari, M. (2004) *Synlett*, 2307–10.
73 Waller, F.J. (1986) *Catalysis Reviews Science and Engineering*, **28**, 1–12.
74 Xu, Y., Gu, W. and Gin, D.L.J. (2004) *Journal of the American Chemical Society*, **126**, 1616–17.
75 Gray, W.K., Smail, F.R., Hitzler, M.G., et al. (1999) *Journal of the American Chemical Society*, **121**, 10711–18.
76 Hickner, M.A., Ghassemi, H., Kim, Y.S., et al. (2004) *Chemical Reviews*, **104**, 4587–611.
77 Kreuer, K.D. (2001) *Journal of Membrane Science*, **185**, 29–39.
78 Peckham, T.J., Schmeisser, J., Rodgers, M. and Holdcroft, S.J. (2007) *Journal of Materials Chemistry*, **17**, 3255–68.
79 Bergbreiter, D.E., Case, B.L., Liu, Y.-S. and Carway, J.W. (1998) *Macromolecules*, **31**, 6053–62.
80 Waller, F.J. (1986) *Polymeric Reagents and Catalysts*, in Symposium Series **308** (ed. W.T. Ford), American Chemical Society, Washington, pp. 42–67.
81 Olah, G.A., Iyer, P.S. and Prakash, G.K.S. (1986) *Synthesis*, 513–31.
82 Connolly, D.J. and Gresham, W.F. (1966) (E.I. Du Pont de Nemours & Co.). U.S. Patent 3.282.875.
83 Childs, R.F. and Mika-Gabala, A. (1982) *Journal of Organic Chemistry*, **47**, 4204–7.
84 Olah, G.A., Prakash, G.K.S. and Sommer, J. (1979) *Science*, **206**, 13–20.
85 Farcasiu, D., Ghenciu, A., Marino, G. and Rose, K.D. (1997) *Journal of the American Chemical Society*, **119**, 11826–31.
86 Olah, G.A., Kaspi, J. and Bukala, J. (1977) *Journal of Organic Chemistry*, **42**, 4187–91.
87 See, for instante, McClure, J.D. and Brandernberger, S.G. (1977) (Shell Oil Company). U.S. Patent 4.065.515.
88 Torok, B., Kiricsi, I., Moinar, A. and Olah, G.A. (2000) *Journal of Catalysis*, **193**, 132–8.
89 Liu, Y.J., Lotero, E. and Goodwin, J.G. (2006) *Journal of Catalysis*, **243**, 221–8.
90 Olah, G.A., Malhotra, R., Narang, S.C. and Olah, J.A. (1978) *Synthesis*, 672–3.
91 Krespan, C.C. (1979) *Journal of Organic Chemistry*, **44**, 4924–9.
92 Konishi, H., Suetsugu, K., Okano, T. and Kiji, J. (1982) *Bulletin of the Chemical Society of Japan*, **55**, 957–8.
93 Prakash, G.K.S., Mattew, T., Mandal, M., Farnia, M. and Olah, G.A. (2004) *Arkivoc*, 103–10.
94 Yamato, T., Hideshima, C., Prakash, G.K.S. and Olah, G.A. (1991) *Journal of Organic Chemistry*, **56**, 3955–7.
95 Olah, G.A., Mathew, T. and Prakash, G.K.S. (1999) *Synlett*, 1067–8.
96 Olah, G.A., Mathew, T. and Prakash, G.K.S. (2001) *Chemical Communications*, 1696–7.
97 Olah, G.A. and Meidar, D. (1979) *New Journal of Chemistry*, **3**, 269–73.
98 Olah, G.A., Meidar, D., Malhotra, R., et al. (1980) *Journal of Catalysis*, **61**, 96–102.
99 Yamato, T., Hideshima, C., Prakash, G.K.S. and Olah, G.A. (1991) *Journal of Organic Chemistry*, **56**, 2089–91.
100 Aleksiuk, O. and Biali, S.E. (1993) *Tetrahedron Letters*, **34**, 4857–60.
101 Kaspi, J., Montgomery, D.D. and Olah, G.A. (1978) *Journal of Organic Chemistry*, **43**, 3147–50.
102 McClure, J.D. (1977) (Shell Oil Company). U.S. Patent 4.041.090.

103 Harmer, M.A., Sun, Q., Vega, A.J., *et al.* (2000) *Green Chemistry*, 1–14.
104 Olah, G.A., Torok, B., Asma, T., *et al.* (1996) *Catalysis Letters*, **42**, 5–13.
105 Armandi, R., Hyde, J.R., Ross, S.K., *et al.* (2005) *Green Chemistry*, **7**, 288–93.
106 Olah, G.A., Prakash, G.K.S., Iyer, P.S., *et al.* (1987) *Journal of Organic Chemistry*, **52**, 1881–4.
107 Yamato, T., Hideshima, C., Miyazawa, A., *et al.* (1990) *Catalysis Letters*, **6**, 345–8.
108 Olah, G.A., Laali, K. and Mehrotra, A.K. (1983) *Journal of Organic Chemistry*, **48**, 3360–2.
109 Nunan, J.G., Klier, K. and Herman, R.G. (1993) *Journal of Catalysis*, **139**, 406–20.
110 Cho, B.R. and Yang, H.J. (1992) *Bulletin of the Korean Chemical Society*, **13**, 586–7.
111 Olah, G.A., Shamma, T. and Prakash, G.K.S. (1997) *Catalysis Letters*, **46**, 1–4.
112 Bucsi, I., Molnar, A., Bartok, M. and Olah, G.A. (1995) *Tetrahedron*, **51**, 3319–26.
113 Bucsi, I., Molnar, A., Bartok, M. and Olah, G.A. (1994) *Tetrahedron*, **50**, 8195–202.
114 Yamato, T., Hideshima, C., Prakash, G.K.S. and Olah, G.A. (1991) *Journal of Organic Chemistry*, **56**, 3192–4.
115 Yamato, T., Hideshima, C., Suehiro, K., Tashiro, M., Prakash, G.K.S. and Olah, G.A. (1991) *Journal of Organic Chemistry*, **56**, 6248–50.
116 Pruckmayr, G. and Weir, R.H. (1978) (E.I. du Pont de Nemours & Co.). U.S. Patent 4.120.903.
117 Olah, G.A., Narang, S.C., Meidar, D. and Salem, G.F. (1981) *Synthesis*, 282–3.
118 Olah, G.A., Husain, A. and Singh, B.P. (1983) *Synthesis*, 892–5.
119 Sartori, G., Ballini, R., Bigi, F., *et al.* (2004) *Chemical Reviews*, **104**, 199–250.
120 Petrakis, K.S. and Fried, J. (1983) *Synthesis*, 891–2.
121 Carlson, R., Gautun, H. and Westerlund, A. (2002) *Advanced Synthesis & Catalysis*, **344**, 57–60.
122 Olah, G.A., Yamato, T., Iyer, P.S. and Prakash, G.K.S. (1986) *Journal of Organic Chemistry*, **51**, 2826–8.
123 Olah, G.A. and Mehrotra, A.K. (1982) *Synthesis*, 962–3.
124 Olah, G.A., Husain, A., Gupta, B.G. and Narang, S.C. (1981) *Synthesis*, 471–2.
125 Casas, C.P., Yamamoto, H. and Yamato, T.(2005) *Journal of Chemical Research-S*, 694–6.
126 Olah, G.A., Keumi, T. and Mediar, D. (1978) *Synthesis*, 929–30.
127 Cho, B.R. and Yang, H.J. (1991) *Bulletin of the Korean Chemical Society*, **12**, 1–9.
128 Kumareswaran, R., Pachamuthu, K. and Vankar, Y.D. (2000) *Synlett*, 1652–4.
129 Liu, Y.J., Lotero, E. and Goodwin, J.G. (2006) *Journal of Catalysis*, **242**, 278–86.
130 Pittman, C.U. and Liang, Y. (1980) *Journal of Organic Chemistry*, **45**, 5048–52.
131 McClure, J.D. and Newman, F.E. (1981) (Shell Oil Co.). U.S. Patent 4.053.522.
132 Olah, G.A. and Ip, W.M. (1988) *New Journal of Chemistry*, **2**, 299–301.
133 Yamato, T., Hideshima, C., Miyazawa, A., *et al.* (1990) *Catalysis Letters*, **6**, 345–8.
134 Joseph, J.K., Jain, S.L. and Sain, B. (2006) *Journal of Molecular Catalysis A–Chemical*, **247**, 99–102.
135 Lin, H., Zhao, Q., Xu, B. and Wang, S. (2007) *Journal of Molecular Catalysis A–Chemical*, **268**, 221–6.
136 Laufer, M.C., Hausmann, H. and Hölderivh, W.(2003) *Journal of Catalysis*, **218**, 315–20.
137 Prakash, G.K.S., Mathew, T., Krishnaraj, S. and Marinez, E.R. (1999) *Applied Catalysis A: General*, **181**, 283–8.
138 Olah, G.A., Arvanaghi, M. and Krishnamurthy, V.V. (1983) *Journal of Organic Chemistry*, **48**, 3359–60.
139 French, L.G. and Charlton, T.P. (1993) *Heterocycles*, **35**, 305–13.
140 Hachoumy, M., Mathew, T., Tongco, E.C., *et al.* (1999) *Synlett*, 363–5.
141 Olah, G.A., Reddy, V.P. and Prakash, G.K.S. (1991) *Synthesis*, 29–30.
142 Wabnitz, T.C., Yu, J.Q. and Spencer, J.B. (2003) *Synlett*, 1070–2.
143 Tsumori, N., Xu, Q., Souma, Y. and Mori, H. (2002) *Journal of Molecular Catalysis A–Chemical*, **179**, 271–7.
144 Heidkum, A., Harmer, M. and Hölderivh, W.F. (1997) *Catalysis Letters*, **47**, 243–6.

145 Heidekum, A., Harmer, M.A. and Hölderivh, W.F. (1999) *Journal of Catalysis*, **181**, 217–22.
146 Olah, G.A., Meidar, D. and Fung, A.P. (1979) *Synthesis*, 270–1.
147 Li, Z. and Ganesan, A. (1998) *Synthetic Communications*, **28**, 3209–12.
148 Yoshida, J., Hashimoto, J. and Kawabata, N. (1981) *Bulletin of the Chemical Society of Japan*, **54**, 309–10.
149 Johnston, R.D., Marston, C.R., Krieger, P.E. and Goe, F.L. (1988) *Synthesis*, 393–4.
150 Menger, F.M. and Chu, C.H. (1981) *Journal of Organic Chemistry*, **46**, 5044–5.
151 Ishiara, K., Hasegawa, A. and Yamamoto, H. (2001) *Angewandte Chemie – International Edition*, **40**, 4077–9.
152 Hon, Y.-S., Lee, C.-F. and Chen, R.-J. and Szu, P.-H. (2001) *Tetrahedron*, **57**, 5991–6001.
153 Hon, Y.-S., Lee, C.-F. and Chen, R.-J. and Huang, Y.-F. (2003) *Synthetic Communications*, **33**, 2829–42.
154 Masaki, Y., Tanaka, N. and Miura, T. (1998) *Tetrahedron Letters*, **39**, 5799–802.
155 Tanaka, N. and Masaki, Y. (1999) *Synlett*, 1960–2.
156 Masaki, Y., Yamada, T. and Tanaka, N. (2001) *Synlett*, 1311–13.
157 Tanaka, N. and Masaki, Y. (1999) *Synlett*, 1277–9.
158 Tanaka, N. and Masaki, Y. (2000) *Synlett*, 406–8.
159 Corain, B., Zecca, M. and Jerábek, K. (2001) *Journal of Molecular Catalysis*, **177**, 3–20.
160 Mukaiyama, T. and Iwakiri, H. (1985) *Catalysis Letters*, 1363–6.
161 Tanabe, K. and Hölderich, W.F. (1999) *Applied Catalysis A: General*, **181**, 399–434.
162 Pathak, V.P. (1993) *Synthetic Communications*, **23**, 83–5.
163 Podrebarac, G.G., Ng, F.T.T. and Rempel, G.L. (1997) *Chemical Engineering Science*, **52**, 2991–3002.
164 Serra-Holm, V., Salmi, T., Maki-Arvela, P., Paatero, E. and Lindfors, L.P. (2001) *Organic Process Research and Development*, **5**, 368–75.
165 Stowell, J.C. and Hauck, H.F. (1981) *Journal of Organic Chemistry*, **46**, 2428–9.
166 Simon, C., Peyronel, J.F., Clerc, F. and Rodriguez, J. (2002) *European Journal of Organic Chemistry*, 3359–64.
167 Konwar, D., Dutta, D.K. and Goswami, B.N. (1998) *Journal of Chemical Research (S)*, 342–3.
168 Jin, T.S., Zhang, J.S., Wang, A.Q. and Li, T.S. (2004) *Synthetic Communications*, **34**, 2611–16.
169 Di Girolano, M. and Marchiona, M. (2001) *Journal of Molecular Catalysis A – Chemical*, **177**, 33–40.
170 Buchmeiser, M.R. (ed.) (2003) Polymeric materials, in *Organic Synthesis And Catalysis*, Wiley-VCH Verlag GmbH, Weinheim.
171 For a general overview of applications of functional polymers, see accompanying chapters in this book.
172 For a recent review on phosphine-functionalized polymers see: Guinó, M. and Hii, K.K. (2007) *Chemical Society Reviews*, **36**, 608–17.
173 Oververger, C.G., Guterl, A.C., Kawakami, Y., et al. (1978) *Pure and Applied Chemistry*, **50**, 309–19.
174 Mancecke, G. and Stork, W. (1978) *Angewandte Chemie – International Edition*, **17**, 657–70.
175 Hierl, M.A., Gamson, E.P. and Klotz, I.M. (1979) *Journal of the American Chemical Society*, **101**, 6020–2.
176 Delaney, E.J., Wood, L.E. and Klotz, I.M. (1982) *Journal of the American Chemical Society*, **104**, 799–807.
177 Vaidya, R.A. and Mathias, L.J. (1986) *Journal of the American Chemical Society*, **108**, 5514–20.
178 Shinkai, S., Tsuji, H., Hara, Y. and Manabe, O. (1981) *Bulletin of the Chemical Society of Japan*, **54**, 631–2.
179 Tomoi, M., Akada, Y. and Kakiuchi, H. (1982) *Makromolekulare Chemie – Rapid Communications*, **3**, 537–42.
180 Deratani, A., Darling, G.D., Horak, D. and Fréchet, J.M.J. (1987) *Macromolecules*, **20**, 767–72.
181 Altava, B., Burguete, M.I., Fraile, J.M., et al. (2000) *Angewandte Chemie – International Edition*, **39**, 1503–6.

182 Altava, B., Burguete, M.I., García-Verdugo, E., *et al.* (1999) *Tetrahedron*, **55**, 12897–906.
183 Corma, A., García, H. and Leyva, A. (2003) *Chemical Communications*, 2806–7.
184 Huang, J.-W. and Shi, M. (2003) *Advanced Synthesis Catalysis*, **345**, 953–8.
185 Guendouz, F., Jacquier, R. and Verducci, J. (1988) *Tetrahedron*, **44**, 7095–108.
186 Menger, F.M. and McCann, D.J. (1985) *Journal of Organic Chemistry*, **50**, 3928–30.
187 Kwong, C.K.-W., Huang, R., Zhang, M., *et al.* (2007) *Chemistry – A European Journal*, **13**, 2369–76.
188 Sharma, K.K. and Asefa, T. (2007) *Angewandte Chemie – International Edition*, **46**, 2879–82.
189 Bergbreiter, D.E. and Li, C. (2003) *Organic Letters*, **5**, 2445–7.
190 Bergbreiter, D.E., Osburn, P.L. and Li, C. (2002) *Organic Letters*, **4**, 737–40.
191 Helms, B., Guillaudeu, S.J., Xie, Y., *et al.* (2005) *Angewandte Chemie – International Edition*, **44**, 6384–7.
192 Leadbater, N.E. and Marco, M. (2002) *Chemical Reviews*, **102**, 3217–74.
193 End, N. and Scöning, K.U. (2004) *Topics in Current Chemistry*, **242**, 241–71.
194 Skouta, R., Varma, R.S. and Li, C.J. (2005) *Green Chemistry*, **7**, 571–5.
195 Zhao, L.J., Kwong, C.K.W., Shi, M. and Toy, P. (2005) *Tetrahedron*, **61**, 12026–32.
196 Stanetty, P. and Kremslehner, M. (1998) *Tetrahedron Letters*, **39**, 811–12.
197 Wang, Y.G., Jiang, H.F., Liu, H.L. and Liu, P. (2005) *Tetrahedron Letters*, **46**, 3935–7.
198 Schuchardt, U., Vargas, R.M. and Gelbard, G. (1996) *Journal of Molecular Catalysis A – Chemical*, **109**, 37–44.
199 Simoni, D., Rondanin, R., Morini, M., *et al.* (2000) *Tetrahedron Letters*, **41**, 1607–10.
200 Simpson, J., Rathbpme, D.L. and Billington, D.C. (1999) *Tetrahedron Letters*, **40**, 7031–3.
201 Ogawa, C., Sugiera, M. and Kobayashi, S. (2003) *Chemical Communications*, 192–3.
202 Bensa, D., Constantieux, T. and Rodriguez, J. (2004) *Synthesis*, 923–7.
203 Flowers, R.A., Xu, X., Timmons, C. and Li, G. (2004) *European Journal of Organic Chemistry*, 2988–90.
204 Kobayashi, N. and Iwai, K. (1978) *Journal of the American Chemical Society*, **100**, 7071–2.
205 Kobayashi, N. and Iwai, K. (1980) *Tetrahedron Letters*, **21**, 2167–70.
206 Sera, A., Takagi, K., Katayama, H., *et al.* (1988) *Journal of Organic Chemistry*, **53**, 1157–61.
207 Inagaki, M., Hiratake, J., Yamamoto, Y. and Oda, J. (1987) *Bulletin of the Chemical Society of Japan*, **60**, 4121–6.
208 Hodge, P., Khoshdel, E., Waterhouse, J. and Frechet, J.M.J. (1985) *Journal of the Chemical Society – Perkin Transactions 1*, 2327–31.
209 Alvarez, R., Hourdin, M.-A., Cavè, C., *et al.* (1999) *Tetrahedron Letters*, **40**, 7091–4.
210 Yamauchi, K., Kinoshita, M. and Imoto, M. (1971) *Bulletin of the Chemical Society of Japan*, **44**, 3186–7.
211 Yamashita, T., Yasueda, H., Miyauchi, Y. and Nakamura, N. (1977) *Bulletin of the Chemical Society of Japan*, **50**, 1532–4.
212 Hermann, K. and Wynberg, H. (1977) *Helvetica Chimica Acta*, **60**, 2208–12.
213 Hafez, A.M., Taggi, A.E., Dudding, T. and Letcka, T. (2001) *Journal of the American Chemical Society*, **123**, 10853–9.
214 France, S., Bernstein, D., Weatherwax, A. and Lectka, T. (2005) *Organic Letters*, **7**, 3009–12.
215 Bernstein, D., France, S., Wolfer, J. and Lectka, T. (2005) *Tetrahedron Asymmetry*, **16**, 3481–3.
216 Danelli, T., Annunziata, R., Benaglia, M., Cinquini, M., *et al.* (2003) *Tetrahedron Asymmetry*, **14**, 461–7.
217 Priem, G., Pelotier, B., Macdonald, S.J.F., *et al.* (2003) *Synlett*, 679–83.
218 Priem, G., Pelotier, B., Macdonald, S.J.F., *et al.* (2003) *Journal of Organic Chemistry*, **68**, 3844–8.
219 Müller, C.A., Hoffart, T., Holbach, M. and Reggelin, M. (2005) *Macromolecules*, **38**, 5375–80.
220 Starks, C.M., Liotta, C.L. and Halpern, M. (1994) *Phase-Transfer Catalysis*, Chapman & Hall, New York.

221 Halpern, M. (ed.) (1997) *Phase-Transfer Catalysis: Perspectives in Mechanism and Syntheses*, in Symposium Series **659**, American Chemical Society, Washington.

222 Regen, S.L. (1975) *Journal of the American Chemical Society*, **97**, 5956–7.

223 Regen, S.L. (1979) *Angewandte Chemie – International Edition*, **18**, 421–9.

224 Mathur, N.K., Narang, C.K. and Williams, R.E. (1980) *Polymers as Aids in Organic Chemistry*, Academic Press, New York, pp. 209–13.

225 Tomoi, M. and Ford, W.T. (1988) Polymeric Phase Transfer Catalysts, in *Syntheses and Separations Using Functional Polymers* (eds D.C. Sherrington and P. Hodge), John Wiley & Sons, Inc., New York, pp. 181–208.

226 Svec, F. (1988) *Pure and Applied Chemistry*, **60**, 377–86.

227 Hodge, P. (1997) *Chemical Society Reviews*, **26**, 417–24.

228 Vaino, A.R. and Janda, K.D. (2000) *Journal of Combinatorial Chemistry*, **2**, 579–96.

229 Cinquini, M., Solonna, S., Molinari, H., et al. (1976) *Chemical Communications*, 394.

230 Tundo, P. (1978) *Synthesis*, 315–16.

231 Regen, S.L. and Besse, J.J. (1979) *Journal of the American Chemical Society*, **101**, 4059–63.

232 Fukunishi, K., Czech, B. and Regen, S.L. (1981) *Journal of Organic Chemistry*, **46**, 1218–21.

233 Montanari, F. and Tundo, P. (1981) *Journal of Organic Chemistry*, **46**, 2125–30.

234 Anelli, P.L., Montanari, F. and Quici, S. (1986) *Journal of Organic Chemistry*, **51**, 4910–14.

235 Altava, B., Burguete, M.I., Frías, J.C., et al. (2000) *Industrial and Engineering Chemistry Research*, **39**, 3589–95.

236 Molinari, H., Montanari, F., Quici, S. and Tundo, P. (1979) *Journal of the American Chemical Society*, **101**, 3920–7.

237 Tomoi, M., Yanai, N., Shiiki, S. and Kakiuchi, H. (1984) *Journal of Polymer Science. Polymer Chemistry Edition*, **22**, 911–26.

238 Tomoi, M., Ogawa, E., Hosokama, Y. and Kakiuchi, H.(1982) *Journal of Polymer Science. Polymer Chemistry Edition*, **20**, 3421–9.

239 Tomoi, M. and Ford, W.T. (1981) *Journal of the American Chemical Society*, **103**, 3821–8.

240 Regen, S.L. (1976) *Journal of the American Chemical Society*, **98**, 6270–4.

241 Balakrishnan, T. and Ford, W.T. (1983) *Journal of Organic Chemistry*, **48**, 1029–35.

242 Pugia, M.J., Czech, A., Czech, B.P. and Bartsch, R.A. (1986) *Journal of Organic Chemistry*, **51**, 2945–8.

243 McKenzie, W.M. and Sherrington, D.C. (1978) *Chemical Communications*, 541–3.

244 Regen, S.L. (1977) *Journal of Organic Chemistry*, **42**, 875–9.

245 Chiles, M.S., Jackson, D.D. and Reeves, P.C. (1980) *Journal of Organic Chemistry*, **45**, 2915–18.

246 Akelah, A. and Sherrington, D.C. (1981) *Chemical Reviews*, **81**, 557–87.

247 Nishikubo, T., Iizawa, T., Shimojo, M., Kato, T. and Shiina, A.(1990) *Journal of Organic Chemistry*, **55**, 2536–42.

248 Jovanovic, S.S., Misic-Vukovic, M.J., Djokovic, D.D. and Bajic, D.S. (1992) *Journal of Molecular Catalysis*, **73**, 9–16.

249 Ford, W.T., Chandran, R. and Turk, H. (1988) *Pure and Applied Chemistry*, **60**, 395–400.

250 Lee, J.-J. and Ford, W.T (1994) *Journal of the American Chemical Society*, **116**, 3753–9.

251 Ford, W.T. (2001) *Reactive and Functional Polymers*, **48**, 3–13.

252 Tamami, B. and Mahdavi, H. (2002) *Tetrahedron Letters*, **43**, 6225–8.

253 Kobayashi, N. and Iwai K. (1981) *Makromolekulare Chemie – Rapid Communications*, **2**, 105–8.

254 Hodge, P., Khoshdel, E. and Waterhouse, J. (1983) *Journal of the Chemical Society – Perkin Transactions 1*, 2205–9.

255 O'Donnell, M.J. (2001) *Aldrichimica Acta*, **34**, 3–15.

256 Chinchilla, R., Mazón, P. and Nájera, C. (2000) *Tetrahedron Asymmetry*, **11**, 3277–81.

257 Chinchilla, R., Mazón, P. and Nájera, C. (2004) *Molecules*, **9**, 349–64.

258 Zhengpu, Z., Yongmer, W., Zhen, W. and Hodge, P. (1999) *Reactive and Functional Polymers*, **41**, 37–43.

259 Thierry, B., Plaquevent, J.-C. and Cahard, D. (2001) *Tetrahedron Asymmetry*, **12**, 983–6.
260 Thierry, B., Perrard, T., Audouard, C., et al. (2001) *Synthesis*, 1742–6.
261 Kim, M.-J., Jew, S., Park, H. and Jeong, D.-S. (2007) *European Journal of Organic Chemistry*, 2490–6.
262 Chiellini, E. and Solaro, R. (1977) *Chemical Communications*, 231–2.
263 Colonna, S., Fornasier, R. and Pfeiffer, U. (1978) *Journal of the Chemical Society – Perkin Transactions 1*, 8–11.
264 Castells, J. and Duñach, E. (1984) *Chemistry Letters*, 1859–60.
265 Kimura, Y. and Regen, S.L. (1982) *Journal of Organic Chemistry*, **42**, 2493–4.
266 Dou, H.J.-M., Gallo, R., Massanally, P. and Metzger, J. (1977) *Journal of Organic Chemistry*, **42**, 4275–6.
267 Kimura, Y., Kirszensztejn, P. and Regen, S.L. (1983) *Journal of Organic Chemistry*, **48**, 385–6.
268 Wakui, T., Xu, W.Y., Chen, C.S. and Smid, J. (1986) *Makromolekulare Chemie – Macromolecular Chemistry and Physics*, **187**, 533–45.
269 Grinberg, S. and Shaubi, E. (1991) *Tetrahedron*, **47**, 2895–902.
270 Annunziata, R., Benaglia, M., Cinquini, M., et al. (2000) *Organic Letters*, **2**, 1737–9.
271 Albanese, D., Benaglia, M., Landini, D., et al. (2002) *Industrial and Engineering Chemistry Research*, **41**, 4928–35.
272 Benaglia, M., Cinquini, M., Cozzi, F. and Tocco, G. (2002) *Tetrahedron Letters*, **43**, 3391–3.
273 For an example of a dendrimer containing 36 ammonium groups on the surface see: Lee, J.J., Ford, W.T., Moore, J.A. and Li, Y. (1994) *Macromolecules*, **27**, 4632–4.
274 Altava, B., Burguete, M.I., García-Verdugo, E., et al. (1999) *Tetrahedron*, **55**, 12897–906.
275 Burguete, M.I., Fraile, J.M., García-Verdugo, E., et al. (2005) *Industrial and Engineering Chemistry Research*, **44**, 8580–7.
276 Bergbreiter, D.E. and Blanton, J.R. (1985) *Journal of Organic Chemistry*, **50**, 5828–33.
277 Grinberg, S., Kas'yanov, V. and Srinivas, B. (1997) *Reactive and Functional Polymers*, **34**, 53–63.
278 Taylor, R.T. (1986) *Polymeric Reagents and Catalysts*, in Symposium Series **308** (ed. W.T. Ford), American Chemical Society, Washington, pp. 132–54.
279 De Nooy, A.E.J., Besemer, A.C. and van Dekkum, H. (1996) *Synthesis*, 1153–74.
280 Miyazawa, T. and Endo, T. (1985) *Journal of Polymer Science. Polymer Chemistry Edition*, **23**, 2487–94.
281 Miyazawa, T. and Endo, T. (1988) *Journal of Molecular Catalysis*, **49**, L31-L34.
282 DF: degree of functionalization; percentage of aromatic rings, for PS-DVB resins, containing the desired fucntionality.
283 Dijksman, A., Arends, I.W.C. and Sheldon, R.A. (2000) *Chemical Communications*, 271–2.
284 Dijksman, A., Arends, I.W.C. and Sheldon, R.A. (2001) *Synlett*, 102–4.
285 Tsubokawa, N., Kimono, T. and Endo, T. (1995) *Journal of Molecular Catalysis A – Chemical*, **101**, 45–50.
286 Fey, T., Fischer, H., Bachmann, S., Albert, K. and Bolm, C. (2001) *Journal of Organic Chemistry*, **66**, 8154–9.
287 But, T.Y.S., Tashino, Y., Togo, H. and Toy, P.H. (2005) *Organic and Biomolecular Chemistry*, **3**, 970–1.
288 Weik, S., Nicholson, G., Jung, G. and Rademann, J. (2001) *Angewandte Chemie – International Edition*, **40**, 1436–9.
289 Jacobson, S.E., Mares, F. and Zambri, P.M. (1979) *Journal of the American Chemical Society*, **101**, 6938–46.
290 Jacobson, S.E., Mares, F. and Zambri, P.M. (1979) *Journal of the American Chemical Society*, **101**, 6946–50.
291 Taylor, R.T. and Flood, L.A. (1983) *Journal of Organic Chemistry*, **48**, 5160–4.
292 Denmark, S.E. and Wu, Z. (1999) *Synlett*, 847–59.
293 Shiney, A., Rajan, P.K. and Sreekumar, K. (1996) *Polymer International*, **41**, 377–81.
294 Boehlow, T.R., Buxton, P.C., Grocock, E.L., et al. (1998) *Tetrahedron Letters*, **39**, 1839–42.

295 Song, C.E., Lim, J.S., Kim, S.C., et al. (2000) *Chemical Communications*, 2415–16.
296 Neimann, K. and Neumann, R. (2001) *Chemical Communications*, 487–8.
297 Yamamoto, Y., Kawano, Y., Toy, P.H. and Togo, H. (2007) *Tetrahedron*, **63**, 4680–7.
298 Julià, S., Masana, J. and Vega, J.C. (1980) *Angewandte Chemie – International Edition*, **19**, 929–31.
299 Julià, S., Guixer, J., Masana, J., et al. (1982) *Journal of the Chemical Society – Perkin Transactions 1*, 1317–24.
300 Colonna, S., Molinari, H., Banfi, S., et al. (1983) *Tetrahedron*, **39**, 1635–41.
301 Banfi, S., Colonna, S., Molinari, H., et al. (1984) *Tetrahedron*, **40**, 5207–11.
302 Ebrahim, S. and Wills, M. (1997) *Tetrahedron Asymmetry*, **8**, 3163–73.
303 Pu, L. (1998) *Tetrahedron Asymmetry*, **9**, 1457–77.
304 Porter, M.J., Roberts, S.M. and Skidmore, J. (1999) *Bioorganic and Medicinal Chemistry*, **7**, 2145–56.
305 Porter, M.J. and Skidmore, S. (2000) *Chemical Communications*, 1215–25.
306 Lauret, C. and Roberts, S.M. (2002) *Aldrichimica Acta*, **35**, 47–51.
307 Itsuno, S., Sakakura, M. and Ito, K. (1990) *Journal of Organic Chemistry*, **55**, 6047–9.
308 Bentley, P.A., Bergeron, S., Cappi, M.W., et al. (1997) *Chemical Communications*, 739–40.
309 Allen, J.V., Drauz, K.-H., Flood, R.W., et al. (1999) *Tetrahedron Letters*, **40**, 5417–20.
310 Adger, B.M., Barkley, J.V., Bergeron, S., et al. (1997) *Journal of the Chemical Society – Perkin Transactions 1*, 3501–7.
311 Bentley, P.A., Bickley, J.F., Roberts, S.M. and Steiner, A. (2001) *Tetrahedron Letters*, **42**, 3741–3.
312 Cappi, M.W., Chen, W.-P., Flood, F.W., et al. (1998) *Chemical Communications*, 1159–60.
313 Berkessel, A., Gasch, N., Glaubitz, K. and Koch, C. (2001) *Organic Letters*, **3**, 3839–42.
314 Bentley, P.A., Cappi, M.W., Flood, R.W., et al. (1998) *Tetrahedron Letters*, **39**, 9297–300.
315 Bentley, P.A., Flood, R.W., Roberts, S.M., et al. (2001) *Chemical Communications*, 1616–17.
316 Ray, P.C. and Roberts, S.M. (2001) *Journal of the Chemical Society – Perkin Transactions 1*, 149–53.
317 Flood, R.W., Geller, T.P., Petty, S.A., et al. (2001) *Organic Letters*, **3**, 683–6.
318 Kelly, D.R., Bui, T.T.T., Caroff, E., et al. (2004) *Tetrahedron Letters*, **45**, 3885–8.
319 Tsogoeva, S.B., Wöltinger, J., Jost, C., et al. (2002) *Synlett*, 707–10.
320 Dijkstra, H.P., Kruithof, C.A., Ronde, N., et al. (2003) *Journal of Organic Chemistry*, **68**, 675–85.
321 Coffey, P.A., Drauz, K.-H., Roberts, S.M., et al. (2001) *Chemical Communications*, 2330–1.
322 Sartori, G., Armstrong, A., Maggi, R., et al. (2003) *Journal of Organic Chemistry*, **68**, 3232–7.
323 Gröger, H. and Wilken, J. (2001) *Angewandte Chemie – International Edition*, **40**, 529–32.
324 Kondo, K., Yamano, T. and Takemoto, K. (1985) *Makromolekulare Chemie – Macromolecular Chemistry and Physics*, **186**, 1781–5.
325 Toy, P.H. and Janda, K.D. (1999) *Tetrahedron Letters*, **40**, 6329–32.
326 Clapham, B. and Cho, C.W. and Janda, K.D. (2001) *Journal of Organic Chemistry*, **66**, 868–73.
327 Cordova, A., Tremblay, M.R., Clapham, B. and Janda, K.D. (2001) *Journal of Organic Chemistry*, **66**, 5645–8.
328 Font, D., Jimeno, C. and Pericàs, M.A. (2006) *Organic Letters*, **8**, 4653–5.
329 Font, D., Bastero, A., Sayalero, S., Jimeno, C. and Pericàs, M.A. (2007) *Organic Letters*, **9**, 1943–6.
330 Kehat, T. and Portnoy, M. (2007) *Chemical Communications*, 2823–5.
331 Giacalone, F., Gruttadauria, M., Marculescu, A.M. and Noto, R. (2007) *Tetrahedron Letters*, 255–9.
332 Gruttadauria, M., Giacalone, F., Marculescu, A.M., et al. (2007) *European Journal of Organic Chemistry*, 4688–98.
333 Szöllosi, G., London, F., Baláspiri, L., et al. (2003) *Chirality*, **15**, S90-S96.
334 Benaglia, M., Celentano, G. and Cozzi, F. (2001) *Advanced Synthesis & Catalysis*, **343**, 171–3.

335 Benaglia, M., Cinquini, M., Cozzi, F., et al. (2002) *Advanced Synthesis & Catalysis*, **344**, 533–42.
336 Benaglia, M., Cinquini, M., Cozzi, F., et al. (2003) *Journal of Molecular Catalysis A–Chemical*, **204**, 157–63.
337 Calderón, F., Fernández, R., Sánchez, F. and Fernández-Mayoralas, A. (2005) *Advanced Synthesis Catalysis*, **347**, 1395–403.
338 Miao, W. and Chan, T.H. (2006) *Advanced Synthesis Catalysis*, **348**, 1711–18.
339 Yang, S.-D., Wu, L.-Y., Yan, Z.-Y., et al. (2007) *Journal of Molecular Catalysis A–Chemical*, **268**, 107–11.
340 Kucherenko, A.S., Struchkova, M.I. and Zlotin, S.G. (2006) *European Journal of Organic Chemistry*, 2000–4.
341 Tang, Z. and Marx, A. (2007) *Angewandte Chemie – International Edition*, **46**, 1–5.
342 Benaglia, M., Celentano, G., Cinquini, M., Puglisi, A. and Cozzi, F. (2002) *Advanced Synthesis Catalysis*, **344**, 149–52.
343 Benaglia, M., Celentano, G., Cinquini, M., et al. (2004) *European Journal of Organic Chemistry*, 567–73.
344 Selkälä, S.A., Tois, J., Pihko, P.M. and Koskinen, A.M.P. (2002) *Advanced Synthesis Catalysis*, **344**, 941–5.
345 Zhang, Y., Zhao, L., Lee, S.S. and Ying, J.Y. (2006) *Advanced Synthesis & Catalysis*, **348**, 2027–32.
346 Ueyanagi, K. and Inoue, S. (1976) *Makromolekulare Chemie – Macromolecular Chemistry and Physics*, **177**, 2807–17.
347 Oku, J. and Inoue, S. (1981) *Chemical Communications*, 229–30.
348 Kim, H.J. and Jackson, W.R. (1992) *Tetrahedron Asymmetry*, **3**, 1421–30.
349 Song, C.E., Chun, Y.J. and Kim, I.O. (1994) *Synthetic Communications*, **24**, 103–9.
350 Shvo, Y., Becker, Y. and Gal, M. (1994) *Chemical Communications*, 2719–20.
351 Sigman, M.S. and Jacobsen, E.N. (1998) *Journal of the American Chemical Society*, **120**, 4901–2.
352 Krattiger, P., McCarthy, C., Pfaltz, A. and Wennemers, H. (2003) *Angewandte Chemie – International Edition*, **42**, 1722–4.
353 Krattiger, P., Kovàsy, R., Revell, J.D., et al. (2005) *Organic Letters*, **7**, 1101–3.
354 Krattiger, P., Kovàsy, R., Revell, J.D. and Wennemers, H. (2005) *QSAR and Combinatorial Science*, **24**, 1158–63.
355 Revell, J.D., Gantenbein, D., Krattiger, P. and Wennemers, H. (2006) *Biopolymers Peptide Science*, **84**, 105–13.
356 Sell, C.S. and Dorman, L.A. (1982) *Chemical Communications*, 629–30.
357 Van der Berg, H.J., Challa, G. and Pandit, U.K. (1989) *Journal of Molecular Catalysis*, **51**, 13–27.
358 Barret, A.G.M., Love, A.C. and Tedeschi, L. (2004) *Organic Letters*, **6**, 3377–80.
359 Karbass, N., Sans, V., Garcia-Verdugo, E., et al. (2006) *Chemical Communications*, 3095–7.
360 Lozano, P., García-Verdugo, E., Piamtongkam, R., et al. (2007) *Advanced Synthesis & Catalysis*, **349**, 1077–84.
361 Burguete, M.I., Galindo, F., García-Verdugo, E., et al. (2007) *Chemical Communications*, 3086–8.
362 Marion, N., Díez-González, S. and Nolan, S.P. (2007) *Angewandte Chemie – International Edition*, **46**, 2988–3000.
363 Burquete, M.I., Erythropel, H., Garciá-Verdugo, E., Luis, S.V. and Sans, V. (2008) *Green Chemistry*, doi: 10.1039/b714977h.
364 Karimian, K., Mohanazadeh, F. and Rezai, S. (1983) *Journal of Heterocyclic Chemistry*, **20**, 1119–21.
365 Zeitler, K. and Mager, I. (2007) *Advanced Synthesis & Catalysis*, **349**, 1851–7.
366 Wahlen, J., De Vos, D.E., Jacobs, P.A. and Alsters, P.L. (2004) *Advanced Synthesis & Catalysis*, **346**, 152–64.
367 Blossey, E.C., Neckers, D.C., Thayer, A.C. and Schaap, A.P. (1973) *Journal of the American Chemical Society*, **95**, 5820–2.
368 Neckers, D.C. (1986) *Polymeric Reagents and Catalysts*, in Symposium Series **308** (ed. W.T. Ford), American Chemical Society, Washington, pp. 107–31.
369 Neckers, D.C. (1988) *Properties of Polymeric Rose Bengals – Polymers as Photochemical Reagents*, in *Synthesis and Separations Using Functional Polymers* (eds D.C. Sherrington and P. Hodge), John Wiley & Sons, Inc., New York, pp. 209–26.

370 Nowakowska, M., Kepczynski, M. and Szczubialka, K. (2001) *Pure and Applied Chemistry*, **73**, 491–5.

371 Nowakowska, M. and Kepczynski, M. (2007) *Biomacromolecules*, **8**, 433–8.

372 Benaglia, M., Danelli, T., Fabris, F., *et al.* (2002) *Organic Letters*, **4**, 4229–32.

373 Mattay, J., Vondenhof, M. and Denig, R. (1989) *Chemische Berichte*, **122**, 951–8.

374 Nowakowska, M. and Kepczynski, M. (1998) *Journal of Photochemistry and Photobiology A: Chemistry*, **116**, 251–6.

375 Faust, D., Funken, K.-H., Horneck, G., *et al.* (1999) *Solar Energy*, **65**, 71–4.

376 Akhavan-Tafti, H., Handley, R.S., Sandison, M.D. and Larkin, R.K. (Lumigen, Inc.). (2003) U.S. Patent 6,545,102 B1.

377 Gryglik, D., Miller, J.S. and Ledakowicz, S. (2004) *Solar Energy*, **77**, 615–23.

378 Motherwell, W.B., Bingham, M.J. and Six, Y. (2001) *Tetrahedron*, **57**, 4663–86.

379 Wulff, G., Sarhan, A. and Zabrocki, K. (1973) *Tetrahedron Letters*, 4239–332.

280 Ramström, O. and Mosbach, K. (1999) *Current Opinion in Chemical Biology*, **3**, 759–64.

381 Wulff, G. (2002) *Chemical Reviews*, **102**, 1–27.

382 Alexander, C., Davidson, L. and Hayes, W. (2003) *Tetrahedron*, **59**, 2025–57.

383 Alexander, C., Andersson, H.S., Andersson, L.I., *et al.* (2006) *Journal of Molecular Recognition*, **19**, 106–80.

384 Cheng, Z., Zhang, L. and Li, Y. (2004) *Chemistry–A European Journal*, **10**, 3555–61.

385 Emgenbroich, M. and Wulff, G. (2003) *Chemistry–A European Journal*, **9**, 4106–17.

386 Visnjevski, A., Schomäcker, R., Yilmaz, E. and Brüggemann, O. (2005) *Catalysis Communications*, **6**, 601–6.

387 Volkmann, A. and Brüggemann, O. (2006) *Reactive and Functional Polymers*, **66**, 1725–33.

388 Kofoed, J. and Reymond, J.-L. (2005) *Current Opinion in Chemical Biology*, **9**, 656–64.

389 Haag, R. and Roller, S. (2003) Polymeric materials, in *Organic Synthesis and Catalysis* (ed. M.R. Buchmeiser), Wiley-VCH Verlag GmbH, Weinheim, pp. 305–44.

390 Lee, J.J., Ford, W.T., Moore, J.A. and Li, Y. (1994) *Macromolecules* **27**, 4632–4.

391 Davis, A.V., Driffield, M. and Smith, D.K. (2001) *Organic Letters*, **3**, 3075–8.

392 Zubia, A., Cossio, F.P., Morao, I., Rieumont, M. and Lopez, X. (2004) *Journal of the American Chemical Society*, **126**, 5243–52.

393 Delort, E., Nguyen-Trung, N.Q., Darbre, T. and Reymond, J.-L. (2006) *Journal of Organic Chemistry*, **71**, 448–4480.

394 Guillena, G., Kreiter, R., van de Coevering, R., *et al.* (2003) *Tetrahedron Asymmetry*, **14**, 3705–12.

395 Kofoed, J., Darbre, T. and Reymond, J.-L. (2006) *Organic and Biomolecular Chemistry*, **4**, 3268–81.

11
Transition Metal Catalysts

Rajiv Banavali, Martin J. Deetz and Alfred K. Schultz

Rohm and Haas Chemicals, LLC, 727 Norristown Road, Spring House, PA 19477-0904, USA

11.1
Introduction

Transition metal loaded organic synthetic resins are employed industrially for numerous synthetic transformations. With increasing cost pressures and environmental restrictions on processes using transition metal catalysts, the need for more recyclable transition metal catalysts intensifies. Organic synthetic resins have been around for more than 60 years and are ideal substrates for transition metal immobilization. They are well poised for chemical modification and the resultant catalysts can be used in either batch or continuous systems.

This chapter will walk through the various forms these catalytic resins take. The catalysts covered in this review fall into three classes, (i) transition metals covalently bonded to the polymer support through an organometallic bond, (ii) transition metals coordinated to the polymer support, typically in ionic form and (iii) transition metal clusters that are formed by precipitating metals into nanoparticles within the polymeric framework. Additionally, this chapter covers the synthetically useful and industrially practiced reactions catalyzed by transition metals loaded onto organic supports and comments on the mechanisms and reusability aspects of the processes [1].

11.2
Synthetic Avenues for Producing Metal Loaded Organic Resins

Transition metals are used as catalysts for various reactions. These same transition metals can be anchored to an organic resin. Resins, as used here, are defined as functional organic synthetic polymers. The transition metal can be introduced into the polymer matrix in two different ways. The transition metal can be included as part of the functional monomer prior to polymerization. This approach presents

The Power of Functional Resins in Organic Synthesis. Judit Tulla-Puche and Fernando Albericio
Copyright © 2008 WILEY-VCH Verlag GmbH & Co. KGaA, Weinheim
ISBN: 978-3-527-31936-7

several drawbacks, including the need for several, often non-efficient, steps, an uncontrolled distribution of the metal on the resin and, potentially, side product of cross-links and/or grafted by-products can also occur. The cross-links will alter the polymer morphology. This topic is further discussed below. The preferred method of introducing the transition metal onto a functional synthetic organic polymer matrix is post-polymerization functionalization (see below) [2].

11.2.1
Covalent Bonding

Covalent bonding refers to the materials made in which the transition metal is bonded directly to the resin through an organometallic bond. Two different approaches can be used to covalently attach metal complexes to polymer supports: (i) synthesis of appropriate functional monomers and their (co)polymerization to form catalytically active polymers (Scheme 11.1) or (ii) attachment of metal complexes to preformed functional polymer supports by chemical reactions. Following these approaches, both soluble and cross-linked chiral polymeric metal complexes can be prepared. An example of an organometallic tin catalyst suitable for transesterification was reported by workers at Rohm and Haas Company [3].

11.2.2
Coordination Complexes

Coordination of transition metals upon a support typically involves binding ionic transition metal species onto a support with ionic ligands capable of binding the transition metal multivalently. The loading is typically accomplished by adding the metal salt to the swollen resin in an appropriate solvent. Often, this is the intermediate species in the precipitation of metal nanoparticles onto a solid support, which will be covered later (Scheme 11.2).

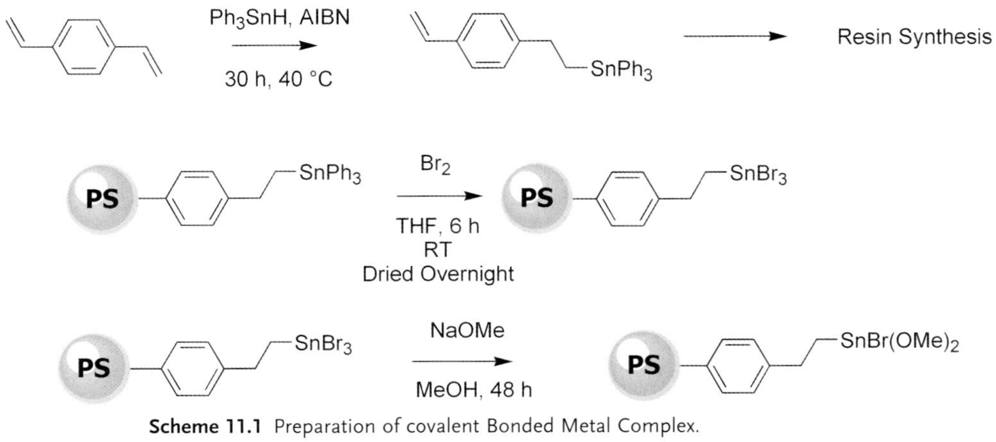

Scheme 11.1 Preparation of covalent Bonded Metal Complex.

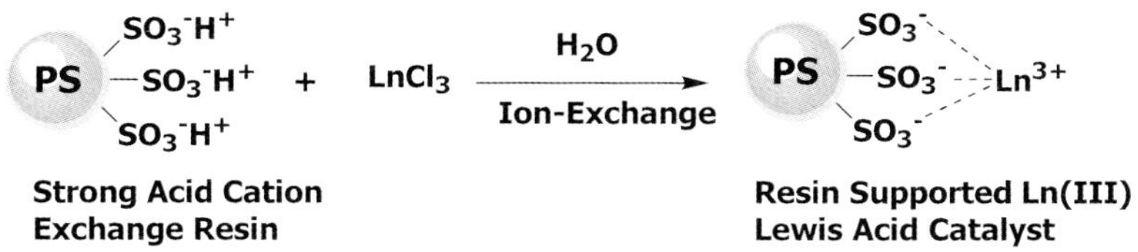

Scheme 11.2 Preparation of coordination complex.

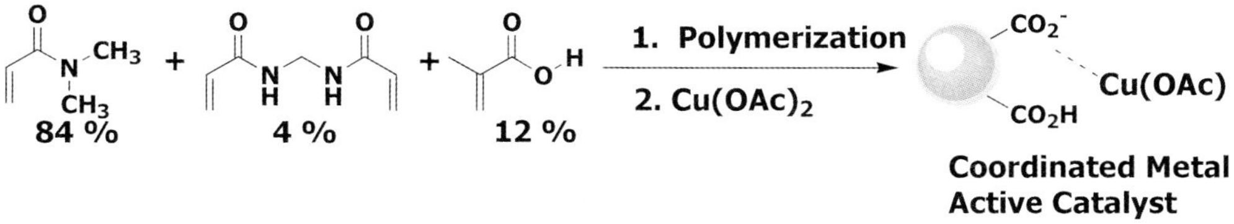

Scheme 11.3 Preparation of coordination complex by Ligand Exchange.

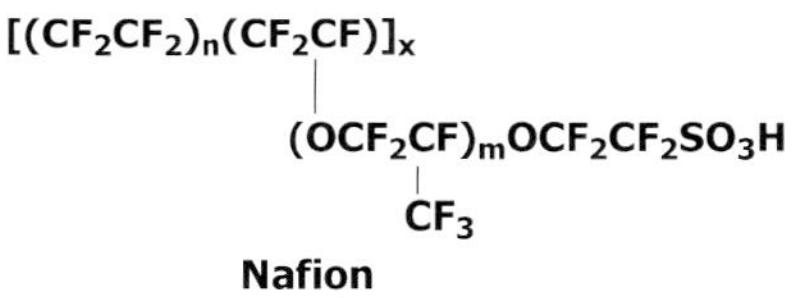

Figure 11.1 Structure of Nafion® resin.

An example of a polymer-supported catalyst was produced from a tailor-made resin based on N,N-dimethylacrylamide with 4 mol% methylene bis(acrylamide) as the cross-linker and 12 mol% methacrylic acid as the functional, metal binding comonomer. Treatment of the resin with a solution of $Cu(OAc)_2$ in methanol resulted in a ligand exchange reaction with partial substitution of the acetates with polymer-bound carboxylate groups (Scheme 11.3) [4]. The use of the catalyst is discussed further below.

Fierro and coworkers [5] reported a very effective catalyst based on Pd^{2+} centers dispersed inside the polymer framework of a commercially available macroreticular resin bearing sulfonic groups (Lewatit K 2621/Hþ) [6].

Another example of coordination of a transition metal into a selected polymer is the use of Nafion resin [7]. Several studies over the past 20 years have demonstrated that the perfluorinated ion-exchange polymer is an excellent support for metal-complex catalysts (Figure 11.1).

11.2.3
Precipitation

This section describes the precipitation of the transition metals within a polymer support. The precipitation process can be random, or much more specific as in the case of formation of nanoparticulate transition metal nanoclusters within a polymer matrix. The synthetic route to making precipitated transition metal complexes involves three steps, namely (i) synthesis of a suitably functionalized polymer; (ii) loading of the polymer with convenient metal nanoparticle precursors; (iii) generation within the polymer of the metal nanoparticles. The first two steps can be condensed to one upon utilization of metal-containing monomers in the polymer synthesis. Furthermore, the third step can be omitted by directly loading the polymer support with pre-formed metal nanoparticles [8]. An example of a transition metal precipitate is shown below. Bergbreiter and coworkers have described the precipitation of palladium onto a functionalized polystyrene backbone. The polymer was produced by grafting the ligands onto pSTY support and functionalization on the grafts [9] (Scheme 11.4).

A proper choice of the reaction medium for the metal loading reaction is very important, especially when nonporous resins are used as polymer supports. Thus, a solvent must be chosen that can solubilize the metal precursor, but which is also capable of swelling the nonporous resin to an appreciable extent; swelling is needed to guarantee accessibility of the reactants to most of the functional groups.

The final step in the preparation of polymer-supported metal nanoparticles is the generation of the nanoparticles within the polymer, which is usually accomplished by reduction of the polymer-bound metal precursors. Often, techniques similar to the preparation of conventional metal catalysts supported on inorganic solids are employed.

Commonly employed reductants include elemental hydrogen, sodium borohydride, hydrazine, alcohols and formaldehyde. When nonporous resin supports are employed, the growth of the metal nanoparticles during reduction becomes limited by the steric restrictions imposed by the three-dimensional polymer network

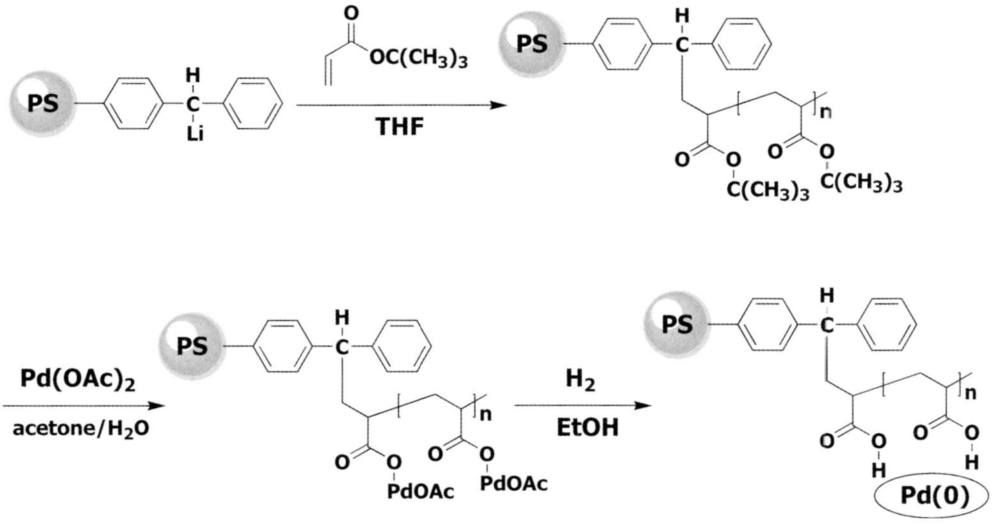

Scheme 11.4 Preparation of nanoparticle Pd(0) Complex.

(Figure 11.2). This allows some level of control on the size of the nanoparticle [10].

The size and size distribution of metal nanoparticles throughout the particles of the support can be controlled by the choice of reducing agent, the reduction procedure and the metal concentration. Some researchers have seen increasing homogeneity of the metal nanoparticle distribution throughout the support with decreasing metal concentration and increasing concentration of reductant [8, 11].

Corain *et al.* [6] have described the polymer support as a soluble macromolecule or a micellar aggregate that "wraps" the metal nanoparticle in solution, thus preventing metal sintering and precipitation. It can also be a resin, that is an insoluble material consisting in a bundle of physically and/or chemically cross-linked polymer chains in which the metal nanoparticles are embedded (Figure 11.2). Thus, soluble cross-linked polymers ("microgels") that can stabilize metal nanoparticles can be prepared; in addition, metal colloids protected by soluble linear polymers have been grafted onto insoluble resin supports to yield insoluble catalysts. This chapter is devoted mainly to metal nanoparticles on insoluble resin supports [8].

The overall preparation scheme of functionalized metal polymer complex is summarized in Figure 11.3.

11.2.4
Microstructural Aspects of Organic Functional Resins

Functional resins are produced in two basic morphological types. Gel-type (microporous) resins are without an appreciable porosity in the dry state; their interior

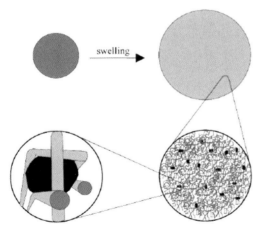

Figure 11.2 Representation of the process used to precipitate Pd(0) nanoparticles in the matrix of a gel-type resin. The process begins with the swelling of the bead with a Pd(II) solution, followed by reduction of the Pd(II) to Pd(0). The metal nanoparticles are not covalently attached, rather mechanically trapped within copolymer matrix. Source: Image taken from Reference [6].

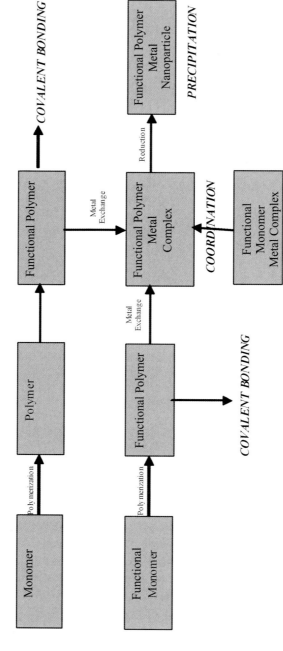

Figure 11.3 Flow diagram for creating functional polymer–metal nanoparticles.

11.2 Synthetic Avenues for Producing Metal Loaded Organic Resins

is accessible only after swelling in the reaction environment. Macroreticular (macroporous) resins contain pores stable even in the dry state, in addition to the micropores generated by the swelling of the polymer skeleton. Spherical beads with diameter usually in the range 0.3–1.25 mm are the most widely applied form, although some types are also supplied as powder with particles smaller than 0.2 mm [12]. Gel-type resins are prepared by the functionalization of the precursor gel-type copolymer, whereas macroreticular resins are prepared from the corresponding macroporous copolymer. The polymers, which can be functionalized, are prepared by the suspension polymerization reaction with the appropriate monomers. Macroreticular copolymers are produced in a suspension polymerization process, in the presence of a porogen. The porogen material is responsible for the generation of the resultant pore structure [13]. Often, macroporous resins contain more cross-linking monomer than gel-type resins.

Gel-type, microporous, resins must swell to expose their catalytically active sites, whereas macroreticular resins have a permanent pore structure (inside these pores, catalytically active sites reside). Pores of the macroreticular resins can be described acceptably in terms of the conventional cylindrical pore model (pore diameter and volume). Pore structure, size, pore volume, and so on have been studied intensively in recent years. Examples of analytical techniques include X-ray microprobe analysis, ESR spectroscopy, NMR, and inverse steric exclusion chromatography (ISEC); the latter yields the best quantitative assessment of the nanomorphology of swollen resins.

In one example [14], millimeter-sized particles of polystyrene were chloromethylated, then functionalized with phosphino groups and, eventually, the phosphinated material was reacted with rhodium(I) complexes. By means of X-ray microprobe analysis (XRMA), the penetration of the reactants towards the core of the particles was estimated from the distribution of chlorine, phosphorus and rhodium atoms in the material as determined by XRMA after each step. The higher the cross-linking degree, the more difficult is the penetration of the reacting species. Table 11.1 below lists the techniques studied for determination of pore structure.

In an example, Corain, Jerabek, Zecca, and coworkers have investigated the preparation of metal palladium catalysts (metal crystallite size circa 2–4 nm) supported on microporous resins. They employed ISEC, ESR and field-gradient spin-echo NMR spectroscopies to assess the quantitative relationships between the nanoscopic morphology (nanostructure) and molecular accessibility in the swollen state. They evaluated these catalysts with different metal loading and cross-linking degrees, in the hydrogenation of cyclohexene (IM) under mild conditions ($T = 25\,°C$; $P = 0.5$ MPa) in methanol, at a Pd concentration equal to 2.5×10^{-4} M. The metal crystallites were dispersed inside amphiphilic microporous resins based on styrene and 2-methacryloxyethylsulfonic acid, cross-linked with methylenebisacrylamide (1–6%, mol) and fully characterized, *inter alia*, by means of the above-mentioned techniques [15]. The use of these resin catalysts is discussed below.

Table 11.1 Physiochemical methods for pore structure analysis.

Method	Information	Reference
Diffraction	Nanoparticle morphology	a
EPR of paramagnetic probes	Accessibility, mobility	b
NMR of confined solvent	Accessibility, mobility	c
ISEC	Size and volume of "pores" and surface area	d
X-Ray microprobe analysis	Distribution of functional groups	e
EPR of paramagnetic probes + ISEC	Swollen state morphology, accessibility, mobility	f

a   Yu, L.B., Chen, D.P. and Wang, P.G. (1997) *Journal of Organic Chemistry*, **62** 3575.
b   1. Yu, Z., Liao, S., Yang, B. and Yu, D. (1997) *Journal of Molecular Catalysis A: Chemical*, **120**, 247. 2. Mdleleni, M.M., Rinker, R.G. and Ford, P.C. (1998) *Inorganic Chimica Acta*, **270**, 345. 3. Parshall, G.W. (1980) *Homogeneous Catalysis*, Wiley-Interscience, New York. 4. Sherrington, D.C. (1998) *Supported Reagents and Catalysts in Chemistry* (eds B.K. Hodnett, A.P. Kybett, J.H. Clark, K. Smith), The Royal Society of Chemistry, Cambridge, p. 220.
c   1. Miller, M.M., Sherrington, D.C. (1995) *Journal of Catalysis*, **152**, 368. 2. Miller, M.M., Sherrington, D.C. (1995) *Journal of Catalysis*, **152**, 377.
d   Jerabek K. (1996) *Cross-Evaluation of Strategies in Size-Exclusion Chromatography*, ACS Symposium Series **635**, (eds M. Potschka, P.L. Dubin), American Chemical Society, Washington, p. 211–24.
e   Li, F., Huang, J., Zou, J., Pan, P., Yuan, G. (2002) *Carbon*, **40**, 2871.
f   Leporini, D., Zhu, X.X., Krause, M., Jeschke, G., Spiess, H.W. (2002) *Macromolecules*, **35**, 3977.

11.3
Reactions Utilizing Transition Metal Catalysts Supported on Organic Resins

11.3.1
Hydrogenation

Metal complexes are often used in hydrogenation reactions. These metal catalysts can be supported on a resin in the form of nanoparticles [8]. Palladium metal nanoparticles are useful as hydrogenation catalysts. They are produced by the metal exchange reaction of Pd(OAc)$_2$ onto an appropriate acid containing resin. Once the Pd(II) is in place, the resultant polymer supported Pd(II) is reduced by NaBH$_4$ [9].

The chosen polymer support influences the reaction as well. Various polymer supports are described above. Studies of the hydrogenation of 2-propen-2-ol with Pd(0) catalysts showed interesting efficiency and selectivity differences dependant on the polymer resin chosen. Zharmagambetova *et al.* [16] studied three different organic polymer supports for Pd(0) nanoclusters and compared them to the inorganic Al$_2$O$_3$. Organic polymers included poly-2-vinlypyridine (PVP), cellulose and styrene-divinyl benzene (STY-DVB) copolymer. The cellulose and STY-DVB resins were functionalized with amino groups for anchoring the Pd. During the hydrogenation of 2-propene-2-ol, a potential side reaction, that is the isomerization of the olefin, leads to the unreactive propanal as a by-product. Table 11.2 summarizes the stoichiometry of the reactions, and the extent of hydrogenation versus isomerization for each catalyst.

The use of PVP as a support material for the Pd(0) enhances selectivity for hydrogenation over the "free" form of the catalyst. Also, the reaction rate is increased dramatically as compared to the catalyst supported on inorganic supports. Clearly, the use of PVP resin as support for Pd(0) enhances efficiency and selectivity.

A shortcoming of organic polymeric supports is temperature stability. Typically, acidic organic functional resins are stable in the 120 °C range, whereas basic organic functional resins are stable to approximately 50–60 °C range. To overcome the temperature sensitivity of organic supports, Corain and coworkers studied isonitrile-functionalized polymer supports [17]. They synthesized the novel isonitrile functional resin shown in Figure 11.4.

Combined TGA-DTA measurements of the catalysts revealed that the polymeric structure of these materials is stable up to 300 °C; this was confirmed by monitoring the morphology of the materials by electron scanning microscopy. To establish that these resins were not leaching the active Ru or Rh catalyst, they were subjected to

Table 11.2 Selectivity differences as a function of resin support.

Resin Support	H_2 uptake	Hydrogenation (%)	Isomerization (%)
None	0.05	67	33
Al_2O_3	0.32	75	25
PVP	1.06	98	2
Cellulose-NH_2	1.00	77	23
Sty/DVB-NH_2	0.01	68	32

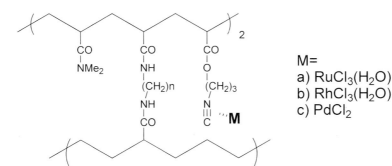

Figure 11.4 Isonitrile-functionalized polymer support.

Table 11.3 Hydrogenation using high-temperature stable catalyst.

Substrate	Conversion (21 °C; 3 atm H_2; 10 h)
Nitrobenzene	0
Phenylacetylene	Styrene (33%); ethylbenzene (67%)
Benzaldehyde	Benzyl alcohol (81%); other reduction products (19%)

five consecutive cycles (10 mg material in 10 mL CH_2Cl_2 at 100 °C for 10 h) and showed an initial loss of 12% of the metal on the first cycle, but no determinable loss thereafter as measured by IR spectroscopy. The palladium catalyst (compound c in Figure 11.4) was used to illustrate catalytic activity. Table 11.3 summarizes the efficiency and selectivity of the hydrogenation of nitrobenzene, phenylacetylene and benzaldehyde.

In another example, nanoclustered Pt(0) catalysts based on cross-linked macromolecular matrixes were evaluated in the hydrogenation of an α,β-unsaturated aldehyde, citral. The monometallic catalysts exhibit remarkable selectivity for geraniol/nerol when 2-3 nm, regularly shaped, spherical metal nanoclusters are deposited on the supports from solutions of solvated platinum atoms prepared by metal vapor synthesis (MVS). The immobilization in the polymer framework of ions of a second metal such as Fe(II), Co(II), or Zn(II) enhances the selectivity of the Pt catalysts by up to more than 90% [18].

Chen et al. [19] have reported very active, stable platinum nanoparticle catalysts prepared by alcohol reduction of $PtCl_6$ using poly(N-isopropylacrylamide) previously grafted on PS microspheres as stabilizing polymer. The observed catalytic activity in the hydrogenation of allyl alcohol was more than five times higher than with Pt/C. Moreover, it was possible to recycle the resin-based catalysts for at least six cycles, whereas Pt/C was not recyclable at all. When comparing the catalytic activity of free and heterogeneous colloidal platinum particles, only a small decrease in the reaction rate was observed.

Sabadie and Germain [20] have investigated the stereoselectivity of polymer-supported metal catalysts in the hydrogenation of 1,2-dimethylcyclohexene. Depending on the pressure of hydrogen, different ratios of cis- and trans-isomers (0.44 and 0.57 at 1.25 and 10 MPa, respectively) of 1,2-dimethylcyclohexanes were obtained over Pd/APSDVB catalysts.

The partial hydrogenation of dienes was successfully carried out by Michalska et al. [21] using palladium supported on heterocyclic polyamides. Under the reaction conditions employed (MeOH, atmospheric pressure, 25 °C) the resin-supported catalyst was able to selectively hydrogenate one of the two double bonds present. Recycling experiments proved the high stability of the used catalysts. For example, in 11 hydrogenation runs with 2-methyl-1,3-pentadiene, which is equivalent to 4300 catalytic cycles per palladium atom, neither loss of activity nor changes in selectivity were observed.

A successful partial hydrogenation of alkynes to alkenes in the presence of other double bonds in the substrate was also reported by Sulman et al. [22]. The authors prepared linalool(LN) (3,7-dimethyl 1,6-octadiene-3-ol) by the selective hydrogenation of dehydrolinalool (3,7-dimethyl-octa-6-ene-1-yne-3-ol) using a Pd/PVP/Al$_2$O$_3$ catalyst. They achieved 99.8% selectivity to LN in toluene at 90 °C and 0.1 MPa by running the reaction under hydrogen limitation (480 min reactor shaking frequency). The catalyst was recycled 20 times and exhibited higher stability than an analogous catalyst prepared from a polystyrene-co-butadiene copolymer deposited on Al$_2$O$_3$.

11.3.2
Bi-functional Catalysis

(−)-(2S,3S)-1-Dimethylamino-3-(3-methoxyphenyl)-2-methylpentan-3-ol hydrochloride was directly dehydroxylated to the new opioidic (−)-(2R,3R)-[3-(3-methoxyphenyl)-2-methylpentyl]dimethylamine hydrochloride over Amberlyst-15 containing Pd. Two reaction steps are involved in this one-pot synthesis: dehydration of the starting tertiary alcohol followed by hydrogenation of the obtained olefin. Dehydration of the alcohol occurs only in the presence of an Amberlyst resin as solid acid catalyst, which also serves as support for the hydrogenation component Pd. Such a relatively simple well-known Pd/Amberlyst system provides excellent results for this complicated one-pot conversion. Other acidic heterogeneous catalysts did not show catalytic performance for this kind of reaction. The various reaction parameters of this one-pot dehydroxylation reaction were optimized using a statistical test design program [23] (Scheme 11.5).

11.3.3
Isomerization of Olefins

Raje and Datta [24] have describe an investigation of the catalytic reactivity and stability of the cationic tetrakis(triethylphosphite)nickel(0) hydride complex

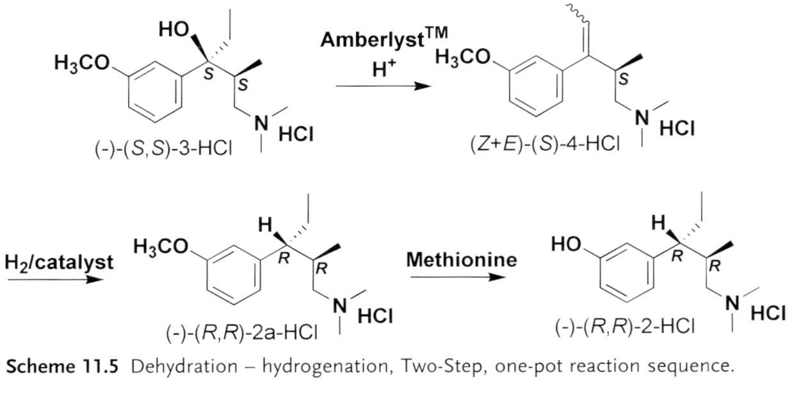

Scheme 11.5 Dehydration – hydrogenation, Two-Step, one-pot reaction sequence.

supported on Amberlyst-15, a sulfonic acid synthetic organic macroreticular resin based on styrene-divinylbenzene copolymer, for the vapor-phase isomerization of 1-butene. The ionic-heterogeneous catalyst was better than its homogeneous counterpart, in terms of activity and, dramatically, superior stability.

11.3.4
Hydrosilylation Reaction

A review by Michalska and Strzelec [25] describes the preparation of polyamides suitable for use as supports for Rh complexes and the activity of the supported catalysts in hydrosilylation reactions. A series of polyamides were synthesized by low-temperature interfacial condensation of 2,5- and 2,6-pyridinedicarboxylic chlorides and aliphatic diamines $H_2N(CH_2)_nNH$, where $n = 2$ and 6, and used as the model supports for immobilization of the Rh(I) complex catalyst (Figure 11.5).

The structural parameters of the complexes were studied in the dry and swollen states by WAXS, SAXS, DSC, nitrogen BET adsorption, ISEC and pycnometry. The original polymer structure changed during complex formation to materials of higher porosity. The relationship between the support structure and catalyst activity and selectivity was studied in model reactions, namely, hydrosilylation of alkenes, dienes and alkynes (Scheme 11.6).

The catalyst activity decreased with increasing polymer crystallinity. A high regioselectivity of the catalyst in the hydrosilylation of alkenes towards formation of the linear products was achieved due to the favorable microporous structure of the polyamide supports with pore size of 10–20 . The stereoselectivity of the reaction can be reversed by a proper choice of donor functions in a polymer support, for example the traditional cis-selectivity of Rh catalysts in hydrosilylation of phenylacetylene was changed to trans-selectivity by use of a 2,5-py instead of a 2,6-py moiety. The polyamide-supported catalysts showed high stability through 6–9 synthesis runs [25].

11.3.5
Aldol Reactions

Lanthanide(III) catalysts supported on ion exchange resins (Ln-resins) were prepared from Dowex, Amberlite, Amberlyst and other cation exchange resins. The

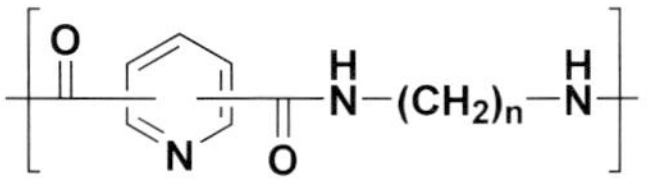

Figure 11.5 Polyamide used for rhodium complexation.

$$Me_2PhSiH + PhHC=CH_2 \longrightarrow \underset{\beta}{Me_2PhSiCH_2CH_2Ph} + \underset{\alpha}{CH_3(SiMe_2Ph)CHPh}$$

Scheme 11.6 Hydrosilylation of Olefins.

amount of lanthanides on resin supports was measured through EDTA titration The lanthanides in aqueous solution exchanged with almost all H⁺ or Na⁺ on the resins to form stable ionic complexes between the lanthanides(III) and the resins. The effects of resin types, resin sizes and lanthanide salts were studied with a reaction of indole and hexanal in aqueous solution and with an aldol reaction of benzaldehyde and silyl enol ether in organic solvents. Among polymeric resins tested, Amberlyst XN-1010 and Amberlyst-15 complexed with lanthanides(III) were the most effective catalysts. The selective Ln-resins were used to catalyze acetalization of aldehydes, aldol reaction of formaldehyde or benzaldehyde in aqueous solution, nucleophilic addition to an imine, allylation of an aldehyde, an aza Diels-Alder reaction and a ring-opening reaction of an epoxide. Also, glycosylation of alcohols using glucosyl fluoride as a donor was also promoted with the Ln-resin. This work demonstrated potential uses of lanthanide(III) catalysts supported on functional organic resins in routine organic reactions [26].

11.3.6
The C–O Coupling Reaction

For the C–O coupling reaction, supporting the metal caused a significant decrease in catalytic activity. This was the case both in dichloromethane and 1,2-dichloroethane and at room temperature as well as at 50 °C. This reduced catalytic activity was due to diffusional limitations to the transport of the reagents through the polymer support. This is often witnessed when using polymer supports. One way to circumvent the negative effects of the polymer support is to use a polymer system with lower cross-linking level. Polymers made with lower cross-linking levels tend to swell more, and this higher swelling "exposes" more catalytic centers to the reactants. However, a control test performed with a fully analogous polymer support with only 2% cross-linking, and consequently with a consistently higher swelling volume, yielded the same result. Enhancing the quantity of catalyst led also, in this case, to a significant improvement in the reaction yield, which, however, never exceeded 50% [4].

In contrast to previous results obtained in related C-N coupling reactions, the polymer supported catalyst was in this case less efficient than the corresponding homogeneous one. Furthermore, attempts to recycle the supported catalyst following separation from the reaction mixture, by washing with 1,2-dichloroethane and drying in air, led to significantly decreased productivity of the catalyst. This decrease could be due to leaching of copper species from the support. However, copper analysis on the reaction mixture of the first reaction cycle after catalyst separation revealed that only 4.7% of the total copper amount in the employed catalyst charge was released. Thus, the decrease in catalytic productivity should be due to decomposition of the active catalyst, presumably upon formation of inactive complex species, or by other fouling mechanisms.

Various cationic metal-complex catalysts have been successfully immobilized on Nafion with little if any leaching occurring during catalysis, and in several instances the supported catalyst has been reused with little loss in activity. While the activity of Nafion supported catalysts depends upon catalyst loading, suggesting diffusion limitations, activities comparable with homogeneous activities have been observed by increasing the dispersion of Nafion. The physical structure and chemical properties of Nafion offer the additional benefits of providing protection to the supported catalyst from deactivation and the potential to increase the activity of the supported catalyst compared with its homogeneous analog. Although diffusion limitations and the high cost of Nafion have potentially limited its application as a metal-complex catalyst support, the development of high surface area Nafion silica nanocomposites provides the opportunity to solve these issues [27].

11.4
Commercial Interest in Transition Metal Catalysts

Metal-loaded organic functional resins are available commercially and are currently being employed as catalysts in a few large-scale industrial processes. Strongly acidic resins are used as active supports for metal palladium in the preparation of bifunctional catalysts that consist of acid as well as hydrogenation-active centers.

11.4.1
Methyl Isobutyl Ketone (MIBK) Synthesis

The commercial production of the industrial solvent methyl isobutyl ketone (MIBK) is based on metal impregnated polymeric resin (Scheme 11.7).

The self-aldolization of acetone by the catalytic acid centers gives 2-methyl-2-hydroxypentan-4-one (diacetone alcohol), which is the intermediate in the process. The intermediate is then dehydrated to the unsaturated oxide. MIBK is then obtained by hydrogenation of the condensation product under pressurized hydrogen in the presence of Amberlyst CH28, which is a commercial Pd(0)-impregnated organic functional resin [28]. This allows the consecutive condensation, dehydration, and hydrogenation of acetone. Acetone conversions varying from 25–50% are obtained at temperatures from 130-150 °C; the MIBK selectivities varied in the range 70–90%. The results indicate that the condensation–dehydration rate-limiting step severely decreases as conversion increases; this was attributed to the rate-inhibiting effect of the water formed in the dehydration reactions on the catalyst [29]. Amberlyst CH10 from Rohm and Haas Company, a 0.3% Pd loaded sulfonic macroreticular resin, is suggested as a tri-functional catalyst for condensation,

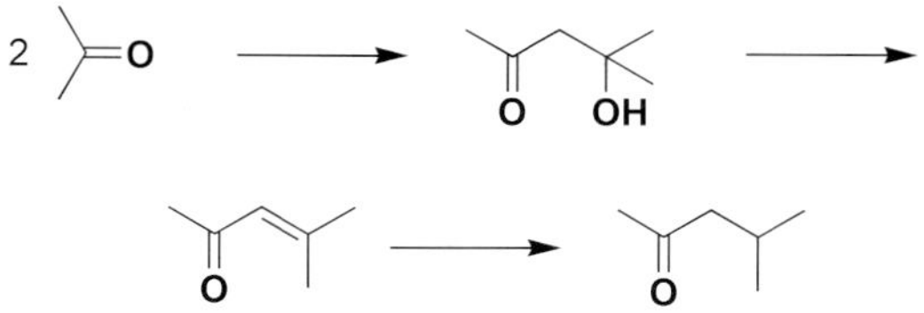

Scheme 11.7 Synthesis of Methyl isobutyl ketone.

dehydration and hydrogenation for *t*-amyl methyl ether (TAME) production [30]. The use of Amberlyst CH10 as a trifunctional catalyst in TAME production is to provide selective hydrogenation of diolefins (e.g. butadiene or piperylenes) in the C5 feed, as well as acid sites for the etherification of the tertiary olefins and the olefin isomerization from linear to tertiary olefins [31, 32].

Similar catalysts based on acidic organic functional resins (Lanxess catalysts K 6333 and VP OC 1063) [28] are employed in industrial heat-exchange units for the reduction of dioxygen level in water from ppm to ppb levels.

11.4.2
Conversion of Glycerol into 1,2-Propanediol

The combination of Ru/C and Amberlyst functional organic resin is effective for the dehydration and hydrogenation (denoted as hydrogenolysis) of glycerol to 1,2-propanediol under mild reaction conditions (393°K). A Ru/C catalyst prepared by using active carbon with a low surface area (approximately 250 $m^2 g^{-1}$) showed better performance than that prepared by using active carbon with a high surface area. In addition, treatment of Ru/C catalysts prepared from $Ru(NO)(NO_3)_3$ at the appropriate temperature enhanced the performance compared to that of the commercially available Ru/C catalysts. This temperature treatment can be influenced by the decomposition of Ru precursor salt and aggregation of Ru metal particles. In addition, the degradation reaction, as a side-reaction, to C1 and C2 compounds of glycerol hydrogenolysis was more structure-sensitive than the hydrogenolysis reaction, and the selectivity of hydrogenolysis was lower on smaller Ru particles. The combination of Ru/C with the Amberlyst resin enhanced the turnover frequency of 1,2-propanediol formation considerably, indicating that 1,2-propanediol can be formed mainly by dehydration of glycerol to acetol catalyzed by Amberlyst and subsequent hydrogenation of acetol to 1,2-propanediol catalyzed by Ru/C [32]. A combination of active carbon supported Ru catalyst with an organic functional resin (Amberlyst-15) exhibited much higher activity in glycerol hydrogenolysis under mild reaction conditions, such as 393°K and 4 MPa H_2, than other metal-acid bifunctional catalyst systems using various zeolites, sulfated zirconia, H_2WO_4, and liquid H_2SO_4 [33].

The authors found that of the various noble metals (Ru/C, Rh/C, Pt/C, and Pd/C) and acid catalysts [an organic functional resin (Amberlyst), H_2SO_4(aq), and HCl(aq)] the combination of Ru/C with Amberlyst resin was most effective in the dehydration + hydrogenation (i.e. hydrogenolysis) of glycerol under mild reaction conditions (393°K, 8.0 MPa). The dehydration of glycerol to acetol is catalyzed by the acid catalysts. The subsequent hydrogenation of acetol on the metal catalysts gives 1,2-propanediol. The activity of the metal catalyst + Amberlyst in glycerol hydrogenolysis can be related to that of acetol hydrogenation over the metal catalysts. Regarding acid catalysts, H_2SO_4(aq.) shows lower glycerol dehydration activity than Amberlyst, and HCl(aq.) strongly decreases the activity of acetol hydrogenation on Ru/C. In addition, the OH group on Ru/C can also catalyze the dehydration of glycerol to 3-hydroxypropionaldehyde, which can then be converted into 1,3-propanediol through subsequent hydrogenation and other degradation products [34].

11.4.3
Oxidation – Synthesis of H_2O_2

Most hydrogen peroxide is produced by the selective hydrogenation of 2-ethylanthraquinone (EAQ) to 2-ethylanthrahydroquinone (EAHQ), followed by treatment with dioxygen; this produces hydrogen peroxide and regenerates 2-ethylanthraquinone. Biffis, Jerabek, and Corain have developed novel catalysts for this process that are based on palladium supported on very lipophilic functional resins and that can promote a chemoselectivity for EAHQ slightly but definitely superior to that provided by an industrial catalyst under identical conditions. This finding demonstrates the potential of variations of the lipophilic/hydrophilic character of the support as a tool for the improvement of the chemoselectivity of resin-based metal catalysts [35].

Two polymers, namely poly(4-vinylpyridine) (PVP) and polyaniline (PANI), were used as supports for palladium catalysts acting in 2-ethyl-9,10-anthraquinone (9,10-EAQ) hydrogenation, a key step in the industrial production of H_2O_2. The nature of PVP and PANI interactions with various chloro complexes of Pd(II) coexisting in $PdCl_2$–H_2O–HCl solutions was studied using IR (mid-IR, far-IR), UV/Vis and XPS spectroscopies. It was found that the type of interactions involving nitrogen atoms of the polymers depended mainly on the acidity of $PdCl_2$ solution. Protonation of polymers (via acid–base reactions) as well as coordination of Pd^{2+} ions by nitrogen atoms of the polymers took place in highly acidic $PdCl_2$ solution (2 M HCl) containing predominantly anionic $[PdCl_4]^{2-}$, $[PdCl_3(H_2O)]^-$ complexes. In the weakly acidic $PdCl_2$ solutions, (0.66×10^{-3} M HCl) containing predominantly electro-neutral $[PdCl_2(H_2O)_2]$ complexes, hydrolysis of the complex proceeded as the main process, resulting in precipitation of palladium oxide on PVP. With PANI, a redox mechanism was involved, resulting in the reduction of some Pd^{2+} to Pd^0, accompanied by partial oxidation of the polymer chain. As a consequence of various mechanisms of polymer reactions with Pd^{2+} ions, the surface morphology of the final catalysts, characterized by XRD and SEM methods was different. The dispersion of palladium in Pd/PVP and Pd/PANI catalysts (1–10 wt.% Pd) influenced the course of EAQ hydrogenation. The presence of large palladium particles promoted reactions leading to the formation of the so-called "degradation products" not capable of hydrogen peroxide formation. Pd/PVP catalysts exhibited higher activity. The selectivity of EAQ hydrogenation in their presence was better than that seen for Pd/PANI catalysts [36].

11.4.4
Hydrogenation of N–O Bonds

The hydrogenation of unsaturated nitrogen-containing groups is another class of reactions in which immobilized catalysts can be employed [37–39]. The reduction

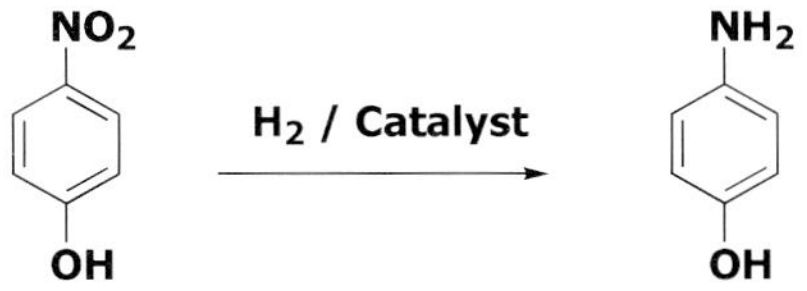

Scheme 11.8 Reduction of N–O species to amines.

$$O_2 + 2 H_2 \xrightarrow{Pd(0)} 2 H_2O$$

Scheme 11.9 Reaction Involved in the Removal of Oxygen from Water.

of *p*-nitrophenol to *p*-aminophenol is a common example used to compare catalysts (Scheme 11.8). The reduction has been successfully carried out by immobilizing palladium on organic supports. Indeed, there are many examples in the literature of the reduction being carried out when the Pd(0) was immobilized on basic [40] or acidic [41] resin supports. Another example involves immobilization of Pd(II) by coordination to a chelating polymer, such as poly(*N*-vinyl-pyrrolidone) [42].

11.4.5
Removal of Oxygen from Water

While not a traditional use for immobilized transition metal catalysis, this section exhibits the incredible diversity of the uses of these materials.

Palladium loaded resins can be utilized for the process of catalytically removing oxygen from boiler feed water (Scheme 11.9). Scientists from Lanxess have reported that their basic polymeric resins can be used in a fixed bed reactor for this application [43].

The commercial deoxygenation catalysts are prepared from anion exchange resins containing Pd(0) nanoparticles. The palladium nanoparticles are generated near the periphery of the beads, allowing for good kinetics and resulting in the need for low catalysts loadings for efficient deoxygenation.

In the above-mentioned applications, the resins are typically used as beads (0.2–1.25 mm diameter), in fixed-bed or suspension reactors or, more frequently, in flow-through reactors. These are typically continuous operations and the reactor sizes are meant to have a production capacity in several hundreds of tons per year. Working temperatures range from room temperature up to about 120 °C. Most resin materials suffer from relatively low mechanical, thermal and chemical stability when compared with the more traditional inorganic materials. For this reason, resin-based catalysts are mainly applied as fixed-beds operating under modest temperature and pressures. However, resin supports do have certain advantages over conventional supports. These result from the fact that, in functional resins, most of the functional groups are embedded inside the polymer matrix, and not simply on the surface of the support particles, as is commonly the case with inorganic supports. These functional groups become accessible when the resin is swollen by a suitable liquid medium having a good compatibility with the polymer. The importance of the proper design of a polymer support that is highly swellable in the reaction solvent is thus critical.

11.5
Conclusions and Future Outlook

In conclusion, over the past 20 years, various transition metals have been successfully immobilized on polymeric resins. The catalytic activity of these transition metal complexes have been shown with little if any leaching occurring during catalysis. In several instances the supported catalyst has been re-used with little loss in activity. This proves that the reactions are truly catalyzed by anchored metal.

Such catalyst resins are now used in the production of many industrially important materials, including solvents such as MIBK, oxygenate additives such as TAME (t-amyl methyl ether), hydrogen peroxide and 1,2 propanediol. In contrast, there is much less use of catalytic ion exchange resins in the commercial production of fine chemicals. The reasons for this might include selectivity aspects, the availability of resins in a shape that is well suited for large reactors and a lack of knowledge with respect to the accessibility and stability of the active sites. The importance and the scope of uses of such catalysts are often limited by diffusional issues and problems of mechanical and thermal stability.

Although swelled resins could allow the diffusion of large molecules and their approach to the reactive sites, the bead form of the resins may induce some kinetic limitations. The use of fibers or membranes might help solve this issue.

Recently, the thermal stability of acidic polymeric resin catalysts has been significantly improved by the introduction of halogen atoms on the benzene rings of the polystyrene matrix. This also increased the acidity of the sulfonic acid groups [44].

Further work is needed to understand and improve these aspects.

Acknowledgments

The authors gratefully acknowledge the generous support of Rohm and Haas Company.

References

1 Corain, B., Jerabek, K. and Kralik, M. (2001) Introduction to a special edition of *Journal of Molecular Catalysis A–Chemical*, **177**.

2 Chemin, A., Deleuze, H. and Maillard, B. (1998) *European Polymer Journal*, **34**, 1395.

3 Jiang, O., McDade, C. and Gross, A. (1996) US5561205.

4 Biffis, A., Gardan, M. and Corain, B. (2006) *Journal of Molecular Catalysis A–Chemical*, **250**, 1.
5 Brieva, G.B., Serrano, E.C., Campos-Martin, J.M. and Fierro, L.G. (2004) *Chemical Communications*, 1184.
6 Burato, C., Centomo, P., Rizzoli, M., et al. (2006) *Advanced Synthesis Catalysis*, **348**, 255.
7 Seen, A. (2001) *Journal of Molecular Catalysis A–Chemical*, **177**, 105.
8 Kralik, M. and Biffis, A. (2001) *Journal of Molecular Catalysis A–Chemical*, **177**, 113.
9 Bergbreiter, D.E. and Zhong, Z. (2005) *Industrial and Engineering Chemistry Research*, **44**, 8616.
10 Biffis, A., D'Archivio, A.A., Jerabek, K., et al. (2000) *Advanced Materials*, **12**, 1909.
11 Kralik, M., Hronec, M., Lora, S., et al. (1995) *Journal of Molecular Catalysis*, **97**, 145.
12 Corain, B., Zecca, M. and Jerabek, K. (2001) *Journal of Molecular Catalysis A–Chemical*, **177**, 3.
13 Kunin, R. (1958) *Ion Exchange Resins*, 2nd edn, John Wiley and Sons Inc., New York.
14 (a) Grubs, R.H. and Sweet, E.M. (1975) *Macromolecules*, **8**, 241. (b) Ro, K.S. and Woo, S.I. (1990) *Journal of Molecular Catalysis*, **61**, 27.
15 Corain, B., D'archivio, A.A., Galantini, L., et al. (1998) *Special Publication – Royal Society of Chemistry*, **216**, 182.
16 Zharmagambetova, A.K., Ergozhin, E.E., Sheludyakov, Y.L., et al. (2001) *Journal of Molecular Catalysis A–Chemical*, **177**, 165.
17 Corain, B., Basato, M., Main, P., et al. (1987) *Journal of Organometallic Chemistry*, **326**, C43.
18 Centomo, P., Zecca, M., Lora, S., et al. (2005) *Journal of Catalysis*, **229**, 283.
19 Chen, C.-W., Serizawa, T. and Akashi, M. (1999) *Chemistry of Materials*, **11**, 1381.
20 Sabadie, J. and Germain, J.-E. (1974) *Bulletin de la Societe Chimique de France*, 1133.
21 Michalska, Z.M., Ostazewski, B. Zientarska, J. and Sobczak, J.W. (1998) *Journal of Molecular Catalysis A–Chemical*, **129** (2–3), 207.
22 Sulman, E., Bodrova, Y., Matveeva, V., et al. (1999) *Applied Catalysis A: General*, **176**, 75.
23 Wissler, M.C., Jagusch, U.-P., Sundermann, B. and Hoelderich, W.F. (2007) *Catalysis Today* **121**, 6.
24 Raje, A.P. and Datta, R. (1992) *Journal of Molecular Catalysis*, **72**, 97.
25 Michalska, Z.M. and Strzelec, K. (2001) *Journal of Molecular Catalysis A–Chemical*, **177**, 89.
26 Yu, L., Chen, D., Li, J. and Wang, P.G. (1997) *Journal of Physical Chemistry*, **62**, 3575.
27 Seen, A.J. (2001) *Journal of Molecular Catalysis A–Chemical*, **177**, 105.
28 Talwalkar, S. and Mahajani, S. (2006) *Applied Catalysis A: General*, **302**, 140.
29 Nicol, W. and Toit, E.L. (2004) *Chemical and Process Engineering*, **43**, 1539.
30 http://www.rohmhaas.com/wcm/ April 2008.
31 Lange, P.M., Martinola, F. and Oecki, S. (1985) *Hydrocarbon Processing*, 51.
32 Miyazawa, T., Koso, S., Kunimori, K. and Tomishige, K. (2007) *Applied Catalysis A: General* **318**, 244.
33 Kusunoki, Y., Miyazawa, T., Kunimori, K. and Tomishige, K. (2005) *Catalysis Communications*, **6**, 645.
34 Miyazawa, T., Kusunoki, Y., Kunimori, K. and Tomishige, K. (2006) *Dalton Transactions*, **29**, 3561, and *Journal of Catalysis*, **240**, 213.
35 Biffis, A., Ricoveri, R., Campestrini, S., et al. (2002) *Chemistry–A. European Journal*, **8**, 2962.
36 Drelinkiewicz, A. and Hasik, M. (2001) *Journal of Molecular Catalysis A–Chemical*, **177**, 149.
37 Augustine, R.L. (1997) *Catalysis Today*, **37**, 419.
38 Blaser, H.U. and Studer, M. (1999) *Applied Catalysis A: General*, **189**, 191.
39 Coq, B. and Figueras, F. (1998) *Coordination Chemistry Reviews*, **178/180**, 1753.
40 Abdullaev, M.G., ANasibulin, A.A. and Klyuev, M.V. (1997) *Russian Journal of Organic Chemistry*, **33**, 1676.
41 Kratky, V., Kralik, M., Hronec, M. and Zecca, M. (2000) *Pet Coal*, **42**, 28.

42 (a) Gao, Y., Wang, F., Liao, S. and Yu, D. (1998) *Reaction Kinetics and Catalysis Letters* **64**, 351. (b) Mani, R., Mahadevan, V. and Srinivasan, M. (1991) *Reactive Polymers*, **14**, 263.

43 Wagner, R. and Lange, P.M. (1989) *Erdöl Erdgas Kohle* **115**, 414.

44 Banavali, R. (2006) New Thermally Stable AmberlystTM Catalyst for the Synthesis of Fine Chemicals. Paper presented at Fifth Tokyo Conference on Advanced Catalytic Science and TechnologyTokyo, Japan.

12
Chiral Auxiliaries on Solid Support

Peter Gaertner and Amitava Kundu

Vienna University of Technology, Institute of Applied Synthetic Chemistry, Getreidemarkt 9/163-OC, A-1060 Vienna, Austria

12.1
Introduction

A conventional linker such as a kind of protecting group bound to a solid support determines only the type of connection of the substrate to the polymer, the reaction conditions that are tolerated during the synthesis sequence and, eventually, the mode of detaching and the functional group present in the product [1–4]. In addition to these features, chiral auxiliaries as enantiomerically pure linker offer a simple approach to chiral compound libraries and thus introduce a further diversity element [5].

The use of resin-bound chiral auxiliaries is an attractive mode of operation not only because of the advantages exhibited by solid phase organic synthesis (SPOS), isolation of the product and possibility of automation, but most importantly because the chiral auxiliary, which is usually quite expensive, can be recovered very easily by simple filtration (Scheme 12.1) [6]. Hence, many chiral auxiliaries used in solution phase have been applied to SPOS.

The first example of polymer supported chiral auxiliaries was published by Kawana in 1972, describing the asymmetric synthesis of α-hydroxy acids [7]. Although some other groups [8, 9] entered this field of research shortly afterwards, only during the 1990s the concept became popular again; the application of different chiral auxiliaries on solid support has been reported, either newly developed ones or modifications of some known for their good induction of stereoselectivity [10–13].

The Power of Functional Resins in Organic Synthesis. Judit Tulla-Puche and Fernando Albericio
Copyright © 2008 WILEY-VCH Verlag GmbH & Co. KGaA, Weinheim
ISBN: 978-3-527-31936-7

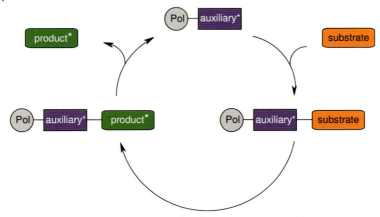

Scheme 12.1 Reaction and recycling of a chiral auxiliary on a solid support.

12.2
Carbohydrate Derived Auxiliaries

The convenience of employing carbohydrate auxiliaries is based on the multiplicity of stereogenic centers and their commercial availability at economical prices. Carbohydrates offer hydroxy functionalities that serve as ideal anchoring groups for the attachment to the resin as well as the connection to the achiral substrate.

Kunz and coworkers reported the application of an immobilized chiral galactosylamine in diastereoselective synthesis of piperidine and amino acid derivates [14]. Preparation of the galactosylamine auxiliary from 1,6-hexandiol and β-D-galactopyranosyl azide (**1**) required a six-step synthesis protocol. The precursor **2** was loaded onto a polymer bound silane. Reduction of the azide function gave the corresponding immobilized amine **3** (Scheme 12.2).

This chiral auxiliary **3** has been employed in the diastereoselective Ugi four-component condensation using the auxiliary's amine, an aldehyde, formic acid, *tert*-butyl isocyanide and zinc chloride at ambient temperature. Fluoride induced cleavage from the polymer gave N-formylated, N-galactosylated amino acid derivates in 58–96% yield. The diastereomeric ratios ranged from 74:26 to 96:4 (Scheme 12.3).

Preparation of piperidine derivates required the conversion of the auxiliary's amine into an imine function. Subsequent Lewis acid promoted Mannich–Michael reaction with Danishefsky's diene gave aliphatic, resin-bound dehydropiperidinones **5**. After cleavage, the yields of the obtained enaminones **6** ranged between 13 and 90% (Scheme 12.4). The diastereomeric ratios varied from 78:22 to 99:1. The authors reported that aliphatic substituted enaminones were formed with lower diastereoselectivity than aromatic products.

Further additions of soft nucleophiles to resin-bound dehydropiperidinones **5** were examined. Conversion with a cuprate reagent, $R_2Cu(CN)Li_2$, in the presence of $BF_3 \cdot OEt_2$ gave access to 2,6-disubstituted piperidinones with high structural

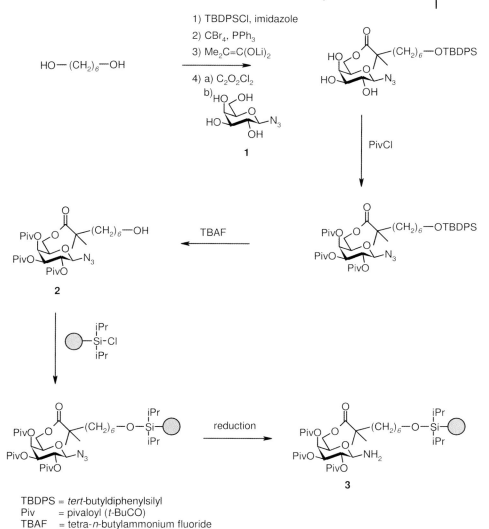

TBDPS = *tert*-butyldiphenylsilyl
Piv   = pivaloyl (*t*-BuCO)
TBAF = tetra-*n*-butylammonium fluoride

Scheme 12.2 Preparation of the carbohydrate derived auxiliary **3**.

diversity (Scheme 12.4). Fluoridolytic detachment from the polymer support has been performed and yielded 38–76% product with diastereomeric ratios of 67:33 to 98:2.

Kawana *et al.* used xylofuranose derivates as chiral auxiliaries [7, 13]. Through its primary hydroxy function, the auxiliary was loaded to the polystyrene resin. Esterification of the immobilized auxiliary **8** gave α-keto esters. Subsequent nucleophilic additions of Grignard reagents afforded resin-bound α-hydroxy esters. Subsequent saponification afforded the chiral α-hydroxy acids **9** (Scheme 12.5) in 18–84% yields and 36–65% enantiomeric excesses. The recovered polymer supported chiral auxiliaries could be reused without decrease of enantioselectivity.

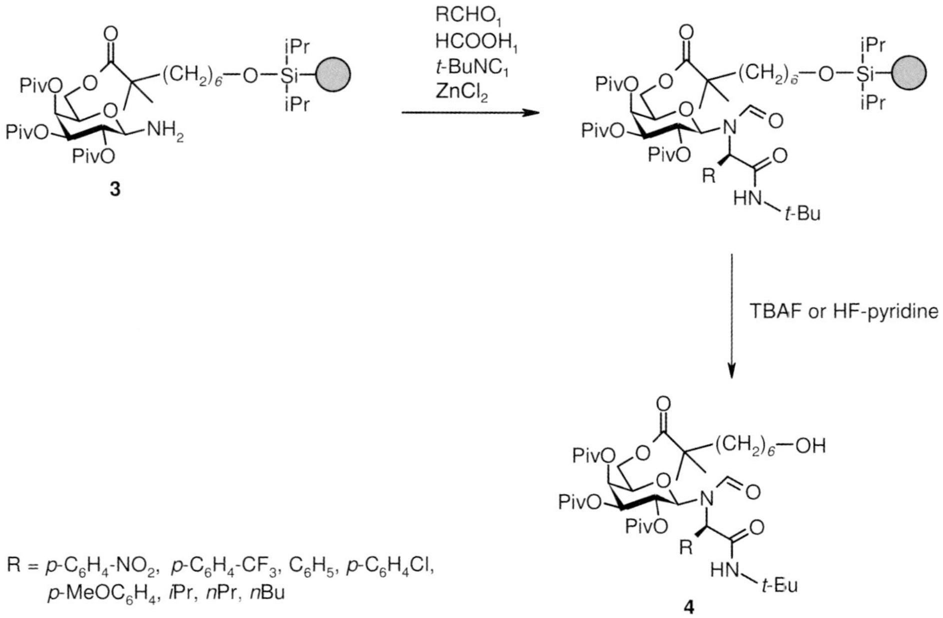
RCHO, HCOOH, t-BuNC, ZnCl2
TBAF or HF-pyridine
3
4
R = p-C6H4-NO2, p-C6H4-CF3, C6H5, p-C6H4Cl, p-MeOC6H4, iPr, nPr, nBu

Scheme 12.3 Ugi four-component condensation.

12.3 Alcohols as Chiral Auxiliaries

Chiral alcohols can easily be coupled to substrates by esterification. The benefit of their use as auxiliaries on solid support is provided by the option of simple cleavage from asymmetric products.

Successful polymer supported stereoselective Diels–Alder reaction was performed using immobilized enantiopure 4-(3-hydroxy-4,4-dimethyl-2-oxopyrrolidin-1-yl)benzoic acid **12** as a chiral auxiliary [15]. The corresponding resin-bound acrylate derivate has been applied as the dienophile **13.** Preparation of the precursor started with the combination of pantolactone **10** and the sodium salt of 4-aminobenzoic acid. Conversion into the corresponding benzyl ester followed. The obtained racemate was esterified with (1*S*)-camphanic acid chloride to a diastereomeric mixture to gain the enantiopure compounds by chromatographic separation. After subsequent saponification of the camphanic acid moiety and hydrolysis of the benzyl ester the (*R*)-enantiomer **11** was coupled to Rink amide resin (Scheme 12.6).

To generate dienophile **13**, the auxiliary **12** was esterified with acrylic acid chloride. The reactions of the resin-bound acrylate ester **13** with isoprene, 2,3-dimethylbutadiene, cyclopentadiene and 1,3-cyclohexadiene were carried out in the presence of $TiCl_4$ and yielded 80–98% of **14** after cleavage from the polymer (Scheme 12.7). Enantiomeric excesses (ee) from 40 to 99% have been reported.

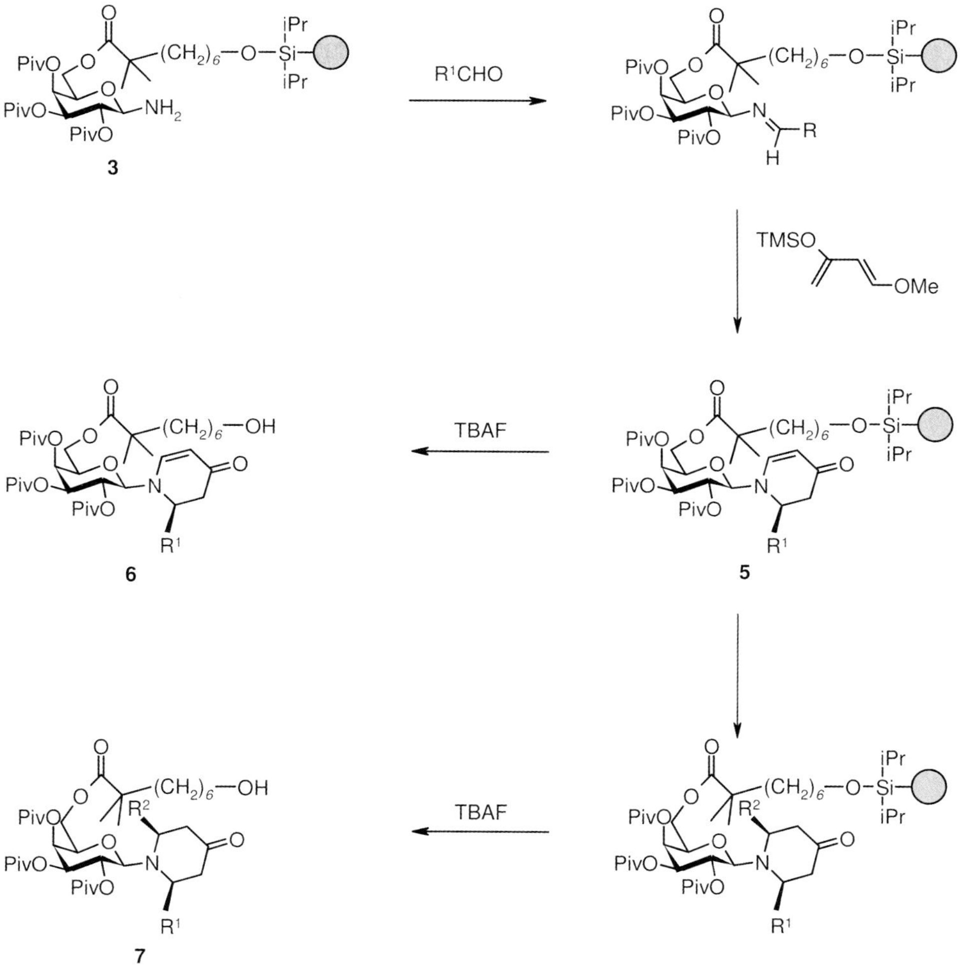

Scheme 12.4 Synthesis of piperidinone derivates.

To increase the low enantiomeric excess of 40% in the stereoselective Diels–Alder reaction with isoprene, the resin-bound auxiliary was modified. After introduction of an aminohexanoic acid spacer to the chiral auxiliary, the corresponding acrylate **15** showed an enhanced ee (70%) on applying the same reaction conditions and yielded 77% (Scheme 12.8).

Previously, the same authors had reported the successful application of a similar solid-supported chiral auxiliary (**16**) derived from (3-hydroxy-4,4-dimethyl-2-oxopyrrolidin-1-yl)acetic acid [13, 16] (Scheme 12.9). Reactions of the auxiliary's hydroxy functionality with aryl ketenes afforded chiral, aryl substituted propionic esters in high yields of up to 100% and ees from 80 to 96%.

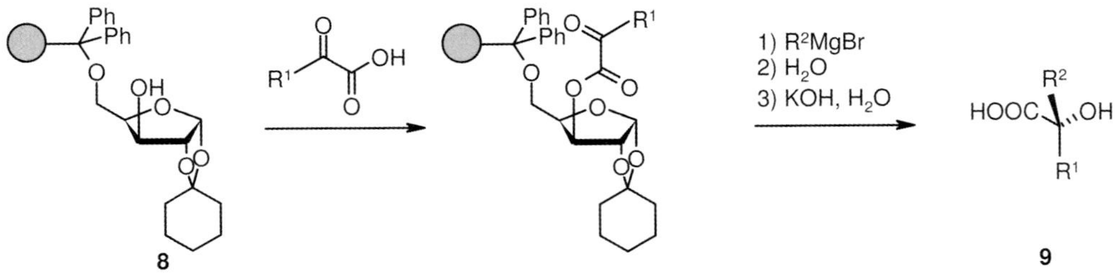

Scheme 12.5 Asymmetric synthesis of α-hydroxy acids.

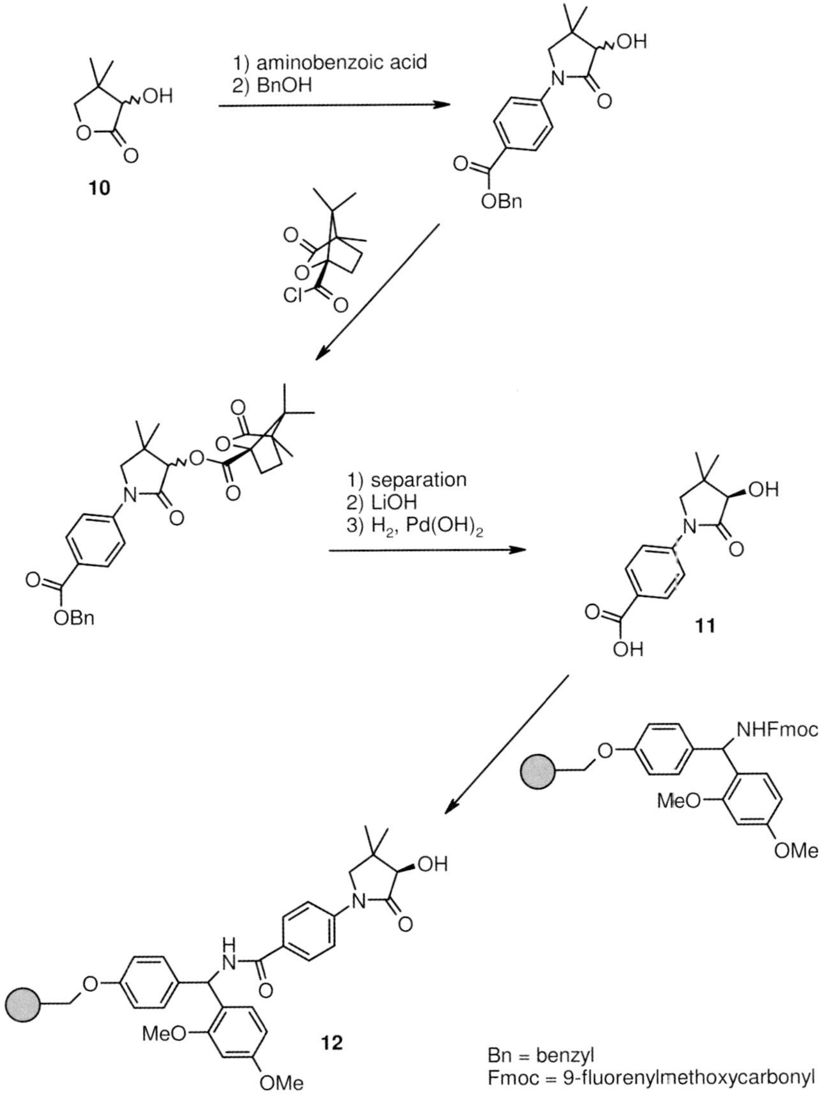

Scheme 12.6 Synthesis of immobilized 4-(3-hydroxy-4,4-dimethyl-2-oxopyrrolidin-1-yl)benzoic acid (**12**).

Bn = benzyl
Fmoc = 9-fluorenylmethoxycarbonyl

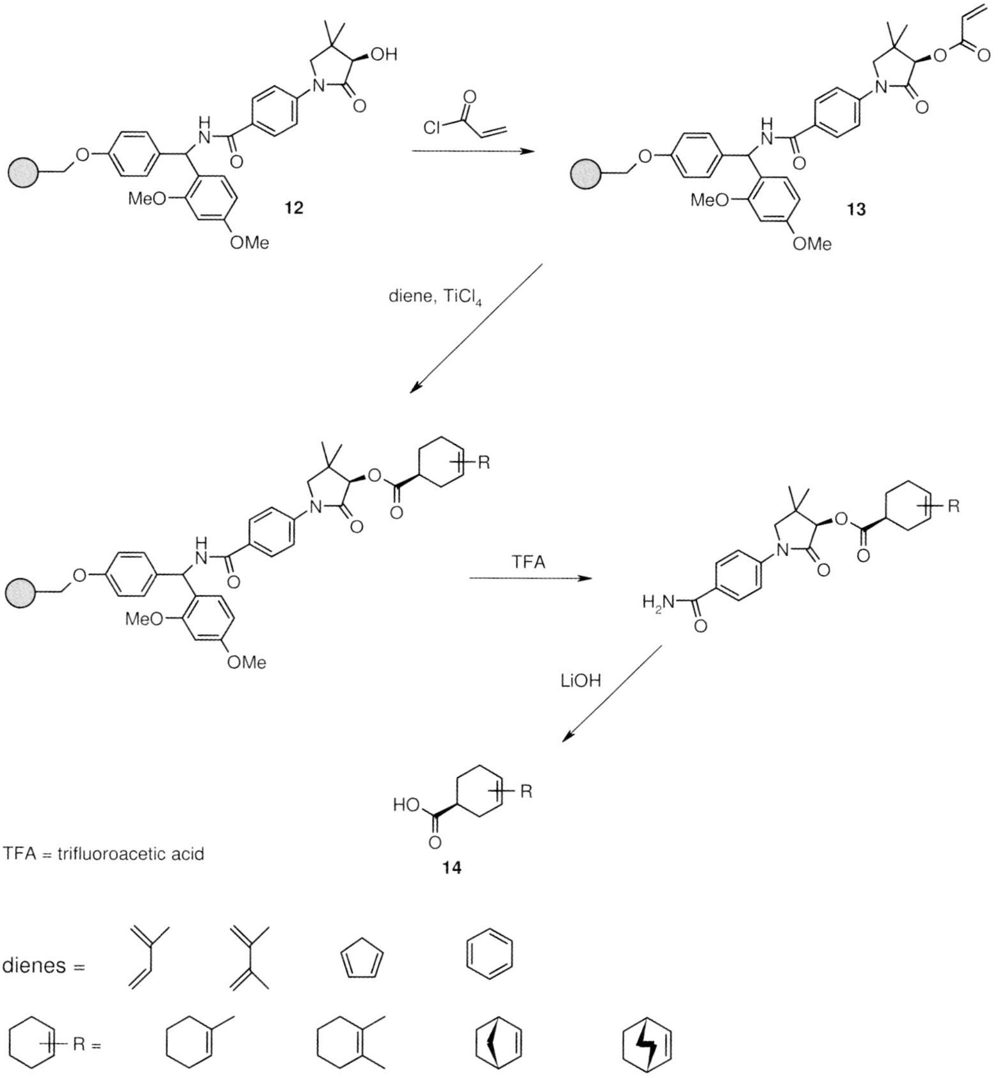

Scheme 12.7 Asymmetric Diels–Alder reaction with different dienes.

Auxiliary **16** has also been used in the asymmetric preparation of β-homoarylglycines [13, 17]. The reactions yielded 63–68% and gave diastereomeric ratios from 85:15 to 93:7.

To achieve a high degree of facial selectivity in asymmetric cycloaddition, the employment of π-shielding in combination with a chiral auxiliary has turned out to be an effective strategy [18]. The capability of asymmetric induction was applied to an immobilized system, designed on the basis of π-shielding. For this purpose

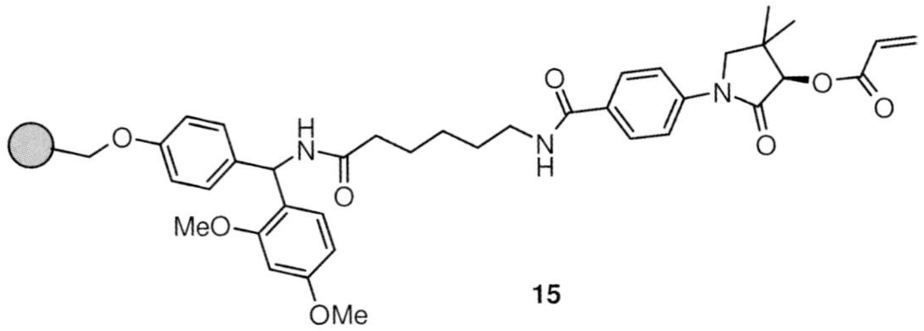

Scheme 12.8 Modified acrylate.

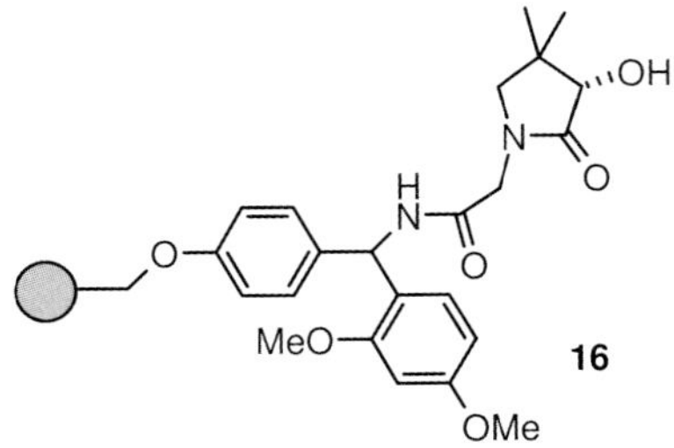

Scheme 12.9 (3-Hydroxy-4,4-dimethyl-2-oxopyrrolidin-1-yl)acetic acid derived chiral auxiliary.

L-proline was reduced to prolinol, and conversion with benzyl chloride into N-benzyl substituted prolinol **17** followed. Subsequent complex formation was achieved by the usage of Cr(CO)$_6$. Irradiation of complex **18** in the presence of polystyrene-diphenylphosphine gave the polymer-supported chiral auxiliary **19** (Scheme 12.10).

Esterification of **19** with acrylic acid chloride made diene **20** available. Subsequent stereoselective Diels–Alder cycloaddition with cyclopentadiene proceeded with complete diastereoselectivity in 55% yield. The asymmetric product **21** was cleaved from the polymer by exposure to light (Scheme 12.11).

A solution phase chiral auxiliary for 1,3-dipolar cycloaddition of isomunchnones with vinyl ethers has been adapted for solid phase synthesis by attaching both enantiomers of the precursor α-hydroxyvaline to benzhydrylamine resin (Scheme 12.12) [13, 19]. The auxiliary **22** was then functionalized by acylation and diazotization to provide diazoimide resin **23**. Rhodium(II)-catalyzed nitrogen extrusion and cycloaddition in the presence of different vinyl ethers afforded, after detachment from the polymer, various bicyclic molecules (**24**) in 49–65% yield and provided high degrees of selectivity (93–95% ee).

12.4
Amine Derived Auxiliaries

Due to facile introduction to the substrate, chiral amines are frequently used as auxiliaries. Their versatility is based on the opportunity of easy transformation into

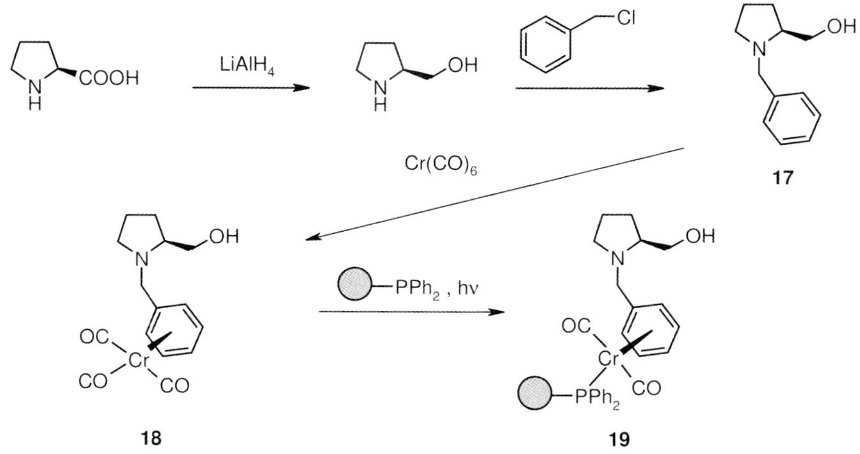

Scheme 12.10 Preparation of polymer supported chiral auxiliary, chromium carbonyl complex **19**.

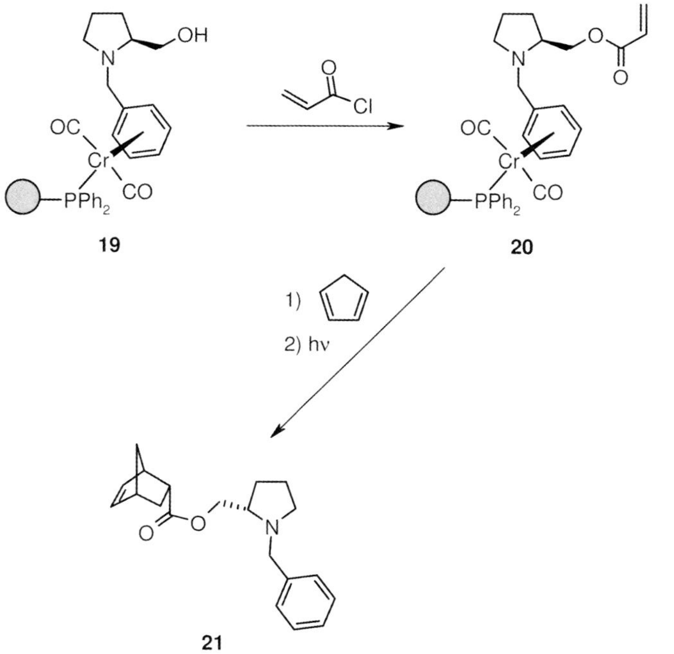

Scheme 12.11 Stereoselective Diels–Alder cycloaddition using auxiliary **19**.

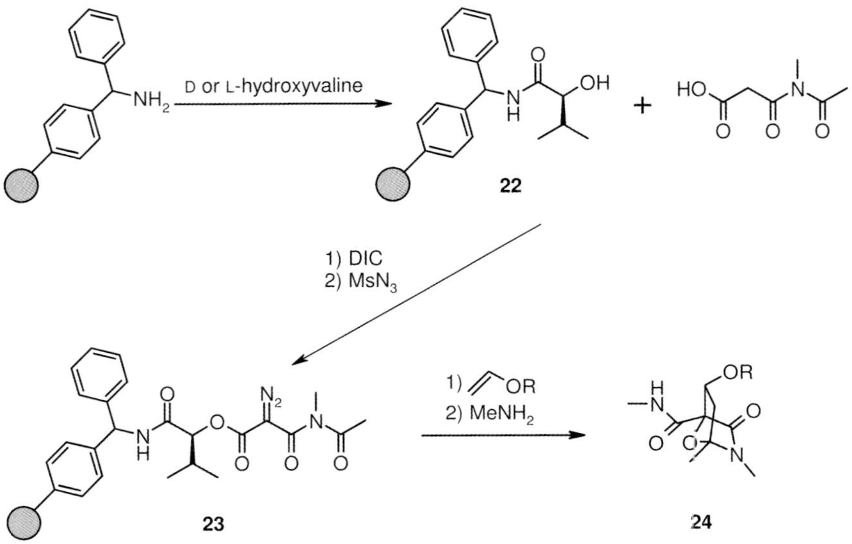

Scheme 12.12 Asymmetric 1,3-dipolar cycloaddition.

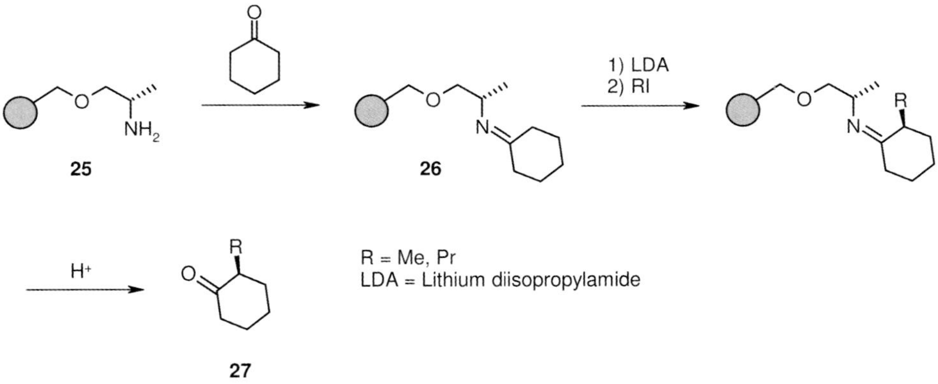

Scheme 12.13 Leznoff's α-alkylation.

imines and hydrazines. However, a second functionality such as an alcohol or acid is required for attachment to the resin.

The first application of a supported chiral amine auxiliary was published by Leznoff et al. in 1979 [8, 13]. The auxiliary **25** was readily prepared by coupling of alaninol to Merrifield resin. Cyclohexanone was attached to the auxiliary **25** by formation of imine **26**. Subsequent α-alkylation using MeI and PrI gave 2-substituted cyclohexanones **27** in 80% yield with ees of 60–95% after hydrolytic cleavage (Scheme 12.13). The recovered chiral auxiliary could be reused without a decrease in stereoselectivity.

Ephedrine and pseudoephedrine auxiliaries have the advantage of being commercially available and cheap. In addition, no further modification is necessary prior to their use.

An immobilized pseudoephedrine has been employed in solid phase asymmetric alkylation [20]. (1*R*,2*R*)-Pseudoephedrine was conveniently attached to Merrifield resin through an ether linkage. The auxiliary **28** was then acylated with propionic anhydride, phenylacetyl chloride or 3-phenylpropionyl chloride to give the corresponding resin-bound pseudoephedrine amides **29**. These amides were deprotonated and subsequently alkylated using benzyl bromide or butyl iodide. Detachment from the auxiliary afforded alcohols **30** and various ketones **31** in 19–65% yield and gave enantiomeric excesses of 80–93% (Scheme 12.14). Alcohols **30** were cleaved using lithium amidotrihydroborate (LiH_2NBH_3) while ketones **31** were derived from reaction of amide **29** with different alkyllithium reagents. Employment of heteroaryllithiums in the cleavage step provided several heteroaromatic ketones. The pseudoephedrine auxiliary **28** could be recovered and reused without significant loss in yield or enantiomeric excess.

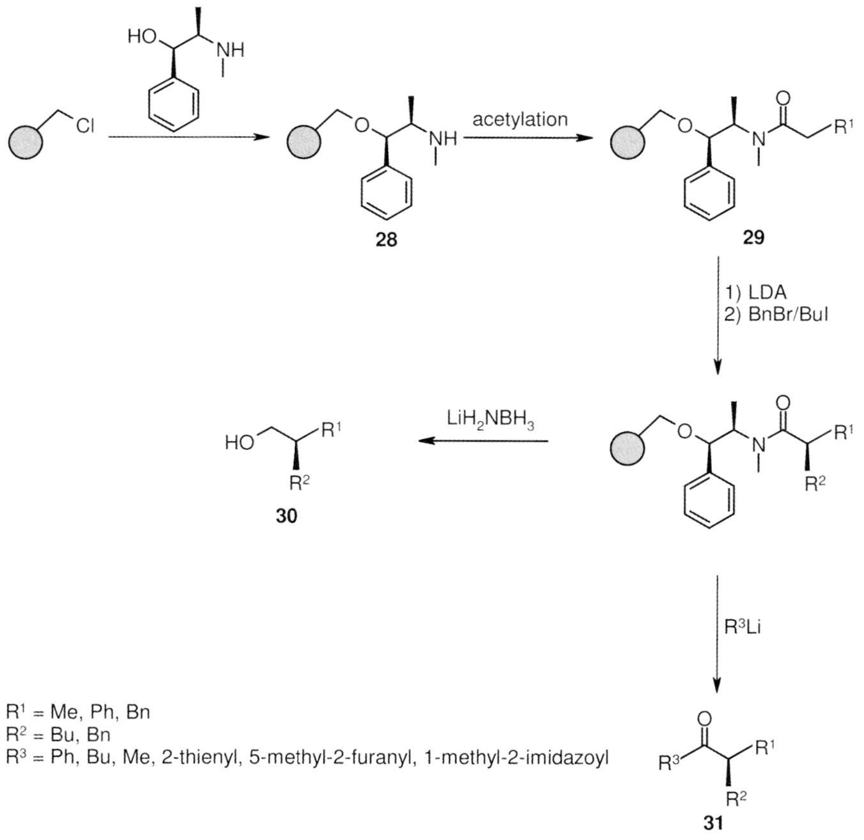
HO
NH
Cl
O
NH
acetylation
O
N
O
R1
28
29
1) LDA
2) BnBr/BuI
LiH2NBH3
HO
R1
R2
30
R3Li
R1 = Me, Ph, Bn
R2 = Bu, Bn
R3 = Ph, Bu, Me, 2-thienyl, 5-methyl-2-furanyl, 1-methyl-2-imidazoyl
R3
31

Scheme 12.14 Preparation of chiral α-alkylated alcohols and ketones.

An ephedrine derived chiral auxiliary **32** on a solid support has been applied in the asymmetric synthesis of γ-butyrolactones [13, 21]. The auxiliary was attached to the resin through the nitrogen. Treatment with α,β-unsaturated acid chlorides afforded the corresponding esters **33**. SmI mediated conversion with aldehydes or ketones gave γ-butyrolactones **34** in the so-called "asymmetric catch-release" reaction (Scheme 12.15). The resin "catches" the reactive intermediate and undergoes stereocontrolled addition. The product is "released" spontaneously from the resin-bound auxiliary **32**. The resulting γ-butyrolactones (**34**) were obtained in 37–73% yield and 70–96% ee.

Asymmetric iodolactonization on solid support was carried out using a C_2-symmetric chiral auxiliary **35** [13, 22]. The prolinol derived precursor was allylated, followed by treatment with iodine and H_2O. The resulting lactone (**36**) was obtained with exclusive trans selectivity and 87% ee (Scheme 12.16).

Polymer-bound chiral hydrazines have been synthesized to furnish α-branched amides (Scheme 12.17) [13, 23]. Enantiopure β-methoxyamines **37** and **38**, derived from readily available hydroxyproline and N,N-dibenzylleucinol, respectively, have been attached to solid support and transformed into the corresponding hydrazine auxiliaries **39** and **40** via several steps. The synthesis of a series of enantiomerically enriched α-branched amides required the coupling of aliphatic and aromatic aldehydes to form hydrazones **41** and **42** followed by the addition of different nucleophiles. Cleavage of the N–N bond of the resulting hydrazines **43** and **44** led to α-branched amines **45**. Further conversion with benzoyl chloride or acetyl chloride furnished amides **46** in yields of 24–51% and 50–83% ee.

Diastereoselective allylation of immobilized chiral imines has been carried out by employing a polymer-bound substrate bearing a pendant chiral auxiliary (Scheme 12.18) [13, 24]. The auxiliary was not directly attached to the resin.

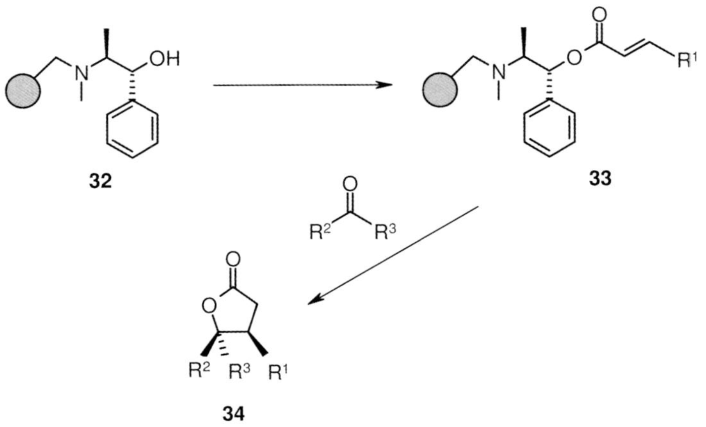

R^1 = H, Me
R^2 = i-Pr, t-Bu, Ph, c-Hex, C_5H_{11}
R^3 = H, Me, Et, Pr

Scheme 12.15 Asymmetric catch-release reaction.

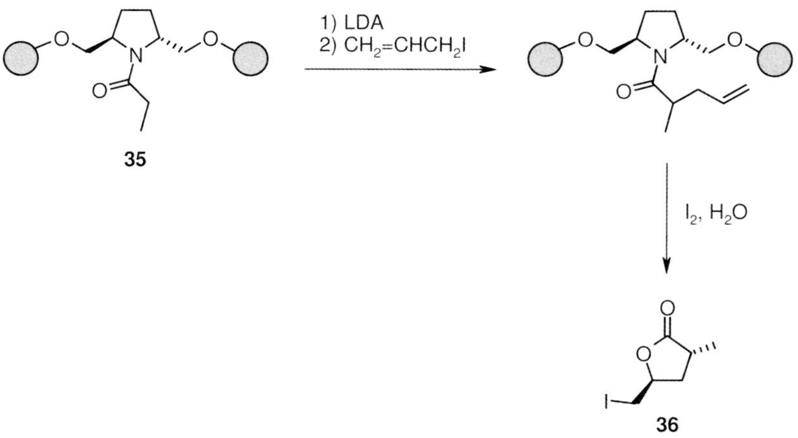

Scheme 12.16 Stereocontrolled iodolactonization on solid support.

Hence, it did not act as a linker between substrate and polymer. Through a sulfonate (OTs) or benzyl ether linkage the enantiopure imine monomers **47** and **48** were polymerized with styrene under radical conditions. The polymer bound imines **49** and **50** were treated with allyl bromide. Cleavage of the auxiliary, followed by detachment from the sulfonate linked solid support, released the desired allylamine **51a** in 95% yield with 99% ee. Cleavage of the benzyl ether linkage afforded product **51b** in 93–96% yield and 99–100% de.

12.5
Oxazolidinones, Oxazolidines and Oxazolines as Auxiliaries

The option of commercial availability and facile introduction and cleavage make these types of heterocycles one of the most popular chiral auxiliaries used in organic synthesis [5]. First introduced by Evans et al. [25], oxazolidin-2-ones are versatile chiral auxiliaries for asymmetric acyl group based transformation. Therefore, there was great interest in immobilizing them on a solid support as linker groups – mainly those derived from L-serine and L-tyrosine, which already offer a second functional group for bonding to the polymer [13, 26–32]. However, for one of these examples [13, 26] it was shown later [13, 33] that the proposed structure cannot be correct and thus the mode of chiral induction and the achieved results have to be questioned.

A recently new designed Wang resin supported Evans' chiral auxiliary (**52**) has been shown to perform Evans' asymmetric alkylation on solid support (Scheme 12.19) [34, 35]. Preparation of the auxiliary started with coupling of Fmoc-piperidine-4-carboxylic acid to Wang resin. Subsequent removal of the Fmoc protection was followed by coupling to N-protected (2R,3S)-3-amino-2-hydroxy-4-phenylbutanoic acid. After deprotection, the amino-alcohol moiety was converted into the oxazolidinone auxiliary **52** using carbonyldiimidazole (CDI).

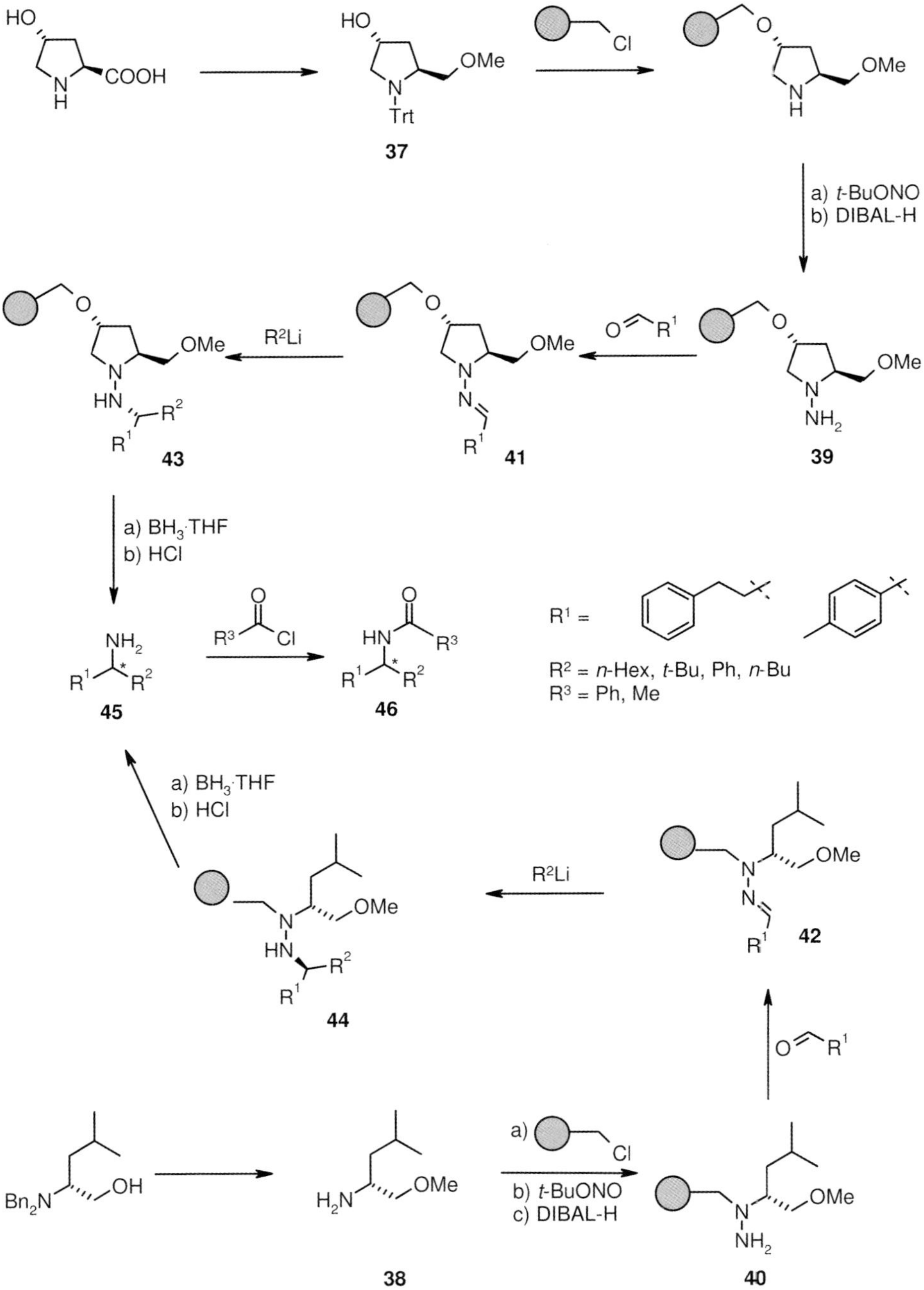

Scheme 12.17 Preparation of chiral hydrazine auxiliaries **39** and **40** and their application.

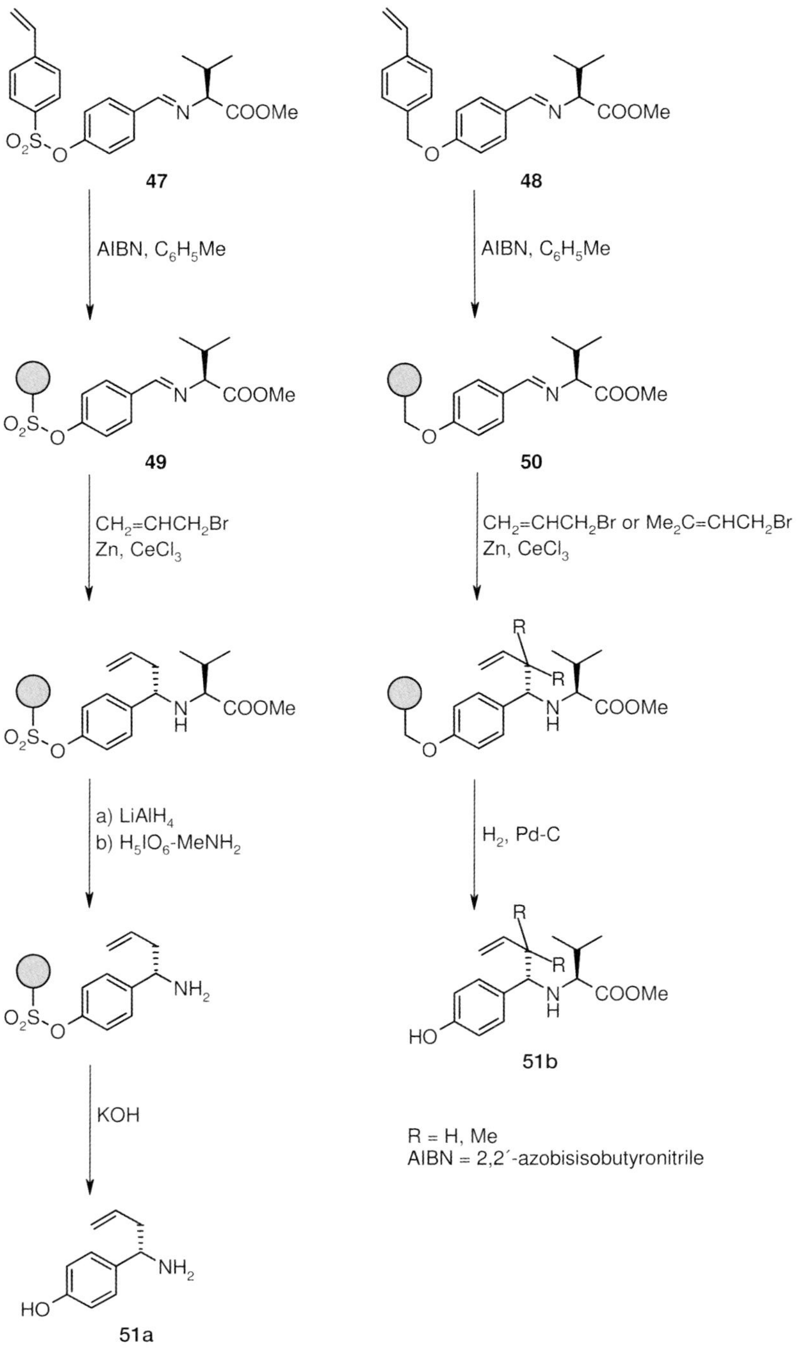

Scheme 12.18 Asymmetric allylation of chiral imines.

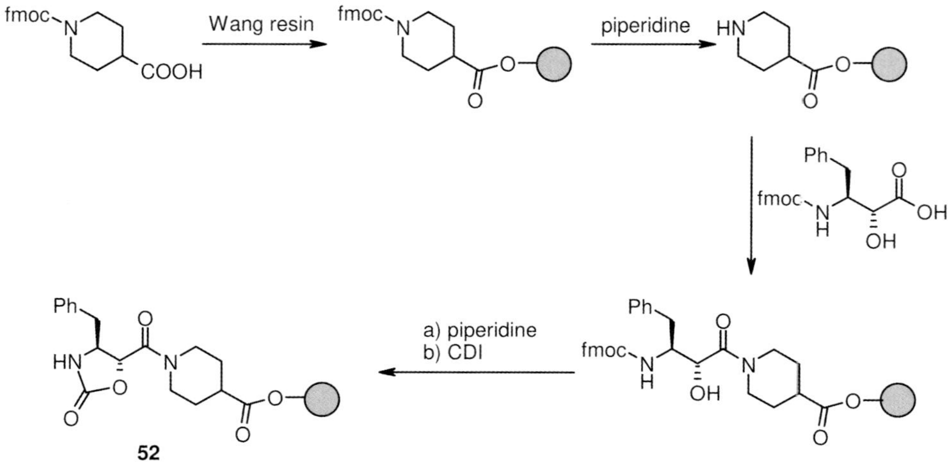

Scheme 12.19 Preparation of an Evans type chiral auxiliary on solid support.

CDI = 1,1´carbonyldiimidazole
fmoc = 9-fluorenylmethoxycarbonyl

Introduction of the auxiliary **52** to the substrate required the acylation of the oxazolidinone nitrogen with various carboxylic acids. Deprotonation and subsequent alkylation with different electrophiles (R^2X) were performed. Through LiOOH mediated hydrolysis, the α-branched carboxylic acids **53** were obtained in 50–70% yield and enantiomeric excesses of 84–97% (Scheme 12.20). The resin-bound chiral auxiliary **52** could be recovered and recycled, thereby maintaining stereoselectivity.

Asymmetric 1,3-dipolar cycloadditions employing diphenylnitrone and mesitonitrile oxide were carried out by applying a tyrosine derived Evans' auxiliary on solid support [13, 31]. Preparation of the chiral linker **56** started with esterification, protection of the amino functionality and subsequent reduction of commercially available L-tyrosine **54**. The obtained alcohol **55** was then cyclized under thermal conditions and has been attached to Wang or Merrifield resin (Scheme 12.21).

The resulting chiral auxiliary **56** was acylated with *trans*-crotonic anhydride. Conversion with 1,3-dipoles and cleavage from the support with $NaBH_4$ afforded the corresponding isoxazolines and isoxazolidines in 12–62% yield with an endo:exo ratio of up to 9:1 and enantiomeric excesses up to 89% (Scheme 12.22).

The immobilized Evans' chiral auxiliary **56** has also been employed in asymmetric α-alkylation of resin-bound propionic amide [13, 27]. Reductive cleavage afforded the α-benzylated propanol (2-methyl-3-phenylpropanol).

An asymmetric Diels–Alder reaction between **56**, featuring a crotonoyl moiety as dienophile, and cyclopentadiene led to the bicyclic ester **57** in 26% yield and 86% ee [13, 30] (Scheme 12.23).

Solid phase aldol and conjugate addition reactions were carried out using auxiliary **56** [13, 29]. The auxiliary was acylated with propionyl chloride and treated with

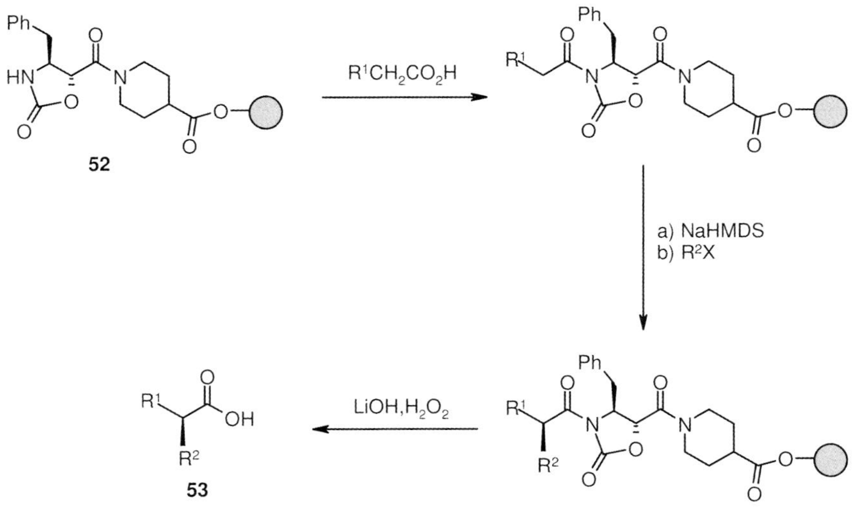

12.5 Oxazolidinones, Oxazolidines and Oxazolines as Auxiliaries

R¹ = Me, Bn, PhO, 2,4-dichlorobenzyl
R² = Me, Et, propargyl, allyl, Bn, 4-BrBn, 4-NO$_2$Bn, 2,4-diClBn
X = Br, I

Scheme 12.20 Selective alkylation using immobilized Evan's chiral auxiliary **52**.

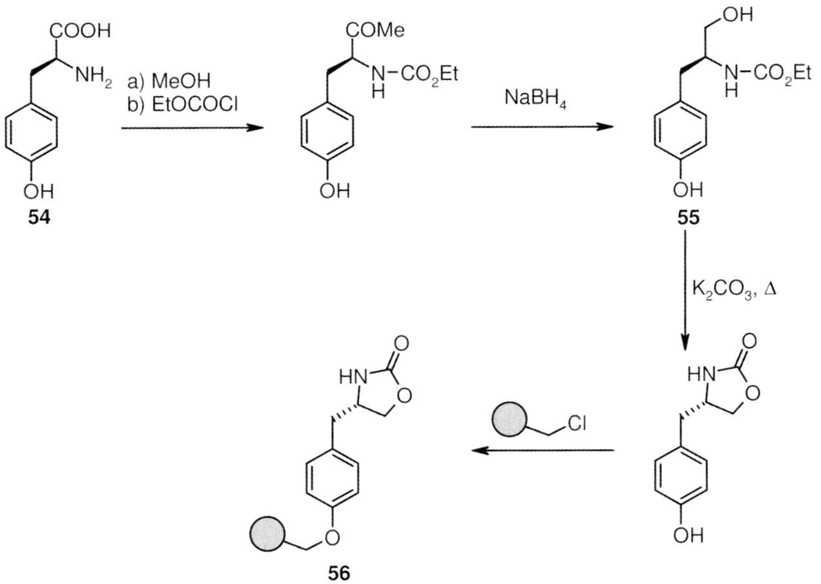

Scheme 12.21 Synthesis of an oxazolidinone chiral auxiliary.

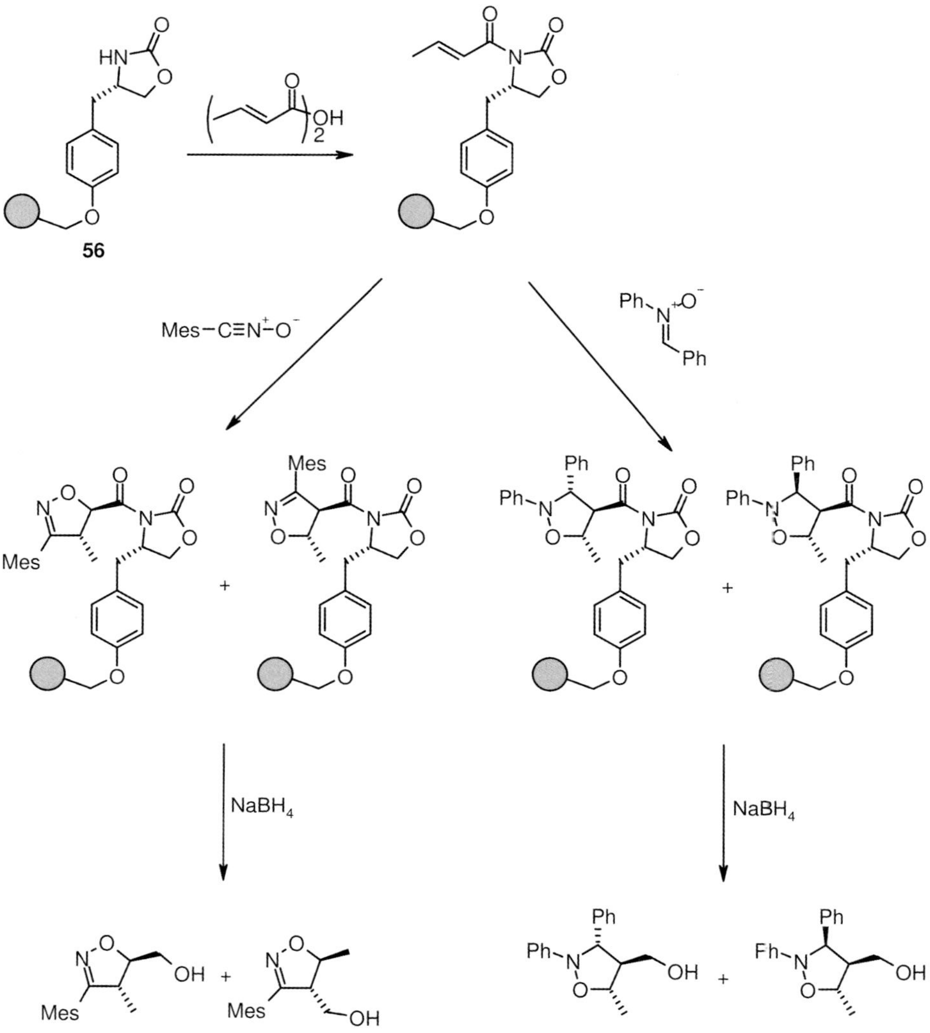

Scheme 12.22 Solid-supported chiral auxiliary in asymmetric 1,3-dipolar cycloadditions.

benzaldehyde to provide the aldol adduct **58**. The β-hydroxy acid product was cleaved from the chiral auxiliary with lithium hydroxide and yielded 63% of product with a diastereomeric excess of >98% (Scheme 12.24). For asymmetric conjugate addition the propionylated Evans' chiral auxiliary was treated with acrylonitrile. Detachment from the resin provided 4-cyano-2-methyl-butyric acid in 52% yield and 78% ee.

A further aldol condensation employing **56** has been reported [13, 28]. The chiral linker **56** was N-acetylated with hydrocinnamoyl chloride and finally treated with isovaleraldehyde. The afforded β-hydroxylated immobilized product was detached

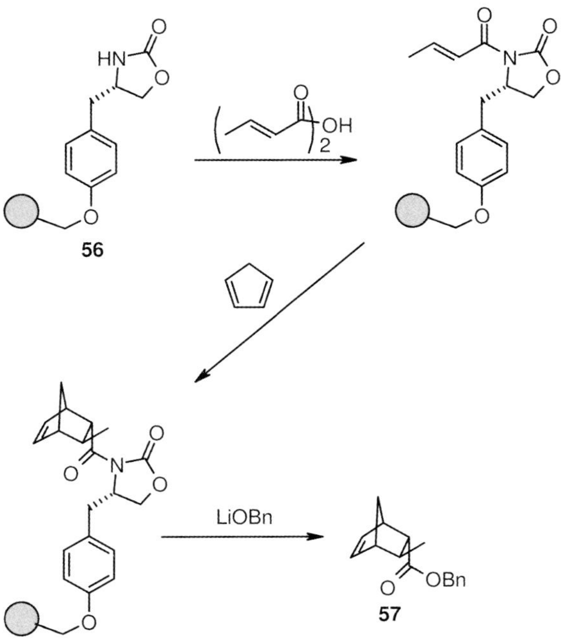

12.5 Oxazolidinones, Oxazolidines and Oxazolines as Auxiliaries

Scheme 12.23 Asymmetric Diels–Alder reaction using oxazolidinone chiral auxiliary **56**.

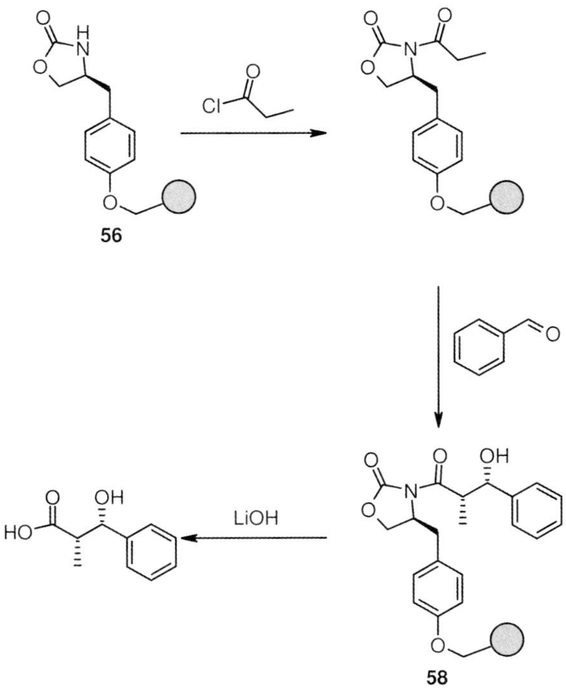

Scheme 12.24 Asymmetric solid phase aldol addition.

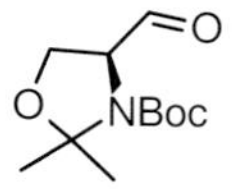

Scheme 12.25 Garner's aldehyde.

from the solid support either with LiOH, to gain 2-benzyl-3-hydroxy-5-methylhexanoic acid, or with NaOMe to provide the corresponding methyl ester. The *syn*-adduct was formed predominantly (20:1).

A polymer-supported version of the Garner aldehyde (Scheme 12.25) was applied in the synthesis of a β-lactam [36]. Owing to its strong electrophilicity the chiral, oxazolidine based Garner aldehyde is a versatile intermediate and often provides excellent asymmetric induction [37].

Preparation of the chiral oxazolidine-aldehyde **64** from D-serine **59** required various steps, including conversion into the silyl ether **60**, reaction with ketone **61** to provide oxazolidine silyl ether **62** and subsequent desilylation of the *tert*-butyldiphenylsiloxy group. Oxidation of the hydroxy functionality and hydrogenolytic cleavage of the benzyl ester then gave the corresponding aldehyde **63**. The acid functionality of **63** was then coupled to aminomethylated Merrifield resin (Scheme 12.26).

Conversion of **64** with benzylamine, followed by phenoxyacetyl chloride gave immobilized β-lactam **65**. The product **66** was detached from the resin by cleavage of the N,O-acetal with trifluoroacetic acid and was obtained in 75% yield (Scheme 12.27).

Stereoselective synthesis of an α-alkylated ester has been carried out employing a polymer supported chiral oxazoline [13, 38]. An oxazoline bound ethyl moiety was alkylated with benzyl chloride to provide, after acid catalyzed detachment from the resin, an α-alkylated propionic acid ester in 43–48% yield and 56% ee.

12.6
Sulfoxide, Sulfinamide and Sulfoximine Auxiliaries

A chiral sulfoxide supported on Merrifield resin was used in asymmetric syntheses of optical active pentenoates by conjugate addition (Scheme 12.28) [13, 39]. The auxiliaries, bearing a phenyl or biphenyl linkage to the polymer, were prepared from *(S)*-diacetone-D-glucosyl methanesulfinate **67** and 4-(*tert*-butyldimethylsiloxy)phenyl- or 4-(*tert*-butyldimethylsiloxy)biphenylmagnesium bromide. Following treatment with LDA (lithium diisopropylamide) and (iodomethyl)trimethylsilane gave the siloxyaryl β-silylethyl sulfoxides **68**. After deprotection, the precursors were coupled to Merrifield resin to afford the chiral sulfoxides **69**. The loaded resins were deprotonated and treated with methyl cinnamate. For cleavage from the polymer support, **70a,b** were heated in benzene to furnish methyl 3-phenyl-5-(trimethylsilyl)pent-4-enoate (**71**) in 48–51% yield with 75–90% ee. Detachment

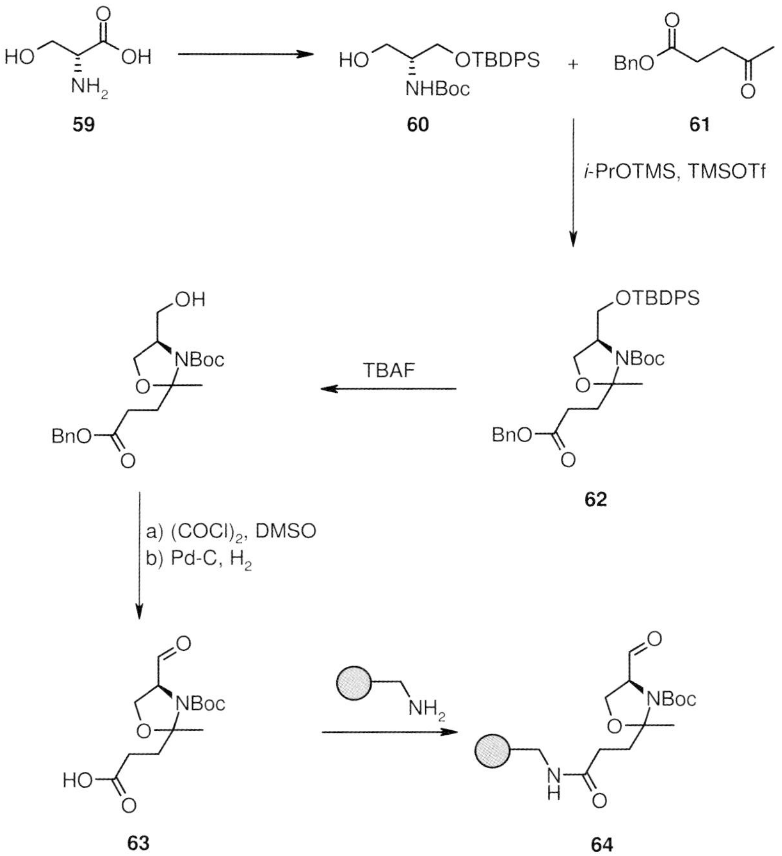

12.6 Sulfoxide, Sulfinamide and Sulfoximine Auxiliaries

Scheme 12.26 Preparation of a chiral oxazolidine aldehyde (**64**) on solid support.

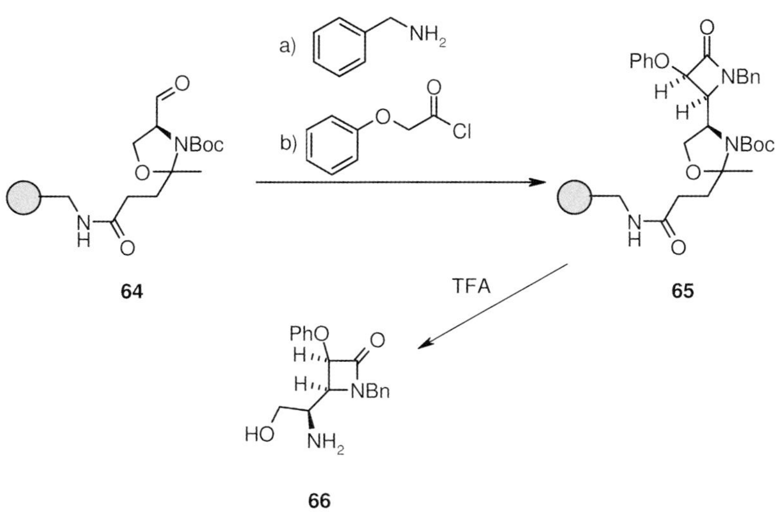

Scheme 12.27 Stereoselective synthesis of β-lactam **66**.

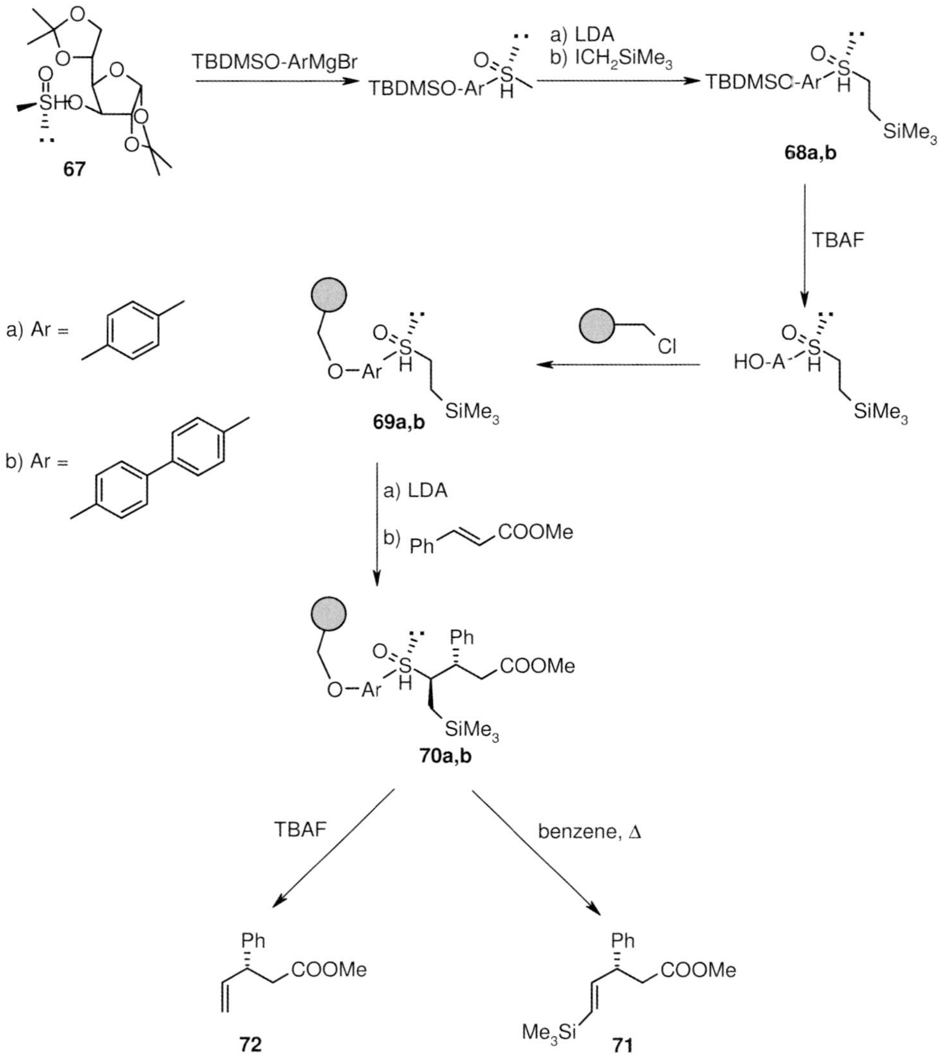

Scheme 12.28 Synthesis and application of a polymer supported chiral sulfoxide.

from **70b** was also performed using TBAF, liberating methyl 3-phenylpent-4-enoate (**72**).

Asymmetric α-alkylation of amines was achieved applying an immobilized chiral sulfonamide (Scheme 12.29) [13, 40]. To synthesize auxiliary **76**, a tertiary alcohol (**73**) was converted into Grignard reagent **74**. Addition of sulfur dioxide and subsequent chlorination furnished the sulfinyl chloride **75**. Reaction with (S)-2-amino-1,1,2-triphenylethanol, followed by reduction, hydroboration and

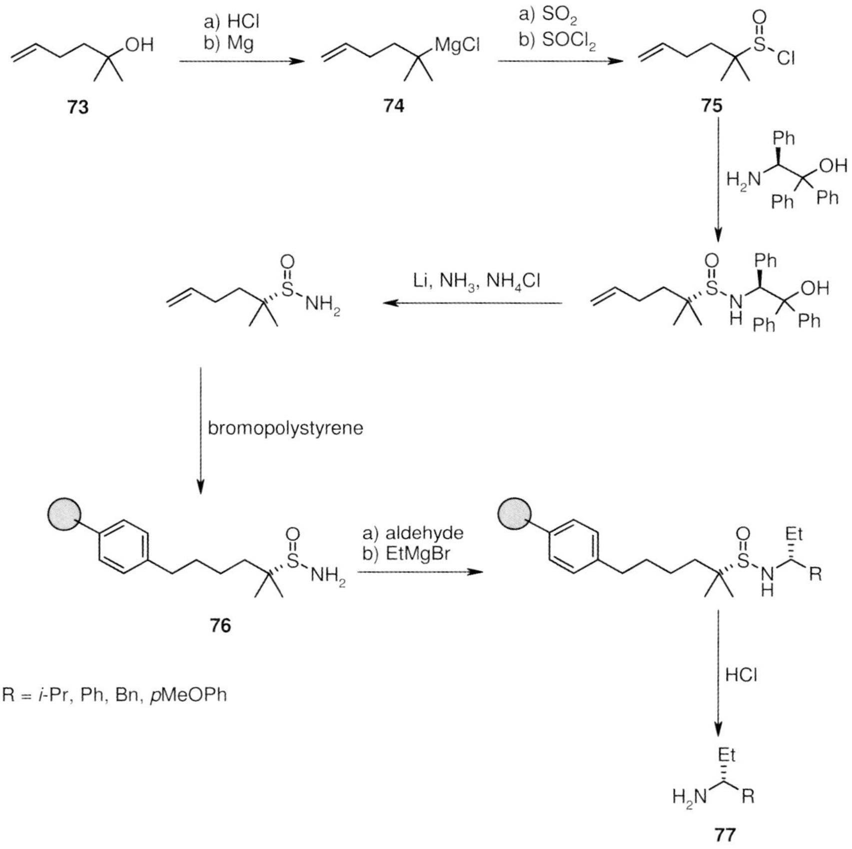

Scheme 12.29 Synthesis and application of a polymer supported chiral sulfonamide.

Suzuki coupling with bromopolystyrene provided the linker **76**. Syntheses of α-branched amines required condensation of **76** with different aldehydes and addition of ethylmagnesium bromide. Detachment of alkylated amines **77** from the resin was accomplished by treatment with HCl in dichloromethane and yielded 90–95% product. Diastereomeric ratios varied between 88:12 and 97:3. This method was also used in the asymmetric formation of pavine and isopavine. These alkaloids were obtained with complete stereoselectivity.

Application of a polymer-supported enantiopure sulfoximine in stereoselective conversion into hydroxysulfones has been reported (Scheme 12.30) [13, 41]. The chiral sulfoximine resin was prepared by coupling the sulfoximine potassium salt **78**, in the presence of tetrabutylammonium bromide (TBAB), to Merrifield resin. Hydroxyalkylation of the polymer-bound sulfoximine **79** at the α-position with benzaldehyde and propanal furnished the hydroxysulfoximine resin **80**. Oxidative cleavage gave hydroxysulfones **81** in 81 and 84% yield and enantiomeric excesses of 24 and 26% ee.

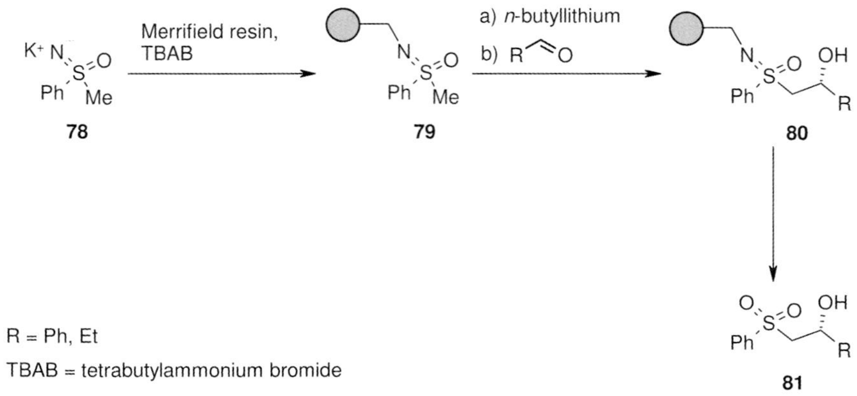

Scheme 12.30 Solid phase preparation of hydroxysulfones.

R = Ph, Et
TBAB = tetrabutylammonium bromide

12.7
Cyclohexanone as a Chiral Auxiliary

A polymer-supported cyclohexanone auxiliary (**84**) has been described in the stereoselective synthesis of α-branched carbonyl compounds [42]. Preparation of the chiral linker **84** started with the conversion of cyclohexanone into 2-(4-bromobenzyl)-1-cyclohexanone via a chiral imine. Subsequently, the substituted cyclohexanone **82** was converted into the boronate **83** and was then coupled to bromopolystyrene (Scheme 12.31).

The resin-bound chiral auxiliary **84** was allowed to react with different vinyllithium reagents. The resulting alcohols **85** were treated with n-BuLi, methyl chloroformate and different cuprate reagents. Finally, the α-alkylated adducts **86** were liberated from the polymer with ozone followed by one of three different workup conditions to provide a primary alcohol, an aldehyde or a carboxylic acid (Scheme 12.32).

12.8
An Enone as a Chiral Auxiliary

Diastereoselective [2+2] photocycloaddition of a polymer-supported cyclic chiral enone with ethylene has been reported (Scheme 12.33) [43]. The auxiliary was derived from (−)-8-(p-methoxyphenyl)menthol (**87**). Protection of the secondary alcohol and demethylation were carried out to give (−)-8-(p-hydroxyphenyl)menthyl acetate (**88**). An alkyl linker was introduced and finally loaded to poly(ethylene glycol) grafted Wang resin. Deprotection of the alcohol functionality was followed by esterification with cyclohexen-3-one-1-carboxylic acid to provide the chiral enone **89**. The photochemical reaction with ethylene was performed by irradiating with light (λ > 280 nm). Trifluoroacetic acid (TFA) or aqueous hydrolysis with

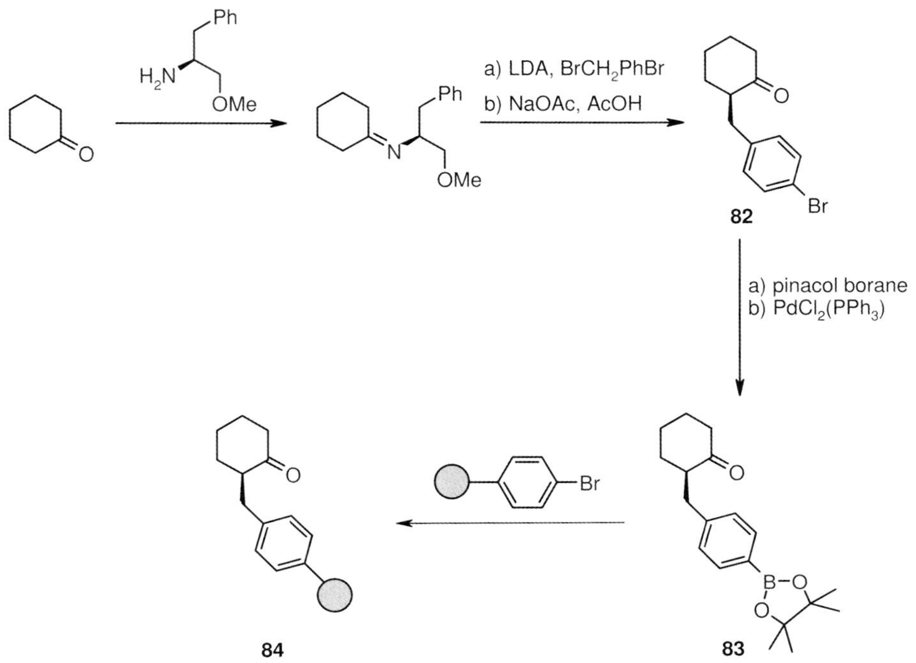

12.8 An Enone as a Chiral Auxiliary

Scheme 12.31 Synthesis of a solid supported cyclohexanone auxiliary.

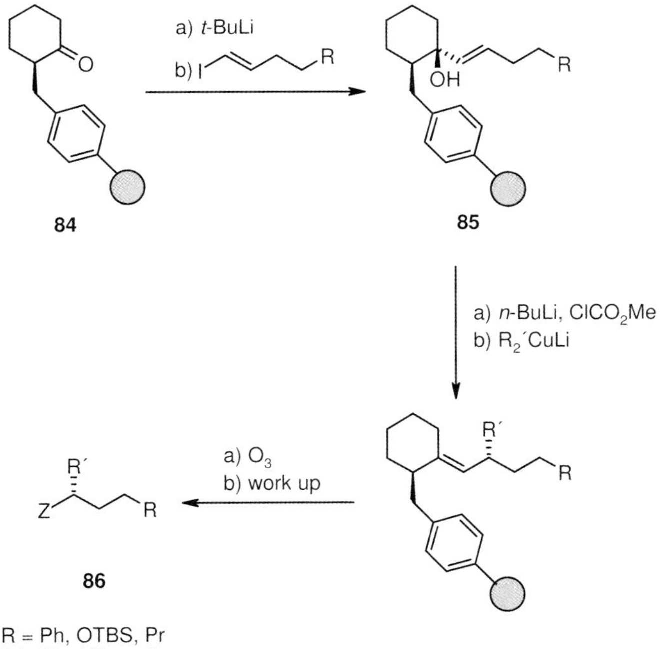

Scheme 12.32 Stereoselective preparation of α-branched carbonyl compounds.

R = Ph, OTBS, Pr
R´ = Ph, t-Bu, s-Bu
Z = CH$_2$OH, CHO, COOH

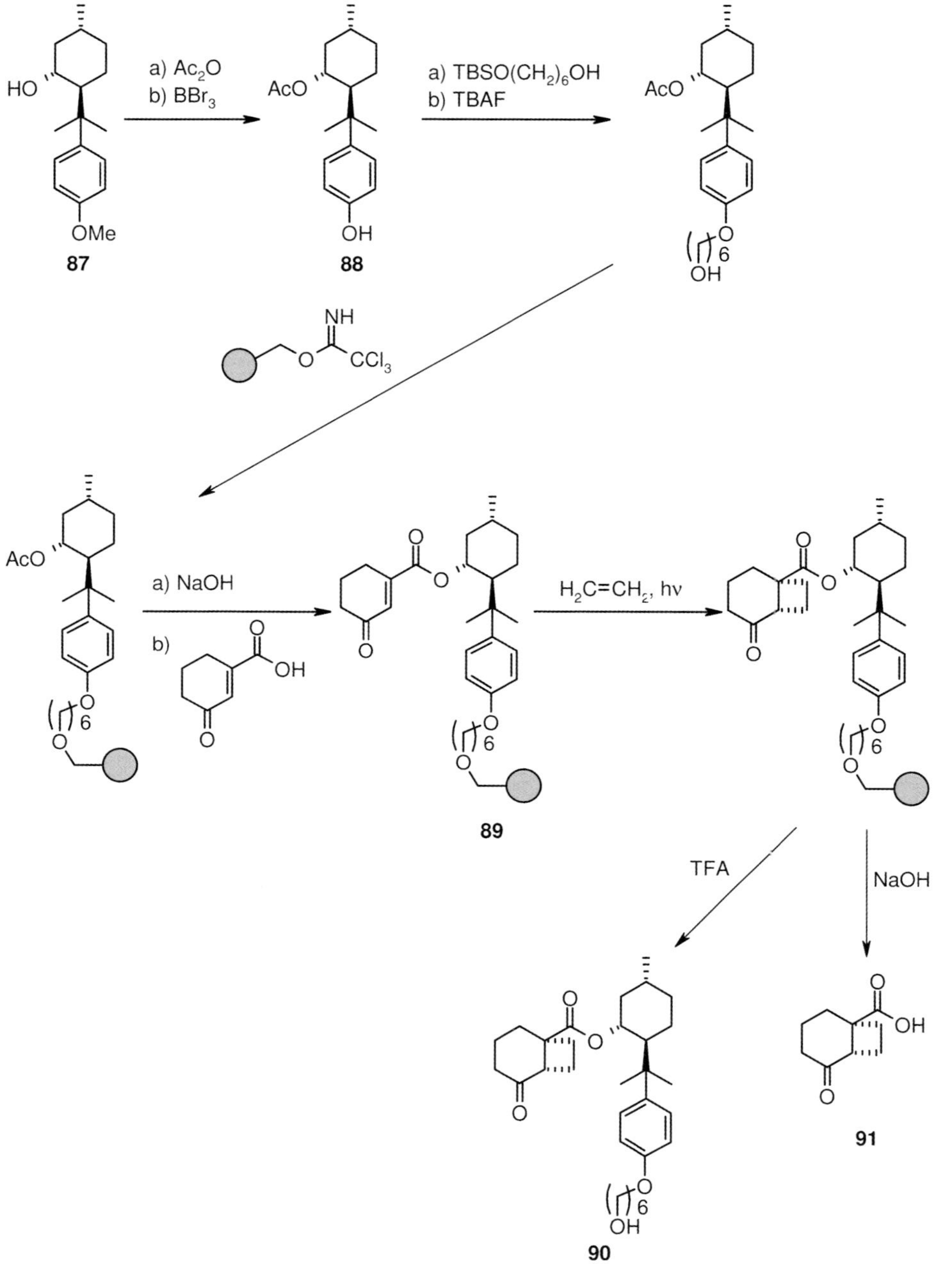

Scheme 12.33 Selective photocycloaddition of an immobilized enone.

NaOH released the cycloaddition products **90** and **91** in yields of up to 86% and 71% de.

12.9
Hydrobenzoin Derived Auxiliaries

m-Hydrobenzoin is easily accessible by treatment of racemic benzoin with LiAlH$_4$ or NaBH$_4$ [44] and can simply be desymmetrized by using either commercially available chiral anhydrolactols **92a** or **92b** as enantioselective protective groups [45], providing an easy approach to both derivatives **93a** and **93b**. They are protected at either the (*S*)- or (*R*)-carbon-center, and allow the synthesis of both possible enantiomers in subsequent auxiliary applications of linkers **94a** and **94b** (Scheme 12.34) [46].

Thus, *m*-hydrobenzoin was selectively protected at the (*R*)-carbon-center by reaction with *exo*-anhydrolactol [(MBE)$_2$O] (**92b**) under acidic conditions [45], and the obtained product **93b** was then attached to commercially available Wang resin, which first had been chloromethylated twice, following literature procedures [47], by standard protocols [48]. Resin **95** was finally deprotected under acidic conditions, leading to the polymer bound auxiliary resin **96**. After esterification with propionic acid, the substrate resin was benzylated. Finally, saponification detached the α-alkylated acid from the support and gave products in 90% yield and in moderate enantiomeric ratios of 28–36%. However, the polymer-supported auxiliary can be reused without loss of stereoselectivity and results obtained in solution were comparable with those achieved on solid support (Scheme 12.35) [46].

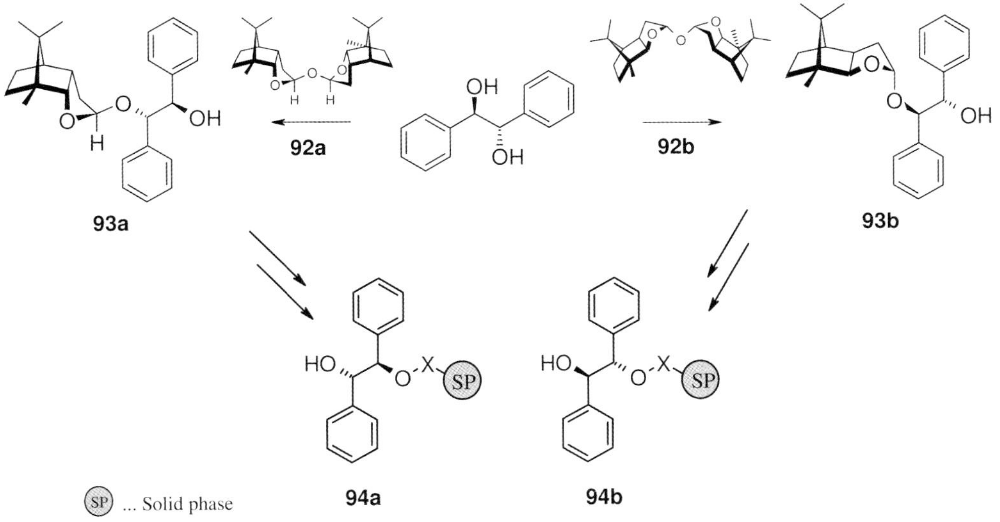

Scheme 12.34 Desymmetrization of *m*-hydrobenzoin.

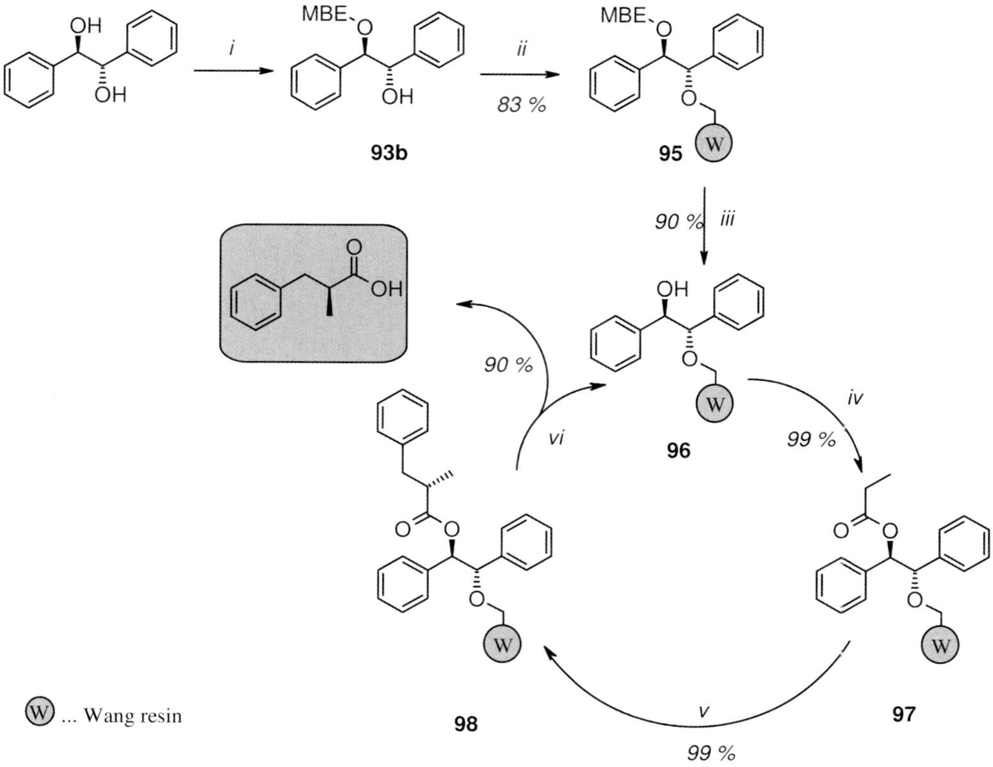

Scheme 12.35 (shown).

Reagents and conditions: (i) (MBE)₂O **92b**, p-TsOH, CH₂Cl₂; (ii) 1. NaH (1.0 equiv.), DMF, 2 h; 2. NaI (cat.), Wang-Cl (Wang-OH, 200-400 mesh, 1.0 mmol/g was purchased from Advanced ChemTech) (0.25 equiv.), 48 h; (iii) PPh₃·HBr (1 equiv.), meOH (60 equiv.), CH₂Cl₂, 48 h; (iv) Propanoic acid (10 equiv.), DIC (10 equiv.), DMAP (1 equiv.), CH₂Cl₂, 48 h; (v) 1. LDA (5 equiv.), THF, -78°C, 90 min.; 2. BnBr (8 equiv.), -78°C, 1 h; (vi) LiOH (10 equiv.), THF:MeOH:H₂O = 10:4:1, 48 h.

Scheme 12.35 Preparation and application of hydrobenzoin auxiliary **96**.

In solution phase test reactions, the introduction of ethylene glycol ether-type sub-linkers has been shown to increase stereoselectivity in the presence of Lewis acid $ZnCl_2$, most probably due to coordination between these additional chelating atoms of the sub-linker and zinc [49]. A modified solid supported hydrobenzoin auxiliary, containing this sub-linker, was prepared by esterification of **93b** with the *tert*-butyl ester of bromoacetic acid. After subsequent reduction the ethylene glycol sub-linker was coupled to chloromethylated Wang resin. To prevent reactions from unloaded resin, and thus dramatically decreased stereoselectivity, the remaining chloromethyl groups were deactivated by conversion into the corresponding iodide followed by Bu_3SnH mediated reduction to obtain a methyl residue, according to literature procedures [8]. Removal of the protecting group released the chiral auxiliary **102** (Scheme 12.36).

For the asymmetric synthesis of *(S)*-mandelic acid **105**, auxiliary **102** was esterified with benzoylformic acid and the resulting polymer bound phenylglyoxylate

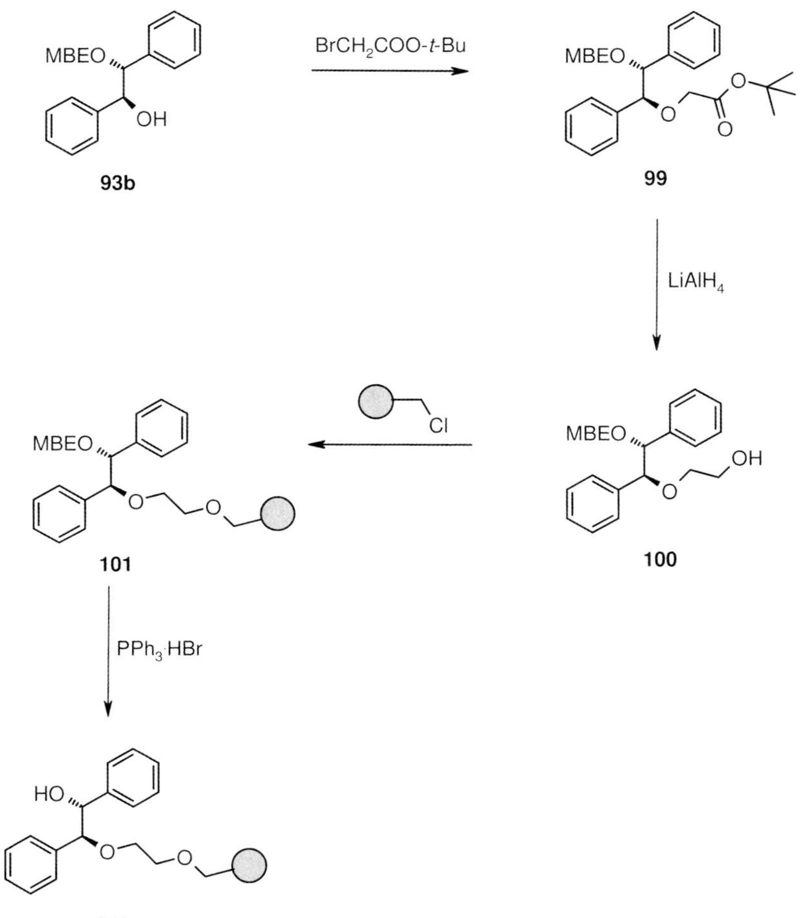

Scheme 12.36 Synthesis of the chiral hydrobenzoin auxiliary **102**.

103 was reduced by L-selectride in the presence of $ZnCl_2$. Saponification released the α-hydroxy acid **105**, which was then transformed into the L-valine methyl ester **106** derivative to determine the diastereomeric ratio. Stereoselectivities were a little bit better for resins in which remaining chloromethyl groups had been deactivated (86% compared to 80% de for the untreated resin) and the yields were about 69–75%. Again, recycling was possible with quite stable stereoselectivities; however, yields varied from 52 to 98% for different runs (Scheme 12.37) [50]. The same procedure was also applied in the asymmetric preparation of an α-hydroxy acid containing a heteroaromatic moiety [51]. 2-Oxo-2-(thien-2-yl)acetic acid was esterified with **102** and subsequently reduced. After detachment of *(S)*-2-hydroxy-2-(thien-2-yl)acetic acid from the resin an enantiomeric access of 58% could be determined.

Considering the good performance of auxiliary **102** in the reduction experiments, it was obvious to test it also for addition of organozinc reagents to immo-

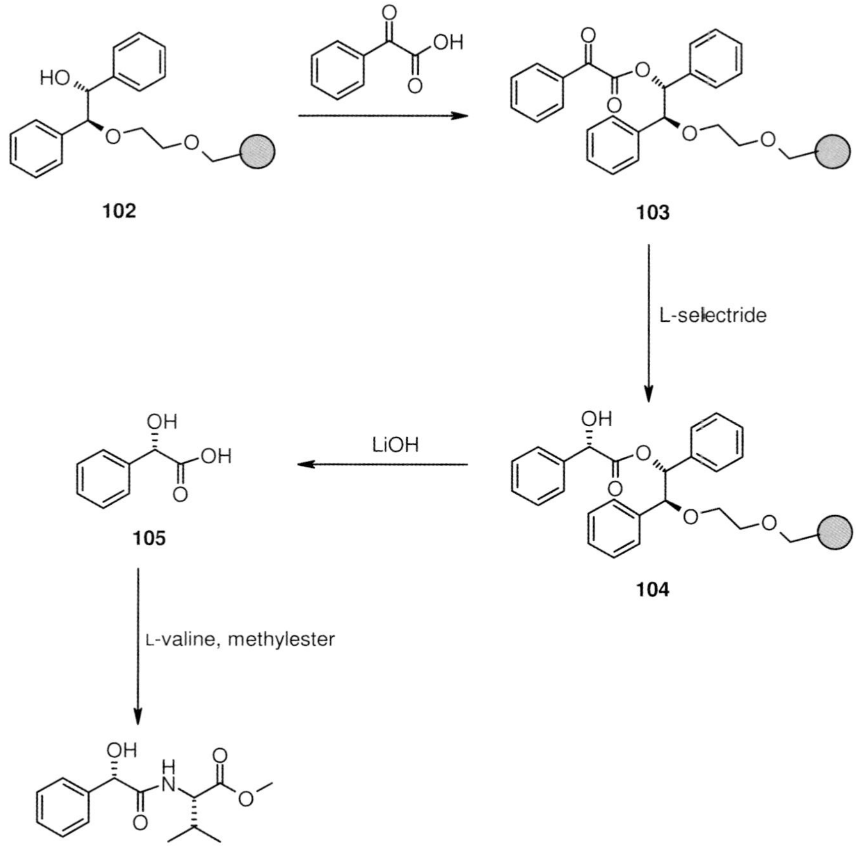

Scheme 12.37 Application of hydrobenzoin auxiliary **102**.

bilized α-keto carboxylic esters (**107**), such as phenylglyoxylates and pyruvates. LiOH mediated saponification released the α-hydroxy acids **109** (Scheme 12.38) [52]. Yields between 75 and 95% and enantiomeric excesses of up to 90% could be obtained.

The applicability of this method was further demonstrated in the stereoselective synthesis of diol **111**, which is the key intermediate in the synthesis of (+)-frontalin (**112**) [53]. Ozonolysis of **111**, obtained from solid phase chemistry, furnished frontalin **112** with 86% ee (Scheme 12.39).

A further mean of making simple changes to the electronic and coordinative properties of the hydrobenzoin auxiliary is the introduction of different substituents into the aryl moieties. The effect on the stereoinductive potential of auxiliaries modified in this manner was investigated first in solution, and the model showing best results was used in the reduction of polymer-bound phenylglyoxylate to mandelic acid (**105**) under the same conditions as described above [54]. Thus, application of auxiliary **113b** (Scheme 12.40) gave enantiomeric excesses from 53 to 70%

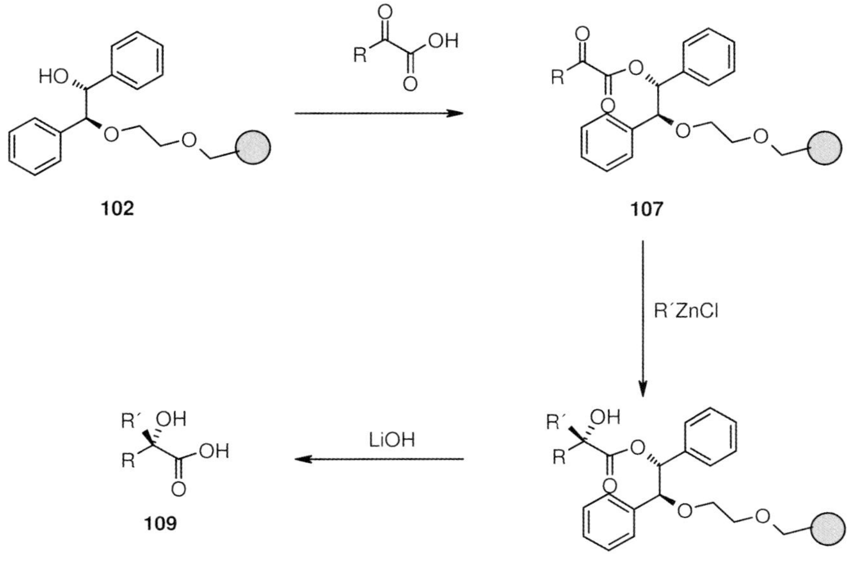

102, 107, 108, 109

R = Ph, Me
R′= *n*-Bu, *i*-Pr, *c*-Hex, Et, Me, Ph

Scheme 12.38 Addition of R'ZnCl to immobilized α-keto carboxylic esters.

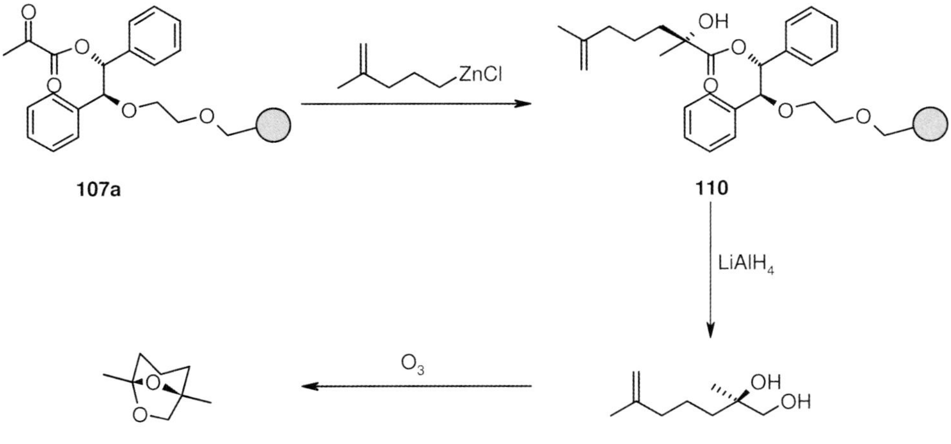

107a, 110, 111, 112

Scheme 12.39 Stereoselective preparation of (+)-frontalin **112** by application of hydrobenzoin auxiliary **102**.

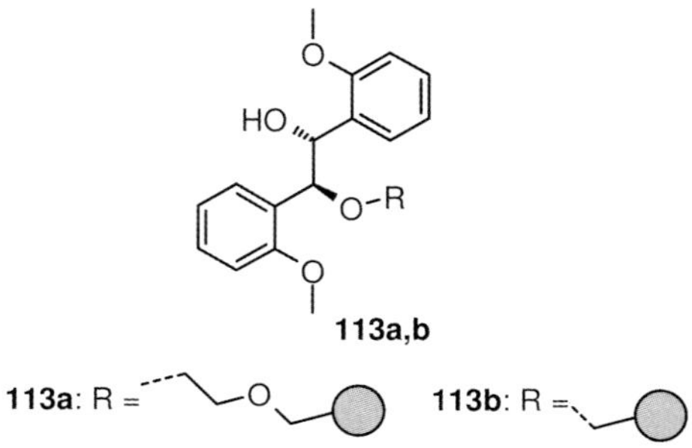

Scheme 12.40 Hydrobenzoin auxiliary 113a,b.

for different recycling runs. However, it was important to achieve a good loading of the resin, because otherwise stereoselectivity dropped (ee 33% for 51% loading). Interestingly, in this case the introduction of a further coordination site in the sub-linker, which had improved stereoselectivities for other auxiliaries, had no positive influence. Auxiliary **113a** provided **105** with 31% ee only. Again deactivation of the remaining chloromethyl groups of the Wang resin improved stereoselectivities to 55% de, as expected.

The stereoselective potential of auxiliaries **113a,b** was additionally investigated in the asymmetric α-alkylation of propionic acid (**114a**: 65% ee and **114b**: 55% ee), addition of an organozinc reagent to an α-keto carboxylic acid (**116a**: 57% ee and **116b**: 62% ee) and a Diels–Alder reaction between acrylic acid and cyclopentadiene (**118b**: *endo/exo* 13.5 : 1; 35% ee) (Scheme 12.41) [55].

12.10
Conclusion

Concerning the expanding necessity for enantiopure compounds, the application of chiral auxiliaries on solid support has increased in recent years. Although some achievements have been reached for selected examples there are still challenges to be overcome. Most of them are not directly connected with chiral auxiliaries on solid support but concern reactions on a solid carrier in general. Altogether they are increasing stereoselectivity, for which the type of used polymeric backbone is another crucial factor [27], monitoring of reactions' yields and selectivities, which is difficult because the covalently attached organic molecules constitute only a small part of the entire molecule [56], and recyclability of auxiliaries, which is always given as the main argument for chiral auxiliaries on solid support but has only been demonstrated rarely. In overcoming these challenges, solid bound chiral auxiliaries will enlarge their applicability in asymmetric synthesis.

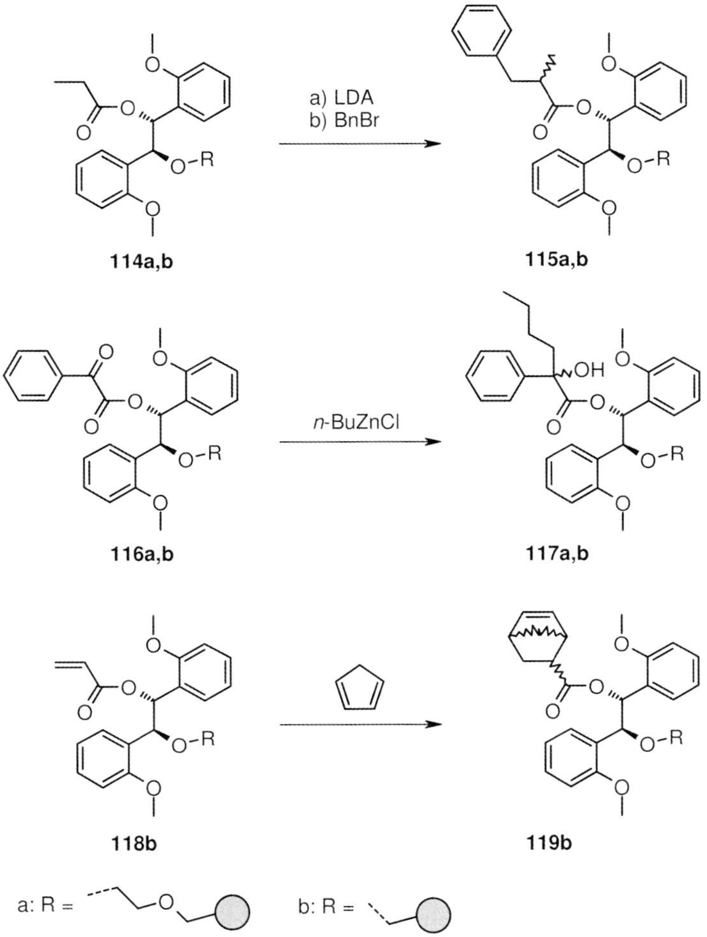

Scheme 12.41 Stereoselective reactions using chiral auxiliary **113a,b**.

References

1 Dörner, B., Steinauer, R. and White, P. (1999) *Chimia*, **53**, 11–17.
2 James, I.W. (1999) *Tetrahedron*, **55**, 4855–946.
3 Guillier, F., Orain, D. and Bradley, M. (2000) *Chemical Reviews*, **100**, 2091–157.
4 Comely, A.C. and Gibson, S.E. (2001) *Angewandte Chemie – International Edition*, **40**, 1012–32.
5 Gnas, Y. and Glorius, F. (2006) *Synthesis*, **12**, 1899–930.
6 Kirchhoff, J.H., Lormann, M.E.P. and Bräse, S. (2001) *Chimica Oggi – Chemistry Today*, **19**, 28–33.
7 (a) Kawana, M. and Emoto, S. (1972) *Tetrahedron Letters*, **48**, 4855–8. (b) Kawana, M. and Emoto, S. (1974) *Bulletin of the Chemical Society of Japan*, **47**, 160–5.
8 (a) Worster, P.M., McArthur, C.R. and Leznoff, C.C. (1979) *Angewandte Chemie*, **91**, 255. (b) McArthur, C.R., Worster, P.M., Jiang, J.-L. and Leznoff, C.C. (1982)

Canadian Journal of Chemistry–Revue Canadienne de Chimie, **60**, 1836–41.
9 Colwell, A.R., Duckwall, L.R., Brooks, R. and McManus, S.P. (1981) *Journal of Organic Chemistry*, **46**, 3097–102.
10 Furman, B., Thürmer, R., Kaluza, Z., et al. (1999) *Angewandte Chemie*, **111**, 1193–5.
11 Gordon, K., Bolger, M., Khan, N. and Balasubramanian, S. (2000) *Tetrahedron Letters*, **41**, 8621–5.
12 Backes, B.J., Dragoli, D.R. and Ellman, J.A. (1999) *Journal of Organic Chemistry*, **64**, 5472–8.
13 Chung, C.W.Y. and Toy, P.H. (2004) *Tetrahedron Asymmetry*, **15**, 387–99.
14 Zech, G. and Kunz, H. (2004) *Chemistry–A European Journal*, **10**, 4136–49.
15 Akkari, R., Calmès, M., Escale, F., et al. (2004) *Tetrahedron Asymmetry*, **15**, 2515–25.
16 Akkari, R., Calmès, M., Mai, N., et al. (2001) *Journal of Organic Chemistry*, **66**, 5859–65.
17 Akkari, R., Calmès, M., Malta, D., et al. (2003) *Tetrahedron Asymmetry*, **14**, 1223–8.
18 Xie, L. and Jones, G.B. (2005) *Tetrahedron Letters*, **46**, 3579–82.
19 Savinov, S.N. and Austin, D.J. (2002) *Organic Letters*, **4**, 1419–22.
20 Hutchinson, P.C., Heightman, T.D. and Procter, D.J. (2004) *Journal of Organic Chemistry*, **69**, 790–801.
21 Kerrigan, N.J., Hutchinson, P.C., Heightman, T.D. and Procter, D.J. (2003) *Chemical Communications*, 1402–3.
22 Price, M.D., Kurth, M.J. and Schore, N.E. (2002) *Journal of Organic Chemistry*, **67**, 7769–73.
23 Enders, D., Kirchhoff, J.H., Köbberling, J. and Pfeiffer, T.H. (2001) *Organic Letters*, **3**, 1241–4.
24 Itsuno, S., El-Shehawy, A.A., Abdelaal, M.Y. and Ito, K. (1998) *New Journal of Chemistry*, **22**, 775–7.
25 Evans, D.A., Bartroli, J. and Shih, T.L. (1981) *Journal of the American Chemical Society*, **103**, 2127.
26 Allin, S.M. and Shuttleworth, S.J. (1996) *Tetrahedron Letters*, **37**, 8023–6.
27 Burgess, K. and Lim, D. (1997) *Chemical Communications*, 785–6.
28 Purandare, A.V. and Natarajan, S. (1997) *Tetrahedron Letters*, **38**, 8777–80.
29 Phoon, C.W. and Abell, C. (1998) *Tetrahedron Letters*, **39**, 2655–8.
30 Winkler, J.D. and McCoull, W. (1998) *Tetrahedron Letters*, **39**, 4935–6.
31 (a) Faita, G., Paio, A., Quadrelli, P., et al. (2000) *Tetrahedron Letters*, **41**, 1265–9. (b) Faita, G., Paio, A., Quadrelli, P., et al. (2001) *Tetrahedron*, **57**, 8313–22.
32 Desimoni, G., Faita, G., Galbiati, A., et al. (2002) *Tetrahedron Asymmetry*, **13**, 333–7.
33 (a) Bew, S.P., Bull, S.D. and Davies, S.G. (2000) *Tetrahedron Letters*, **41**, 7577–81. (b) Bew, S.P., Bull, S.D., Davies, S.G., et al. (2002) *Tetrahedron*, **58**, 9387–401.
34 Kotake, T., Hayashi, Y., Rajesh, S., et al. (2005) *Tetrahedron*, **61**, 3819–33.
35 Kotake, T., Rajesh, S., Hayashi, Y., et al. (2004) *Tetrahedron Letters*, **45**, 3651–4.
36 Wills, A.J., Krishnan-Ghosh, Y. and Balasubramanian, S. (2002) *Journal of Organic Chemistry*, **67**, 6646–52.
37 Liang, X., Andersch, J. and Bols, M. (2001) *Journal of the Chemical Society–Perkin Transactions 1*, 2136.
38 Colwell, A.R., Duckwall, L.R., Brooks, R. and McManus, S.P. (1981) *Journal of Organic Chemistry*, **46**, 3097–102.
39 Nakamura, S., Uchiyama, Y., Ishikawa, S., et al. (2002) *Tetrahedron Letters*, **43**, 2381–3.
40 Dragoli, D.R., Burdett, M.T. and Ellman, J.A. (2001) *Journal of the American Chemical Society*, **123**, 10127–8.
41 Hachtel, J. and Gais, H.-J. (2000) *European Journal of Organic Chemistry*, 1457–65.
42 Spino, C., Gund, V.G. and Nadeau, C. (2005) *Journal of Combinatorial Chemistry*, **7**, 345–52.
43 Shintani, T., Kusabiraki, K., Hattori, A., et al. (2004) *Tetrahedron Letters*, **45**, 1849–51.
44 (a) Cullis, P.M. and Lowe, G. (1981) *Journal of the Chemical Society–Perkin Transactions 1*, 2317–21. (b) Yamada, M., Horie, T., Kawai, M., et al. (1997) *Tetrahedron*, **53**, 15685–90.
45 (a) Noe, C.R., Knollmüller, M., Steinbauer, G. and Völlenkle, H. (1985) *Chemische Berichte*, **118**, 1733–45. (b) Noe, C.R., Knollmüller, M., Steinbauer, G. and Völlenkle, H. (1985) *Chemische Berichte*, **118**, 4453–8. (c) Noe, C.R., Knollmüller,

M., Steinbauer, G., et al. (1988) *Chemische Berichte*, **121**, 1231–9.
46 Gärtner, P., Schuster, C. and Knollmüller, M. (2004) *Letters in Organic Chemistry*, **1**, 249–53.
47 Nugiel, D.A., Wacker, D.A. and Nemeth, G.A. (1997) *Tetrahedron Letters*, **38**, 5789–90.
48 Meerwein, H. (1965) Herstellung aliphatischer und alicyclischer Äther durch Alkylierung von Alkoholen bzw. Alkoholaten mit Alkylhalogeniden, in *Methoden der Organischen Chemie (Houben-Weyl)*, Vol. VI/3 (ed. E. Müller), Thieme, Stuttgart, pp. 24–32.
49 Schuster, C., Bröker, J., Knollmüller, M. and Gärtner, P. (2005) *Tetrahedron Asymmetry*, **16**, 2631–47.
50 Schuster, C., Knollmüller, M. and Gärtner, P. (2005) *Tetrahedron Asymmetry*, **16**, 3211–23.
51 Schuster, C. (2004) Meso-hydrobenzoin derivatives as new chiral auxiliaries and enantiomerically pure linkers for stereoselective synthesis on solid support. Institute of Applied Synthetic Chemistry-Vienna University of Technology. PhD Thesis.
52 Schuster, C., Knollmüller, M. and Gärtner, P. (2006) *Tetrahedron Asymmetry*, **17**, 2430–41.
53 Whitesell, J.K. and Buchanan, C.M. (1986) *Journal of Organic Chemistry*, **51**, 5443–5.
54 Bröker, J., Knollmüller, M. and Gärtner, P. (2006) *Tetrahedron Asymmetry*, **17**, 2413–29.
55 Bröker, J. (2004) Aryl substituted meso-hydrobenzoins as new chiral auxiliaries and enantiomerically pure linkers for stereoselective synthesis on solid support. Institute of Applied Synthetic Chemistry-Vienna University of Technology. PhD Thesis.
56 Cironi, P., Alvarez, M. and Albericio, F. (2004) *QSAR & Combinatorial Science*, **23**, 61–8.

13
Immobilized Enzymes in Organic Synthesis
Jesper Brask

Novozymes A/S, Krogshoejvej 36, DK-2880 Bagsvaerd, Denmark

13.1
Introduction

Catalysis is one of the most important and rapidly expanding areas in modern organic chemistry. Catalytic reactions can be achieved by either chemocatalysis or biocatalysis. The former field is dominated by transition metal catalysis, whereas in biocatalysis the use of isolated enzymes dominates over whole-cell transformations.

It has been a widespread assumption that enzymes are fragile molecules that only work in aqueous environments. However, over the last 20 years numerous reports have appeared that feature a wide range of enzymes used for organic synthesis in non-aqueous environments. The discovery that many enzymes retain their catalytic activity in non-aqueous media is often attributed to Klibanov, despite a few much earlier reports [1]. A non-aqueous system may be required for a given transformation due to solubility properties or to drive a reaction equilibrium (such as lipases working in the synthesis/acylation direction) and can even lead to advantages such as enhanced thermostability or altered substrate selectivity.

Whereas biocatalysis previously was a last option that was only looked into when all other synthetic methods had failed, it is now a discipline well integrated into classical organic synthesis in the pharma-, agro-, and fine chemical industries [2]. An example from the latter group is laboratory chemicals producer Fluka, which has reported over 100 biocatalytic processes in routine production [3]. Biocatalysis can offer outstanding chemo-, regio- and/or enantioselectivities under mild reaction conditions. It is, hence, often used to create chirality, for example, in the pharma industry [4].

In the present chapter, emphasis will be on practical, laboratory-scale application of enzymes for organic synthesis. The benefits of using immobilized enzymes will be highlighted, together with ideas on how to immobilize and apply immobilized enzymes.

The Power of Functional Resins in Organic Synthesis. Judit Tulla-Puche and Fernando Albericio
Copyright © 2008 WILEY-VCH Verlag GmbH & Co. KGaA, Weinheim
ISBN: 978-3-527-31936-7

13.1.1
Enzyme Classification and Availability

According to the Enzyme Commission (EC) system, enzymes are classified according to their function and substrates. The code numbers contain four elements, with the first number indicating one of six main divisions (classes) and the following numbers indicating subclasses. The six main divisions are:

1. Oxidoreductases: these enzymes catalyze redox reactions. Examples are oxidases that catalyze oxidation of a substrate by reducing molecular oxygen (O_2), and peroxidases that reduce H_2O_2. Laccases (EC 1.10.3.2) are oxidases that catalyze the oxidation of (poly)phenolic substrates. Reductases and dehydrogenases (EC 1.1.1) catalyze the reduction of carbonyls, using NADH/NADPH cofactors. Catalases (EC 1.11.1.6) catalyze the decomposition of H_2O_2 to O_2 and H_2O.
2. Transferases: are enzymes transferring a group (e.g. a methyl or glycosyl group) from one compound to another.
3. Hydrolases: these enzymes catalyze the hydrolytic cleavage of C–O, C–N, C–C and a few other bonds. Examples of subclasses are 3.1 Esterases (including 3.1.1.3 Lipases, described in more detail below), 3.2 Glycosidases and 3.4 Peptidases. This is, industrially, the most important class.
4. Lyases: enzymes cleaving, typically, C–C, C–O or C–N bonds by elimination, typically leaving double bonds.
5. Isomerase: an industrially important example is glucose isomerase (EC 5.3.1.18), which catalyzes the isomerization of D-glucose to D-fructose.
6. Ligases: these enzymes catalyze the joining of two molecules coupled with the hydrolysis of ATP or similar. The bonds formed are often high-energy bonds.

Researchers in academia and companies without internal microbiology and/or molecular biology facilities rely on partnering or the commercial availability of enzymes. Fortunately, many enzymes are available either directly from the large producers (e.g. Novozymes, Gencor – mostly bulk quantities) or through distributors or smaller specialized biocatalysis companies (Amano, BioCatalytics, etc.). The Sigma-Aldrich conglomerate has a significant portfolio within biocatalysis, including immobilized enzymes such as Novozym 435 and Amano PS-C lipases. Some companies (e.g. BioCatalytics) offer screening kits containing the most common enzymes within a given class (lipases, reductases, etc.). A comprehensive list of close to 1000 commercially available enzymes has been collected, together with EC number, synonyms and brand names, origin, scale of production, and producer/distributor contact information [5].

13.1.2
Popular Enzymes for Biocatalysis

13.1.2.1 Lipases
Lipases have become the most versatile class of biocatalysts in organic synthesis due to their broad acceptance of diverse substrates and their tolerance of a wide

range of organic solvents. They have been implemented both in hydrolysis and synthesis (acylation) reactions, controlled by the water activity, often with chemo-, regio- or enantioselective control [6]. Not only can lipases accept esters in hydrolysis and transesterification reactions, or synthesize them from alcohol and carboxylic acid, other nucleophiles are also tolerated, allowing amides, thioesters, etc. as acceptable substrates or targets. On the other hand, serine-hydrolase peptidases can often also catalyze the same range of reactions (i.e. not restricted to amide substrates).

Recently, even examples of lipase-catalyzed Michael additions and aldol condensations have appeared [7]. These are dramatic examples of catalytic promiscuity, that is, the ability of an enzyme to catalyze more distinctly different chemical transformations [8]. Often such activities are explained in terms of the active site offering a scaffold in which substrates adopt favorable conformations and/or reactants are brought together in a desired geometry. Accordingly, after being observed in wild-type enzymes, these "side activities" can often be enhanced in site-directed variants, in which residues in or close to the active site are mutated.

Activated acyl donors such as vinyl esters, halogenated methyl or ethyl ester, oxime esters or carboxylic acid anhydrides are often preferred for acylation of alcohols. With vinyl esters, acylation is in general fast and quantitative, as the equilibrium is driven by the release of vinyl alcohol, which will spontaneously tautomerize to volatile acetaldehyde. Amine nucleophiles will typically react spontaneously with these acyl donors, which hence cannot be used for resolution of amines. Vinyl acetate and ethyl acetate are common reagents (and often also solvents) for enzymatic kinetic resolution of alcohols and amines, respectively [9].

13.1.2.2 Oxidoreductases

Together with enantioselective hydrolysis/acylation reactions, enantioselective ketone reductions dominate biocatalytic reactions in the pharma industry [10]. In addition, oxidases [11] have found synthetic applications, such as in enantioselective Baeyer–Villiger reactions [12] catalyzed by, for example, cyclohexanone monooxygenase (EC 1.14.13) or in the TEMPO-mediated oxidation of primary alcohols to aldehydes, catalyzed by laccases [13]. Hence, the class of oxidoreductases is receiving increased attention in the field of biocatalysis. Traditionally they have been perceived as "difficult" due to cofactor requirements etc, but recent examples with immobilization and cofactor regeneration seem to prove the opposite.

The use of baker's yeast for selective reductions has a long history, while the use of isolated enzymes is more recent. Dehydrogenases and reductases require a nicotinamide cofactor (NADH or NADPH), from which a hydride is transferred to the substrate carbonyl. Enzymes from different species have been classified according to their selectivity (hydride transfer to *si*- or *re*-face of the carbonyl) [14]. The cofactors to be used together with isolated enzymes are commercially available (e.g. from Sigma-Aldrich), but are for most applications too costly to use in stoichiometric amounts. However, cofactor *in situ* regeneration can be

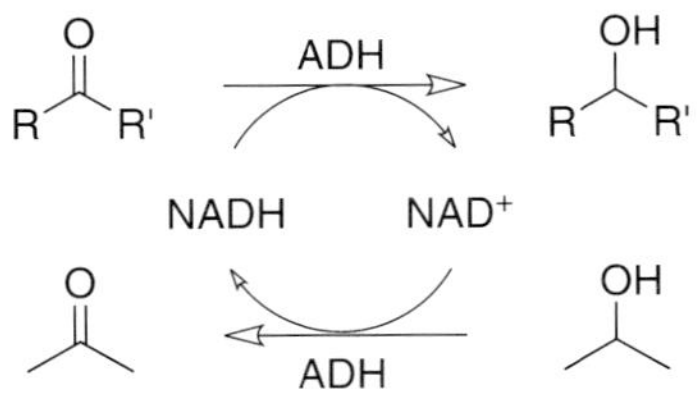

Figure 13.1 Alcohol dehydrogenase (ADH) cofactor recycling with 2-propanol/acetone.

obtained photochemically or electrochemically, as well as biocatalytically through reduction/oxidation of, for example, acetone/2-propanol [15] (Figure 13.1).

The use of oxidoreductases in solution clearly dominates over immobilized applications. Use of immobilized whole cells (e.g. baker's yeast) [16] is, however, well described, and reports have also appeared claiming increased stability and activity of isolated horse liver alcohol dehydrogenase and other oxidoreductases immobilized on agarose [17] or salt crystals [18] (protein-coated microcrystals, PCMC [19]). Furthermore, immobilization of oxidoreductases on surfaces has been studied more intensively for the development of biosensors.

13.1.2.3 Nitrile-Converting Enzymes

The two enzyme classes nitrile hydratases (RCN + H_2O → $RCONH_2$) and nitrilases (RCN + $2H_2O$ → RCOOH + NH_3) actually belong to two distant groups in the EC system, with the hydratases being classified as lyases (EC 4.2.1.84) and nitrilases as hydrolases (EC 3.5.5.1). Microorganisms that produce a nitrile hydratase also seem to produce amidases, which enable them to convert nitriles into carboxylic acids in a two-step reaction. Actually, amidase side-activity can be a problem with commercial nitrile hydratase preparations (if the target structure is the amide). Nitrilases, however, hydrolyze the nitrile without the formation of a free amide intermediate.

There are numerous reported applications for organic synthesis [20], including examples with chemo- regio-, and/or stereoselective control, as well as the industrial large-scale conversion of acrylonitrile into acrylamide (Section 13.3.2). In the early 1990s Novozymes offered a whole-cell immobilized *Rhodococcus* sp. preparation containing nitrile hydratase and amidase activity under the name "Novo SP361" [21]. This experimental product is, however, not in the current product range.

13.1.3
Solvents and Stability

For lipases, as for other more fragile enzymes, not all organic solvents are equally well tolerated. High concentrations of water-miscible organic solvents (e.g. DMF, DMSO) can lead to denaturation, whereas good enzymatic performance is often observed in hydrophobic solvents (e.g. hexane, toluene), in which the enzymes are suspended [22]. For hydrolysis reactions in water/buffer, low concentrations of

DMSO can be tolerated as a cosolvent, e.g., to suspend an insoluble crystalline substrate.

Water activity is very important not only to control the equilibrium of a hydrolysis/acylation reaction but also for general enzyme stability. Most enzymes require a shell of surrounding water and under totally anhydrous conditions (use of molecular sieves, high vacuum, etc.) many enzymes, including lipases, will be inactive. In certain cases the activity can be improved by adding a small amount of water to a suspension of enzyme in organic medium. The effect has been described as water being a molecular lubricant that improves the flexibility of the enzyme. Water activity can be controlled by incubating enzyme and reaction mixture in a desiccator over saturated salt solutions [23].

Further, the ionization state of the enzyme should be correct before it is exposed to an organic medium. This is done by adjusting the pH (in aqueous solution) to the pH-optimum (or pH 7 if unknown) of the enzyme. When transferred to organic media, the enzyme will generally keep this ionization state (called "pH memory"). Alternatively, soluble or insoluble buffers can be used in organic solution [24].

13.1.4
Predicting Hydrolase Enantiopreference

Lipases often show excellent enantioselectivity in kinetic resolution of secondary alcohols, whereas primary and tertiary alcohols in general are more challenging. What has become known as "Kazlauskas' rule" predicts which enantiomer reacts faster in the acylation of secondary racemic alcohols. The model is represented in Figure 13.2. If the bigger group (L) has priority over the medium group (M) according to Prelog's rule, the (R)-alcohol will react faster. The same stereopreference is observed for the hydrolysis of esters, in this case the (R)-alcohol is obtained as product. This empirical rule was originally proposed [25] based on results from 130 esters of secondary alcohols hydrolyzed by cholesterol esterase and lipases from *Pseudomonas cepacia* and *Candida rugosa,* but is now accepted as broadly applicable to other lipases as well [26]. Interestingly, subtilisins (bacterial serine endo-peptidases) commonly show the opposite enantiopreference. The rule further implies that better enantioselectivities will be obtained on substrates with a large size difference between M and L substituents. The rule is also valid for the isosteric primary amines (M-CHNH$_2$-L).

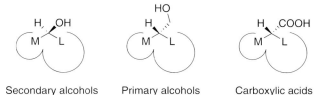

Figure 13.2 Kazlauskas enantiopreference for lipases.

Prediction of enantiopreference towards primary alcohols and carboxylic acids can be more complex. Kazlauskas and coworkers suggested the models outlined in Figure 13.2 based on substrate mapping studies of lipases from *Pseudomonas cepacia* (toward primary alcohols with a chiral carbon in the β-position) [27] and *Candida rugosa* (toward carboxylic acids with a chiral carbon in the α-position) [28]. The model for primary alcohols is only reliable if there is no oxygen substituent on the chiral carbon. To explain the opposite enantiopreference observed with primary and secondary alcohols, it was suggested that the CH_2 in the primary alcohol is accommodated as a kink, allowing a similar position of the alcohol oxygen for both substrates.

Such models can be of value for initial suggestions to enantiopreference, but the complex interactions between lipase, substrate and reaction medium often surprise, and experimental data (in a screening of enzymes and conditions) are still essential [29].

13.2
Immobilized Enzymes

A simple way of applying enzymes in a non-aqueous reaction is through addition of lyophilized enzyme powder. With a hydrophobic solvent, this will create an enzyme suspension and, if other conditions (substrate, temperature, etc.) are acceptable, it will probably work. For a quick screening of a range of enzymes it may even be the most optimal setup. It is, however, well known that many enzymes lose activity upon lyophilization (which to some degree can be prevented by the use of lyoprotectants). Another concern is that enzyme dust is potentially allergenic if inhaled.

Alternatively, if introduction of water in the system is not considered a problem (e.g. if water is used as a solvent or cosolvent in a homogenous or heterogeneous, two-phase, system), the enzyme can be added as an aqueous formulation. Advantages are also reported on the use of reversed micelles (detergent micelles containing aqueous enzyme in organic media). In any case, for all reaction conditions, the use of an immobilized enzyme preparation should be considered as outlined below.

Briefly, the advantages of immobilization are:

- Possibility of recycling
- The enzyme can easily be removed from the product or reaction mixture
- Often improved enzyme stability (towards organic solvents and high temperature)
- Easy handling and dosing
- Improved shelf stability (Novozym 435 is stable at room temperature for years)
- Possibility of packing columns (fixed bed reactors) for batch or continuous operation.

Disadvantages, on the other hand, include: [30]

- Loss of absolute activity due to the immobilization process
- Diffusion limitations lowering reaction rates
- Cost of carrier and extra work in immobilizing the enzyme if not commercially available
- Cannot be used together with immobilized or insoluble substrates.

Numerous methods exist for enzyme immobilization and the field is often reviewed [31].

13.2.1
Immobilization without Carrier

Many applications rely on enzymes being retained by membranes, aggregated by cross-linking, or immobilized by encapsulation. These techniques are often simple and inexpensive, but typically also generate a poorly defined immobilized enzyme. The immobilization can involve isolated enzymes or whole cell preparations. Sweetzyme IT, an immobilized glucose isomerase produced by Novozymes is an example of the latter, in which the cells are cross-linked by glutaraldehyde (GA) and extruded to produce dry, solid particles [32].

13.2.1.1 Cross-linking
The use of cross-linked enzyme crystals (CLEC) was developed in the early 1990s [33]. The technology involves the cross-linking of enzyme microcrystals with bifunctional reagents, typically GA. The result is a solid microporous material that is insoluble in both organic and aqueous media, allowing separation by filtration or simple settling/decantation. The problem is, however, often the crystallization step. Sheldon and coworkers, therefore, introduced cross-linked enzyme aggregates (CLEA) in which enzyme in solution is simply brought to aggregate before the cross-linking step [34]. Aggregation can be induced from addition of $(NH_4)_2SO_4$, poly(ethylene glycol) or alcohols. Both CLEC and CLEA products are close to 100% pure protein and hence typically have very high activity per weight. However, as enzymes typically are expensive, the cost per weight is equally high and so, with the unavoidable loss in recovery from a batch process, reuse can be problematic from an economy point of view. CLECs were previously commercialized by Altus, whereas CLEAs are now available from CLEA Technologies.

Despite reports that larger polyaldehydes (e.g. periodate-oxidized polysaccharides) result in higher activity of the cross-linking enzymes [35], GA continues to be the cross-linking agent of choice. Applications involve direct cross-linking of enzyme in solution, CLEC/CLEA type immobilizations, cross-linking of enzyme adsorbed on a carrier, as well as being a bifunctional linker between a carrier and enzyme. GA reacts with amines (primarily from lysine residues) but the chemistry involved is surprisingly complex [36]. Hence cross-linking procedures have developed largely through empirical methods. Historically, it has been proposed that the cross-linking ability was due to Schiff base formation. Owing to the lability of such linkages some protocols suggest reducing the putative Schiff bases with

NaBH$_4$. There is, however, little evidence that such reduction has any effect. In fact it is well known that GA immobilized proteins are stable and the linkages cannot be hydrolyzed. Furthermore, the literature supports the idea that GA has unique characteristics that render it a far more effective cross-linking agent than other dialdehydes. Commercial GA solutions, pH 3–4, are multicomponent mixtures, containing hydrated aldehydes, cyclic hemiacetals and polymeric structures of cyclic acetals. Under acidic conditions these acetals can react with amines to create the corresponding aminals. However, cross-linking is often performed at pH 7–8, where GA undergoes intermolecular aldol condensations, resulting in α,β-unsaturated oligomeric aldehydes. Schiff bases formed with these aldehydes would be stabilized due to conjugation. Further, it has been suggested that the Schiff bases cyclize into quaternary pyridinium structures. Another possibility is that amines add to unsaturated aldol condensation products in a Michael addition.

13.2.1.2 Entrapment

Entrapment is another immobilization method not relying on resin carriers as the enzyme is entrapped in natural or synthetic gel matrices. Natural gels include alginate and κ-carrageenan. Dropping an aqueous solution of enzyme and alginate or κ-carrageenan into a gelling agent (e.g. CaCl$_2$ for alginate and KCl or κ-carrageenan) will produce water-insoluble beads containing entrapped enzyme. An often-used synthetic gel is polyacrylamide. The enzyme, acrylamide and a cross-linker (e.g. N,N'-methylenebisacrylamide) are polymerized by addition of an initiator (e.g. potassium persulfate) and a stimulator (e.g. 3-dimethylaminopropionitrile). A specific form of entrapment immobilization is the so-called sol–gel process. Addition of acid or base catalyst to a solution of Si(OR)$_4$ and enzyme leads to hydrolysis (liberation of ROH) and cross-linking of the formed Si-monomers into an amorphous SiO$_2$-network, entrapping the enzyme [37].

13.2.2
Immobilization on a Carrier

An enzyme can be immobilized on/in a resin carrier either by adsorption (by hydrophobic, electrostatic or other forces) or it can be covalently linked to the resin. Carrier materials used for immobilization in biocatalysis include natural, synthetic, organic, inorganic, porous and non-porous materials. The main advantage compared to immobilization without a carrier is in general a better defined immobilized enzyme, as particle size, pore size, porosity, hydrophobicity and so on is pre-determined from the choice of carrier. However, the carrier cost is often significant.

Loading (milligrams of enzyme per gram of carrier) is an important parameter for enzymatic activity. Initially, an activity versus loading plot would be approximately linear, but it quickly curves and a point can be reached after which higher loading no longer gives better activity, but in some cases actually the opposite. This can often be correlated to the surface area of the carrier and be explained in terms of monolayer versus multilayer structures. For industrial preparations, it is obviously important to find the optimal loading in a compromise between activity and cost of carrier and enzyme. For initial laboratory-scale experiments, finding this exact point is typically of less importance.

For many applications it is important to know what the loading actually is (e.g. to compare activity of different enzymes immobilized to the same loading). Loading determination of an immobilized enzyme is often possible by methods such as quantitative amino acid analysis or active site titration. The latter has the advantage that only active enzyme is quantified. Typically, however, loading is determined indirectly during the immobilization process by measuring enzyme concentration in the aqueous solution before and after incubation with the carrier. The concentration of a pure enzyme in aqueous buffer can be determined from the absorption at 280 nm. The specific absorption can be calculated from the amino acid sequence of the enzyme. For less pure enzyme solutions, the total protein content can be determined with Bradford, BCA or other assays.

13.2.2.1 Adsorption

Lipases act in nature on oil–water interfaces and often have hydrophobic domains on their surface. Hence, immobilization by adsorption to a hydrophobic carrier is often a simple and effective way to immobilize lipases. A wide range of different hydrophobic support materials is commercially available, including synthetic acrylic, divinylbenzene-styrene or polypropylene polymers. An example of the latter is Accurel MP 1000, which is available from Membrana. Novozym 435 is immobilized on Lewatit VP OC 1600, a divinylbenzene-cross-linked poly(methyl methacrylate) resin produced by Lanxess (previously Bayer).

An alternative strategy is to immobilize by electrostatic binding. Enzymes contain both positive and negative surface charges, and both cationic and anionic ion-exchange resins have been used for enzyme immobilization. However, many lipases have slightly acidic pI-values (pI 5–6), meaning that their net charge will be negative at pH 7. Hence, cationically charged resins (anion exchangers) will in these cases be preferred. Macroporous synthetic ion-exchange resins, also used for water treatment and chromatography, are good choices (trademarks Duolite, Amberlite, Lewatit, etc.). Alternatively, an ion-exchange resin can be generated from an otherwise inert carrier material by coating with polyethyleneimine (PEI), functionalization with diethylaminoethanol (DEAE), etc. A potential disadvantage of electrostatic adsorption is that the ion-exchange carriers by nature contain functional groups that can change the microenvironment around the enzyme or even react with substrate, etc. Lipozyme RMIM is the Novozymes trade name for *Rhizomucor miehei* lipase immobilized on Duolite A568, a weak anion exchange resin from Rohm and Haas.

In general, adsorption immobilization is an extremely easy process as the resin is simply mixed with the enzyme in aqueous buffer and gently stirred/shaken, typically over night, followed by filtering and drying. However, a potential problem with the method is leaching of the enzyme from the support in the application. This limits the number of reuses and enzyme contamination in the reaction product can cause problems. The degree of leaching depends much on the solvent used in the application, and can be from close to zero to almost quantitative. Neat DMSO will denature most enzymes, but is very effective in stripping enzyme adsorbed to a hydrophobic carrier. Hence, it can be used to determine loading or the amount of non-covalently linked enzyme.

13.2.2.2 Covalent Binding

Like cross-linking procedures, covalent binding to a resin typically utilizes accessible amino groups or carboxylic acids exposed on the enzyme. Fortunately, these common residues (Glu, Asp, Lys) are in general found on the surface (whereas more hydrophobic residues are in the interior). If not directly involved in the catalytic mechanism, the residues can be used for immobilization without significantly affecting the catalytic activity. However, as with other immobilization methods, some loss of specific activity can be expected due to distortion of the structure (loss of mobility), shielding of the active site and so on.

Traditionally in biochemistry (e.g. for affinity chromatography), enzymes have been covalently linked to carbohydrates such as agarose and dextran (Sepharose, Sephadex). The most widely used activation method is the cyanogen bromide (CNBr) method, yielding isourea and imidocarbonate functionalities that react with amines on the enzyme to produce N-substituted carbamate linkages.

Within biocatalysis, Eupergit C is one of the most popular resins for enzyme immobilization on a laboratory as well an industrial scale. Oxidoreductases, transferases, hydrolases and lyases have all been successfully immobilized [38]. The porous, epoxy-activated acrylic beads (100–250 µm diameter) are produced by Degussa (Röhm GmbH) and distributed by Sigma-Aldrich in research quantities. A standard immobilization procedure is simply to dissolve enzyme in aqueous buffer (typically pH 7–8), add Eupergit beads and leave the suspension at room temperature or 4 °C for 12–72 h. After initial adsorption to the hydrophobic resin, enzyme nucleophiles (primarily Lys side chains at neutral or alkaline pH) react with the epoxy groups to form stable linkages. If a non-leaking immobilized enzyme preparation is the aim, extensive washing is needed to remove adsorbed enzyme that has not reacted.

The literature contains numerous other examples of resin-to-enzyme coupling strategies, including activated esters (e.g. NHS-esters), vinylsulfones, maleimides (e.g. to react with Cys thiols), aldehydes (e.g. from GA), azides (e.g. for "click chemistry"), thiols (for disulfide linkages), carboxylic acids or primary amines (e.g. for carbodiimide mediated couplings), etc. The enzyme can react with the functional groups directly on the carrier, or through bifunctional linkers. Further, excess reactive groups can be capped using small molecules with desired properties to modify the microenvironment around the immobilized enzymes.

However, even though immobilization by covalent binding to a carrier is superior to most other immobilization strategies, the very high cost of the activated carriers, regrettably, often limits the use in large-scale applications.

13.3
Applications

A range of excellent and recent reviews can be found, in which the use of enzymes within specific branches or disciplines of organic chemistry is highlighted. These include biocatalysis in carbohydrate chemistry [39], polymer chemistry [40] and for protecting group manipulations [41]. The present chapter is focused on immobilized enzymes. Hence, as an appetizer, a few selected applications with Novozym 435 are presented below, followed by a short subsection discussing industrial-scale applications of immobilized enzymes.

13.3.1
Novozym 435 – A Versatile Immobilized Biocatalyst

The single most used lipase for biocatalysis is probably the *Candida antarctica* B-lipase (CALB) [42]. It is commercialized by Novozymes in liquid formulation as well as in immobilized form under the trade name Novozym 435 (previously SP 435). CALB has high activity on a wide range of substrates (it has some problems with very bulky substrates), often with outstanding selectivities. Formulated as Novozym 435 it is stable up to approx. 90 °C in solvents such as toluene (or solvent-free reaction mixtures). The A-lipase (CALA), currently only commercially available in liquid form, has attractive properties too, including even better thermostability and higher activity on sterically hindered substrates [43].

An early application is the preparation of 6-*O*-acylglucopyranosides (sugar esters) (Scheme 13.1). CALB was found to chemoselectively esterify the primary 6-OH with fatty acids. Ethyl D-glucopyranoside was initially selected due to improved solubility in neat fatty acid [44], but reactions with unmodified mono- and disaccharides suspended in acetone, *tert*-butyl alcohol or DMSO-mixtures have also been reported [45]. The products have found applications as non-ionic tensides.

Another early discovery was that CALB accepts H_2O_2 as nucleophile to produce peroxycarboxylic acids from esters or carboxylic acids (perhydrolysis activity can also be found in other serine hydrolases) [46, 47]. The *in situ* formed peracid can subsequently be used to epoxidize an alkene by (non-enzymatic) Prileshajev epoxidation. Hence, oleic acid incubated with CALB and H_2O_2 will produce 9,10-epoxyoctadecanoic acid [48]. Other alkenes can be epoxidized by H_2O_2 and a catalytic amount of carboxylic acid (and CALB) (Scheme 13.2) [49].

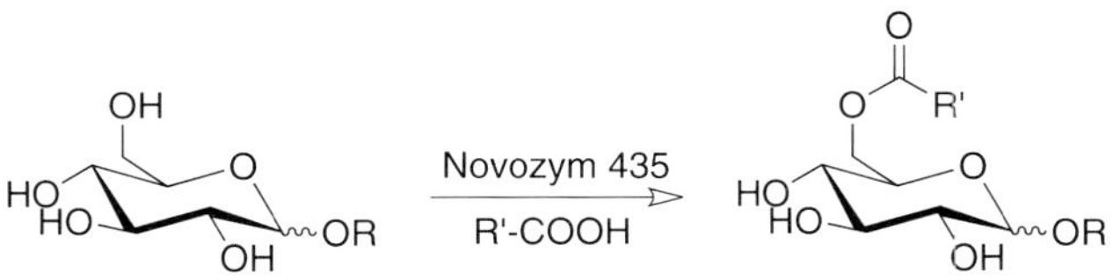

Scheme 13.1 Chemoselective synthesis of 6-*O*-acylglucopyranoside.

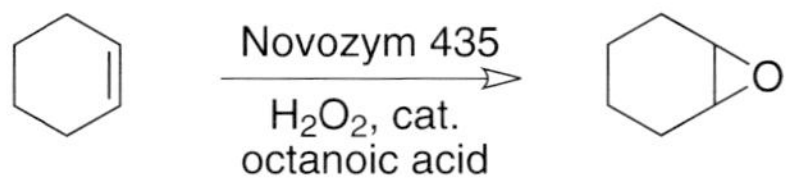

Scheme 13.2 Epoxidation by *in situ* perhydrolysis.

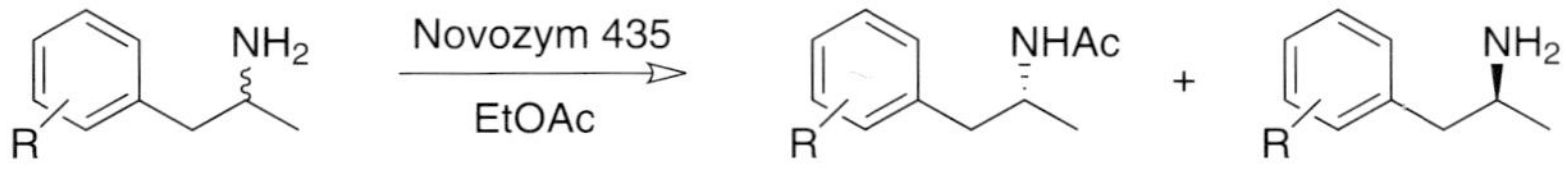

Scheme 13.3 Enantioselective acetylation of β-substituted isopropylamines in EtOAc.

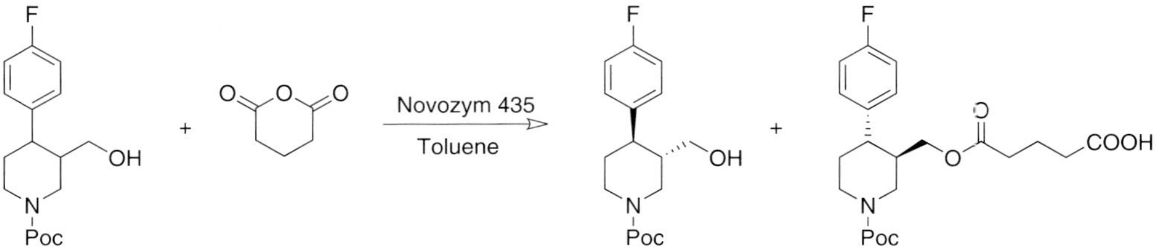

Scheme 13.4 Kinetic resolution by acylation of a primary alcohol.

Examples of kinetic resolutions with lipases are numerous [9]. Impressive enantioselectivities are often obtainable with secondary alcohols, e.g., in acetylations with vinyl acetate, or in hydrolysis of the racemic ester. Likewise, the corresponding amines can be resolved, e.g. by enantioselective acetylation with EtOAc as both acyl donor and solvent. This has been demonstrated by Gotor and coworkers using Novozym 435 [50]. The reaction (Scheme 13.3) follows Kazlauskas' selectivity. In fact an impressive range of CALB (Novozym 435) catalyzed transformations on nitrogenated compounds have been collected in a recent review article [51].

Another example by Gotor and coworkers is kinetic resolution by acylation of a primary alcohol [52]. With cyclic anhydrides as acyl donors the reaction product (now bearing a carboxylic acid) could easily be separated from the mixture by alkaline extraction. The non-reacting enantiomer is an intermediate in the synthesis of (−)-paroxetine, a selective inhibitor of 5-hydroxytryptamine reuptake (Scheme 13.4).

By definition, a maximum 50% yield can be obtained through kinetic resolution of racemates. In some applications, a 100% yield is obtainable through meso ("the meso trick") or prochiral substrates. Researchers from Schering-Plough used Novozym 435 for desymmetrization of a 2-substituted-1,3-propanediol intermediate in the synthesis of an antifungal agent (Scheme 13.5) [53]. Details were reported for scale-up to pilot plant, giving >95% ee with <2% diol remaining.

Alternatively, the resolution reaction can be combined with rapid *in situ* racemization of the substrate in dynamic kinetic resolution processes. Combinations of

Scheme 13.5 Desymmetrization by acetylation of a 2-substituted-1,3-propanediol.

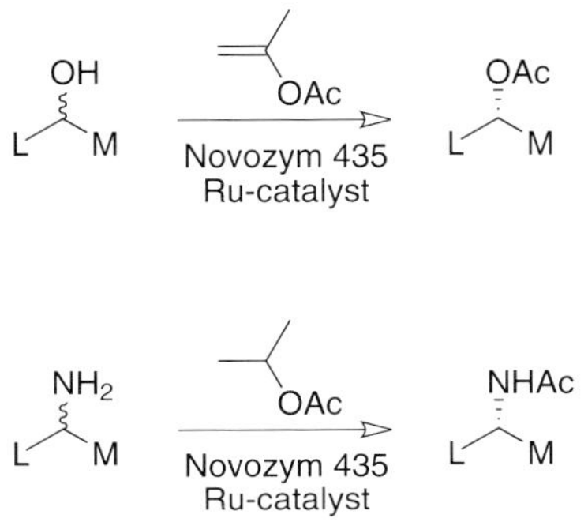

Scheme 13.6 Dynamic kinetic resolutions. L and M refer to "large" and "medium" size substituents, respectively.

biocatalysis (lipase-catalyzed resolution) and metal-catalyzed racemization have been demonstrated by Bäckvall and others [54]. Hence, using Novozym 435 for acylation together with Ru(II)-catalysts for racemization, dynamic kinetic resolutions of a range of secondary alcohols [55] and primary amines [56] have been reported with yields up to 99% and ee generally exceeding 99% (Scheme 13.6).

13.3.2
Immobilized Enzymes in Industrial-Scale Biocatalysis

Applications of immobilized enzymes in industrial processes have previously been reviewed [57]. Large-scale processes for food applications include production of high-fructose corn syrup with immobilized glucose isomerase, and interesterification of triglycerides (to modify their melting point) with immobilized lipase. In addition, the dipeptide sweetener aspartame (H-Asp-Phe-OMe) can be prepared with immobilized thermolysin (by enantiospecific coupling of Z-L-Asp-OH and H-DL-Phe-OMe in EtOAc). Within the pharma- and chemical industries, acrylamide is produced on a large scale using immobilized nitrile hydratase, while 6-aminopenicillanic acid (6-APA) is prepared with immobilized penicillin amidase (Scheme 13.7).

13.4
Concluding Remarks

Enzymatic processes in fields such as food and leather manufacturing have ancient roots, yet application of enzymes for organic synthesis is still a relatively young

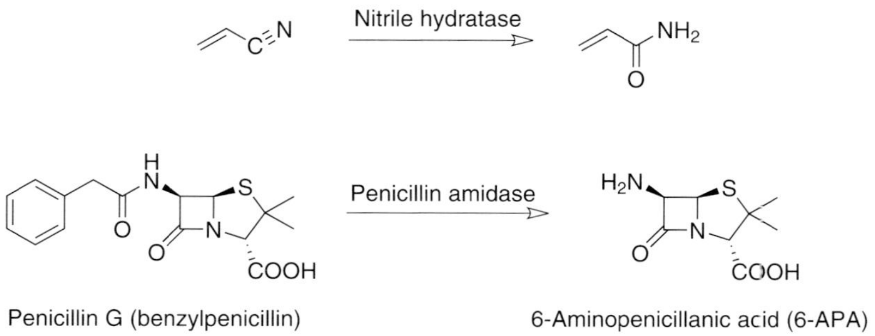

Scheme 13.7 Industrial-scale synthesis of acrylamide and 6-APA by immobilized enzymes.

Acrylonitrile → (Nitrile hydratase) → Acrylamide

Penicillin G (benzylpenicillin) → (Penicillin amidase) → 6-Aminopenicillanic acid (6-APA)

and much developing discipline [58]. Development of an enzymatic process can still be a very significant task, involving high-throughput screening and generation of optimized enzyme variants. In other cases, the conversion is straightforward, for which commercially available enzymes can be applied like any other standard shelf chemical. Immobilized enzymes in particular have the advantage of being easy ready-to-use tools in the repertoire of organic chemistry.

References

1 (a) Klibanov, A.M. (1990) *Accounts of Chemical Research*, **23**, 114–20. (b) Roberts, S.M., Turner, N.J., et al. (1995) *Introduction to Biocatalysis Using Enzymes and Micro-Organisms*, Cambridge University Press, Cambridge.
2 (a) Straathof, A.J.J., Panke, S. and Schmid, A. (2002) *Current Opinion in Biotechnology*, **13**, 548–56. (b) Liese, A. and Filho, M.V. (1999) *Current Opinion in Biotechnology*, **10**, 595–603. (c) Schulze, B. and Wubbolts, M.G. (1999) *Current Opinion in Biotechnology*, **10**, 609–15.
3 Wohlgemuth, R. (2004) Large-scale application of biocatalysis in the asymmetric synthesis of laboratory chemicals, in *Asymmetric Catalysis on Industrial Scale: Challenges, Approaches and Solutions* (eds H.U. Blaser and E. Schmidt), Wiley-VCH Verlag Gmbh, Weinheim, pp. 309–19.
4 (a) Patel, R.N. (2006) *Current Opinion in Drug Design and Development*, **9**, 741–64. (b) Gotor-Fernández, V., Brieva, R. and Gotor, V. (2006) *Journal of Molecular Catalysis B: Enzymatic*, **40**, 111–20.
5 Rasor, P. (2002) Tabular survey of commercially available enzymes, in *Enzyme Catalysis in Organic Synthesis*, 2nd edn, Vol. 3 (eds K. Drauz and H. Waldmann), Wiley-VCH Verlag Gmbh, Weinheim, pp. 1461–518.
6 (a) Ghanem, A. (2007) *Tetrahedron*, **63**, 1721–54. (b) Ghanem, A. and Aboul-Enein, H.Y. (2005) *Chirality*, **17**, 1–15.
7 (a) Svedendahl, M., Hult, K. and Berglund, P. (2005) *Journal of the American Chemical Society*, **127**, 17988–9. (b) Torre, O., Gotor-Fernández, V., Alfonso, I., et al. (2005) *Advanced Synthesis Catalysis*, **347**, 1007–14.
8 Bornscheuer, U.T. and Kazlauskas, R.J. (2004) *Angewandte Chemie–International Edition*, **43**, 6032–40.
9 Hanefeld, U. (2003) *Organic & Biomolecular Chemistry*, **1**, 2405–15.
10 Panke, S. and Wubbolts, M. (2005) *Current Opinion in Chemical Biology*, **9**, 188–94.
11 (a) Burton, S.G. (2003) *Trends Biotechnology*, **21**, 543–9. (b) Burton, S.G. (2003) *Current Organic Chemistry*, **7**, 1317–31. (c) Niedermeyer, T.H.J., Mikolasch, A.

and Lalk, M. (2005) *Journal of Organic Chemistry*, **70**, 2002–8.
12 (a) Clouthier, C.M., Kayser, M.M. and Reetz, M.T. (2006) *Journal of Organic Chemistry*, **71**, 8431–7. (b) Hilker, I., Wohlgemuth, R., Alphand, V. and Furstoss, R. (2005) *Biotechnology and Bioengineering*, **92**, 702–10.
13 Fabbrini, M., Galli, C., Gentili, P. and Macchitella, D. (2001) *Tetrahedron Letters*, **42**, 7551–3.
14 (a) Nakamura, K., Yamanaka, R., Matsuda, T. and Harada, T. (2003) *Tetrahedron Asymmetry*, **14**, 2659–81. (b) Kroutil, W., Mang, H., Edegger, K. and Faber, K. (2004) *Current Opinion in Chemical Biology*, **8**, 120–6.
15 Nakamura, K. and Matsuda, T. (2006) *Current Organic Chemistry*, **10**, 1217–46.
16 Howarth, J., James, P. and Dai, J. (2001) *Tetrahedron Letters*, **42**, 7517–19.
17 Bolivar, J.M., Wilson, L., Ferrarotti, S.A., et al. (2006) *Journal of Biotechnology*, **125**, 85–94.
18 Kreiner, M. and Parker, M.-C. (2005) *Biotechnology Letters*, **27**, 1571–7.
19 Kreiner, M., Moore, B.D. and Parker, M.C. (2001) *Chemical Communications*, 1096–7.
20 (a) Mylerová, V. and Martínkova, L. (2003) *Current Organic Chemistry*, **7**, 1279–95. (b) Schulze, B. (2002) Hydrolysis and formation of C–N bonds: hydrolysis of nitriles, in *Enzyme Catalysis in Organic Synthesis*, 2nd edn, Vol. 2 (eds K. Drauz and H. Waldmann), Wiley-VCH Verlag GmbH, Weinheim, pp. 699–715.
21 Crosby, J., Moiliet, J., Parratt, J.S. and Turner, N.J. (1994) *Journal of the Chemical Society – Perkin Transactions 1*, 1679–87.
22 Klibanov, A.M. (2001) *Nature*, **409**, 241–6.
23 Adlercreutz, P. (2000) Biocatalysis in non-conventional media, in *Applied Biocatalysis*, 2nd edn (eds A.J.J. Straathof and P. Adlercreutz), Taylor & Francis, London, pp. 295–316.
24 (a) Harper, N., Dolman, M., Moore, B.D. and Halling, P.J. (2000) *Chemistry – A European Journal*, **6**, 1923–9. (b) Partridge, J., Halling, P.J. and Moore, B.D. (2000) *Journal of the Chemical Society – Perkin Transactions 2*, 465–71.
25 Kazlauskas, R.J., Weissfloch, A.N.E., Rappaport, A.T. and Cuccia, L.A. (1991) *Journal of Organic Chemistry*, **56**, 2656–65.
26 (a) Cygler, M., Grochulski, P., Kazlauskas, R.J., et al. (1994) *Journal of the American Chemical Society*, **116**, 3180–6. (b) Kazlauskas, R.J. and Weissfloch, A.N.E. (1997) *Journal of Molecular Catalysis B: Enzymatic*, **3**, 65–72. (c) Ema, T., Yamaguchi, K., Wakasa, Y., et al. (2003) *Journal of Molecular Catalysis B: Enzymatic*, **22**, 181–92. (d) Ema, T. (2004) *Current Organic Chemistry*, **8**, 1009–25.
27 Weissfloch, A.N.E. and Kazlauskas, R.J. (1995) *Journal of Organic Chemistry*, **60**, 6959–69.
28 Ahmed, S.N., Kazlauskas, R.J., Morinville, A.H., et al. (1994) *Biocatalysis*, **9**, 209–25.
29 Berglund, P. (2001) *Biomolecular Engineering*, **18**, 13–22.
30 Castro, G.R. and Knubovets, T. (2003) *Critical Reviews in Biotechnology*, **23**, 195–231.
31 (a) Cao, L. (ed.) (2005) *Carrier-bound Immobilized Enzymes: Principles, Application and Design*, Wiley-VCH Verlag GmbH, Weinheim. (b) Guisan, J.M. (ed.) (2006) Immobilization of enzymes and cells, in *Methods in Biotechnology*, 2nd edn, Vol. 22, Humana Press Inc., Totowa. (c) Lalonde, J. and Margolin, A. (2002) Immobilization of enzymes, in *Enzyme Catalysis in Organic Synthesis*, 2nd edn, Vol. 1 (eds K. Drauz and H. Waldmann), Wiley-VCH Verlag GmbH, Weinheim, pp. 163–84. (d) Cao, L. (2005) *Current Opinion in Chemical Biology*, **9**, 217–26. (e) Tischer, W. and Wedekind, F. (1999) *Topics in Current Chemistry*, **200**, 95–126. (f) Rasor, P. (2000) Immobilized enzymes, in enantioselective organic synthesis, in *Chiral Catalyst Immobilization and Recycling* (eds D.E. De Vos, I.F.J. Vankelecom and P.A. Jacobs), Wiley-VCH Verlag GmbH, Weinheim, pp. 97–122. (g) Woodley, J.M. (1993) Solid supports and catalysts in organic synthesis, in *Immobilized Biocatalysts* (ed. K. Smith), Horwood, Chichester, pp. 254–71.

32 Bhosale, S.H., Rao, M.B. and Deshpande, V.V. (1996) *Microbiological Reviews*, 280–300.
33 (a) St.Clair, N.L. and Navia, M.A. (1992) *Journal of the American Chemical Society*, **114**, 7314–16. (b) Govardhan, C.P. (1999) *Current Opinion in Biotechnology*, **10**, 331–5. (c) Roy, J.J. and Abraham, T.E. (2004) *Chemical Reviews*, **104**, 3705–21.
34 Cao, L., Van Rantwijk, F. and Sheldon, R.A. (2000) *Organic Letters*, **2**, 1361–4.
35 (a) Schoevaart, R., Siebum, A., van Rantwijk, F., Sheldon, R. and Kieboom, T. (2005) *Starch/Stärke*, **57**, 161–5. (b) Matero, C., Palomo, J.M., van Langen, L.M., van Rantwijk, F. and Sheldon, R.A. (2004) *Biotechnology and Bioengineering*, **86**, 273–6.
36 (a) Walt, D.R. and Agayn, V.I. (1994) *Trends in Analytical Chemistry*, **13**, 425–30. (b) Migneault, I., Dartiguenave, C., Bertrand, M.J. and Waldron, K.C. (2004) *BioTechniques*, **37**, 790–802.
37 (a) Reetz, M.T. (1997) *Advanced Materials*, **9**, 943–54. (b) Pierre, A.C. (2004) *Biocatalysis and Biotransformation*, **22**, 145–70.
38 (a) Katchalski-Katzir, E. and Kraemer, D.M. (2000) *Journal of Molecular Catalysis B: Enzymatic*, **10**, 157–76. (b) Boller, T., Meier, C. and Menzler, S. (2002) *Organic Process Research & Development*, **6**, 509–19.
39 Hamilton, C.J. (2004) *Natural Product Reports*, **21**, 365–85.
40 (a) Gross, R.A., Kumar, A. and Kalra, B. (2001) *Chemical Reviews*, **101**, 2097–124. (b) Kobayashi, S., Uyama, H. and Kimura, S. (2001) *Chemical Reviews*, **101**, 3793–818.
41 Kadereit, D. and Waldmann, H. (2001) *Chemical Reviews*, **101**, 3367–96.
42 (a) Uppenberg, J., Öhrner, N., Norin, M., et al. (1995) *Biochemistry*, **34**, 16838–51. (b) Anderson, E.M., Larsson, K.M. and Kirk, O. (1998) *Biocatalysis and Biotransformation*, **16**, 181–204.
43 De María, P.D., Carboni-Oerlemans, C., Tuin, B., Bargeman, G., van der Meer, A. and van Gemert, R. (2005) *Journal of Molecular Catalysis B: Enzymatic*, **37**, 36–46.
44 (a) Adelhorst, K., Björkling, F., Godtfredsen, S.E. and Kirk, O. (1990) *Synthesis*, 112–15. (b) Kirk, O., Björkling, F., Godtfredsen, S.E. and Larsen, T.O. (1992) *Biocatalysis*, **6**, 127–34.
45 (a) Ferrer, M., Soliveri, J., Plou, F.J., et al. (2005) *Enzyme and Microbial Technology*, **36**, 391–8. (b) Chaiyaso, T. H-kittikun, A. and Zimmermann, W. (2006) *Journal of Industrial Microbiology and Biotechnology*, **33**, 338–42.
46 Bernhardt, P., Hult, K. and Kazlauskas, R.J. (2005) *Angewandte Chemie – International Edition*, **44**, 2742–6.
47 (a) Björkling, F., Godtfredsen, S.E. and Kirk, O. (1990) *Journal of the Chemical Society D – Chemical Communications*, 1301–3. (b) Björkling, F., Frykman, N., Godtfredsen, S.E. and Kirk, O. (1992) *Tetrahedron*, **48**, 4587–92.
48 Orellana-Coca, C., Törnvall, U., Adlercreutz, D., Mattiasson, B. and Hatti-Kaul, R. (2005) *Biocatalysis and Biotransformation*, **23**, 431–7.
49 (a) Moreira, M.A., Bitencourt, T.B. and Nascimento, M.G. (2005) *Synthetic Communications*, **35**, 2107–14. (b) Jarvie, A.W.P., Overton, N. and St Pourcain, C.B. (1998) *Chemical Communications*, 177–8.
50 González-Sabín, J., Gotor, V. and Rebolledo, F. (2002) *Tetrahedron Asymmetry*, **13**, 1315–20.
51 Gotor-Fernández, V., Busto, E. and Gotor, V. (2006) *Advanced Synthesis Catalysis*, **348**, 797–812.
52 De Gonzalo, G., Brieva, R., Sánchez, V.M., et al. (2003) *Journal of Organic Chemistry*, **68**, 333–3336.
53 Morgan, B., Dodds, D.R., Zaks, A., et al. (1997) *Journal of Organic Chemistry*, **62**, 7736–43.
54 Pàmies, O. and Bäckvall, J.-E. (2003) *Chemical Reviews*, **103**, 3247–61.
55 Martín-Matute, B., Edin, M., Bogár, K., Kaynak, F.B. and Bäckvall, J.-E. (2005) *Journal of the American Chemical Society*, **127**, 8817–25.
56 Paetzold, J. and Bäckvall, J.E. (2005) *Journal of the American Chemical Society*, **127**, 17620–1.
57 (a) Katchalski-Katzir, E. (1993) *Trends Biotechnology*, **11**, 471–8. (b) End, N. and Schöning, K.-U. (2004) *Topics in Current Chemistry*, **242**, 273–317.
58 Bornscheuer, U.T. and Buchholz, K. (2005) *Engineering in Life Sciences*, **5**, 309–23.

Part Four Resins for Solid Phase Synthesis

14
Acid-Labile Resins

Peter D. White

Merck Chemicals, Padge Road, Beeston, Nottinghamshire NG9 2JR, UK

14.1
Introduction: Linker Design

The vast majority of linker systems used in solid phase synthesis undergo product cleavage by acidolysis (for some recent reviews see references [1, 2]). The reason for this preference is partly because most linkers were originally designed for use in peptide synthesis, where the use of acid-cleavable linkers and side-chain protecting groups are favored, but is also due to the ease with which the properties of such linkers can be fine-tuned to suit particular applications.

In most linker systems, the product cleavage reaction follows an S_N1-type mechanism, involving heterolytic fission of a carbon–heteroatom bond. The rate of this reaction depends on the stability of the carbocation produced, the nature of leaving group and the basicity of the heteroatom. It therefore follows that a molecule anchored by a good leaving group to a given linker will be cleaved under milder acidolysis conditions than one attached by a poor leaving group. Inversely, a molecule anchored to a linker that generates a highly stabilized cation will cleave under milder conditions than when anchored to a linker that gives a less stabilized cation.

Most of the acid-labile linkers in common usage are based on benzylic structures, since the stability of the cation formed on cleavage can be easily modulated by the introduction of *p*- or *o*-electron-donating or -withdrawing substituents to the aromatic ring. Substitution of the benzylic protons with additional aryl groups gives rise to benzhydryl and trityl structures offering further stabilization. Linkers have also been based on structures that give aromatic cations on cleavage, such as xanthine [3, 4] and cyclohexatriene [5], and those that exploit the β-silyl effect (Figure 14.1) [6, 7].

Tables 14.1 and 14.2 show the comparative ease of cleavage of common linker types and leaving groups encountered in solid phase synthesis. By taking into account the relative positions of linker and leaving groups within these tables, the

The Power of Functional Resins in Organic Synthesis. Judit Tulla-Puche and Fernando Albericio
Copyright © 2008 WILEY-VCH Verlag GmbH & Co. KGaA, Weinheim
ISBN: 978-3-527-31936-7

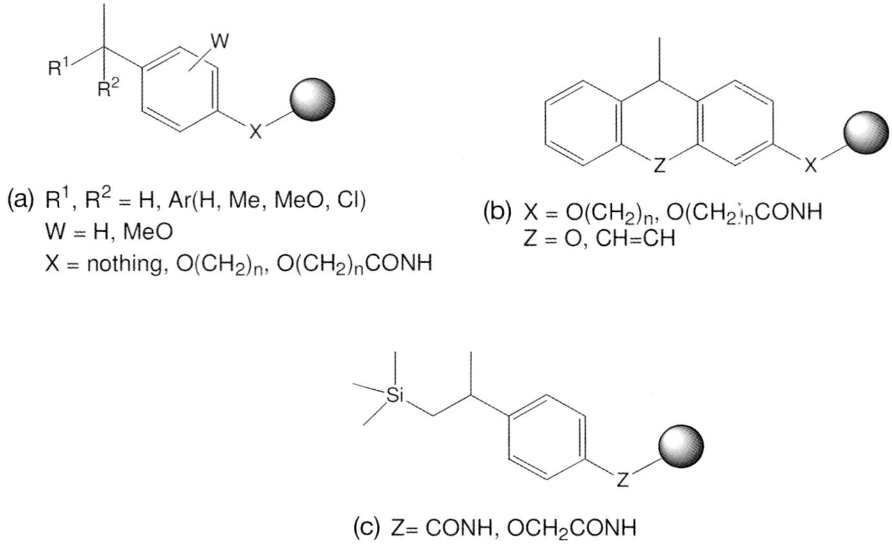

(a) R^1, R^2 = H, Ar(H, Me, MeO, Cl)
W = H, MeO
X = nothing, $O(CH_2)_n$, $O(CH_2)_n CONH$

(b) X = $O(CH_2)_n$, $O(CH_2)_n CONH$
Z = O, CH=CH

(c) Z= CONH, OCH_2CONH

Figure 14.1 Design templates for common classes of acid-labile linkers. (a) Benzyl, benzhydryl and trityl; (b) xanthinyl and cyclohepta(di)trienyl; (c) β-silyl.

comparative acid sensitivities of different linkers and leaving group combinations can be rationalized.

Notably, the relative orders are based on empirical observation and those of close neighbors may change, depending on the nature of the leaving group and actual experimental conditions.

There are two principle ways in which an acid-labile linker is attached to the solid support. First, it is synthesized directly on the solid phase, with part of the polymer, typically a phenyl ring, becoming an integral part of the linker structure. Such linkers have been termed "integral linkers" [1]. Second, the linker is preformed in solution and attached to the support usually via an ether or amide bond. The advantage of the former approach is that it is cost efficient since it avoids the need for isolation and purification of intermediates, with all reactions being performed directly on the solid phase. However, control of synthesis can be difficult, leading to problems in obtaining reproducible substitutions and batch-to-batch consistency. As linkers have become more sophisticated, the trend has been towards using preformed linkers as their use gives precise control over resin functionalization and provides confidence regarding the integrity of the linker.

In resins where the linker is anchored by an ether bond, the ether plays an essential role as an electron-donating substituent, thereby enhancing the acid sensitivity of the linker. In many cases, such as the Wang [8] (**1**) and Sasrin [10] linkers (**3**), this ether is benzylic as it derives from chloromethyl polystyrene (**8**). With such resins, there is always the risk that during the cleavage reaction the benzylic ether may also cleave, resulting in a product modified or contaminated

Table 14.1 Common linker types ordered by stability of incipient carbocation.

Increasing stability of carbocation[a]	Linker type

[a] Relative orders are based on empirical observation and those of close neighbors may change depending on the nature of the leaving group and actual experimental conditions.

14 Acid-Labile Resins

Table 14.2 Leaving groups ordered by ease of cleavage by acidolysis.

	Leaving group
Easy	O$_2$CR
	Oar
	OR
	NRSO$_2$R
	NHSO$_2$R
	NRCOX
	NHCOX
	NHAr
	NHR
	NHC(=NH)NHR
Hard	SR

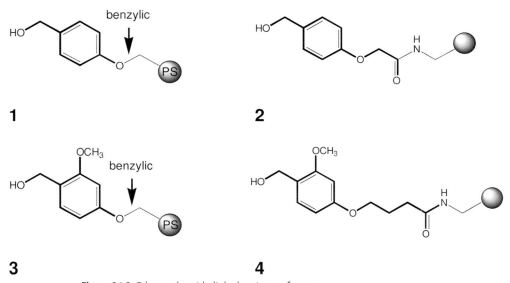

Figure 14.2 Ether and amide linked variants of some hydroxymethylphenoxy linkers. Wang (**1**) [8]; HMPA (**2**) [9]; Sasrin (**3**) [10]; HMPB (**4**) [11].

by the linker. These problems can be overcome by anchoring the linker via an acid-stable alkyl ether. This can be achieved, for instance, by employing hydroxyethylpolystyrene as the base support, or by incorporating an alkylcarboxylic spacer into the base linker structure (Figure 14.2). The latter approach has been extensively adopted, as such linkers are carboxylic acids and so can be easily linked to amino-functionalized polymer supports via formation of an amide bond [12]. The acid sensitivity of such linkers can be fine-tuned by varying the length of the spacer between the electron-donating ether oxygen and the electron-withdrawing carbonyl group of the spacer. For example, protected peptide amides are cleaved twice as fast with 1% TFA in DCM from Barany's XAL linker with a 4-methylene spacer **5** than from the corresponding linker with only a single methylene spacer **6**. Ether and amide linked variants exist for most common linker types.

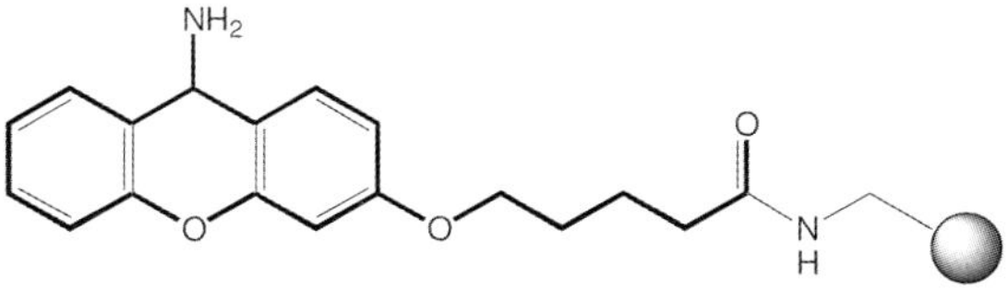

5

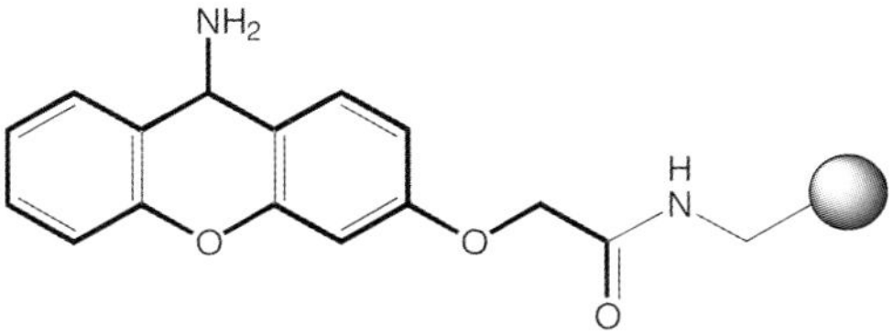

6

Linkers can be used in two distinct modes in solid phase synthesis. They can be used simply as polymer-supported protecting groups, in which case the functional groups loaded on and cleaved from the resin are the same. Alternatively, the linker performs the role of a reagent such that there is a functional group transformation between resin loading and cleavage. Classic examples are amide-forming linkers such as the XAL linkers **5**, **6** or Rink amide **7** [13], which is used to produce peptide amides by Fmoc solid phase peptide synthesis, where a carboxylic acid is loaded onto the resin but a carboxamide is released (Figure 14.3). Table 14.4 below shows commonly used linkers arranged by the functional group released on acidolysis.

14.2
Linker Types

14.2.1
Benzyl-Based Linkers

The simplest benzyl-based acid-labile linker is a polymer-supported benzyl group. It is most commonly encountered in the form of chloromethyl polystyrene (**8**),

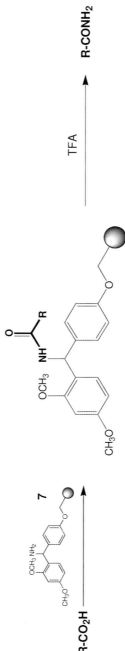

Figure 14.3 Rink amide linker **7** acting as an ammonia equivalent.

also known as Merrifield resin [14], which was developed for Boc/Bzl-based solid phase peptide synthesis.

Merrifield resin is most frequently used as a simple solid-phase benzyl protecting group and is loaded by displacement of the resin-bound chloride by the appropriate nucleophile (Figure 14.4). Carboxylic acids are attached most frequently by heating the resin in DMF with the appropriate caesium salt of the carboxylic acid in the presence of KI [15]. The addition of a phase transfer catalyst such as 18-crown-6 is often helpful [16]. Phenols may also be attached in a similar manner. Alcohols are attached using the corresponding alkoxide generated *in situ* with either NaH or KH [17, 18]. Amines can be loaded simply by heating the resin with excess reagent.

The esters derived from Merrifield resin are only cleaved by strong acids such as HF, TFMSA/TFA, TMSOTf/TFA, HBr/TFA and TMSBr/TFA (Table 14.4). Weaker acids like TFA may also be used but the reaction is sluggish and requires catalysis with reagents like soft nucleophiles such as thioanisole. Similar methods should also work for alcohols and phenols, whereas thiols and amines can not generally be cleaved from this resin by acidolysis.

In the Boc/Bzl method of solid phase synthesis, where the growing peptide chain is subjected to repeated treatments with TFA for Boc group removal, this lack of total stability to TFA has serious implications. Not only can it cause

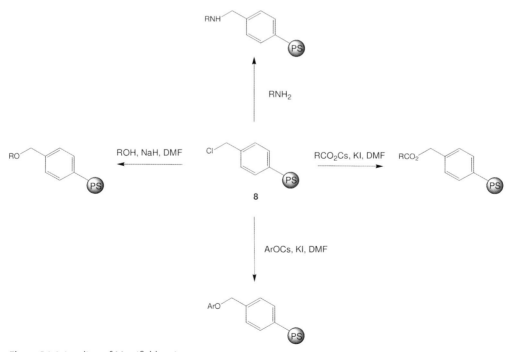

Figure 14.4 Loading of Merrifield resin.

leaching of product from the peptide support but, more seriously, chain termination by trifluoroacetylation [19]. This problem is overcome in the PAM linker **9** [20], a preformed variation of the simple benzyl ester linker incorporating an electron-withdrawing acetyl group. This spacer group not only increases the acid stability of the benzyl ester but also provides a means of attachment of the linker to the solid support. Esters derived from this linker are over 100× more stable to TFA/DCM (1:1) than those obtained from Merrifield-type resins.

9

The hazardous nature of HF, and the delirious effects of prolonged TFA exposure on peptide chains, has led several groups to investigate alternative protecting group strategies to Boc/Bzl for SPPS ([21] and references therein). These endeavors led to the development of several alkoxybenzyl-based linkers from which the products could be cleaved under milder conditions. The original was the *p*-alkoxybenzyl alcohol linker of Wang **1**, which was designed originally for the synthesis of protected peptide fragments in conjunction with N^α-Bpoc protection [8], and is produced from Merrifield resin by reaction with the *p*-hydroxybenzyl alcohol. Other variations of this linker, such as the previously mentioned HMPA (**2**) [9] and the analogous 3(4-hydroxymethylphenoxy)propionic acid linker [22], have been developed for Fmoc/*t*Bu-based SPPS incorporating alkyl carboxylic acid spacers to allow ready attachment to amino-functionalized polymers.

Esters, ethers and hydroxamates derived from these linkers can be cleaved with mild acids such as TFA or HCl in dioxane (Table 14.4). If TFA is used for release of alcohols, trifluoroacetates can be formed [2], but these can be easily hydrolyzed later with aqueous base. Simple amines cannot be cleaved from this type of linker by acidolysis. Thiols may be cleaved with HF.

Carboxylic acids can be attached to these linkers using methods of ester bond formation such as carbodiimide/DMAP [23] and acid chloride/base. For the loading of N-protected-α-amino acids in particular, an array of different methods has been developed to minimize enantiomerization and dipeptide formation during the esterification reaction. These include the use of MSNT/N-methylimidazole [24], mixed anhydrides generated with 2,6-dichlorobenzoyl chloride [25], esters of 2,5-diphenyl-2,3-dihydro-3-oxo-4-hydroxythiophene [26] and acid fluorides [27]. Phenols and N-protected hydroxylamines have been immobilized using the Mitsunobu reaction [28, 29]. The latter are particularly useful for the preparation of hydroxamates [29, 30].

Reaction with trichloroacetonitrile and DBU can also be used to generate a resin-bound trichloroacetimidate (**10**), which can be used to immobilize alcohols

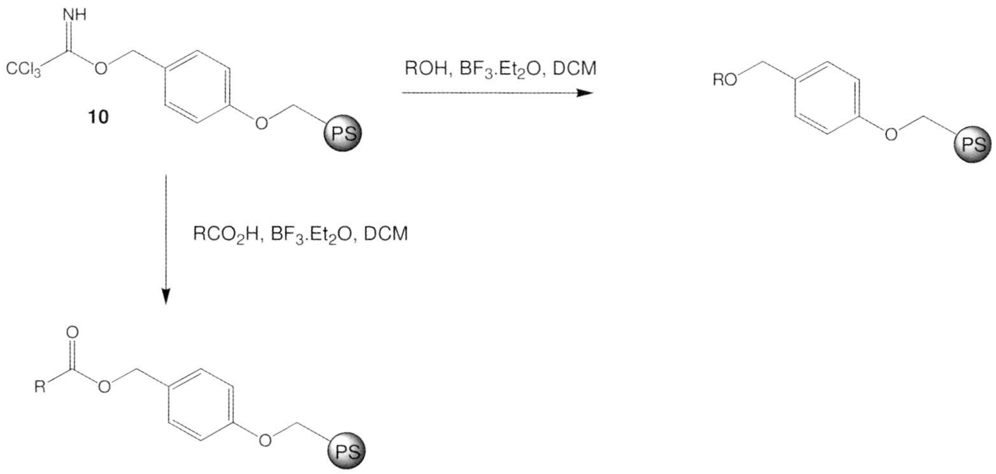

Figure 14.5 Loading of alcohols and acid onto trichloroacetimidate Wang resin.

[31], carboxylic, phosphonic, sulfonic and thioacids under acidic conditions (Figure 14.5) [32]. Furthermore, the hydroxyl group can be converted into halogen [33–36] or sulfonic ester derivatives [35, 36], allowing attachment of functionalities such as amines, thiols, phenols and alcohols by nucleophilic displacement in a similar manner as previously described for Merrifield resin. The bromide derived from Wang resin is particularly useful as it has good reactivity but is sufficiently stable for long-term storage [34, 36]. In addition, N-protected-α-amino acids can be coupled to this resin in the presence of base at room temperature [37], presumably without loss of optical integrity [38].

Amines derived from *p*-alkoxybenzyl-type linkers, despite not being acid cleavable, still have synthetic utility. Anilines anchored to Wang resin, once converted into carboxamides or sulfonamides by reaction with the appropriate electrophile, can be cleaved with TFA (Figure 14.6) [34]. Sulfonamides of aliphatic amines may also be cleaved with TFA [36]. Stronger acids are generally required for acyl derivatives of amines but this can cause cleavage of the linker benzylic ether bond, leading to formation of *p*-hydroxybenzylated by-products.

For the reversible immobilization of amines to such linkers, the most frequently used strategy is conversion of the hydroxymethyl linkers into activated carbonic esters (Figure 14.7) [8, 39, 40]. These react with primary and secondary amines or hydrazines to form immobilized carbamates or carbazates analogous to the classical benzyloxycarbonyl protecting group. Amines and hydrazides attached by this strategy to Wang resin can be cleaved with TFA [8, 40].

The addition of further *o*-methoxy groups to the *p*-alkoxybenzyl alcohol framework naturally results in linkers with enhanced acid sensitivities. One of the first linkers of this type was 2-(4-hydroxy-3-methoxyphenoxy)acetic acid (**11**) developed by Sheppard and colleagues for the Fmoc solid phase synthesis of protected peptide acids for fragment condensation [41]. Treatment of peptidyl resins with

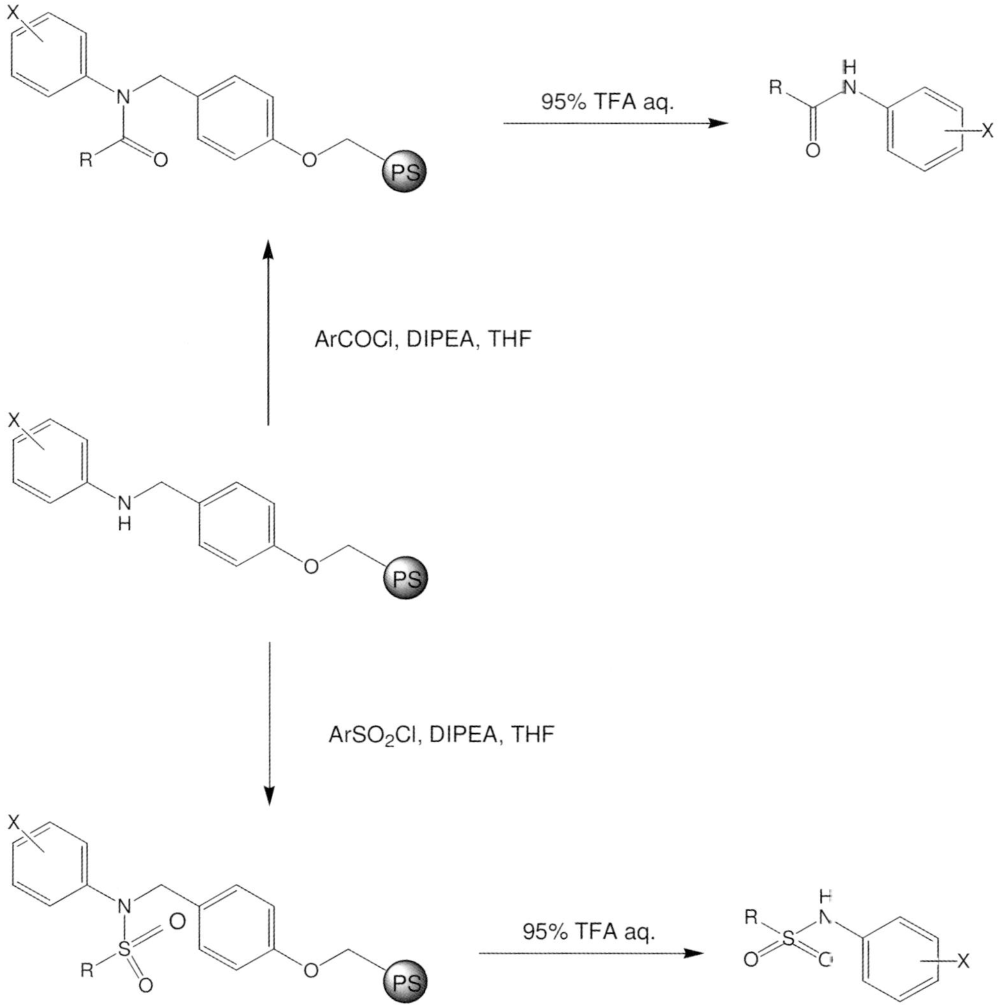

Figure 14.6 Synthesis of anilides on Wang resin.

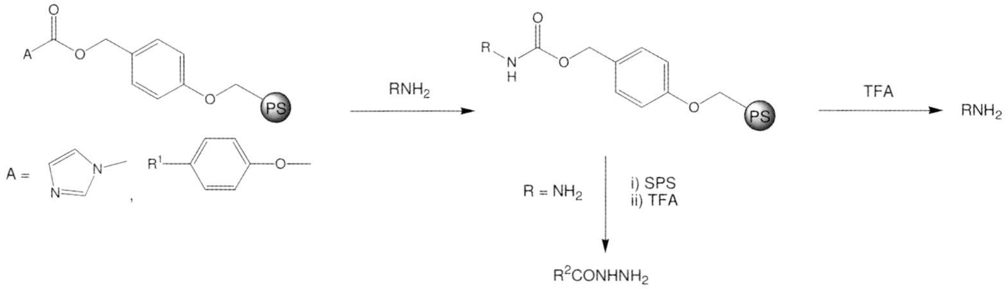

Figure 14.7 Solid phase synthesis (SPS) using immobilized carbamates and carbazates on Wang resin.

1% TFA in DCM cleaved the peptide acid from the resin whilst retaining tBu/Boc-based side-chain protection. Riniker [11] later developed a version of this linker with enhanced acid sensitivity by insertion of two extra methylene groups into the linker spacer **4**. The Sasrin linker **3**, developed by Mergler [10], is also based on the same dialkoxybenzyl alcohol structure, but in this case the linker is attached directly to Merrifield resin via an ether bond.

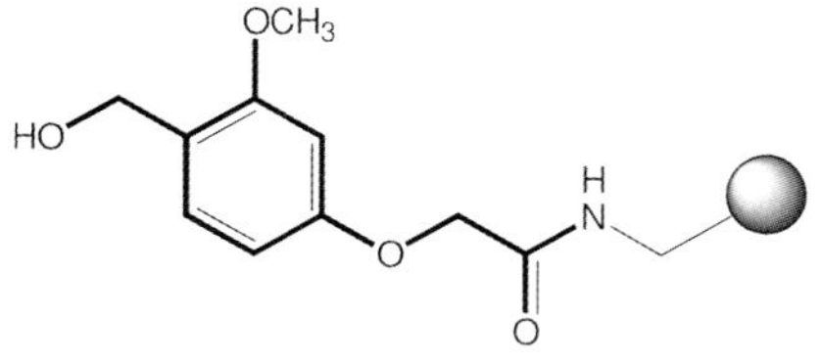

11

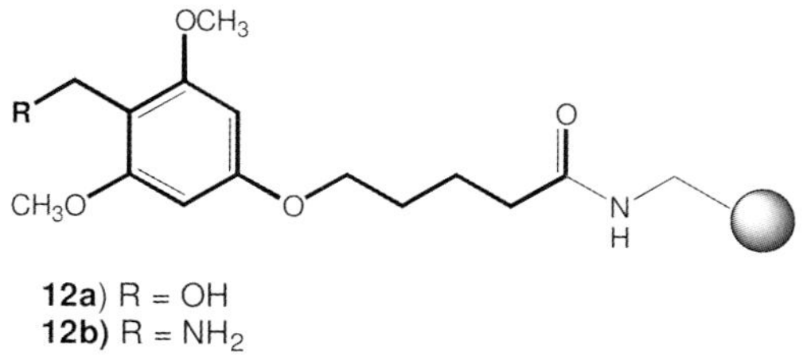

12a) R = OH
12b) R = NH$_2$

12

These linkers can also be converted into bromo and chloro derivatives as previously described for *p*-hydroxymethylphenoxy linkers **1** and **2** [33] and used in similar applications [38, 42].

The most acid labile linker in the series, the HAL linker **12a**, was described by Albericio and Barany [43]. Here, the linker ester bond is extremely easily cleaved with dilute acids, even being labile to AcOH. The amine form, known as the PAL linker **12b**, is one of the standard linkers for the synthesis of peptide amides [44]. Acylation of this amino functionality under standard coupling conditions results in a polymer-supported secondary amide. Owing to the high stability of the nascent trialkoxybenzyl cation, treatment with moderately strong acids like TFA is sufficient to cleave the linker-amide C–N bond, liberating the primary carboxamide, in an identical manner to that previously described for Rink amide linker **7** in Figure 14.3.

One important variation of this linker is Barany's BAL linker **13**, which contains a formyl group [45–51]. Primary amines [45] and O-protected hydroxylamines [52] are attached to this linker by reductive amination. The resulting secondary amines can be acylated and the corresponding secondary amides or hydroxamates cleaved with TFA. Anilines [53] and enamines [54] can be cleaved without needing to be acylated.

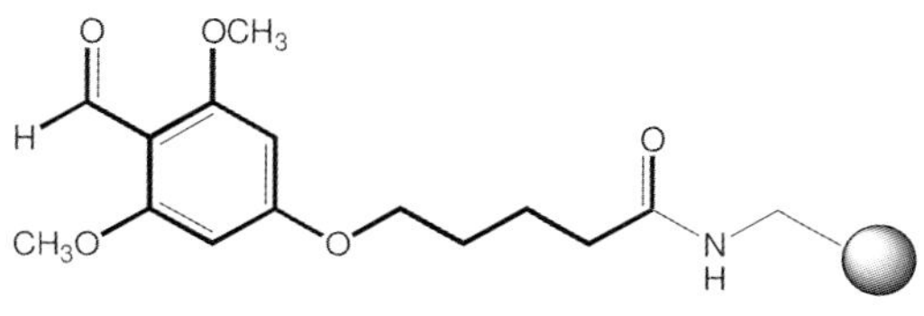

13

14 Acid-Labile Resins

Barany and coworkers have exploited this approach to provide peptides bearing a wide range of C-terminal modifications, including esters [45], N-alkylamides [45], alcohols [45], aldehydes [45, 49, 51] and hydrazides [48] (Figure 14.8). By employing amino acid allyl esters, selective deprotection on the solid phase provided a route to cyclic peptides [45] and base-sensitive C-terminal peptide modifications, such as thioesters and *p*-nitroanilides (Figure 14.9) [47]. Variants of the BAL linker have been described where the core 4-formyl-3,5-dimethoxyphenoxy group has been attached directly **14** [55] or through an ethoxy spacer **15** [56] to Merrifield resin, or anchored to bromoethylpolystyrene **16** [57].

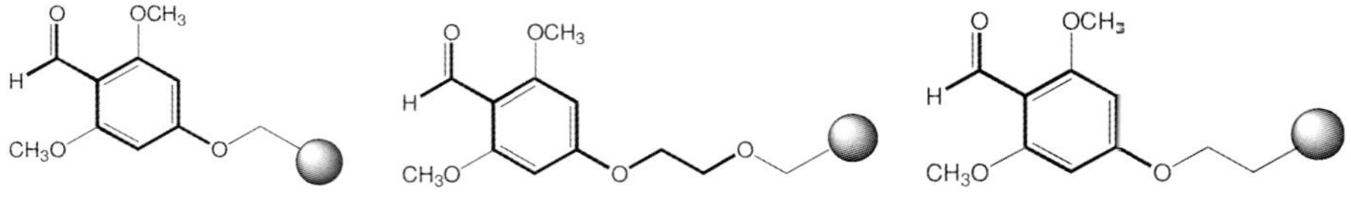

14   **15**   **16**

When the resin-leaving group is a secondary amide or similar derivative, 3-dialkoxybenzyl-based linkers are sufficiently electron rich to allow product release

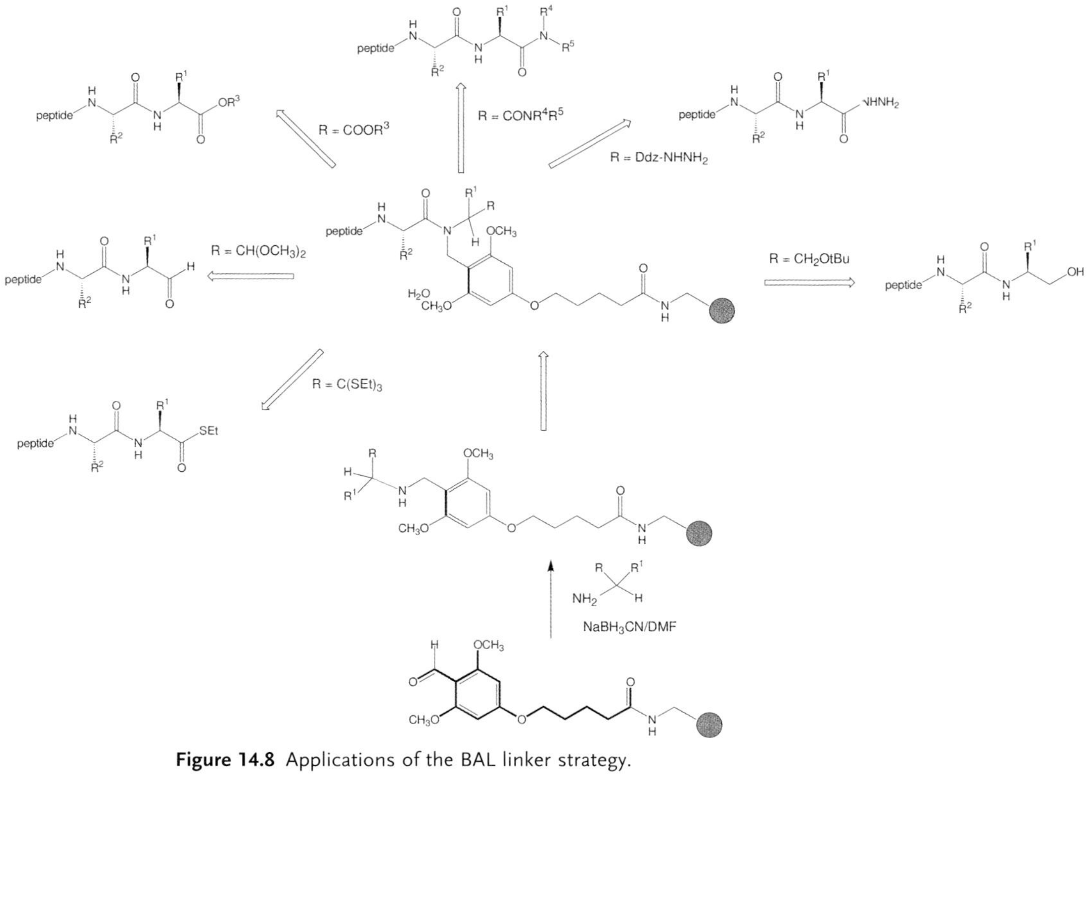

Figure 14.8 Applications of the BAL linker strategy.

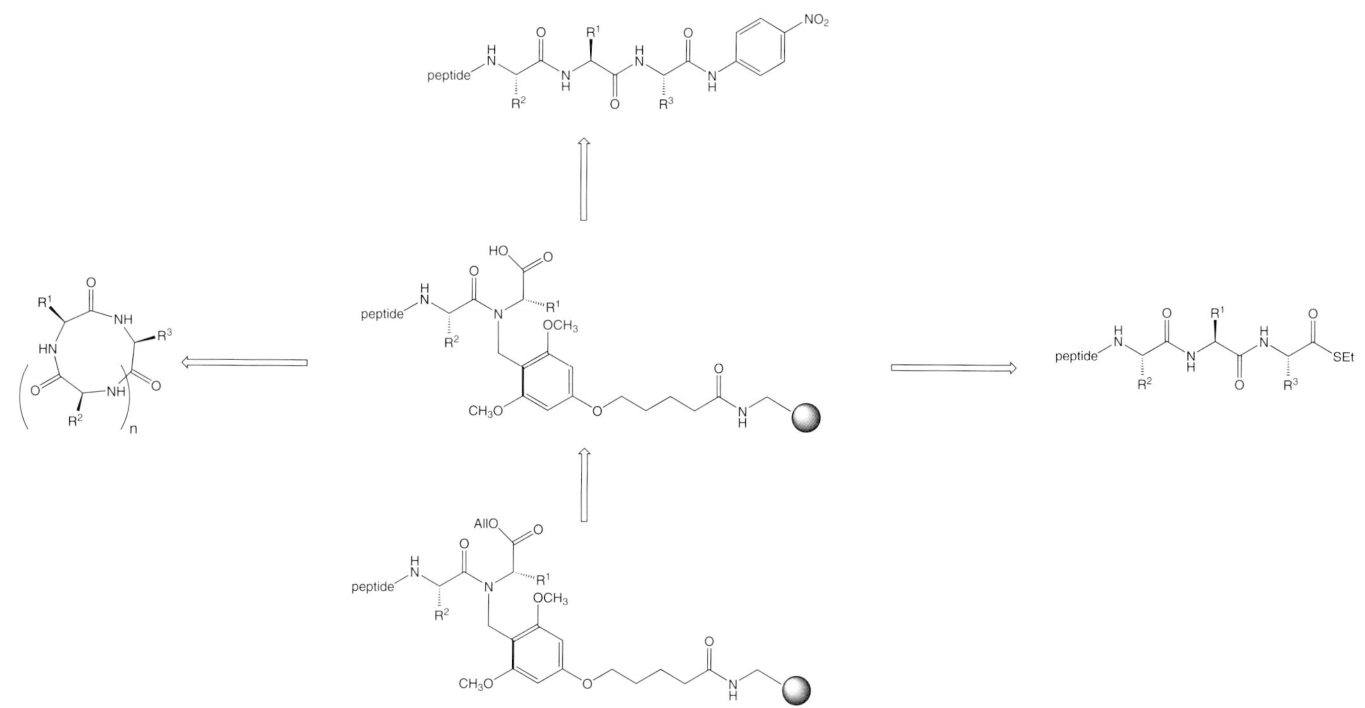

Figure 14.9 Synthesis of cyclic peptides, thioesters and *p*-nitroanilides using the BAL linker.

with TFA. Consequently, numerous formyl linkers have been developed for use in solid phase organic synthesis to prepare secondary carboxamides, sulfonamides, ureas, carbamates and nitrogen-containing heterocycles (Figure 14.10) [58], and peptide ester, alcohols, aldehydes, and N-alkylamides [59].

In a study that evaluated various formyl resins in the synthesis of furanopyrazines, FMPB resin **18** was found to perform the best. This was ascribed to its enhanced acid stability and low steric hindrance compared to BAL-type linkers [60].

One interesting variation of this kind of linker is the formylindole linker of Estep (**20**) [61]. It is used in exactly the same manner as other formylalkoxybenzyl-type linkers and has properties similar to the BAL linker. The low steric hindrance of the indole group compared to the o-alkoxybenzyl moiety means acylation of secondary amines prepared on the former linker is easier.

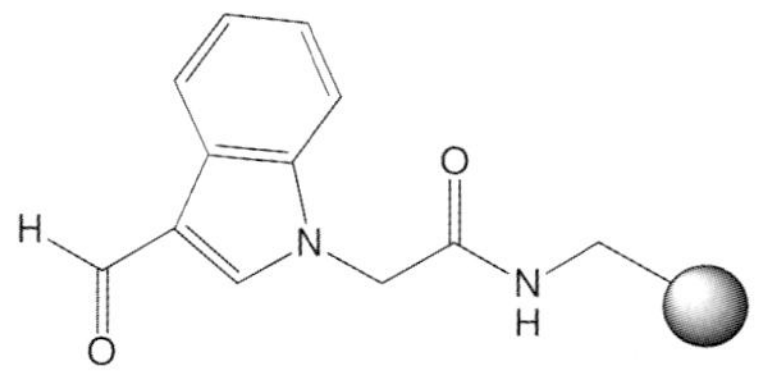

20

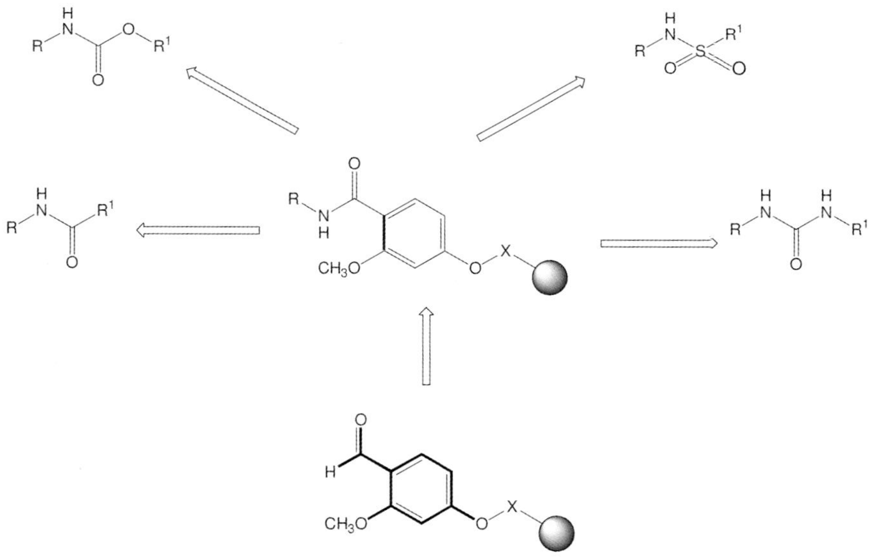

17: X = CH$_2$
18: X = (CH$_2$)$_3$CONHCH$_2$
19: X = (CH$_2$)$_2$

Figure 14.10 Application of 4-formyl-3-methoxyphenoxy linkers.

14.2.2
Benzhydryl-Based Linkers

The simplest benzhydryl linker is the integral benzhydrylamine (BHA) (**21**) developed by Marshall for the preparation of peptide amides by Boc SPPS [62]. Matsueda enhanced the acid sensitivity of the linker by introduction of a *p*-methyl group (MBHA, **22**) [63]. Nevertheless, both linkers require HF for primary amide release, although in the case of MBHA cleavage can also be effected with TFMSA and HBr/TFA. Diphenyldiazomethane resin **23** [64] derived from the keto intermediate used in the preparation of BHA resin reacts cleanly with carboxylic acids to form supported esters (Figure 14.11). Alcohols can also be loaded to this resin in the presence of a Lewis acid. During the loading process, the resin loses its purple color, providing a visual indication of completeness of reaction. Both alcohols and acids can be released with TFA. Carboxylic acids attached to the hydroxy form of the *p*-methylbenzhydryl linker can also be cleaved with TFA.

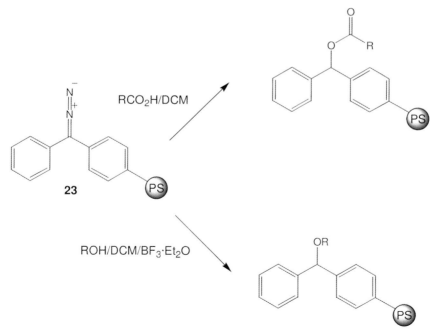

Figure 14.11 Use of diphenyldiazomethane (**23**) resin.

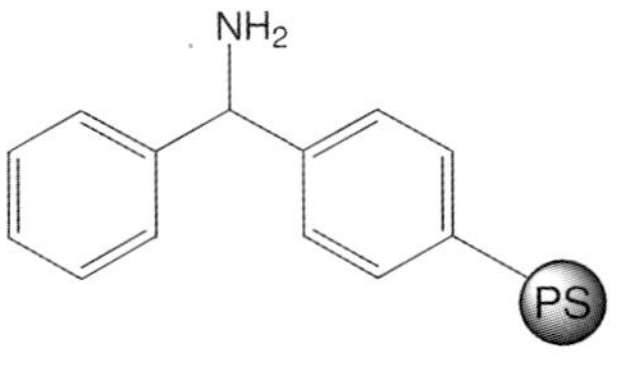

21, 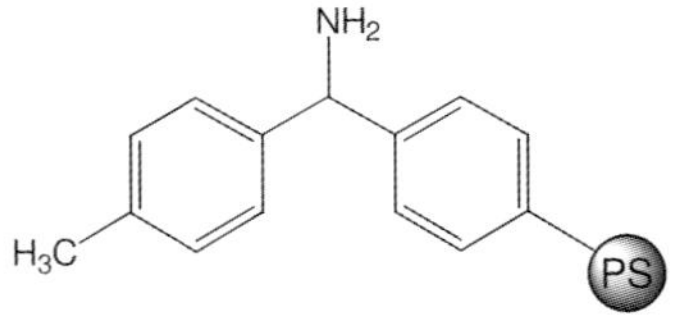 **22**

A wide range of benzhydryl-based linkers bearing additional electron-donating functionalities has also been developed for Fmoc SPPS of peptide amides, the most popular of which are the eponymous Rink amide (**7**) [13] and Knorr linkers **24** [65]. Primary amides prepared on this resin can be cleaved with TFA in the same manner as previously described for the PAL linker.

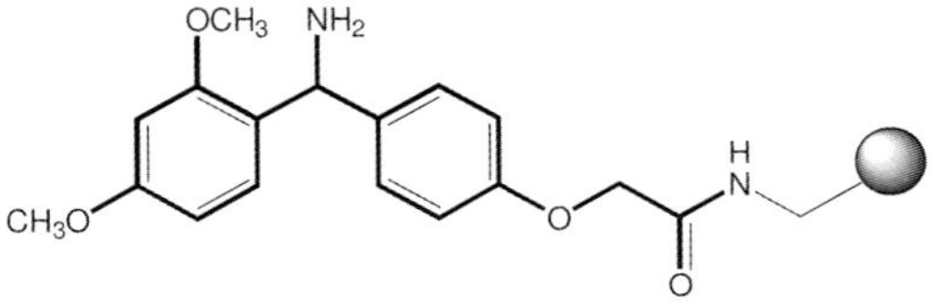

24

The hydroxyl version of the Rink amide linker, known as the Rink acid resin (**25**), was developed as a tool for the preparation of protected peptide fragments [13]. The peptide-linker ester bond is labile to extremely weak acids, such as HOBt or acetic acid, allowing peptides bearing t-butyl-based side-chain protection to be cleaved intact. Conversion of the hydroxyl group into chloride [66] or trifluoroacetyl [67] provides linkers that have been used for immobilization of various nucleophiles, including alcohols, N-protected hydroxylamines, phenols, purines, amines, anilines and thiols [66–68]. The stability of the cation derived from this linker is such that even thiols and amines can be cleaved from this linker with TFA (Figure 14.12).

Less electron-rich benzhydryl-based systems such as the MAMP linker **26** have been used for the preparation of secondary amides, following immobilization of primary amines and acylation [69]. Products are released by treatment with TFA. The hydroxyl version of the same linker system has been used to immobilize Fmoc-amino acids for SPPS. Release of products was effected using 1% TFA [70].

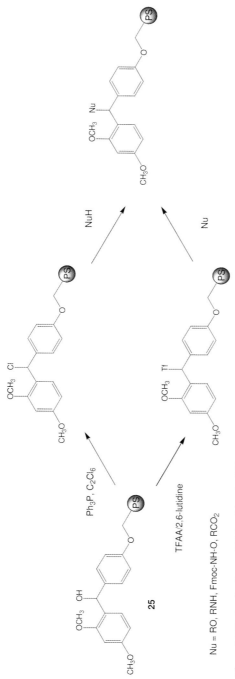

Nu = RO, RNH, Fmoc-NH-O, RCO$_2$

Figure 14.12 Applications of Rink acid resin (**25**).

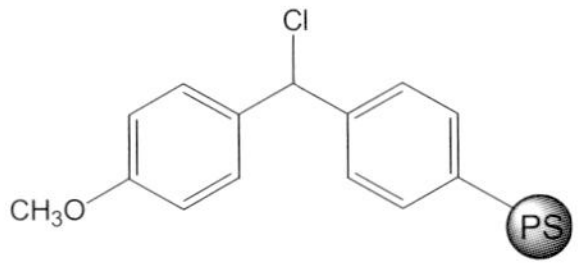

26

The α-mercaptobenzhydryl linker **27** [71] was developed for the synthesis of thioacids by Boc synthesis. Acylation of the thiol group with Boc-amino acids leads to a resin-bound thioester. Subsequent peptide chain extension and then treatment with HF cleaves the ester-linker bond, releasing the C-terminal peptide thioacid. Unfortunately, owing to the nucleophile lability of thioesters, this approach cannot be applied to Fmoc-based methods.

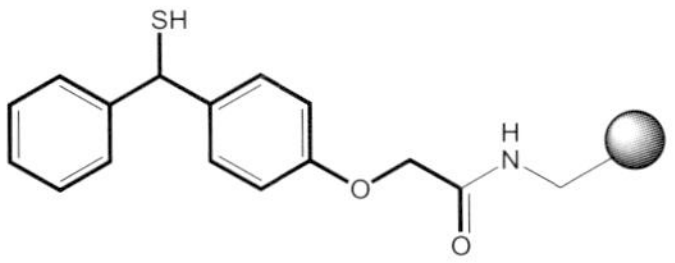

27

14.2.3
Trityl-Based Linkers

The use of trityl-based linkers in solid phase synthesis was pioneered by Leznoff [72] and Fréchet [73] in the 1960s and 1970s. A remarkable array of trityl-based linkers of differing acid sensitivities has been developed over the years (Table 14.3). They can be used to protect and immobilize almost any heteroatom-containing functional group, including amines, since the bulk of the trityl group prevents alkylation or acylation of the amine nitrogen. Nucleophilic functionalities like acids, alcohols, phenols, thiols, imidazoles, guanidines, anilines and amines are loaded via an S_N1 reaction with the appropriate polymer-supported trityl chloride in the presence of base, whilst amides are attached under acid-catalyzed conditions by reaction with the appropriate trityl alcohol. In all cases, product release can be achieved by treatment with varying concentrations of TFA in DCM, with thiols, guanidines and amides being the hardest functionalities to cleave.

Of those in common usage, the 2-chlorotrityl **28** [74] and 4-carboxytrityl **29** [75] linkers give the least stabilized cations and are suitable for immobilization of carboxylic acids. They are ideal for Fmoc/tBu-based SPPS as their use avoids many of the side reactions that occur with standard benzyl-based linkers. Firstly, racemization does not occur during loading of the resin with the C-terminal residue [79], as is the case with esterification to hydroxy-functionalized resins. Secondly, the bulky trityl cation does not cause alkylation side reactions with nucleophilic amino-acid side-chains. Thirdly, cysteine does not undergo racemization [80, 81]

Table 14.3 Selection of trityl linkers ordered according to increasing acid sensitivity. Ease of functional group release is indicated: 1: too labile; 2: <5% TFA in DCM; 3: >5% TFA in DCM; 4: >50% TFA in DCM; 5: not cleaved.

Amine: 2	Amine: 2	Amine: 1	Amine: 1	Amine: 1
Alcohol: 2	Alcohol: 2	Alcohol: 2	Alcohol: 1	Alcohol: 1
Carboxylic acid: 2	Carboxylic acid: 1	Carboxylic acid: 1	Carboxylic acid: 1	Carboxylic acid: 1
Guanidine: 4	Guanidine: 4	Guanidine: 4	Guanidine: 3	Guanidine: nd
Amide: 5	Amide: 4	Amide: 4	Amide: 3	Amide: 2
Thiol: 4	Thiol: 4	Thiol: 3	Thiol: 2	Thiol: nd

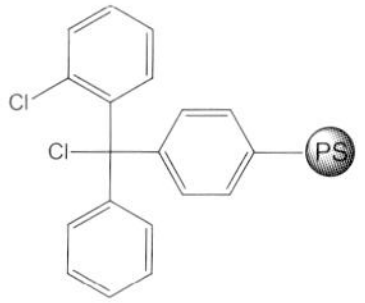

28 [74]

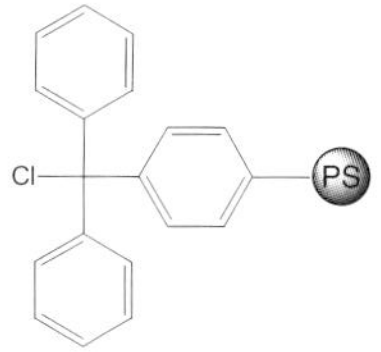

31

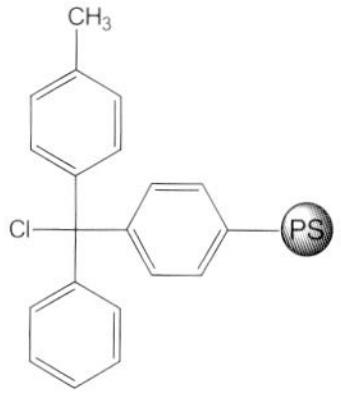

32

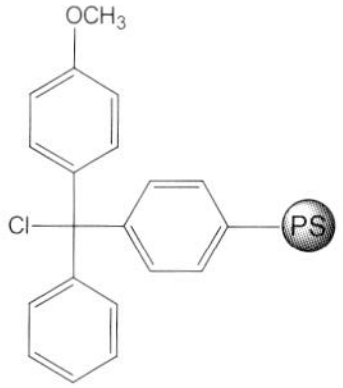

33

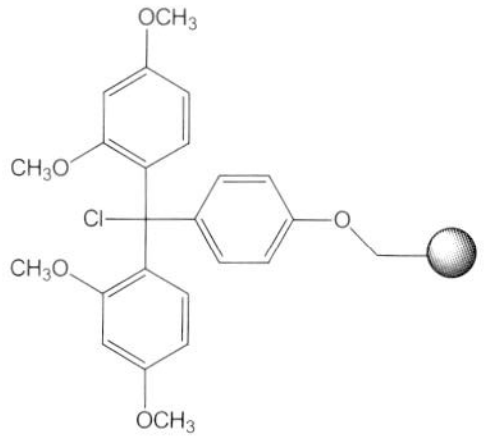

35 [78]

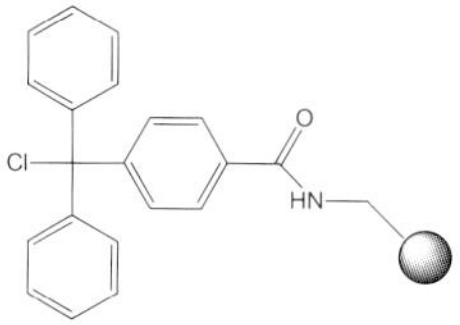

29 [75]

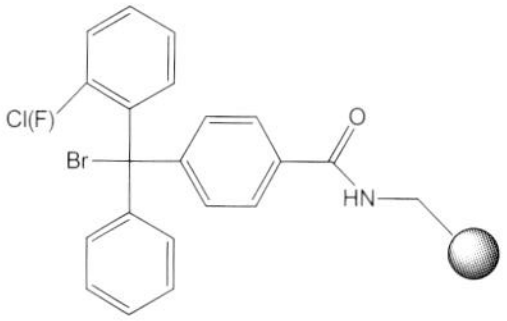

34 [77]

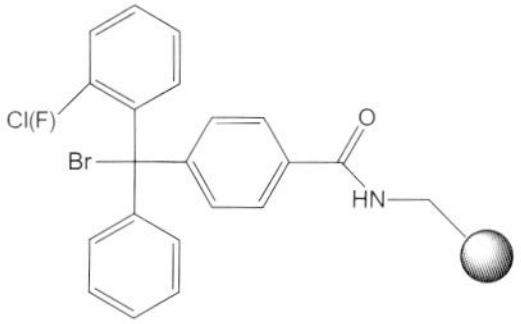

30 [76]

or β-piperidinylalanine [82] formation during chain extension, as is the case when loaded as the C-terminal residue on Wang and similar resins. Finally, dipeptides containing proline do not undergo diketopiperazine formation [83]. A further benefit of such linkers is that the resin-ester bond is easily cleaved with mild acidic reagents such as TFE/DCM or HFIP/DCM, providing fully protected peptide fragments [84, 85].

Trityl amine [78], -thiol, -hydrazine [86] and -hydroxylamine [87] linkers are sufficiently nucleophilic to be acylated or alkylated. Such supports are useful tools for the synthesis of amides, thiols, hydrazides and hydroxamates.

14.2.4
Cyclic Linkers

Several linkers have been developed that rely on the formation of highly stabilized aromatic carbocations. The most frequently used are the eponymous Sieber amide linker **36** [3] and Barany's 3-XAL linker **6** [4]. Both are based on a 3-methoxyxanthine scaffold, which owing to the highly stabilized nature of the xanthenium ion can provide primary amides on treatment with 1% TFA in DCM, making them excellent tools for the synthesis of protected peptide carboxamides. The Sieber amide resin has also been used to prepare secondary amides via reductive alkylation of the amino group, acylation of the resultant amine and cleavage with dilute TFA [88]. Brill et al. [67] have effected transamination of trifluoroacetylated Sieber amide resin in good yield. This approach offers considerable potential for the immobilization of amines on this support.

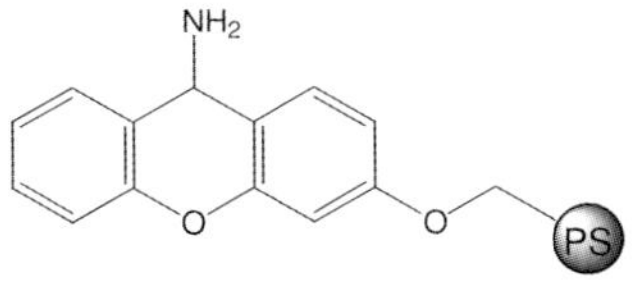

36

Ramage et al. [89, 90] have reported TFA-labile linkers for the synthesis of peptide amides **37**, hydrazides **38** and semicarbazones **39** based on dibenzocycloheptadiene scaffold.

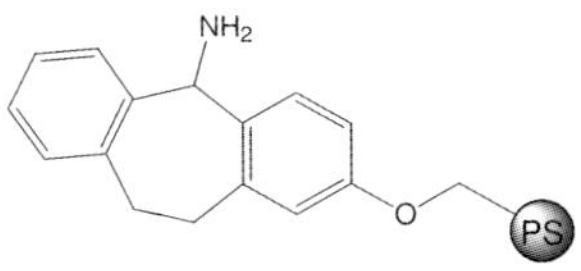

37

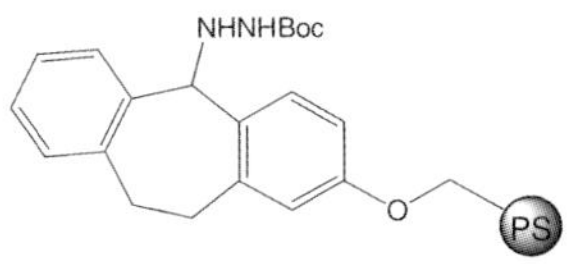

38

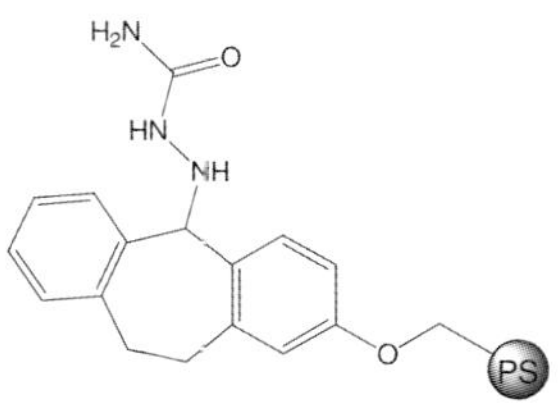

39

Similar linkers based on the dibenzocycloheptadiene **40** and dibenzocycloheptatriene **41** structures have also been described by Noda *et al.* [91]. Primary amides could be cleaved from **40** and **41** with 25% TFA with a half-life of 4 and 2 min, respectively.

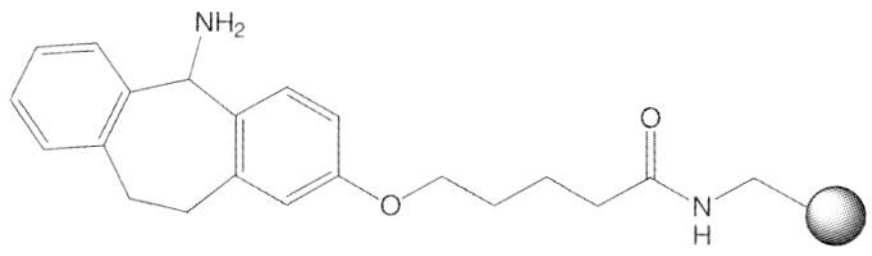

40

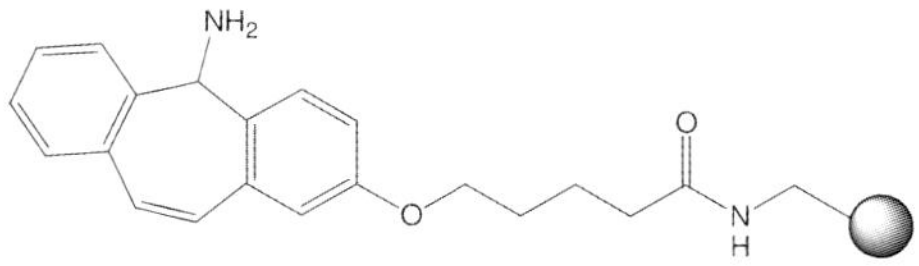

41

14.2.5
Silyl-Based Linkers

Several acid-labile linker systems have been developed that rely on the property of silicon to stabilize the formation of cations on a carbon two atoms away. For instance, linkers **42** [92] and **43** [93] release acids and amides upon treatment with TFA, respectively.

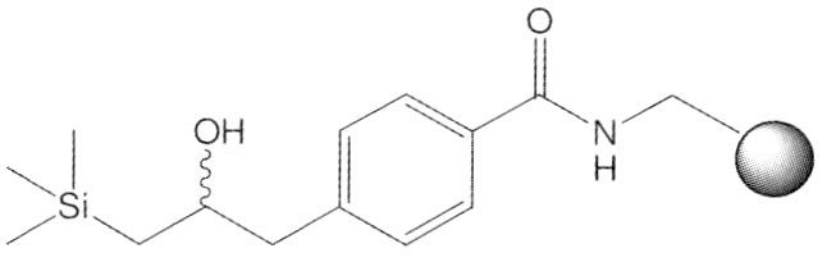

42

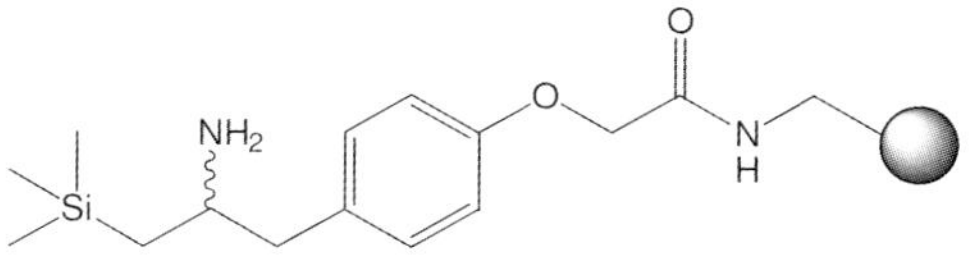

43

Protiodesilylation has been exploited by many researchers as a mechanism for releasing arenes directly from the linker by cleavage of a C–Si bond (Figure 14.13) [94–99].

The approach generally only works well for electron-rich arenes, although for electron-deficient ones fluoride ion mediated cleavage can be used. Resins **44** [98] and **45** [99] are commercially available. They can be converted into the reactive halide or triflate form by treatment with HCl in ether or TFMSA/DCM. Attach-

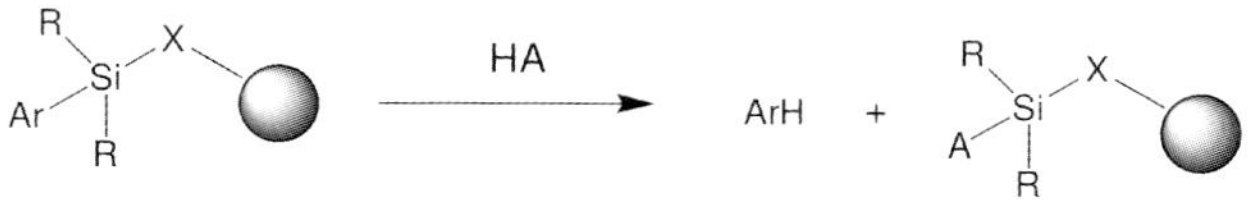

Figure 14.13 Design template for linkers cleaved by protiodesilylation; X is a spacer unit; R is alkyl.

ment of the arene is usually achieved by displacement of halide or triflate from the silyl linker with the appropriate aryl carbanion.

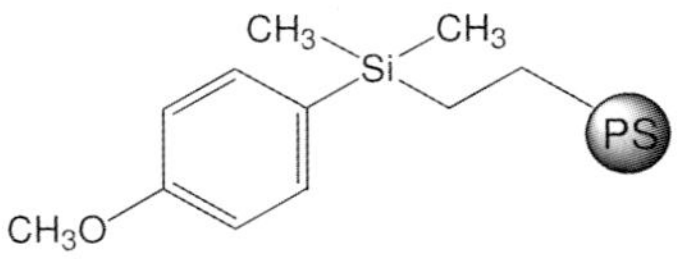

44

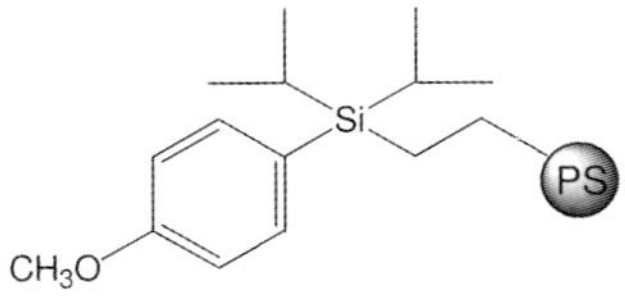

45

14.2.6 Acetal/Aminal-Type Linkers

Linkers of this type fall into two categories, depending on whether it is the carbonyl or alcohol/amine component of the acetal/aminal that is tethered to the solid phase. A classic example of the former type is the DHP linker **46** (Figure 14.14) [100]. This linker has been used extensively for the solid phase immobilization of primary and secondary alcohols [100–102], but also works well for phenols [103], indoles [104], purines [105] and tetrazoles [106]. Loading is typically achieved by treating the resin with excess substrate in the presence of pyridinium *p*-toluenesulfonate (PPTS) or *p*-toluenesulfonic acid (pTsOH). Hydroxylic compounds are generally cleaved from this linker with acidic cocktails containing water or alcohols: 95% TFA aq., TFA/DCM/MeOH, *p*-TsOH/*n*-BuOH/DCE, PPTS/*n*-BuOH/DCE. The use of TFA can occasionally lead to formation of the trifluoroacetate. For indoles, tetrazoles and purines the addition of hydroxylic scavengers to the cleavage mixture is not required, although in some cases it leads to higher yields.

For diols, aldehyde linkers of the general structure **47** have been used [107–109]. Cleavage can be effected with TFA mixtures containing hydroxylic scavengers (Figure 14.15).

For the solid phase synthesis of aldehydes and ketones, various linkers based on diols **48** have been used (Figure 14.16) [110–112]. Loading of this type linker with carbonyl compound is generally achieved by treatment of the resin with pTsOH in benzene with azeotropic removal of water. Products are released upon treatment with HCl or pTsOH in aqueous dioxane.

An interesting recent development has been the threonine-based 1,2-aminoalcohol linker **49** [113, 114]. Fmoc-protected amino aldehydes can be attached to this linker via formation of an oxazolidine. Following peptide synthesis, the

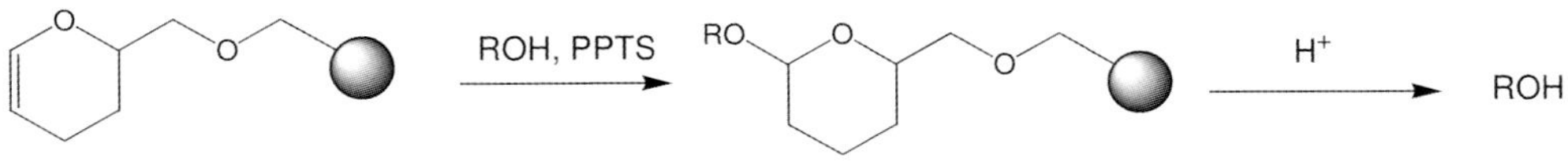

46

Figure 14.14 SPS of alcohols using the DHP linker.

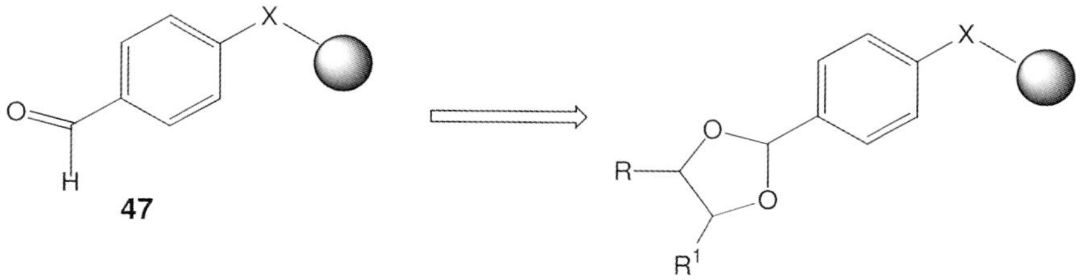

Figure 14.15 Immobilization of diols using polymer-supported aldehydes.

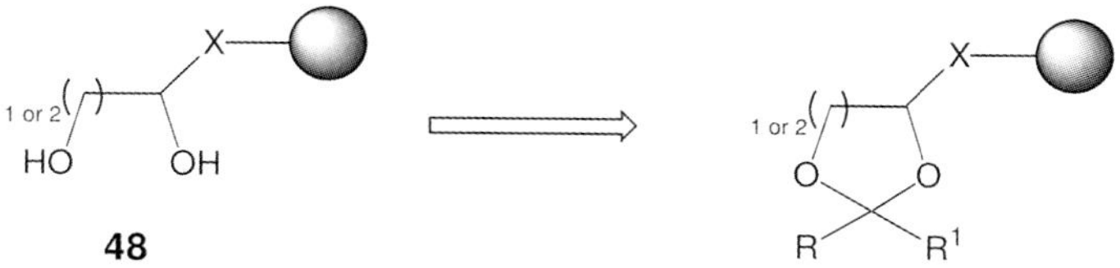

Figure 14.16 Immobilization of carbonyl compounds using polymer-supported diols.

side-chain protecting groups can be removed with the peptide still attached to the resin by treatment with TFA, before the desired peptide aldehyde is released with MeCN/water/TFA 60:40:0.1 (Figure 14.17).

14.3
The Cleavage Reaction

The choice of acid used for releasing the product from the solid support is largely determined by the nature of the linker used, the immobilized functional group and the experiment design. The acids selected to effect product cleavage are usually volatile liquids for ease of removal, such as anhydrous HF, TFA, TMSBr/TFA, TMSOTf/TFA and HBr/TFA. As a general guide, functionalities attached to benzyl-based linkers without alkoxy substituents tend to require strong acid-like anhydrous HF or TFMSA for release, whereas with trityl, alkoxy-substituted benzyl and benzhydryl-based systems weaker acids like TFA or even AcOH can be used. The requirements for commonly encountered functional group linker combinations are given in Section 14.4.

As previously mentioned, product cleavage generally involves an S_N1 reaction that generates a highly reactive cation on the solid phase. This cation can potentially react with nucleophilic functionalities in the target molecule, leading to reattachment of the product to the solid phase. Depending on the nature of linker and nucleophile, the reaction can be irreversible and in such situations will lead to reduced product yield. Benzyl and benzhydryl linkers are most problematic in this regard. The nucleophiles most at risk are electron-rich arenes (e.g. pyrroles, indoles, anisoles, phenols), thiols and thioethers. With oxygen and nitrogen nucleophiles, the reactions are either reversible or are inhibited by protonation, and so are not of concern. To avoid such problems, it is normal practice to add

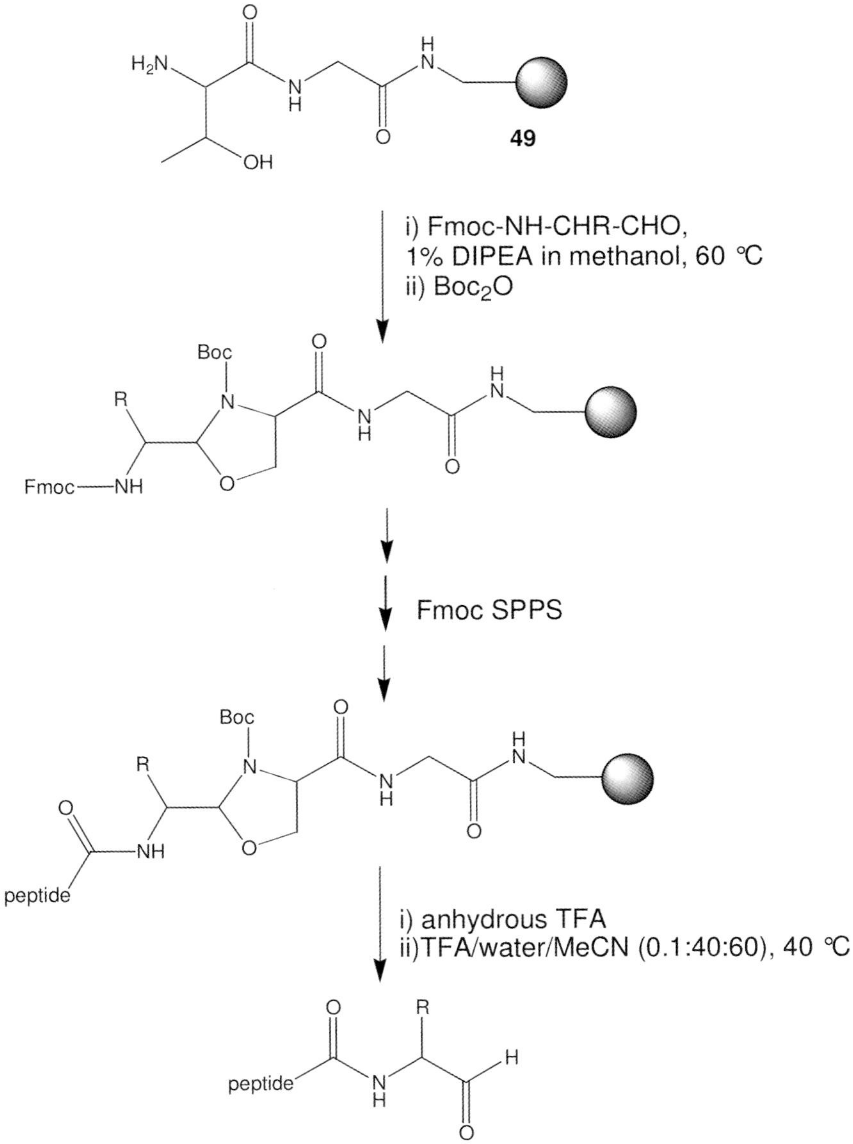

Figure 14.17 Synthesis of peptide aldehydes on an oxazolidine linker.

additional reagents to the cleavage to scavenge these reactive cations. The most effective scavengers are sulfur-based compounds, since they have a high affinity for carbon electrophiles, their reactions are generally irreversible and their low basicity (pK_a of −7 to −9) [115]. This means that they are compatible with strong acids like HF and TMFSA. The most frequently used are ethanedithiol [116] and dimethyl sulfide [117]. Thioanisole is a useful scavenger for catalyzing slow cleavage reactions, which by their nature have more S_N2 character and so can be accelerated by a push–pull mechanism [118].

Water is a commonly used scavenger with TFA-based cleavage cocktails as it is volatile and non-toxic; however, it should not be used with strong or hydrolysable acids like TFMSA or TMSBr. Furthermore, in some circumstances, its use may result in product hydrolysis, for example hydroxamates to carboxylic acids [2].

The use of silanes such as triethylsilane (TES) and triisopropylsilane (TIPS) has found favor for use in TFA-based cleavage cocktails [119]. They are particularly effective scavengers for linkers that give highly stabilized cations, such as trityl and alkoxylbenzylbenzhydryl. This is because with standard nucleophilic scavengers the reactions tend to be reversible, whereas with silanes the cation is reduced irreversibly to the parent hydrocarbon. Whilst TES is a better scavenger than TIPS, the use of latter is generally preferred as it gives less side reactions owing to its greater bulk. Typical side reactions of silanes include conversion of indoles into indolines and reduction of those substrates that can form stabilized cations in TFA, such as tertiary alcohols and electron-rich aldehydes and ketones.

For small molecule synthesis, linker-functional group combinations that are labile to TFA tend to be selected. In the absence of nucleophilic functionalities such as those described above within the target molecule, simple mixtures of TFA and dichloromethane tend to be used for release of the product from the support as this mixture is volatile and so makes work up of the reaction a simple matter of evaporation.

For peptides, the use of scavengers is almost mandatory, owing to the range of potentially reactive functionalities within the peptide and the large number of cation-generating protecting groups employed for masking of amino-acid sidechains. With peptides made by the t-Boc SPPS strategy, very strong acids, such as HF or TFMSA, are required to cleave the product from the resin and for side-chain deprotection. HF cleavage is generally the preferred approach as this method gives products of higher purity and excess HF can be easily removed by evaporation. Nevertheless, careful control of reaction and temperature is necessary to avoid side reactions. For peptides that do not have problematic amino acid residues, such as tryptophan, methionine and cysteine, mixtures of HF/anisole can be used. However, when combinations of these amino acids are present procedures such as Tam's low-high HF [121] are preferred. This involves treating the peptidyl resin with HF/DMS/p-cresol/p-thiocresol at 0 °C. Under these conditions, the mechanism of the reaction becomes predominately S_N2, thereby reducing by-products arising from alkylation side reactions. Once the reaction is complete, the HF and DMS are evaporated and the reaction vessel is recharged with HF so that the final HF concentration is 90% to allow cleavage to proceed under high HF conditions. Boc SPPS deprotection procedures are nicely reviewed in Reference [115].

For cleavage from the resin and side-chain deprotection of peptide made by the Fmoc methodology, originally complex cocktails of scavenger containing a wide variety of scavengers were advocated, such as Reagent R [122] or Reagent K [123]. With the introduction of Pmc [124] and Pbf [125] side-chain protection for arginine residues and N^{in}-Boc protection for tryptophan [126], the need for such complex mixtures has been largely eliminated. In practice, simple non-odorous mixtures such as TFA/TIPS/water are used to cleave the vast majority of peptides produced by Fmoc SPPS. For an historical perspective of the cleavage cocktails employed in Fmoc SPPS see Reference [21].

14.4
Functional Group and Linker Combinations

Table 14.4 shows commonly used linker and functional group combinations. Notably, the cleavage conditions provided are mostly taken from the literature and

Table 14.4 Cleavage conditions for common linker and functional group combinations.

Released	Attached	Linker	Cleavage conditions
RCO$_2$H	RCO$_2$H		HF, TFMSA/TFA, TMSOTf/TFA, HBr/TFA [115, 127]
			HF, TFMSA/TFA, TMSOTf/TFA, HBr/TFA [115, 127]
			95% TFA [9, 70]
			1% TFA in DCM [10, 41, 70]
			AcOH/DCM/TFE (1:2:7) [85]; 0.1% TFA in DCM [43]; HFIP/DCM (3:7), TFE/DCM (2:8) [85] (not tested with HAL and Rink acid linkers)

14.4 Functional Group and Linker Combinations

Table 14.4 Continued.

Released	Attached	Linker	Cleavage conditions
ROH/ArOH	ROH/ArOH	(dihydropyran-based linker)	95% TFA aq. [100], TFA/DCM/MeOH, p-TsOH/n-BuOH/DCE [100], PPTS/n-BuOH/DCE [101]
		(trichloroacetimidate benzyl linker)	3–50% TFA [31]
		(4-bromomethylphenoxy linker)	
		(2,4-dimethoxybenzhydryl linker, Cl(Tf))	5% TFA in DCM [66]
		(bis-trityl chloride linker with amide)	1–50% TFA in DCM [127]
		(2-chlorotrityl chloride linker)	1–5% TFA in DCM [127]
RCONHOH	RCO₂H	(2-chlorotrityl-O-NH₂ linker)	5% TFA in DCM [87]
		(aminooxymethyl phenoxy linker)	70% TFA in DCM [29]
RCHO	RCHO	(diol linker)	HCl/dioxane/water [110]
		(threonine–glycine amide linker)	MeCN/water/TFA 60/40/0.1 [113] AcOH/water/DCM/MeOH [114]

Table 14.4 Continued.

Released	Attached	Linker	Cleavage conditions
1. RCONH$_2$ 2. RNHCONH$_2$ 3. ROCONH$_2$	1. RCO$_2$H 2. RNCO 3. ROCOCl		HF, TFMSA, HBr/TFA [115, 127]
			90–95% TFA [13, 44]
			1% TFA in DCM [3]
RCONH$_2$	RCONH$_2$		95% TFA [128]
1. RCONHR^1 2. RNHCONHR^1 3. ROCONHR^1	1 (i) R^{1}NH$_2$ 1 (ii) RCO$_2$H 2 (i) R^{1}NH$_2$ 2 (ii) RNCO 3 (i) R^{1}NH$_2$ 3 (ii) ROCOCl		1–5% TFA in DCM [58, 61, 129]
RCONHNH$_2$	(i) NH$_2$NH$_2$ (ii) RCO$_2$H		95% TFA [9, 86]
R(C=NH)NH$_2$	R(C=NH)NH$_2$		50% TFA [130]

Table 14.4 Continued.

Released	Attached	Linker	Cleavage conditions
RNHC(=NH)NH₂	(i) RNH₂ (ii) (BocHN)₂CS		50% TFA [61]
	RNHC(=NH)NH₂		95% TFA [128]
RNH₂, ArNH₂			50% TFA [40, 127]
			1–5% TFA in DCM [66, 127]
1,2- and 1,3-Diols	1,2- and 1,3-Diols		90% TFA aq. [107]
RR¹CO	RR¹CO		HCl/dioxane/water [110]
RSO₂NH₂	RSO₂Cl		HF, TFMSA, HBr/TFA
			20% TFA in DCM [131]
			1% TFA in DCM
RSO₂NHR¹	(i) R¹NH₂ (ii) RSO₂Cl		5–95% TFA [34, 36, 58, 61]

Table 14.4 Continued.

Released	Attached	Linker	Cleavage conditions
RSH	RSH		5% TFA in DCM [66, 127]
RBr	RSH		5% TFA [127]
ArH	ArLi		HF, TFA (only if Ar is electron-rich aryl) [94–99]

are only intended as a guide. In practice, lower concentrations of acid may also suffice for the release of a given functionality as will acids of comparable strength. In almost all cases, with the exception of some functionalities attached to trityl resins, product release may also be achieved using higher concentrations of acid. However, with resins containing benzylic linkages, acids stronger than TFA should not be used, to avoid linker cleavage.

List of Abbreviations

Boc	t-butoxycarbonyl
Bpoc	1-biphenyl-1-methyl-ethoxycarbonyl
BHA	benzhydrylamine
Bzl	benzyl
DBU	diaza(1,3)bicyclo[5.4.0]undecane
DCE	dichloroethane
DCM	dichloromethane
DMAP	N,N-dimethylaminopyridine
DMS	dimethyl sulfide
HMPB	4-(4-hydroxymethylphenoxy)butyric acid
FMPB	4-(4-formyl-2-methoxyphenoxy)butyric acid

Fmoc   fluoren-9-ylmethoxycarbonyl
HMPA   4-hydroxymethylphenoxyacetic acid
HFIP   1,1,1,3,3,3-hexafluoroisopropanol
MBHA   4-methoxybenzhydrylamine
MSNT   1-(mesitylene-2-sulfonyl)-3-nitro-1H-1,2,4-triazole
Pbf   2,2,4,6,7-pentamethyldihydrobenzofuran-5-sulfonyl
Pmc   2,2,5,7,8-pentamethylchroman-6-sulfonyl
PPTS   pyridinium p-toluenesulfonate
TES   triethylsilane
TFA   trifluoroacetic acid
TFMSA   trifluoromethanesulfonic acid
TMSBr   trimethylsilyl bromide
TMSOTf   trimethylsilyl trifluoromethanesulfonate
TIPS   triisopropylsilane

References

1 Guillier, F., Orain, D. and Bradley, M. (2000) *Chemical Reviews*, **100**, 2091–158.
2 James, I.W. (1999) *Tetrahedron*, **55**, 4855–946.
3 Sieber, P. (1987) *Tetrahedron Letters*, **28**, 2107–10.
4 Han, Y.X., Bontems, S.L., Hegyes, P., et al. (1996) *Journal of Organic Chemistry*, **61**, 6326–39.
5 Noda, M., Yamaguchi, M., Ando, E., et al. (1994) *Journal of Organic Chemistry*, **59**, 7968–75.
6 Chao, H.G., Bernatowicz, M.S., Reiss, P.D., et al. (1994) *Journal of the American Chemical Society*, **116**, 1746–52.
7 Chao, H.G., Bernatowicz, M.S. and Matsueda, G.R. (1993) *Journal of Organic Chemistry*, **58**, 2640–4.
8 Wang, S.-S. (1973) *Journal of the American Chemical Society*, **95**, 1328–33.
9 Atherton, E., Logan, J.C. and Sheppard, R.C. (1981) *Journal of the Chemical Society–Perkin Transactions 1*, 538–46.
10 Mergler, M., Tanner, R., Gosteli, J. and Grogg, P. (1988) *Tetrahedron Letters*, **29**, 4005–8.
11 Riniker, B., Florsheimer, A., Fretz, H., et al. (1993) *Tetrahedron*, **49**, 9307–20.
12 Atherton, E., Clive, D.L.J. and Sheppard, R.C. (1975) *Journal of the American Chemical Society*, **97**, 6585–6585.
13 Rink, H. (1987) *Tetrahedron Letters*, **28**, 3787–90.
14 Merrifield, R.B. (1964) *Biochemistry*, **3**, 1385–90.
15 Gisin, B.F. (1973) *Helvetica Chimica Acta*, **56**, 1476–82.
16 Sharma, S. and Pasha, S. (1997) *Bioorganic & Medicinal Chemistry Letters*, **7**, 2077–80.
17 McArthur, C.R., Worster, P.M., Jiang, J.L. and Leznoff, C.C. (1982) *Canadian Journal of Chemistry–Revue Canadienne de Chimie*, 1836–41.
18 Stones, D., Miller, D.J., Beaton, M.W., et al. (1998) *Tetrahedron Letters*, **39**, 4875–8.
19 Kent, S.B.H., Mitchell, A.R., Engelhard, M. and Merrifield, R.B. (1979) *Proceedings of the National Academy of Sciences of the United States of America*, **76**, 2180–4.
20 Mitchell, A.R., Erickson, B.W., Ryabtsev, M.N., et al. (1976) *Journal of the American Chemical Society*, **98**, 7357–62.
21 Fields, G.B. and Noble, R.L. (1990) *International Journal of Peptide and Protein Research*, **35**, 161–214.
22 Albericio, F. and Barany, G. (1985) *International Journal of Peptide and Protein Research*, **26**, 92–7.
23 2006 Novabiochem Catalog, pp. 2.36.
24 Blankemeyermenge, B., Nimtz, M. and Frank, R. (1990) *Tetrahedron Letters*, **31**, 1701–4.
25 Sieber, P. (1987) *Tetrahedron Letters*, **28**, 1647–6150.

26 Kirstgen, R., Sheppard, R.C. and Steglich, W. (1987) *Journal of the Chemical Society D – Chemical Communications*, 1870–1.
27 Green, J. and Bradley, K. (1993) *Tetrahedron*, **49**, 4141–6.
28 Hamper, B.C., Dukesherer, D.R. and South, M.S. (1996) *Tetrahedron Letters*, **37**, 3671–4.
29 Floyd, C.D., Lewis, C.N., Patel, S.R. and Whittaker, M. (1996) *Tetrahedron Letters*, **37**, 8045–8.
30 Richter, L.S. and Desai, M.J. (1997) *Tetrahedron Letters*, **38**, 321–2.
31 Hanessian, S. and Xie, F. (1998) *Tetrahedron Letters*, **39**, 733–6.
32 Phoon, C.W., Oliver, S.F. and Abell, C. (1998) *Tetrahedron Letters*, **39**, 7959–62.
33 Mergler, M., Nyfeler, R. and Gosteli, J. (1989) *Tetrahedron Letters*, **30**, 6741–4.
34 Raju, B. and Kogan, T.P. (1997) *Tetrahedron Letters*, **38**, 4965–8.
35 Nugiel, D.A., Wacker, D.A. and Nemeth, G.A. (1997) *Tetrahedron Letters*, **38**, 5789–90.
36 Ngu, K. and Patel, D.V. (1997) *Tetrahedron Letters*, **38**, 973–6.
37 Corbett, J.W., Graciani, N.R., Mousa, S.A. and DeGrado, W.F. (1997) *Bioorganic & Medicinal Chemistry Letters*, **7**, 1371–6.
38 Mergler, M., Nyfeler, R. and Gosteli, J. (1989) *Tetrahedron Letters*, **30**, 6745–8.
39 Dixit, D.M. and Leznoff, C.C. (1977) *Journal of the Chemical Society D – Chemical Communications*, 798–9.
40 Hauske, J.R. and Dorff, P. (1995) *Tetrahedron Letters*, 1589–92.
41 Sheppard, R.C. and Williams, B.J. (1982) *International Journal of Peptide and Protein Research*, **20**, 451–4.
42 Barlaam, B., Koza, P. and Berriot, J. (1999) *Tetrahedron*, **55**, 7221–32.
43 Albericio, F. and Barany, G. (1991) *Tetrahedron Letters*, **32**, 1015–18.
44 Albericio, F., Kneibcordonier, N., Biancalana, S., et al. (1990) *Journal of Organic Chemistry*, **55**, 3730–43.
45 Jensen, K.J., Alsina, J., Songster, M.F., et al. (1998) *Journal of the American Chemical Society*, **120**, 5441–52.
46 Alsina, J., Jensen, K.J., Albericio, F. and Barany, G. (1999) *Chemistry – A European Journal*, **5**, 2787–95.
47 Alsina, J., Yokum, T.S., Albericio, F. and Barany, G. (1999) *Journal of Organic Chemistry*, **64**, 8761–9.
48 Royo, M., Fresno, M., Frieden, A., et al. (1999) *Reactive and Functional Polymers*, **41**, 103–10.
49 Guillaumie, F., Kappel, J.C., Kelly, N.M., Barany, G. and Jensen, K.J. (2000) *Tetrahedron Letters*, **41**, 6131–5.
50 Alsina, J., Yokum, T.S., Albericio, F. and Barany, G. (2000) *Tetrahedron Letters*, **41**, 7277–80.
51 Kappel, J.C. and Barany, G. (2005) *Journal of Peptide Science*, **11**, 525–35.
52 Ngu, K. and Patel, D.V. (1997) *Journal of Organic Chemistry*, **62**, 7038–9.
53 Gray, N.S., Kwon, S. and Schultz, P.G. (1997) *Tetrahedron Letters*, **38**, 1161–4.
54 Gordeev, M.F., Patel, D.V. and Gordon, E.M. (1996) *Journal of Organic Chemistry*, **61**, 924–8.
55 Boojamra, C.G., Burow, K.M. and Ellman, J.A. (1995) *Journal of Organic Chemistry*, **60**, 5742–3.
56 Vergnon, A.L., Pottorf, R.S. and Player, M.R. (2004) *Journal of Combinatorial Chemistry*, **6**, 91–8.
57 Grimstrup, M. and Zaragoza, F. (2002) *European Journal of Organic Chemistry*, 2953–60.
58 Fivush, A.M. and Willson, T.M. (1997) *Tetrahedron Letters*, **38**, 7151–4.
59 Doerner, B. and White, P. (1999) in Peptides 1998, *Proceedings of the 25th European Peptide Symposium* (eds B. Sandor and F. Hudecz), Akademiai Kiado, Budapest, pp. 90.
60 Fernández, E., Garcia-Ochoa, S., Huss, S., et al. (2002) *Tetrahedron Letters*, **43**, 4741–5.
61 Estep, K.G., Neipp, C.E., Stramiello, L.M.S., et al. (1998) *Journal of Organic Chemistry*, **63**, 5300–1.
62 Pietta, P.G. and Marshall, G.R. (1970) *Chemical Communications*, 650–1.
63 Matsueda, G.R. and Stewart, J.M. (1981) *Peptides*, **2**, 45.
64 Mergler, M., Dick, F., Gosteli, J. and Nyfeler, R. (1999) *Tetrahedron Letters*, **40**, 4663–4.
65 Bernatowicz, M.S., Daniels, S.B. and Koster, H. (1989) *Tetrahedron Letters*, **30**, 4645–8.

66 Garigipati, R.S. (1997) *Tetrahedron Letters*, **38**, 6807–10.
67 Brill, W.K.D., Schmidt, E. and Tommari, R.A. (1998) *Synlett*, 906–8.
68 Brill, W., Riva-Toniolo, K., D. and C. (2001) *Tetrahedron Letters*, **42**, 6515–18.
69 Brown, D.S., Revill, J.M. and Shute, R.E. (1998) *Tetrahedron Letters*, **39**, 8533–6.
70 Barlos, K., Gatos, D., Hondrelis, J., Matsoukas, J., Moore, G.J. and Schäfer, W. (1989) *Liebigs Annalen Der Chemie*, 951–5.
71 Canne, L.E., Walker, S.M. and Kent, S.B.H. (1995) *Tetrahedron Letters*, **36**, 1217–20.
72 Fyles, T.M. and Leznoff, C.C. (1976) *Canadian Journal of Chemistry–Revue Canadienne de Chimie*, **54**, 935–42.
73 Fréchet, J.M.J. and Haque, K.E. (1975) *Tetrahedron Letters*, **16**, 3055–6.
74 Barlos, K., Gatos, D., Kallitsis, J., *et al.* (1989) *Tetrahedron Letters*, **30**, 3943–6.
75 Bayer, E., Clausen, N., Goldammer, C., Henkel, B., Rapp, W. and Zhang, L. (1994) Peptides: chemistry, structure & biology, in *Proceedings of the 13th American Peptide Symposium* (eds S.H. Robert and A.S. John), ESCOM, Leiden, pp. 156–8.
76 Zikos, C.C. and Ferderigos, N.G. (1994) *Tetrahedron Letters*, **35**, 1767–8.
77 Van Vliet, A. and Tesser, G. (1996) in *Innovation and Perspectives in Solid Phase Synthesis and Combinatorial Libraries* (ed. R. Epton), Mayflower Scientific, Birmingham, pp. 545–8.
78 Meisenbach, M. and Voelter, W. (1997) *Chemistry Letters*, 1265–6.
79 Barlos, K., Chatzi, O., Gatos, D. and Stavropoulos, G. (1991) *International Journal of Peptide and Protein Research*, **37**, 513–20.
80 Atherton, E., Hardy, P.M., Harris, D.E. and Mattews, B.H. (1990) in Peptides 1990, in *Proceedings of the 21st European Peptide Symposium* (eds E. Giralt and D. Andreu), ESCOM, Leiden, pp. 243–4.
81 Fujiwara, Y., Akaji, K. and Kiso, Y. (1994) *Chemical and Pharmaceutical Bulletin*, **42**, 724–6.
82 Lukszo, J., Patterson, D., Albericio, F. and Kates, S.A. (1996) *Letters in Peptide Science*, **3**, 157–66.
83 Barlos, K., Gatos, D., Kallitsis, J., *et al.* (1989) *Tetrahedron Letters*, **30**, 3943–6.
84 Bollhagen, R., Schmiedberger, M., Barlos, K. and Grell, E. (1994) *Journal of the Chemical Society D–Chemical Communications*, 2559–60.
85 Barlos, K. and Gatos, D. (2000) Convergent Peptide Synthesis, in *Fmoc Solid Phase Peptide Synthesis: A Practical Approach* (eds W.C. Chan and P.D. White), Oxford University Press, Oxford, pp. 215–28.
86 Stravropoulos, G., Gatos, D., Magafa, V. and Barlos, K. (1996) *Letters in Peptide Science*, **2**, 315–18.
87 Mellor, S.L., McGuire, C. and Chan, W.C. (1997) *Tetrahedron Letters*, **38**, 3311–14.
88 Chan, W.C. and Mellor, S.L. (1995) *Journal of the Chemical Society D–Chemical Communications*, 1475–7.
89 Ramage, R., Irving, S.L. and McInnes, C. (1993) *Tetrahedron Letters*, **34**, 6599–602.
90 Patterson, J.A. and Ramage, R. (1999) *Tetrahedron Letters*, **40**, 6121–4.
91 Noda, M., Yamaguchi, M., Ando, E., *et al.* (1994) *Journal of Organic Chemistry*, **59**, 7968–75.
92 Chao, H.G., Bernatowicz, M.S., Reiss, P.D., Klimas, C.E. and Matsueda, G.R. (1994) *Journal of the American Chemical Society*, **116**, 1746–52.
93 Chao, H.G., Bernatowicz, M.S. and Matsueda, G.R. (1993) *Journal of Organic Chemistry*, **58**, 2640–4.
94 Finkelstein, J.A., Chenera, B. and Veber, D.F. (1995) *Journal of the American Chemical Society*, **117**, 11999–20000.
95 Plunkett, M.J. and Ellman, J.A. (1995) *Journal of the American Chemical Society*, **117**, 3306–7.
96 Newlander, K.A., Chenera, B., Veber, D.F., Yim, N.C.F. and Moore, M.L. (1997) *Journal of Organic Chemistry*, **62**, 6726–32.
97 Hone, N.O., Davies, S.G., Devereux, N.J., Taylor, S.L. and Baxter, A.D. (1998) *Tetrahedron Letters*, **39**, 897–900.
98 Woolard, F.X., Paetsch, J. and Ellman, J.A. (1997) *Journal of Organic Chemistry*, **62**, 6102–3.
99 Hu, Y. and Porco, J.A. Jr. (1998) *Journal of Organic Chemistry*, **63**, 4518–21.
100 Thompson, L.A. and Ellman, J.A. (1994) *Tetrahedron Letters*, 9333–6.

101. Liu, G. and Ellman, J.A. (1995) *Journal of Organic Chemistry*, **60**, 7712–13.
102. Wallace, O.B. (1997) *Tetrahedron Letters*, **38**, 4939–42.
103. Pearson, W.H. and Clark, R.B. (1997) *Tetrahedron Letters*, **38**, 7669–72.
104. Smith, A.L., Stevenson, G.I., Swain, C.J. and Castro, J.L. (1998) *Tetrahedron Letters*, **39**, 8317–20.
105. Nugiel, D. and Cornelius, L. (1997) *Journal of Organic Chemistry*, **62**, 201–3.
106. Yoo, S.E., Seo, J.S., Yi, K.Y. and Gong, Y.D. (1997) *Tetrahedron Letters*, **38**, 1203–6.
107. Fréchet, J.M. and Pellé, G. (1975) *Journal of the Chemical Society D – Chemical Communications*, 225.
108. Hanessian, S. and Huynh, H.K. (1999) *Tetrahedron Letters*, **40**, 671–4.
109. Wu, Y.T., Hsieh, H.P., Wu, C.Y., et al. (1998) *Tetrahedron Letters*, **39**, 1783–4.
110. Leznoff, C.C. and Wong, J.Y. (1973) *Canadian Journal of Chemistry – Revue Canadienne de Chimie*, **51**, 3756–64.
111. Leznoff, C.C. and Greenberg, S. (1976) *Canadian Journal of Chemistry – Revue Canadienne de Chimie*, **54**, 3824–9.
112. Hodge, P. and Waterhouse, J. (1983) *Journal of the Chemical Society – Perkin Transactions 1*, 2319–23.
113. Ede, N.J., Eagle, S.N., Wickham, G., et al. (2000) *Journal of Peptide Science*, **6**, 11–18.
114. Sorg, G., Thern, B., Mader, O., Rademann, J. and Jung, G. (2005) *Journal of Peptide Science*, **11**, 142.
115. Tam, J.P. (1988) *Macromolecular Sequencing and Synthesis: Selected Methods and Applications*, Alan R. Liss, Inc, pp. 153–84.
116. Chang, C., Felix, A.M., Jimenez, M.H. and Meienhofer, J. (1980) *International Journal of Peptide and Protein Research*, **15**, 485–94.
117. Kitagawa, K., Kitade, K., Kiso, Y., et al. (1980) *Chemical and Pharmaceutical Bulletin*, **28**, 926–31.
118. Kiso, Y., Ukawa, K. and Akita, T. (1980) *Journal of the Chemical Society D – Chemical Communications*, 101–2.
119. Pearson, D.A., Blanchette, M., Baker, M.L. and Guindon, C.A. (1989) *Tetrahedron Letters*, **30**, 2739–43.
120. Tam, J.P., Heath, W.F. and Merrifield, R.B. (1983) *Journal of the American Chemical Society*, **105**, 6442–55.
121. Hudson, D. (1989) in *Peptides 1988, Proceedings of the 20st European Peptide Symposium* (eds G. Jung and E. Bayer), Walter de Gruyter & Company, Berlin, pp. 211–13.
122. King, D.S., Fields, C.G. and Fields, G.B. (1990) *International Journal of Peptide and Protein Research*, **36**, 255–66.
123. Ramage, R., Green, J. and Blake, A.J. (1991) *Tetrahedron*, **47**, 6353–70.
124. Carpino, L.A., Shroff, H. Triolo, S.A., et al. (1993) *Tetrahedron Letters*, **34**, 7829–32.
125. White, P. (1992) Peptides: chemistry & biology, in *Proceedings of the 12th American Peptide Symposium* (eds J.A. Smith and J.E. Rivier), ESCOM, Leiden, pp. 537–8.
126. Deegan, T.L., Gooding, O.W., Baudart, S. and Porc, J.A. Jr. (1997) *Tetrahedron Letters*, **38**, 4973–6.
127. Novabiochem 2006 Catalog, pp. 2.17–2.22.
128. Bernhardt, A., Drewello, M. and Schutkowski, M. (1997) *Journal of Peptide Research*, **50**, 143–52.
129. Brown, E.G. and Nuss, J.M. (1997) *Tetrahedron Letters*, **38**, 8457–60.
130. Roussel, P., Bradley, M., Matthews, I. and Kane, P. (1997) *Tetrahedron Letters*, **38**, 4861–4.
131. Beaver, K.A., Siegmund A.C. and Spear, K.L. (1996) *Tetrahedron Letters*, **37**, 1145–8.

15
Base/Nucleophile-Labile Resins

Francesc Rabanal

Universitat de Barcelona, Departament de Química Orgànica, Facultat de Química, Martí i Franquès, 1, 08028 Barcelona, Spain

15.1
Introduction

Nucleophile- and base-labile linkers and resins are based on classical organic chemistry reactions. The first type releases organic compounds generally by nucleophilic substitution on bound carboxylic acid derivatives while the second involves an elimination reaction promoted by a base. Linkers may be viewed as immobilized protecting groups on a solid support. Hence, they have to be labile to the desired reagents, the chosen nucleophile or base in this case, but be stable to the synthetic assembly conditions of the organic compound, which may include reagents that are either nucleophilic, basic or both. A clear example is the solid phase synthesis of peptides, either by the Boc/Bn or Fmoc/tBu protocols, which generally includes the presence of nucleophiles such as HOBt or related compounds (HOAt, HOSu, etc.) and bases such as tertiary amines (in the Boc/Bn synthesis) or secondary amines like piperidine (in the Fmoc/tBu synthesis). The present chapter gives an overview of different nucleophile- and base-labile linkers and resins. Their reactivity/stability will be rationalized according to their chemical properties, in particular the pK_a of the linker or functional group of the resin.

15.2
Nucleophile-Labile Resins

The cleavage of organic compounds from nucleophile-labile resins usually relies on the addition–elimination chemistry at the carbonyl group of the carboxylic acid derivative (i.e. an ester or thioester) that mediates the linkage between the assembled compound and the linker-resin. The overall reaction is a nucleophilic substitution that involves the release into the solution of the compound of interest with the attacking nucleophile generally incorporated. The linker-resin generally acts

The Power of Functional Resins in Organic Synthesis. Judit Tulla-Puche and Fernando Albericio
Copyright © 2008 WILEY-VCH Verlag GmbH & Co. KGaA, Weinheim
ISBN: 978-3-527-31936-7

as the leaving group, although the reverse approach is also feasible (Section 15.2.1.2). A typical example of nucleophile-labile resins is hydroxybenzylic resins or linkers (e.g. hydroxymethylpolystyrene), where the organic compound is attached to the linker-resin via an ester bond. In this common case, the linker is displaced as an alkoxide, the conjugated base of the alcohol (Figure 15.1).

Alkoxides are not good leaving groups. It is well known that the ability of a leaving group to act as such correlates well with the strength of the conjugated base, which is measured by the acidity or pK_{aH} of the acid (in this case the alcohol) [1]. The lower the pK_{aH} is, the better the leaving group. The pK_{aH} of a typical alcohol such as methanol is 15.7. Benzylic alcohols are slightly more acidic, with a pK_{aH} in the order of 14–15, due to the presence of the aromatic ring. The presence of electron-withdrawing groups in the phenyl ring reduces the pK_{aH} to below 14, such as with the Nbb resin (Figure 15.2), thus making it a better leaving group and more reactive to nucleophiles. Linkers based on the α-hydroxyacetic acid (glycol acid derivatives) or 2-hydroxyethers show pK_{aH}s of around 14–15 [2].

The fact that many nucleophile-labile linkers are alcohols despite not being good leaving groups reflects the compromise between reactivity towards cleavage and stability to the chemical conditions of the synthetic assembly, where some nucleophiles may be present. Other type of linkers have also been used, such as phenol-based linkers (with a pK_{aH} around 10), aliphatic-thiol and thiophenol-based linkers ($pK_{aH} \approx 8$–9) and N-hydroxy-based linkers (oxime $pK_{aH} \approx 8$). These linkers are more reactive to nucleophiles but they also show some limitations, such as excessive lability to secondary amines. Hence, they are hardly compatible with the Fmoc/tBu synthesis of peptides due to premature cleavage by the repetitive use of piperidine, although some alternatives have been proposed (Section 15.2.2.1). The structure of these linkers is displayed in Figure 15.2.

Classical reactions involving nucleophiles such as saponification (⁻OH as the nucleophile), aminolysis (with amines; also ammonia in ammonolysis reactions), transesterification (alkoxides, ⁻OR) and others (hydrazinolysis, hydroxamic acid synthesis, etc.) have been adapted to solid phase and used to obtain, for instance, carboxylic acids, amides and esters. Internal or intramolecular nucleophilic attack has been employed to obtain cyclic products such as lactones, lactams (including cyclic peptides) and a great variety of heterocycles (hydantoins, diketopiperazines, benzodiazepinones, etc.).

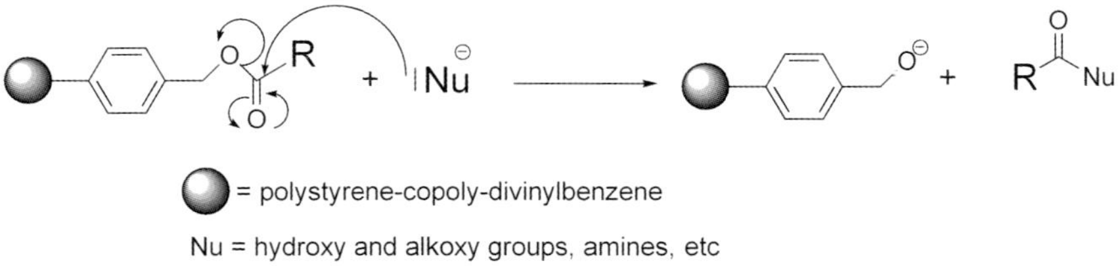

● = polystyrene-copoly-divinylbenzene

Nu = hydroxy and alkoxy groups, amines, etc

Figure 15.1 General representation of a nucleophile-labile resin (example with an ester bound to hydroxymethylpolystyrene resin). A simplified addition–elimination mechanism is shown.

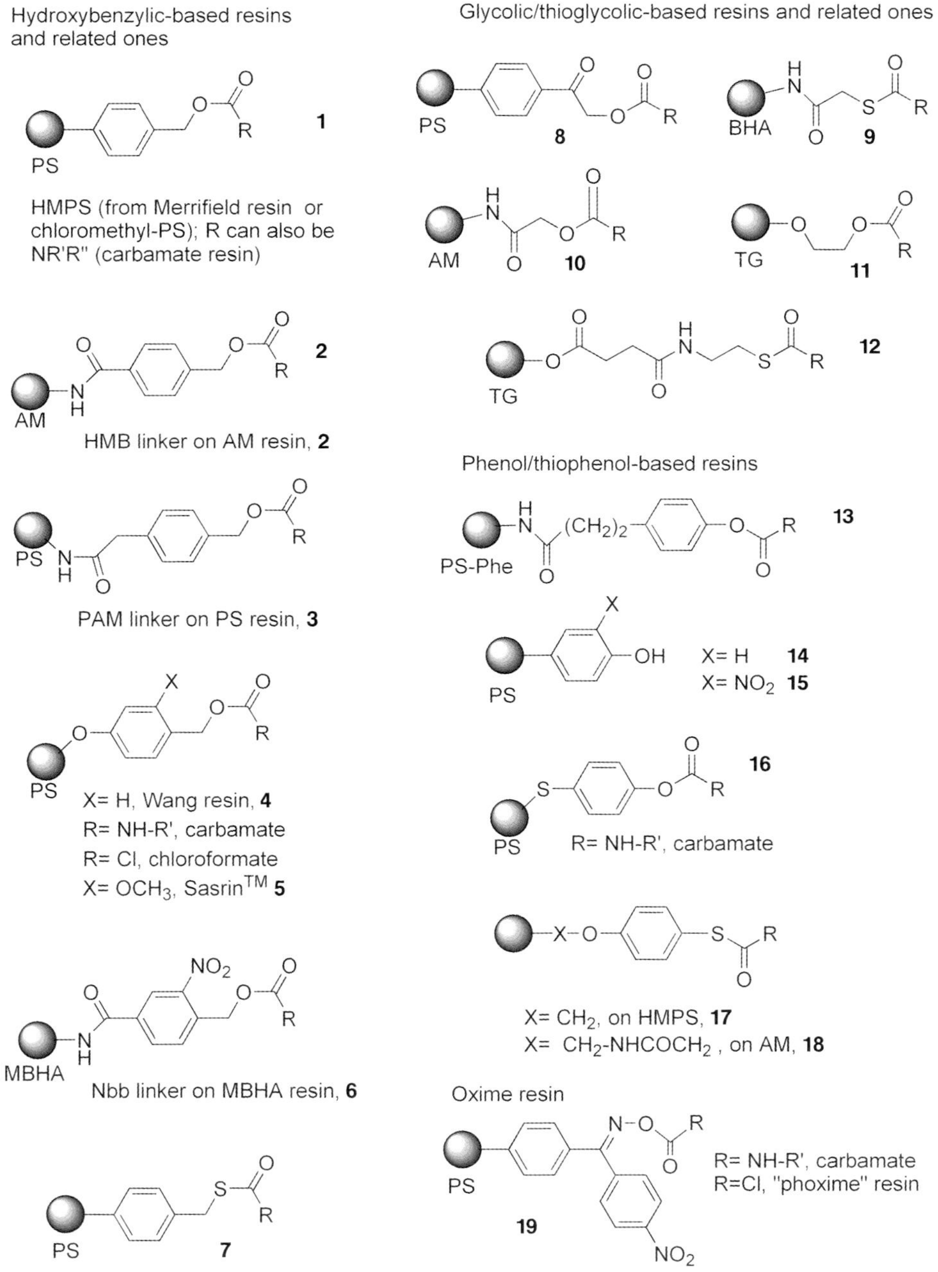

15.2 Nucleophile-Labile Resins

Hydroxybenzylic-based resins and related ones

1 HMPS (from Merrifield resin or chloromethyl-PS); R can also be NR'R'' (carbamate resin)

2 HMB linker on AM resin

3 PAM linker on PS resin

4 X= H, Wang resin
R= NH-R', carbamate
R= Cl, chloroformate
5 X= OCH$_3$, Sasrin™

6 Nbb linker on MBHA resin

7

Glycolic/thioglycolic-based resins and related ones

8, **9**, **10**, **11**, **12**

Phenol/thiophenol-based resins

13

14 X= H
15 X= NO$_2$

16 R= NH-R', carbamate

17 X= CH$_2$, on HMPS
18 X= CH$_2$-NHCOCH$_2$, on AM

Oxime resin

19 R= NH-R', carbamate
R= Cl, "phoxime" resin

Figure 15.2 Nucleophile labile resins and linkers. An ester/thioester derivative is shown (R-CO-Nu will be the released product, R is generally alkyl or aryl, unless otherwise indicated). Linkers are also displayed anchored to suitably functionalized resins. Carbamates and chloroformates are also shown in the cases reported (see text). Some linkers and resins are labile to other cleavage conditions such as Merrifield, HMPS, Wang and Sasrin (acidolysis) or Nbb (photolysis).

Below, the usefulness of nucleophile- and base-labile linkers and resins is reviewed in the preparation of various organic compounds. First, intermolecular nucleophilic displacement (saponification, transesterification and amminolysis and related reactions) are analyzed. Finally, the obtention of cyclic products by intramolecular nucleophilic reactions is reviewed.

15.2.1
Intermolecular Nucleophilic Displacement

15.2.1.1 Hydroxy Functionalized Linkers and Resins and Related Ones

Nucleophilic detachment of compounds from resins has been associated with solid phase chemistry since its origins. In the article in which B. R. Merrifield describes the first synthesis of a tetrapeptide on a solid support in 1963, basic hydrolysis of the ester bond (with 0.2 M NaOH aqueous ethanolic solution) was used as the method to cleave the peptide acid from a nitrated version of the classical chloromethylated polystyrene resin [3]. Generally, the preparation of carboxylic acids from ester-bound compounds to hydroxy-containing linkers (Figure 15.2) requires a source of hydroxide ion (NaOH, LiOH, Bu$_4$NOH) in aqueous or organic solvents or in a mixture of both. Reaction times tend to be long (12–48 h) and, sometimes, heating is required. For instance, benzoic acids bound to linker **1** have been cleaved with LiOH·H$_2$O (5 equiv) in MeOH/H$_2$O/THF under reflux (18–42 h) [4] while benzofuran carboxylic acids have been released from the 2-hydroxyether resin **11** with aqueous NaOH (1 M) in isopropanol (8 h at 50 °C) [5]. Leznoff described the complete removal of cyclohexencarboxylic acids with Bu$_4$NOH [6]. Milder saponifications were described by Ueki and coworkers, where peptides were detached from esterified PAM resins with TBAF·3H$_2$O in 30 min at room temperature [7]. Peptides have been also cleaved from Nbb resin using TBAF·3H$_2$O (8 equiv. in acetonitrile, 40 min) or aqueous LiOH (15 equiv, 10 min, rt) [8].

Hydroxy linkers bearing an electron-withdrawing group such as an α-carbonyl (oxyacyl resins, phenacyl type-linkers or glycolic acid derivatives) are also suitable for the nucleophilic release of ester bound compounds. The glycolamidic ester linker **10** has been successfully used to synthesize peptides by the Fmoc/tBu strategy. It is compatible with the repetitive piperidine treatment but peptides can be cleaved with dilute NaOH solutions, ammonia or alkoxides [9]. Esters bound to oxyacyl resins **8** have been reported to be cleavable by thiolysis, saponification, ammonia, hydrazine and potassium cyanide (complexed with dicyclohexyl-18-crown-6) [10, 11].

The release and obtention of esters by transesterification is another classical procedure to detach esterified compounds using, in this case, an alcohol or an alkoxide as the nucleophile. Care must be taken that water is excluded otherwise the acid will be obtain as a by-product [4]. Typical cleavage conditions involved the use of the desired alcohol in the presence of a non-nucleophilic base such as a tertiary amine (if primary or secondary amines are used, there is the risk of obtaining also the amide) [12] or the alkoxide dissolved in the corresponding alcohol or in an inert solvent (e.g. THF). Many examples have been described, employing unhindered primary alcohols such as methanol, ethanol or benzyl alcohol. Peptide esters bound to Merrifield resin have been cleaved by alcoholysis using tertiary bases [12] and methyl esters of cyclohexenecarboxylic acids have been released with MeOH in the presence of K_2CO_3 [13]. Sodium methoxide (in MeOH:THF, reflux for 20 h) has also been used to detach methyl benzoates and related aromatic methylesters from the same resin [14, 15]. Methyl esters of indolecarboxylic acids have been obtained using the HMB linker (**2**) by cleavage with MeOH and triethylamine at 50 °C [16, 17]. Likewise, Wang linker (**4**) has been cleaved with NaOMe (2 equiv.) in MeOH:THF (in the preparation of tetrahydro-1,4-benzodiazepine-2-ones) [18] and HMPS bound esters clipped with NaOMe in refluxing THF (in the synthesis of 3,4-dihydro-2*H*-pyrans) [19]. Catalysis with KCN and NEt_3 in MeOH and benzene has been used to detach nitrosoacetals prepared by cycloaddition reactions [20]. Likewise, peptide methyl esters have prepared from Nbb resin (**6**) by KCN-MeOH or DIEA-MeOH treatments [8]. DBU/LiBr has also been used with various alcohols (MeOH, EtOH, *i*PrOH, BzlOH, etc.; Me_3COH gave very low yields) to detach peptides from Wang [21] and Sasrin [22] resins (**5**) and oligomeric thioureas and peptides from PAM (**3**) resin [21–23].

Phenolic resins **13** and **14** have also been used successfully in peptide synthesis to prepare Met-enkephalin and LH-RH analogs. The ester-bound peptides have been released both by saponification (with 1 M NaOH in aqueous DMF, 45 min) and transesterification (DMAE:DMF, 1:1, 24–48 h) [24, 25].

Amides can also be prepared by nucleophilic detachment of esterified compounds to hydroxy linkers or resins using amines or ammonia. First attempts to obtain amides by solid phase were reported for peptide amides in the 1960s. Oxytocin and bradykinin synthesis were carried out by the Boc/Bn procedure on Merrifield resin and cleaved with NH_3/MeOH (9 M, 17 h) or in liquid ammonia [26, 27]. Other nitrogen nucleophiles were also used to detach oxytocin, such as methylamine (7 M in MeOH, 2 h), hydroxylamine (10 M in MeOH, 2 h) and hydrazine hydrate (50% in MeOH, 4 h) [28]. LH-RH analogs have been detached with neat ethylamine, butylamine, ethylenediamine and hydrazine [29, 30]. Later, the HMB linker (**2**) was proposed to prepare peptide amides following an Fmoc/tBu protection strategy. Notably, Fmoc/tBu chemistry uses piperidine as the reagent for the repetitive removal of Fmoc. However, the fact that it is a secondary amine prevents the peptide-bound benzyl ester from being prematurely released during the SPPS. The HMB linker bound to polydimethylacrylamide resin has thus been used to prepare Substance P. The peptide was cleaved with ammonia in MeOH (at 0 °C in 160 min) [31]. More recently, the HMB linker has been used to prepare constrained glicidol amino acids and tri- and tetracyclic scaffolds obtained by Pictet–Spengler reaction on indoles, furans and thiophene heterocycles. Cleavage was afforded with 0.1 M aqueous NaOH or ethylamine [32–34].

Similarly, peptide amides have been prepared with the Nbb resin. The steric hindrance of the different amines seemed to play a role in the cleavage as the yields were very good for primary amines but dropped for secondary ones [8].

Phenol-type resins **13–15** have also been used to prepare LH-RH, enkephalin and bradykinin peptide analogs. Peptides were successfully cleaved by ammonolysis (NH_3 in MeOH–DMF or DMF, 5–18 h) or aminolysis with protected amino acid or peptide esters bearing the free α-amino group [24, 25, 35]. Marshall's mercaptophenol resin (**16**) has been used to prepare libraries of piperazine-2-carboxamides (consisting of 22 different amines) and 6-carboxybenzopyran-4-ones. Cleavage was performed using an excess of appropriate amines in pyridine for 24 to 48 h [36, 37]. Tetrahydro-β-carboline-3-carboxamides have also been similarly synthesized and released [38].

The oxime resin (polymer bound *p*-nitrobenzophenone oxime, **19**) was introduced in 1980 for the preparation of protected peptides by aminolysis with amino acid esters (1.2–4 equiv, 4–18 h) and hydrazinolysis (0.5 M anhydrous hydrazine in CH_2Cl_2 : MeOH). Alquilamides and arylamides of peptides have also been prepared with the oxime resin. Primary amines cleaved peptides in good to excellent yields. However, the yields dropped to 60% with diethylamine, a hindered secondary amine. Aromatic amines gave very low yields, except aniline, which furnishes an excellent yield in the presence of 2% of acetic acid [39–43]. The catalytic effect of acetic acid in aminolysis is well known. Other nucleophiles, such as hydroxylamine (O-TBDMS protected) have been employed to obtain hydroxamic acids [44] and hydrogen sulfur equivalents ($Me_3Si-S-SiMe_3$) to prepare protected peptide thioacids [45].

The oxime resin has also been proven to be labile to hydroxy-containing compounds such as *N*-hydroxypiperidine, *N*-hydroxysuccinimide and *N*-hydroxybenzotriazole at 50–70 °C [46]. The corresponding active esters of peptides were transacylated to amino containing resins [47]. Another N-hydroxy resin, N-hydroxysuccinimide active ester resin, has been described for the synthesis of amide products. Bound esters were successfully cleaved even with secondary amines in high yields and purities [48].

The α-hydroxyacrylic acid has been used to develop linkers for the preparation of Boc/Bn protected peptides that are cleaved by means of a conjugated nucleophilic attack. This linker is stable to TFA (trifluoroacetic acid) (55% in CH_2Cl_2) and DIEA (50% in DMF). Peptides were detached with 5–10% piperidine or morpholine [49].

Sulfur analogs of some of the described hydroxy linkers and resins (Figure 15.2) have been employed to synthesize resin-bound thioesters that can then be cleaved with alcohols, amines and organometallic reagents to furnish esters, amides, ketones, aldehydes and alcohols [50]. Secondary amines react sluggishly with linker **9** and, therefore, more reactive thioester-linker derivatives were developed, such as **17** and **18**. Yields for a hindered amine such as Pro-OMe were in the order of 60–70%. Thiol-containing PS-DVB resin **7** has been used for the obtention of β-hydroxyacids by hydrolysis with 0.2 M NaOH in aqueous dioxane [51].

15.2.1.1.1 **Carbonic Acid Derivatives on Resin** Carbamate derivatization of some common hydroxy resins (i.e. hydroxymethylpolystyrene, Wang or the oxime resins) has also been described for the synthesis of ureas, sulfonamides and dihydropyridones [52]. Thus, the oxime resin (**19**) has been treated either with isocyanates or triphosgene (to obtained the corresponding chloroformate, the so-called "phoxime" resin) followed by reaction with an amine to afford in each case the carbamate on resin (Figure 15.3). Thermolytic treatment of such a carbamate resin with additional amines in toluene at 80 °C overnight released the corresponding ureas in solution [53]. A similar approach has been followed with the mercaptophenol resin **16** [54]. Likewise, treatment of Wang resin (**4**) with *p*-nitrophenylchloroformate (or phosgene) provided the corresponding active carbonate (or chloroformate) that upon treatment with amines afforded the subsequent carbamates on resin (**4a**, Figure 15.3). Proton uptake with a base (LHMDS) followed by reaction with sulfonyl chloride afforded the resin bound sulfonamides that were cleaved with LiOH/H_2O–THF and/or NaOMe/THF in excellent purity and good overall yield [55]. Other carbonic acid derivatives such as polysubstituted aminosulfonyl ureas [56] and guanidines [57] have been similarly prepared by cleavage with amines.

15.2.1.1.2 **Miscellanea** A methodology to prepare α-substituted-β-hydroxy acids and esters has been introduced in solid phase based on an Evans' oxazolidinone-based linker to produce enantiospecific aldol condensations (Figure 15.4). Acids and esters were released by treatment with LiOH and H_2O_2 in THF (at −20 °C) or NaOMe in THF, respectively [58, 59]. Diels–Alder adducts of oxazolidinone-bound crotonates have also been detached with LiOCH$_2$Ph [60].

Resins based on Meldrum's acid chemistry (resin bound cyclic malonic acid ester) have been described for the synthesis of thiophene derivatives (Figure 15.5). Release and heterocycle formation was performed in NaOMe/MeOH at reflux for

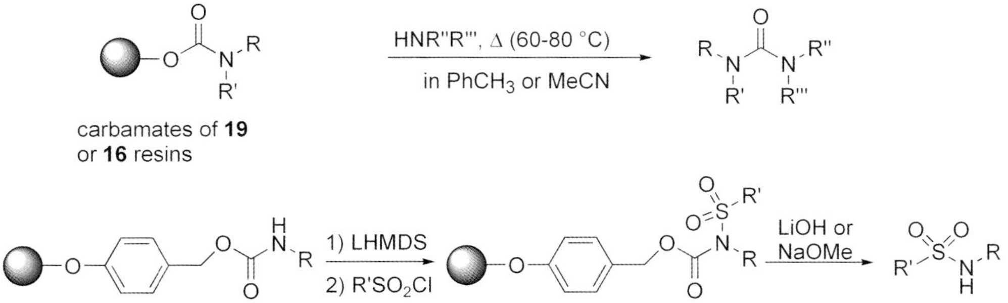

Figure 15.3 Urea and sulfonamide synthesis from carbamate derivatized oxime (**19**), thiophenol (**16**) and Wang (**4**) resins.

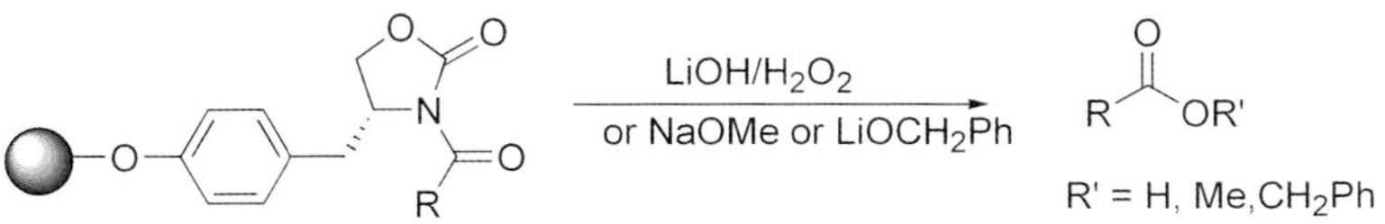

Figure 15.4 Evans' oxazolidinone based linker.

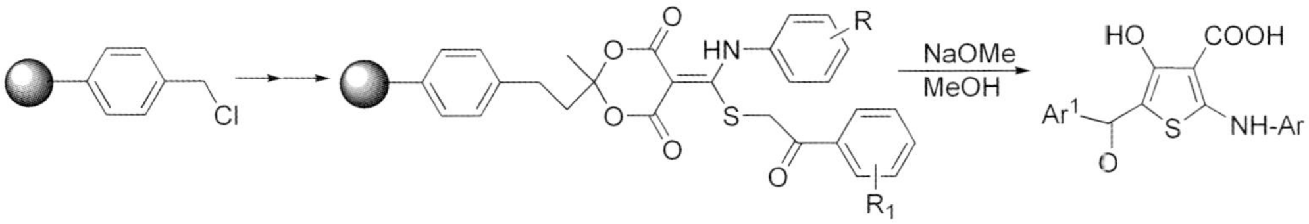

Figure 15.5 Synthesis of thiophene derivatives on resin-bound cyclic malonic esters.

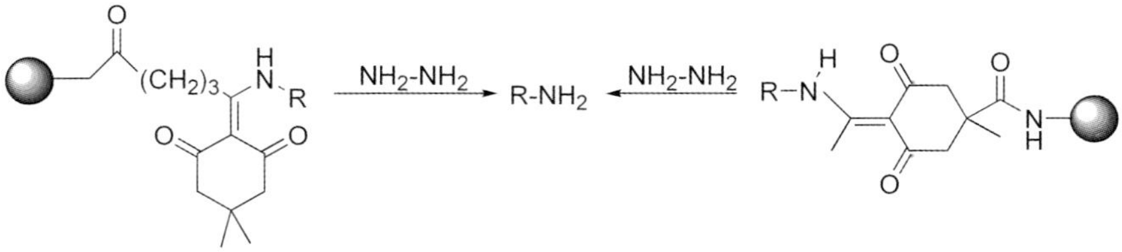

Figure 15.6 Dimedone-based linkers furnish amines upon treatment with hydrazine.

4 h with good yields and purities. Benzothiazoles were similarly obtained by acidolysis (HClO$_4$ in MeCN, reflux, 4 h) [61].

Some specific linkers have been designed to be cleaved by hydrazine due to the high nucleophilicity and ambident character of this reagent. Inspired by the Gabriel and Ing-Manske procedure for the synthesis of amines, phthalimide linkers were developed to generate a library phthalhydrazides with three points of diversity [62]. Likewise, the particular reactivity of dimedone-based structures with hydrazine has also been exploited to construct some linkers based on this chemistry. Based on the Dde amino protecting group [Dde = 1-(4,4-dimethyl-2,6-dioxocyclohexylidene)ethyl], Bycroft introduced 2-acyldimedone linkers to anchor primary amines (Figure 15.6). The linker can be cleaved with 2% hydrazine hydrate and as it is fairly stable to 20% piperidine/DMF (12% loss after 6 h at rt) it is amenable to the Fmoc/tBu synthesis of peptides. It can also be cleaved with *n*-propylamine [63]. Another version of a Dde-based linker that is bound to the solid support through a carboxylic group placed instead of one of the 4-methyl groups of dimedone has been developed [64].

Formamidine-based linkers have also been described for the preparation of amine derivatives upon cleavage with hydrazine (although other reagents were also possible, such as KOH/MeOH, LiAlH$_4$ and ZnCl$_2$/EtOH) [65].

15.2.1.2 Carboxy- and Sulfonic-Based Linkers and Resins

Nucleophilic displacement reactions in the solid phase using a reverse approach (e.g. the carboxy acid derivative on resin, and target product as the leaving group) has been undertaken to prepare alcohols and phenols by alcoholysis, saponification, aminolysis or ammonolysis (Figure 15.7).

The benzylic ester resin **20** derived from Merrifield resin was described in the 1970s by Leznoff and coworkers for the selective synthesis of monotrityl ethers of symmetrical aliphatic diols, phenols and porphyrins [66–68]. More recently, the

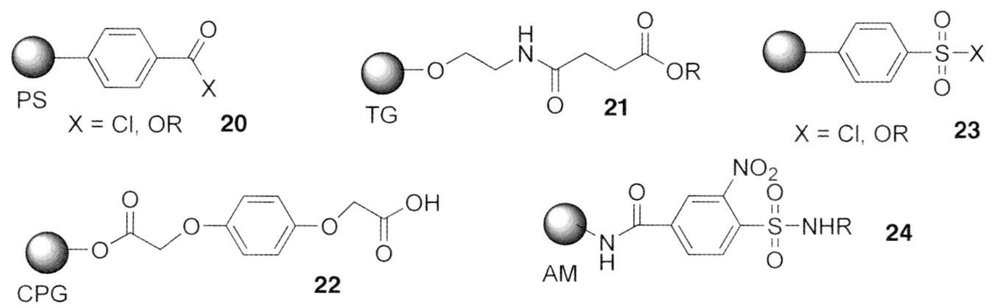

Figure 15.7 Carboxy- and sulfonic-based resins and linkers.

use of this resin has been extended to combinatorial chemistry techniques to prepare diazines by inverse electron demand Diels–Alder reactions [69], polyisoxazolines (by 1,3-dipolar addition and selenide oxidation/elimination) [70], isoxazole and isoxazolines (by an [3 + 2] cycloaddition) [71] and Pauson–Khand cycloaddition products [72]. Cleavage conditions included NH_3 in water–dioxane, NaOH in dioxane, $NaOCH_3$ in THF/MeOH, K_2CO_3 in MeOH/THF or $(n\text{-}C_4H_9)_4NOH$ in, H_2O/THF.

Diacid-based linkers, such as the succinic linker **21**, have been described to prepare alcohols. The procedure involves the esterification of the starting alcohol with succinic anhydride and DMAP to yield the hemiester that is anchored to an amino containing-resin by means of an amide bond. The bound alcohol is then elaborated and finally released with a nucleophile. Oligosaccharides have been assembled following this approach and released with aqueous ammonia or sodium methoxide in methanol–dioxane [73, 74]. Peptide alcohols have also been prepared with the succinic linker on BHA resin and released by treatment with NH_3 in MeOH for 72–96 h or hydrazine in DMF for 24 h [75]. Similarly, hydroquinone-O,O'-diacetic acid (linker **22**) has been used to link nucleosides to polystyrene or CPG supports. Cleavage of oligonucleotides was carried out with aqueous ammonia [76]. Other diacids with a similar function have also been described [77].

Sulfonyl resins (**23**) have been developed to prepare indoles via a palladium-catalyzed cyclization. Cleavage was carried out with TBAF with excellent yields and purities (85–100%) [78]. Likewise, a library of bivalent ligands (including guanidine, pyridinium and carboxylic and sulfonic acids constituents) for a protein receptor was prepared on nitrobenzenesulfonamide resin **24**. Cleavage was achieved with sodium sulfide [79].

A similar approach, but mechanistically different, has been performed with sulfonate resins. Taking advantage of the good leaving group properties of sulfonates, a sulfonate resin (Figure 15.8) was developed from chloromethyl resin. The resin was compatible with Grignard, Wittig and Suzuki reactions as well as with $NaBH_4$ reduction and reductive aminations. Subsequent reaction of the elaborated bound molecule with nucleophiles such as primary and secondary amines, thiolates and imidazole furnished the corresponding alkylation products [80, 81]. Oxazolidinones have also been prepared by cyclization cleavage [82].

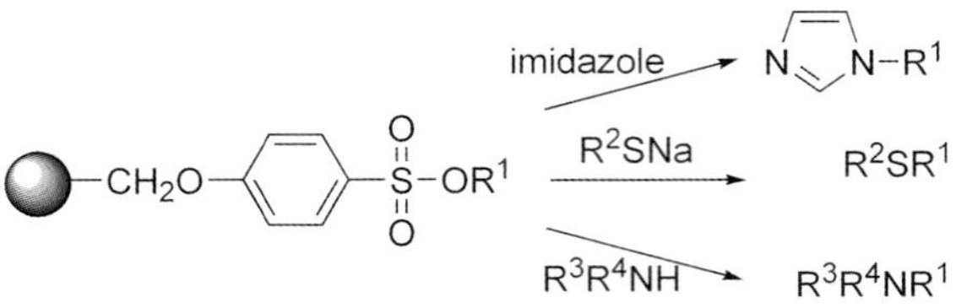

Figure 15.8 Arylsulfonate resins and their cleavage by imidazole, thiolates and primary and secondary amines. The sulfonate resin acts as the leaving group in the nucleophilic displacement reaction.

15.2.2
Intramolecular Nucleophilic Displacement: Cleavage by Cyclization

If the nucleophilic attack to release the target bound compound is carried out by a nucleophilic group within the same molecule (intramolecular attack) then cyclization with concomitant release may take place. The rate of such cyclizations depends on several factors such as the number and type of substituents, the attacking group and, particularly, on the size of the ring formed. Five- and six-membered rings are the most favored, as expected, although some seven-membered rings (benzodiazepines) and large rings such as cyclic peptides are also accessible. The main advantage of the cleavage by cyclization is purity as only the molecule that contains the nucleophilic group will undergo the intramolecular reaction and be released, even if the reaction that served to introduce or unmask the nucleophile proceeds with a low yield. The first solid phase intramolecular cleavage reactions by nucleophilic attack were described in peptide synthesis: diketopiperazine formation and cyclic peptide preparation. The first was a side reaction and reduced the yield of peptide synthesis; the second provided a route to prepare cyclic macrolactam biologically active peptide molecules. Later, with the introduction of combinatorial chemistry and molecular scaffold development, many other heterocyclic structures followed. Next, nucleophilic cyclization strategies in solid phase are overviewed for the most common compounds.

15.2.2.1 Cyclic Peptides

Many strategies have been developed for the synthesis of cyclic peptides. Some of them involved cyclization with simultaneous release from the solid support. The most nucleophile-labile resins have been described for this purpose, such as the oxime resin **19** and the thiol-based resin **12**. The peptide sequence is generally assembled by the Boc/Bn protection scheme due to the lability of the anchoring bond to nucleophiles such as piperidine. However, the Fmoc/tBu assembly of peptides has been described on thiol-functionalized resin **12** by using buffered solutions of DBU (80 mM) and HOBt (74 mM) in DMF to deblock the Fmoc group. This reagent has been demonstrated to minimally affect the stability of the thioester bond linking the assembled peptide to the resin. A cyclic decapeptide, tyrocidine A, has been synthesized following this method. Cyclization occurs spontaneously in aqueous ammonia and no intermolecular ammonolysis products were observed [83–85]. The same authors described a safety-catch acylsulfonamide resin that rendered cyclic peptides by intramolecular nucleophilic cyclization and used it to prepare a library of 192 members [86, 87]. The oxime resin has also been used to prepare cyclic peptides. The sequence is assembled following the Boc/Bn methodology. The N-terminal Boc is then deprotected, the N-terminal ammonium salt is neutralized with DIEA (1.5–2 equiv) and cyclization generally takes place with detachment typically in 24 h (yields are strongly dependent on the sequence, around 50–90% may be achieved) [88–90].

15.2.2.2 Diketopiperazines

Diketopiperazine formation has long been described as a side reaction in peptide synthesis. It occurs after deprotection or neutralization of the α-amino group at the dipeptide stage and reduces the overall yield of the synthesis. However, diketopiperazine structures have also been found in natural products with therapeutic properties and hence they have been used as a scaffolds to design new potential drugs [38, 91, 92]. A typical example would be the synthesis of indolyl diketopiperazine alkaloids. Access to these compounds may be achieved by Pictet–Spengler reaction of L-tryptophan bound to hydroxymethylpolystyrene resin with aldehydes. Fmoc amino acids were then coupled and final Fmoc deprotection resulted in cyclative release to yield alkaloids in 50–99% yields (Figure 15.9) [93, 94].

15.2.2.3 Hydantoins

Two main routes have been described for the preparation of hydantoins: cyclization of urea-derivatized amino acids and of carbamate-bound amino acid carboxamides (Figure 15.10) [95, 96]. In the first route, a bound amino acid is treated with an isocyanate to obtain the urea that is subsequently cyclized and released by acid (HCl) or base (tertiary amines) catalysis upon warming (60–100 °C) [95, 97]. The second route employs a carbamate linkage to attach the amino acid through the amino

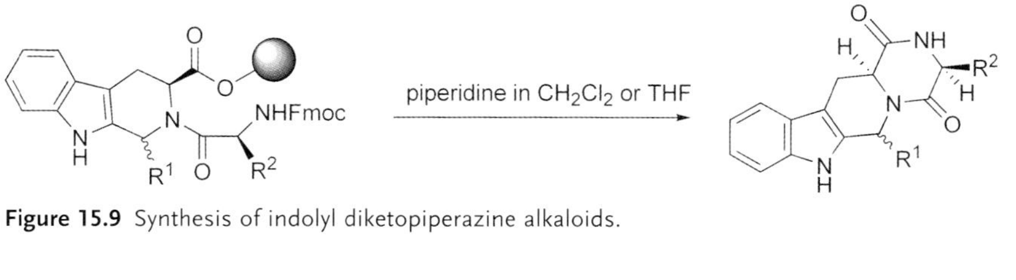

Figure 15.9 Synthesis of indolyl diketopiperazine alkaloids.

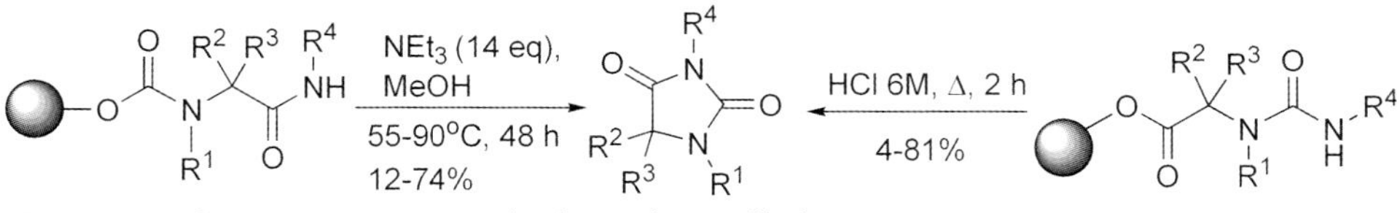

Figure 15.10 The two main strategies for the synthesis of hydantoins.

group. After amidation, cyclative cleavage is promoted by bases at 55–90 °C [96]. The hydantoins are usually obtained with high purity but yields are very variable.

Thiohydantoins have been similarly prepared following the first route but using isothiocyanates [98]. Sulfahydantoins have been also prepared using sulfamoyl chloride [99]. The "phoxime" resin (phosgenated oxime resin, **19**, see Section 15.2.1.1.1) has been used to synthesize 3-aminohydantoins from the oxime carbamates [100].

15.2.2.4 Benzodiazepinones

Cyclative cleavage to obtain the benzodiazepine ring has been described for 1,4-benzodiazepine-2-ones and 1,4-benzodiazepine-2,5-diones. 2-Aminobenzophenone imines were reacted with amino acids esterified to HMPS resin, and subsequent acid catalysis with TFA at 60 °C for 24 h promoted the cyclization with simultaneous cleavage (Figure 15.11) to yield benzodiazepinones [95]. A similar approach consists of anchoring Fmoc-anthranilic acid to an amino acid bound onto Wang resin. Cyclorelease is afforded by base treatment [101]. Another method involves loading of Wang resin with fumaric diacid; o-aminobenzyl alcohols are then coupled to the resulting hemiacid. Mesylation of the alcohol followed by nucleophilic displacement by amines leaves the compound ready for detachment by transesterification with NaOMe and simultaneous cyclization to yield benzodiazepinones with three points of diversity in excellent purity and good overall yield [18].

Many other heterocycles have been prepared following intramolecular nucleophilic cleavage strategies, like pyrazolones [102, 103], benzimidazoles [104], indoles [105], quinazoline-2,4-diones [106], quinolinones [107], tetramic acids [108], oxazolidinones [82] and lactones [109, 110].

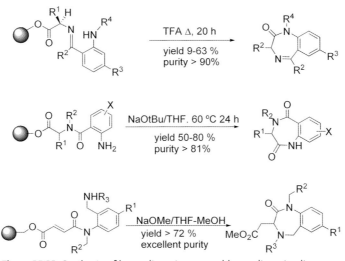

Figure 15.11 Synthesis of benzodiazepinones and benzodiazepinediones.

15.3
Base-Labile Linkers and Resins

Base-labile linkers are designed to release the target molecule by a β-elimination reaction promoted by a base reagent (amines, DBU, hydroxide ion, etc.). The general structure of this family of linkers consists of a resonance electron-withdrawing group, which enhances the acidity of a neighboring α-hydrogen, and a leaving group in a β position (Figure 15.12). The general mechanism is supposed to be an E1cb (unimolecular elimination, E1, with formation of a stable anion of the conjugate base, cb) [111–113]. Examples of electron-withdrawing groups used are fluorene ($pK_{AH} \approx 22.6$ in DMSO), sulfone ($pK_{AH} \approx 30$ in DMSO), and the carbonyl of an ester group ($pK_{AH} \approx 30$ in DMSO) [2].

Several fluorene-based linkers have been developed (Figure 15.13). In the 1980s, 9-(hydroxymethyl)fluorene-4-carboxylic acid was introduced for the synthesis of

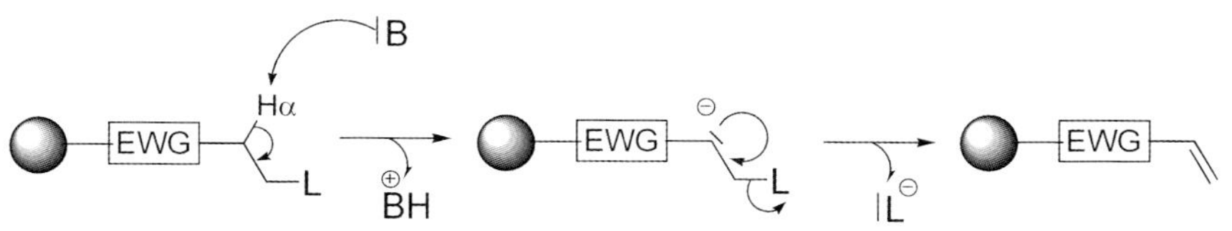

EWG = electron-withdrawing group: sulfone, fluorene (Hα imbedded in the fluorene), nitrophenyl, carbonyl, etc.

Figure 15.12 Generic representation of a base-labile resin and the β-elimination reaction. An electron-withdrawing group (EWG) increases the acidity of an α-proton. The presence of a base (B) induces the elimination reaction with the organic compound of interest behaving as the leaving group (L).

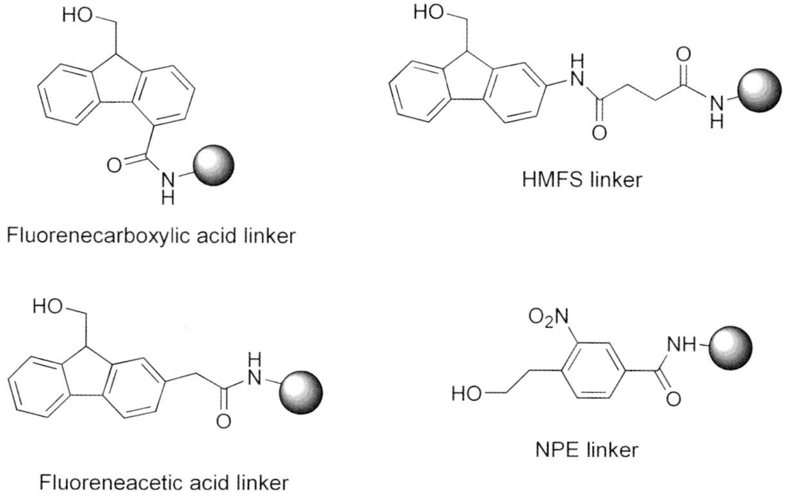

Fluorenecarboxylic acid linker

HMFS linker

Fluoreneacetic acid linker

NPE linker

Figure 15.13 Fluorene-based linkers and nitrophenylethyl (NPE) linker.

protected peptides. Derived from the Fmoc or Fm protecting groups, it was too base-labile and some unexpected peptide loss was observed due to tertiary amines (used in the peptide assembly process) or amines from the decomposition of the DMF [114]. Later, a version that contained a methylene group between the fluorene nucleus and the carboxamide group, 9-(hydroxymethyl)-2-fluoreneacetic acid, was designed to increase stability. However, there was still evidence for partial undesired cleavage in DMF and by the free amino functions of some amino acids [115].

Finally, the presence of an N-acylamido electron-donating group in the HMFS linker provided the desired stability in addition to preserving the lability to cyclic secondary bases such as piperidine (pK_{AH} = 11.1) [116, 117]. Even the weaker base morpholine (pK_{AH} = 8.3) cleaved quantitatively the protected peptide (20% in DMF, 1–2 h) a fact that reduced side reactions [118, 119]. The HMFS linker has been successfully used in combinatorial chemistry [120], oligonucleotide synthesis [121], peptide-oligonucleotide hybrids [122, 123] and in the preparation of large proteins following a convergent approach, such as the *Aequorea* green fluorescent protein (238 amino acids) that involved the preparation and assembly of 26 fully protected peptides [124].

The NPE linker is based on a design similar to that of fluorene-based linkers [125, 126]. Here, the acidity of the α-hydrogens of the benzylic position is enhanced by the presence of two electron-withdrawing groups (nitro and carboxamide carbonyl groups). This linker has been described for the solid phase synthesis of oligonucleotides (via a carbonate bond), oligonucleotide 3'-phosphates and Boc/Bn protected peptides (via an ester bond). The NPE resin was stable to tertiary amines like triethylamine (40% in pyridine, which is used to remove the 2-cyanoethyl phosphate protecting group) and DIEA (5% in DMF, which is used to neutralize the ammonium salt after Boc removal in peptide synthesis). Oligonucleotides are generally cleaved with DBU (0.5 M in pyridine) or concentrated ammonia (16 h, 55 °C) [125] and peptides with DBU (0.1 M in dioxane, 2 h, rt) or piperidine (20% in DMF, 2 h, rt) [126].

The classical Hofmann elimination reaction (which dates back to 1851) has been adapted to the solid phase in combination with the Michael addition. The REM resin, called this way because the resin linker is REgenerated after product cleavage and functionalized by means of a Michael addition, has been developed to prepare arrays of tertiary amines. The procedure involves acylation of hydroxymethylpolystyrene with acrylic chloride to furnish the acrylate on resin. Then, a secondary amine, whose substituents offer two potential sites of diversity, is bound by Michael addition. Quaternization of the amine with an alkyl halide (or reductive amination) introduces another site of diversity and activates the linker to release the amine by a Hofmann elimination with DIEA (Figure 15.14) [127–129]. Additionally, the use of a second basic resin has been described as a source reagent to promote the elimination [130, 131].

A similar approach has been described for vinylsulfone-type resins [132, 133]. Vinyl sulfones are stable to a wider range of chemical conditions than the REM benzyl ester analogs (Figure 15.15). Resins **25** and **26** showed moderate stability

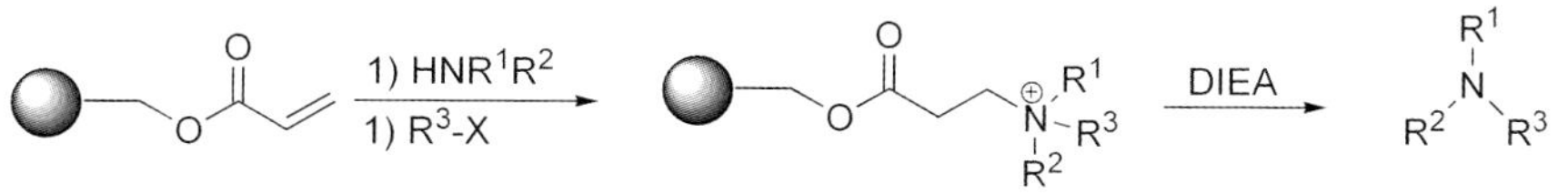

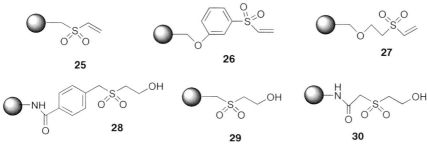

Figure 15.14 Preparation of tertiary amines with the REM linker, based on the Hofmann elimination reaction.

Figure 15.15 Sulfone-based resins. Resins **25–27** are used as alternatives to REM linkers. Resin **28** is based on the safety-catch principle while resins **29** and **30** were used to synthesize peptides.

to acids and offered high stability to nucleophiles, including Grignard reagents. Tertiary amines were prepared in moderate to good yields. Linker **27** was similarly used to prepare tetrahydroquinolines with yields ranging from 25 to 100% [133, 134].

The safety-catch principle was applied to design linker **28**. Aminoarylsulfon-amides were prepared on it [by means of 4-(chlorosulfonyl)phenyl-isocyanate and amines] and detached with 10% ammonia in 2,2,2-trifluoroethanol [135]. Last but not least, sulfone linkers have also been designed to synthesize peptides. Linker **29** was designed in the 1970s to synthesize peptides like ACTH (5–10) following a Boc/Bn protection strategy. Detachment was carried out with NaOH in MeOH:-dioxane and was very fast (66% in 3 min) [136, 137]. These conditions were described not to induce epimerization of the peptide. Linker **30** was also used to prepare Leu-enkephalin analogs and release was achieved under similar conditions [138]. Similar approaches have been described for the synthesis of dehydroalanine and dehydropeptides from cysteine bound to Merrifield resin [139] and N-benzylpiperidin-4-ones [140].

Finally, a solid phase version of the Reissert reaction has been developed for combinatorial chemistry purposes, and differently substituted isoquinolines have been prepared (Figure 15.16). Starting from polymer-supported benzoyl chloride, isoquinoline was bound and treatment with trimethylsilyl cyanide (TMSCN), followed by alkylation, yielded resin **31**. After performing synthetic transformations (carbonylation etc.), Reissert hydrolysis was induced with aqueous NaOH in THF.

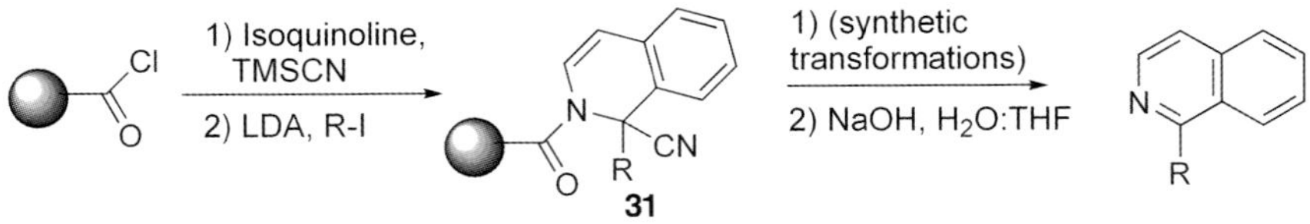

Figure 15.16 Reissert reaction for the preparation of isoquinolines.

Isoxazolinoisoquinoline heterocycles and several C1-alkylated isoquinolines were thus obtained [141].

In summary, many chemical organic reactions based on nucleophilic displacement or elimination by bases have been adapted to solid phase procedures to take advantage of this methodology. The most general ones have described in this chapter. Some specific linkers and resins designed *ad hoc* for particular reactions have not been treated here (Wittig, Wittig–Horner, Weinreb amide synthesis and others). The reader can access them through more general reviews [142–146].

Acknowledgments

Funding by MEC and FEDER (CTQ 2007–65615/BQU) is greatly acknowledged.

List of Abbreviations

AM	aminomethylpolystyrene
BHA	benzhydrylamine resin
Bn	benzyl
Boc	*tert*-butoxycarbonyl
tBu	*tert*-butyl
CPG	controlled pore glass
DBU	1,8-diazabicyclo[5.4.0]undec-7-ene
DMAE	N,N-dimethylaminoethanol
DMSO	dimethyl sulfoxide
Fmoc	fluorenylmethoxycarbonyl
HMB	4-hydroxymethylbenzoic
HMPS	hydroxymethylpolystyrene
MBHA	4-methylbenzhydrylamine resin
Nbb	3-nitrobenzamidobenzyl
Nu	nucleophile
PAM	4-(hydroxymethyl)phenylacetamidomethyl
PS	polystyrene
rt	room temperature
SPPS	solid phase peptide synthesis

TBAF	tetrabutylammonium fluoride
TBDMS	*tert*-butyldimethylsilyl
TG	TentaGel resin
THF	tetrahydrofuran
TMSCN	trimethylsilylcyanide

References

1 Clayden, J., Greeves, N., Warren, S. and Wothers, P. (2001) Chapter 12, in *Organic Chemistry*, Oxford University Press, UK.
2 Bordwell pKa table Avialable at http://www.chem.wisc.edu/areas/reich/pkatable/index.htm (accessed 11.04.2008)
3 Merrifield, R.B. (1963) *Journal of the American Chemical Society*, **85**, 2149–54.
4 Chamoin, S., Houldsworth, S. and Snieckus, V. (1998) *Tetrahedron Letters*, **39**, 4175–8.
5 Fancelli, D., Fagnola, M.C., Severino, D. and Bedeschi, A. (1997) *Tetrahedron Letters*, **38**, 2311–14.
6 Goldwasser, J.M. and Leznoff, C.C. (1978) *Canadian Journal of Chemistry – Revue Canadienne de Chimie*, **56**, 1562–8.
7 Ueki, M., Kai, K., Amemiya, M., *et al.* (1988) *Journal of the Chemical Society D – Chemical Communications*, 414–5.
8 Nicolás, E., Clemente, J., Ferrer, T., *et al.* (1997) *Tetrahedron*, **53**, 3179–94.
9 Baleux, F., Calas, B. and Mery, J. (1986) *International Journal of Peptide and Protein Research*, **28**, 22–8.
10 Tam, J.P., Cunnigham-Rundles, W.F., Erickson, B.W. and Merrifield, R.B. (1977) *Tetrahedron Letters*, **18**, 4001–4.
11 Mizoguchi, T., Shigezane, K. and Takamura, N. (1970) *Chemical and Pharmaceutical Bulletin*, **18**, 1465–74.
12 Beyerman, H.C., Hindriks, H. and De Leer, E.W.B. (1968) *Journal of the Chemical Society D – Chemical Communications*, 1668.
13 Leznoff, C.C. and Goldwasser, J.M. (1977) *Tetrahedron Letters*, 1875–8.
14 Kang, S.K., Kim, J.S., Yoon, S.K., Lim, K.H. and Yoon, S.S. (1998) *Tetrahedron Letters*, **39**, 3011–12.
15 Frenette, R. and Friesen, R.W. (1994) *Tetrahedron Letters*, **35**, 9177–80.
16 Hutchins, S.M. and Chapman, K.T. (1996) *Tetrahedron Letters*, **37**, 4869–72.
17 Cheng, Y. and Chapman, K.T. (1997) *Tetrahedron Letters*, **87**, 1497–500.
18 Bhalay, G., Blaney, P., Palmer, V.H. and Baxter, D. (1997) *Tetrahedron Letters*, **38**, 8375–8.
19 Tietze, L.F., Hippe, T. and Steinmetz, A. (1996) *Synlett*, 1043–4.
20 Kuster, G.J. and Scheeren, H.W. (1998) *Tetrahedron Letters*, **39**, 3613–16.
21 Seebach, D., Thaler, A., Baser, D. and Ko, S.Y. (1991) *Helvetica Chimica Acta*, **74**, 1102–18.
22 Mergler, M. and Nyfeler, R. (1994) Innovation and perspectives in solid phase synthesis. Peptides, proteins and nucleic acids, in *Biological and Biomedical Applications* (ed. R. Epton), Mayflower Worldwide Ltd., Oxford, UK, pp. 559–602.
23 Smith, J., Liras, J.L., Schneider, S.E. and Anslyn, E.V. (1996) *Journal of Organic Chemistry*, **61**, 8811–18.
24 Hudson, D., Kenner, G.W., Sharpe, R. and Szelke, M. (1979) *International Journal of Peptide and Protein Research*, **14**, 177–85.
25 Rivaille, P., Gautron, J.P., Castro, B. and Milhaud, G. (1980) *Tetrahedron*, **36**, 3413–19.
26 Beyerman, H.C., Boers-Boonekamp, C.A.M. and Maasen van den Brink, H. (1968) *Recueil Des Travaux Chimiques Des Pays-Bas*, **87**, 275–73.
27 Takashima, H., DuVigneaud, V. and Merrifield, R.B. (1968) *Journal of the American Chemical Society*, **90**, 1323–5.
28 Beyerman, H.C. and Maasen van den Brink-Zimmermannovà, H. (1968) *Recueil Des Travaux Chimiques Des Pays-Bas*, **87**, 1196–200.

29 Coy, D.H., Coy, E.J., Schally, A.V. and Vilchez-Martinez, J.A. (1975) *Journal of Medicinal Chemistry*, **18**, 275–77.
30 Haviv, F., Palabrica, C.A., Bush, E.N., et al. (1989) *Journal of Medicinal Chemistry*, **32**, 2340–4.
31 Atherton, E., Logan, C.J. and Sheppard, R.C. (1981) *Journal of the Chemical Society–Perkin Transactions 1*, 538–46.
32 Nielsen, T.E. and Meldal, M. (2004) *Journal of Organic Chemistry*, **69**, 3765–73.
33 Diness, F., Beyer, J. and Meldal, M. (2007) *Chemistry–A European Journal*, **12**, 8056–66.
34 Danieli, E., Trabocchi, A., Menchi, G. and Guarna, A. (2005) *European Journal of Organic Chemistry*, 4372–81.
35 Fridkin, M., Patchornik, A. and Katchalski, E. (1968) *Journal of the American Chemical Society*, **90**, 2953–7.
36 Breitenbucher, J.G. and Johnson, C.R. (1998) *Tetrahedron Letters*, **39**, 1295–8.
37 Breitenbucher, J.G. and Hui, H.C. (1998) *Tetrahedron Letters*, **39**, 8207–10.
38 Fantauzzi, P.P. and Yager, K.M. (1998) *Tetrahedron Letters*, **39**, 1291–4.
39 DeGrado, W.F. and Kaiser, E.T. (1980) *Journal of Organic Chemistry*, **45**, 1295–300.
40 DeGrado, W.F. and Kaiser, E.T. (1982) *Journal of Organic Chemistry*, **47**, 3258–61.
41 Lansbury, P.T., Hendrix, J.C. and Coffman, A.I. (1989) *Tetrahedron Letters*, **30**, 4915–18.
42 Kaiser, E.T., Mihara, H., Lafforet, G.A., et al. (1989) *Journal of the American Chemical Society*, **243**, 187–92.
43 Voyer, N., Lavoie, A., Pinette, M. and Bernier, J. (1994) *Tetrahedron Letters*, **35**, 355–8.
44 Golebiowski, A. and Klopfenstein, S. (1998) *Tetrahedron Letters*, **39**, 3397–400.
45 Schwabacher, A.W. and Maynard, T.L. (1993) *Tetrahedron Letters*, **34**, 1269–70.
46 Sasaki, T., Findeis, M.A. and Kaiser, E.T. (1991) *Journal of Organic Chemistry*, **56**, 3159–68.
47 Hamuro, Y., Scialdone, M.A. and DeGrado, W.F. (1999) *Journal of the American Chemical Society*, **121**, 1636–44.
48 Shao, H., Zhang, Q., Goodnow, R., Chen, L. and Tam, S. (2000) *Tetrahedron Letters*, **41**, 4257–60.
49 Eggenweiler, H.M., Clausen, N. and Bayer, E. (1998) Peptides 1996, in *Proceedings of the 24th European Peptide Symposium* (ed. R. Ramage and R. Epton), Mayflower Scientific, Kingswinford, UK, pp. 359–60.
50 Vlattas, I., Dellureficio, J., Duna, R., et al. (1997) *Tetrahedron Letters*, **38**, 7321–4.
51 Kobayashi, S., Hachiya, I. and Yasuda, M. (1996) *Tetrahedron Letters*, **37**, 5569–72.
52 Chen, C., McDonald, I.A. and Muñoz, B. (1998) *Tetrahedron Letters*, **39**, 217–20.
53 Scialdone, M.A., Shuey, S.W., Soper, P., et al. (1998) *Journal of Organic Chemistry*, **63**, 4802–7.
54 Dressman, B.A., Singh, J. and Kaldor, S.W. (1998) *Tetrahedron Letters*, **39**, 3631–4.
55 Raju, B. and Kogan, T.P. (1997) *Tetrahedron Letters*, **38**, 3373–6.
56 Fitzpatrick, L.J. and Rivero, R.A. (1997) *Tetrahedron Letters*, **38**, 7479–82.
57 Dodd, D.S. and Wallace, O.B. (1998) *Tetrahedron Letters*, **39**, 5701–4.
58 Allin, S.M. and Shuttleworth, S.J. (1996) *Tetrahedron Letters*, **37**, 8023–6.
59 Purandare, A.V. and Natajaran, S. (1997) *Tetrahedron Letters*, **38**, 8777–80.
60 Winkler, J.D. and McCoull, W. (1998) *Tetrahedron Letters*, **39**, 4935–6.
61 Huang, X. and Tang, J. (2003) *Tetrahedron*, **59**, 4851–6.
62 Nielsen, J. and Rasmussen, P.H. (1996) *Tetrahedron Letters*, **37**, 3351–4.
63 Chabra, S.R., Khan, A.N. and Bycroft, B.W. (1998) *Tetrahedron Letters*, **39**, 3585–8.
64 Bannwarth, W., Huebscher, J., Barner, R. and Bycroft, B.W. (1996) *Bioorganic & Medicinal Chemistry Letters*, **13**, 1525–8.
65 Furth, P.S., Reitman, M.S. and Cook, A.F. (1997) *Tetrahedron Letters*, **38**, 5403–6.
66 Leznoff, C.C. and Wong, J.Y. (1972) *Canadian Journal of Chemistry–Revue Canadienne de Chimie*, **50**, 2892–3.
67 Leznoff, C.C. and Dixit, D.M. (1977) *Canadian Journal of Chemistry–Revue Canadienne de Chimie*, **55**, 3351–5.

68 Leznoff, C.C. and Svirskaya, P.I. (1978) *Angewandte Chemie – International Edition in English*, **17**, 947.
69 Panek, J.S. and Zhu, B. (1996) *Tetrahedron Letters*, **37**, 8151–4.
70 Kurth, M.J., Randall, L.A. and Takenouchi, K. (1996) *Journal of Organic Chemistry*, **61**, 8755–61.
71 Kantorowski, E. and Kurth, M.J. (1997) *Journal of Organic Chemistry*, **62**, 6797–903.
72 Schore, N.E. and Najdi, S.D. (1990) *Journal of the American Chemical Society*, **112**, 441–2.
73 Adinolfi, M., Barone, G., Napoli, L., *et al.* (1996) *Tetrahedron Letters*, **37**, 5007–10.
74 Zhu, T. and Boons, G.J. (1998) *Angewandte Chemie – International Edition in English*, **37**, 1898–900.
75 Swistok, J., Tilley, J.W., Danho, W., *et al.* (1989) *Tetrahedron Letters*, **30**, 5045–448.
76 Pon, R.T. and Yu, S. (1997) *Tetrahedron Letters*, **38**, 3327–30.
77 Berteina, S. and DeMesmaeker, A. (1998) *Tetrahedron Letters*, **39**, 5759–62.
78 Zhang, H.C., Ye, H., Meretto, A.F., *et al.* (2000) *Organic Letters*, **2**, 89–92.
79 Pattarawarapan, M., Chen, J., Steffensen, M. and Burgess, K. (2001) *Journal of Combinatorial Chemistry*, **3**, 102–16.
80 Rueter, J.K., Nortey, S.O., Baxter, E.W., *et al.* (1998) *Tetrahedron Letters*, **39**, 975–8.
81 Baxter, E.W., Rueter, J.K., Nortey, S.O. and Reitz, A.B. (1998) *Tetrahedron Letters*, **39**, 979–82.
82 Ten Holte, P., Van Esseveldt, B.C.J., Thijs, L. and Zwanenburg, B. (2001) *European Journal of Organic Chemistry*, 2965–9.
83 Bu, X., Xie, G., Law, C.W. and Guo, Z. (2002) *Tetrahedron Letters*, **43**, 2419–22.
84 Bu, X., Wu, X., Xie, G. and Guo, Z. (2002) *Organic Letters*, **4**, 2893–5.
85 Bu, X., Wu, X., Ng, N.L.J., *et al.* (2004) *Journal of Organic Chemistry*, **69**, 2681–5.
86 Qin, C., Zhong, X., Ng, N.L.J., *et al.* (2004) *Tetrahedron Letters*, **45**, 217–20.

87 Qin, C., Bu, X., Zhong, X., *et al.* (2004) *Journal of Combinatorial Chemistry*, **6**, 398–406.
88 Ösapay, G., Profit, A. and Taylor, J.W. (1990) *Tetrahedron Letters*, **31**, 6121–4.
89 Ösapay, G. and Taylor, J.W. (1990) *Journal of the American Chemical Society*, **112**, 6046–51.
90 Mihara, H., Yamabe, S., Niidome, T., *et al.* (1995) *Tetrahedron Letters*, **36**, 4837–40.
91 Kowalski, J. and Lipton, M.A. (1996) *Tetrahedron Letters*, **37**, 5839–40.
92 Li, W.R. and Peng, S.Z. (1998) *Tetrahedron Letters*, **39**, 7373–6.
93 Van Loevezijn, A., van Maarseveen, J.H., Stegman, K., *et al.* (1998) *Tetrahedron Letters*, **39**, 4737–40.
94 Wang, H. and Ganesan, A. (1999) *Organic Letters*, **1**, 1647–9.
95 De Witt, S.H., Kiely, J.S., Stankovic, C.J., *et al.* (1993) *Proceedings of the National Academy of Sciences of the United States of America*, **90**, 6909–13.
96 Dressman, B.A., Spangle, L.A. and Kador, S.W. (1996) *Tetrahedron Letters*, **37**, 937–40.
97 Kim, S.W., Ahn, S.Y., Koh, J.S., *et al.* (1997) *Tetrahedron Letters*, **38**, 4603–6.
98 Matthews, J. and Rivero, R.A. (1997) *Journal of Organic Chemistry*, **62**, 6090–2.
99 Albericio, F., Garcia, J., Michelotti, E.L., Nicolás, E. and Tice, M.C. (2000) *Tetrahedron Letters*, **41**, 3161–3.
100 Hamuro, Y., Marshall, W.J. and Scialdone, M.A. (1999) *Journal of Combinatorial Chemistry*, **1**, 163–72.
101 Mayer, J.P., Zhang, J., Bjergarde, K., *et al.* (1996) *Tetrahedron Letters*, **37**, 8081–4.
102 Tietze, L.F. and Steinmetz, A. (1996) *Synlett*, **39**, 667–8.
103 Tietze, L.F. and Steinmetz, A. (1997) *Bioorganic & Medicinal Chemistry Letters*, **7**, 1303–6.
104 Huang, W. and Scarborough, R.M. (1999) *Tetrahedron Letters*, **40**, 2665–8.
105 Macleod, C., Hartley, R.C. and Hampercht, D.W. (2002) *Organic Letters*, **4**, 75–8.
106 Gouilleux, L., Fehrentz, J.A., Winternitz, F. and Martinez, J. (1996) *Tetrahedron Letters*, **37**, 7031–4.
107 Sim, M.M., Lee, C.L. and Ganesan, A. (1998) *Tetrahedron Letters*, **39**, 6399–402.

108 Matthews, J. and Rivero, R.A. (1998) *Journal of Organic Chemistry*, **63**, 4808–10.
109 Gouault, N., Cupif, J.F., Sauleau, A. and David, M. (2000) *Tetrahedron Letters*, **41**, 7293–7.
110 Le Hete, C., David, M., Carreaux, F., et al. (1997) *Tetrahedron Letters*, **38**, 5153–6.
111 Larey, F.A. and Sumdberg, R.Z.(1990) in *Advanced Organic Chemistry, Part A: Structure and Mechanisms* 3rd edition Plenum Press, New York, pp. 368–73.
112 More O'Ferral, R.A. (1970) *Journal of the Chemical Society B – Physical Organic*, 274–7.
113 More O'Ferral, R.A. and Slae, S. (1970) *Journal of the Chemical Society B – Physical Organic*, 260–8.
114 Mutter, M. and Bellof, D. (1984) *Helvetica Chimica Acta*, **67**, 2009–16.
115 Ling, Y.Z., Ding, S.H., Chu, J.Y. and Felix, A.M. (1990) *International Journal of Peptide and Protein Research*, **35**, 95–8.
116 Rabanal, F., Giralt, E. and Albericio, F. (1992) *Tetrahedron Letters*, **33**, 1775–8.
117 Rabanal, F., Giralt, E. and Albericio, F. (1995) *Tetrahedron*, **51**, 979–82.
118 Rabanal, F., Pastor, J.J., Nicolás, E., et al. (2000) *Tetrahedron Letters*, **42**, 8093–6.
119 Albericio, F., Cruz, M., Debéthune, L., et al. (2001) *Synthetic Communications*, **31**, 225–32.
120 Pastor, J.J., Fernandez, I., Rabanal, F. and Giralt, E. (2002) *Organic Letters*, **4**, 3831–3.
121 Robles, J., Beltran, M., Marchán, V., et al. (1999) *Tetrahedron*, **55**, 13251–64.
122 Marchán, V., Pulido, D., Pedroso, E. and Grandas, A. (2006) *European Journal of Organic Chemistry*, 958–63.
123 Marchán, V., Debéthune, L., Pedroso, E. and Grandas, A. (2004) *Tetrahedron*, **60**, 5461–9.
124 Nishiuchi, Y., Inui, T., Nishio, H., et al. (1998) *Proceedings of the National Academy of Sciences of the United States of America*, **95**, 13549–54.
125 Robles, J., Fernández-Forner, D., Albericio, F., Giralt, E. and Pedroso, E. (1991) *Tetrahedron Letters*, **32**, 1511–14.
126 Albericio, F., Giralt, E. and Eritja, R. (1991) *Tetrahedron Letters*, **32**, 1515–18.
127 Morphy, J.R., Rankovic, Z. and Rees, D.C. (1996) *Tetrahedron Letters*, **32**, 3209–12.
128 Brown, A.R., Rees, D.C., Rankovic, Z. and Morphy, J.R. (1997) *Journal of the American Chemical Society*, **119**, 3288–95.
129 Morphy, J.R., Rankovic, Z. and York, M. (2002) *Tetrahedron Letters*, **43**, 6413–15.
130 Yamamoto, Y., Tanabe, K. and Okonogi, T. (1999) *Chemistry Letters*, 103–4.
131 Ouyang, X., Armstrong, R.W. and Morphy, M.M. (1998) *Journal of Organic Chemistry*, **63**, 1027–32.
132 Kroll, F.E.K., Morphy, R., Rees, D. and Gani, D. (1997) *Tetrahedron Letters*, **38**, 8573–6.
133 Heinonen, P. and Lönnberg, H. (1997) *Tetrahedron Letters*, **38**, 8569–72.
134 Aktar, M., Kroll, F.E.K. and Gani, D. (2000) *Tetrahedron Letters*, **38**, 4487–91.
135 Garcia-Echeverria, C. (1997) *Tetrahedron Letters*, **38**, 8933–4.
136 Tesser, G.I., Buis, J.W.A.R.M., Wolters, E.T.M. and Bothé-Helmes, E.G.A.M. (1976) *Tetrahedron*, **32**, 1069–72.
137 Buis, J.W.A.M., Tesser, G.I. and Nivard, R.J.F. (1976) *Tetrahedron*, **32**, 2321–5.
138 Katti, S.B., Misra, P.K., Haq, W. and Mathur, K.B. (1992) *Journal of the Chemical Society D – Chemical Communications*, 843–4.
139 Yamada, M., Miyahima T. and Horikawa, H. (1998) *Tetrahedron Letters*, **39**, 289–92.
140 Barco, A., Nenetti, S., De Rissi, C., et al. (1998) *Tetrahedron Letters*, **39**, 7591–4.
141 Lorsbach, B.A., Bagdanoff, J.T., Miller, R.B. and Kurth, M.J. (1998) *Journal of Organic Chemistry*, **63**, 2244–50.
142 James, I.W. (1999) *Tetrahedron*, **39**, 4855–946.
143 Guillier, F., Orain, D. and Bradley, M. (2000) *Chemical Reviews*, **100**, 2091–157.
144 Blaney, P., Grigg, R. and Sridharan, V. (2002) *Chemical Reviews*, **102**, 2607–24.
145 Blackburn, C. (1998) *Biopolymers*, **47**, 311–51.
146 Krchnak, V. and Holladay, M.W. (2002) *Chemical Reviews*, **102**, 61–91.

16
Safety-Catch and Traceless Linkers in Solid Phase Organic Synthesis

Matthias Sebastian Wiehn, Nicole Jung, and Stefan Bräse

University of Karlsruhe (TH), Institute for Organic Chemistry, D-76131 Karlsruhe, Germany

16.1
Introduction

The design of a chemical synthesis of a complex structure often requires the careful consideration of protective group strategies [1]. Although many – and very successful – attempts have been made to reduce or even avoid protective groups [2, 3], the broad range of reaction conditions, the combination of reagents and the sensitivity of building blocks require the capping of functional groups.

Translated to solid phase chemistry, this would mean the careful selection of the appropriate linkage of the substrates to the solid support. An ever increasing number of various linker types is now available, most of them used very successfully in the synthesis of complex molecules on solid supports [4–7].

In this chapter, we present two special types of linkers, the so-called safety-catch linkers and traceless linkers.

16.2
Safety-Catch Linkers

A safety-catch linker is defined as "a linker which is cleaved by performing two different reactions instead of the normal single step, thus providing better control over the timing of compound release" [8]. The "safety-catch" principle consists of a linker system that is inert throughout all operations of the synthesis and has to be converted before the cleavage step from its stable form into an activated one that is labile towards the cleavage conditions.

Since "safety-catch" means the activation of the linker prior to cleavage, such a system can be applied for monodirectional, such as traceless linkers, or multi-functional linkers [9] as well as for cleavage-cyclization strategies. Table 16.1 gives an overview of the safety-catch linker types known to date. Slight differences

The Power of Functional Resins in Organic Synthesis. Judit Tulla-Puche and Fernando Albericio
Copyright © 2008 WILEY-VCH Verlag GmbH & Co. KGaA, Weinheim
ISBN: 978-3-527-31936-7

Table 16.1 Overview of safety-catch linker types.

Structure	Activation	Cleavage	Reference
2, 5 Sulfonamide linker Kenner- and Ellman-type (alkylation)	Alkylation with CH_2N_2, $TMSCHN_2$ or ICH_2CN; allylation with Pd(0); alkylation with N,N',O-trialkyl-isoureas	Nucleophiles (hydroxide, amines, hydrazine, thiols)	[10–17]
10 Reversed Kenner linker (alkylation)	Alkylation with MeI	NH_3/MeOH	[16, 18]
13 (oxidation)	Oxidation with H_2O_2 or mCPBA	Nucleophiles (hydroxides, amines)	[19–21]
17 Also oxazoles instead of pyrimidines (oxidation)	Oxidation with mCPBA or oxone	Nucleophilic aromatic substitution with amines or azides	[22–28]
20 (oxidation)	Oxidation with mCPBA	NH_4OH, base to give β-elimination (DBU, Me_2NH, NaOtBu)	[29–32]
21 (oxidation)	Oxidation with mCPBA	Base to give β-elimination (DBU, Me_2NH)	[32, 33]
24 (oxidation)	Oxidation with tBuOOH/CSA	Pummerer rearrangement with TFAA and cleavage with Et_3N or $Et_3N/NaBH_4$	[34]
28 (oxidation)	Oxidation to sulfoxide/selenoxide with H_2O_2 or H_2O_2/HFIP	Pericyclic elimination (spontaneous or by heating)	[35–38]

Table 16.1 Continued.

Structure	Activation	Cleavage	Reference
31	Oxidation with mCPBA	Bases (KOtBu, LDA)	[39]
32 SCAL linker	Me₃SiBr/PhSMe	TFA/(EtO)₂(PS)SH or TFA	[40, 41]
35 DSA linker	SiCl₄/PhSMe/ PhOMe/EDT	TFA	[42]
36 DSB linker	SiCl₄/PhSMe/ PhOMe/EDT	TFA	[43]
37 HMPPA linker	Reduction with TMSBr/EDT	TFA	[44]
38 TRAM linker	N-deprotection and quaternization with RX	Base (DIEA) to give β-elimination	[45]
39 REM linker	Alkylation	Base (DIEA) to give β-elimination	[46]

Table 16.1 Continued.

Structure	Activation	Cleavage	Reference
40 (aryl with NO$_2$ → reduction; acyloxy with R)	Reduction with SnCl$_2$/HCl	TFA	[47]
43 (resin–O–aryl(NO$_2$)–CH$_2$OR; reduction and tosylation)	Reduction with SnCl$_2$ and tosylation	TFA	[48]
44 (aryl hydrazide; oxidation)	Oxidation with NBS (N-bromosuccinimide) or Cu(OAc)$_2$	Nucleophiles (amines, alcohols)	[49–54]
45 (hydrazide with aryl–R; oxidation)	Oxidation with NBS or Cu(OAc)$_2$	Nucleophiles (amines, alcohols)	[55, 56]
46 (dihydroquinoline with Ph, N–C(O)R; oxidation)	Oxidation with DDQ, CAN or Ph$_3$CBF$_4$	Nucleophiles (amines, H$_2$O)	[57]
50 (benzylic OH with phenyl, O–C(O)R; dehydration)	Dehydration with TFA	Amines	[58]
54 (TG resin, Gln side chain ester to p-hydroxybenzyl–OC(O)R; NBoc proline; deprotection)	Boc-removal with TFA	Phosphate buffer	[59–61]
57 (CPG resin, succinate ester, NHAlloc, phosphoramidite with CN; deprotection)	Alloc-removal with Pd(0)	Et$_3$N/NH$_3$	[62]

Table 16.1 Continued.

Structure	Activation	Cleavage	Reference
61 TG resin	Boc-removal with TFA	Phosphate buffer	[63]
62	Boc-removal with TFA	Spontaneous	[64]
63	N-deprotection with $Na_2S_2O_4/H^+$	Et_3N	[65]
64	Deprotection with TBAF	TFA	[66]
65 X = O, NH	Penicillin G acylase, pH 7, MeOH	Spontaneous or by heating	[67, 68]
66	Deprotection with TFA	Amines	[69–72]
69	Deprotection with TFA	Spontaneous	[64]

Table 16.1 Continued.

Structure	Activation	Cleavage	Reference
70 (protection)	N-Protection with Boc	Hydrolysis with LiOH/ H_2O_2 or alcoholysis with MeOH/MeONa	[73]
71 (acylation)	Reaction with benzyl isocyanate to give urea	iPr$_2$NH to give resin bound hydantoin and release of the alcohol	[74]
75 Dpr(Phoc) linker (basic hydrolysis)	NaOH to give isocyanate intermediate	Nucleophiles (NaOH or NH$_3$ in iPrOH)	[75–77]
80 (complexation)	Complexation with Co$_2$(CO)$_8$	TFA	[78]
83 (removal of dithioketal)	Removal of dithioketal with Hg(ClO$_4$)$_2$ or H$_5$IO$_6$	Irradiation ($h\nu$)	[79–81]
86 (hydrolysis)	Acidic hydrolysis	Irradiation ($h\nu$)	[82]
87 (hydrolysis)	Hydrolysis with PPTS	Nucleophiles (amines, hydroxides, alcohols)	[83]

concerning the attachment of the linker to the resin are not considered (wavy lines indicate several possible linkage strategies; for a detailed description the literature has to be consulted).

The first safety-catch linker was developed by Kenner et al. in 1971 [10]. Starting from sulfonated polystyrene-divinylbenzene copolymer (**1**), successive treatment with chlorosulfonic acid and aqueous ammonia led to the polymer-bound sulfonamides. In a next step activated N-protected amino acids were coupled with the sulfonamide functionality, yielding N-acylsulfonamides (**2**). These are stable to strong anhydrous acids such as TFA or HBr/AcOH as well as to strongly nucleophilic reagents and aqueous alkali since basic attack ionizes the acidic sulfonamide group. Kenner et al. activated the linker by N-alkylation with diazomethane (Scheme 16.1). The resulting N-acyl-N-alkyl-sulfonamides **3** can be cleaved easily through nucleophilic attack on the carbonyl carbon by hydrolysis, aminolysis or hydrazinolysis to give the corresponding carboxylic acids, amides or hydrazides (**4**) [10, 15].

The concept of Kenner's sulfonamide safety-catch linker was adapted by Backes and Ellman, who created further derivatives of the linker and worked on the activation methods [11–14]. It was shown that N-alkylation can also be achieved by reaction with ICH_2CN, $TMSCHN_2$ or by transformation with O,N,N'-trialkylureas as Link et al. demonstrated more recently [16]. Another example for the activation of Ellman's sulfonamide linker is the palladium-catalyzed allylation by Kiessling et al. who used thiols (**8**) as nucleophilic cleaving reagents, yielding the corresponding thioesters (**9**) [17] (Scheme 16.2).

In 2001, Maclean et al. modified the Kenner safety-catch strategy to obtain secondary sulfonamides **10**. Using this reversed Kenner linker, the carboxylic acid remains attached on the solid support while the sulfonamide unit **12** is released into the solution [18] (Scheme 16.3).

The oxidation of sulfides to sulfoxides or sulfones, as well as the reverse reduction strategy, is a well-known method for the design of a safety-catch protocol. As nucleophilic displacement is facilitated if electron-withdrawing groups are placed

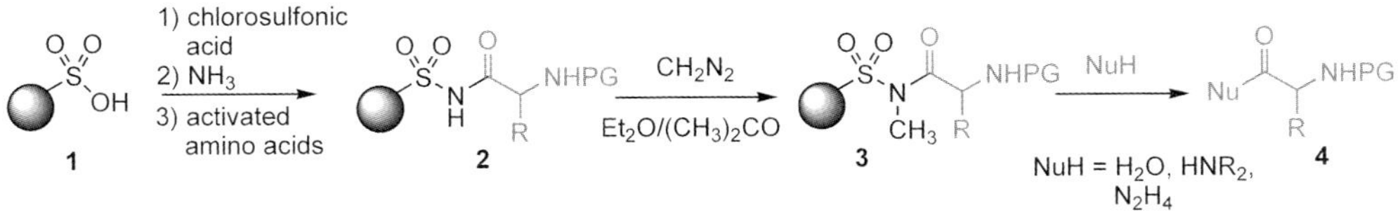

Scheme 16.1 Preparation and use of Kenner's safety-catch linker.

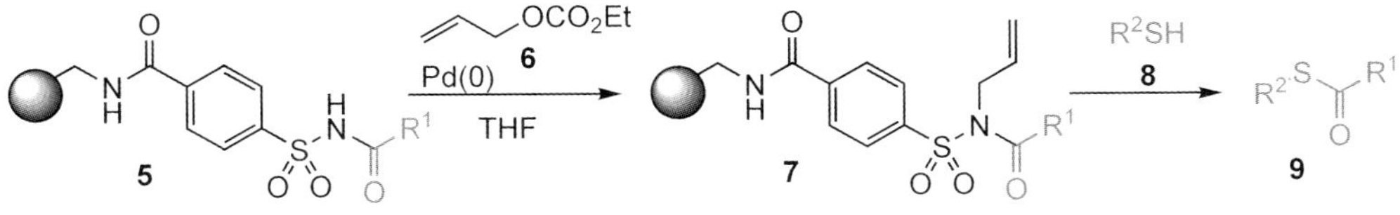

Scheme 16.2 Solid-phase synthesis of thioesters using Ellman's linker by Kiessling et al.

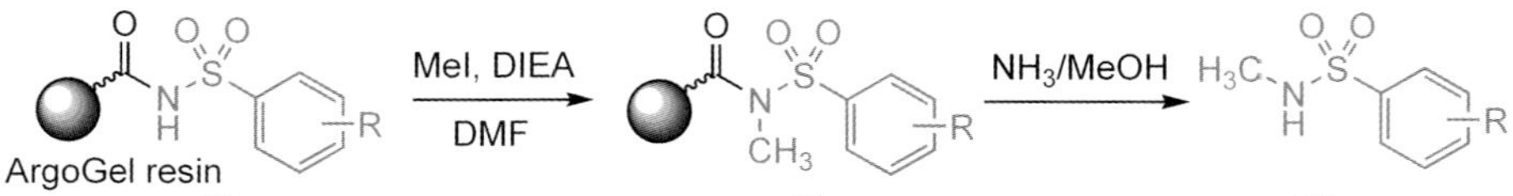

Scheme 16.3 Maclean's reversed Kenner linker yielding sulfonamides.

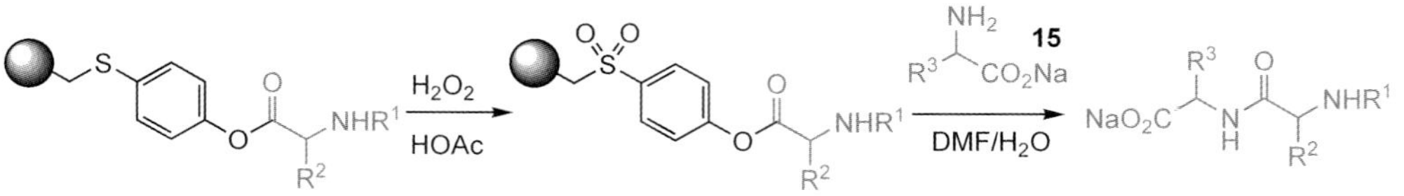

Scheme 16.4 Sulfide safety-catch linker by Marshall et al.

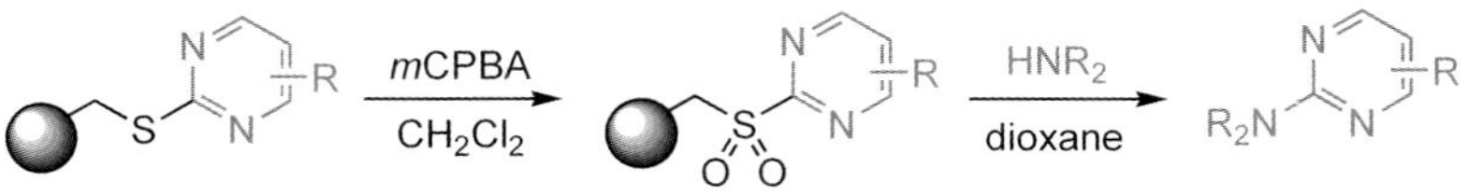

Scheme 16.5 Sulfide safety-catch linker by Villalgordo et al.

in an ortho- or para-position of aromatic linkers, a resin-bound thioether can be activated by oxidation to enable the nucleophilic cleavage of a phenoxide moiety. Marshall has anchored amino acids onto the phenolsulfide linker (**13**). Oxidative activation with H_2O_2 to give the sulfone **14** allowed the displacement of the ester moiety with amino acids, yielding dipeptides **16** [20] (Scheme 16.4). A conceptually similar approach was devised by Dressman et al., who produced different ureas by oxidation of the safety-catch linker with mCPBA and subsequent cleavage of resin-bound carbamates with amines [21].

Villalgordo et al. [22, 23] as well as Gayo and Suto [25] developed a strategy to cleave pyrimidines from the solid support. After oxidation of the thioether-linkage **17**, aromatic substitution of the sulfonyl unit was performed with different N-nucleophiles as amines and azides to give free amino- or azido-pyrimidines **19** (Scheme 16.5). To demonstrate the stability of the linker, the resin-bound derivatives were subjected to different reactions such as saponification, ester reduction, acid chloride formation or Mitsunobu alkylation. A similar approach was presented later on by Hwang and Gong in the SPOS of 2-aminobenzoxazoles [26].

The oxidation of sulfide-linkers to the corresponding sulfones also offers the possibility of β-elimination [29–33]. Compounds can be loaded as amines, esters or carbamates on the resin. After activation of the linker with mCPBA – and in the case of amines previous quaternization of the N-atom – β-elimination with NaOtBu, NH_4OH or N-bases like dimethylamine or DBU affords the free amines **23** (Scheme 16.6), carboxylic acids or the carbamic acids, which decompose spontaneously to the corresponding amines [32]. Although they do not contain sulfur-linkages, the

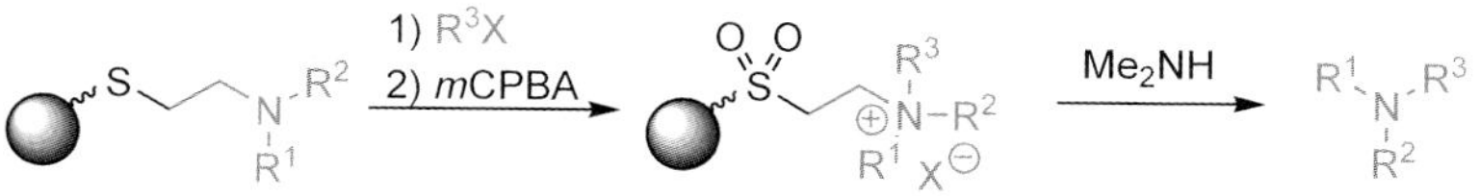

Scheme 16.6 Activation and subsequent cleavage by β-elimination on Wade's safety-catch linker.

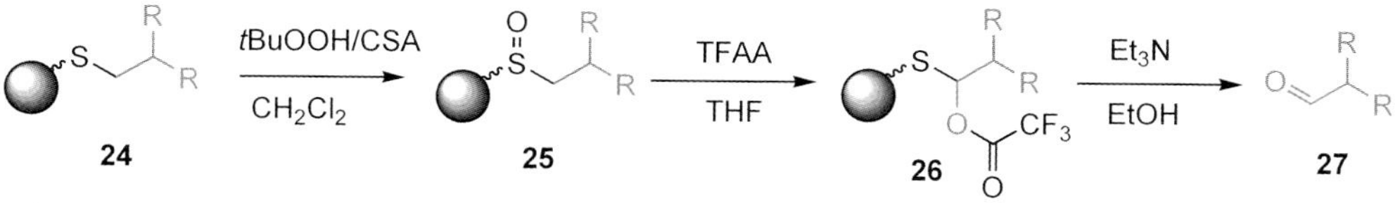

Scheme 16.7 Li's safety-catch system including a Pummerer rearrangement.

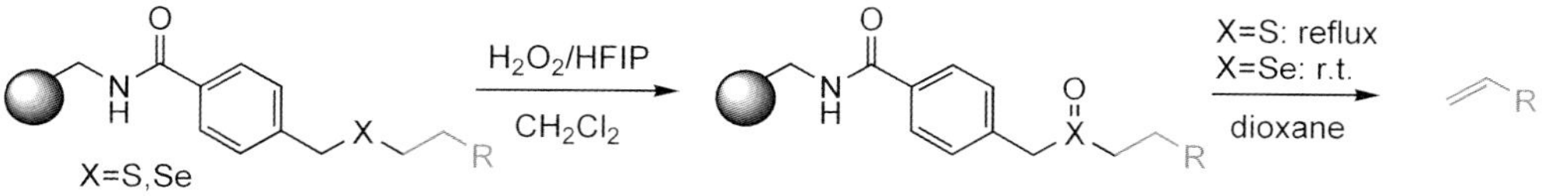

Scheme 16.8 Sulfide and selenide safety-catch linker by Bradley et al.

TRAM linker by Belshaw et al. [45] and the REM resin by Morphy et al. [46] should be mentioned here, as they are built on a quaternization strategy of amines as activation step and subsequent β-elimination by basic treatment as well. While the REM resin releases tertiary amines from the solid support, the use of the TRAM linker yields Michael acceptors.

A special activating and cleavage protocol of a sulfide safety-catch linker was developed by Li et al. [34]. The linker was generated by binding S-protected p-hydroxythiophenol to Merrifield resin. After removal of the protecting group, different electrophiles were attached. Activation was achieved via selective oxidation to the sulfoxide **25** with tBuOOH/CSA; subsequent Pummerer rearrangement using trifluoroacetic anhydride (TFAA) led to trifluoroacetoxythioacetals **26**, which were cleaved by treatment with Et$_3$N in ethanol to give aldehydes **27** (Scheme 16.7). Alternatively, the corresponding alcohols could be obtained when employing, additionally, NaBH$_4$ in the cleavage step.

The groups of Bradley and Nicolaou investigated the last safety-catch approach via activation by sulfur oxidation that should be described herein [35–39]. The linker **28** was activated by reaction with H$_2$O$_2$ or H$_2$O$_2$/HFIP to give the sulfoxide **29**, which undergoes a pericyclic elimination reaction, releasing olefins **30** from the solid support (Scheme 16.8). A similar concept could be adapted to the corresponding selenium linkers, as Nicolaou demonstrated in the solid-phase semi-synthesis of vancomycin [37]. The required conditions for the cleavage step strongly depend on the nature of the linker – the selenoxide already eliminates at room temperature whereas the sulfoxide must be heated under reflux in dioxane to yield the desired olefins.

Based on the dependence in heterolytic benzyl-oxygen or benzyl-nitrogen cleavage of the electronic character of ortho- and para-substituents, several safety-catch systems were developed. Pátek and Lebl introduced a safety-catch acid-labile (SCAL) linker for the synthesis of primary amides [40] (Scheme 16.9). The linker in its oxidized form 32 is extremely stable towards acids (TFA, thioanisole/TFA) and bases (aq. NaOH, piperidine). The system is activated by reduction of both sulfoxide groups to the corresponding sulfides 33 either with $PPh_3/TMSCl/CH_2Cl_2$ or with $(EtO)_2P(S)SH/DMPU$; acidolytic cleavage is then accomplished by treatment with TFA to give the target molecules 34.

Similar strategies for solid-phase peptide synthesis were followed by Kiso and coworkers in the design of the DSB (4-(2,5-dimethyl-4-methylsulfinylphenyl)-4-hydroxybutanoic acid)- and the DSA (4-(4-methoxyphenyl-aminomethyl)-3-methoxyphenylsulfinyl-6-hexanoic acid) linker [42, 43] as well as by Undén and Erlandsson in the design of HMPPA (3-(4-hydroxymethylphenylsulfanyl) propanoic acid) linker [44]. The first amino acid is attached via an ester bond to the DSB linker. After peptide synthesis, activation and cleavage are performed in a one-pot procedure via reductive acidolysis using $SiCl_4/PhSMe/PhOMe/EDT$ and TFA to give free peptide acids. On the conceptually similar DSA linker, amino acids are connected via an ester bond, yielding the free peptide amides after activation and cleavage with TFA.

The concept of transforming an electron-withdrawing substituent of an aromatic system into an electron-donating one and, therefore, facilitating an acidolytic cleavage of the benzylic carbon–heteroatom bond was also utilized by Albericio et al. [47] (Scheme 16.10) as well as by Takahashi et al. [48] Both groups activated

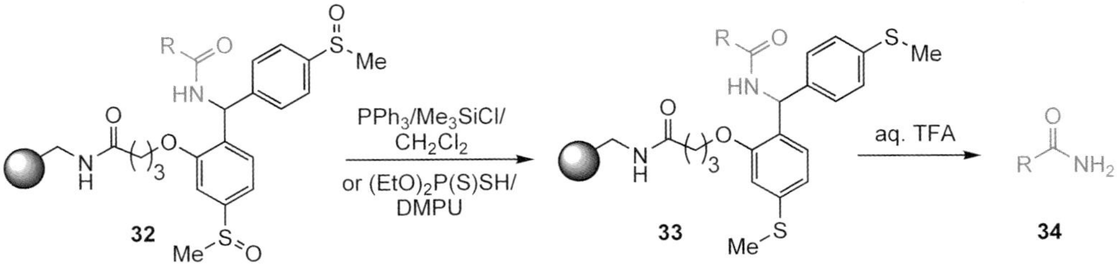

Scheme 16.9 Safety-catch acid labile (SCAL) linker by Pátek and Lebl.

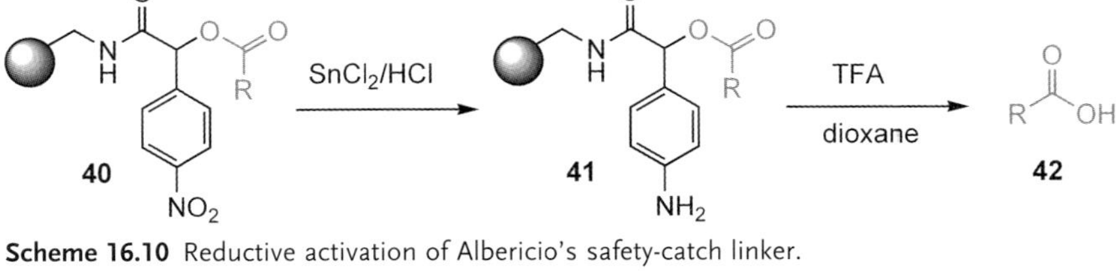

Scheme 16.10 Reductive activation of Albericio's safety-catch linker.

the linker **40** by reduction of nitro-substituents to amines using $SnCl_2$. This step enabled the cleavage with TFA, yielding peptide acids **42** or alcohols, respectively.

Wieland established a safety-catch linker based on the benzyl hydrazide functionality **44**. Oxidation with NBS or $Cu(OAc)_2$ gives a reactive diazene derivative, which can be cleaved by nucleophiles like amines or alcohols to produce the corresponding acyl derivatives under elimination of elemental nitrogen [49]. Analogically to Maclean's reversed Kenner safety-catch linker, Waldmann *et al.* used an inverse benzyl hydrazide linker **45** to obtain aryl derivatives after cleavage while leaving the acyl unit on the solid support [55]. A novel safety-catch approach also founded on nitrogen oxidation as the activation step was shown by Mioskowski *et al.*, who generated the resin-bound N-acyl dihydroquinoline linker **46** [57]. This linker system is treated with oxidative reagents like DDQ, CAN or triphenylcarbenium tetrafluoroborate, yielding the activated N-acyl quinolinium salt **47** that can be cleaved with nucleophiles like amines or hydroxides to give amides **49** or carboxylic acids, respectively (Scheme 16.11).

Besides the benzyl hydrazide linker **44**, Wieland *et al.* introduced a 2,2-diphenyl-2-hydroxyethyl-ester derived resin [58]. This linker type can be activated by acid-catalyzed dehydration (aq. TFA) to give the reactive enol ester **51**, which affords amides **53** upon treatment with amines (Scheme 16.12).

A very useful approach in safety-catch strategy is the application of linkers bearing a protected nitrogen functionality. After activation by removal of the protecting group, the free N-atom attacks intramolecularly a neighboring carbonyl functionality of the linker under neutral or basic conditions and, thereby, releases the target structures from the solid support. One example of this type is linker **54** (Scheme 16.13) by Bradley *et al.* [59, 60], which was shown to be cleavable via modulation of the pH. After removal of the Boc group, the deprotected N-atom attacks under neutral to weakly basic conditions the ester group, thus forming a

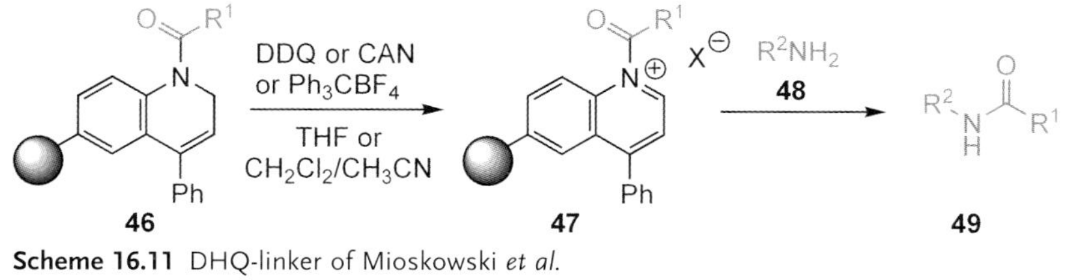

Scheme 16.11 DHQ-linker of Mioskowski *et al.*

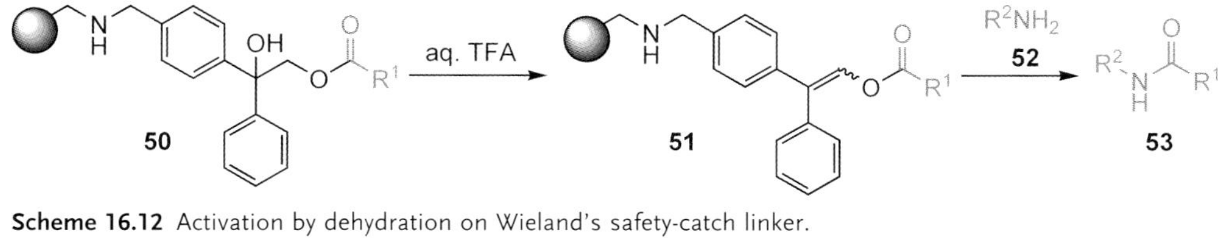

Scheme 16.12 Activation by dehydration on Wieland's safety-catch linker.

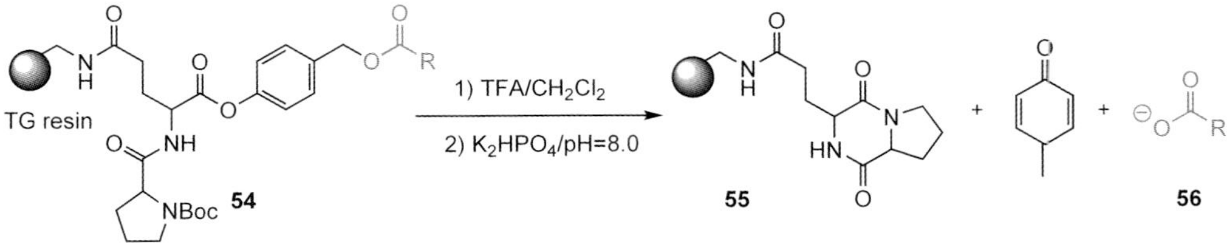

Scheme 16.13 Safety-catch linker by Bradley et al.

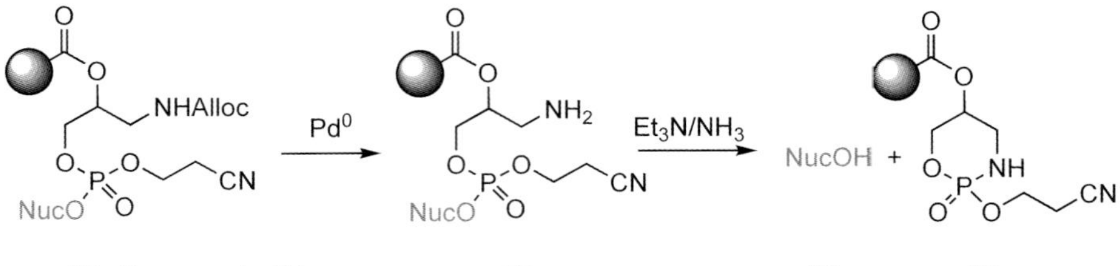

Scheme 16.14 A safety-catch palladium activated linker by Lyttle et al.

resin-bound diketopiperazine **55** and releasing a cyclohexadienone and the desired peptidic acids **56** via an 1,6-elimination process. Later, the same group modified the nature of the linker system in such a way that the reactive quinone by-product remains on the bead after the cleavage step [61].

Another instance for this linker class is the safety-catch linker by Lyttle, which was developed for the synthesis of nucleic acids on solid supports [62]. Starting from a resin carrying an Alloc-protected amino group fragment, conventional phosphoramidite chemistry was carried out to build up the desired immobilized nucleotide **57**. Removal of the Alloc group via palladium catalysis under neutral conditions produces a polymer-bound intermediate **58** with a free amino functionality that can intramolecularly attack activated phosphonates and liberate the nucleotide **59** from the solid support (Scheme 16.14). More examples of safety-catch linkers that use the deprotection of an N-functionality as the activation step are listed in Table 16.1 (resins **61–65**) [63–68].

The concept of activation of a linker by the removal of protecting groups is not only common in combination with amines but also in case of phenolic linker systems. The groups of Bourne and Plé developed safety-catch linkers that could be activated by deprotecting the phenol group of linker **66** [69–72]. By that transformation, the nucleophilic attack of amine groups is enabled and results in the release of cyclic amider from the solid support (Scheme 16.15). Parang et al. introduced a similar safety-catch strategy for the synthesis of carbohydrates and nucleoside monophosphates [64].

An example of a safety-catch approach relying on the introduction of a protecting group to activate the linker for cleavage is the benzamide linker system **70** by Hulme et al. [73] The inert benzamide carbonyl is activated by Boc-protection of

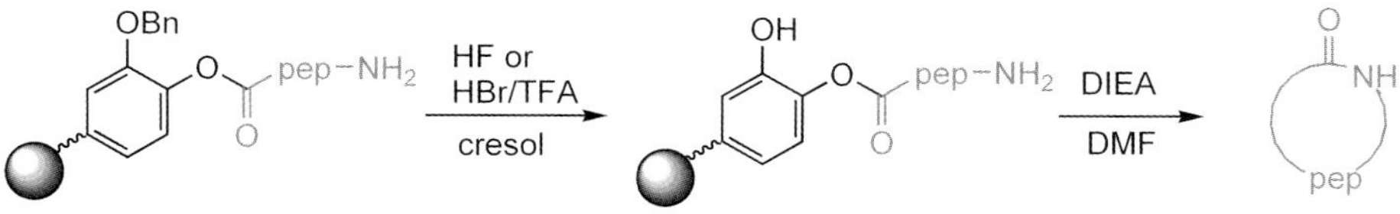

Scheme 16.15 Safety-catch linker by Bourne et al.

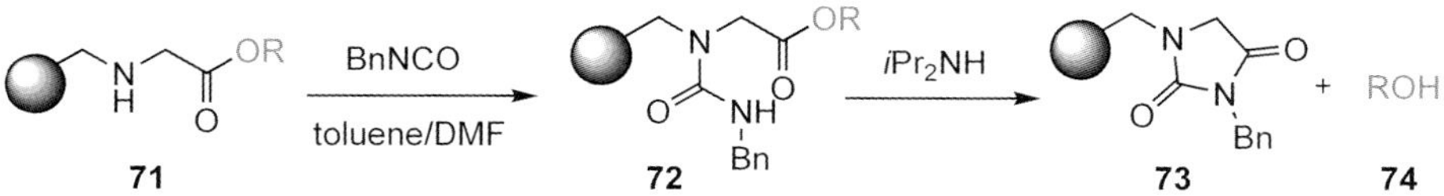

Scheme 16.16 Safety-catch linker by Raghavan and Rajender.

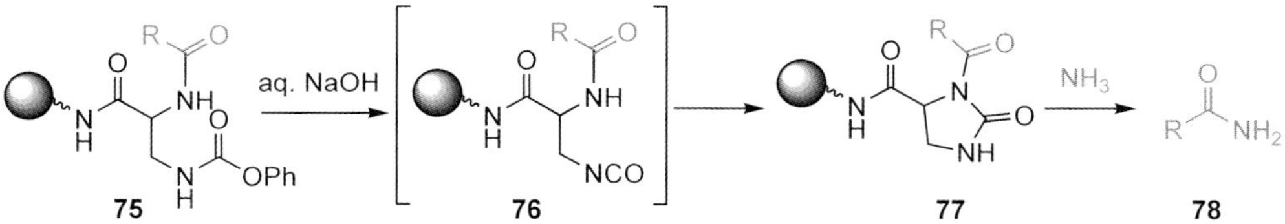

Scheme 16.17 Dpr(Phoc) linker by Pascal et al.

the N-atom. The amide is now sensitive to hydrolytic (aq. LiOH/H_2O_2) or alcoholic cleavage (MeONa/MeOH), yielding the free acids or esters, respectively. Raghavan and Rajender added benzyl isocyanate to convert the amino function of the glycine type linker **71** into the corresponding urea **72** [74]. Under basic conditions, this intermediate cyclizes easily to resin-bound hydantoin (**73**) and thus releases alcohols **74** from the solid support (Scheme 16.16).

Pascal et al. have reported the Dpr(Phoc) linker (Dpr = L-2,3-diaminopropionic acid, Phoc = phenyloxycarbonyl), which releases – similar to the safety-catch approaches shown before – the target structures from the resin after an intramolecular cyclization step [75, 76]. The linker system (**75**) is a derivative of β-aminophenyloxycabonyl-2,3-diaminopropionic acid and shows a high stability under neutral and acidic conditions. Pretreatment under mild alkaline conditions affords an isocyanate intermediate (**76**) that is attacked by a neighboring amide to give the activated cyclic urea **77** (Scheme 16.17). Cleavage with NaOH/iPrOH/H_2O or NH_3/iPrOH gives the peptide acids or the corresponding peptide amides.

An interesting activation methodology of a safety-catch linker was developed by Fürst and Rück-Braun [78] (Scheme 16.18). They investigated a propargyl based linker strategy with complexation of the alkyne unit as the activation step. Propargyl alcohol as well as propargyl chloroformate linkers were used to generate carboxylic acids and amines. The propargyl unit **80** is stable under acidic conditions. After treatment with $Co_2(CO)_8$ to build Co-alkyne complexes, cleavage is readily performed using TFA to give the desired target structures.

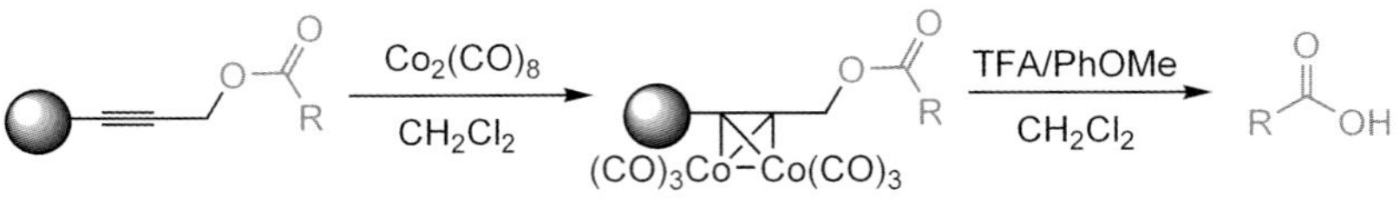

Scheme 16.18 Activation via alkyne complexation on the propargyl linker by Fürst and Rück-Braun.

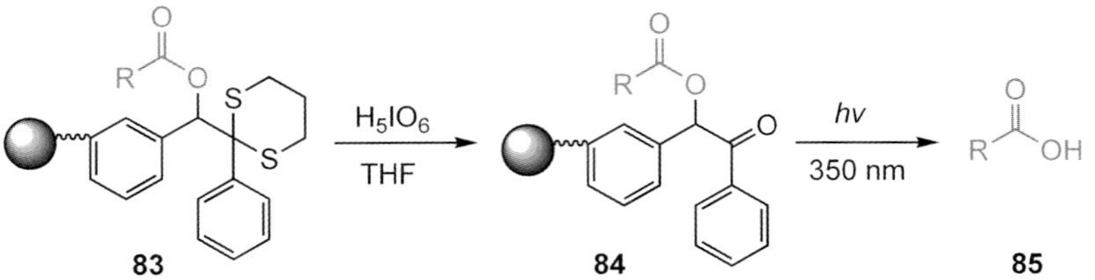

Scheme 16.19 A photocleavable safety-catch linker by Balasubramanian et al.

The last group of safety-catch systems described herein is the class of photocleavable safety-catch linkers. These linkers have aroused much interest as they offer an additional dimension of orthogonality. Balasubramanian et al. have developed a dithiane-protected 3-alkoxybenzoin linker (**83**) that can be activated by removal of the dithioacetal unit using $Hg(ClO_4)_2$ or H_5IO_6 [79–81]. Subsequent cleavage procedure by irradiation yields carboxylic acids **85** (Scheme 16.19). A conceptually similar linker containing an acetal unit for the synthesis of oligonucleotides was shown by Belshaw et al. [82] Deprotection of an acetal-protected carbonyl functionality was also utilized as activation step by Ladlow et al., who achieved cleavage not by irradiation but by nucleophilic attack to the intramolecularly formed and activated amide to give acids, esters or amides [83].

To cope with the demands for the synthesis of diverse libraries of small organic molecules, peptide mimetics and peptides, an approach to a broad arsenal of linkers having specific requirements of particular SPOS strategy is important. In this respect, the benefit of safety-catch linkers offering the great advantage of extended orthogonality further expands the potential for use of a large catalog of reactions and synthetic protocols.

16.3
Traceless Linkers

A traceless linker is a linker "which leaves no residue on the compound after cleavage, i.e. is replaced by a hydrogen" [8]. Traceless linking is nowadays considered as "leaving no functionality," meaning that cleavage leads exclusively to the formation of a C–H bond (or a C–C bond) on the original position of attachment, giving alkanes, alkenes, alkynes or arenes. A broadening of this definition to OH

16.3 Traceless Linkers

or NH groups is not useful, because otherwise every linker derived from polymeric protecting groups would have to be regarded as a traceless linker.

In terms of designing a traceless linker, one has to start from a heteroatom–carbon bond, which is labile towards protogenolytic, hydrogenolytic or hydridolytic cleavage. Since most heteroatom–carbon single bonds are less stable than a carbon–carbon bond, traceless linkers can be synthesized based on nearly all heteroatoms. However, the enthalpies of C–X bonds are only relevant for homolytic bond scission. Many linkers are cleaved heterolytically, and the kinetic stability towards heterolytic bond cleavage is decisive in these cases. Table 16.2 gives an overview of traceless linkers.

Table 16.2 Overview of traceless linkers.

Structure	Cleavage	Possible structures achievable	References
88	NaBH$_4$ (alkanes) or bases (alkenes)	Alkenes **89**, alkanes **90**	[84]
TG resin **91**	Photolytic cleavage	Methylarenes **92**	[85, 86]
PEG resin **93** (R = H, R = CF$_3$)	H$_2$/Raney nickel	Alkanes **95**	[87, 88]
PEG resin **94**	Na/Hg	Alkanes **95**	[89]
97	Boronic acids	Biphenylmethyl derivatives **98**	[90]
99 (n=0,1,2)	SmI$_2$	Alkanes (oxindoles/tetrahydroquinolones)	[91, 92]

Table 16.2 Continued.

Structure	Cleavage	Possible structures achievable	References
100	Pd(OAc)$_2$/dppp/Et$_3$N/ HCO$_2$H	Arenes 101	[93]
102	Mg/HgCl$_2$ (alkanes), Et$_3$N (alkenes)	Butyrolactones 103, butenolides 104	[94]
105	RMgX	Olefins	[95]
106	Pd(OAc)$_2$/dppp/Et$_3$N/ HCO$_2$H (arenes) or boronic acids (biphenyls)	Arenes, biphenyls	[96]
107	Grignard reagents using Ni(0)-catalyst	Biphenyls	[97]
108	Bu$_3$SnH/AIBN (alkanes) H$_2$O$_2$ or mCPBA (alkenes)	Alkanes 111, alkenes 112	[98–103]
113 Trialkylsilyl linker	TFA, HCl, TBAF, CsF or HF	Arenes 115	[104–111]
116 Dialkylsilyloxy linker	TBAF	Arenes	[112–114]
117	TFA	arenes	[105, 115]

Table 16.2 Continued.

Structure	Cleavage	Possible structures achievable	References
119 (nBu, nBu Sn linker)	Pd(PPh$_3$)$_4$/toluene	Cyclic alkenes **121**	[116]
122 T1 linker	HCl/THF, HSiCl$_3$ or DMF	Arenes	[117–124]
123 T1 linker	HCl/THF or DMF/THF or HSiCl$_3$/CH$_2$Cl$_2$ or H$_3$PO$_2$/Cl$_2$HCCO$_2$H	Arenes **124, 125**	[121, 125]
126 Arylhydrazide linker	Oxidation with NBS or Cu(OAc)$_2$, cleavage with nucleophiles (MeOH)	Arenes **127**	[55, 56]
128	NaOH/MeOH	Methylarenes **129**	[126, 127]
130	Pd(OAc)$_2$	Methylarenes **131**	[128, 129]
133 (R=CH$_2$NR$_2$, CN)	AcOH or TFA and spontaneous decarboxylation	Ketone derivatives **135**	[130–134]
136	HCl aq. or TMSCl/NaI/ MeCN/dioxane and spontaneous decarboxylation	Pyrimidine derivatives	[135]

Table 16.2 Continued.

Structure	Cleavage	Possible structures achievable	References
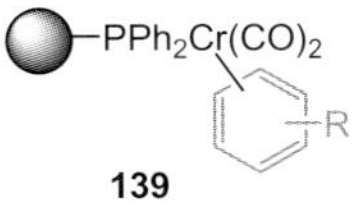 139 (–PPh$_2$Cr(CO)$_2$–R)	Pyridine, I$_2$/THF or CH$_3$CN/THF	Arenes 140	[136, 137]
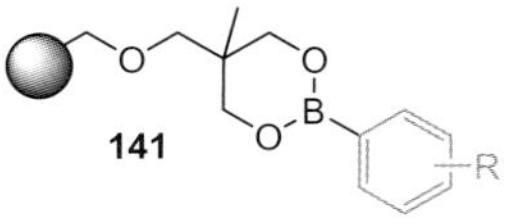 141 (boronate ester)	Ag (NH$_3$)$_2$NO$_3$	Arenes	[138]
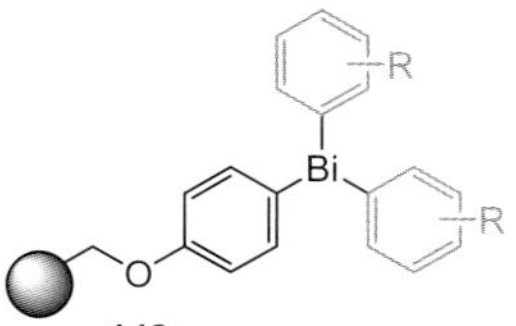 142 (bismuth aryl)	TFA	Arenes	[139]
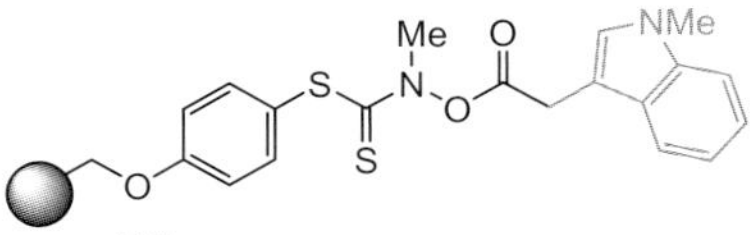 143 (thiocarbamate indole)	Photolytic cleavage	Methylarenes	[140]

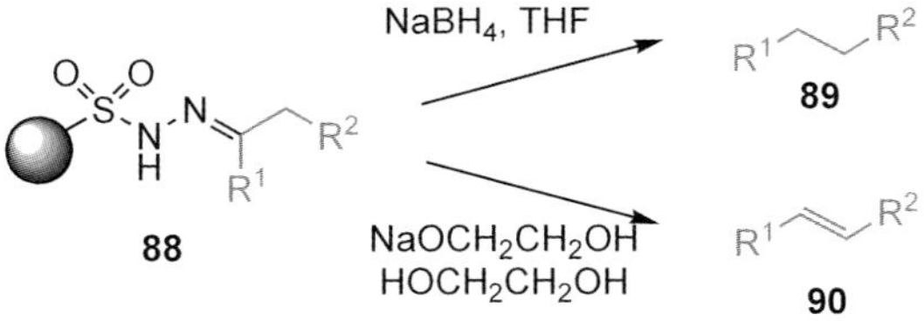

Scheme 16.20 The first traceless linker by Kamogawa et al.

The first traceless linker was developed by Kamogawa and coworkers in 1983 [82]. Starting from a polymer-bound sulfonylhydrazine, sulfonylhydrazone resin 88 was formed by reaction with ketones or aldehydes. The cleavage step was conducted either by reduction with sodium borohydride or lithium aluminium hydride to yield alkanes 89 or by treatment with base to give the corresponding alkenes 90 in a Bamford–Stevens reaction (Scheme 16.20). This work was a pioneering approach in the field of traceless linkers.

One of the very early investigations dealing with traceless linkers was published in 1994 by Sucholeiki and describes the use of thioethers 91, attached via an aromatic core that enhances photolytic cleavage. Irradiation at 350 nm affords the hydrocarbon **92** [85, 86] (Scheme 16.21). The linker has not been fully explored and is limited in the range of functionalized arenes, since, for example, phenyl instead of biphenyl residues result in the formation of disulfides.

The aryl sulfide linker system shown in Scheme 16.22 was reported by Janda *et al.* in 1996 [87, 88]. After attachment of alkyl bromides to the benzenethiol linker to give resin **93**, cleavage was achieved by desulfurization under quite harsh conditions with Raney-nickel, yielding the new hydrocarbon product **95**. Alternatively, the linker can be oxidized with *m*CPBA or $KHSO_5$ to the corresponding sulfone **94** (Table 16.2); subsequent treatment with 5% sodium/mercury furnishes the parent hydrocarbon in high yields [89].

Wagner and coworkers developed a traceless sulfur-based linker by applying the chemistry of benzylsulfonium salts [90]. Various benzyl bromides were attached to a thiol linker synthesized in four steps starting from Merrifield resin. Reaction of the resulting resin **96** with triethyloxonium tetrafluoroborate gave the solid supported sulfonium salt **97**, which could be cleaved by reaction with boronic acids to afford biphenylmethyl derivatives **98** (Scheme 16.23). The alkylation can be regarded as an activation step and the linker as a safety-catch system.

Similar concepts describing the formation of novel C–C bonds in the cleavage step were introduced by Holmes *et al.* [96] and Park *et al.* [97] Both groups utilized aryl sulfonyloxy linkers that were cleaved by reaction with boronic acids or Grignard reagents to give biphenyl derivatives. Wustrow *et al.* introduced a

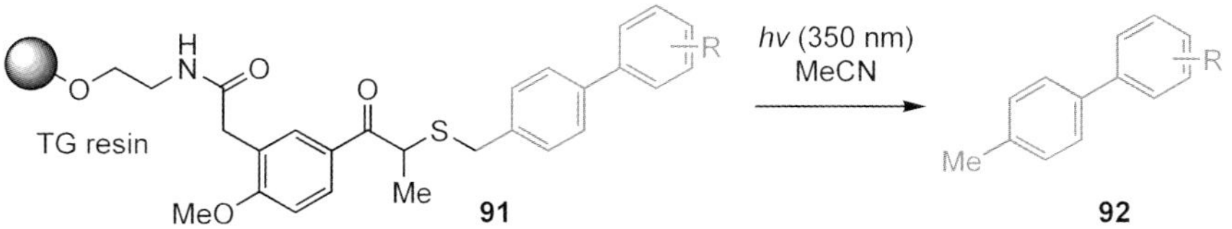

Scheme 16.21 Photolabile traceless linkage by Sucholeiki *et al.*

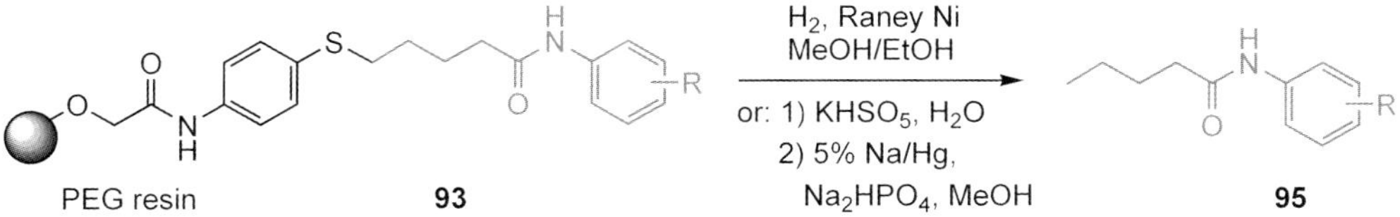

Scheme 16.22 Traceless linker for alkanes according to Janda *et al.*

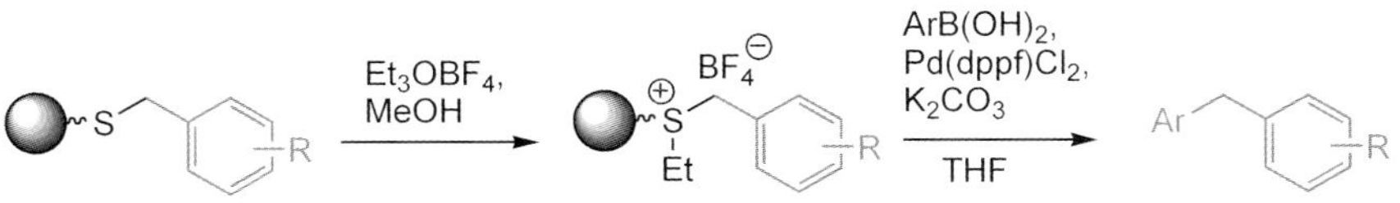

Scheme 16.23 Traceless sulfur linker by Wagner *et al.*

sulfonyloxy linker for the synthesis of arenes **101** [93]. The target molecules were gained by application of reductive cleavage conditions using a Pd(0) catalyst (Scheme 16.24).

Recently, Sheng et al. reported the synthesis of butenolides and butyrolactones using the traceless sulfone linker **102** [94]. After synthesis of substituted dihydrofuran-2(3H)-ones on solid supports, cleavage with Mg/HgCl$_2$ in EtOH/THF afforded butyrolactones **103** while treatment with Et$_3$N in CH$_2$Cl$_2$ gave the corresponding butenolides **104** (Scheme 16.25). Other examples for traceless sulfone linkers were given by Procter et al. [91] and Yu et al. [92] who cleaved the target molecules from the resin under reductive conditions using SmI$_2$ and by Kurth et al. who performed Grignard reactions during the cleavage step, thus forming cyclobutylidene-containing structures via SPOS [95].

Selenium has proved to be a useful element for traceless SPOS as well. The selenium–carbon bond is prone to undergo homolytic cleavage to give radicals. Ruhland et al. [98] as well as Nicolaou et al. [100, 101] independently developed efficient methods for the preparation of selenium-containing supports to obtain traceless linkers. Starting from polystyrene, various steps, including selenation with selenium powder or dimethyl diselenide, gave rise to selenium resins that were alkylated to give selenoethers **110**. The traceless cleavage affording alkanes **111** was performed by reduction with tributyltin hydride while the corresponding alkenes **112** could be obtained using mild oxidizing reagents (Scheme 16.26). The linker class was also applied in the synthesis of heterocycles like γ-lactones [99], benzopyrans [102] or indulines [103].

The most common anchors for traceless linkage of arenes are based on silyl linkers due to the well-known protodesilylation of the Si–aryl bond. The first traceless silyl linkers were described independently by Ellman et al. and Veber et al. in 1995 [104, 106]. Starting from resin-bound arylstannane **113**, Ellman has shown the advantages of this type of detachment in the synthesis of a diverse benzodiazepine library (Scheme 16.27). Improvements in the chemoselectivity of the cleav-

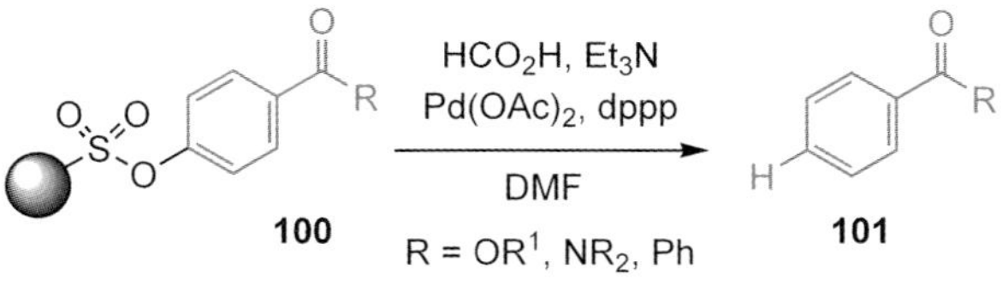

Scheme 16.24 Cleavage–hydrogenation reaction with a sulfonyloxy linker according to Wustrow et al.

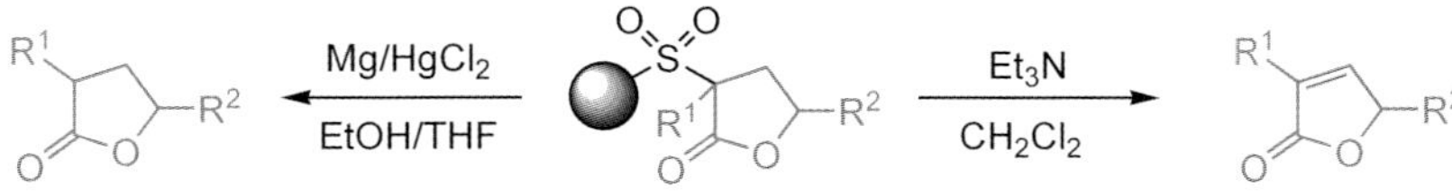

Scheme 16.25 Cleavage to give alkyl and alkenyl structures using a sulfone linker by Sheng et al.

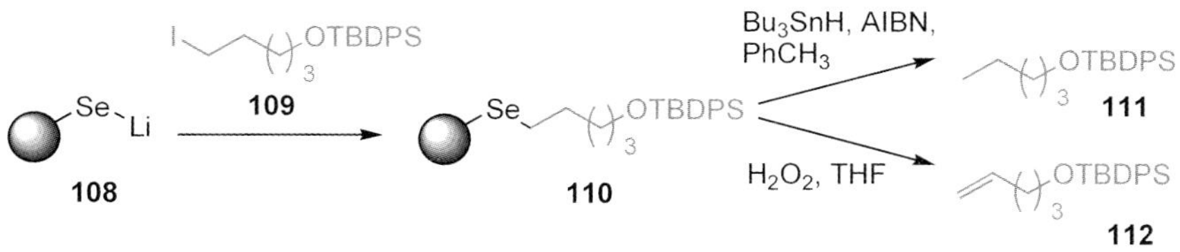

Scheme 16.26 Selenium linker yielding alkanes or alkenes by Nicolaou et al.

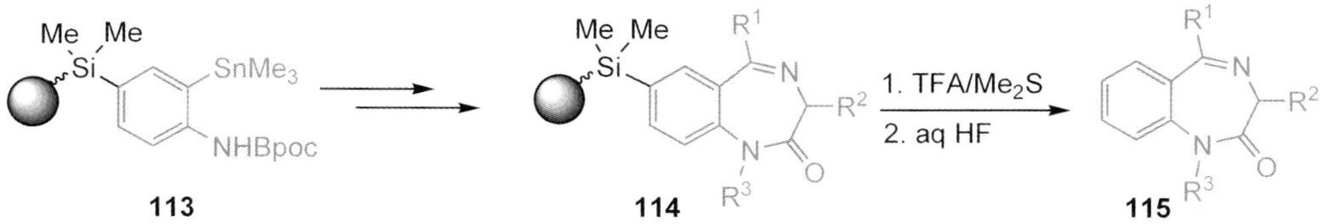

Scheme 16.27 Synthesis of a benzodiazepine library with the aid of a silyl linker by Ellman et al.

age step (the silyl linker produces a substantial amount of the silyl arene upon cleavage) were accomplished using a germanium linker **117**, which is more labile towards acids [105, 115]. Other trialkylsilyl or dialkylsilyloxy linkers have also been used to facilitate loading, synthesis and detachment from solid supports [107–114]. Furthermore, silicon linkers can be used for the immobilization of allylsilanes, which release alkenes in a traceless fashion [141].

Tin hydride reagents are versatile tools for the functionalization of alkenes and alkynes. Based on this concept, Nicolaou and coworkers developed a polymer-bound tin hydride (**118**) that reacts via Pd-catalyzed hydrostannylation (or nucleophilic attack on the tin chloride with a vinyl lithium) with alkynes to give alkenylstannanes [116]. After further transformation to derivatives **119**, the resin-bound substrates undergo proteolytic traceless cleavage to yield unsubstituted alkenes **120**. Alternatively, the stannane can be employed for intramolecular Stille coupling to produce macrolactones **121** in the cleavage step (Scheme 16.28).

The use of nitrogen as the element of linkage in traceless linkers has centered around the chemistry of diazonium compounds [121, 125]. The synthetic utility of the T1 triazene linker as a traceless anchor for arenes has been demonstrated by Bräse et al. using short reaction sequences. Thus, cinnamic esters were synthesized in a sequence starting from the iodoarene resin **123**. Heck coupling with acrylates using palladium catalysis affords an immobilized cinnamate. This can be detached either directly or after a sequence of transformations to yield allylic amine **125** in a traceless fashion. Either trichlorosilane [118] or an HCl/THF mixture [125] can be used to give products **124** and **125** in high yields and without further purification (Scheme 16.29).

Waldmann's safety-catch hydrazide linker is also part of the class of traceless linkers [55]. Starting from hydrazide resin **126**, which is converted into an activated species by oxidation with $Cu(OAc)_2$, the molecules are cleaved by the addition of nucleophiles like amines to give arenes **127** (Scheme 16.30).

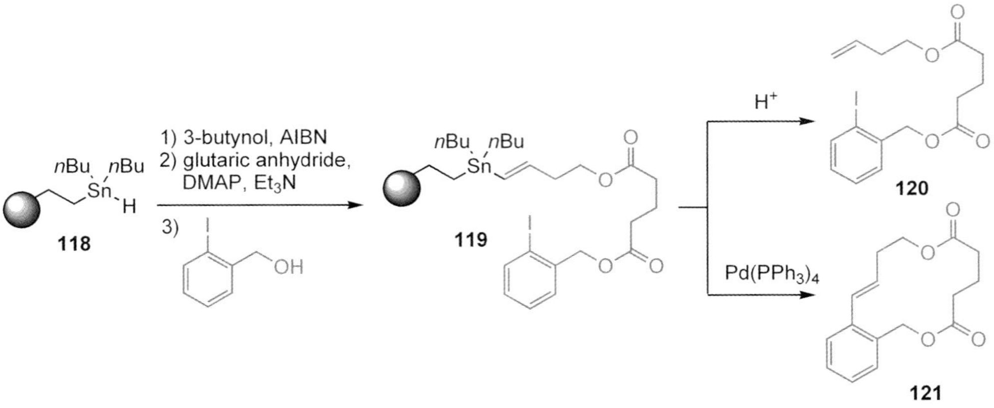

Scheme 16.28 Stannane linker according to Nicolaou et al.

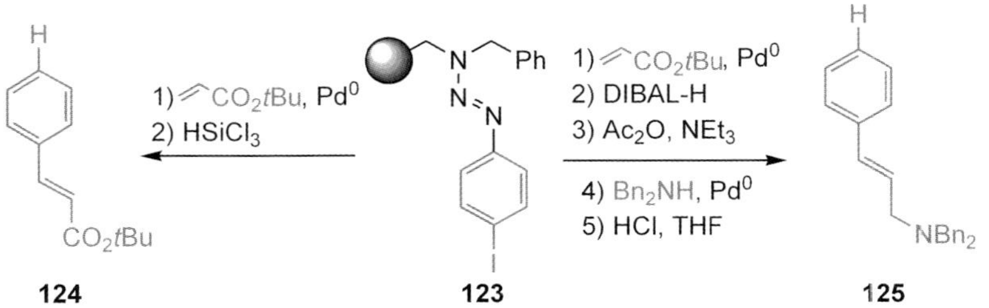

Scheme 16.29 T1 linker for traceless cleavage by Bräse et al.

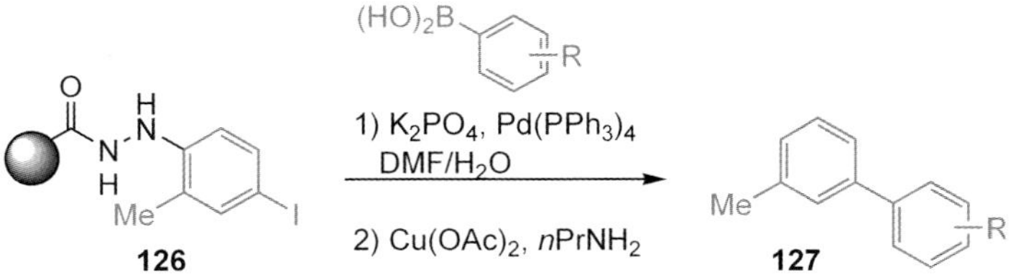

Scheme 16.30 Hydrazide linker according to Waldmann et al.

The phosphorous–carbon bond in phosphonium salts is readily cleavable by the aid of a base in the absence of aldehydes. Hence, the polymer-bound phosphonium salt **128** offers a direct access to methylarenes **129** (Scheme 16.31). An interesting feature of this linker is the fact that carbonyl compounds can be olefinated and this leads to a cleavage–olefination linker system [126, 127].

Moreover, methylarenes can be generated by cleavage from polystyrene resins using homogeneous palladium catalysis by either formate reduction [128] as shown in Scheme 16.32 or under an atmosphere of hydrogen [129].

Decarboxylation of appropriately substituted alkanes [130–134] and arenes [135] is also a traceless method that has been used to generate hydrocarbons. Since the

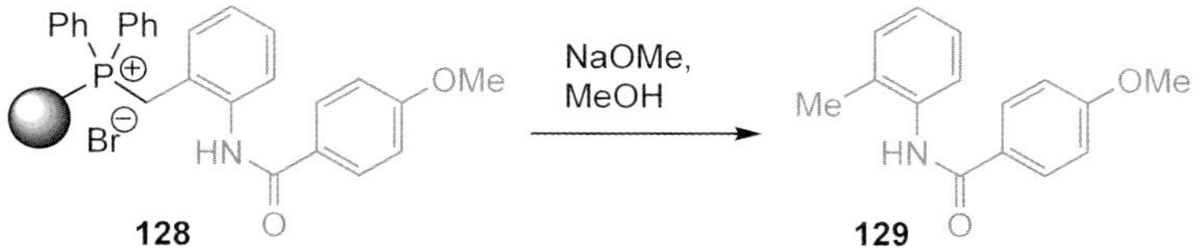

Scheme 16.31 Phosphonium linker for methylarenes by Hughes et al.

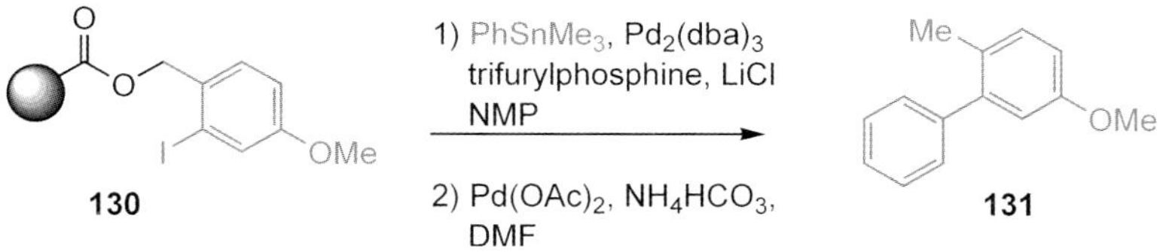

Scheme 16.32 Synthesis of methylarenes on solid supports by Sucholeiki.

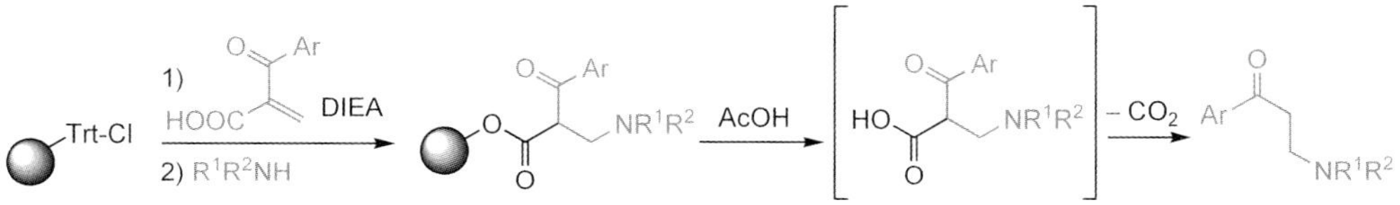

Scheme 16.33 Synthesis of β-aminoketones by a decarboxylative strategy according to Hoeg-Jensen et al.

neighboring group effect is essential, the target structures that can be achieved are limited. Scheme 16.33 shows an example of this strategy. Hoeg-Jensen et al. succeeded in the synthesis of β-aminoketones **135** via decarboxylation in the cleavage step [130]. Starting from trityl chloride resin **132**, the group generated, within a two-step procedure, resin-bound compounds **133** that subsequently can be cleaved in glacial acetic acid to give the desired β-aminoketones under elimination of CO_2.

An interesting traceless linker using chromium carbonyl complexes of aromatic compounds has been investigated by Peplow et al. [136] Chromium carbonyl complex **138** was photochemically loaded onto polystyrene phosphine resin **137** to give the solid-supported complex **139**. After modification of the substrate, traceless cleavage was achieved by heating in pyridine to give arene **140** (Scheme 16.34). Rigby et al. reported a similar π-arene chromium linker for the synthesis of tertiary alcohols and esters [137].

Other notable traceless approaches are the arylboronic acid linker **141** by Carboni et al. [138], the bismuthane linker **142** by Ruhland et al. [139] and the photolabile thiohydroxamic acid linker **143** by Routledge et al. [140]

Moreover, cyclative cleavage is an often used technique that has been developed into a powerful tool for traceless solid-phase synthesis. One example of particular

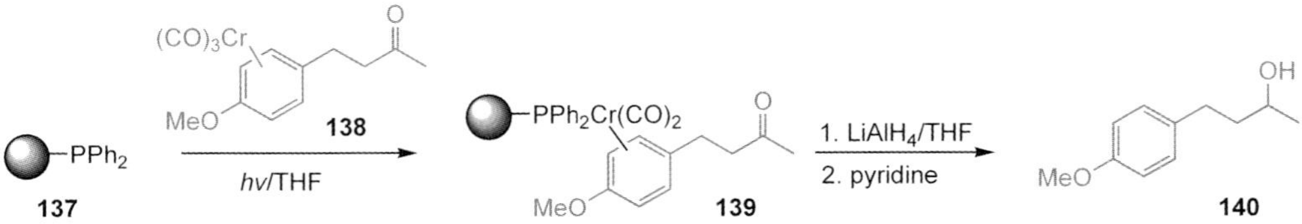

Scheme 16.34 Arylic chromium carbonyl complex as traceless linker by Peplow et al.

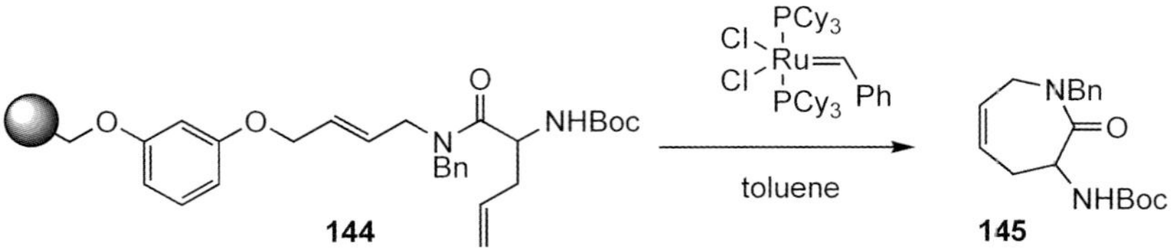

Scheme 16.35 Ring-closing metathesis used for traceless cleavage by Maarseveen et al.

interest is the ring-closing metathesis first used by Maarseveen et al. for the SPOS of lactams [142]. The RCM reaction was achieved with Grubbs ruthenium benzylidene catalyst to release the seven-membered protected lactam **145** (Scheme 16.35). Further examples for cyclative traceless cleavage procedures include, for example, cycloaddition reactions [141]. For further reading, see the corresponding literature (for reviews see References [144–147]).

16.4
Conclusions

The chemistry of traceless linkers is a fast emerging field in the intensive investigated area of Solid-Phase Organic Synthesis. Although some confusion about the definition or classification has been related with this linker type and, therefore, a careful designation has to be made, it is clear that this anchoring mode plays an important role in the design and syntheses of drug-like molecules.

List of Abbreviations

Ac	acetyl
AIBN	azobis(isobutyro)nitrile
Ala	alanine
Alloc	allyloxycarbonyl
Bn	benzyl
Boc	tert-butyloxycarbonyl

Bpoc	2-(*para*-biphenyl-2-yl)isopropyloxycarbonyl
CAN	ceric ammonium nitrate
CPG	controlled pore glass
CSA	camphorsulfonic acid
DBU	1,8-diazabicyclo[5.4.0]undec-7-ene
DDQ	dichlorodicyanobenzoquinone
DIBAL-H	diisobutylaluminium hydride
DIEA	diisopropylethylamine
DMAP	dimethylaminopyridine
DMF	dimethylformamide
DMPU	N,N'-dimethylpropylene-urea
dppf	1,3-bis(diphenylphosphino)ferrocene
dppp	1,3-bis(diphenylphosphino)propane
Dpr(Phoc)	L-2,3-diaminopropionic acid (phenyloxycarbonyl)
DSA	4-(4-methoxyphenyl-aminomethyl)-3-methoxyphenylsulfinyl-6-hexanoic acid
DSB	4-(2,5-dimethyl-4-methylsulfinylphenyl)-4-hydroxybutanoic acid
EDT	ethanedithiol
HFIP	hexafluoroisopropanol
HMPPA	3-(4-hydroxymethylphenylsulfanyl)propanoic acid
LDA	lithium diisopropylamide
mCPBA	*meta*-chloroperbenzoic acid
NBS	*N*-bromosuccinimide
Nu	nucleophile
PEG	poly(ethylene glycol)
Pep	peptide
PG	protecting group
Pipoc	*N*-piperidinooxycarbonyl
PPTS	*para*-pyridinumtoluene sulfonic acid pyridinium-*para*-Aorylate
RCM	ring closure metathesis
REM	regenerative Michael acceptors
SCAL	safety-catch acid-labile
SPOS	Solid Phase Organic Synthesis
TBAF	tetrabutylammonium fluoride
TBDPS	*tert*-butyldiphenylsilyl
Tf	trifluoromethylsulfonate
TFA	trifluoroacetic acid
TFAA	trifluoroacetic anhydride
TG	TentaGel
THF	tetrahydrofuran
TMS	trimethylsilyl
Tos	tosyl
TRAM	traceless release of acrylamides

References

1 Kocienski, P.J. (2005) *Protecting Groups*, Thieme, Stuttgart.
2 Hoffmann, R.W. (2006) *Synthesis*, 21, 3531–41.
3 Baran, P.S., Maimone, T.J. and Richter, J.M. (2007) *Nature*, 446, 404.
4 Jung, N., Encinas, A. and Bräse, S. (2006) *Current Opinion in Drug Discovery & Development*, 9, 713.
5 Bräse, S. and Dahmen, S. (2002) Rinkers for Solid-phane Synthesis, in *Handbook of Combinatorial Chemistry*, Vol. 1 (eds K.C. Nicolaou, R. Hanko and W. Hartwig), Wiley-VCH Verlag GmbH, Weinheim, pp. 59–169.
6 Knepper, K., Gil, C. and Bräse, S. (2004) Rinkers for solid-phase Synthesis. *Highlights in Bioorganic Chemistry* (eds C. Schmuck and H. Wennemers), Wiley-VCH Verlag GmbH, Weinheim, pp. 449–484.
7 Scott, P.J.H. and Steel, P.G. (2006) *European Journal of Organic Chemistry*, 10, 2251.
8 Maclean, D., Balchuim, J.J., Juanov, V.T., Kato, Y., Shaw, A., Schneider, P., Gordon, E.M. (1999) *Pure Appl. Chem.*, 71, 2349.
9 Jung, N., Wichin, M.S., Bräse, S. (2007) *Top. Curr. Chem.*, 278, 1.
10 Kenner, G.W., McDermott, J.R. and Sheppard, R.C. (1971) *Journal of the Chemical Society, Chemical Communications* 636.
11 Backes, B.J. and Ellman, J.A. (1999) *Journal of Organic Chemistry*, 64, 2322.
12 Shin, Y., Winans, K.A., Backes, B.J., et al. (1999) *Journal of the American Chemical Society*, 121, 11684.
13 Backes, B.J., Virgilio, A.A. and Ellman, J.A. (1996) *Journal of the American Chemical Society*, 118, 3055.
14 Backes, B.J. and Ellman, J.A. (1994) *Journal of the American Chemical Society*, 116, 11171.
15 Heidler, P. and Link, A. (2005) *Bioorganic & Medicinal Chemistry*, 13, 585.
16 Zohrabi-Kalantari, V., Heidler, P., Larsen, T. and Link, A. (2005) *Organic Letters*, 7, 5665.
17 He, Y., Wilkins, J.P. and Kiessling, L.L. (2006) *Organic Letters*, 8, 2483.
18 Maclean, D., Hale, R. and Chen, M. (2001) *Organic Letters*, 3, 2977.
19 Marshall, D.L. and Liener, I.E. (1970) *Journal of Organic Chemistry*, 35, 867.
20 Flanigan, E. and Marshall, D.L. (1970) *Tetrahedron Letters*, 21, 2403.
21 Dressman, B.A., Singh, U. and Kaldor, S.W. (1998) *Tetrahedron Letters*, 39, 3631.
22 Obrecht, D., Abrecht, C., Grieder, A. and Villalgordo, J.M. (1997) *Helvetica Chimica Acta*, 80, 65.
23 Chucholowski, A., Masquelin, T., Obrecht, D., et al. (1996) *Chimia*, 50, 525.
24 Font, D., Heras, M. and Villalgordo, J.M. (2003) *Journal of Combinatorial Chemistry*, 5, 311.
25 Gayo, L.M. and Suto, M.J. (1997) *Tetrahedron Letters*, 38, 211.
26 Hwang, J.Y. and Gong, Y.-D. (2006) *Journal of Combinatorial Chemistry*, 8, 297.
27 Lebreton, S., Newcombe, N. and Bradley, M. (2003) *Tetrahedron*, 59, 10213.
28 Ding, S., Gray, N.S., Ding, Q., et al. (2002) *Journal of Combinatorial Chemistry*, 4, 183.
29 Timar, Z. and Gallagher, T. (2000) *Tetrahedron Letters*, 41, 3173.
30 Garcia-Echeverria, C. (1997) *Tetrahedron Letters*, 38, 8933.
31 Miranda, L.P., Lubell, W.D., Halkes, K.M., et al. (2002) *Journal of Combinatorial Chemistry*, 4, 523.
32 Wade, W.S., Yang, F. and Sowin, T.J. (2000) *Journal of Combinatorial Chemistry*, 2, 266.
33 Fridkin, G. and Lubell, W.D. (2005) *Journal of Combinatorial Chemistry*, 7, 977.
34 Tai, C.-H., Wu, H.-C. and Li, W.-R. (2004) *Organic Letters*, 6, 2905.
35 Russell, H.E., Luke, R.W.A. and Bradley, M. (2000) *Tetrahedron Letters*, 41, 5287.
36 Nicolaou, K.C., Winssinger, N., Hughes, R., et al. (2000) *Angewandte Chemie (International Edition in English)*, 39, 1084.
37 Nicolaou, K.C., Cho, S.Y., Hughes, R., et al. (2001) *Chemistry – A European Journal*, 7, 3798.
38 Wang, Y.-G., Xu, W.-M. and Huang, X. (2007) *Journal of Combinatorial Chemistry*, 9, 513.

39 Nicolaou, K.C., Snyder, S.A., Bigot, A. and Pfefferkorn, J.A. (2000) *Angewandte Chemie (International Edition in English)*, **39**, 1093.
40 Pátek, M. and Lebl, M. (1991) *Tetrahedron Letters*, **32**, 3891.
41 Brust, A. and Tickle, A.E. (2007) *Journal of Peptide Science*, **13**, 133.
42 Kimura, T., Fukui, T., Tanaka, S., et al. (1997) *Chemical & Pharmaceutical Bulletin*, **45**, 18.
43 Kiso, Y., Fukui, T., Tanaka, S., Kimura, T. and Akaji, K. (1994) *Tetrahedron Letters*, **35**, 3571.
44 Erlandsson, M. and Undén, A. (2006) *Tetrahedron Letters*, **47**, 5829.
45 Ciolli, C.J., Kalagher, S. and Belshaw, P.J. (2004) *Organic Letters*, **6**, 1891.
46 Brown, A.R., Rees, D.C., Rankovic, Z. and Morphy, J.R. (1997) *Journal of the American Chemical Society*, **119**, 3288.
47 Isidro-Llobet, A., Alvarez, M., Burger, K., et al. (2007) *Organic Letters*, **9**, 1429.
48 Ohno, H., Tanaka, H. and Takahashi, T. (2004) *Synlett*, **508**,
49 Wieland, T., Lewalter, J. and Birr, C. (1970) *Liebigs Annalen der Chemie*, **740**, 31.
50 Semenov, A.N. and Gordeev, K.Y. (1995) *Journal of Peptide & Protein Research*, **45**, 303.
51 Berst, F., Holmes, A.B. and Ladlow, M. (2003) *Organic & Biomolecular Chemistry*, **1**, 1711.
52 Peters, C. and Waldmann, H. (2003) *Journal of Organic Chemistry*, **68**, 6053.
53 Camarero, J.A., Hackel, B.J., De Yoreo, J.J. and Mitchell, A.R. (2004) *Journal of Organic Chemistry*, **69**, 4145.
54 Berst, F., Holmes, A.B., Ladlow, M. and Murray, P.J. (2000) *Tetrahedron Letters*, **41**, 6649.
55 Stieber, F., Grether, U. and Waldmann, H. (1999) *Angewandte Chemie (International Edition in English)*, **38**, 1073.
56 Millington, C.R., Quarrell, R. and Lowe, G. (1998) *Tetrahedron Letters*, **39**, 7201.
57 Arseniyadis, S., Wagner, A. and Mioskowski, C. (2004) *Tetrahedron Letters*, **45**, 2251.
58 Wieland, T., Birr, C. and Fleckenstein, P. (1972) *Liebigs Annalen der Chemie*, **756**, 14.
59 Orain, D. and Bradley, M. (2000) *Molecular Diversity*, **5**, 25.
60 Atrash, B. and Bradley, M. (1997) *Chemical Communications*, 1397.
61 Chitkul, B., Atrash, B. and Bradley, M. (2001) *Tetrahedron Letters*, **42**, 6211.
62 Lyttle, M.H., Hudson, D. and Cook, R.M. (1996) *Nucleic Acid Research*, **24**, 2793.
63 Hoffmann, S. and Frank, R. (1994) *Tetrahedron Letters*, **35**, 7763.
64 Ahmadibeni, Y. and Parang, K. (2005) *Journal of Organic Chemistry*, **70**, 1100.
65 Xiao, X.-Y., Nova, M.P. and Czarnik, A.W. (1999) *Journal of Combinatorial Chemistry*, **1**, 379.
66 Scicinski, J.J., Congreve, M.S. and Ley, S.V. (2004) *Journal of Combinatorial Chemistry*, **6**, 375.
67 Grether, U. and Waldmann, H. (2000) *Angewandte Chemie (International Edition in English)*, **39**, 1629.
68 Grether, U. and Waldmann, H. (2001) *Chemistry–A European Journal*, **7**, 959.
69 Bourne, G.T., Golding, S.W., McGeary, R.P., et al. (2001) *Journal of Organic Chemistry*, **66**, 7706.
70 Ravn, J., Bourne, G.T. and Smythe, M.L. (2005) *Journal of Peptide Science*, **11**, 572.
71 Horton, D.A., Severinsen, R., Kofod-Hansen, M., et al. (2005) *Journal of Combinatorial Chemistry*, **7**, 421.
72 Beech, C.L., Coope, J.F., Fairley, G., et al. (2001) *Journal of Organic Chemistry*, **66**, 2240.
73 Hulme, C., Peng, J., Morton, G., et al. (1998) *Tetrahedron Letters*, **39**, 7227.
74 Raghavan, S. and Rajender, A. (2002) *Chemical Communications*, 1572.
75 Sola, R., Saguer, P., David, M.L. and Pascal, R. (1993) *Journal of the Chemical Society, Chemical Communications*, 1786.
76 Sola, R., Mery, J. and Pascal, R. (1996) *Tetrahedron Letters*, **37**, 9195.
77 Pascal, R. and Sola, R. (1997) *Tetrahedron Letters*, **38**, 4549.
78 Fürst, M. and Rück-Braun, K. (2002) *Synlett*, 1991.
79 Routledge, A., Abell, C. and Balasubramanian, S. (1997) *Tetrahedron Letters*, **38**, 1227.

80 Lee, H.B. and Balasubramanian, S. (1999) *Journal of Organic Chemistry*, **64**, 3454.
81 Cano, M., Ladlow, M. and Balasubramanian, S. (2002) *Journal of Organic Chemistry*, **67**, 129.
82 Flickinger, S.T., Patel, M., Binkowski, B.F., et al. (2006) *Organic Letters*, **8**, 2357.
83 Li, X., Abell, C. and Ladlow, M. (2003) *Journal of Organic Chemistry*, **68**, 4189.
84 Kamogawa, H., Kanzawa, A., Kadoya, M., et al. (1983) *Bulletin of the Chemical Society of Japan*, **56**, 762.
85 Sucholeiki, I. (1994) *Tetrahedron Letters*, **35**, 7307.
86 Forman, F.W. and Sucholeiki, I. (1995) *Journal of Organic Chemistry*, **60**, 523.
87 Jung, K.W., Zhao, X.-Y. and Janda, K.D. (1996) *Tetrahedron Letters*, **37**, 6491.
88 Jung, K.W., Zhao, X.-Y. and Janda, K.D. (1997) *Tetrahedron*, **53**, 6645.
89 Zhao, X.-Y., Jung, K.W. and Janda, K.D. (1997) *Tetrahedron Letters*, **38**, 977.
90 Vanier, C., Lorge, F., Wagner, A. and Mioskowski, C. (2000) *Angewandte Chemie (International Edition in English)*, **39**, 1679.
91 McAllister, L.A., Turber, K.L., Brand, S., et al. (2006) *Journal of Organic Chemistry*, **71**, 6497.
92 Xie, J., Sun, J., Zhang, G., et al. (2007) *Journal of Combinatorial Chemistry*, **9**, 566.
93 Jin, S., Holub, D.P. and Wustrow, D.J. (1998) *Tetrahedron Letters*, **39**, 3651.
94 Sheng, S.-R., Xu, L., Zhang, X.-L., et al. (2006) *Journal of Combinatorial Chemistry*, **8**, 805.
95 Cheng, W.-C., Halm, C., Evarts, J.B., et al. (1999) *Journal of Organic Chemistry*, **64**, 8557.
96 Pan, Y., Ruhland, B. and Holmes, C.P. (2001) *Angewandte Chemie (International Edition in English)*, **40**, 4488.
97 Cho, C.-H., Park, H., Park, M.-A., et al. (2005) *European Journal of Organic Chemistry*, 3177.
98 Ruhland, T., Anderson, K. and Pederson, H. (1998) *Journal of Organic Chemistry*, **63**, 9204.
99 Fujita, K., Watanabe, K., Oishi, A., et al. (1999) *Synlett*, 1760.
100 Nicolaou, K.C., Pfefferkorn, J.A., Cao, G.-Q., Kim, S. and Kessabi, J. (1999) *Organic Letters*, **1**, 807.
101 Nicolaou, K.C., Pfefferkorn, J.A. and Cao, G.-Q. (2000) *Angewandte Chemie (International Edition in English)*, **39**, 734.
102 Nicolaou, K.C., Pfefferkorn, J.A. and Cao, G.-Q. (2000) *Angewandte Chemie (International Edition in English)*, **39**, 739.
103 Nicolaou, K.C., Roecker, A.J., Pfefferkorn, J.A. and Cao, G.-Q. (2000) *Journal of the American Chemical Society*, **122**, 2966.
104 Plunkett, M.J. and Ellman, J.A. (1995) *Journal of Organic Chemistry*, **60**, 6006.
105 Plunkett, M.J. and Ellman, J.A. (1997) *Journal of Organic Chemistry*, **62**, 2885.
106 Finkelstein, J.A., Chenera, B. and Veber, D.F. (1995) *Journal of the American Chemical Society*, **117**, 11999.
107 Woolard, F.X., Paetsch, J. and Ellman, J.A. (1997) *Journal of Organic Chemistry*, **62**, 6102.
108 Merluzzi, V.J., Hargrave, K.D., Labadia, M., et al. (1990) *Science*, **250**, 1411.
109 Curtet, S. and Langlois, M. (1999) *Tetrahedron Letters*, **40**, 8563.
110 Lee, Y. and Silverman, R.B. (2001) *Tetrahedron*, **57**, 5339.
111 Han, Y., Walker, S.D. and Young, R.N. (1996) *Tetrahedron Letters*, **37**, 2703.
112 Boehm, T.L. and Showalter, H.D.H. (1996) *Journal of Organic Chemistry*, **61**, 6498.
113 Briehn, C.A., Kirschbaum, T. and Bäuerle, P. (2000) *Journal of Organic Chemistry*, **65**, 352.
114 Harikrishnan, L.S. and Showalter, H.D.H. (2000) *Tetrahedron*, **56**, 515.
115 Spivey, A.C., Diaper, C.M., Adams, H. and Rudge, A.J. (2000) *Journal of Organic Chemistry*, **65**, 5253.
116 Nicolaou, K.C., Winssinger, N., Pastor, J. and Murphy, F. (1998) *Angewandte Chemie (International Edition in English)*, **37**, 2534.
117 Bräse, S. and Dahmen, S. (2000) *Chemistry – A European Journal*, **6**, 1899.
118 Lormann, M., Dahmen, S. and Bräse, S. (2000) *Tetrahedron Letters*, **41**, 3813.
119 Schunk, S. and Enders, D. (2000) *Organic Letters*, **2**, 907.

120 Bräse, S. and Schroen, M. (1999) *Angewandte Chemie (International Edition in English)*, **38**, 1071.
121 Bräse, S. (2004) *Accounts of Chemical Research*, **37**, 805.
122 Bräse, S., Dahmen, S. and Lormann, M.E.P. (2003) *Methods in Enzymology*, **369**, 127.
123 Zimmermann V., Avemaria F. and Bräse S. (2007) *Journal of Combinatorial Chemistry*, **9**, 200.
124 Dahmen, S. and Bräse, S. (2000) *Angewandte Chemie (International Edition in English)*, **39**, 3681.
125 Bräse, S., Enders, D., Köbberling, J. and Avemaria, F. (1998) *Angewandte Chemie (International Edition in English)*, **37**, 3413.
126 Hughes, I. (1996) *Tetrahedron Letters*, **37**, 7595.
127 Slade, M.A., Phillips, M.A. and Berger, J.G. (2000) *Molecular Diversity*, **4**, 215.
128 Sucholeiki, I. (Solid-phase Sciences Corporation) (1997) [US5684130].
129 Pavia, M.R., Whitesides, G., Hangauer, D.G. and Hediger, M.E. (Sphinx Pharmaceuticals Corporation) (1995) [WO PCT.95/04277].
130 Garibay, P., Nielsen, J. and Hoeg-Jensen, T. (1998) *Tetrahedron Letters*, **39**, 2207.
131 Sim, M.M., Lee, C.L. and Ganesan, A. (1998) *Tetrahedron Letters*, **39**, 6399.
132 Sim, M.M., Lee, C.L. and Ganesan, A. (1998) *Tetrahedron Letters*, **39**, 2195.
133 Zaragoza, F. (1997) *Tetrahedron Letters*, **38**, 7291.
134 Hamper, B.C., Gan, K.Z. and Owen, T.J. (1999) *Tetrahedron Letters*, **40**, 4973.
135 Cobb, J.M., Fiorini, M.T., Godard, C.R., Theoclitou, M.-E. and Abell, C. (1999) *Tetrahedron Letters*, **40**, 1045.
136 Gibson, S.E., Hales, N.J. and Peplow, M.A. (1999) *Tetrahedron Letters*, **40**, 1417.
137 Rigby, J.H. and Kondratenko, M.A. (2001) *Organic Letters*, **3**, 3683.
138 Pourbaix, C., Carreaux, F., Carboni, B. and Deleuze, H. (2000) *Chemical Communications*, 1275.
139 Rasmussen, L.K., Begtrup, M. and Ruhland, T. (2006) *Journal of Organic Chemistry*, **71**, 1230.
140 Horton, J.R., Stamp, L.M. and Routledge, A. (2000) *Tetrahedron Letters*, **41**, 9181.
141 Schuster, M., Lucas, N. and Blechert, S. (1997) *Chemical Communications*, 823.
142 Van Maarseveen, J.H., Hartog, J.A.J., Engelen, V., *et al.* (1996) *Tetrahedron Letters*, **37**, 8249.
143 Gowravaram, M.R. and Gallop, M.A. (1997) *Tetrahedron Letters*, **38**, 6973.
144 Guillier, F., Orain, D. and Bradley, M. (2000) *Chemical Reviews*, **100**, 2091.
145 Blaney, P., Grigg, R. and Sridharan, V. (2002) *Chemical Reviews*, **102**, 2607.
146 James, I.W. (1999) *Tetrahedron*, **55**, 4855.
147 Gil, C. and Bräse, S. (2004) *Current Opinion in Chemical Biology*, **8**, 230.

17
Photolabile and Miscellaneous Linkers/Resins

Soo Sung Kang and Mark A. Lipton

Purdue University, Department of Chemistry, 560 Oval Drive, West Lafayette, IN 47907-2084 USA

17.1
Introduction

As the interest in synthesizing complex and acid-sensitive molecules on solid supports has grown, the need for new methods of detaching molecules from synthesis resins has also increased. To supplement the original methods of acid- and base-catalyzed deprotection, other methodologies have been explored and developed. One of the most appealing ideas has been to use photochemical detachment of substrate from resin as an extra dimension of protecting group orthogonality in solid phase synthesis.

Photolabile linkers, in which long-wave UV radiation (300–360 nm) is used to detach an organic functional group – usually a carboxylic acid, a carbamate (a latent amine), an alcohol, or a phosphate – from a polymer or glass surface, provide an alternative to the classic acidic or basic cleavage protocols used in solid phase organic synthesis. Ideally, photochemical deprotection should proceed with high functional group selectivity, low amounts of side reactions and involve no separation from reactants.

17.2
Types of Photolabile Linker

17.2.1
o-Nitrobenzyl as a Photolabile Group

The introduction of photolabile linkers for use in solid phase synthesis has its origins in the development of photolabile protecting groups in peptide synthesis. An early breakthrough was Scofield's [1] report in 1962 of the photochemical deprotection of benzyloxycarbonylglycine (Cbz-Gly) to afford glycine upon its irradiation with UV light. An outgrowth of that discovery was the development of the

The Power of Functional Resins in Organic Synthesis. Judit Tulla-Puche and Fernando Albericio
Copyright © 2008 WILEY-VCH Verlag GmbH & Co. KGaA, Weinheim
ISBN: 978-3-527-31936-7

o-nitrobenzyl carbamate as a modification of the Cbz protecting group, also by Schofield [2]. Woodward [3] later used the 6-nitroveratryloxycarbonyl (NVOC) and 2-nitrobenzyloxycarbonyl (NBOC) groups (Figure 17.1) in a study of the structure-based effects of carbamates on photolysis rate and efficiency. NVOC was found to have better photo-release properties than NBOC when they were treated with UV irradiation (>320 nm) for 1–12 h.

The likely pathway of photochemical cleavage is a Norrish type II cleavage reaction (Scheme 17.1). During the photoisomerization, the activated oxygen of the nitro group abstracts a γ-hydrogen from the benzylic position to produce a quinone methide-like intermediate, which subsequently rearranges to afford an o-nitrosobenzaldehyde and thereby releases the carboxylic acid product.

17.2.1.1 o-Nitrobenzyl Linker

The solid phase synthesis of peptides using a photolabile nitrobenzyl linker was first reported by Rich [4] in 1973. He used a nitro-functionalized Merrifield resin (1) and attached a Boc-protected amino acid by refluxing it with base. The peptide chain was synthesized using Merrifield's procedure [5] and the synthesized peptide was cleaved from the resin using UV irradiation (350 nm) in methanol (Scheme 17.2).

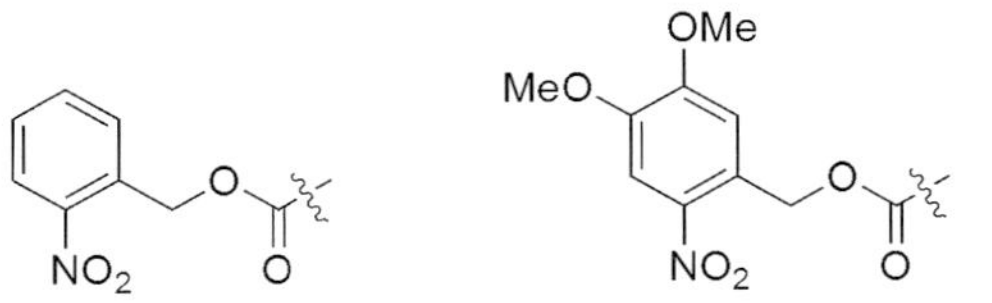

Figure 17.1 The 6-nitroveratryloxycarbonyl (NVOC) and 2-nitrobenzyloxycarbonyl (NBOC) groups.

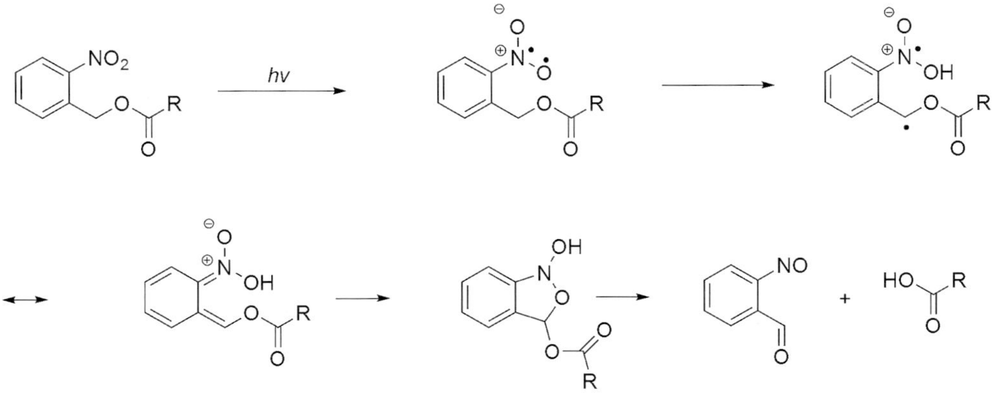

Scheme 17.1 Norrish type II photochemical cleavage of NBOC.

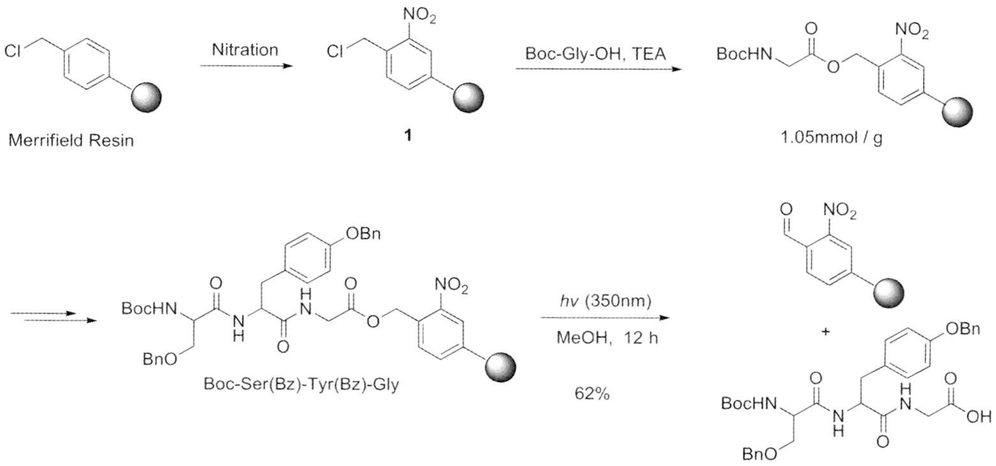

Scheme 17.2 First solid phase synthesis of peptides using a photolabile nitrobenzyl linker.

17.2.1.2 4-Bromomethyl-3-Nitrobenzoic Acid Derived Linker

After Rich introduced the general concept of o-nitro functionalized photolabile linkers for peptide synthesis, its substructures, especially the benzylic methylene, the meta or para positions of the aromatic ring, and the types of resin became targets of modification to improve (i) swelling in diverse solvents, (ii) the rate and efficiency of photocleavage, (iii) bond formation between the linker and substrates and (iv) suppression of side reactions.

Because Rich's first photolabile resin [4] (**1**) has poor swelling efficiency in chloroalkane solvents, its application was limited to the synthesis of smaller peptides. The low extension property of resin **1** reduces penetration of the solvents and reagents, and thereby reduces the reaction rate [6] of larger protected peptides. The synthesis of a tripeptide using resin **1** resulted in a 62% overall yield, but the tetrapeptide Boc-Leu-Arg(Tos)-Pro-Gly was obtained in only 32% yield. To alter the swelling properties of **1**, Rich synthesized 4-bromomethyl 3-nitrobenzoic acid and connected it to an aminomethyl PS resin by amide bond formation. This new linker [7] (**2**) had an acceptable loading yield (0.3 mmol g^{-1}) and improved swelling properties in all commonly used solvents, and was used to synthesize Gly-His(Bzl)-Trp-Ser(Bzl)-Tyr(Bzl)-Gly-Leu-Arg(Tos)-Pro-Gly in 64% yield. Tam [8] also made a tridecapeptide, Boc-Asn-Lys(ClZ)-Tyr(BrZ)-Thr(Bzl)-Thr(Bzl)-Glu(Bzl)-Tyr(BrZ)-Ser(Bzl)-Ala-Ser(Bzl)-Val-Lys(ClZ)-Gly, in 54% yield using the same linker (Figure 17.2).

Albericio introduced bromomethyl-NBB-resin (**3**), in which 4-bromomethyl-3-nitrobenzoic acid is linked to a benzhydrylamine resin, and used it in the synthesis of Boc-Cys(Acm)-Asn-Cys(Acm)-Lys(Z)-Ala-Pro [9] in 68% yield, and Boc-Cys(Acm)-Asn-Cys(Acm)-Lys(Z)-Ala-Pro-Glu(Bzl)-Thr(Bzl)-Ala-Leu-Cys(Acm)-Ala [10] in 89% yield. However, the NBB resin was unstable to base: the peptide and linker bonds were not completely stable to treatment with piperidine, so the synthesis of

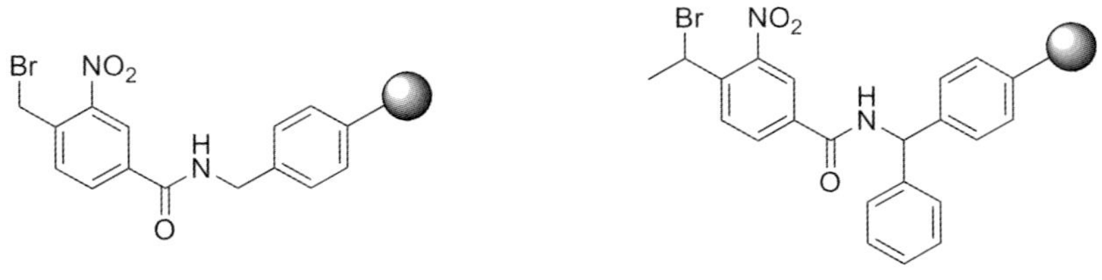

Figure 17.2 Early resins used in the synthesis of oligopeptides.

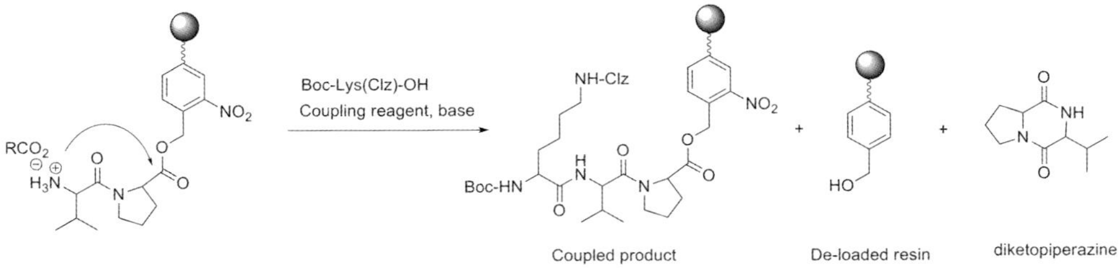

Scheme 17.3 The formation of diketopiperazine (DKP) not only reduces the total yield but can also lead to increased impurities due to the de-loaded benzyl alcohol's involvement in side reactions.

longer peptides using **3** and an Fmoc-based strategy was discouraged by the authors [11]. Another drawback of **3** was a propensity for diketopiperazine (DKP) formation during the third amino acid coupling.

DKP formation is a universally undesired side reaction in peptide synthesis regardless of the peptide synthesis strategy. DKP formation is sequence dependent [12, 13], happens during both acid- [12] and base- [13] catalyzed procedures, and seems to be favored by the electron-withdrawing character of aromatic rings in the resin. DKP formation is troublesome not only because it reduces total yield but it can also increase the impurities by the de-loaded benzyl alcohol's involvement in other side reactions on the resin (Scheme 17.3). If the first amino acid is a secondary amine or bulky, DKP formation was often more than 40% of the total product [13] on the NBB resin.

An Fmoc synthesis protocol is likewise not recommended when using the nitrobenzyl linker because of DKP formation during Fmoc deprotection. Gilralt [14] suggested the Bop coupling reagent as a means of suppressing DKP formation and compared it with Suzuki's [15] and standard methods (Table 17.1). Such methods involve minimizing the time of base contact with the resin after acidic deprotection of a Boc group.

Table 17.1 Comparison of coupling agents used to suppress the formation of diketopiperazine (DKP).

Coupling procedure for the 3rd amino acid	3rd	Dipeptide	Formation of DKP (%)
Standard[a]	Boc-Lys(Clz)-OH	Boc-Ala-Phe-	59
Suzuki[b]	Boc-Lys(Clz)-OH	Boc-Ala-Phe-	0
BOP DIEA[c]	Boc-Lys(Clz)-OH	Boc-Ala-Phe-	5
BOP, HOBt, DIEA[c]	Boc-Lys(Clz)-OH	Boc-Ala-Phe-	2
Standard	Boc-Lys(Clz)-OH	Boc-Ala-Phe-	91
Suzuki	Boc-Lys(Clz)-OH	Boc-Ala-Phe-	8
BOP, HOBt, DIEA	Boc-Lys(Clz)-OH	Boc-Ala-Phe-	9

a Boc deprotection with TFA, neutralization with 5% DIEA and addition of the amino acid with DCC.
b Boc deprotection with 4N HCl in dioxane and addition of NMM salt of the amino acid with DCC.
c Boc deprotection with TFA and addition of coupling reagents in DMF.

17.2.1.3 Hydroxynitrobenzyl and Nitroveratryl Linkers

Other variations of the nitrobenzyl linker are the nitroveratryl (**4**) and hydroxynitrobenzyl (**5**) linkers, which are characterized by one or two alkoxy substituents on the nitrobenzene ring. The increased electron density at the benzylic position in these linkers improves photochemical cleavage efficiency through stabilization of radical intermediates in photolysis and also retards diketopiperazine formation during the coupling of the third amino acid in peptide synthesis.

It was Zehavi [16] who first used the nitroveratryl linker (**4**) in 1973 to synthesize an oligosaccharide using a photolabile solid phase strategy. Starting with isovanillin, **4** was prepared by nitration, alkylation with the Merrifield polystyrene resin and reduction of the aldehyde. Photolysis using UV irradiation (>320 nm) in dioxane produced product in 13% yield after 32 h of irradiation (Scheme 17.4).

The hydroxynitrobenzyl linker **5** was designed by Nicolaou [17] for the combinatorial synthesis of oligosaccharides. The single sugar photolysis test (Scheme 17.5) resulted in a 95% cleavage yield after 8 h of irradiation, and the heptasaccharide product was obtained in 20% overall yield.

Amines can also be anchored to a photolabile linker through carbamate bonds. During the early development of solid phase synthesis methodology, Woodward [18] investigated the cleavage efficiency of various amino acids from nitrobenzyl derivatives. His study showed that nitroveratryl derivatives were more efficient than monomethoxynitrobenzyl derivatives (Table 17.2) and that side reactions were totally avoided by the addition of sulfuric acid, hydrazine, hydroxylamine hydrochloride or semicarbazide hydrochloride to scavenge the reactive nitrosobenzaldehyde side product.

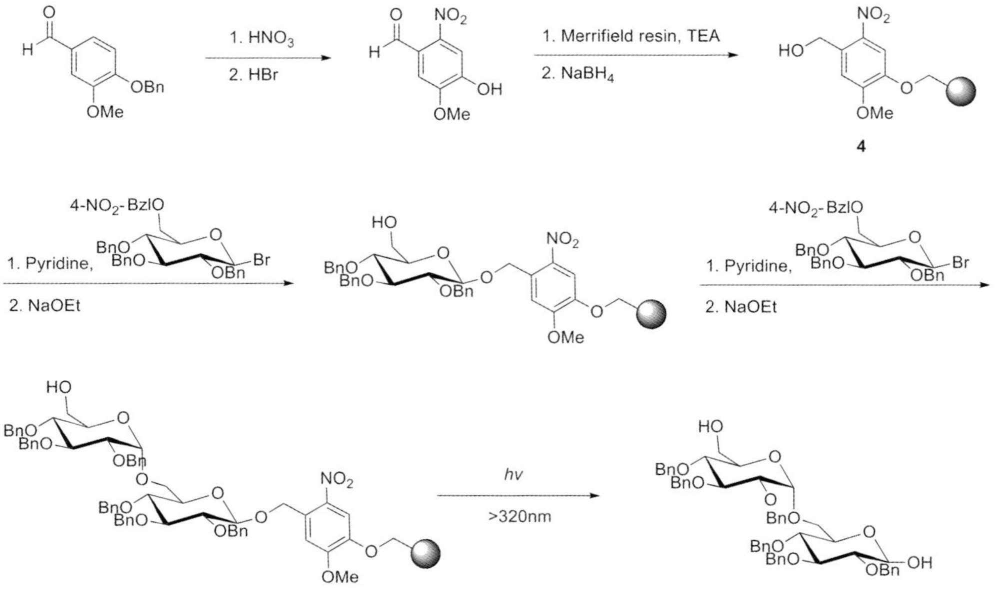

Scheme 17.4 An early example of the synthesis of an oligosaccharide using nitroveratryl linker (**4**).

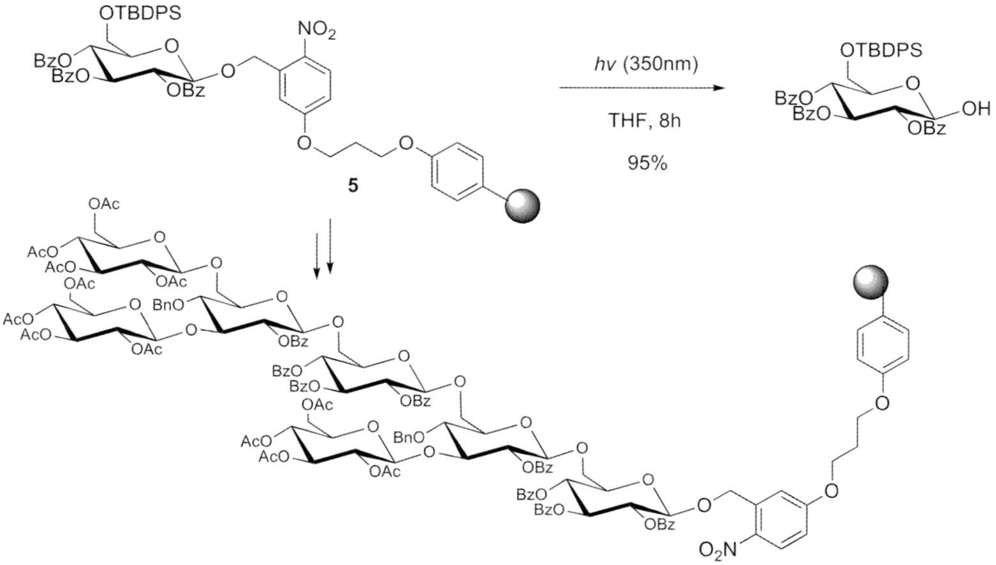

Scheme 17.5 Cleavage of a sugar from hydroxynitrobenzyl linker **5**.

Table 17.2 Cleavage efficiency of various amino acids from nitrobenzyl derivatives.

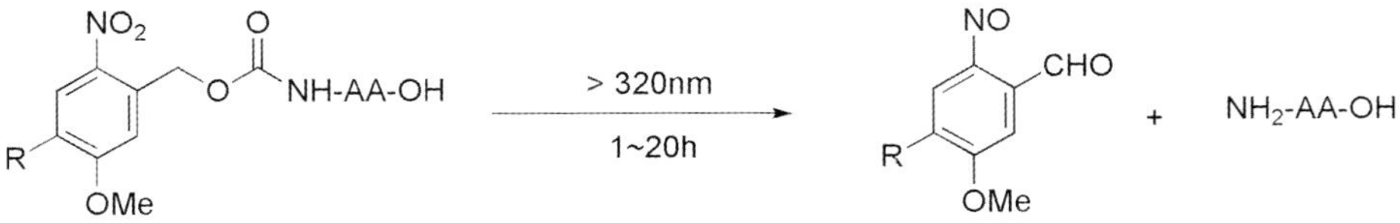

R	AA	Yield (%)
MeO	Ala	80
MeO	Phe	35
H	Ala	35
H	Phe	17
MeO	Val	>99[a]
H	Ala	>99[a]

a 5 eq of H_2SO_4 or 10 eq of $H_2NNHCONH_2 \cdot HCl$.

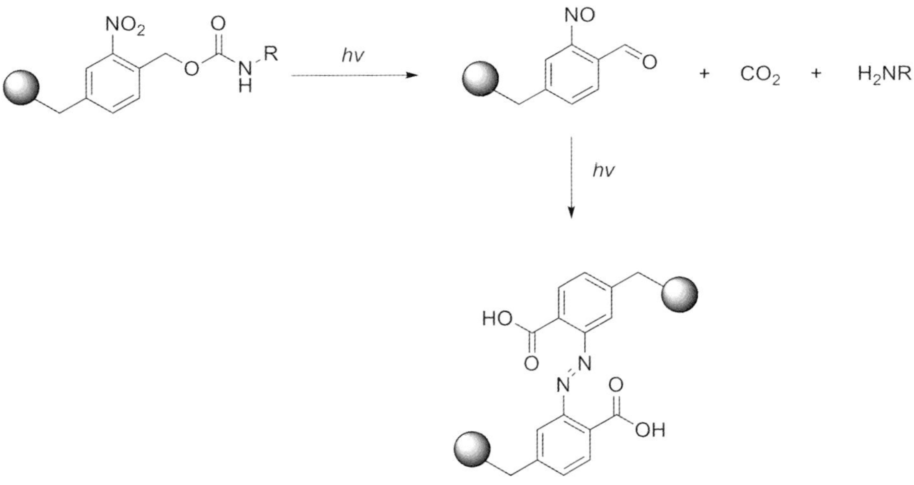

Scheme 17.6 Production and reaction of unwanted nitrosobenzaldehydes associated with the use of nitrobenzyl linkers.

The production of nitrosobenzaldehydes is a third possible drawback associated with the use of nitrobenzyl linkers. The reactive aldehyde can react with an amine in the released product, thereby substantially decreasing the yield of photolysis. Additionally, the nitroso group can engage in a second photoreaction through intermolecular diazene formation (Scheme 17.6). The produced diazene is not only an internal filter ($\lambda_{max} = 325$ nm) [19], but also cross-links the resin, which slows down photocleavage and eventually blocks product release from the resin.

In addition to Woodward's scavenger treatment [18], the use of an α-substituted nitrobenzyl linker was used to solve the problem of aldehyde condensation, since the ketone photolysis byproduct is substantially less reactive than an aldehyde. On the basis of the accepted cleavage mechanism, though, photochemical deprotection should still be feasible with an α-substituted nitrobenzyl linker.

17.2.1.4 α-Substituted Nitrobenzyl Linkers

Even though the efficiency of deprotection of α-substituted nitrobenzyl carbamates as an amine protecting group had been published by Schofield [2] (6) in 1966, and later by Woodward [18] (7), it took 20 years more for α-substituted nitrobenzyl linkers to be applied to solid phase synthesis to decrease photo-byproduct formation (Figure 17.3).

Pillai developed the 2'-nitrobenzhydryl PS resin [20] (8, NBH resin) as an α-substituted nitrobenzyl linker. The NBH resin was prepared from o-nitrobenzoyl chloride and 1% cross-linked PS resin by Friedel–Crafts acylation and ketone reduction (Scheme 17.7). Photolysis of single amino acids from the NBH resin

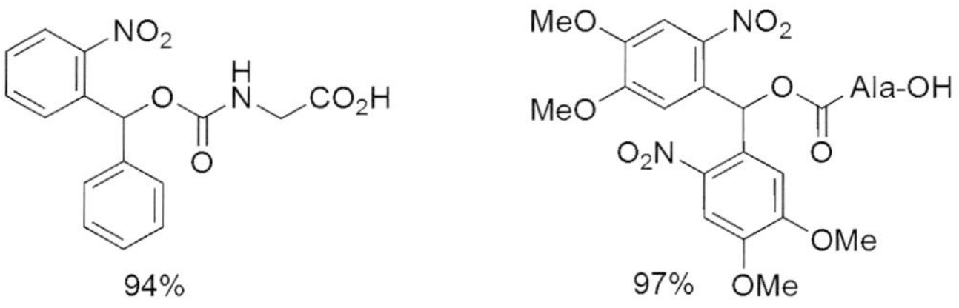

Figure 17.3 Early examples of α-substituted nitrobenzyl linkers used to decrease photo-byproduct formation.

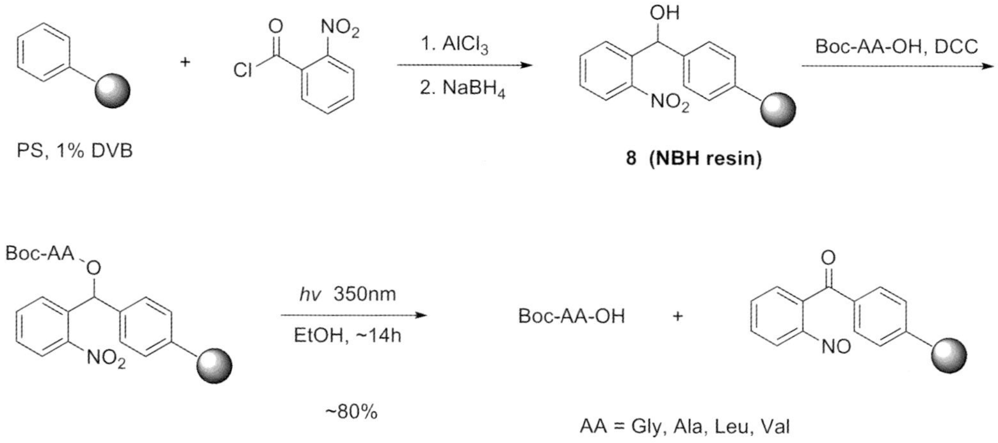

Scheme 17.7 Preparation of, and photolysis of single amino acids from, NBH resin.

17.2 Types of Photolabile Linker

was greater than 78%, and the tetrapeptide Boc-Leu-Ala-Gly-Val was detached in 58% yield after 24 h of irradiation. Pillai's other approach used the o-nitro-α-methylbromobenzyl resin [21] **9** (Figure 17.4), but the high loading capacity (5.6 mmol g^{-1}) could reduce the efficiency of coupling reactions.

Geysen [22] prepared a new α-substituted linker (**10**, ANP) from 3-amino-3-(2-nitrophenyl)propionic acid. This ANP resin, distinguished by its α-substituted junction to the resin, showed good photolytic activity (80% yield) after 20 h of irradiation at 365 nm. The drawback of the ANP linker – unwanted β-elimination of the amide – was rectified by Schreiber [23] by substituting the 3-amino-2,2-dimethyl-3-(2-nitrophenyl)propionic acid-derived linker **11**. Photolysis in 4:1 H$_2$O–MeOH resulted in 86% amide release after 4 h of irradiation.

Harren [24] provided another example of an α-substituted photolabile linker (**12**) and anchored it to an aminopropylsiloxane-grafted controlled pore glass (amino CPG). It had a longer tether and sterically less bulky amine. An Fmoc-Gly photolysis test resulted in a 77% yield after 2 h of irradiation at 350 nm (Scheme 17.8).

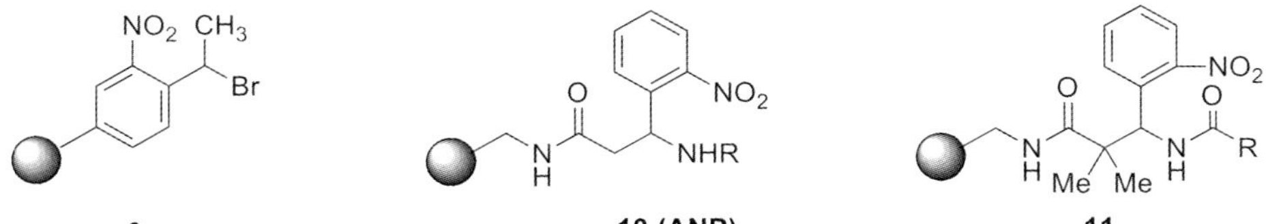

Figure 17.4 More recent examples of α-substituted linkers.

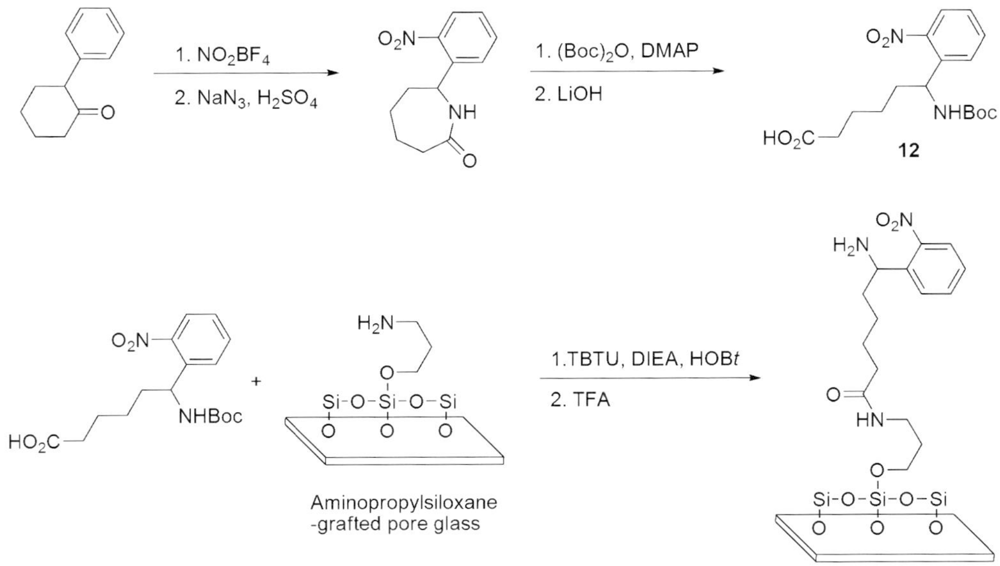

Scheme 17.8 Preparation and anchoring to controlled pore glass of α-substituted photolabile linker **12**.

17.2.1.5 Comparison of o-Nitrobenzyl Linkers

Holmes [25] prepared a series of known nitrobenzyl-derived photolabile linkers and examined their rates of photolytic cleavage in various solvents (Table 17.3). When irradiated at 365 nm, o-nitrobenzyl linker **13** showed the slowest rate of cleavage in all solvents; α-methyl substituted linkers (**15–17**) were also cleaved faster than the non-substituted linker **14** in all solvents. Increasing the length of tether (**16**) between the resin and linker slightly improved reaction rates in organic solvents, but diminished them in aqueous buffer.

The bathochromic and hyperchromic shifts of n→π* or π→π* absorbances (Figure 17.5) of the aryl rings caused by the introduction of two alkoxy groups correlated well with the increased photolability of nitroveratryl linkers during 365 nm irradiation.

Table 17.3 Half-lives (min) during 365 nm UV irradiation.

Linker	Solvent		
	Aqueous[a]	MeOH	Dioxane
13	14.1	362	36.8
14	12.9	16.2	4.0
15	1.7	8.8	0.6
16	0	3.9	0.5
14	0.7	0.5	0.2

a Saline phosphate buffer.

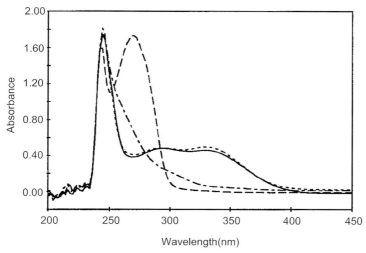

Figure 17.5 Bathochromic and hyperchromic shifts of n→π* or π→π* absorbances of nitrobenzyl-derived photolabile linkers 13 (—); 14 (----); 15 (– -), and phenacyl (– –).

A similar comparison was made by Geysen [26], in which the photocleavage yields of **18**, **19**, and his new linker **20** were monitored at various times (Table 17.4). Surprisingly, his new linker **20** showed similar or slightly better results than the nitrobenzyl linker **18** and the α-methyl nitroveratryl linker **19** when irradiated at 350 nm.

It would appear from these data that the α-hydrogen abstraction process is more important to the yield of photolytic cleavage than the process of nitro group photoexcitation in the overall photolysis procedure.

17.2.2
Functionalized Phenylacyl Linkers

Another useful photoreleasable functionality is the phenacyl [PhC(O)CH$_2$-] group. The aryl carbonyl easily undergoes n→π* or π→π* excitation upon 300–350 nm UV radiation, and then rearranges to release an α-substituent. Depending on the substituents on the aryl ring, various photolytic pathways with different efficiencies are possible. The parent phenacyl group undergoes direct solvolysis and bond scission [27], whereas photolysis of the 4-hydoxyphenacyl group involves a spirocyclopropanone [28] as an intermediate. Benzoin [29] (an α-phenyl-substituted phenacyl group) and o-methylphenacyl [30] groups, in contrast, undergo a hydrogen abstraction process from neighboring groups and cleavage during rearrangement.

Table 17.4 Photocleavage yields of linkers **18–20** at various times.

Linker	Yield (%)		
	3 h	6.5 h	23 h
18	12	19	43
19	8	30	82
20	19	56	87

Table 17.5 Effect of the nature of the leaving group and the substitution of the phenyl ring.

R	R'	Photolysis time (h)	Yield (%)
H	4'-Methoxy	9	10
H	3'-Methoxy	11	88
H	3',5'-Dimethoxy	1	94
3,4-Dimethoxy	3',4'-Dimethoxy	2	75
2,3-Dimethoxy	2',3'-Dimethoxy	4	76

17.2.2.1 Benzoin Linker

Early studies of the photoreactivity of benzoin groups were conducted by Sheehan [31] in 1964. Sheehan [29] found that the photolysis reaction was critically dependent upon the nature of leaving group and sensitive to the substitution of the phenyl group (Table 17.5). In particular, methoxy groups placed at the meta positions of the benzyl ring were essential for the fast and efficient cleavage of various functional groups with 350 nm irradiation.

The photolysis mechanism can be rationalized as a diradical process starting from an n→π* transition, the mechanistic details of which are not clear [9, 32, 33]. The highly efficient process of 3′,5′-dimethoxybenzoin (DMB) photolysis can be rationalized by Wan's [34] suggestion (Scheme 17.9), involving a transient absorption (485 nm) assigned to the cyclohexadienyl cation **22**. He proposed that a charge transfer interaction of the electron-rich dimethoxyphenyl ring with the electron deficient oxygen of an n→π* singlet excited ketone formed an intramolecular exciplex **21**, which rearranges to **22**, leading then to the benzofuran product.

However, the chromophore in the DMB linker proved to be too sensitive to light, so a dithiane-protected structure, introduced by Chan [35] and Balasubramanian, was used to employ a "safety-catch" strategy. Using this strategy, peptides could be anchored to the linker by a C-terminal ester bond [36] or by an N-terminal carbamate bond [37] (Scheme 17.10). In a Fmoc-alanine releasing test, the dithiane was deprotected using methyl triflate or periodic acid, and photocleavage resulted in a 92% yield of Fmoc-Ala after 2 h.

17.2.2.2 Phenacyl Linker

The photoinduced cleavage of a simple phenacyl group was first observed by Reese [38] in 1962 during a study of the Fries rearrangement. This photoreactivity was used as a protecting group for carboxylic acids by Sheehan [39], who suggested that the reaction mechanism was a simple radical scission of the carbon–oxygen bond followed by hydrogen atom abstraction from solvent (Scheme 17.11).

The *p*-methoxyphenacyl ester shows more efficient photolysis than that of the parent phenacyl ester, which can be rationalized by the added stabilization of radical intermediates and a bathochromic shift of the carbonyl chromophore through additional π-electron donation. α-Substitution appears to be important for

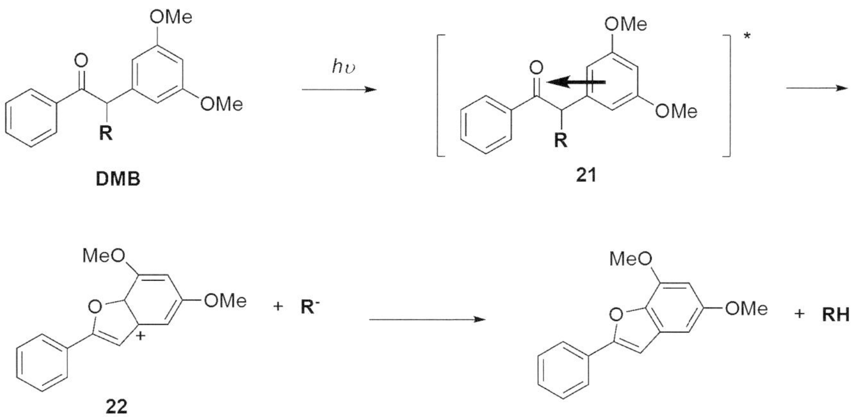

Scheme 17.9 Photolysis of 3′,5′-dimethoxybenzoin (DMB).

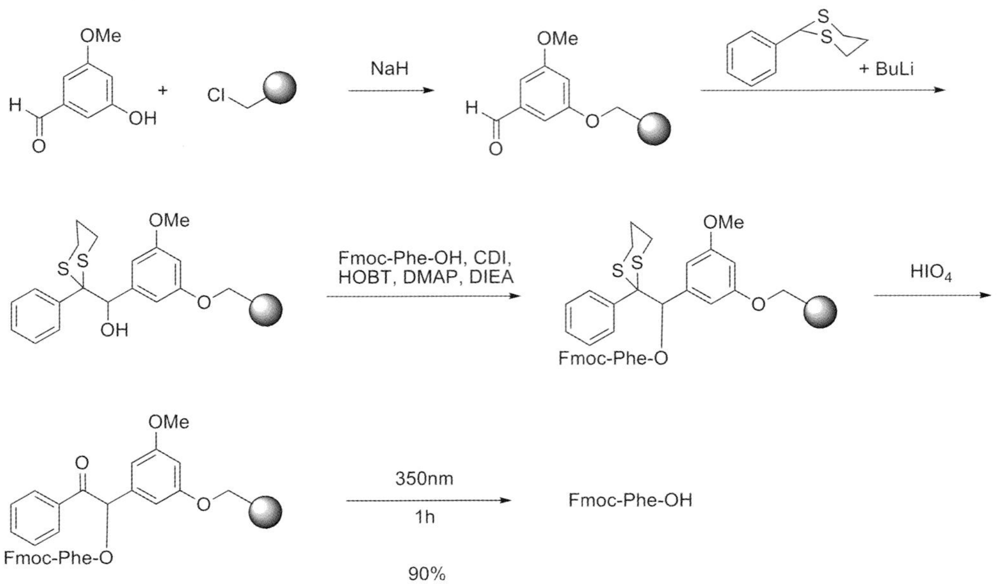

Scheme 17.10 Example of a "safety-catch" strategy, using a dithiane-protected structure.

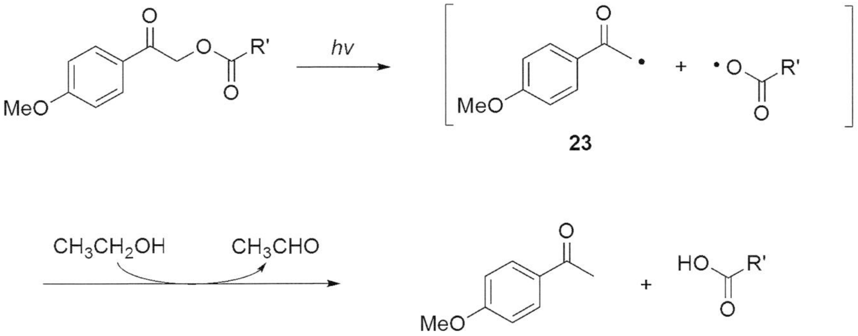

Scheme 17.11 Photoinduced cleavage of a simple phenacyl group.

good phenacyl group reactivity from Sheehan's experiments [39]. α-Methyl-substituted phenacyl esters have similar reactivity to *p*-methoxy-substituted phenacyl esters when treated with 320 nm UV radiation (Table 17.6).

Wang [40] transferred the known photoreactivity of the phenacyl group, especially α-methyl substituted, to a PS resin in 1976 for peptide synthesis (Scheme 17.12). The resin **26** is easily available by Friedel–Crafts acylation of PS, and its photolytic cleavage of a tetrapeptide appeared to be superior to that of the *o*-nitrobenzyl linker under similar reaction conditions. The α-methylphenacyl linker gave

Table 17.6 Comparison of the reactivity of α-methyl-(**25**) and p-methoxy-(**24**) substituted phenacyl esters.

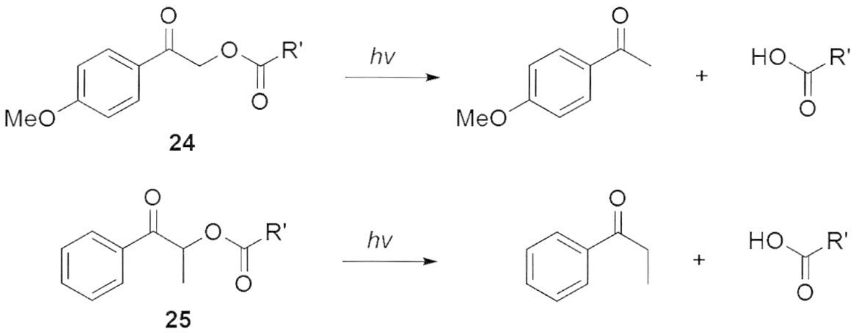

Substrate	Solvent	Irradiation time (h)	Yield (%)
Boc-Ala-24	Dioxane	17	82
Boc-Ala-24	EtOH	6	93
Boc-Ala-25	EtOH	6	94
Boc-Ala-25	Dioxane	6	96

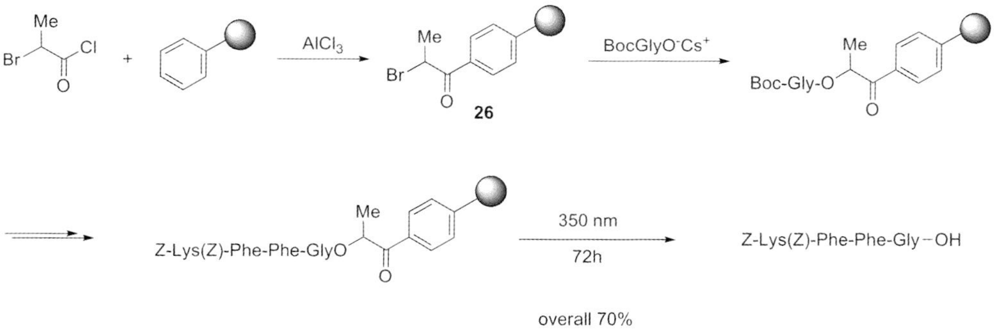

Scheme 17.12 Attachment of a phenacyl group to a PS resin and subsequent use in peptide synthesis.

a 70% overall yield, whereas the o-nitrobenzyl linker gave a 40% overall yield of the same tetrapeptide sequence.

Mutter devised the [4-(2-bromopropionyl)phenoxy]acetic acid linker [41] (**27**, PPOA), which presumably has the mechanistic advantages of both a p-alkoxy aryl substituent and α-methyl substitution (Figure 17.6). Unfortunately, synthesis of the pentapeptide Boc-Tyr(Bzl)-Gly-Gly-Phe-Leu using the PPOA linker resulted in only a 14% yield. It was realized that PPOA is only partially compatible with nucleophiles such as hydrazine, NH_3 and Triton B.

Figure 17.6 Examples of α-methylphenacyl linkers.

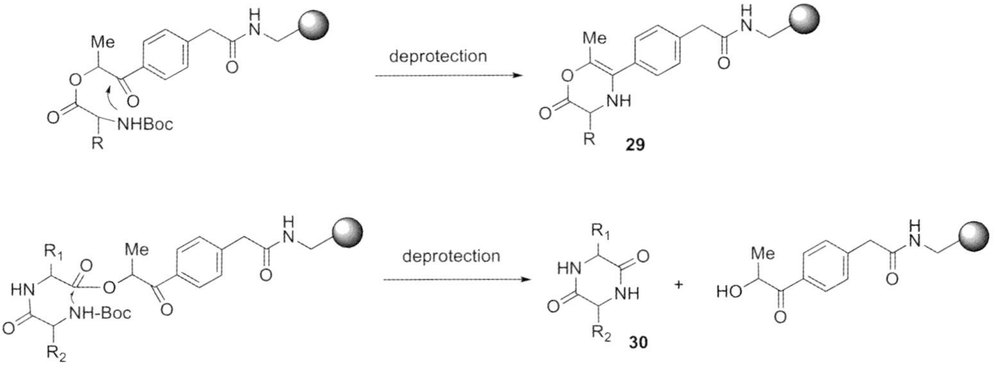

Scheme 17.13 Intramolecular cyclization of phenacyl linkers to form oxazinone **29** or diketopiperazines **30**.

Tjoeng synthesized the closely related α-methylphenacyl linker [42] **28** using (2-chloropionyl-4-phenyl)acetic acid as a starting material. As discussed with nitrobenzyl linkers (Section 17.2.1), amino acids attached to phenacyl linkers can undergo intramolecular cyclization to form an oxazinone **29** or diketopiperazine [43] **30** (Scheme 17.13). In this instance, however, intramolecular cyclization occurred at insignificant levels, resulting in ~75% yields of Boc-Asp(Obzl)-Val-Tyr(Bzl)-Val-Glu(OBzl)-OH and Z-Arg(Z,Z,)-Lys(Z)-Asp(OBzl)-Val-Tyr(Bzl)-OH after cleavage with 350 nm UV radiation.

Recently, Belshaw [44] reported a safety-catch photolabile linker based on the phenacyl group that permits selective photorelease of oligonucleotides from a solid substrate. The safety-catch concept is based on masking the carbonyl group as a dimethyl ketal that renders the linker photoinert during oligonucleotide synthesis (Scheme 17.14).

17.2.2.3 o-Methylphenacyl Linker

Photoenolization [45], one of the well-documented [46] photochemical rearrangements of o-alkylphenacyl groups (Scheme 17.15) can be used for the deprotection of alcohols [47], carboxylic acids [48] and phosphates [49]. Klan chose the 2,5-dimethylphenacyl (DMP) chromophore [30] as a photoremovable protecting group, and its photolysis with >280 nm UV radiation results in an almost quantitative yield of deprotection.

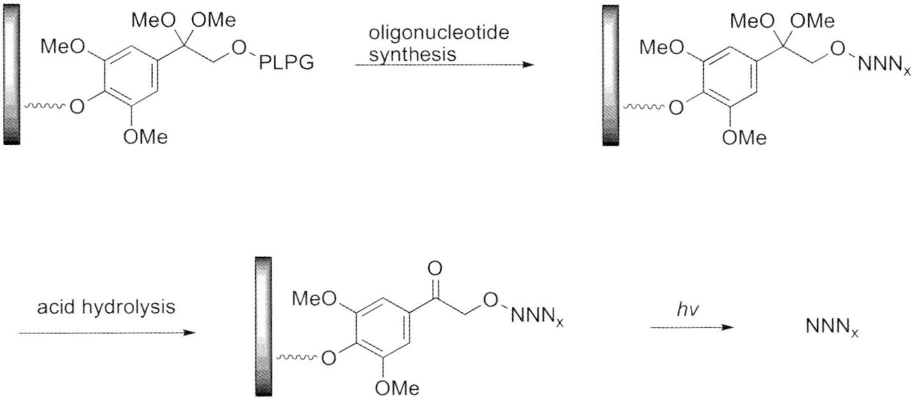

Scheme 17.14 Use of a safety-catch photolabile linker based on the phenacyl group for the selective photorelease of oligonucleotides.

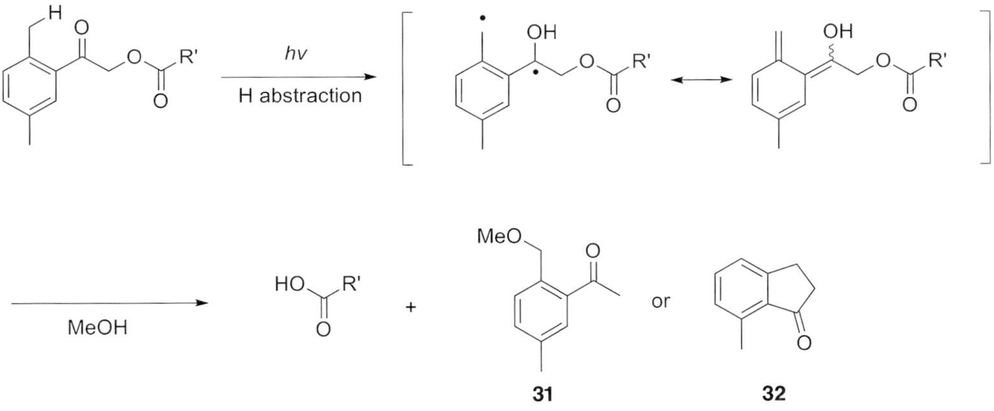

Scheme 17.15 Photoenolization of o-alkylphenacyl groups.

A solid-phase version of the DMP linker using p-methylphenylacetic acid **33** was reported by Wang [50]; it afforded good yields of deprotection of various carboxylic acids within 8 h using 280–366 nm UV radiation (Scheme 17.16).

Giese has reported [51, 52] that the triplet state of an excited phenacyl ketone can undergo intramolecular hydrogen atom abstraction from an attached peptide, resulting in photocyclization of the phenacyl group with peptides, and cyclization with peptide backbone. One of the attractions of the DMP linker is that their fast photoenolization will probably help to minimize unwanted hydrogen atom abstraction from an attached peptide.

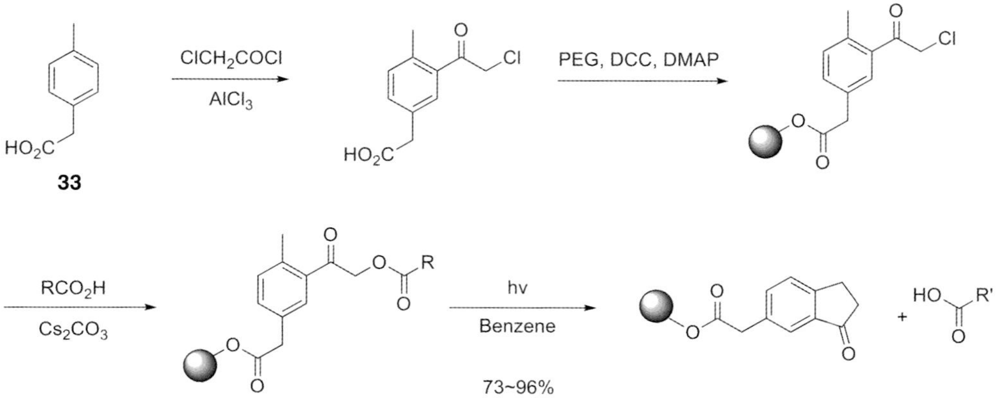

Scheme 17.16 A solid-phase version of the DMP linker and its use in deprotection.

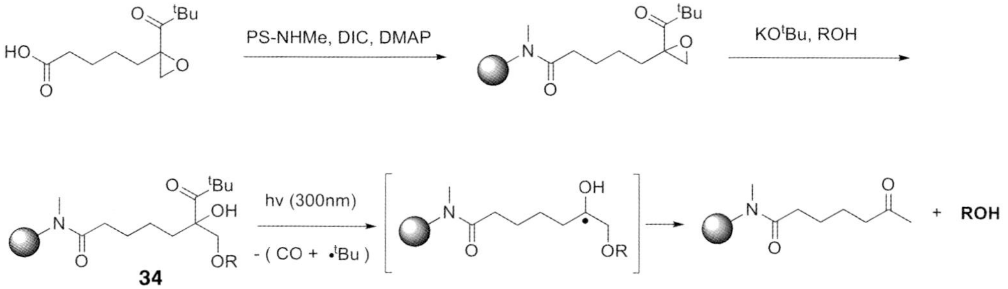

Scheme 17.17 Incorporation of the pivaloyl motif in the photolabile linker **34** and subsequent use in the solid phase synthesis of alcohols.

17.2.3
Pivaloyl Linker

During their investigation of glyceryl radicals [53], Giese and coworkers found the pivaloyl motif [54] as a radical precursor, initiating a two-step photolytic cleavage process (Scheme 17.17). Giese employed this motif in the design of a novel photolabile linker (**34**) for solid phase synthesis of carboxylic acids [55] and alcohols [56]. The cleavage reaction was solvent independent [55, 56], affording cleavage yields of ~90% and purities of ~90%[55]. The photolysis half-life was 2 min at >320 nm and 0.5 min at >305 nm [55]. The linker was quite stable to harsh acids and bases [55, 56], and a study comparing it [55] with the nitroveratryl linker (**16**) showed that **34** released its substrate ~4× faster.

Giese also suggested a solid phase peptide synthesis strategy using both the photolabile linker **34** and the NVOC protecting group to avoid the use of common

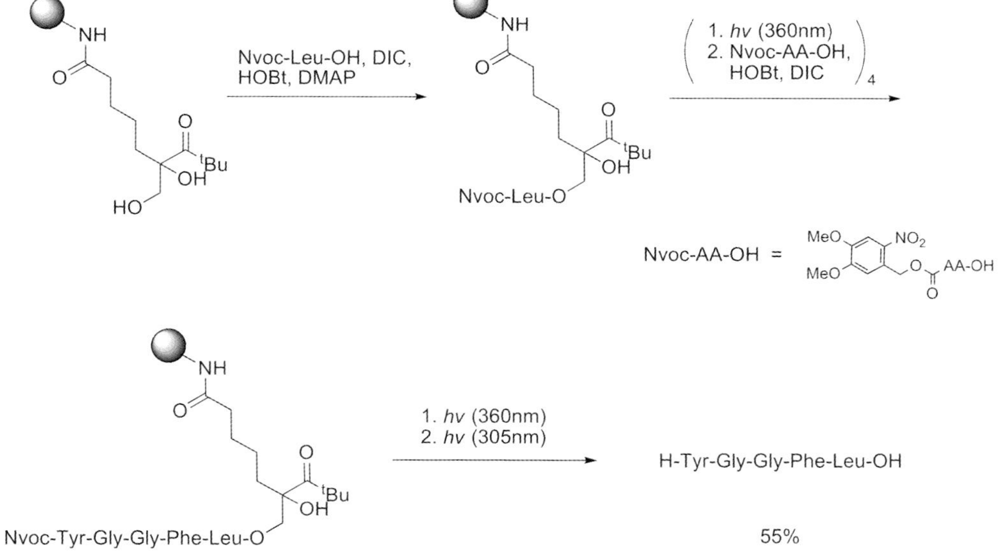

Scheme 17.18 Use of both photolabile linker **34** and the NVOC protecting group in the solid phase synthesis of peptides.

deprotection reagents [57]. Because the chromophores of the pivaloyl group (305 nm) and the nitroveratryl group (360 nm) have distinct absorbances, they could selectively deprotect either the N-terminal NVOC or the C-terminal linker (Scheme 17.18).

17.2.4
Miscellaneous Photolabile Linkers

Homolytic photocleavage of the C–S bond of benzyl sulfide [58, 59] has been adapted for use in linkers for solid-phase organic reactions. Cleavage was substrate-dependent: photolytic scission of site "b" happened only to structures such as **37**, which contain a biphenyl group. Trimethylphenyltin **36** underwent a palladium-catalyzed Stille coupling with the resin-bound aryl iodide **35** to give, after photolysis, a biphenyl (Scheme 17.19).

Transition metal π-complexes can be used in photolabile linkers. The π-complex **40** was formed from the chromium arene complex **39** and polymer-supported triphenylphosphine (**38**) [60] (Scheme 17.20). The complex **40** tolerated LiAlH$_4$ reduction and acetylation, but released the product **41** in 70% yield when oxidized with air under the influence of UV irradiation.

Photoactivation of thiohydroxamic esters initiates a rapid decarboxylative radical rearrangement discovered by Barton [61]; it has been used [62] in an application as a photolabile linker (Scheme 17.21). Coupling of indole-3-acetic acid **43** and

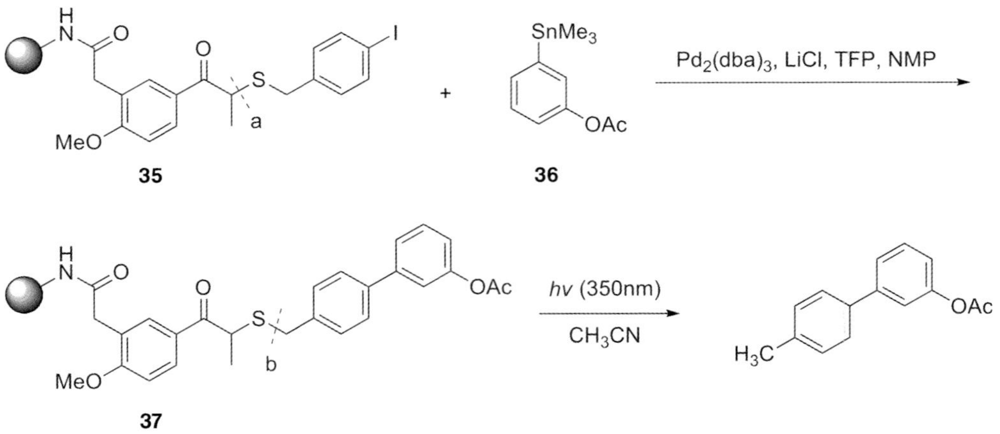

Scheme 17.19 Homolytic photocleavage of the C–S bond of benzyl sulfide.

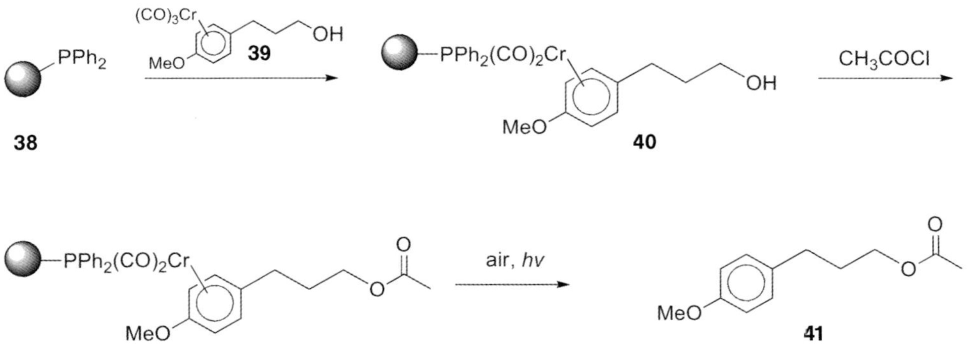

Scheme 17.20 Example of the use of a transition metal π-complexes in photolabile linkers.

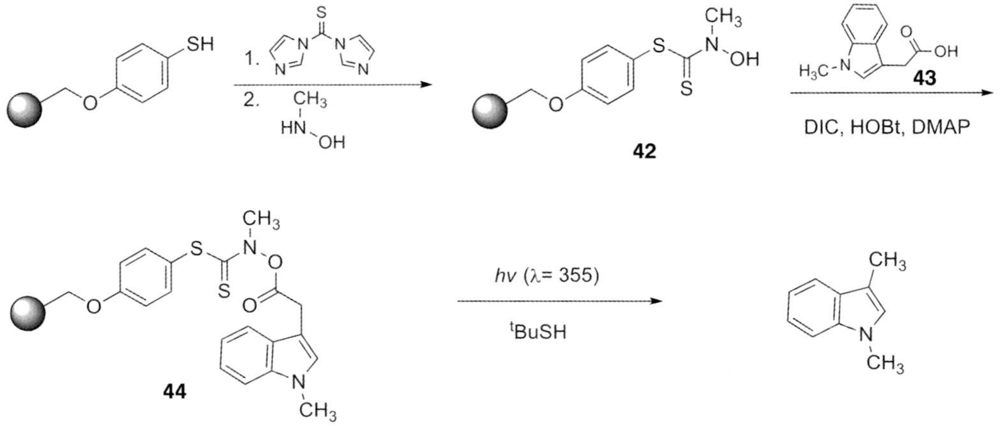

Scheme 17.21 Deployment of thiohydroxamic esters in decarboxylation.

linker **42** using DIC, DMAP and HOBt produced resin-bound substrate **44**, which could then be photochemically cleaved from the linker with 350 nm UV radiation.

The application of a triazene motif to the development of a photolabile linker was reported by Enders in 2004 [63]. The reaction of a secondary amine with polymer-bound diazonium salt **45** produced the triazene moiety **46**. It was stable to a wide range of non-acidic conditions, suitable for other organic reactions. Photocleavage was carried out with 355 nm UV laser irradiation (3ω Nd-YAG) in methanol–diethyl ether (Scheme 17.22).

17.3
Miscellaneous Linkers

17.3.1
1,3-Dithiane Linker

Thioacetals have been widely used as protecting groups for the carbonyl functionality and are stable to a wide range of reaction conditions. They were first employed as solid phase anchors by Bertini [64] (Scheme 17.23). He noted that polymeric

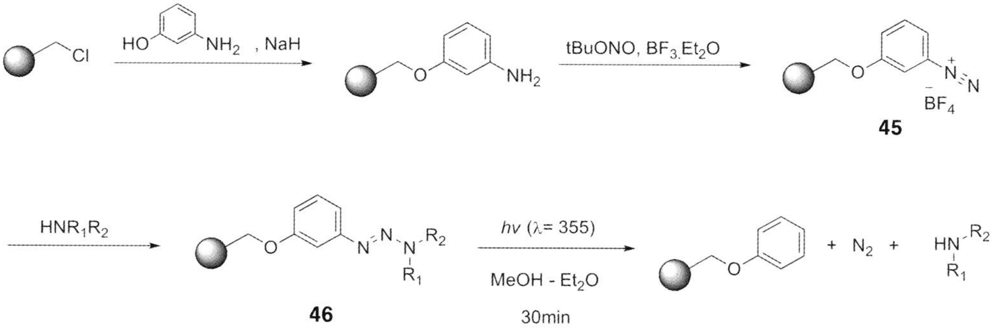

Scheme 17.22 Application of a triazene motif to the development of a photolabile linker.

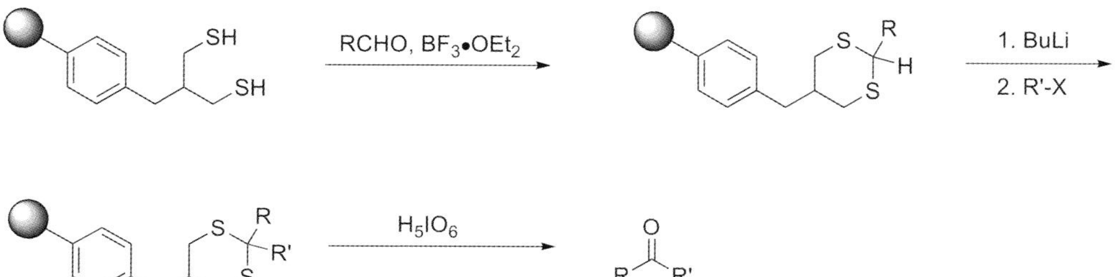

Scheme 17.23 Synthesis of anchored thioacetals and their use as protecting group for the carbonyl functionality.

thioacetal reagents offer two main advantages: an easy workup and the avoidance of trace amounts of volatile sulfhydryl compounds.

Bertini used copolymerization of a dithio-functionalized styrene with normal styrene to produce a dithiol-functionalized resin, whereas Künzer [65] simplified the preparation of a 1,3-dithiol linker by employing commercially available (±)-α-lipoic acid as a starting material (Scheme 17.24).

17.3.2
p-Alkoxybenzyl Linker

The solid phase version of the PMB protecting group, the *p*-alkoxybenzyl linker ("Wang resin"), provides a simple way to link alcohols [66] or amines [67] with a functionalized Merrifield resin. This linker was initially detached by treatment with 20% TFA in CH_2Cl_2. Porco's study [66] focused on an alternative means of detachment of alcohols because TFA cleavage produced trifluoroacetate esters as a significant byproduct (~25%). He rationalized the use of DDQ for oxidative cleavage and a mixed-bed ion exchange resin scavenger (Amberlyst A-26) to facilitate the removal of excess DDQ and DDQH (Scheme 17.25).

To increase the stability of *p*-alkoxybenzyl resins to acids, Porco and Kusumoto introduced the *p*-acylaminobenzyl linker [68], which they found to be suitable for anchoring glycosides for oligosaccharide synthesis (Scheme 17.26).

17.3.3
Alkene-Functionalized Linker

Alkenes can be considered as a class of protected carbonyl compounds since they can be converted into carbonyls by ozonolysis. Hall [69] demonstrated a practical strategy for solid phase synthesis of peptide aldehydes using an olefinic linker, which was constructed using Wittig chemistry. After normal peptide synthesis using the alkene linker, ozonolysis of the linker and subsequent workup with

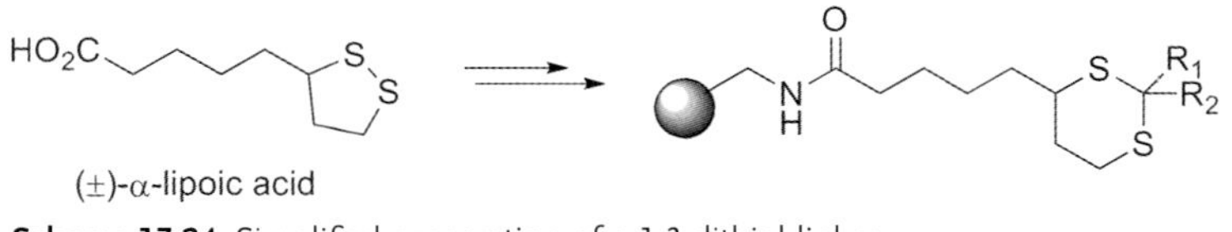

Scheme 17.24 Simplified preparation of a 1,3-dithiol linker.

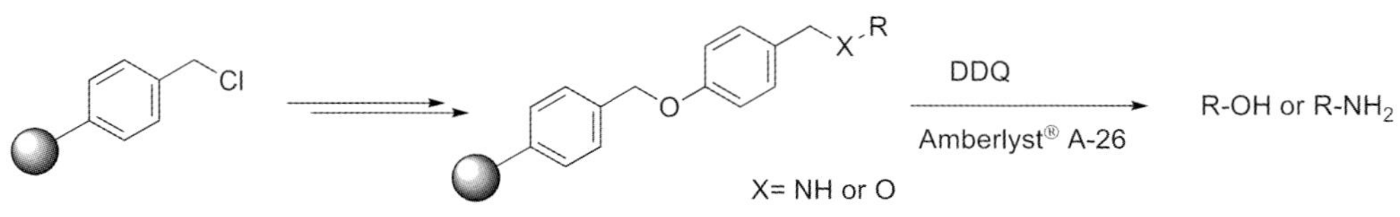

Scheme 17.25 Attachment and of the *p*-alkoxybenzyl linker and subsequent cleavage.

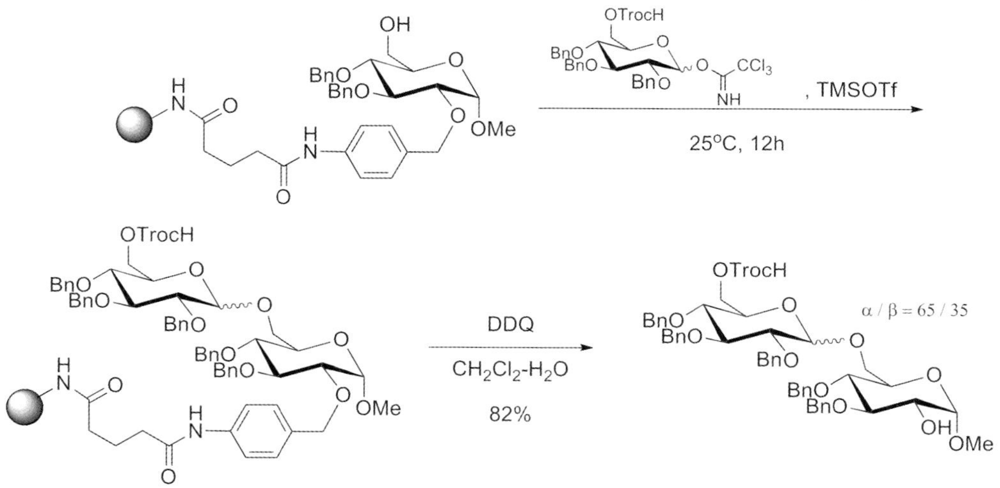

Scheme 17.26 Use of the *p*-acylaminobenzyl linker in anchoring glycosides for oligosaccharide synthesis.

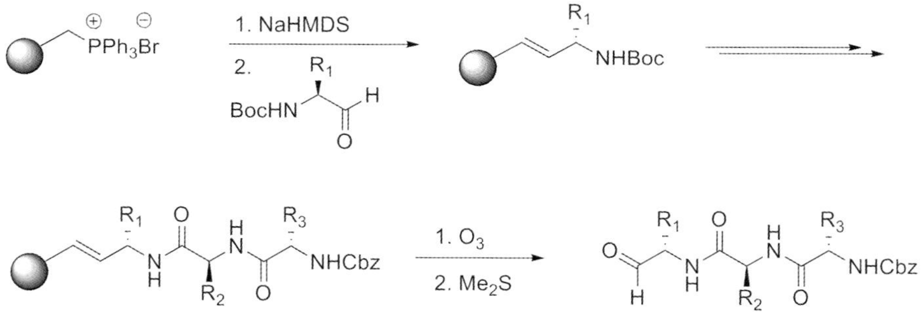

Scheme 17.27 Solid phase synthesis of peptide aldehydes using an olefinic linker.

dimethyl sulfide affords the peptide aldehydes (Scheme 17.27). A library of 27 tripeptide aldehydes was synthesized using a 3 × 3 × 3 array.

17.4 Conclusion

Photolabile linkers have evolved to meet the need for additional degrees of protecting group orthogonality during on-resin organic synthesis. With the development of high-throughput screening techniques, resin-based bioactivity tests and the synthesis of arrays of small molecules on microchips, the importance of photolabile linkers has grown beyond its original purpose. Today, photolabile linkers are regarded as an important alternative to the limitations of the more commonly used acid- and base-labile linkers.

List of Abbreviations

Acm	acetamido
ANP	3-amino-3-(2-nitrophenyl)propionic acid
Bn	benzyl
Boc	t-butyloxycarbonyl
Bop	benzotriazol-l-yl-oxy-tris(dimethylamino)phosphonium hexafluorophosphate
BrZ	2-bromobenzyloxycarbonyl
Bzl	benzyl
Cbz	carbobenzyloxy (benzyloxycarbonyl)
ClZ	2-chlorobenzyloxycarbonyl
CPG	aminopropylsiloxane-grafted controlled pore glass
DCC	dicyclohexylcarbodiimide
DDQ	2,3-dichloro-5,6-dicyano-1,4-benzoquinone
DIC	N,N'-diisopropylcarbodiimide
DIEA	N,N-diisopropylethylamine
DKP	diketopiperazine
DMAP	4-dimethylaminopyridine
DMB	3′,5′-dimethoxybenzoin
DMF	dimethylformamide
DMP	2,5-dimethylphenacyl
Fmoc	(9-fluorenylmethyl) carbamate
HOBt	1-hydroxybenzotriazole
NBB	(3-nitrobenzamido)benzyl
NBH	2-nitrobenzhydryl
NBOC	2-nitrobenzyloxycarbonyl
NMM	N-methylmorpholine
NMP	N-methylpyrrolidone
NVOC	6-nitroveratryloxycarbonyl
PEG	poly(ethylene glycol)
PPOA	[4-(2-bromopropionyl)phenoxy]acetic acid
PS	polystyrene
TBDPS	tetrabutyldiphenylsilyl
TBTU	O-(benzotriazol-1-yl)-N,N,N',N'-tetramethyluronium tetrafluoroborate
Tos	p-toluenesulfonyl
Z	benzyloxycarbonyl

References

1 Schofield, P. and Barltrop, J.A. (1962) *Tetrahedron Letters*, **3**, 679.
2 Schofield, P., Barltrop, J.A. and Plant, P.J. (1966) *Journal of the Chemical Society, Chemical Communications*, 822.
3 Woodward, R.B., Patchornik, A. and Amit, B. (1970) *Journal of the American Chemical Society*, **92**, 6333.
4 Rich, D. and Gurwara, S.K. (1973) *Journal of the Chemical Society, Chemical Communications*, **17**, 610.

5 Merrifield, R.B. (1964) *Biochemistry*, **3**, 1385.
6 Hoffman, J.M. and Breslow, R. (1972) *Journal of the American Chemical Society*, **94**, 2110.
7 Rich, D.H. and Gurwara, S.K. (1975) *Journal of the American Chemical Society*, **97**, 1575.
8 Tam, J.P., Voss, C.R., Whitney, D.B., et al. (1983) *International Journal of Peptide and Protein Research*, **22**, 2042.
9 Albericio, F., Girant, E., Pedroso, E., et al. (1982) *Tetrahedron*, **38**, 1193.
10 Albericio, F., Granier, C., Labbe-Jullie, C., et al. (1984) *Tetrahedron*, **40**, 4313.
11 Albericio, F., Giralt, E. and Lloyd-Williams, P. (1993) *Tetrahedron*, **49**, 11065.
12 Khosla, M.C., Smeby, R.R. and Bumpus, F.M. (1972) *Journal of the American Chemical Society*, **94**, 4721.
13 Menifield, R.B. and Gisin, B.F. (1972) *Journal of the American Chemical Society*, **94**, 3102.
14 Giralt, E., Gairi, M., Williams, P.L. and Albericio, F. (1990) *Tetrahedron Letters*, **31**, 7363.
15 Suzuki, K., Nitta, K. and Endo, N. (1975) *Chemical & Pharmaceutical Bulletin*, **23**, 222.
16 Zehavi, U. and Patchornik, A. (1973) *Journal of the American Chemical Society*, **95**, 5673.
17 Nicolaou, K.C., Winssinger, N., Pastor, J. and Deroose, F. (1997) *Journal of the American Chemical Society*, **119**, 449.
18 Patchornik, A., Amit, B. and Woodward, R.B. (1970) *Journal of the American Chemical Society*, **92**, 6333.
19 Griess, P. (1877) *Chemische Berichte*, **10**, 1868.
20 Pillai, V.N.R. and Ajayaghosh, A. (1987) *Journal of Organic Chemistry*, **52**, 5714.
21 Pillai, V.N.R. and Ajayaghosh, A. (1988) *Tetrahedron*, **44**, 6661.
22 Brown, B.B., Wagner, D.S. and Geysen, H.M. (1995) *Molecular Diversity*, **1**, 4.
23 Schreiber, S.L. and Sternson, S.M. (1998) *Tetrahedron Letters*, **39**, 7451.
24 Ryba, T.D. and Harran, P.G. (2000) *Organic Letters*, **2**, 851.
25 Holmes, C.P. (1997) *Journal of Organic Chemistry*, **62**, 2370.
26 Geysen, H.M., Rodebaugh, R. and Fraser-Reid, B. (1997) *Tetrahedron Letters*, **38**, 7653.
27 Sheehan, J.C. and Umezawa, K. (1973) *Journal of Organic Chemistry*, **38**, 3771.
28 Given, R.S., Conrad, P.G.II , Weber, J.F.W. and Kandler, K. (2000) *Organic Letters*, **2**, 1545.
29 Sheehan, J.C. and Wilson, R.M. (1964) *Journal of the American Chemical Society*, **86**, 5277.
30 Klan, P., Zabadal, M. and Heger, D. (2000) *Organic Letters*, **2**, 1569.
31 Sheehan, J.C. and Davies, G.D. (1964) *Journal of Organic Chemistry*, **29**, 2006.
32 Peach, J.M., Pratt, A.J. and Snaith, J.S. (1995) *Tetrahedron*, **51**, 10013.
33 Givens, R.S., Athey, P.S., Matuszewski, B., Kueper, L.W. III , Xue, J. and Fister, T. (1993) *Journal of the American Chemical Society*, **115**, 6001.
34 Shi, Y., Corrie, J.E.T. and Wan, P. (1997) *Journal of Organic Chemistry*, **62**, 8278.
35 Chan, S.I. and Rock, R.S. (1996) *Journal of Organic Chemistry*, **61**, 1526.
36 Balasubramanian, S. and Lee, H. (1999) *Journal of Organic Chemistry*, **64**, 3454.
37 Balasubramanian, S., Routledge, A. and Abell, C. (1997) *Tetrahedron Letters*, **38**, 1227.
38 Reese, B. and Anderson, J.C. (1962) *Tetrahedron Letters*, **2**, 1.
39 Sheehan, J.C. and Umezawa, K. (1973) *Journal of Organic Chemistry*, **38**, 3771.
40 Wang, S.-S. (1976) *Journal of Organic Chemistry*, **41**, 3258.
41 Bellof, D. and Mutter, M. (1985) *Chimia*, **39**, 317.
42 Heavner, G.A. and Tjoeng, F. (1983) *Journal of Organic Chemistry*, **48**, 355.
43 Tam, J.P., Merrifield, R.B. and Tjoeng, F.S. (1979) *International Journal of Peptide and Protein Research*, **14**, 262.
44 Belshaw, P.J., Flickinger, S.T., Patel, M., et al. (2006) *Organic Letters*, **8**, 2357.
45 Wagner, P.J. and Chen, C. (1976) *Journal of the American Chemical Society*, **98**, 239.
46 Sammes, P.G. (1976) *Tetrahedron*, **32**, 405.
47 Klan, P., Literak, J. and Wirz, J. (2005) *Photochemical & Photobiological Sciences*, **4**, 43.
48 Banerjee, A. and Falvey, D.E. (1997) *Journal of Organic Chemistry*, **62**, 6245.
49 Pelliccili, A.P., Pospisil, T. and Wirz, J. (2002) *Photochemical & Photobiological Sciences*, **1**, 920.
50 Wang, Y., Du, L. and Zhang, S. (2005) *Tetrahedron Letters*, **46**, 3399.

51 Giese, B., Wyss, C., Batra, R., *et al.* (1996) *Angewandte Chemie (International Edition in English)*, **35**, 2529.
52 Giese, B., Sauer, S., Schumacher, A. and Barbosa, F. (1998) *Tetrahedron Letters*, **39**, 3685.
53 Giese, B., Muller, S.N., Batra, R., *et al.* (1997) *Journal of the American Chemical Society*, **119**, 2795.
54 Bamford, C.H. and Norrish, R.G.W. (1935) *Journal of the Chemical Society*, 1504.
55 Giese, B. and Peukert, S. (1998) *Journal of Organic Chemistry*, **63**, 9045.
56 Giese, B. and Glatthar, R. (2000) *Organic Letters*, **2**, 2315.
57 Giese, B., Kessler, M., Glatthar, R. and Bochet, C.C. (2003) *Organic Letters*, **5**, 1179.
58 Sucholeiki, I. (1994) *Tetrahedron Letters*, **35**, 7307.
59 Sucholeiki, I. and Forman, F.W. (1995) *Journal of Organic Chemistry*, **60**, 523.
60 Gibson, S.E., Hales, N.J. and Peplow, M.A. (1999) *Tetrahedron Letters*, **40**, 1417.
61 Barton, D.H.R., Crich, D. and Potier, P. (1985) *Tetrahedron Letters*, **26**, 5943.
62 Routledge, A., Horton, J.R. and Stamp, L.M. (2000) *Tetrahedron Letters*, **41**, 9181.
63 Enders, D., Rijksen, C., Köbberling, E.B., *et al.* (2004) *Tetrahedron Letters*, **45**, 2839.
64 Bertini, V., Lucchesini, F., Pocci, M. and De Munno, A. (1998) *Tetrahedron Letters*, **39**, 9263.
65 Künzer, H. and Huwe, C. (1999) *Tetrahedron Letters*, **40**, 683.
66 Porco, J., Deegan, T., Gooding, O. and Baudart, S. (1997) *Tetrahedron Letters*, **38**, 4973.
67 Kobayashi, S. and Aoki, Y. (1998) *Tetrahedron Letters*, **39**, 7345.
68 Kusumoto, S., Fukase, K., Nakai, Y., Egusa, K. and Porco, J. (1999) *Synlett*, 1074.
69 Hall, B. and Sutherland, J. (1998) *Tetrahedron Letters*, **39**, 6593.

Part Five Solid Phase Synthesis of Biomolecules. The State of the Art (from the Resin Point of View)

18
Peptides

Judit Tulla-Puche[1] and Fernando Albericio[1,2]

[1]Institute for Research in Biomedicine, Barcelona Science Park, 08028 Barcelona, Spain.
[2]University of Barcelona, Department of Organic Chemistry, 08028 Barcelona, Spain.

18.1
Introduction

Several pharmaceutical analysts predict that peptides as pharmaceutical agents are entering their golden age [1, 2]. The irruption of GnRH agonists and antagonists (Leuprolide, Goserelin, Buserelin, Triptorelin, Nafarelin, Cetrorelix, Ganirelix, etc.), Somatostatin analogs (Octeotride, Lanreotride, etc.), Desmopressin and Calcitonins was decisive for the rapid growth of the peptide market in the early 1990s. In the late 1990s, this growth slowed down reflecting the maturation of the market. However, the large number of peptide-based new chemical entities (NCE) introduced in the early 2000s has accelerated the development of this market, from US$5.9 billion in 2004 to 7.94 billion in 2007, which represents a two-figure annual growth rate. The new products have also implied an expansion of the therapeutic fields targeted by peptides, which have traditionally been reserved to oncology, cardiovascular diseases (Bivalirudin, Eptifibatide), CNS-neurobiology (Ziconotide, Taltirelin), immunology (Glatiramer), infection (Enfuvirtide, Thymalfasin), metabolism (Exenatide, Pramlintide) and reproduction (Atosiban). Of these, the non-interferon and non-steroidal immunomodulator Glatiramer, administered to decrease the frequency of relapses in patients with Relapsing-Remitting Multiple Sclerosis, has become a blockbuster with a market value of over US$1200 million. Furthermore, Exenatide, the first non-insulin anti-diabetic drug, is at the level of sales of Desmopressin and Calcitonins.

Core peptide technology is receiving contributions not only from advances in basic research but also from breakthroughs in peptide drug delivery and peptide drug manufacturing and improved knowledge of the pharmacokinetics of these compounds. In both basic research and peptide manufacturing, the solid phase approach developed by Merrifield [3] has been a cornerstone. Nowadays, and thanks to that methodology, it is possible to synthesize millions of peptides in a

The Power of Functional Resins in Organic Synthesis. Judit Tulla-Puche and Fernando Albericio
Copyright © 2008 WILEY-VCH Verlag GmbH & Co. KGaA, Weinheim
ISBN: 978-3-527-31936-7

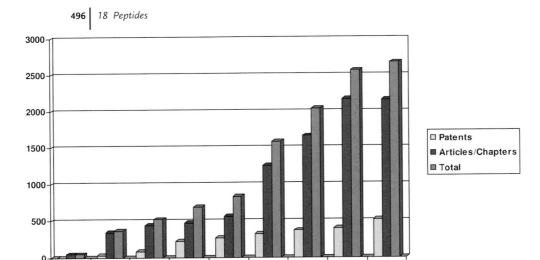

Figure 18.1 Number of articles/chapters and patents devoted to "solid phase peptide synthesis" in the period 1963–2007.

combinatorial mode [4], as well as in multi-kg scales of just one peptide in the amount required for an active pharmaceutical ingredient (API) [5].

One indication of the impact of solid-phase methodology on peptide-based research is the term "solid phase peptide synthesis", which has appeared more than 13 000 times in the scientific literature (SciFinder Scholar 2006) since the advent of the technique in 1963. The number of articles/chapters and patents devoted to this technique have grown constantly over the years (Figure 18.1). Regarding the latter, patents filed have risen from 84 (1973–1977) to 220 (1978–1982) and more than 500 in 2003–2007. This increase reflects the peptide boom in the early 1990s and the >25% growth between the periods 1998–2002 and 2003–2007. These observations augur a promising future for peptides as pharmaceutical agents.

This chapter is divided into two parts. The first briefly discusses common methodological tools while the second describes examples of recent and representative peptide syntheses that illustrate the state-of-the-art.

18.2
Methodological Remarks

Solid phase peptide synthesis[6, 7] (SPPS) is now an established methodology for the synthesis of small- to medium-sized peptides. Since the first reports on this technique in the early 1960s [3] several improvements in protecting groups, coupling reagents and solid supports have allowed access to a broad range of peptides in a routine and automatic manner. However, the synthesis of complex peptides containing heterodetic bonds, (poly)cyclic structures or a large number of residues requires the coming together of sophisticated strategies. A brief summary of solid supports, protecting groups and coupling reagents is described below. The linkers will be discussed in each section.

18.2.1
Solid Supports

The solid support was crucial for the development of SPPS. In the early years, several support materials were tested, such as cellulose, poly(vinyl alcohol) or ion-exchange resins, but polystyrene soon became the polymer of choice because it shows good swelling properties and a good level of substitution can be achieved. Nevertheless, the hydrophobic nature of polystyrene (PS) has limitations in the synthesis of highly hydrophobic peptides or in sequences that tend to aggregate, the so-called "difficult sequences." Attention turned to the development of more hydrophilic supports and resins. Addition of poly(ethylene glycol) (PEG) was evaluated because of its amphiphilic properties, which would permit solvation in both polar and non-polar solvents. Thus, PEG-PS supports, which bear both a hydrophobic PS core and hydrophilic PEG chains, were developed independently by Zalipsky, Albericio and Barany [8] (PEG-PS) and Bayer and Rapp [9] (TentaGel). The benefits of these resins for the assembly of long peptides prompted the appearance of other supports, such as Champion I and II (NovaGel) [10] and ArgoGel [11], and also more hydrophilic PEG resins where a small amount of PS or polyamide [12], or acrylate with polymerizable vinyl groups [13] had been added. Recent years have witnessed the development of resins such as SPOCC [14] and ChemMatrix (CM) [15]. Compared to the earlier PEG-containing resins, these supports are 100% formed by primary ether bonds and thus show improved chemical stability and can reach higher loadings, comparable to those of PS resins.

18.2.2
Protecting Groups

The first twenty years of peptide chemistry were dominated by *tert*-butyloxycarbonyl (Boc) as temporary protecting group for the N^α-amino function and benzyl-type protecting groups as permanent protecting groups for side-chains. The main drawback of the Boc/Bzl strategy is the use of HF for the final cleavage step, which hampers the application of this methodology to large-scale synthesis. With the introduction of the N^α-(9-fluorenylmethoxycarbonyl) (Fmoc) [16] as temporary protecting group, *t*-butyl-type groups were employed for side-chain protection, thereby allowing the use of the milder trifluoroacetic acid (TFA) for the final detachment of the peptide from the resin. Several other N^α-amino protecting groups have since appeared, such as allyloxycarbonyl (Alloc) [17, 18], p-nitrobenzyloxycarbonyl [19] (pNZ), and 6-nitroveratryloxycarbonyl (Nvoc), all of which are

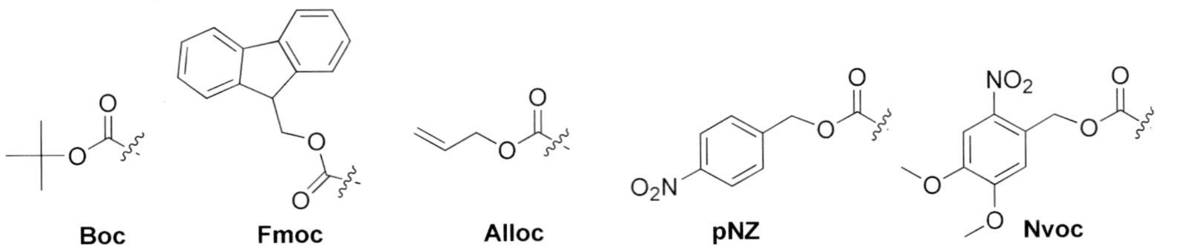

Figure 18.2 N$^\alpha$-Amino protecting groups in peptide synthesis.

orthogonal to the Boc and Fmoc groups, and allow for the synthesis of cyclic, branched and dendrimeric peptides (Figure 18.2).

18.3
Coupling Reagents

Coupling reagents have also experienced a great evolution [20]. Initially, carbodiimides [21] were the reagents of choice for amide formation in combination with 1-hydroxybenzotriazole (HOBt) [22] as additive. N,N'-Diisopropylcarbodiimide (DIPCDI) soon became increasingly more common for SPPS than N,N'-dicyclohexylcarbodiimide (DCC) due to the solubility of the former's urea in DMF (dimethylformamide), which implies that a filtration step is not required (Figure 18.3A). Despite their lower cost, carbodiimides have been exceeded by aminium and phosphonium salts because these salts show higher coupling efficiency combined with a low risk of racemization. 1-[bis(Dimethylamino)methylene]-1H-benzotriazolium hexafluorophosphate 3-oxide (HBTU) and 1-[bis(dimethylamino)methylene]-1H-benzotriazolium tetrafluoroborate 3-oxide (TBTU) [23] in combination with HOBt are used routinely, while 1-[bis(dimethylamino)methylene]-6-chloro-1H-benzotriazolium hexafluorophosphate 3-oxide (HCTU) [24] and 1-[bis(dimethylamino)methylene]-1H-1,2,3-triazolo-[4,5-b]pyridinium hexafluorophosphate 3-oxide (HATU) [25], derived from Cl-HOBt and 1-hydroxy-7-azabenzotriazole (HOAt), respectively, have shown to be superior, but are somewhat less utilized due to their higher cost. The amino acid fluoride can be formed *in situ* by means of tetramethylfluoroformadinium hexafluorophosphate (TFFH) (Figure 18.3B) [26]. In cyclizations, which occur at a lower rate, phosphonium salts [benzotriazol-1-yl-oxytris(pyrrolidino)phosphonium hexafluorophosphate (PyBOP) [27] and 7-azabenzotriazol-1-yl-oxytris(pyrrolidino)phosphonium hexafluorophosphate (PyAOP)][28] are preferred, since guanylation of the amino terminus is minimized in comparison to aminium salts (Figure 18.3C). Recently, a novel proton acceptor coupling reagent has proved to be superior to the ones previously described. The oxygen in the carbocation moiety confers more solubility to the reagent. Furthermore, it enhances coupling yields and decreases racemization, thereby allowing the use of one equivalent of base (Figure 18.3D) [29].

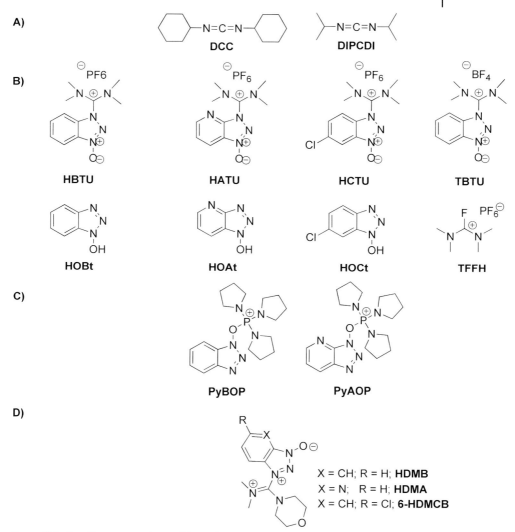

Figure 18.3 Coupling reagents in peptide synthesis: (A) carbodiimides; (B) aminium salts; (C) phosphonium salts; (D) proton acceptor-based aminium salts.

18.4
Synthesis of Long Peptides

Two general strategies are envisioned for the assembly of large peptides: (a) convergent approaches which include coupling of protected segments on (i) solid-support (classically named convergent) or (ii) in solution (hybrid); and (b) stepwise approaches using PEG-based resins and/or additional tools such as

N-2-hydroxy-4-methoxybenzyl (Hmb) for backbone protection, pseudoprolines and o-acyl isopeptides.

18.4.1
Convergent Approaches

Limitations in the assembly of large peptides arise from incomplete couplings or deprotection. These can be due to the presence of consecutive hindered amino acids in the sequence or to aggregation of the growing chain in the case of hydrophobic sequences. Consequently, stepwise synthesis has been limited to around 50 cycles. As stated above, convergent synthesis may include the assembly of peptide fragments on solid-support followed by an on-resin fragment coupling (solid phase fragment coupling) [30–34] or the use of solid-phase for the synthesis of protected fragments and the final assembly of the fragments in solution (hybrid strategy). In both cases, the synthesis of protected fragments is required. The construction of these segments call for linkers or resins that maintain the side-chain protecting groups intact upon cleavage. This is the case of allyl, and photo-labile, nucleophilic or low-acid labile linkers or resins, where the cleavage conditions are orthogonal to side-chain protecting removal conditions. Of these, low-acid labile linkers are preferred, since handling of the cleavage step is easier, and good recoveries and purities are obtained. To obtain C-terminal acid peptides, HMPB linker [35], Sasrin resin [36, 37] and 2-chlorotrityl resin [38] (2-CTC, Barlos resin) can be used (Figure 18.4). 2-CTC resin has been used extensively because of its wide availability and lower cost. Synthetically, 2-CTC resin offers other advantages, most importantly the minimization of diketopiperazine (DKP) formation as a result of the steric hindrance that the trityl group imposes [39], low racemization on incorporating the first amino acid [40] and the possibility of being recycled [41–43]. For the synthesis of protected peptide amides, the Sieber amide resin [44] is used. In all these supports, protected fragments can be released by low-acidic cocktails of 1–2% TFA in CH_2Cl_2. When necessary, 2-CTC resin can be cleaved even in milder conditions by using trifluoroethanol (TFE)–CH_2Cl_2 and hexafluoroisopropanol (HFIP)–CH_2Cl_2 mixtures [45].

Before starting a convergent strategy two points require careful examination: (i) choice of the C-terminal amino acid of each segment to prevent undesirable racemization in the fragment-coupling step (Gly and Pro are preferred); (ii) length of protected segments. These variables may have a major impact on the purity of the protected peptide, and also on its solubility. A low solubility may not only hamper fragment-coupling but also characterization and purification of the protected fragment itself.

18.4.1.1 Hybrid Approaches: The Synthesis of Enfuvirtide
In a landmark synthesis, Trimeris-Roche manufactured the 36-residue fusion inhibitor Enfuvirtide (T-20 or Fuzeon) (Figure 18.5A) on a multi-ton scale by using a hybrid solid-solution approach [46, 47]. For this purpose, the peptide was divided

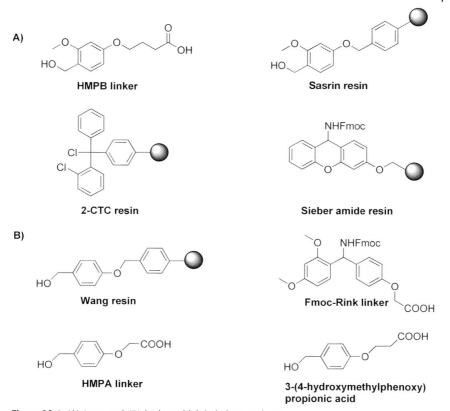

Figure 18.4 (A) Low- and (B) high-acid labile linkers and resins.

in three fragments of 9–16 residues in length, and each was constructed on 2-CTC resin using an Fmoc/t-Bu strategy (Figure 18.5B). For the stepwise elongation of each fragment, HBTU and HOBt or Cl-HOBT were used as coupling reagents, and after cleavage from the resin with 1% TFA in CH_2Cl_2, the protected fragments were isolated by precipitation, obtaining purities greater than 90%. The remaining steps were carried out in solution. H-Phe-NH_2 was coupled to the C-terminal fragment to provide the protected Fmoc-1-36-NH_2 amide peptide, which was coupled, after Fmoc removal, to the middle fragment. Condensation reactions were carried out in DMF, and the intermediates were isolated by precipitation following addition of H_2O. After coupling of the N-terminal fragment, precipitation and global side-chain deprotection, the crude T-20 was reached with 75% purity by HPLC. Final purification by preparative HPLC afforded the pure peptide in approximately 30% overall yield, much higher than the 6–8% yield reached using a stepwise approach. A similar approach was undertaken for the second-generation fusion inhibitor T-1249, a 39-aa peptide that shares 60% homology with T-20.

A)

Ac-Tyr¹-Thr-Ser-Leu-Ile-His-Ser-Leu-Ile-Glu-Glu-Ser-Gln-Asn-Gln-Gln-Glu-Lys-Asn-Glu-Gln-Glu-Leu-Leu-Glu-Leu-Asp-Lys-Trp-Ala-Ser-Leu-Trp-Asn-Trp-Phe³⁶-NH₂

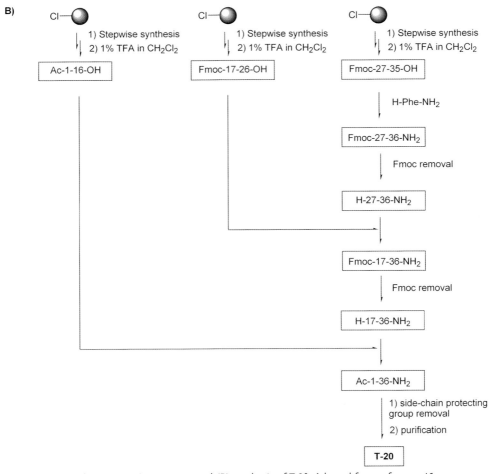

Figure 18.5 (A) Sequence and (B) synthesis of T-20. Adapted from reference 46.

18.4.1.2 Solid Phase Fragment Coupling: Synthesis of Hirudin

Barlos *et al.* [48] assembled a 65-residue thrombin inhibitor variant of Hirudin (HV1) (Figure 18.6) by solid-phase fragment coupling using 2-CTC resin. To prevent racemization, fragments with Gly at the C-terminus were chosen. Thus, the protein was dissected into six fragments that varied in size from the resin-bound 23-aa C-terminal fragment to a small seven-residue fragment. The selected

H-Val^1-Val^2-Tyr-Thr-Asp-Cys-Thr-Glu-Ser-Gly^{10}-Gln^{11}-Asn-Leu-Cys-Leu-Cys-Glu-Gly^{18}-Ser^{19}-Asn-Val-Cys-Gly-Gln-Gly^{25}-Asn^{26}-Lys-Cys-Ile-Leu-Gly-Ser-Asp-Gly^{34}-Glu^{35}-Lys-Asn-Gln-Cys-Val-Thr-Gly^{42}-Glu^{43}-Gly-Thr-Pro-Lys-Pro^{48}-Gln^{49}-Ser-His-Asn-Asp-Gly^{54}-Asp^{55}-Phe-Glu-Glu-Ile-Pro-Glu-Glu-Tyr(SO$_3$)-Leu-Gln^{65}-OH

Figure 18.6 Sequence of Hirudin (HV1).

fragments were Glu^{43} to Gln^{65}; Glu^{35} to Gly^{42}; Asn^{26} to Gly^{34}; Ser^{19} to Gly^{25}; Gln^{11} to Gly^{18}; and Val^1 to Gly^{10}. Fragments were constructed using conventional Fmoc SPPS using *t*-Bu for Glu, Thr, Tyr, Asp, and Ser; Boc for Lys; Trt for Gln, Asn, and His; and methoxytrityl (Mmt) for Cys in a first strategy; Mmt for Cys^{6,14,22,39} and acetamidomethyl (Acm) for Cys^{16,28} were used in a second strategy. The peptides were elongated with DIPCDI/HOBt and protected fragments were cleaved using a TFE–CH$_2$Cl$_2$ (3:7) mixture.

At this step, the purity and solubility of the protected fragments was evaluated. Middle fragments were obtained in >95% purity and were soluble in 0.1 M DMSO. In contrast, the N- and C-terminal fragments were not obtained in sufficient purity and showed poor solubility. Thus, the N-terminal fragment was further dissected in three segments (Glu^{43} to Pro^{48}; Gln^{49} to Gly^{54}; and Asp^{55} to Gln^{65}). Moreover, solubility of the C-terminal fragment improved considerably by omitting Val^1. As for the fragment containing Cys^{16}, protection with Acm showed better solubility and reactivity than with Mmt, so the two-step oxidative folding was used.

With the eight segments in hand, condensation was carried out overnight at room temperature with 0.05–0.1 M solutions in DMSO (dimethyl sulfoxide) in a threefold excess also using DIPCDI/HOBt as coupling reagents. Even after extensive washings with DMF and DMSO, insoluble C-components remained, and multiple couplings of each fragment with less than equimolar amounts were assayed instead. Condensations were checked after 2h by the Kaiser test, and when the coupling was incomplete a new condensation was performed. Finally, Boc-Val-OH was coupled to resin-bound HV1(2-65). Using this strategy, the crude [Cys^{16,28}(Acm)]-HV1 was obtained in 40% purity after release of the resin and global deprotection. After purification and folding in a two-step procedure with Tris-HCl buffer (pH 8.5) in the presence of β-mercaptoethanol (200 μM) followed by reaction with I$_2$/AcOH, the final folded Hirudin was obtained in a 12% overall yield.

18.4.2
Stepwise Approaches

To overcome aggregation, stepwise elongation of large peptides requires not only potent coupling reagents but also additional tools. In recent years, beneficial results have been obtained by means of PEG resins, pseudoprolines (ΨPro) and O-acyl isopeptide or depsipeptide methodology.

18.4.2.1 PEG Resins

As explained previously, the amphiphilic nature of PEG allows aggregation of the growing chain to be minimized. Thus, totally PEG-based resins such as SPOCC [14] and ChemMatrix [15] have been applied with success to the synthesis of complex peptides. These resins have the advantage of showing good chemical stability and a good level of substitution. Several linkers can be incorporated, thereby allowing the synthesis of a range of C-terminal modified peptides.

18.4.2.1.1 Example: The Synthesis of β-Amyloid (1-42) Using CM Resin [49]

For the assembly of β-amyloid (1-42) (Figure 18.7), the linker 3-(4-hydroxymethylphenoxy)propionic acid (AB linker) was coupled to aminomethyl CM resin (0.58 mmol g^{-1}-resin) using HBTU–HOBt–N,N-diisopropylethylamine (DIEA)–DMF (3:3:3:9) for 90 min. The first amino acid by ester formation was introduced with two couplings of Fmoc-L-Ala-OH–DIPCDI–HOAt–DMAP (5:5:5:0.5) in CH$_2$Cl$_2$ and the potential remaining free hydroxyls were capped by acetylation with Ac$_2$O–DIEA in DMF. At this stage, the peptidyl-resin was introduced in an ABI synthesizer and elongation of the peptide was continued automatically using HBTU/HOBt/DIEA as coupling system. The final cleavage and global side-chain deprotection was performed with TFA–TIS–H$_2$O–1,2-ethanedithiol (EDT) (95:2:2:1) for 90 min and the peptide was lyophilized. To determine the purity of the final crude product, the peptide was dissolved in neat TFA to obtain the monomeric form. It was then evaporated, redissolved in HFIP and evaporated again. This last step was repeated twice more and an aliquot dissolved in HFIP was injected into the analytical HPLC to obtain a final purity of 91%.

H-Asp-Ala-Glu-Phe-Arg-His-Asp-Ser-Gly-Tyr-Glu-Val-His-His-Gln-Lys-Leu-Val-Phe-Phe-Ala-Glu-Asp-Val-Gly-Ser-Asn-Lys-Gly-Ala-Ile-Ile-Gly-Leu-Met-Val-Gly-Gly-Val-Val-Ile-Ala-OH

Figure 18.7 Sequence of Aβ(1-42) peptide.

18.4.2.2 Pseudoprolines

To disrupt hydrophobic interactions and to maximize coupling and deprotections yields, several tools have been applied such as DMSO [50] or "magic mixtures" [51], the addition of chaotropic salts [52], or the introduction of Pro residues [53]. More general has been the application of Hmb [54, 55] for backbone amide protection and the introduction of ψPro dipeptides [56–58]. In ψPro dipeptides, the C-terminal amino acid is an oxazolidine-protected Ser, Thr or Cys. In elongating a peptide, ψPros prevent aggregation of the growing chain in a similar way as Pro. The ψPros are introduced as Fmoc-protected dipeptides to prevent incomplete couplings over the oxazolidine moiety. Once the peptide elongation is completed, TFA treatment results in cleavage of the oxazolidine moiety, thereby recovering the natural amino acids (Scheme 18.1). When comparing this strategy with Hmb protection, the introduction of pseudoprolines has proved superior [59]. By using pseudoprolines, otherwise inaccessible peptides have been assembled [58, 60–63].

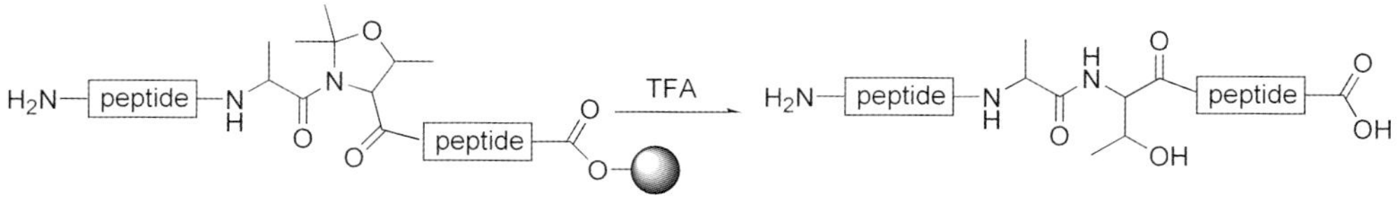

Scheme 18.1 TFA-mediated cleavage of pseudoprolines to render the corresponding dipeptide.

H-Lys-Cys-Asn-Thr-Ala-Thr-Cys-Ala-Thr-Gln-Arg-Leu-Ala-Asn-Phe-Leu-Val-His-Ser-Ser-Asn-Asn-Phe-Gly-Ala-Ile-Leu-Ser-Ser-Thr-Asn-Val-Gly-Ser-Asn-Thr-Tyr-NH$_2$

Figure 18.8 Sequence of Amylin.

18.4.2.2.1 Example: Synthesis of Human IAPP Using Pseudoprolines [61] The 37-mer Human islet amyloid polypeptide (IAPP) or Amylin (Figure 18.8) bears a very hydrophobic sequence and has proved very difficult to assemble because of low coupling yields and incomplete deprotections. To examine several strategies, the synthesis of the fragment 8–37 was undertaken. In a first strategy, assembly of this fragment was carried out by recoupling at β-branched amino acids and amino acids right after them, but the desired product could not be purified because of the presence of overlapping peaks. In a second approach, double couplings were used at all positions. Although a slight improvement was observed, it was not enough to obtain the pure product. Only the introduction of pseudoprolines at three positions, Fmoc-Ala^8-Thr^9($\Psi^{Me,Me}$)Pro-OH, Fmoc-Ser^{19}-Ser^{20}($\Psi^{Me,Me}$)Pro-OH and Fmoc-Leu^{27}-Ser^{28}($\Psi^{Me,Me}$)Pro-OH, resulted in a major peak. Synthesis was carried out on a PAL-PEG-PS resin with recouplings at pseudoprolines and β-branched positions, and at residues after either of them. The synthesis was repeated using the same conditions for the full-length Amylin, and the crude product also showed a major peak with the correct mass, which was of sufficient purity to allow disulfide formation without a previous purification. The final oxidized Amylin was purified to obtain the peptide in >90% purity.

H-Ser^1-Pro-Tyr-Ser-Ser-Asp-Thr-Thr-Pro-Cys-Cys-Phe-Ala-Tyr-Ile-Ala-Arg-Pro-Leu-Pro-Arg-Ala-His-Ile-Lys-Glu-Tyr-Phe-Tyr-Thr-Ser-Gly-Lys-Cys-Ser-Asn-Pro-Ala-Val-Val-Phe-Val-Thr-Arg-Lys-Asn-Arg-Gln-Val-Cys-Ala-Asn-Pro-Glu-Lys-Lys-Trp-Val-Arg-Glu-Tyr-Ile-Asn-Ser-Leu-Glu-Met-Ser^{68}-OH

Figure 18.9 Sequence of RANTES.

18.4.2.2.2 Example: The Synthesis of RANTES Using CM Resin in Combination with Pseudoprolines [63] Several attempts at assembling the 68-residue RANTES (Figure 18.9) by standard stepwise Fmoc SPPS have been pursued. These include the change of coupling reagents from HBTU to the more powerful HATU, and manual and automatic conditions. All of these failed to deliver the final peptide. Pseudoprolines were then introduced in the synthesis at key positions. These included two ΨPro at the β-turn positions to destabilize β-sheet formation, and two ΨPros at the C-terminal and at the N-terminal, respectively. An automatic synthesis using the four pseudoprolines on PS support was carried out, but, even though HATU was used as coupling reagent, only the 45-residue fragment was obtained. In contrast, when the same synthesis was performed on ChemMatrix resin, a main peak was obtained, which allowed for the purification of RANTES with a final purity of 90%.

The first amino acid was introduced on a Wang–ChemMatrix resin by reaction with Fmoc-Ser(^{t}Bu)-OH, MSNT, NMI (N-methylimidazole) and DIEA in CH_2Cl_2, followed by an acetylation step with Ac_2O–DIEA in DMF. Fmoc quantification gave a loading of 0.42 mmol g^{-1}. Peptide chain elongation was carried out on an automatic synthesizer using HATU as coupling reagent and Fmoc-protected amino acids. The ΨPro dipeptides Fmoc-Asn(Trt)-Ser($\psi^{Me,Me}$Pro)-OH, Fmoc-Val-Thr($\psi^{Me,Me}$Pro)-OH, Fmoc-Tyr(^{t}Bu)^{29}-Thr(ψMe,MePro)^{30}-OH and Fmoc-Asp(O^{t}Bu)-Thr(ψMe,MePro)-OH were used to introduce residues $N^{63}S,^{64}$ $Val^{42}T,^{43}$ $Y^{29}T^{30}$ and $D^6T,^7$ respectively. Detachment from the resin and side-chain deprotection was achieved by treatment with reagent K (TFA–phenol–H_2O–thioanisole–1,2-ethanedithiol, 82.5:5:5:5:2.5). After precipitation and lyophilization, characterization by HPLC indicated a 31% purity, which increased to 90% after purification by reversed-phase HPLC.

18.4.2.3 O-Acyl Isopeptide

A method that has recently received considerable attention for the synthesis of "difficult sequences" is the so-called O-acyl isopeptide or depsipeptide method [64–67]. This technique involves the conversion of an O-acyl isopeptide into its peptide counterpart under physiological conditions (Scheme 18.2). Previously, O-acyl prodrug analogues had shown enhanced solubility compared to their parent drugs, probably as a result of the presence of the free amino group [68, 69]. This was also the case with peptides, where sequences that were very difficult to assemble by standard stepwise synthesis because of high hydrophobicity gave the O-acyl isoform in good yields and purities. The presence of the ester is believed to change the secondary structure of the peptide. In fact, O-acyl isopeptides or "switch" peptide transitions to their peptide counterpart have been studied by circular dichroism (CD), revealing controlled induction or reversal of secondary structure and self-assembly of small peptides [67, 70, 71].

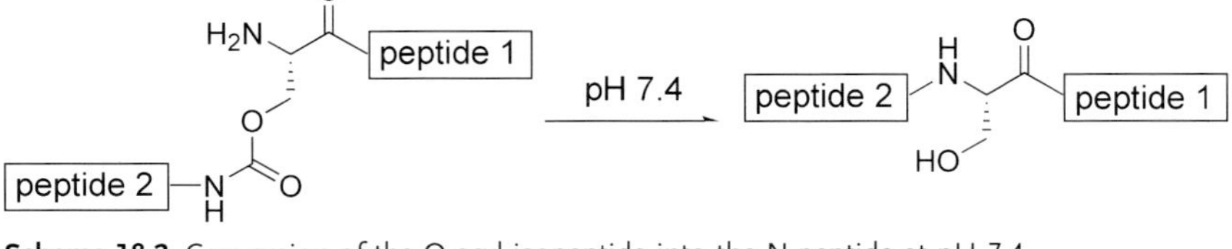

Scheme 18.2 Conversion of the O-acyl isopeptide into the N-peptide at pH 7.4.

In a first report, Sohma and coworkers [72] synthesized the highly hydrophobic peptide Ac-Val-Val-Pns-Val-Val-NH$_2$ (Pns, phenylnorstatine) on a Rink amino methyl resin and obtained only a 6.9% yield after purification. Analysis of side products indicated incomplete Fmoc deprotection and incomplete acetylation arising from aggregation. In contrast, when the branched ester was constructed on the same resin and then converted into the parent peptide in phosphate-buffered saline (PBS) at pH 7.4, the yield increased to 54%, although a small amount of racemized product (3.2%) was observed. Similar results were obtained when using Ser in the place of Pns.

To further the scope of this methodology, application to the synthesis of longer peptides, such as Aβ(1-42), on a 2-CTC resin was undertaken [73] (see below). Recently, this peptide has also been assembled using a photocleavable Nvoc group [74]. The protecting group was completely removed by photolysis under the conditions used for the O to N acyl shift, thereby obtaining the final peptide in a clean manner.

To prevent racemization, which was one of the limitations of the method (in some cases levels around 20% were reached), pre-formed isodipeptide units, such as Boc-Thr(Fmoc-Val)-OH, were introduced [75]. On the basis of these previous results, a completely convergent approach to suppress racemization was also developed [76]. Thus, for a given peptide, an N-terminal fragment, bearing a C-terminal O-acyl isopeptide, was coupled to a C-terminal fragment. Owing to the presence of the urethane-protected Ser/Thr residue, oxazolone formation, and therefore racemization, is avoided.

The influence of the solid support has been addressed in only a few studies. While assembly of the first O-acyl isopeptides was carried out on a Rink AM resin, Sohma and coworkers reported high amounts (20%) of fragment Aβ1-25 in the synthesis of Aβ1-42 peptide with a Tentagel A resin using the AB linker, which attributed to cleavage of the ester by the use of a high content TFA cleavage cocktail. Since treatment of the protected peptide released from 2-CTC resin with the same cleavage cocktail in solution did not result in formation of the by-product, the authors attributed this result to cleavage of the ester moiety while the isopeptide is resin-bound. However, Coin and coworkers [77] pointed out that the by-product originated in the acylation of the uncapped free hydroxyls of the solid support, since in 2-CTC the non-reacted sites are routinely capped with MeOH and they are non-reactive towards the acylating species formed during O-acylation.

The same authors also found extensive DKP formation when attempting to assemble an analog of the WW domain FBP28, [Asn[15]]FBP28-NH$_2$ with the aid of two depsi units. Analysis at each step showed that DKP formation was dependent on the nature of the two amino acids following the depsi bond and, probably, also on their position within the sequence. Therefore, to minimize DKP formation, if there are several Ser of Thr residues in the sequence, careful attention should be paid to the two amino acid residues at the N-terminal side next to the isopeptide site. To overcome this side reaction, the group Bsmoc, which can be removed by only using a 2% piperidine, was employed [78, 79]. In an optimized version, the peptide was synthesized on a Rink-TentaGel resin with a capping step performed after loading the first amino acid and after the introduction of each depsi unit. The method is amenable to automation, and results in yields and purities were comparable with the pseudoproline technique. Recently, this methodology has been used to synthesize a peptide–polymer conjugate [80], which self-assembles with the formation of microstructures upon recovery of the native peptide backbone by O to N acyl migration.

18.4.2.3.1 Example: Synthesis of β-Amyloid (1-42) by the o-Acyl Isopeptide Method
[71]. Aβ(1-42) has two Ser that could potentially serve for O-acylation. Ser^{26} was chosen because of the adjacent Gly at position 25, which prevents epimerization during ester formation. Thus, fragment Aβ(27-42) was constructed on 2-CTC resin following standard Fmoc procedures. Next, Boc-Ser-OH was coupled using DIPCDI–HOBt in DMF, and the following amino acid was introduced by ester formation by means of DIPCDI–DMAP in CH_2Cl_2. Elongation was continued and the final crude O-acyl isopeptide was released from the resin by treatment with TFA–m-cresol–thioanisole–H_2O for 90 min. It was then precipitated and lyophilized. After treatment with TFA–H_2O in the presence of NH_4I and dimethyl sulfide, and subsequent purification, the isopeptide was obtained in 34% yield and 96% purity. The enhanced solubility of this isopeptide (100-fold higher) compared to the final product allowed a straightforward purification by preparative HPLC. This TFA salt was found to be stable for long periods, at 4 °C. The purified product was then subjected to O-N intramolecular acyl migration in PBS at pH 7.4. Quantitative conversion was achieved and the final Aβ(1-42) was obtained with a purity of over 95%.

18.5
Native Chemical Ligation

Although several other chemical ligations have been developed [81–83], native chemical ligation [84] (NCL) has become the most widely used, and is an important tool for the chemical synthesis of proteins. Numerous applications have shown the tremendous impact of this methodology, which has been the object of several reviews [85–87]. In NCL, an N-terminal Cys reacts with a C-terminal thioester under physiological conditions at pH 7 to provide a cyclic intermediate that rearranges to give the native amide bond (Scheme 18.3). The reaction is chemoselective and allows the use of unprotected peptides, including internal free Cys. A key part of NCL is the preparation of the peptide thioesters. Owing to the lability of this moiety to nucleophiles such as piperidine, their synthesis was addressed firstly by Boc chemistry, by using the pre-loaded mercaptopropionic linker developed by Hojo and coworkers [88] (Figure 18.10). In this approach, Boc-Gly-SCH_2CH_2COOH, obtained from Boc-Gly-ONp and HS-CH_2CH_2COOH, was loaded on a PAM or MBHA resin. The peptide was then elongated using Boc chemistry, and finally cleaved with HF to obtain the peptide thioester amide [89]. In a simplified and more general approach, Dawson and coworkers [90] coupled Boc-Leu-OH to a 4-

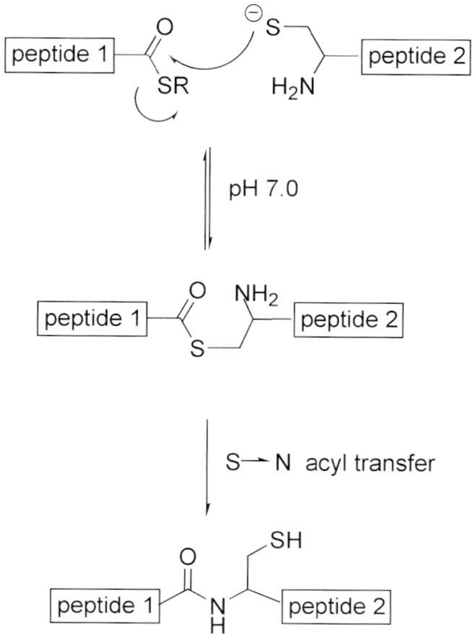

Scheme 18.3 Native chemical ligation. Adapted from reference [84].

(hydroxymethyl)phenylacetamidomethyl (PAM) or MBHA resin. After cleaving the Boc group, the protected STrt-mercaptopropionic acid was coupled to obtain the starting resin TAMPAL (trityl-associated mercaptopropionic acid leucine). After cleaving the Trt group, elongation of the peptide was continued and final cleavage gave the C-terminal mercaptopropionic acid-leucine (MPAL) thioester peptide. This approach has been widely used to prepare a range of peptide thioesters by Boc chemistry. Other linkers based on Hojo's linker have also been prepared, such as the 4-(α-mercaptobenzyl)phenoxyacetic acid DCHA (dicyclohexyl amine) salt [91], which is pre-loaded in solution with Boc-amino acid succinimide esters. Recently, the S-Trt-mercaptophenylacetic acid linker, loaded on a Gly-MBHA resin has been applied to obtain peptide-α-thiophenylesters [92]. This methodology allows the preparation of reactive peptide thioesters with a β-branched amino acid at the C-terminal (Val, Ile, Thr), which otherwise are very difficult to exchange by trans-thioesterification in aqueous solution. One limitation of Boc chemistry is the lability of certain bonds, such as the glycosidic bond, to the final cleavage. Thus, in preparing glycoproteins or phospholipids, the milder Fmoc/tBu strategy is preferred.

To overcome the lability of the thioester moiety to piperidine, numerous strategies for the preparation of peptide thioesters by Fmoc chemistry have been attempted. These include:

(a) The use of modified cleavage cocktails for the removal of the Fmoc group, such as 2% (w/v) HOBt in hexamethyleneimine/1-methylpyrrolidine/1-methyl-2-pyrrolidinone/DMSO (1.6:20:40:40) [93] and the use of DBU/HOBt [94].

Boc chemistry

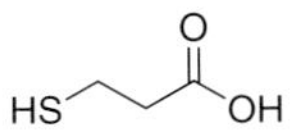

3-mercaptopropanoic acid

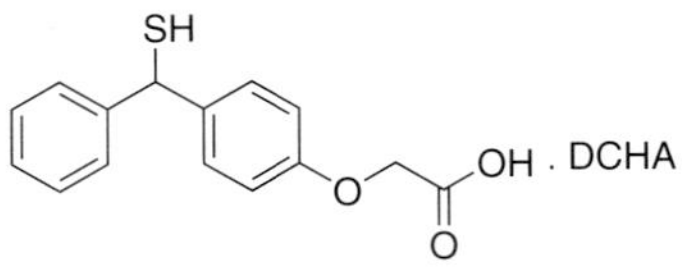

4-α-mercaptobenzyl)phenoxyacetic acid, DCHA salt

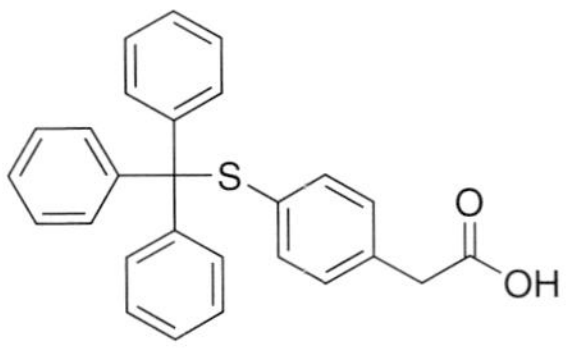

S-Tritylmercaptophenylacetic acid

Fmoc chemistry

a) safety-catch linkers

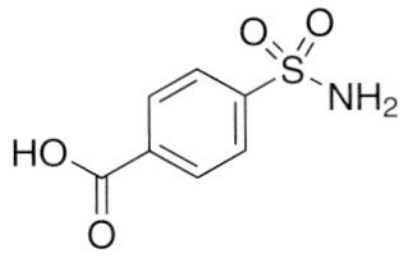

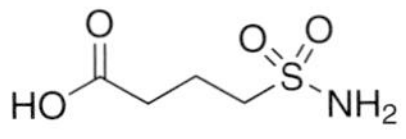

4-sulfamylbutyryl
or 3-carboxypropanesulfonamide

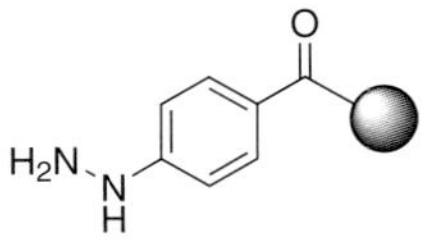

4-Fmoc-hydrazinobenzoyl AM resin

b) BAL

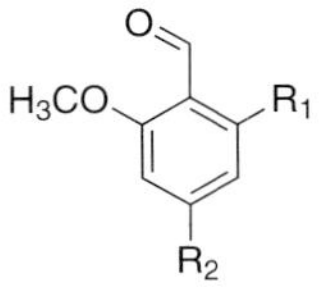

p-BAL: R_1 = -OCH$_3$; R_2 = -O-(CH$_2$)$_4$-COOH
o-BAL: R_1 = -O-(CH$_2$)$_4$-COOH; R_2 = -OCH$_3$

c) O to S acyl shift

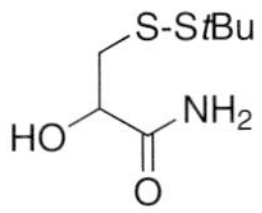

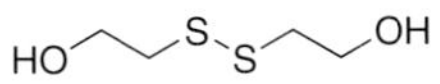

3-tert-butyldisulfanyl-2-hydroxypropionamide

(mercaptocarboxyethylester linker)

Figure 18.10 Linkers for the preparation of peptide thioesters.

(b) The use of trityl-type resins that allow the release of the protected peptide, which is then converted into the peptide thioester in solution [95–98]; For instance, Biancalana et al. [96] used isothiouronium salts like S-phenyltetramethylisothiouronium chloride (CCTU) to obtain phenyl thioesters in solution, from protected peptide acid precursors. By using isothiouronium salts no hydrolysis is detected; however, this method should be avoided in cases where racemization may be an issue.

(c) Cleavage of usual resins, such as PAM and 4-hydroxymethylbenzoic (HMBA) resins, with an alkylaluminum thiolate [99, 100];.

(d) The use of safety-catch type linkers that are first activated and then cleaved by the appropriate thiolate nucleophile, such as Kenner's acylsulfonamide safety-catch linker [101], which is completely stable to basic or strongly nucleophilic conditions. Upon activation by methylation with diazomethane, the acylsulfonamide is converted into the corresponding N-methyl acylsulfonamide, which can then be cleaved with hydroxide or nucleophilic amines. In a first modification of the activation protocol, Ellman and coworkers [102] loaded the 4-carboxybenzenesulfonamide linker on an amino methyl resin and the activation step was carried out with iodoacetonitrile with DIEA in DMSO or NMP (N-methyl pyrrolidone). This gave the N-cyanomethyl acylsulfonamide derivative, which proved to have enhanced reactivity towards nucleophilic displacement because of the electron-withdrawing properties of the cyano group. Taking into account this modification, the same group developed an aliphatic acylsulfonamide linker, 3-carboxypropanesulfonamide [103], which gave better loading yields with no racemization, by using PyBOP and DIEA at $-20\,°C$ in $CHCl_3$, thereby overcoming one of the limitations of the Kenner's linker. In a first application of the 3-carboxypropanesulfonamide linker for the preparation of peptide thioesters [104], the synthesis of the O-linked glycoprotein diptericin was undertaken. After peptide elongation was complete, the linker was activated by cyanomethylation and cleaved with benzyl mercaptan. The glycopeptide thioester moiety was finally deprotected with reagent K, purified and used in NCL to obtain the final glycoprotein. At the same time, Pessi and coworkers [105] found that addition of a catalytic amount of NaSPh (sodium thiophenate) after activation with TMS-CHN$_2$ (trimethylsilyldiazomethane) dramatically improved the yields. To obtain a more stable peptide aliphatic α-thioester, NaSPh was used in conjunction with $(CH_2)_2$-$COOC_2H_5$ (ethyl-3-mercaptopropionate). This approach worked equally well in distinct types of resin, such as amino methyl PS, PEG-PS or TentaGel. Similarly, Biancalana et al. [96] obtained good yields of a peptide thioester bearing a bulky [Asn(Trt)] amino acid at the C-terminal. Quaderer and coworkers [106] found that addition of LiBr to the cleavage step increased yields substantially by facilitating the solubilization of the resin-bound peptide. Recently, Mende and coworkers [107] constructed a peptide on Ellman's safety-catch linker but added a cyclization linker at the N-terminal. Following macrocyclization, thiolysis released acetylated truncated deletion peptide thioesters while the desired peptide thioester was obtained in good purities by cleavage with TFA. The safety-catch linker has several limitations, including the potential alkylation of several sensitive groups, such as the imidazole ring of His [103]. Migration of the thiobenzyl moiety to the Glu side-chain, and hydrolysis of the peptide thioester followed by cyclization, have also been reported [96].

Camarero et al. [108] used the hydrazine safety-catch linker to prepare peptide thioesters. After assembling the peptide using standard Fmoc protocols, the fully protected peptide resin was activated by mild oxidation with N-bromosuccinimide (NBS) in the presence of pyridine, forming a reactive acyl diazene that was then cleaved with an α-amino acid S-alkyl thioester such as H-AA-SEt, where AA is Gly or Ala. After TFA deprotection, peptide thioesters were obtained in good yields. Although the oxidation step did produce racemization, and other sensitive amino acids such as Tyr(tBu) and Trp(Boc) were not affected, Met and Cys presented some problems. Met was completely oxidized, and a reductive cleavage was required. For Cys, the Cys(Trt) derivative should be avoided and use of Cys(Npys) or Cys(S-StBu) is recommended instead.

(e) The use of an acid-labile BAL linker [109, 110]: the BAL linker allows the preparation of C-terminal-modified peptides. The aldehyde linker is coupled to an amino methyl resin, (e.g. PEG-PS) and the first amino acid (AA-OAllyl hydrochloride salt) is introduced by reductive amination by treatment with NaBH$_3$CN in DMF. The second residue is introduced as Ddz-AA-OH to minimize DKP formation and, following cleavage of the Ddz group, the peptide is elongated by standard Fmoc SPPS, except for the N-terminal amino acid, which is introduced as Boc-AA-OH. The allyl group is then removed with Pd(PPh$_3$)$_4$/PhSiH$_3$ and H-AA-SR·HCl is coupled with HATU/DIEA. Final treatment with TFA releases the unprotected peptide thioester.

(f) A BAL strategy wherein the C-terminal residue is a base-stable trithioester that is then converted into the thioester moiety upon treatment with TFA [111]: using the o-BAL linker, this approach has been optimized with Gly trithioortho ester (H$_2$NCH$_2$C-(SEt)$_3$·HCl), which can be previously synthesized in solution or directly in solid-phase by treating the resin-bound Gly(OMe)-o-BAL with AlMe$_3$ and EtSH in CH$_2$Cl$_2$. The latter approach gave lower yields, and in both cases peptide thioesters were obtained in good purities. One advantage of this approach is that the trithioortho ester is not susceptible to nucleophilic attack, and DKP formation is avoided. However, racemization studies have not been carried out.

(g) A side-chain anchoring approach:[96, 112–115] Here the first amino acid, with its C-terminal acid protected (e.g. allyl), is anchored through its side-chain to conventional linkers or resins. The peptide is elongated, the allyl protection removed and the thioester introduced (as pre-formed amino acid thioester or directly with PhSH or ethyl 3-mercaptopropionate). Final treatment with TFA detaches the peptide thioester from the resin.

(h) An O to S acyl shift: Botti and coworkers [116] reasoned that peptide thioesters could be obtained from a peptide-C$^\alpha$carboxyethylester precursor bearing a free thiol in the β position. This would be in equilibrium with its thioester isomer through an O to S acyl shift via a five-membered ring intermediate, which would then undergo NCL in the presence of an N-terminal Cys fragment. To obtain the key mercaptocarboxyethyl ester linker, Fmoc-Cys(S*t*Bu)-OH was coupled to Rink-PEGA resin [117] and, after Fmoc removal, reaction with KNO_2 under aqueous acidic conditions generates the resin-bound 3-*tert*-butyldisulfanyl-2-hydroxypropionamide as a racemic mixture. After elongation of the peptide and TFA cleavage, the peptide-COOH-(CH_2SS*t*Bu)-$CONH_2$ is subjected to NCL and in the presence of mild reducing reagents, such as phosphines or appropriate nucleophiles such as thiophenol, the free thiol is slowly released and the O to S shift takes place, followed by capture by the N-terminal Cys fragment. One limitation of this approach is the presence of a certain amount of hydrolyzed ester as a result of the reaction with KNO_2. In a similar way, Tofteng *et al.* [118] used the linker 2,2-dithiodiethanol loaded on 2-CTC resin to synthesize peptide esters, which, upon reduction conditions, would undergo NCL in the presence of an N-terminal Cys. However, ester hydrolysis was found to be crucial when chaotropic salts (such as Gdn·HCl buffers) were used.

(i) An N to S acyl shift: peptide thioesters were prepared by Ollivier and coworkers [119] by introducing S-TIS-mercaptoethanol under Mitsunobu conditions on the safety-catch 3-carboxysulfonamide linker once elongation of the peptide was completed. Removal of the triisopropyl protecting group of the thiol gives an intermediate that rearranges spontaneously as a result of an intramolecular N,S-acyl shift that involves a cyclic five-membered ring. Final treatment with TFA affords the peptide thioester, which can still be attached to the solid support (amino methyl PS) or can be released in solution (Rink PEG-PS), depending on whether one wishes to perform NCL in the solid- or solution-phase. Although this approach has the advantage of being insensitive to the bulkiness of the amino acid directly attached to the sulfonamide linker, its main drawback is the incomplete Mitsunobu alkylation. Aimoto and coworkers [120, 121] used the *N*-4,5-dimethoxy-2-mercaptobenzyl (Dmbb) group as an auxiliary attached to the backbone amide bond, which upon treatment with TFA undergoes a N to S acyl shift to obtain a peptide thioester intermediate that is treated with excess of thiol to obtain a more stable peptide thioester. The use of aqueous TFA solutions accelerates the N,S-acyl shift and also suppresses the acid-catalyzed cleavage of the Dmmb group. The peptide thioester is obtained with minimum epimerization. This approach is also useful for the preparation of thioesters with a C-terminal bulky amino acid. Ohta and coworkers [122] used an S-protected oxazolidinone linker, which was loaded onto a Rink amino methyl resin. Elongation was carried out by Fmoc chemistry but using the milder Aimoto's cocktail for cleavage of the Fmoc group to prevent cleavage of the amide bond linkage. Cleavage from the resin yielded the S-protected peptide oxazolidinone. The final peptide thioester was obtained by removing the S-MBzl protecting group in solution and subsequent N,S acyl transfer. Nagaike and coworkers [123] used a proline derivative bearing a trityl-protected thiol group to obtain peptide thioesters by an N,S-acyl transfer followed by a microwave-assisted on-resin thioester exchange. Fmoc-Gly-Pro(CH$_2$STrt) was coupled to amino PEGA resin, which was chosen to facilitate the final thioesterification in aqueous solution. After elongation of the peptide chain, thioester exchange was performed in aqueous 10–40% aqueous 3-mercaptopropionic acid (MPA), depending on the substrate, in a microwave (150 W at 80 °C) (Figure 18.11).

To extend the scope of NCL, efforts have been made to introduce modifications so as not to depend on a Cys site. Cys residues can be selectively desulfurized to Ala [124], even in the presence of Cys(Acm) and Met [125]. Homocysteine and selenocysteine also serve as ligation sites, and can then be converted into Met [126] and into Ala or didehydroalanine, respectively [127]. Recently, ligation at Phe sites has also been accomplished [128]. The use of auxiliaries that are removed after ligation has been extensively studied [129–134]. NCL has also been applied to SPPS [135, 136], and to the synthesis of cyclic peptides [137–139]. In another extension of the technique, proteins have been prepared by tandem ligation [135, 140], and expressed protein ligation [141]. Recently, NCL has been applied to the synthesis of glycoproteins [104, 142–144].

A new and elegant extension of NCL is the kinetically controlled ligation (KCL), developed by Kent and coworkers [145]. In this strategy, NCL is performed in a convergent manner. That is, half the protein is assembled by traditional sequential C to N ligations, while the other half is carried out by sequential ligations in the N to C direction. Correct assembly of the fragments is based on the lower reactivity of the thioalkylesters compared to thiophenylesters. Cys protection in middle fragments is accomplished with 1,3-thiazolidine-4-carboxylic acid (Thz) (Scheme 18.4).

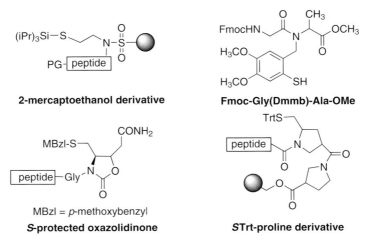

Figure 18.11 Auxiliaries used in the N to S to acyl transfer to construct peptide thioesters.

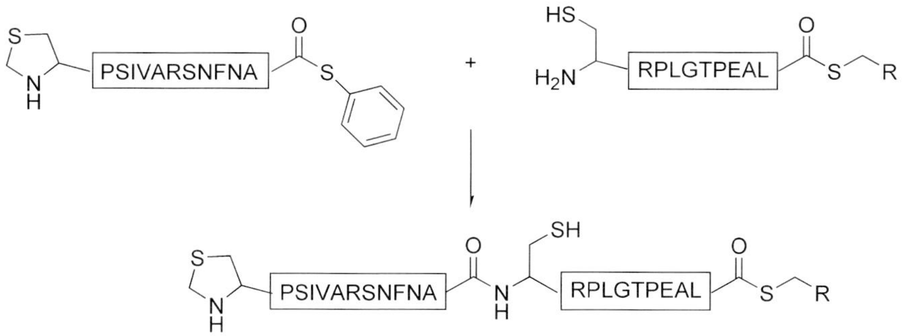

18.5 Native Chemical Ligation | 515

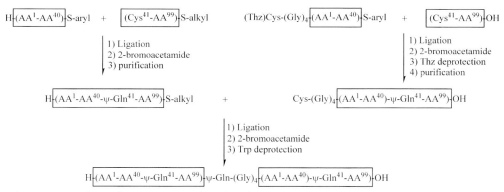

Scheme 18.4 Example of kinetically controlled ligation.

Scheme 18.5 Synthesis of a 203 amino acid covalent dimer, HIV-1 protease.

18.5.1
Example: Synthesis of a 203 Amino Acid Covalent Dimer, HIV-1 Protease [146]

Analogues of HIV-1 protease, an enzyme consisting of a homodimer of two polypeptide chains (each 99-residues long), were constructed by kinetically controlled chemical ligation. To obtain asymmetric analogues, the two monomers were covalently attached through a short linker. The 99-amino acid peptide was split in two fragments: the $(AA^1\text{-}AA^{40})$-thioarylester and the Cys^{41}-$(AA^{42}\text{-}AA^{99})$ thioalkylester (Scheme 18.5). The fragments were prepared by Boc chemistry on a $-OCH_2$-PAM resin or on a $HSCH_2CH_2CO$-Leu-OCH_2-PAM resin using the following side-chain protecting groups: Tos for Arg, Xan for Asn, OcHex for Asp and Glu, 4-CH_3Bzl for Cys, 2-Cl-Z for Lys, Bzl for Ser and Thr, and Br-Z for Tyr. The N-terminal Cys of the middle fragments were protected with the Thz group. After cleavage and purification, conversion of the $(AA^1\text{-}AA^{40})$-thioalkylester into the $(AA^1\text{-}AA^{40})$-thioarylester occurred by reaction with thiophenol in a denaturing aqueous buffer. Next, kinetically controlled reaction was performed at pH 6.8 in a 200 mM phosphate buffer containing 6 M Gdn·HCl. Reactions gave almost exclusively the desired ligated product. Before quenching the reaction, the Thz group was then converted into Cys by addition of methoxyamine hydrochloride. Peptide-α-thiolactones may be formed, but can be driven to the desired ligated thioalkyl ester product by addition of mercaptoethanosulfonate sodium salt (MESNA) into the reaction. After the ligation, Cys was converted into ψ-Gln by reaction with 2-bromoacetamide. The second part of the dimer, which includes a short linker, was constructed by NCL from fragments Thz-Gly$_4$-(AA^1-AA^{40})-($^\alpha$thioalkylester) and Cys^{41}-(AA^{42}-AA^{99}). Following alkylation of Cys to ψ-Gln, and Thz deprotection, fragments (A^1-A^{99})-($^\alpha$thioarylester) and Cys-Gly$_4$-(A^1-A^{99}) were joined by NCL to provide, after alkylation of the Cys residue and removal of Trp-protecting groups, the final polypeptide. After purification by reversed-phase HPLC, the dimer was obtained in a 6.7% overall yield, and was finally folded.

18.6
Cyclic Peptides

Given the potential use of cyclic peptides as therapeutic agents [147–149], the synthesis of these compounds has been the object of extensive studies since the first development of SPPS. Cyclization increases proteolytic resistance and may result in enhanced biological activity compared to their linear counterparts [150, 151]. Cyclic peptides consist of distinct types of linkage: (i) in the most common type, the N- and C-termini are joined ("head-to-tail"); (ii) a side-chain is linked to the C- or N-terminus; (iii) two side-chains are joined (side-chain-to-side-chain). The linkage is usually an amide bond but can also be a disulfide or another type of functionality.

Cyclization can be performed both in solution and in solid-phase [149, 152–154]. Nevertheless, the former must be carried out under high dilution conditions because of the risk of dimerization and oligomerization. As a result of site isolation, intramolecular cyclization is favored on solid-phase, and thus large amounts of solvents are not required. Moreover, coupling reagents can be washed away, thereby releasing the final cyclized peptide in good purity.

To obtain "head-to-tail" cyclic peptides on the solid-phase, the N- and C-termini must be free and not linked to the resin. Two strategies are commonly used: (i) a side-chain anchoring approach and (ii) a backbone amide linker.

In side-chain anchoring the amino acid side-chain is linked to the solid support and the C- and N-termini are orthogonally protected. Once chain elongation is finished, deprotection of both ends and subsequent cyclization and cleavage delivers the final cyclized product. Numerous amino acids have been used for side-chain anchoring, including Asx/Glx [155–161], Lys/Orn [158, 162], Ser/Thr [159, 163], Tyr [159, 163, 164], His [165, 166] and Cys [167, 168], on the usual supports and linkers for peptide synthesis (Figure 18.12).

As for BAL anchoring the first amino acid, with its C-terminal group protected and the N-terminal free, is loaded on the BAL linker by reductive amination. After

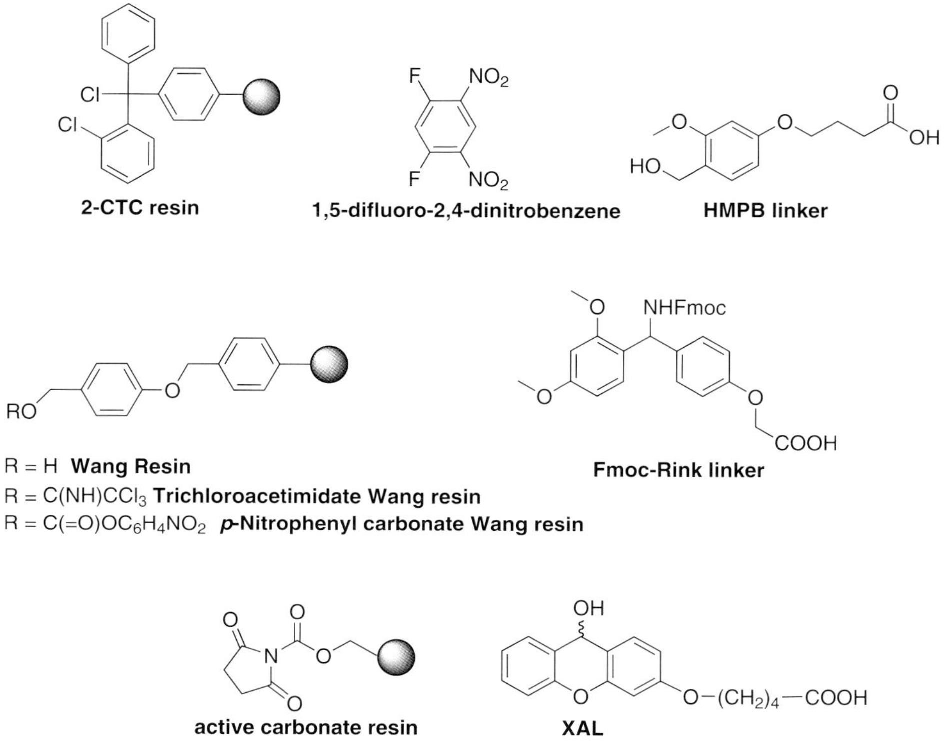

Figure 18.12 Linkers and resins for the synthesis of cyclic peptides by a side-chain anchoring approach.

elongation of the peptide chain, the cyclic peptide is constructed as described above for the side-chain anchoring approach.

18.6.1
Example: Solid Phase Synthesis of Argadin by a Side-Anchoring Approach [169]

Argadin, a potent chitinase inhibitor, has been synthesized in the solid phase by using a side-chain anchoring approach. Argadin is a cyclic pentapeptide that contains two unnatural amino acids: a derivatized Arg and a sensitive glutamic acid-derived hemiaminal, which are constructed from Orn and Glu precursors, respectively. As shown in Scheme 18.6, Fmoc-His-OAllyl was loaded onto 2-CTC resin through its side-chain, and elongation was carried out by an Fmoc/tBu strategy except for Orn, where Dde was used as side-chain protecting group. After the linear peptide was constructed, treatment with Pd(PPh$_3$)$_4$/PhSiH$_3$ followed by piperidine removed the Allyl and the N-terminal Fmoc group, respectively. Next, cyclization was carried out using PyBOP as coupling reagent, and, following Dde removal with hydrazine, Orn was converted into the derivatized Arg on the solid

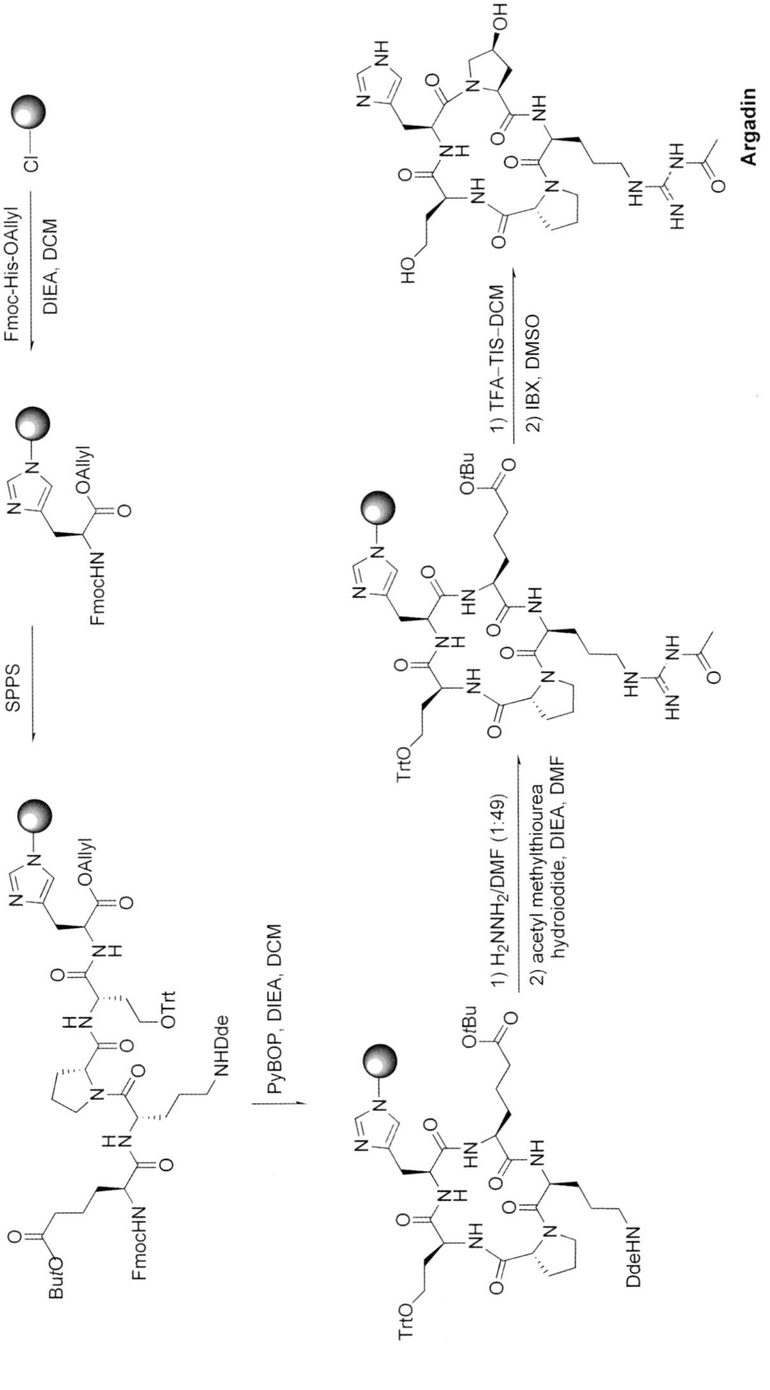

Scheme 18.6 Synthesis of the cyclic peptide Argadin.

phase. Finally, treatment with TFA released the unprotected peptide, which was obtained in a 10% yield after purification. The pure peptide was converted into Argadin by oxidative cyclization of glutamic acid by means of iodoxybenzoic acid (IBX) in DMSO.

18.7 Depsipeptides

Synthesis of depsipeptides by the Fmoc/tBu strategy presents some limitations because of the potential lability of the ester bond in response to piperidine exposure, and to the formation of DKP after the deprotection of the second amino acid after the ester bond.

Spengler and coworkers [170] developed a protocol that allows automatic synthesis of linear depsipeptides. The protocol modifies the pre-programmed modules for Boc chemistry in the following way: (i) the α-hydroxy acid is coupled unprotected by pre-activation with HOBt; (ii) O-acylation is performed with DCC and DMAP, without the presence of HOBt; when two α-hydroxy acids are introduced consecutively, the second is THP-protected; (iii) following O-acylation, a capping step (Ac_2O/DIEA) is not performed to prevent acyl transfer; (iv) the THP-protecting group can be removed under the same conditions used for the Boc group; (v) the remaining couplings are realized using *in situ* neutralization protocols and capping steps.

18.7.1
Example: Synthesis of H-L(1)EAKLKELEAKλ(12)AALEAKLKELEAKL-OH (L12λ)

This depsipeptide was synthesized on a Pam resin, following the machine-assisted protocol described above, and introducing leucic acid in place of leucine at position 12. The cycles shown in Table 18.1 were performed.

Table 18.1 Cycles performed in a depsipeptide synthesis (see text for details).

Cycle	Repetitions	Modules[a]	Description
1	1	c	Washing and swelling of the unprotected resin H-Leu-PAM
2	1	aibcde	Coupling of the first amino acid (K25) to the resin
3–15	13	agcbcde	Standard couplings of A24 until λ12 (incl.)
16	1	AGcdeffffff	Acylation with the symmetric anhydride of K11, prolonged coupling
17–26	10	aGcbcde	Standard couplings without capping
27	1	fGc	Final wash of the resin without N-terminal deprotection

a "a" (activation of the amino acid in the activation vessel, 17 min); "b" (deprotection of the resin in the reaction vessel with TFA, 16 min); "c" (DCM-wash of the resin, 1.5 min); "d" (neutralization of the resin with DIEA and washing, 4.5 min); "e" (transfer of the amino-acid HOBt ester from the activation vessel to the reaction vessel, 5 min); "f" (coupling under shaking, 20 min); "g" (DMSO addition to the reaction vessel and capping of unreacted resin with Ac_2O/DIEA, 16.5 min); "i" (wait without shaking, 15 min); "A" (addition of HOBf suppressed; formation or symmetric anhydride with DCC) "G" (DMSO and DIEA addition to the reaction vessel, no capping, 11.5 min).

However, some other depsipeptides require additional tools. Many depsipeptides of natural origin possess complex architectures such a cyclic or a bicyclic skeleton and may include unnatural or N-methylated amino acids in their structure. In these cases, a manual synthesis combined with careful optimization of each step is required. In this regard, syntheses of callipeltin B [171] and oxathiocoraline [172] have recently been described.

Spengler and coworkers [173] have used *p*-nitromandelic acid (Pnm) as a safety-catch linker for the synthesis of depsipeptides by Boc/Bzl chemistry. The hexafluoroacetone (HFA)-protected Pnm is loaded on the resin with concomitant deprotection of the hydroxy unit. After the depsipeptide is constructed, reduction of the nitro group with 6 M $SnCl_2$/1.6 mM HCl gives the p-aminomandelic (Pam) linker, which can be cleaved in acidic media (5% TFA in dioxane, 50 °C, 1 h), thereby obtaining the target depsipeptides in good yields and purities (Scheme 18.7).

18.8
Click Chemistry

Click chemistry [174] involves the use of reactions that fulfill certain conditions: they are high-yielding reactions that use readily available reagents. Moreover, these reactions are performed in "green" solvents like water, and allow easy isolation of products. In this context, several groups of reaction fit in this profile, such as: (i) cycloaddition reactions; (ii) nucleophilic ring-opening reactions; (iii) carbonyl chemistry of the non-aldol type; and (iv) addition to carbon–carbon multiple bonds

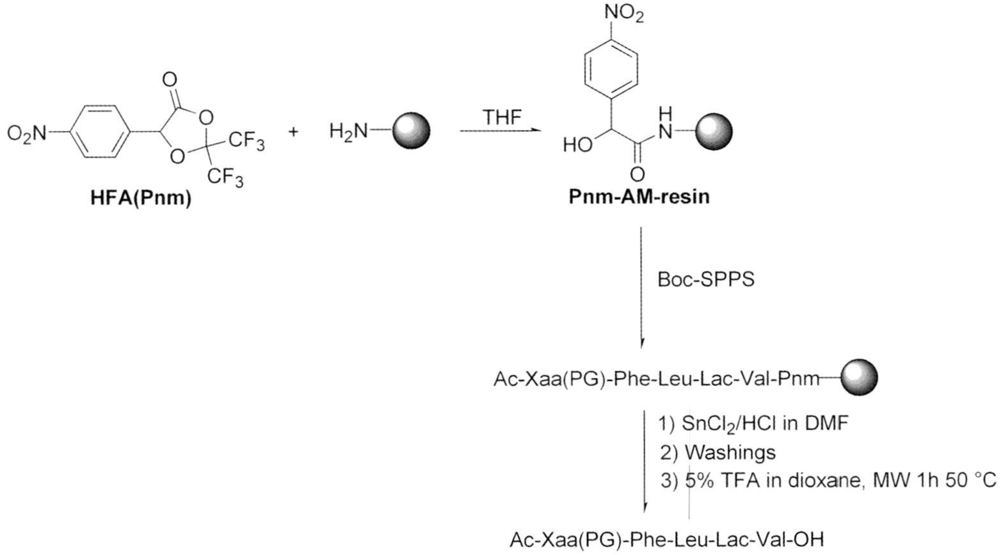

Scheme 18.7 Synthesis of depsipeptides using the Pnm linker.

[175]. Click chemistry is finding applications in the drug discovery field and also in bioconjugation. The reaction that has probably been most employed for click chemistry is the Cu(I)-catalyzed Huisgen 1,3-dipolar cycloaddition of azides and alkynes. Several recent reports have applied this reaction on resin-bound peptides: Tornøe and coworkers [176] synthesized peptidotriazoles on highly hydrophilic resins PEGA$_{800}$ and SPOCC using the HMBA linker. Several resin-bound tripeptides capped with propargylic acid were reacted with alkyl and aryl azides using CuI in DIEA. Owing to site-isolation and to the high solvation of the intermediates on PEG resins, which increase coupling efficiency, alkyne cross-couplings were avoided. Franke and coworkers [177] prepared scaffolded peptides by adding protected azidoacetylated peptides, previously assembled on 2-CTC resin, to an N-terminal alkynyl peptide bound to TentaGel S Ram resin. Gopi and coworkers [178] synthesized peptidotriazole conjugates by introducing an azidoproline at position 6 of the fusion inhibitor peptide RINNIPWSEAMM. The azidopeptide was also elongated in a PEG-containing resin (PAL-PEG-PS) and was then reacted with several alkynes while on the solid phase. One of these peptide conjugates was found to be two orders of magnitude more potent than the parent peptide.

List of Abbreviations

AB linker	3-(4-hydroxymethylphenoxy)propionic acid
Acm	acetamidomethyl
Alloc	allyloxycarbonyl
API	active pharmaceutical ingredient
Boc	*tert*-butyloxycarbonyl
CD	circular dichroism
CM	ChemMatrix
2-CTC	2-chlorotrityl
DCC	N,N'-dicyclohexylcarbodiimide
DCHA	dicyclohexyl amine
DIEA	N,N-diisopropylethylamine
DIPCDI	N,N'-diisopropylcarbodiimide
DKP	diketopiperazine
Dmbb	N-4,5-dimethoxy-2-mercaptobenzyl
EDT	1,2-ethanedithiol
Fmoc	N^{α}-(9-fluorenylmethoxycarbonyl)
HATU	1-[bis(dimethylamino)methylene]-1H-1,2,3-triazolo-[4,5-b]pyridinium hexafluorophosphate 3-oxide
HBTU	1-[bis(dimethylamino)methylene]-1H-benzotriazolium hexafluorophosphate 3-oxide
HCTU	1-[bis(dimethylamino)methylene]-6-chloro-1H-benzotriazolium hexafluorophosphate 3-oxide
HFA	hexafluoroacetone
HFIP	hexafluoroisopropanol

HMBA	4-hydroxymethylbenzoic acid
HOAt	1-hydroxy-7-azabenzotriazole
HOBt	1-hydroxybenzotriazole
IAPP	Human islet amyloid polypeptide
IBX	iodoxybenzoic acid
KCL	kinetically controlled ligation
MESNA	mercaptoethanosulfonate sodium salt
MPA	3-mercaptopropionic acid
MPAL	mercaptopropionic acid-leucine
Mmt	methoxytrityl
MSNT	1-(2-mesitylenesulfonyl)-3-nitro-1H-1,2,4-triazole
NBS	N-bromosuccinimide
NCE	new chemical entities
NCL	native chemical ligation
NMI	N-methylimidazole
NMP	N-methyl pyrrolidone
Nvoc	6-nitroveratryloxycarbonyl
pNZ	p-nitrobenzyloxycarbonyl
Pnm	p-nitromandelic acid
Pam	p-aminomandelic
PAM	4-(hydroxymethyl)phenylacetamidomethyl
PBS	phosphate-buffered saline
PEG	poly(ethylene glycol)
Pns	phenylnorstatine
ΨPro	pseudoprolines
PS	polystyrene
PyAOP	7-azabenzotriazol-1-yl-oxytris(pyrrolidino)phosphonium hexafluorophosphate
PyBOP	benzotriazol-1-yl-oxytris(pyrrolidino)phosphonium hexafluorophosphate
SPPS	solid-phase peptide synthesis
TAMPAL	trityl-associated mercaptopropionic acid leucine
TBTU	1-[bis(dimethylamino)methylene]-1H-benzotriazolium tetrafluoroborate 3-oxide
TFA	trifluoroacetic acid
TFE	trifluoroethanol
TFFH	tetramethylfluoroformadinium hexafluorophosphate
Thz	1,3-thiazolidine-4-carboxylic acid

References

1 Research and Markets, Ireland (2006) Peptides 2006 – New Applications in Discovery, Manufacturing, and Therapeutics.
2 Frost & Sullivan. 2007 Peptides Annual Report.
3 Merrifield, R.B. (1963) *Journal of the American Chemical Society*, **85**, 2149–54.

4 Dooley, C.T. and Houghten, R.A. (1998) Synthesis and screening of positional scanning combinatorial libraries, in Methods, in Molecular Biology, Vol. 87: *Combinatorial Peptide Library Protocols* (ed. S. Cabilly), Humana Press, Totowa, NJ, pp. 13–24.
5 Bruckdorfer, T., Marder, O. and Albericio, F. (2004) *Current Pharmaceutical Biotechnology*, **5**, 29–43.
6 Barany, G. and Merrifield, R.B. (1979) Solid-phase peptide synthesis, in *The Peptides*, Vol. 2 (eds E. Gross and J. Meienhofer), Academic Press, New York, pp. 1–284.
7 Lloyd-Williams, P., Albericio, F. and Giralt, E. (1997) *Chemical Approaches to the Synthesis of Peptides and Proteins*, CRC, Boca Raton, FL.
8 Zalipsky, S., Albericio, F. and Barany, G. (1985) Preparation and use of an Aminoethyl Polyethylene Ctycol-crosslinked Polystyrene Graft Resin Support for solid-phase peptide synthesis. In Peptides: structure and function, Proceedings of the Ninth American Peptide Symposium, (eds. L. M. Deber, V.J. Hruby, K.D. Doppler), Piene Chemical Co., Rockford, IL, pp. 257–60.
9 Bayer, E. and Rapp, W. (1986) *Chemistry of Peptides and Proteins*, **3**, 3–8.
10 Adams, J.H., Cook, R.M., Hudson, D., et al. (1998) *Journal of Organic Chemistry*, **63**, 3706–16.
11 Gooding, O.W., Baudart, S., Deegan, T.L., et al. (1999) *Journal of Combinatorial Chemistry*, **1**, 113–22.
12 Meldal, M. (1992) *Tetrahedron Letters*, **33**, 3077–80.
13 Kempe, M. and Barany, G. (1996) *Journal of the American Chemical Society*, **118**, 7083–93.
14 Meldal, M. and Miranda, L.P. (2003) Matrix for Solid-Phase Organic Synthesis. WO 2003031489.
15 Côté, S. (2005) New Polyether Based Monomers, Crosslinkers, and Highly Crosslinked Amphiphile Polyether Resins. WO 2005012277.
16 Carpino, L.A. and Han, G.Y. (1972) *Journal of Organic Chemistry*, **37**, 3404–9.
17 Guibé, F., Dangles, O. and Balavoine, G. (1986) *Tetrahedron Letters*, **27**, 2365–8.
18 Dangles, O., Guibé, F., Balavoine, G., et al. (1987) *Journal of Organic Chemistry*, **52**, 4984–93.
19 Isidro-Llobet, A., Guasch-Camell, J., et al. (2005) *European Journal of Organic Chemistry*, 3031–9.
20 Albericio, F., Chinchilla, R., Dodsworth, D. and Nájera, C. (2001) *Organic Preparations and Procedures International*, **33**, 203–303.
21 Rich, D.H. and Singh, J. (1979) The carbodiimide method, in *The Peptides: Analysis, Synthesis, Biology*, Vol. 1 (eds E. Gross and J. Meinhofer), Academic Press, New York, pp. 241–61.
22 König, W. and Geiger, R. (1970) *Chemische Berichte*, **103**, 788–98.
23 Dourtoglou, V.J., Ziegler, C. and Gross, B. (1978) *Tetrahedron Letters*, **19**, 1269–72.
24 Marder, O., Shvo, Y. and Albericio, F. (2002) *Chimica Oggi*, **20** (7/8), 37–41.
25 Carpino, L.A. (1993) *Journal of the American Chemical Society*, **115**, 4397–8.
26 Carpino, L.A. and El-Faham, A. (1995) *Journal of the American Chemical Society*, **117**, 5401–2.
27 Coste, J., Le-Nguyen, D. and Castro, B. (1990) *Tetrahedron Letters*, **31**, 205–8.
28 Albericio, F., Cases, M., Alsina, J., et al. (1997) *Tetrahedron Letters*, **38**, 4853–6.
29 El-Faham, A. and Albericio, F. (2007) *Organic Letters*, **9**, 4475–7.
30 Marshall, G.R. and Merrifield, R.B. (1965) *Biochemistry*, **4**, 2394–401.
31 Ommen, G.S. and Anfinsen, C.B. (1968) *Journal of the American Chemical Society*, **90**, 6571–2.
32 Lloyd-Williams, P., Albericio, F. and Giralt, E. (1993) *Tetrahedron*, **49**, 11065–133.
33 Dalcol, I., Rabanal, F., Ludevid, M.-D., et al. (1995) *Journal of Organic Chemistry*, **60**, 7575–81.
34 Barlos, K. and Gatos, D. (1999) *Biopolymers: Peptide Science*, **51**, 266–78.
35 Riniker, B., Floersheimer, A., Fretz, H., et al. (1993) *Tetrahedron*, **49**, 9307–20.
36 Mergler, M., Tanner, R., Gosteli, J. and Grogg, P. (1988) *Tetrahedron Letters*, **29**, 4005–8.

37 Mergler, M., Nyfeler, R., Tanner, R., et al. (1988) *Tetrahedron Letters*, **29**, 4009–12.
38 Barlos, K., Gatos, D., Kallitsis, J., et al. (1989) *Tetrahedron Letters*, **30**, 3943–6.
39 Rovero, P., Vigano, S., Pegoraro, S. and Quartara, L. (1996) *Letters in Peptide Science*, **2**, 319–23.
40 Barlos, K., Chatzi, O., Gatos, D. and Stavropoulos, G. (1991) *International Journal of Peptide and Protein Research*, **37**, 513–20.
41 Barlos, K. and Knipp, B. (2004) PCT Int. Appl. WO 2004056883, A2 20040708, CAN 141:106893, AN 2004:546536.
42 Bohling, J.C. and Zabrodski, W.J. (2004) Eur. Pat. Appl. EP 1391447, A1 20040225, CAN 140:183593, AN 2004:157495.
43 García-Martin, F., Bayó-Puxan, N., Cruz, L.J. and Albericio, F. (2007) *QSAR & Combinatorial Science*, **26**, 1027–35.
44 Sieber, P. (1987) *Tetrahedron Letters*, **28**, 2107–10.
45 Bollhagen, R., Schmiedberger, M., Barlos, K. and Grell, E. (1994) *Journal of the Chemical Society, Chemical Communications*, **22**, 2559–260.
46 Bray, B.L. (2003) *Nature Reviews. Drug Discovery*, **2**, 587–93.
47 Schneider, S.E., Bray, B.L., Mader, C.J., et al. (2005) *Journal of Peptide Science*, **11**, 744–53.
48 Goulas, S., Gatos, D. and Barlos, K. (2006) *Journal of Peptide Science*, **12**, 116–23.
49 García-Martin, F., Quintanar-Audelo, M., et al. (2006) *Journal of Combinatorial Chemistry*, **8**, 213–20.
50 Hyde, C., Johnson, T. and Sheppard, R.C. (1992) *Journal of the Chemical Society, Chemical Communications*, 1573–5.
51 Zhang, L., Goldammer, C., Henkel, B., et al. (1994) 'Magic mixture', a powerful solvent system for solid-phase synthesis of 'difficult sequences', in *Innovation and Perspectives in Solid Phase Synthesis* (ed. R. Epton), Mayflower, Birmingham, pp. 711–16.
52 Seebach, D., Thaler, A. and Back, A.K. (1989) *Helvetica Chimica Acta*, **72**, 857–67.
53 Toniolo, C., Bonora, G.M. and Mutter, M. (1981) *Makromolekulare Chemie*, **182**, 2007–14.
54 Simmonds, R.G. (1996) *International Journal of Peptide and Protein Research*, **47**, 36–41.
55 Zeng, W., Regamey, P.-O., Rose, K., et al. (1997) *Journal of Peptide Research*, **49**, 273–9.
56 Mutter, M., Nefzi, A., Sato, T., et al. (1995) *Peptide Research*, **8**, 145–53.
57 Wöhr, T., Wahl, F., Nefzi, A., et al. (1996) *Journal of the American Chemical Society*, **118**, 9218–27.
58 White, P., Keyte, J.W., Bailey, K. and Bloomberg, G. (2004) *Journal of Peptide Research*, **10**, 18–26.
59 Sampson, W.R., Patsiouras, H. and Ede, N.J. (1999) *Journal of Peptide Research*, **5**, 403–9.
60 Keller, M. and Miller, A.D. (2001) *Bioorganic & Medicinal Chemistry Letters*, **11**, 857–9.
61 Abedini, A. and Raleigh, D.P. (2005) *Organic Letters*, **7**, 693–6.
62 Cremer, G.-A., Tariq, H. and Delmas, A.F. (2006) *Journal of Peptide Science*, **12**, 437–42.
63 García-Martin, F., White, P., Steinauer, R., et al. (2006) *Biopolymers: Peptide Science*, **84**, 566–75.
64 Sohma, Y., Sasaki, M., Hayashi, Y., et al. (2004) *Chemical Communications*, 124–5.
65 Sohma, Y., Yoshiya, T., Taniguchi, A., et al. (2007) *Biopolymers: Peptide Science*, **88**, 253–62.
66 Carpino, L.A., Krause, E., Sferdean, C.D., et al. (2004) *Tetrahedron Letters*, **45**, 7519–23.
67 Mutter, M., Chandravarkar, A., Boyat, C., et al. (2004) *Angewandte Chemie (International Edition in English)*, **43**, 4172–8.
68 Oliyai, R. and Stella, V.J. (1993) *Annual Review of Pharmacology and Toxicology*, **33**, 521–44.
69 Kimura, T., Ohtake, J., Nakata, S., et al. (1995) *Peptide Chemistry*, **32**, 157–60.
70 Dos Santos, S., Chandravarkar, A., Mandal, B., et al. (2005) *Journal of the American Chemical Society*, **127**, 11888–9.

71 Tuchscherer, G., Chandravarkar, A., Camus, M.-S., et al. (2006) *Biopolymers: Peptide Science*, **88**, 239–51.
72 Sohma, Y., Sasaki, M., Hayashi, Y., Kimura, T., Kiso, Y. (2004) *Chemical Communications*, 124–125.
73 Sohma, Y., Hayashi, Y., Kimura, M., et al. (2005) *Journal of Peptide Science*, **11**, 441–51.
74 Taniguchi, A., Sohma, Y., Kimura, M., et al. (2006) *Journal of the American Chemical Society*, **128**, 696–7.
75 Sohma, Y., Taniguchi, A., Skwarczynski, M., et al. (2006) *Tetrahedron Letters*, **47**, 3013–17.
76 Yoshiya, T., Sohma, Y., Kimura, T., Hayashi, Y. and Kiso, Y. (2006) *Tetrahedron Letters*, **47**, 7905–9.
77 Coin, I., Dölling, R., Krause, E., et al. (2006) *Journal of Organic Chemistry*, **71**, 6171–7.
78 Carpino, L.A., Philbin, M., Ismail, M., et al. (1997) *Journal of the American Chemical Society*, **19**, 9915–16.
79 Carpino, L.A., Ismail, M., Truran, G.A., et al. (1999) *Journal of Organic Chemistry*, **64**, 4324–38.
80 Hentschel, J. and Borner, H.G. (2006) *Journal of the American Chemical Society*, **128**, 14142–9.
81 Rose, K. (1994) *Journal of the American Chemical Society*, **116**, 30–3.
82 Liu, C.-F. and Tam, J.P. (1994) *Proceedings of the National Academy of Sciences of the United States of America*, **91**, 6584–8.
83 Köhn, M. and Breinbauer, R. (2004) *Angewandte Chemie (International Edition in English)*, **43**, 3106–16.
84 Dawson, P.E., Muir, T.W., Clark-Lewis, I. and Kent, S.B.H. (1994) *Science*, **266**, 776–9.
85 Dawson, P.E. and Kent, S.B.H. (2000) *Annual Review of Biochemistry*, **69**, 923–60.
86 Macmillan, D. (2006) *Angewandte Chemie (International Edition in English)*, **45**, 7668–772.
87 Muralidharan, V. and Muir, T.W. (2006) *Nature Methods*, **3**, 429–38.
88 Hojo, H., Kwon, Y., Kakuta, Y., et al. (1993) *Bulletin of the Chemical Society of Japan*, **66**, 2700–6.
89 Camarero, J.A., Cotton, G.J., Adeva, A. and Muir, T.W. (1998) *Journal of Peptide Research*, **51**, 303–16.
90 Hackeng, T.L., Griffin, J.H. and Dawson, P.E. (1999) *Proceedings of the National Academy of Sciences of the United States of America*, **96**, 10068–73.
91 Canne, L.E., Walter, S.M. and Kent, S.B.H. (1995) *Tetrahedron Letters*, **36**, 1217–20.
92 Bang, D., Pentelute, B.L., Gates, Z.P. and Kent, S.B. (2006) *Organic Letters*, **8**, 1049–52.
93 Li, X., Kawakami, T. and Aimoto, S. (1998) *Tetrahedron Letters*, **39**, 8669–72.
94 Clippingdale, A.B., Barrow, C.J. and Wade, J.D. (2000) *Journal of Peptide Science*, **6**, 225–34.
95 Futaki, S., Sogawa, K., Maruyama, J., et al. (1997) *Tetrahedron Letters*, **38**, 6237–40.
96 Biancalana, S., Hudson, D., Songster, M.F. and Thompson, S.A. (2001) *Letters in Peptide Science*, **7**, 291–7.
97 Eggelkraut-Gottanka, R., Klose, A., Beck-Sickinger, A.G. and Beyermann, M. (2003) *Tetrahedron Letters*, **44**, 3551–4.
98 Mezo, A.R., Cheng, R.P. and Imperiali, B. (2001) *Journal of the American Chemical Society*, **123**, 3885–91.
99 Swinnen, D. and Hilvert, D. (2000) *Organic Letters*, **2**, 2439–42.
100 Sewing, A. and Hilvert, D. (2001) *Angewandte Chemie (International Edition in English)*, **40**, 3395–6.
101 Kenner, G.W., McDermott, J.R. and Sheppard, R.C. (1971) *Journal of the Chemical Society, Chemical Communications*, 636–7.
102 Backes, B.L., Virgilio, A.A. and Ellman, J.A. (1996) *Journal of the American Chemical Society*, **118**, 3055–6.
103 Backes, B.L. and Ellman, J.A. (1999) *Journal of Organic Chemistry*, **64**, 2322–30.
104 Shin, Y., Winans, K.A., Backes, B.J., et al. (1999) *Journal of the American Chemical Society*, **121**, 11684–9.
105 Ingenito, R., Bianchi, E., Fattori, D. and Pessi, A. (1999) *Journal of the American Chemical Society*, **121**, 11369–74.

106 Quaderer, R. and Hilvert, D. (2001) *Organic Letters*, **3**, 3181–4.
107 Mende, F. and Seitz, O. (2007) *Angewandte Chemie (International Edition in English)*, **46**, 4577–80.
108 Camarero, J.A., Hackel, B.J., de Yoreo, J.J. and Mitchell, A.R. (2004) *Journal of Organic Chemistry*, **69**, 4145–51.
109 Jensen, K.J., Alsina, J., Songster, M.F., et al. (1998) *Journal of the American Chemical Society*, **120**, 5441–52.
110 Alsina, J., Yokum, T.S., Albericio, F. and Barany, G. (1999) *Journal of Organic Chemistry*, **64**, 8761–9.
111 Brask, J., Albericio, F. and Jensen, K.J. (2003) *Organic Letters*, **5**, 2951–3.
112 Tulla-Puche, J. and Barany, G. (2004) *Journal of Organic Chemistry*, **69**, 4101–7.
113 Tulla-Puche, J., Getun, I.V., Alsina, J, et al. (2004) *European Journal of Organic Chemistry*, 4541–4.
114 Gross, C.M., Lelièvre, D., Woodward, C.K. and Barany, G. (2005) *Journal of Peptide Research*, **65**, 395–410.
115 Wang, P. and Miranda, L.P. (2005) *International Journal of Peptide Research and Therapeutics*, **11**, 117–23.
116 Botti, P., Villain, M., Manganiello, S. and Gaertner, H. (2004) *Organic Letters*, **6**, 4861–4.
117 Meldal, M. (1997) *Methods in Enzymology*, **289**, 83–104.
118 Tofteng, A.P., Jensen, K.J. and Hoeg-Jensen, T. (2007) *Tetrahedron Letters*, **48**, 2105–7.
119 Ollivier, N., Behr, J.-B., El-Mahdi, O., Blanplain, A. and Melnyk, O. (2005) *Organic Letters*, **7**, 2647–50.
120 Kawakami, T., Sumida, M., Nakamura, K., et al. (2005) *Tetrahedron Letters*, **46**, 8805–7.
121 Nakamura, K., Mori, H., Kawakami, T., Hojo, H., Ankara, Y. and Aimoto, S. (2007) *International Journal of Peptide Research and Therapeutics*, **13**, 191–202.
122 Ohta, Y., Itoh, S., Shigenaga, A., Shintaku, S., et al. (2006) *Organic Letters*, **8**, 467–70.
123 Nagaike, F., Onuma, Y., Kanazawa, C., et al. (2006) *Organic Letters*, **8**, 4465–8.
124 Yan, L.Z. and Dawson, P.E. (2001) *Journal of the American Chemical Society*, **123**, 526–33.
125 Brad, L., Pentelute, B.L. and Kent, S.B.H. (2007) *Organic Letters*, **9**, 687–90.
126 Tam, J.P. and Yu, Q. (1998) *Biopolymers*, **46**, 319–27.
127 Quaderer, R. and Hilvert, D. (2002) *Chemical Communications*, 2620–1.
128 Crich, D. and Banerjee, A. (2007) *Journal of the American Chemical Society*, **129**, 10064–5.
129 Canne, L.E., Bark, S.J. and Kent, S.B.H. (1996) *Journal of the American Chemical Society*, **118**, 5891–6.
130 Low, D.W., Hill, M.G.M., Carrasco, M.R., et al. (2001) *Proceedings of the National Academy of Sciences of the United States of America*, **98**, 6554–9.
131 Kawakami, T., Akaji, K. and Aimoto, S. (2001) *Organic Letters*, **3**, 1403–5.
132 Offer, J., Boddy, C.N.C. and Dawson, P.E. (2002) *Journal of the American Chemical Society*, **124**, 4642–6.
133 Tchertchian, S., Hartley, O. and Botti, P. (2004) *Journal of Organic Chemistry*, **69**, 9208–14.
134 Tawakami, T., Tsuchiya, M., Nakamura, K. and Aimoto, S. (2005) *Tetrahedron Letters*, **46**, 5533–6.
135 Canne, L.E., Botti, P., Simon, R.J., et al. (1999) *Journal of the American Chemical Society*, **121**, 8720–7.
136 Johnson, E.C.B., Durek, T. and Kent, S.B.H. (2006) *Angewandte Chemie (International Edition in English)*, **45**, 3283–7.
137 Zhang, L. and Tam, J.P. (1997) *Journal of the American Chemical Society*, **119**, 2363–70.
138 Camarero, J.A., Pavel, J. and Muir, T.W. (1998) *Angewandte Chemie (International Edition in English)*, **37**, 347–9.
139 Beligere, G.S. and Dawson, P.E. (1999) *Journal of the American Chemical Society*, **121**, 6332–3.
140 Tam, J.P., Yu, Q. and Yang, J.-L. (2001) *Journal of the American Chemical Society*, **123**, 2487–94.
141 Muir, T.W. (2003) *Annual Review of Biochemistry*, **72**, 249–89.
142 Miller, J.S., Dudkin, V.Y., Lyon, G.J., et al. (2003) *Angewandte Chemie (International Edition in English)*, **4**, 431–4.
143 Warren, J.D., Miller, J.S., Keding, S.J. and Danishefsky, S.J. (2004) *Journal of the American Chemical Society*, **126**, 6576–8.
144 Mezzato, S., Schaffrath, M. and Unverzagt, C. (2005) *Angewandte Chemie (International Edition in English)*, **44**, 1650–4.

145 Bang, D., Pentelute, B.L. and Kent, S.B.H. (2006) *Angewandte Chemie (International Edition in English)*, **45**, 3985–8.
146 Torbeev, V.Y. and Kent, S.B.H. (2007) *Angewandte Chemie (International Edition in English)*, **46**, 1667–70.
147 Kates, S.A., Solé, N.A., Albericio, F. and Barany, G. (1994) Solid-Phase Synthesis of Cyclic Peptides, in *Peptides: Design, Synthesis and Biological Activitys* (eds C. Basava and G.M. Anantharamaiah), Birkhauser, Berlin, pp. 39–58.
148 Wipf, P. (1995) *Chemical Reviews*, **95**, 2115–34.
149 Davies, J.S. (2003) *Journal of Peptide Science*, **9**, 471–501.
150 Kessler, H. (1982) *Angewandte Chemie (International Edition in English)*, **21**, 512–23.
151 Rizo, J. and Gierasch, L.M. (1992) *Annual Review of Biochemistry*, **61**, 387–418.
152 Blackburn, C. and Kates, S.A. (1997) *Methods in Enzymology*, **289**, 175–98.
153 Lambert, J.N., Mitchell, J.P. and Roberts, K.D. (2001) *Journal of the Chemical Society, Perkin Transactions 1*, **5**, 471–84.
154 Gilon, C., Mang, C., Lohof, E., et al. (2002) Synthesis of Cyclic Peptides, in Houben-Weyl: Methods of organic chemistry, vol E22b: Synthesis of Peptides and Peptidomimetics. (eds M. Goodman, A.M. Felix, L. Moroder and C. Toniolo), Georg Thieme Verlag, Stuttgart and New York, pp. 461–542.
155 McMurray, J.S. (1991) *Tetrahedron Letters*, **32**, 7679–82.
156 Rovero, P., Quartara, L. and Fabbri, G. (1991) *Tetrahedron Letters*, **32**, 2639–42.
157 Trzeciak, A. and Bannwarth, W. (1992) *Tetrahedron Letters*, **33**, 4557–60.
158 Kates, S.A., Solé, N.A., Johnson, C.R., et al. (1993) *Tetrahedron Letters*, **34**, 1549–52.
159 Romanovskis, P. and Spatola, A.F. (1998) *Journal of Peptide Research*, **52**, 356–74.
160 Valero, M.L., Giralt, E. and Andreu, D. (1999) *Journal of Peptide Research*, **53**, 56–67.
161 Büttner, F., Norgren, A.S., Zhang, S., et al. (2005) *Chemistry – European Journal* **11**, 6145–58.
162 Alsina, J., Rabanal, F., Giralt, E. and Albericio, F. (1994) *Tetrahedron Letters*, **35**, 9633–5.
163 Alsina, J., Chiva, C., Ortiz, M., et al. (1997) *Tetrahedron Letters*, **38**, 883–6.
164 Cabrele, C., Langer, M. and Beck-Sickinger, A.G. (1999) *Journal of Organic Chemistry*, **64**, 4353–61.
165 Isied, S.S., Kuehn, C.G., Lyon, J.M. and Merrifield, R.B. (1982) *Journal of the American Chemical Society*, **104**, 2632–4.
166 Alcaro, M.C., Orfei, M., Chelli, M., et al. (2003) *Tetrahedron Letters*, **44**, 5217–19.
167 Rietman, B.H., Smulders, R.H.P.H., Eggen, I.F., et al. (1994) *International Journal of Peptide and Protein Research*, **44**, 199–206.
168 Barany, G., Han, Y., Hargittai, B., et al. (2004) *Biopolymers: Peptide Science*, **71**, 652–66.
169 Dixon, M.J., Andersen, O.A., van Aalten, D.M.F. and Eggleston, I.A. (2006) *European Journal of Organic Chemistry*, 5002–6.
170 Spengler, J., Koksch, B. and Albericio, F. (2007) *Biopolymers: Peptide Science*, **88**, 823–8.
171 Krishnamoorthy, R., Vazquez-Serrano, L.D., Turk, J.A., et al. (2006) *Journal of the American Chemical Society*, **128**, 15392–3.
172 Tulla-Puche, J., Bayó-Puxan, N., Moreno, J.A., et al. (2007) *Journal of the American Chemical Society*, **129**, 5322–3.
173 Isidro-llobet, A., Álvarez, M., Burger, K., et al. (2007) *Organic Letters*, **9**, 1429–32.
174 Kolb, H.C., Finn, M.G. and Sharpless, K.B. (2001) *Angewandte Chemie (International Edition in English)*, **40**, 2004–21.
175 Kolb, H.C. and Sharpless, K.B. (2003) *Drug Discovery Today*, **8**, 1128–37.
176 Tornøe, C.W., Christensen, C. and Meldal, M. (2002) *Journal of Organic Chemistry*, **67**, 3057–64.
177 Franke, R., Doll, C. and Eichler, J. (2005) *Tetrahedron Letters*, **46**, 4479–82.
178 Gopi, H.N., Tirupula, K.C., Baxter, S., et al. (2006) *ChemMedChem*, **1**, 54–7.

19
Oligonucleotides and Their Derivatives
Dmitry A. Stetsenko

19.1
Polymer Supported Oligonucleotide Synthesis: An Overview and History

19.1.1
Introduction to Nucleic Acid Synthesis: Never-Ending Quest for Excellence

It is difficult to imagine modern molecular biology without a constant supply of good quality synthetic oligonucleotides. In fact, it is difficult to ascertain how much molecular biology owes to the seemingly simple question of providing reliable custom oligonucleotide synthesis service. It took over three decades and the efforts of many brilliant nucleic acid chemists to answer this question [1]. Today the modern state-of-the-art oligonucleotide synthesis can justly claim it furnishes the most efficient solid-phase chemistry ever designed, attaining stepwise coupling yields in excess of 99.8% [2].

This chapter highlights the most crucial milestones in the history of, and the most important elements in, the polymer-supported synthesis of oligonucleotides. The chemical synthesis of oligonucleotides has been the subject of many excellent reviews on every stage of its genesis. Early development is dealt with in a comprehensive manner [3, 4]; see also some personal accounts by the patriarchs of nucleic acid chemistry from both sides of the Atlantic [5, 6]. Important advances in the phosphotriester and phosphite triester methodologies and the various aspects of solid phase synthesis are covered in the subsequent input [7]. The phosphoramidite method is expertly treated in Marvin Caruthers' contributions [8] and in a series of Serge Beaucage's reviews [9]. The backbone analogs are described in detail in the antisense papers [10]. Oligonucleotide conjugates and derivatives obtainable by solid phase synthesis have also been reviewed [11]. Finally, several books on the subject are available [12].

The Power of Functional Resins in Organic Synthesis. Judit Tulla-Puche and Fernando Albericio
Copyright © 2008 WILEY-VCH Verlag GmbH & Co. KGaA, Weinheim
ISBN: 978-3-527-31936-7

19.1.2
Dawn of the Oligonucleotide Synthesis

Chemical synthesis of oligodeoxyribonucleotides was initiated by Michelson and Todd [13] more than half a century ago. The seminal 1955 paper [13] described the first synthesis of a dinucleoside phosphate, TpT. This pioneering synthesis anticipated several notable advances that would be exploited by other researchers only decades later. These are: (i) protection of the phosphate group; (ii) 3' to 5' direction of the oligonucleotide chain assembly; (iii) 3'-phosphonylation by an H-phosphonate monoester with the help of a coupling reagent (diphenyl phosphorochloridate); and (iv) *in situ* activation of an H-phosphonate diester to the reactive phosphorylating agent during the internucleoside phosphate bond formation (Scheme 19.1). Unfortunately, the protecting group employed for the internucleoside phosphate (benzyl) proved to be too labile for the efficient elongation, deprotection conditions were too harsh, isolation procedures limited, the yields low, and therefore the scheme was applied for the stepwise oligomerization of nucleotides neither by the authors nor by other researchers. However, Todd's legacy [15] eventually found its renaissance later in the development of the phosphotriester and H-phosphonate methods.

19.1.3
Phosphodiester and Phosphotriester Methods: A Slow Maturity

From the late 1950s oligonucleotide synthesis has become dominated for almost two decades by the method introduced by Khorana and coworkers [16] and dubbed the phosphodiester method (Scheme 19.2, method I) [9]. Its central dogma requires the phosphate groups to be left completely unprotected, which renders them sus-

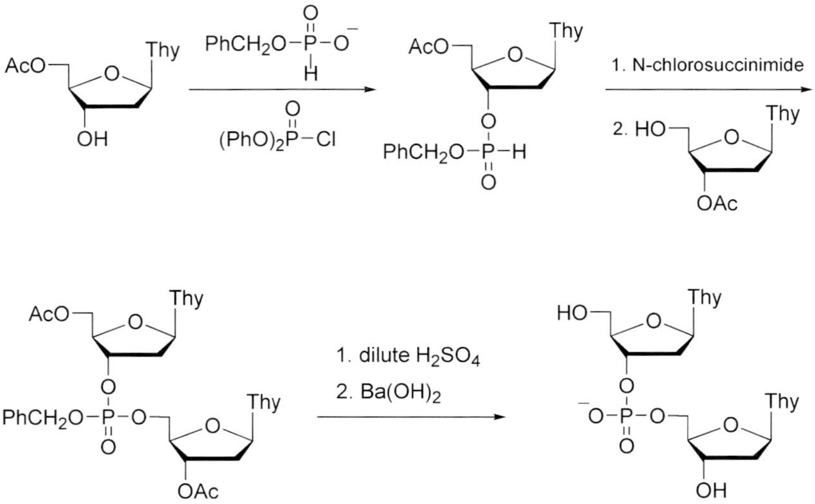

Scheme 19.1 The Michelson and Todd first synthesis of a dinucleoside phosphate, TpT.

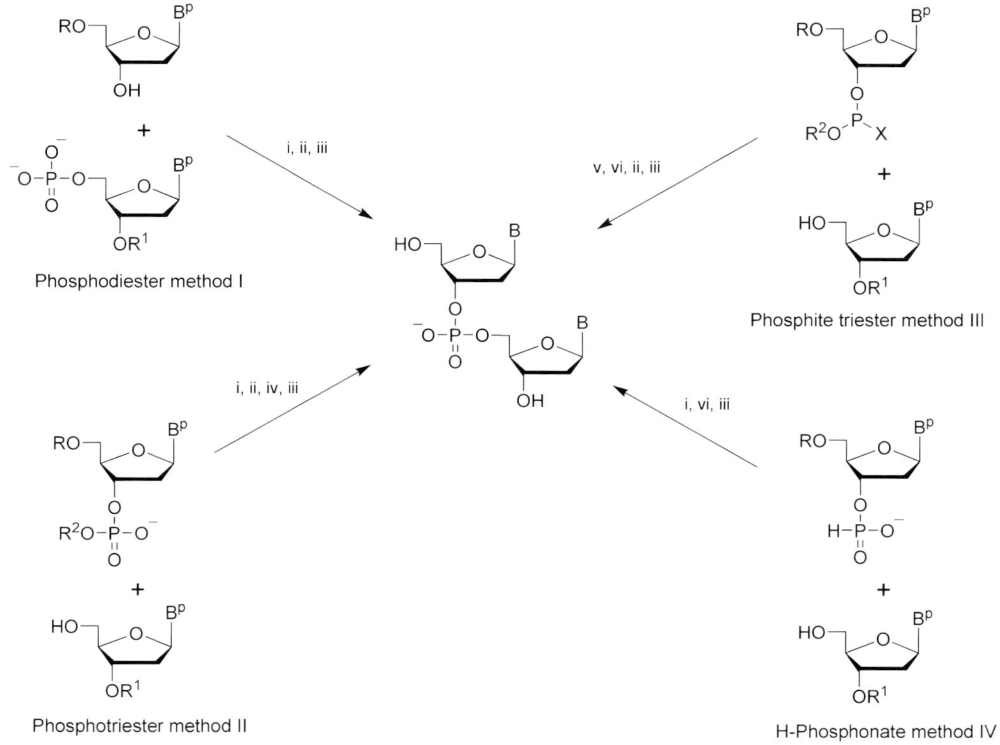

Scheme 19.2 Methods for polymer-supported oligonucleotide synthesis. Dinucleoside phosphate formation in the common methods for oligonucleotide synthesis: phosphodiester (**I**), phosphotriester (**II**), phosphite triester (**III**) and H-phosphonate (**IV**).
B = adenine (Ade), cytosine (Cyt), guanine (Gua), thymine (Thy), uracil (Ura), or other natural or unnatural nucleobase
^{p} = nucleobase protecting group, e.g. benzoyl for Ade and Cyt or isobutyryl for Gua, Thy and Ura are normally left unprotected
R = 5′-protecting group, e.g. 4,4′-dimethoxytrityl (DMTr)
R^1 = 3′-protecting group, e.g. acetyl, or a linkage to the polymer support
R^2 = phosphorus protecting group, e.g. o-chlorophenyl for **II** or β-cyanoethyl for **III**
X = active group at P(III), e.g. Cl in the phosphorochloridite or N(i-C$_3$H$_7$)$_2$ in the phosphoramidite variant, the latter requires *in situ* activation, e.g. by tetrazole
Conditions: (i) coupling reagent, e.g. TPS-Cl (alone for **I** or in the presence of N-methylimidazole for **II** or pivaloyl chloride for **IV**; (ii) 5′-deprotection, e.g. 3% trichloroacetic acid in dichloromethane for DMTr; (iii) N-deprotection and detachment from polymer support, e.g. concentrated aqueous ammonia at 55 °C, it removes also the β-cyanoethyl group from the phosphate; (iv) phosphate protecting group removal, e.g. tetramethylguanidine and p-nitrobenzaldoxime treatment for o-chlorophenyl; (v) activation, e.g. tetrazole; (vi) oxidation, e.g. iodine in aqueous pyridine.

ceptible to numerous side-reactions during the phosphodiester bond formation [17], and makes isolation and purification of the oligomers extremely time- and labor-consuming [18]. Nevertheless, the method gained widespread acceptance in the 1960s and allowed for the major advances in molecular biology that culminated in elucidation of the genetic code [19] and the first total synthesis of a gene [20].

However, from the chemist's point of view the most important contribution of Khorana's group to the nucleic acid chemistry is the development of a complete set of nucleoside protecting groups that is still being commonly used today. Firstly, the ammonia-labile acyl groups have been adopted for the exocyclic amino groups of the nucleobases, for example, benzoyl for adenine [14] and cytosine [21], or isobutyryl for guanine [22], and methods for their introduction [23] and deprotection [24] have been designed. Secondly, Khorana's legacy includes the mild acid-labile p-methoxy-substituted trityl family for the 5′-protection [25], particularly 4,4′-dimethoxytrityl (DMTr) group; deprotection can be assessed spectrophotometrically [26]. Typically, the phosphodiester synthesis run in the 5′-3′ direction [9, 24], where the 3′-acylated 5′-nucleotides were condensed with the 5′-tritylated nucleosides or di- or trinucleotide blocks (Scheme 19.2, method 1). The coupling step was promoted by a condensing agent, initially N,N′-dicyclohexylcarbodiimide (DCC) [22] or trichloroacetonitrile [5a] and later mesitylenesulfonyl chloride (MS-Cl) [27] or 2,4,6-triisopropylbenzenesulfonyl chloride (TPS-Cl) [28]. The latter two reagents have been adopted by another, more progressive approach that was developed a little later and evolved in parallel to the then dominant phosphodiester method [9, 24], namely, the phosphotriester method [4].

The phosphotriester method envisages the protection of the internucleoside phosphate by a suitable protecting group (Scheme 19.2, method II). This may require a separate deprotection step at the end but the charge neutral and nonpolar nature of the phosphotriesters simplifies their isolation and purification by column chromatography quite dramatically. In the 1960s the method was first promoted by Letsinger [29, 30] and then by Eckstein [26, 31], Reese [32] and in the following two decades it was perfected by Narang [33–36], Reese [4, 37, 38], Sekine and Hata [39–41], J. H. van Boom [42–44], Efimov [45–48] and others. The main improvements over the years culminated in the introduction of better phosphorus protecting groups: from the 2-cyanoethyl [29, 30] and 2,2,2-trichloroethyl [26, 31] to 2-chlorophenyl [49], 4-chlorophenyl [33–36] or phenylthio group [50], more powerful coupling agents, primarily arenesulfonyl azolides [34, 35], for example, 1-mesitylenesulfonyl-3-nitro-1,2,4-triazole (MSNT) [37, 51], and related compounds [38, 52], and the nucleophilic catalysts, for example, 1-methylimidazole [45, 46] and others [38, 47, 48].

It is within the framework of the phosphotriester method that the polymer-supported synthesis of oligonucleotides was born [53]. It was pioneered by Letsinger, who published, independently, his first solid-phase procedure [54] in the same year as Merrifield [55]. Letsinger and Mahadevan have successfully demonstrated [54] the first attachment of a nucleoside to an insoluble organic polymer, "popcorn" polystyrene [56], the first solid-phase phosphorylation procedure and the first phosphotriester condensation on solid support (Scheme 19.3), which can be repeated in a stepwise 5′-to-3′ fashion to obtain longer oligonucleotides [57]. The protecting groups involved were the acid-labile trityl for the 5′-hydroxyl and the base-labile 2-cyanoethyl for the phosphate that was cleaved together with the anchorage to the polymer achieved by acylation of the exocyclic amino group of dC by chlorocarbonyl polystyrene [56]. This limitation has been removed later by adopting the ester linkage as the anchor to either the nucleoside 3′- [58] or 5′-

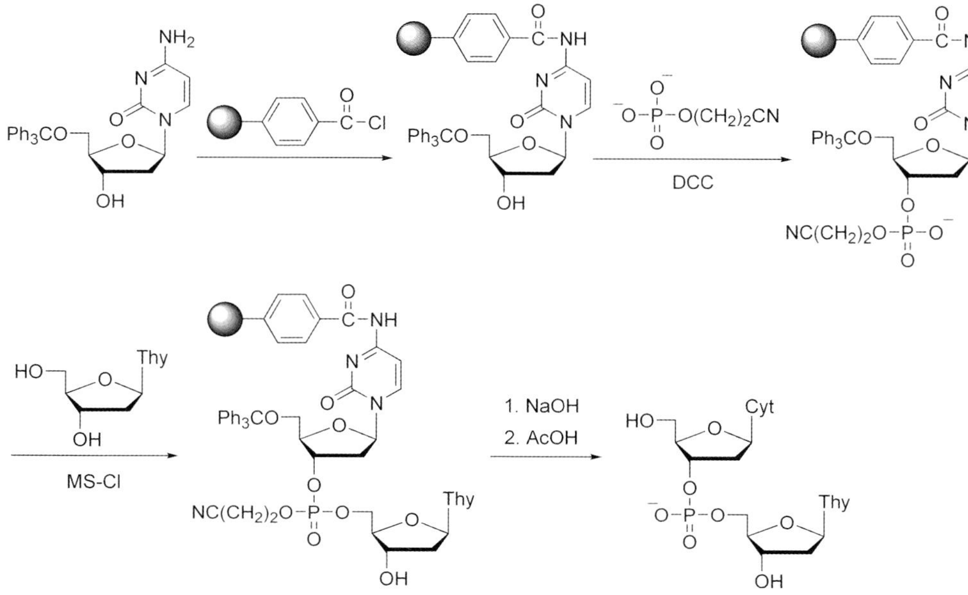

Scheme 19.3 The Letsinger and Mahadevan first polymer-supported synthesis of a dinucleoside phosphate, CpT.

hydroxyl [59]. However, the phosphorylation by a polymer-bound phosphodiester proved to be slow, low-yielding and sensitive to sterical hindrance even when compared with the phosphodiester method [58].

Immediately after the first solid-phase oligonucleotide synthesis was published, other research groups began to probe its potential with the polymer supports, both soluble and cross-linked, using the most popular strategy available at the time, the phosphodiester method. However, their efforts were hampered by the inherently inadequate stepwise yield the method can offer and unbearably long time needed to assemble even tri- or tetranucleotides. The imperfect chemistry continued to foil all attempts at a working solid-phase assembly until the phosphotriester method matured [4]. From the late 1970s, the power of an optimized phosphotriester synthesis and the availability of polymer supports approbated in the solid-phase peptide synthesis have made possible a big leap in the development of polymer-supported techniques.

Still, the time necessary for the solid-phase assembly of a medium-sized oligonucleotide (20–25 nucleotides long) was several days, and the stepwise yields of 90–95% were lower than required for the successful isolation of, for example, a 100-mer. The situation was made worse by the side reactions of U, T and G modification by the condensing agents in the phosphotriester synthesis [60]. Therefore, nucleic acid chemists kept looking for other techniques to speed up and simplify the procedure, and to obtain higher yields. The opportunity was soon offered by the application of the P(III) chemistry that quickly evolved into the solid-phase method of choice. With the advent of the phosphite triester method in its phosphoramidite modification, the phosphotriester chemistry has been firmly relegated to a subsidiary role in oligonucleotide synthesis.

19.1.4
Phosphite Triester and Phosphoramidite Methods: A Breakthrough

Although the higher reactivity of P(III) compounds has long been recognized, their use in nucleic acid chemistry was sporadic [61, 62] until 1975 when Letsinger et al. [63] described a successful approach to oligonucleotide synthesis via the 3'-phosphorochloridite derivatives (Scheme 19.2, method III). The phosphorodichloridites used to obtain the above reacted with 5'-tritylated nucleosides very quickly even at −78 °C, and the most important side-reaction was the formation of a symmetrical 3'-3' dinucleoside phosphite because of the extreme reactivity of the phosphorochloridite synthons. Indeed, in solution the 3',5'-internucleoside phosphite triester was produced within minutes, and the obligatory oxidation step was conveniently performed under either aqueous (iodine) [63] or anhydrous (m-chloroperbenzoic acid) [64, 65] conditions. Letsinger's original choice of 2-chlorophenyl or 2,2,2-trichloroethyl phosphorus protecting group [63] was soon disfavored, and the simple methyl group [66] became a phosphate protection of choice because of the ease of its removal by either thiophenoxide [66] or tert-butylamine [67]. Despite their high sensitivity to moisture and very limited storage capacity, the advantages of fast coupling and good overall yields the nucleoside phosphorochloridites offered have been immediately picked up for solid-phase assembly of DNA [65, 68–70] and RNA [71–74]. Additionally, the internucleoside phosphite triester can be converted into a host of oligonucleotide analogues, and the method was applied successfully for the preparation of a spectrum of P-modified oligonucleotides: phosphorothioates [75, 76] phosphoroselenoates [76], phosphoramidates [76, 77] and methylphosphonates [76, 78]. The problem of high reactivity of the monomers was alleviated by using hindered tertiary alcohol phosphorus protecting group (2,2,2-trichloro-1,1-dimethylethyl) [79, 80] or converting them into more stable yet still highly reactive tetrazolides [81, 82]. The latter derivatives have been used successfully in the promising approach to solid phase synthesis of oligonucleotides by the phosphite triester method [82].

However, even the pre-synthesized tetrazolides were still too reactive to be routinely used in the automated solid phase synthesis. The major advance that solved the problem was made in 1981, when Beaucage and Caruthers, who were experimenting with the nucleoside 3'-phosphoramidite derivatives [83] following some previous Russian work on the phosphorus(III)-amino compounds [84], discovered that these otherwise pretty stable compounds can be rapidly and very efficiently coupled to a solid-supported nucleoside in the presence of a mildly acidic nucleophilic catalyst, tetrazole [85]. This discovery combined with the already existing solid-phase assembly layout [68, 69, 82] paved the way for the very rapid expansion of the polymer-supported oligonucleotide synthesis, which has been summarized in the timely book edited by Gait [12a].

Initially, Beaucage and Caruthers employed methoxy N,N-dimethylamino phosphoramidites [85, 86], which were the least stable representatives of the family. Later, Caruthers [87] and others [88–90] found that more sterically hindered N-alkyl derivatives, particularly N,N-diisopropylamino [88, 90–92], or more electron-deficient, for example, morpholino [89, 90–92], derivatives give generally better yields and, therefore, should be preferred for reproducible results in the solid-supported synthesis of longer oligonucleotides. Another notable improvement has been made by Köster *et al.* [90–92] who combined the ammonia-labile β-eliminable 2-cyanoethyl group with the phosphoramidite chemistry, thus removing an extra step of demethylation from the solid phase synthesis protocol and transforming oligodeoxyribonucleotide deprotection into a single ammonia treatment. Soon the 2-cyanoethoxy *N,N*-diisopropylamino phosphoramidites superseded their morpholino counterparts, mainly because of the difficult purification of the phosphitylating reagent for the preparation of the latter, so that in the last two decades the former have been used in the vast majority of the syntheses. Methoxy *N,N*-diisopropylamino phosphoramidites were also gradually phased out because of the danger of thymine methylation [93] and the extra demethylation step required [65, 94], and have been resurrected only in the last decade in a novel chemistry for RNA synthesis [95]. Methylphosphonamidites are being routinely used for the preparation of methylphosphonate analogues [86, 96].

The improvements in the phosphoramidite method made over the time addressed a few important aspects of the chemistry. First, after the mechanism of the tetrazole-catalyzed phosphoramidite activation has been established and the role of tetrazole as acid and nucleophilic catalyst clarified [97], the search began for more efficient and hence more acidic and/or nucleophilic activators, the criteria for the selection of the right activator being its pK_a and sufficient solubility in anhydrous acetonitrile – the mainstay solvent for the phosphoramidite coupling. Apart from the originally proposed tetrazole (**1**) [85], the most popular activators (Figure 19.1) consist of all three isomeric 5-nitrophenyltetrazoles (**2–4**), *o*- [98, 99], *m*- [100] and particularly the *p*-isomer (*p*-NPT) [101, 102], and the 5-alkylthio derivatives, 5-ethylthiotetrazole (**5**), ETT) [103] and 5-benzylthiotetrazole (**6**), (BTT) [104], the latter two are especially good for sterically hindered RNA phosphoramidites [103c, 104]. Another widely employed promoter for both DNA and RNA solid phase synthesis is the somewhat less acidic but more nucleophilic 4,5-dicyanoimidazole (**7**) (DCI) [105].

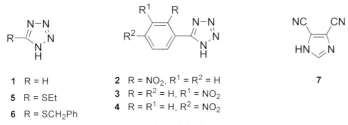

Figure 19.1 Common activators for solid-phase phosphoramidite coupling: tetrazole (**1**), 2-nitrophenyltetrazole (**2**), 3-nitrophenyltetrazole (**3**), 4-nitrophenyltetrazole (**4**), 5-ethylthiotetrazole (**5**), 5-benzylthiotetrazole (**6**) and 4,5-dicyanoimidazole (**7**).

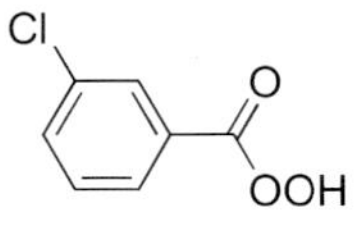

Figure 19.2 Alternative oxidation reagents for polymer-supported phosphite triester synthesis: *m*-chloroperbenzoic acid (**8**), diacetoxyiodobenzene (**9**), 10-camphorsulfonyloxaziridine (**10**), dimethyldioxirane (**11**), ethylmethyldioxirane (**12**).

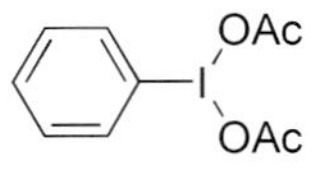

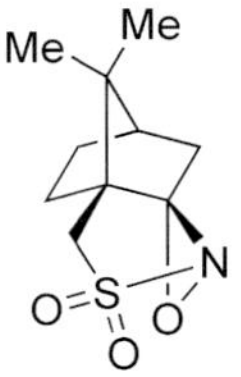

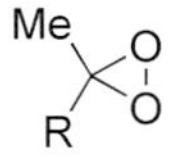

The internucleoside phosphite triester generated during the phosphoramidite coupling is labile under acidic detritylation step and therefore has to be oxidized on solid phase to the stable phosphotriester. This is routinely performed by iodine in aqueous pyridine solution [63, 85], but the requirement for strictly anhydrous conditions during the condensation step prompted a search for non-aqueous oxidants very early in the development. The most frequently used oxidizing agent that works under non-aqueous conditions in solid phase synthesis is *tert*-butyl hydroperoxide (TBHP) [106]. Other useful but less common oxidants (Figure 19.2) include *m*-chloroperbenzoic acid (**8**) [64, 65], diacetoxyiodobenzene (**9**) [107], tetrabutylammonium periodate [107], bis(trimethylsilyl)peroxide in the presence of trimethylsilyl triflate [108], 10-camphorsulfonyloxaziridine (**10**) [109], dimethyldioxirane (**11**) [110] and ethylmethyldioxirane (2-butanone peroxide) (**12**) [111].

Another important procedure is solid-phase sulfurization of the internucleotide phosphite triester to produce phosphorothioate analogues [75, 76]. This was originally accomplished by elemental sulphur [112], but due to low solubility of the latter in the common organic solvents it has been largely replaced by a selection of sulfurizing reagents (Figure 19.3): bis(phenylacetyl) disulfide (PADS) (**13**) [113], tetraethylthiuram disulfide (TETD) (**14**) [114], dibenzoyl tetrasulfide (**15**) [115], bis(*O,O*-diisopropoxyphosphinothioyl) disulfide (**16**) [116], bis(benzyltriethylammonium) tetrathiomolybdate (**17**) [117], arenesulfonyl disulfides (**18**) [118], diethyldithiocarbonate disulfide (DDD) (**19**) [119] and several sulfur-containing heterocycles, notably Beaucage's reagent (3*H*-1,2-benzodithiole-3-one 1,1-dioxide) (**20**) [120] and 3-ethoxy-1,2,4-dithiazoline-5-one (EDITH) (**21**) [121], and other compounds [122]. The aim here is to get as a high sulfurization efficiency as possible to prevent concomitant oxidation; it was shown that >99.8% sulfurization efficiency can be achieved [112c, 113f].

19.1.5
H-Phosphonate Method: Encouraging Diversity

Todd uncovered the potential of nucleoside H-phosphonates for oligonucleotide synthesis in the very early days of nucleic acid chemistry [15]. However, this knowledge laid dormant for almost three decades, while other methods flourished.

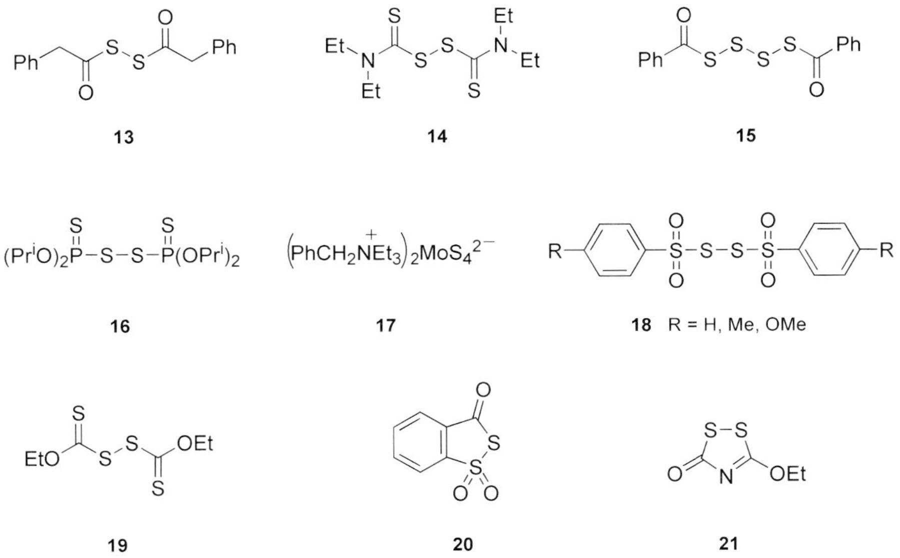

Figure 19.3 Sulfurization reagents for the synthesis of oligonucleotide phosphorothioates: bis(phenylacetyl) disulfide (**13**), tetraethylthiuram disulfide (**14**), dibenzoyl tetrasulfide (**15**), bis(O,O-diisopropoxyphosphinothioyl) disulfide (**16**), bis(benzyltriethylammonium) tetrathiomolybdate (**17**), arenesulfonyl disulfides (**18**), diethyldithiocarbonyl disulfide (**19**), Beaucage's reagent (**20**), 3-ethoxy-1,2,4-dithiazoline-5-one (**21**).

The only exception was an attempt at the 5′-to-3′ solid-phase oligonucleotide synthesis using nucleoside 3′-H-phosphonate monoesters, polymer-bound through a trityl-type linker, and oxidative coupling conditions (HgCl$_2$) [123]. The H-phosphonates were produced by phosphitylation of the polymer-bound N-unprotected nucleoside with phosphorus trichloride followed by hydrolysis by aqueous pyridine [61c]. However, despite some advanced features [124] the stepwise yields were comparable with those in the contemporary phosphodiester method – that is they were too low for the method to be truly competitive.

In the mid-1980s, Garegg *et al.* [125] and, slightly later and independently, Froehler and Matteucci [126] resurrected the H-phosphonate condensation. They discovered that the 5′-tritylated nucleoside 3′-H-phosphonate monoester salts can be efficiently activated in the presence of pyridine by chlorophosphates, for example, diphenyl phosphorochloridate (Todd's reagent) [125], and hindered acyl chlorides, for example, pivaloyl chloride [126], and coupled with the 5′-unprotected nucleosides with ease (Scheme 19.2, method IV). The H-phosphonate condensation proved to be fast and high yielding, the resulting internucleoside H-phosphonate diester is reasonably stable to the mildly acidic conditions of the 5′-detritylation and, therefore, the corresponding solid-phase version of the chemistry has been rapidly put to use for DNA [126–128] and RNA synthesis [129–134]. The pros of the H-phosphonate method include its relatively inexpensive reagents, a single global oxidation step at the end, shorter synthesis cycle and less stringent requirements for anhydrous conditions than for the phosphoramidite method. The cons, however, are significant enough to make the H-phosphonate method come second best after the phosphoramidite method in terms of the product yield. Precious percents are lost because of the basic hydrolysis of the internucleoside H-phosphonate during the final oxidation step [135, 136], and due to the "self-capping" of the support-bound nucleoside 5'-hydroxyl group by pivaloyl chloride during coupling [126, 137]. The H-phosphonate diester was shown to form acylphosphonate during the common acetylation capping step [137]; therefore, the capping step was regularly ignored [126] or a special H-phosphonate monoester capping reagent used [135, 137]. Pivaloyl chloride, which is unstable when stored in pyridine solution, can be replaced with very stable and efficient crystalline 1-adamantanecarbonyl chloride as a condensing reagent [137].

An additional reason for the inferior performance of the H-phosphonate method compared to the phosphoramidite lies in its mechanism, which makes the synthesis very sensitive to pre-activation of the H-phosphonate monoesters with the conventional coupling agents, which must be avoided [138]. However, the main appeal of the H-phosphonate approach lies in the ease it offers for the synthesis of backbone analogues modified at the phosphorus atom: phosphorothioates [139], phosphotriesters [126] and especially phosphoramidates [126, 140, 141]. Now it is rarely used for routine oligonucleotide synthesis but very often for the preparation of variously modified analogues. Further developments in the H-phosphonate method have been made in the last decade by Polish [142] and Japanese researchers [143].

19.2
Supports for the Polymer-Supported Synthesis of Oligonucleotides

19.2.1
Polystyrene (PS) Resins

19.2.1.1 "Popcorn" Polystyrene
While Merrifield experimented with the polystyrene cross-linked with 2% and later 1% DVB [55], Letsinger adopted the "popcorn" polystyrene [54] which relies on less cross-linking agent (0.05–0.2%) to achieve complete insolubility in common organic solvents. All of his early oligodeoxyribonucleotide syntheses by the phosphotriester method were performed on this type of support [53, 58, 59]. Other researchers have used popcorn PS as well in the syntheses by the phosphodiester method [144] and, later, more advanced phosphotriester method [145]. However, with the widespread use of low cross-linked PS and the advent of polyacrylamide and especially silica gel supports (Section 19.2.3), use of popcorn polystyrene in oligonucleotide synthesis has practically ceased.

19.2.1.2 Low Cross-linked Polystyrene
Low cross-linked polystyrene resins were first adopted for oligodeoxyribonucleotide synthesis within the framework of the phosphodiester method [146–150].

However, the shortcomings of the chemistry, notably low stepwise yields (40–80%), have limited the applicability of the solid-phase techniques compared with solution-phase syntheses. One of the unavoidable problems was the striking disparity in the properties between the non-polar polystyrene matrix and the polar ionic phosphodiester oligonucleotide chain, which led to incomplete couplings due to functional group inaccessibility. Overall, despite all the efforts of many researchers and some local successes like the use of uncross-linked insoluble isotactic polystyrene [151], phosphodiester syntheses on polystyrene resins were practically limited to the level of tri-and tetranucleotides.

With the advent of the phosphotriester method, gel-type PS resins were evaluated for stepwise and block syntheses of short- and medium-length oligonucle-

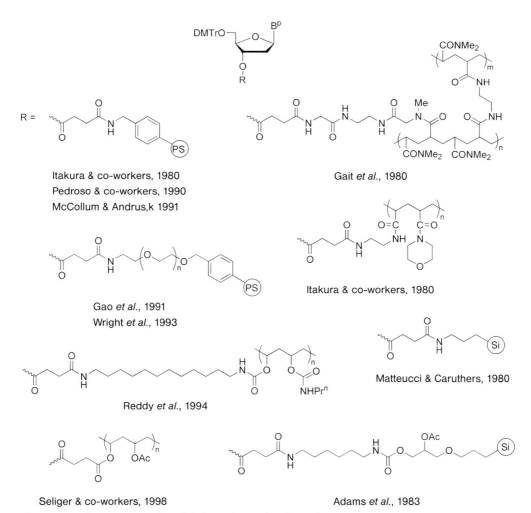

Figure 19.4 Polymer supports for solid-phase oligonucleotide synthesis.

otides with considerable success by Itakura and coworkers (Figure 19.4) [152], for combinatorial synthesis [153] and for the preparation of methylphosphonate analogues [154]. These were usually performed on aminomethyl polystyrene (1–2% DVB) with succinate linker for 3′-to-5′ synthesis [152–154], or, similarly, on a trityl-PS resin for 5′-to-3′ assembly [155]. However, later studies revealed especially favorable properties of rigid inorganic carriers, for example, silica gel and controlled pore glass (CPG), which soon relegated polystyrene resins to an auxiliary role until the appearance of macroporous PS.

Later, a widespread interest in bulk DNA analog synthesis for antisense research prompted Pedroso and coworkers [156] to re-evaluate the suitability of 1% cross-linked PS resin for large-scale oligonucleotide assembly (Figure 19.4). These Spanish researchers have demonstrated that, after optimization of the phosphoramidite protocol, a low cross-linked high-loaded PS resin is a very suitable carrier for preparative oligonucleotide synthesis. A further advantage of the PS resin is the possibility of monitoring the coupling efficiency by gel-phase ^{31}P NMR [157].

19.2.1.3 Highly Cross-linked (Macroporous) Polystyrene

Interest in highly cross-linked (20–30% DVB) macroporous polystyrene arose in the early days of oligonucleotide synthesis [158]. It has been argued that this type of support allows for faster reaction kinetics than the low cross-linked gel-type polystyrene. Modified with a trityl-type linker, it has been tested by Köster and Cramer [159] in the phosphodiester method. The limitations of the chemistry hampered its use for oligomers longer than a few nucleotides.

With the advent of more efficient synthetic methods, macroporous polystyrene supports have been re-evaluated by McCollum and Andrus [160] for the phosphoramidite scheme (Figure 19.4). Since this time, highly cross-linked PS resins have remained at the forefront of oligonucleotide synthesis, being preferred by many as the support of choice for regular [161] as well as non-standard applications [162], for example, the synthesis of peptide–oligonucleotide conjugates [162a,b].

19.2.1.4 Polystyrene–Poly(ethylene glycol) (PEG-PS) Composite Supports

Widespread success of the poly(ethylene glycol) grafted polystyrene supports (PEG-PS, TentaGel) [163] in solid-phase peptide synthesis has been mirrored in oligonucleotide synthesis, prompted by their swelling in the common solvents for DNA synthesis, particularly acetonitrile, good mechanical stability under continuous-flow conditions and high nucleoside loading (0.2–0.4 mmol g^{-1}), which is especially suitable for large-scale synthesis. The use of TentaGel is documented for both H-phosphonate method by Gao *et al.* [164] and phosphoramidite chemistry by Andrus and others (Figure 19.4) [165]. For the best results the synthetic protocol has to be optimized to allow for longer washing times; however, it remains the support of choice for some special applications [166].

19.2.1.5 Linear Polystyrene Grafted onto Other Polymer

Uncross-linked polystyrene grafted onto polytetrafluoroethylene (PTFE, Teflon) was adopted for oligonucleotide synthesis according to the phosphodiester scheme by Potapov *et al.* [167] To produce their support, the authors γ-irradiated PTFE granules in the presence of gaseous styrene, and equipped the linear polystyrene chains grafted onto the Teflon surface with monomethoxytrityl linker for the attachment of the first nucleoside via its 5′-hydroxyl (up to 50–70 µmol g^{-1}). The results of the semi-automated solid-phase oligonucleotide assembly developed were among the best for any polymer-supported phosphodiester DNA synthesis, although the attempts to adapt the support for the phosphotriester method failed [152]. However, the phosphodiester chemistry was outdated, and this original support was eclipsed by the advent of silica gel carriers few years later.

The potential of grafted polymers remained unrecognized until the use of PS-PTFE support was resurrected in the 1990s under the phosphoramidite method [168]. The aminomethylated and loaded with nucleoside 3′-succinate from 12.5 to 48 µmol per g-PS-Teflon support has been employed successfully in the automated synthesis of oligodeoxyribonucleotides as long 143-mer, with stepwise efficiency of up to 99.8%, comparing favorably with such popular carriers as silica beads CPG-500 and macroporous polystyrene Primer Support (Pharmacia).

19.2.2
Polyacrylamide (PA) Resins

Polyacrylamide support was first introduced into solid phase peptide synthesis employing the then novel Fmoc group by Atherton and Sheppard in the 1970s [169]. The authors argued that a polar polymer similar in nature to the main peptide synthesis solvent, DMF, would be inherently more suitable for solid phase assembly of polypeptides than the non-polar hydrophobic polystyrene resin. The same argument has been applied to contemporary oligonucleotide synthesis by the phosphodiester scheme [170]. Independently, Köster prepared a similar resin [171] and used its partially hydrolyzed carboxyl form for oligonucleotide synthesis by the phosphodiester method, adopting for the first time a capping procedure for the truncated sequences by acetylation [172]. A similar chemistry with Khorana's phenylisocyanate capping was employed by Gait and Sheppard in their initial synthesis by the phosphodiester method on polydimethylacrylamide support with aminohexyl side-chains [173], which has been adapted to run on a peptide synthesizer [174], while a Danish group came up with polyacryloylmorpholine partially substituted with piperazino groups [175]. Later, Itakura and coworkers successfully adapted a partly hydrolyzed commercially available polyacryloylmorpholine resin for use with the progressive phosphotriester method (Figure 19.4) [176, 177]. However, the best results were obtained when the first nucleoside was attached via the succinyl linker onto the ethylenediamine-treated polydimethylacrylamide resin (Figure 19.4) [178, 179] or Enzacryl Gel K-2 [177]. The polydimethylacrylamide support has been utilized by Gait *et al.* [180] in the automated solid phase oligonucleotide synthesis by the phosphotriester method on a peptide synthesizer.

However, the polyacrylamide resin beads are compressible and thus more suitable for batch-wise rather than continuous-flow solid phase synthesis. To make the support withstand pressure and thus perform well under continuous-flow conditions, a composite carrier has been designed where polydimethylacrylamide has been physically entrapped within the pores of a macroporous silica matrix, kieselguhr [181]. The support was tested in the Fmoc peptide synthesis with success and then adapted for oligonucleotide assembly by the phosphotriester method [182]. Although the peptide synthesis carrier (Pepsyn K) has been sold until recently, a comparative study soon revealed the kieselguhr-polyamide support to be inferior in its performance in oligonucleotide synthesis compared to a silica-based polymer (CPG) [183]. Nevertheless, polyacrylamide supports continued to be used for oligonucleotide synthesis by the phosphotriester route for special applications, for example, synthesis of cyclic oligonucleotides [184].

19.2.3
Silica Gel Supports

Silica gel was first tried as a solid support for oligonucleotide synthesis as early as in 1972 by Köster [185]. Several linkages were explored (5'-silyl ether, 5'-silyl phosphate and polymer-bound trityl), and the results were typical for the phosphodiester scheme adopted. The true potential of the rigid inorganic carriers was uncovered only when faster and higher-yielding methods for oligonucleotide synthesis were developed. Inspired by the known applications of surface-modified silica gel [186], Matteucci and Caruthers [68] employed macroporous (300 Å) HPLC grade aminopropyl silica gel (20 μm) in conjunction with the Letsinger's phosphorochloridite method [63] and the then novel succinyl linker to the 3'-hydroxyl, and demonstrated this inorganic polymer is a particularly efficient carrier for solid-phase DNA synthesis. Other researchers quickly followed suit, and their studies resulted, especially after the advent of the most efficient phosphoramidite solid-phase method, in a true renaissance of the rigid macroporous supports at the expense of the soft gel-type resins like Merrifield polystyrene or polyacrylamide.

Silica gel and controlled pore glass (CPG) were used in polymer-supported oligonucleotide synthesis for virtually any application. Every method of the phosphodiester bond formation has been adapted to run on silica gel support, including the phosphotriester [187] and H-phosphonate methods [126–129]. The long-chain alkyl amine CPG (LCAA-CPG) [188] first used by Adams et al. [88] (Figure 19.4) has become the most often used carrier for routine small-scale automated oligonucleotide synthesis (500 and 1000 Å pore size) as well as a base support for the preparation of various chemically-modified oligonucleotides and their analogues [189], while even larger pore size silica beads, for example, 3000 Å, have been proposed for special applications like enzymatic synthesis [190].

19.2.4
Miscellaneous Polymers, Surface-Modified Materials and Composite Supports

Many other polymers have been tried as supports for solid phase oligonucleotide synthesis. Even before the first use of polyacrylamide for oligonucleotide synthesis, poly-L-lysine had been proposed for the preparation of di- and trinucleotides by Khorana's method [191]. Poly-L-lysine (MW 80 000) was cross-linked with 1,6-hexamethylenediisocyanate, acylated by *p*-aminobenzoic acid to provide an aromatic phosphoramidate linker, loaded with pT (15% conversion) and subjected to a few rounds of phosphodiester synthesis (43% stepwise yield). The approach proved to be a disappointment and was not pursued further until the introduction of polyacrylamide resins.

Polyhydroxylated supports, for example, polysaccharides and poly(vinyl alcohol), seemed to be a better choice because of their availability and widespread use in chromatography. In the early days, Köster described the preparation of a Sephadex LH 20 support with uridine 5′-phosphate linker [192] which it was argued could be useful for chemo-enzymatic synthesis. However, the functionalization procedure was tedious, and no synthesis data were provided. Another insoluble polysaccharide, cellulose, has been used by Crea and Horn for oligonucleotide synthesis by the block phosphotriester method in the 3′-to-5′ direction [193]. Again the linker was uridine 5′-phosphate attached to hydroxyl group on the polymer by the phosphotriester linkage [193b]. Cellulose also furnished the material for the paper filter discs used in an early version of multiple oligonucleotide synthesis, where the reagents were passed through a column stacked with filter discs, each of them growing a different immobilized oligonucleotide sequence [194]. The method, which was thought of as quite attractive at the time, proved to be difficult to automate and required manual sorting of the filters, therefore got out of practice when more advanced automated synthesizers become available.

More promise has been shown by poly(vinyl alcohol) and its copolymers. Reddy *et al.* [195] have described the use of Toyopearl (Fractogel), which is a methacrylate and vinyl alcohol copolymer widely used in gel permeation chromatography. The abundant hydroxyl groups of the polymer have been reacted with 1,1′-carbonyldiimidazole followed by aliphatic diamine, which furnished a convenient linker for nucleoside 3′-hemisuccinate attachment (Figure 19.4). The authors showed Toyopearl HW-65F can be functionalized with amine up to 0.3 mmol g^{-1} and nucleoside succinate up to 125 µmol g^{-1}, which makes the polymer an attractive option for large-scale oligonucleotide synthesis. Good results for the automated phosphoramidite DNA assembly of up to 120 nucleotides long were reported with the average stepwise yield of 98.5% [195]. Later, Seliger and coworkers [196] employed a partial saponification of poly(vinyl acetate) (Merckogel OR 1000000) to unmask enough hydroxyl groups for the subsequent esterification of nucleoside 3′-succinate (Figure 19.4). The resulting high-loaded support measured up to 368 µmol g^{-1} of nucleoside and showed >98% stepwise yield in the standard automated phosphoramidite synthesis of oligodeoxyribonucleotides up to 20-mer.

Several inherently non-porous materials have been tested for oligonucleotide assembly. It has been argued that the surface-modified non-porous materials have an advantage over the porous for the preparation of extra-long oligonucleotides (>100 nucleotides), where the synthesis on the latter will come to a halt when the inner volume of the pores is filled by the growing oligomeric chains.

The concept has been put to test by Seliger et al. [197], who picked non-porous silica gel microbeads (1.5 µm diameter) used as an HPLC stationary phase, modified their surface by 3-aminopropyl triethoxysilane followed by 5′-DMTr-nucleoside 3′-succinate 4-nitrophenyl ester and obtained a support with a loading of 2–3 µmol g^{-1}. The synthetic performance of the microbeads in the automated solid phase phosphoramidite synthesis was comparable to a wide-pore CPG (1400 Å) with >99% average stepwise yield for 149 cycles. The problem with the surface-modified silica gel microbeads, though, is their low loading and small particle size, which makes them leak through and clog column filters.

Oligonucleotide synthesis on surface-modified polypropylene strips has found application in DNA array technology [198]. The modification has been performed by radical bromination (NBS/AIBN) followed by nucleophilic displacement by diamine for nucleoside 3′-succinate attachment or aminoalcohol for direct phosphoramidite condensation. However, the loading, although sufficient for hybridization studies, is too low for even the minimum scale oligonucleotide synthesis. An approach has been described to circumvent the problem [199]. The authors employed macroporous polyethylene discs (Porex X-4920, 30 µm nominal pore size), surface aminated by a gaseous ammonia plasma process to a uniformly low amino group content. The aminated polyethylene discs were then treated with a low (0.2% DVB) cross-linked polystyrene colloid with ca. 50% chloromethyl groups followed by displacement of the chlorine with triethylene glycol diamine for the anchor group amplification. Indeed, the nucleoside loading after the 5′-DMTr-3′-succinate attachment increased from 0.07 µmol g^{-1} for the starting amino polyethylene discs to 9–12 µmol g^{-1} for the polystyrene-grafted discs, which is close to macroporous polystyrene (12 µmol g^{-1}). The discs compared favorably in the automated phosphoramidite synthesis of a 15-mer DNA phosphorothioate, with >98% average coupling efficiency.

An interesting hybrid support for oligonucleotide with magnetic properties has been described [200] that is composed of an inner polystyrene core (4.5 µm) coated first by an iron salt and then by silica gel to prevent leaching of iron. The beads were then aminopropylated as usual and loaded with nucleoside 3′-succinate either directly or after incorporation of a tetraethylene glycol spacer via the corresponding phosphoramidite up to 20–27 µmol g^{-1}. The magnetic support performed comparably to the standard CPG in the automated oligonucleotide synthesis.

Recently, a silica gel nanoparticle–poly(ethylene glycol) composite support has been described [201]. Aminopropylated colloidal particles of silica gel (average size 230 nm, up to 14 µmol per g of amine) were cross-linked into aggregates by PEG-2000 diisocyanate, modified with 2,2′-sulfonyldiethanol phosphoramidite to provide a base-cleavable linker and subjected to 19 cycles of the automated DNA synthesis. Average coupling yield was up to 98%, and 30–70% of the aggregate was lost during acidic detritylation, so further experimentation is required for the successful use of the nanoparticle-based support for oligonucleotide synthesis.

19.3
Linkers and Anchor Groups for Solid Phase Oligonucleotide Synthesis

19.3.1
Linkers for the Synthesis of 3′- or 5′-Unmodified Oligonucleotides

19.3.1.1 Acid-Labile Trityl Linkers

The dominant current scheme for solid phase oligonucleotide synthesis runs in the 3′-to-5′ direction, that is it employs the starting nucleoside anchored to a polymer support via its 3′-hydroxyl and the monomeric units containing 3′-phosphorus, most often phosphoramidites, although the synthesis could be performed in the opposite, 5′-to-3′, direction by using the appropriate monomers and supports, which may be required for special applications. However, in the early days of the phosphodiester chemistry, the common way of oligonucleotide assembly on solid phase was from 5′-to-3′ employing 5′-nucleotides. This scheme required a 5′-linkage to a polymeric carrier, which was usually achieved via a trityl-type linker, either monoalkoxy or dialkoxy-substituted, normally attached to a polystyrene carrier (Figure 19.5). This kind of linkers has proved to be quite versatile and found its use later for combinatorial organic synthesis of small molecules on solid support.

The most common trityl anchor was monomethoxytrityl (Figure 19.5), employed in the phosphodiester method on PS–1% DVB by Melby and Strobach [146], on macroporous PS by Köster and Cramer [159] and on PTFE grafted polystyrene by Potapov *et al.* [167] A dimethoxytrityl linker was used by Köster and Cramer for the phosphodiester chemistry on popcorn PS [144] and by Belagaje and Brush [155] for their original adaptation of the phosphotriester method for the synthesis in 5′-to-3′ direction (Figure 19.5). Köster has also described a trityl anchor linked to a silica gel support [185]. Similar linkers have been exploited for liquid-phase oligonucleotide synthesis (Section 19.4).

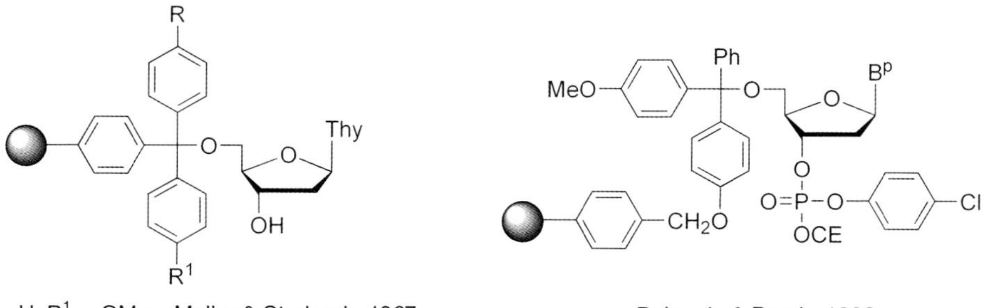

R = H, R¹ = OMe   Melby & Strobach, 1967
R = R¹ = OMe   Köster & Cramer, 1972
R = R¹ = H   Köster, 1972

Belagaje & Brush, 1982

Figure 19.5 Trityl linkers used for the 5′-attachment of a starting nucleoside to a polymer support.

Recently, a polymer-bound DMTr group has been applied by Mihaichuk et al. [124c] for the 5'-to-3' synthesis on PS using either phosphoramidite or H-phosphonate chemistry.

19.3.1.2 Base-Labile Acyl Linkers

A consensus scheme for the protection of nucleoside building blocks during oligonucleotide synthesis, which was been established in the early days of Khorana, employs base-labile acyl groups to protect the exocyclic amino groups of the nucleobases. This makes it natural to design a base-labile linker to join the nucleoside to a polymer support for solid phase synthesis (Figure 19.6). This is best performed by esterification of the 3'- or 5'-hydroxyl group to a carboxyl group linked to support, although historically the first anchor described was an exocyclic amide [53]. First, benzoate ester linkages based on 4-carboxypolystyrene were reported by Letsinger et al. [58, 202] for the 3'-hydroxyl attachment for the synthesis in the 3'-to-5' direction, and by Shimidzu and Letsinger [59] for the 5'-hydroxyl attachment for the 5'-to-3' oligonucleotide assembly, respectively. Later, Yip and Tsou [151] and Ogilvie and Kroeker [150] employed succinylated polystyrene, a polymer-supported equivalent of β-benzoylpropionyl group originally developed by Letsinger and coworkers [203] for solution phase oligonucleotide synthesis via hydrazine deprotection (Figure 19.6).

However, the truly versatile base-labile ester-type linker for the synthesis of oligonucleotides with unmodified 3'-end was found when the succinyl anchor was applied in 1980 independently by Dobrynin et al. [204] for low cross-linked polystyrene, Matteucci and Caruthers [68] for DNA and Ogilvie and coworkers [69, 72] for RNA synthesis on silica gel and Itakura and coworkers for polyacrylamide resin [177], and very swiftly adopted by others [178, 205] for the virtually every kind of a polymer support (Figure 19.6). For its introduction, the 5',N-protected nucleosides are normally treated with succinic anhydride to afford the 3'-succinate hemiesters [177, 206], which sometimes are converted further into the activated succinate esters for the subsequent acylation of the amino support [152, 177], or used directly with a suitable peptide coupling reagent [206]. Alternatively, a support-bound succinamic acid can be esterified with the 3'-unprotected nucleoside in the presence of a condensing agent, for example, a carbodiimide [207]. Variations of the attachment method involve the use of succinyl chloride [205], succinylbis-1,2,4-triazolide [208], isocyanate-mediated nucleoside 3'-hemisuccinate incorporation [209] and nucleophilic substitution of chloromethylated PS with 3'-succinate [205].

A succinyl anchor is commonly cleaved by concentrated aqueous ammonia at ambient temperature for 1–2 h. It is possible to largely preserve the integrity of the succinyl linker and retain most of the N,P-deprotected oligonucleotide on the support by using ethanolamine deprotection [210]. The main problem with the succinate ester attached to the primary amino group on polymer is its propensity for base-catalyzed cyclization into the corresponding succinimide [211], similarly to the aspartimide formation in the Fmoc peptide synthesis. This unwanted side-reaction can be prevented by joining the succinate to the secondary amine, for

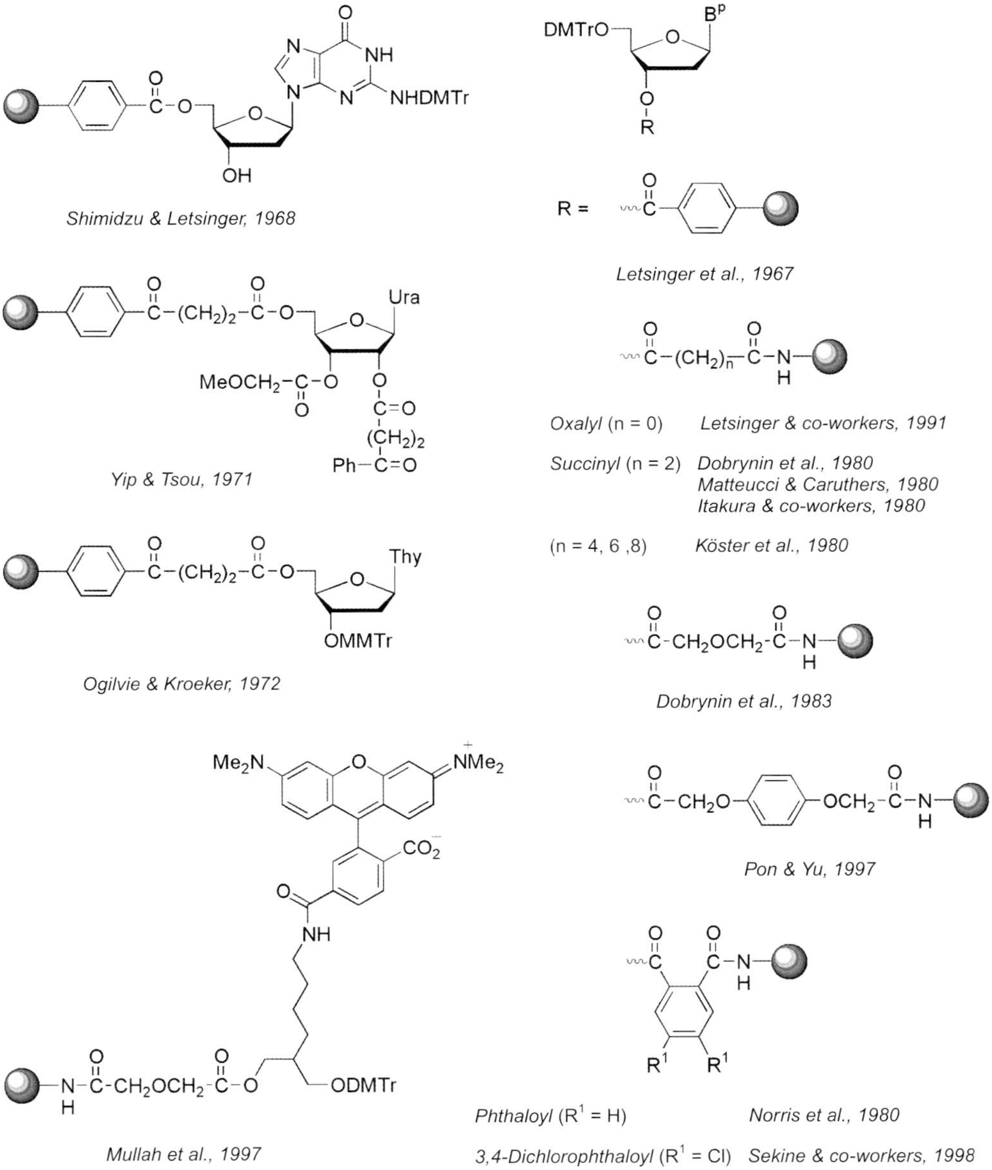

19.3 Linkers and Anchor Groups for Solid Phase Oligonucleotide Synthesis | 547

Figure 19.6 Acyl linkers for solid-phase oligonucleotide synthesis (see text for details).

example, sarcosine (N-methylglycine) [212] or 1,6-bis-methylaminohexane [213]. This makes the succinyl anchor withstand a strongly basic reagent, for example, DBU treatment required for the NPE/NPEOC chemistry [214]. An alternative solution is *in situ* O-silylation of the succinamide by a strong silylating agent, for example, N,O-bis(trimethylsilyl)acetamide (BSA) [215].

The relative stability of the succinyl anchor has prompted researchers to look for other linkers that would be more labile and cleavable under milder basic conditions and that could preserve fragile chemical groups within an oligonucleotide. The one most easily attacked by nucleophiles, the oxalyl anchor, has been described by Letsinger and coworkers (Figure 19.6) [216]. The support-bound nucleoside 3′-oxalate ester can be cleaved in less than 5 min under a range of conditions, for example, dilute (5–20%) ammonia in methanol, n-propylamine in dichloromethane (1 : 5 v/v) or 40% trimethylamine in methanol, yet is stable (<3% cleavage after 14 h) to a hindered secondary or tertiary amines like diisopropylamine, triethylamine or pyridine [216] but not DBU [218]. This enables isolation of fully N-protected oligonucleotides; however, the phosphate 2-cyanoethyl group is at least partly cleaved under these conditions, and so the H-phosphonate chemistry looks more applicable [217]. Attachment of a 5′,N-protected nucleoside to a polymer, LCAA-CPG or macroporous aminomethyl PS, via the oxalyl anchor, can be conveniently performed with in situ prepared oxalylbis-1,2,4-triazolide [217] or commercially available 1,1′-oxalyldiimidazole [219]. The linker is recommended for the preparation of extremely base-sensitive oligonucleotide analogues [219]; however, it is considered to be too labile for routine synthesis [220].

The next lower homolog of the succinyl, the malonyl anchor, was chosen by Guzaev and Lönnberg [221] to attach an original base-labile linker for the preparation of 3′-phosphorylated oligonucleotides (Section 19.3.2.4, Figure 19.11). It is intermediate in its lability between succinyl and oxalyl linkages, being cleaved virtually quantitatively at ambient temperature by concentrated aqueous ammonia in 20 min, 0.05 M methanolic K_2CO_3 in 3 h, and ethylenediamine in ethanol (1 : 1 v/v) in less than 10 min [221].

Higher homologous dicarboxylic acid linkers (adipoyl, suberoyl and sebacoyl) were explored briefly by Köster et al. [205] for the attachment of 5′-DMTr-deoxythymidine to hydroxymethyl PS resin via the corresponding acid chloride (Figure 19.6). No particular advantage over the succinyl anchor was apparent.

Another labile anchoring group, diglycolyl, was first applied by Dobrynin et al. [223] for a solid-phase phosphotriester synthesis on silica gel support, and later adopted by Mullah et al. [222] for the automated solid-phase assembly of ammonia-sensitive tetramethylrhodamine (TAMRA) labeled oligonucleotides useful for real-time PCR (Figure 19.6). The dye-labeled primers were quantitatively cleaved in the synthesizer by the mixture of tert-butylamine–methanol–water (1 : 1 : 2 v/v/v) after 45 min at ambient temperature. However, its rate of deprotection was still deemed too slow by Pon and Yu, who has come up with the next generation of extra-labile anchor based on hydroquinone-O,O′-diacetic acid, the Q-linker [220, 224]. The main advantage of the Q-linker over the oxalyl anchor [216] is its greater stability to storage at room temperature, while it can be cleaved under pretty much the same conditions as the latter albeit slightly more slowly, for example, by 0.05 M methanolic K_2CO_3 or the 1 : 1 mixture of 40% aqueous methylamine–ammonia (AMA) in 1 min, tert-butylamine–methanol–water (1 : 1 : 2 v/v/v) in 10 min, saturated ammonia in methanol or 1 M TBAF in THF in 15 min, 5% methanolic ammonia in 1 h [220]. Despite its ease of deprotection by nucleophiles, the Q-linker is relatively resistant to 20% v/v piperidine in DMF, 0.5 M DBU in pyridine or 1:1 v/v triethylamine in ethanol (5–15% cleavage after 1 h) [220]. These favorable properties prompted its commercial popularity [225]. A disadvantage of the Q-linker is the somewhat tedious preparation of the nucleoside 3′-hemiester derivatives [220, 226]. Particular applications of the Q-linker include reusable solid-phase carriers for large-scale oligonucleotide preparation obtained via esterification of the first nucleoside to a hydroxyl polymer [226], oligonucleotide assembly on underivatized amino supports [227] and tandem synthesis via linker phosphoramidites incorporating cleavable ester bonds [228].

Aromatic carboxylic acid linkers have also been employed. The phthaloyl anchor was described by Norris *et al.* [229] in conjunction with phosphotriester synthesis on polyacryloylmorpholine support (Figure 19.6). The authors prepared deoxythymidine 3′-phthalate monoester with phthalic anhydride, attached it to the piperazine-modified support [175] by DCC-mediated coupling and reported the successful detachment of oligomers during standard ammonia deprotection. Later, Brown *et al.* [212] demonstrated only slightly faster solution-phase cleavage of the 3′-phthaloyl-linked thymidine in 10% DBU in dichloromethane ($\tau_{1/2}$ 3.5 min) and slower ammonolysis (35 min) than for the corresponding succinyl analog (5 min with DBU). These results were mirrored by a Spanish group [218] with the respective DMTr-thymidine 3′-phthaloyl LCAA-CPG support (65% cleavage in 1.5 h with 0.5 M DBU in pyridine). Therefore, the phthalate anchor was used only rarely for standard oligonucleotide synthesis, for example, by Efimov *et al.* [230] for the preparation of oligonucleotide 3′-PEG conjugates, when they found that the corresponding succinyl-anchored PEG is too sensitive to the usual conditions of the phosphoramidite oligonucleotide synthesis. However, the DBU-mediated cyclization release of 3′-phthaloyl nucleosides has been exploited by Sekine and coworkers [231] in the synthesis of oligodeoxyribonucleotides containing N4-acetyl deoxycytidine residues by the H-phosphonate scheme. The 3,4-dichlorophthaloyl anchor (Figure 19.6) designed by the Japanese researchers can be cleaved almost quantitatively after 5 min treatment with 10% DBU in acetonitrile.

19.3.1.3 Carbonic Acid Linkers, Carbamate and Carbonate

In parallel with the search for the linkers more labile than succinyl, Sproat and Brown [232] have described a carbamate anchor that is significantly more stable to the standard concentrated ammonia deprotection (Figure 19.7). The authors have argued that the succinate linkage is not stable enough to the conditions of the phosphotriester oligonucleotide synthesis, particularly to amine impurities in the main solvent, pyridine. The replacement of an ester bond with an N-aryl carbamate bond has indeed resulted in a marked stabilization of the linkage to polymer support, which now required 36–48 h of ammonia treatment at 56 °C. However, with the wholesale switch from the phosphotriester to the phosphoramidite chemistry in the late 1980s, the too robust carbamate linker has found very scarce use [183].

Following the success of NPE/NPEOC chemistry [214], Eritja *et al.* [233] have developed an NPE linker to CPG support based on 4-(2-hydroxyethyl)-3-nitroben-

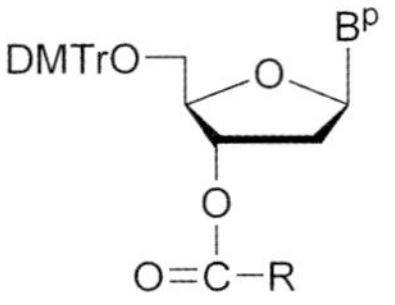

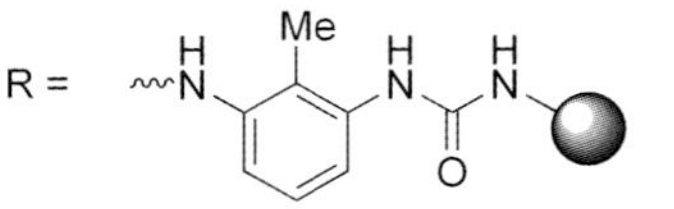

Sproat & Brown, 1985

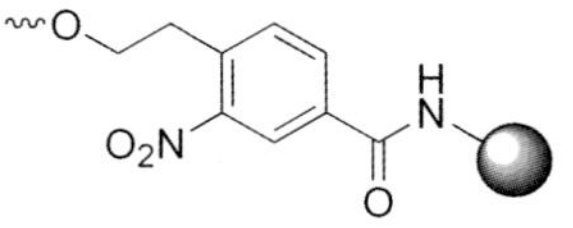

Eritja *et al.*, 1991

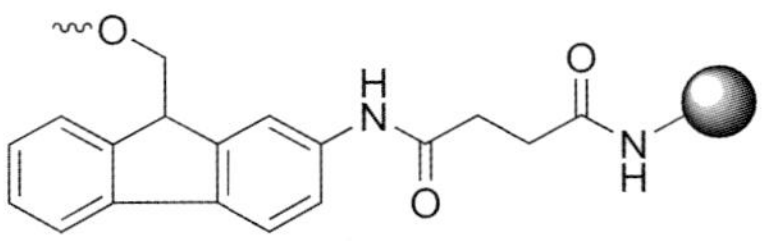

Aviñó *et al.*, 1996

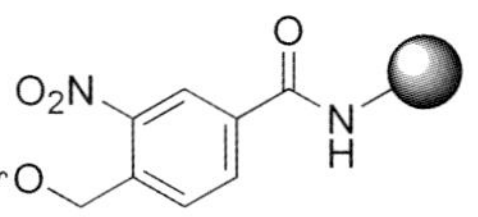

Greenberg & Gilmore, 1994

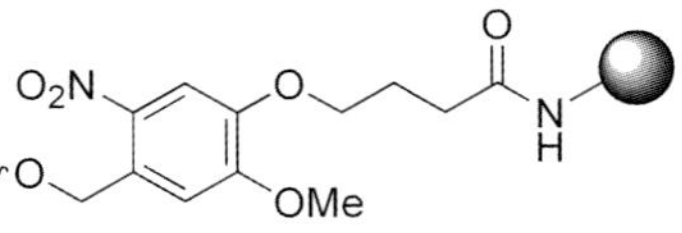

Venkatesan & Greenberg, 1996

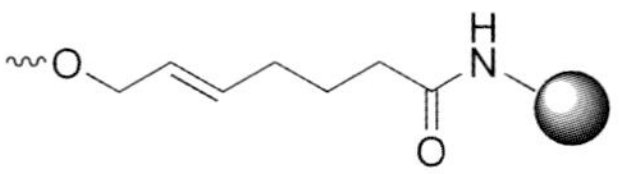

Greenberg *et al.*, 1998

Figure 19.7 Carbamate and carbonate linkers for the synthesis of 3′-hydroxyl oligonucleotides.

zoic acid for use with phosphoramidite chemistry (Figure 19.7). The first nucleoside was attached via the 3′-carbonate linkage, while for the preparation of 3′-phosphorylated oligonucleotides the direct coupling of the corresponding phosphoramidites onto the NPE support can be performed. The 3′-carbonate NPE anchor is cleaved by 0.5 M DBU in pyridine in less than 1 h, by ammonia (5 h, 55 °C) or 20% piperidine in DMF (3 h, ambient temperature); the 3′-phosphate can be liberated from NPE support by standard ammonia treatment. The NPE linker has been used primarily for the preparation of the oligonucleotides containing sensitive nucleobases [234], with DBU-mediated deprotection and phosphoramidite or H-phosphonate chemistry, [235] as well as for the synthesis of peptide–oligonucleotide conjugates [236].

A similar Fmoc-inspired anchor [237] (Figure 19.7) has also been applied for oligonucleotide synthesis following its success in peptide chemistry, for example, for the preparation of oligonucleotides via the NPE scheme [218] and the synthesis of peptide–oligonucleotide conjugates [238].

Photolabile 3′-carbonate linkers for the synthesis of fully-protected oligonucleotide fragments have been described by Greenberg and coworkers. These are based on the 2-nitrobenzyl [239] and 2-nitroveratryl group [240] (Figure 19.7), the latter being photolyzed more efficiently. The photolysis conditions have been optimized to minimize the thymidine photodimer formation and involve irradiation at 365 nm for 3 h in the presence of a transilluminator [240]. The yield of the photolytic cleavage is somewhat moderate (67–82.5%). The nitroveratryl anchor has been used for the synthesis of oligonucleotides with deprotection under non-nucleophilic conditions [242] as well as for the preparation of various 3′-functionalized oligonucleotide derivatives, for example, amino or carboxylic (Sections 19.3.3.1, 19.3.3.3).

Greenberg et al. have also developed a 3′-carbonate-linked allyl-type anchor for the synthesis of oligonucleotides with 3′-hydroxyl and palladium(0)-mediated cleavage (Figure 19.7) [241]. Oligomers were detached quantitatively from the linker by a 1 h treatment at 55 °C with 8 mol equiv of Pd_2dba_3 in the presence of 40 mol equiv of 1,2-bis(diphenylphosphino)ethane (DIPHOS) and 60 mM tetrabutylammonium formate with the slight excess (ca. 0.5 equiv) of formic acid as a proton source. The authors found that a small amount of water facilitates the reaction. Notably, the phosphate 2-cyanoethyl protecting groups are cleaved by the Pd(0) reagent.

19.3.1.4 Silyl, Silanediyl and Disiloxanediyl Linkers

Application of the silyl-type linkers for solid-phase peptide synthesis goes back two decades [243]. The concept of a fluoride-labile silyl ether anchor has been put to use for oligonucleotide synthesis as well. The diisopropylsilanediyl group has been employed by Routledge et al. [244] as a 3′-linker in conjunction with a hydroxyl CPG support for the preparation of N-benzoylated oligodeoxyribonucleotides (Figure 19.8). Cleavage from the polymer has been effected by a 1 min treatment with 1 M TBAF in THF while the phosphate 2-cyanoethyl groups required 2 h at 50 °C for their removal with the same reagent. The same anchor has been applied by Sekine and coworkers [245] for the synthesis of N4-alkoxycarbonylcytosine containing oligomers on hydroxymethyl ArgoPore PS resin.

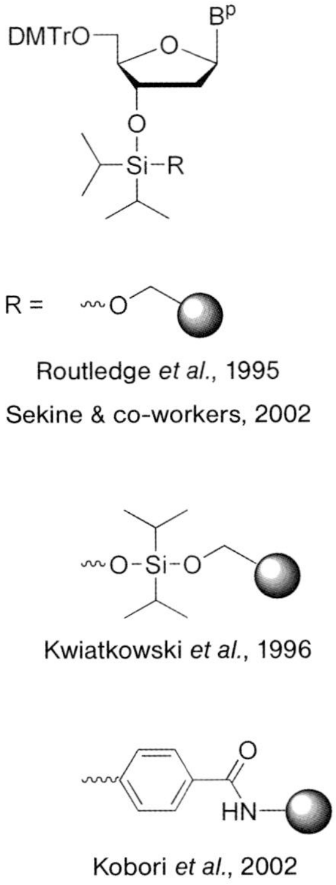

Figure 19.8 Silyl, silanediyl and disiloxanediyl linkers for oligonucleotide synthesis.

Routledge et al., 1995
Sekine & co-workers, 2002

Kwiatkowski et al., 1996

Kobori et al., 2002

However, the Japanese group has noted partial cleavage of the diisopropylsilanediyl linkage under the acidic conditions needed for the DMTr group removal [245, 247]. To circumvent this nuisance, Kobori et al. [247] have prepared a highly cross-linked polystyrene-supported phenyldiisopropylsilyl ether linker that proved to be completely stable to detritylation and used it successfully for oligonucleotide synthesis without N-protection by O-selective phosphoramidite chemistry [246] and pyrophosphate formation on solid phase [248]. The anchor can be cleaved under almost neutral conditions by 1 M TBAF–AcOH in THF (90% release after 1 h) or 0.2 M triethylamine trihydrofluoride in the presence of 0.4 M triethylamine for 4 h.

A disiloxanediyl linker for the preparation of extra-pure oligonucleotides, which is based on the Markiewicz 1,1,3,3-tetraisopropyldisiloxane-1,3-diyl protecting group [249], has been described [250]. The authors have noted its stability to detritylating reagent (<5% decomposition after 36 h) and concentrated aqueous ammonia in dioxane (1 : 1 v/v) treatment (<2% cleavage after 6 h at 65 °C). Oligonucleotides can be released from the disiloxanediyl support by 0.5 M TBAF in THF for 4 h at ambient temperature or the same reagent in DMF for 30 min at 65 °C. The linker has been used successfully for the synthesis of oligodeoxyribonucleotides free from depurinated sequences. Preparation of the disiloxanediyl CPG was quite laborious; however, attachment to a hydroxyalkyl polystyrene support should be quite straightforward [250].

19.3.2 Linkers for the Synthesis of Phosphorylated, Thiophosphorylated or Other Related Oligonucleotides

19.3.2.1 Phosphoramidate and Phosphorothiolate Linkages

The phosphoramidate group is one of the most venerable linkages for the solid phase synthesis of phosphorylated oligonucleotides, being put forward by Blackburn *et al.* [147] as early as in 1967 as an acid-labile linkage between the 5′-phosphate group of an oligonucleotide and the Merrifield-type aminophenoxymethyl polystyrene resin under the phosphodiester synthetic scheme (Figure 19.9). However, the conditions for its protic acid hydrolysis (80% acetic acid, 72 h at 80 °C) were too harsh for the gentle purine deoxyribonucleosides, which prompted the authors to look for an aliphatic phosphoramidate as an alternative. Later Ohtsuka *et al.* [253] found a way to break the aromatic phosphoramidate bond to the resin by diazotization with isoamyl nitrite in the mixture of pyridine and acetic acid (1 : 1 v/v) without damaging glycosidic linkages or N-protecting groups (96% release after 4 h treatment). A similar benzidine phosphoramidate linker has been adopted by Markiewicz and Wyrzykiewicz [251] for the phosphoramidite chemistry on aminopropyl CPG support with isoamyl nitrite cleavage followed by ammonolysis (Figure 19.9).

The phosphoramidate linkage in its aliphatic variant has been exploited by Gryaznov and Letsinger [252] for solid phase synthesis of oligonucleotide 3′-phosphates (Figure 19.9). After phosphate 2-cyanoethyl group removal by prior ammonolysis the phosphoramidate required milder conditions to break down (80% acetic acid, 4 h at ambient temperature) that are tolerable for the purine

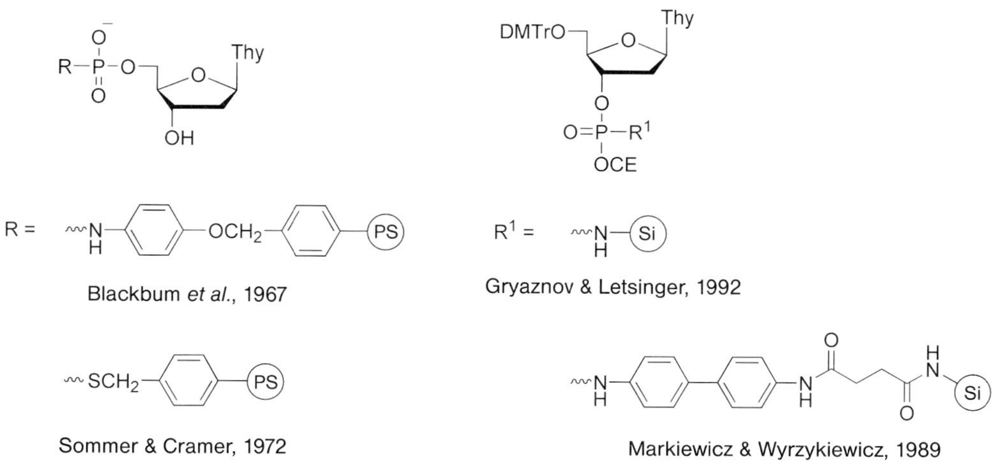

Figure 19.9 Phosphoramidate and phosphorothiolate linkers for oligonucleotide synthesis.

deoxyribonucleosides. A more recent elegant example [254] involves solid phase synthesis of a 5′-terminal "cap" structure of U1 snRNA when the phosphoramidate linkage between the 3′-end of a triribonucleotide and aminomethylated macroporous PS has been cleaved by 24 h treatment with acetic acid followed by the removal of the 3′-terminal phosphate group with alkaline phosphatase.

The phosphorothiolate linkage was used by Sommer and Cramer [255] for the thymidine 5′-phosphorothioate attachment to chloromethylated PS resin (5% cross-linking) in the early days of phosphodiester chemistry (Figure 19.9). In DMF at 60 °C after 6 h of reaction with a stoichiometric amount of the nucleoside, a substitution level of about 35% of the initial chlorine has been obtained. The phosphorothiolate linkage between the oligonucleotide and the resin was split by treatment with iodine solution (5 mg mL^{-1}) in 75% aqueous pyridine at ambient temperature for 22 h.

19.3.2.2 Base-Labile Linkers Based on β-Elimination

Historically, the first base-labile β-eliminable phosphate linker, 2-pyridylethyl phosphate on polystyrene support (5% DVB), was described by Freist and Cramer [149] within the framework of the phosphodiester method (Figure 19.10). However, its difficult introduction and harsh conditions of cleavage (2 M sodium methoxide in 50% methanolic pyridine) have severely limited its usefulness for oligonucleotide synthesis. A few years later β-arylthioethyl groups [256] and their oxidized β-eliminable arenesulfinyl-[256a] and arenesulfonylethyl [256b–d] counterparts were proposed by Narang et al. [256a] and later by Khorana and coworkers [256b,d] for the 5′-phosphate protection in the phosphodiester oligonucleotide synthesis in solution. The principle of a safety catch linker based on a thioether convertible into sulfone was applied shortly after for the soluble polymer-supported synthesis (see Section 19.4) [258]. A similar linker was then reported by Gait and Sheppard [173, 174] for their solid-phase DNA assembly on a polyamide support under the same synthetic scheme (Figure 19.10). With the demise of the phosphodiester method the safety catch approach has been largely abandoned because the respective β-sulfonylethyl groups were found to be stable to the conditions of both phosphotriester [259] and phosphoramidite synthesis [257]. This prompted an adaptation of a related anchoring group by Efimov et al. [260] (Figure 19.10) for the rapid phosphotriester synthesis of 3′-phosphorylated oligodeoxyribonucleotides on a polystyrene resin (1% DVB, 1.25 mmol-Cl g^{-1}). The linker has been introduced into the polymer by nucleophilic substitution of chloromethyl groups by sodium salt of mercaptoethanol in aqueous dioxane for 5–6 h at 80 °C (over 95% substitution) followed by tungstate-promoted oxidation by hydrogen peroxide, and then conventional phosphorylation up to the loading of 0.4 mol g^{-1}. A phosphate derivative can be detached by treatment with 30% triethylamine in dioxane for 2–3 h at ambient temperature [260].

A solid-phase equivalent of the Narang's safety catch thioether convertible into sulfoxide group [256a], 2-(4-carboxyphenylthio)ethanol (Camet), has been proposed by Felder et al. [262] for use with the phosphotriester chemistry on a polydimethylacrylamide support (Figure 19.10). The synthesis was conducted on

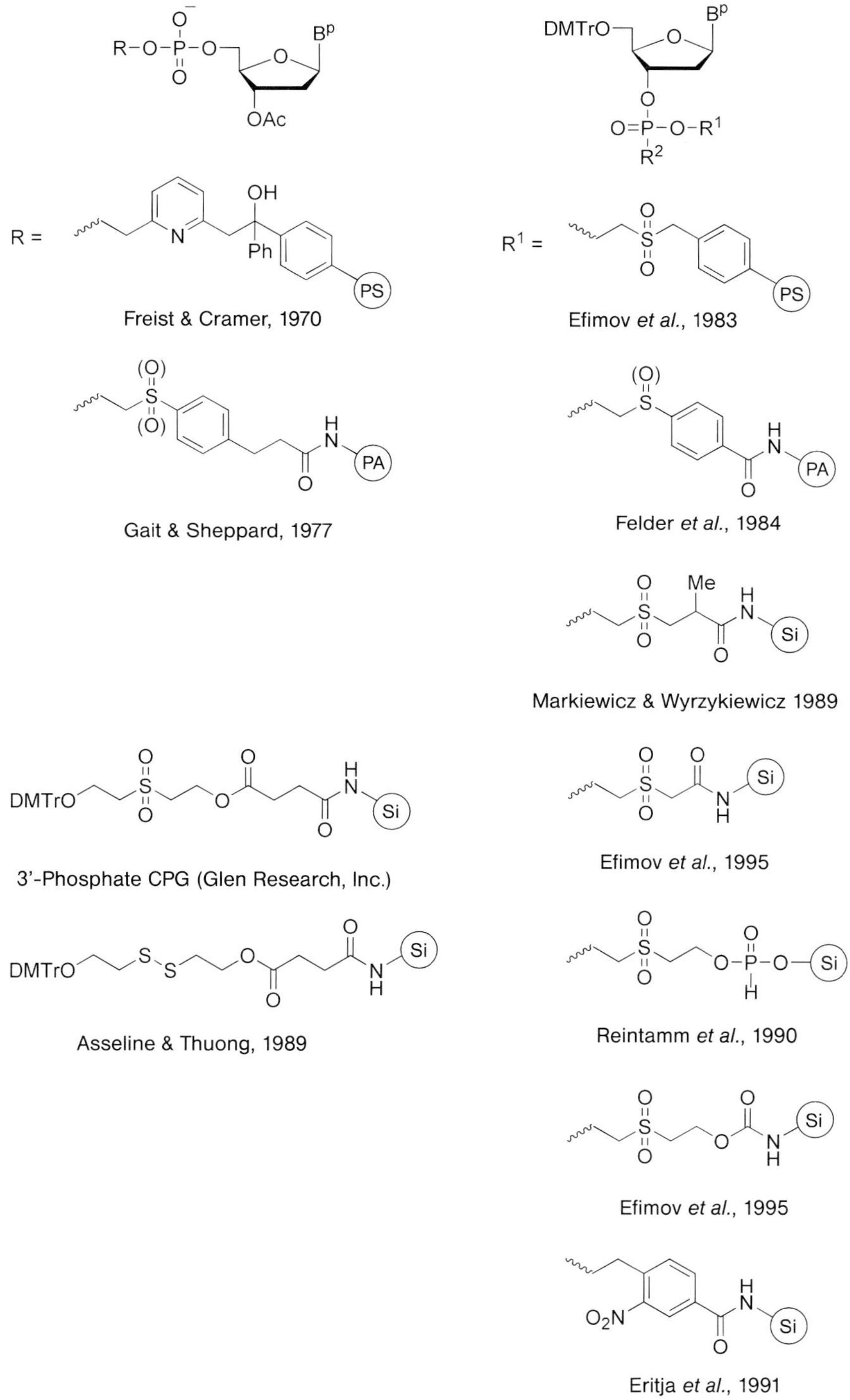

Figure 19.10 Base-labile linkers for the synthesis of oligonucleotide 3′- or 5′-phosphates or phosphorothioates.

a thioether-linked support followed by on-resin 4-chlorophenyl phosphate protecting group deprotection by the oximate reagent [37]; the linker was then transformed into the sulfoxide form by 0.4 M sodium metaperiodate in aqueous dioxane. The oxidized anchor was cleaved by 2 M sodium methoxide in methanol–pyridine (1:1 v/v) for 16 h, just like the patriarch base-labile linker [149]. The authors pointed out that exhaustive oxidation to sulfone is unnecessary because of the negative inductive effect of the 4-carboxamido group [263]. Later a sulfone linker analogous to that described by Gait and Sheppard [173, 174] was employed by Kamaike et al. in their liquid-phase phosphotriester synthesis on soluble cellulose acetate (Section 19.4).

Following the publication of a 5′ phosphoramidite chemical phosphorylation reagent based on β-sulfonylethyl group [257c], similar anchors have been developed for the phosphoramidite solid phase synthesis of 3′-phosphates on silica supports by Markiewicz and Wyrzykiewicz [251] and Efimov et al. [264] The H-phosphonate method has also been harnessed for the preparation of 3′-phosphorylated oligonucleotides via the β-sulfonylethyl linkers by Reintamm et al. [261] and Efimov et al. [264] In these cases [251, 261, 264] the β-eliminative cleavage of the anchor has been accomplished during the final ammonia deprotection. An LCAA-CPG support incorporating 2,2′-sulfonyldiethanol succinate is commercially available from Glen Research, Inc. (Figure 19.10). The latter is used most often for the synthesis of 3′-phosphorylated oligonucleotides. 3′-Thiophosphorylated oligomers can be potentially synthesized on the supports described by substituting sulfurization for oxidation during the first coupling step.

The β-eliminable NPE anchor (Figure 19.10) has also been proposed by Eritja et al. [233] for the synthesis of 3′-phosphates by the phosphoramidite method on LCAA-CPG support (Section 19.3.1.3).

19.3.2.3 Reduction Linkers Based on Disulfide Bond

The anchor based on 2,2′-dithiodiethanol (Figure 19.10) has been designed by Asseline and Thuong [263] for the automated phosphoramidite assembly of 3′-phosphorothioate oligonucleotides. It has been prepared by the reaction of the 4-nitrophenyl ester of mono-dimethoxytritylated 2,2′-dithiodiethanol succinate with aminopropyl silica gel Fractosil 500 (55 µmol g^{-1}). The oligonucleotides can be released by treatment with 0.1 M dithiothreitol (DTT) in concentrated aqueous ammonia (16 h at ambient temperature), which causes reduction of the disulfide bond and base-catalyzed elimination of thiirane from β-mercaptoethylphosphate. The 3′-phosphorothioate group can be used for further chemical modification, for example, by alkylation [265]. The support may also afford heterobifunctional oligonucleotides, for example, with 3′-phosphorothioate and 5′-aminoalkyl group and so on [266]. The same support, albeit synthesized in a slightly different way, has been used for the assembly of 3′-phosphates by Gupta et al. [267].

Kumar et al. [268] have described a similar support based on 3-mercaptopropyl CPG (55 µmol g^{-1}). It has been treated with 2,2′-dithiobis(5-nitropyridine) followed by the O-dimethoxytritylated mercaptoethanol to furnish the unsymmetrical disulfide linker suitable for the preparation of 3′-phosphorylated oligonucleotides with deprotection by 50 mM DTT in concentrated aqueous ammonia for 16 h at 55 °C.

19.3.2.4 Miscellaneous Phosphate Linkers

A palladium(0)-labile allyl linker based on 9-hydroxy-10-undecenoic acid for the preparation of 3′-phosphorylated DNA and RNA (Figure 19.11) on a high-loaded (170 μmol g^{-1}) TentaGel resin has been reported by Zhang and Jones [269]. The advantage of the anchor is that desilylation of RNA by TBAF or triethylamine trihydrofluoride can be performed on a solid phase, thus simplifying isolation of the final product. Analogously, on-resin ammonia treatment will destroy any depurinated DNA oligomers, thereby removing faulty sequences.

Guzaev and Lönnberg [221] have proposed an original diethyl 2,2-bis(hydroxymethyl)malonate anchor (Figure 19.11) on LCAA-CPG (up to 60 μmol g^{-1}) that is cleavable under basic conditions via a retroaldol reaction followed by β-elimination. Formation of the 3′-phosphate takes about 20 min in concentrated ammonia and progressively longer under milder conditions: 3 h in 0.05 M potassium carbonate in methanol and 10 h in 50% ethanolic ethylenedi-

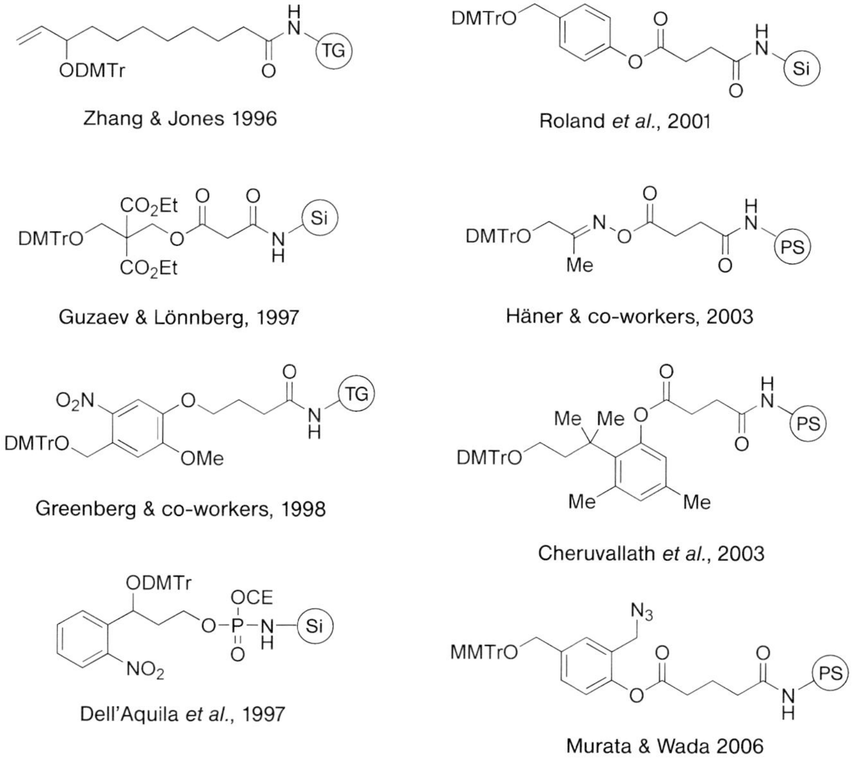

Figure 19.11 Various linkers for the synthesis of oligonucleotide 3′-phosphates and their analogues.

amine. The authors have demonstrated the applicability of their support for the synthesis of methyl phosphotriesters and methylphosphonate analogues.

Greenberg and coworkers [270] have used their 2-nitroveratryl photolinker (Section 19.3.1.3) attached to either LCAA-CPG or TentaGel (Figure 19.11) to synthesize 3′-phosphorylated oligonucleotides via the phosphoramidite chemistry and cleavage by photolysis. The oligomers assembled on TentaGel can be deprotected by ammonia treatment on solid phase and were shown to hybridize with their complementary sequences with high fidelity. The duplexes formed can be then detached from support by photolysis. A similar photolabile anchor based on 1-(2-nitrophenyl)-1,3-propanediol linked to LCAA-CPG by the ammonia-stable phosphoramidate bond (Figure 19.11) was described by Dell'Aquila et al. [271] and used for the preparation of unmodified oligodeoxyribonucleotide 3′-phosphates as well as their methyl phosphotriester analogues.

A simple ammonia-labile linker (Figure 19.11) has been prepared by Roland et al. [272] from 4-hydroxybenzyl alcohol, dimethoxytritylated at the aliphatic hydroxyl and attached through the phenol to the succinylaminopropyl CPG (60–90 µmol per g of DMTr). After the oligonucleotide assembly, the 3′-phosphate or phosphorothioate is produced after succinate hydrolysis followed by the elimination of quinone methide. The support has been used to make a library of 3′-thiophosphorylated dinucleotides.

An α-hydroxyacetone oxime anchor (Figure 19.11) has been developed by Häner and coworkers [273] for the phosphoramidite RNA synthesis. The 2′-protected oligoribonucleotides are detached from the respective macroporous polystyrene support (initial loading 30 µmol g^{-1}) as the 3′-phosphates during 30 min of concentrated aqueous ammonia–ethanolic methylamine (1:1 v/v) treatment at 65 °C. The authors have postulated β-elimination with the formation of 2-nitrosopropene as a plausible mechanism for 3′-phosphate liberation.

Cheruvallath et al. [274] employed a conformationally locked 3,5-dimethyl-2-(1,1-dimethyl-3-hydroxypropyl)phenyl succinate linker (Figure 19.11) on highly crosslinked PS (Pharmacia HL30 Primer Support, 92 µmol g^{-1}) for the preparation of oligonucleotide 3′-phosphorothioate monoesters by the phosphoramidite chemistry. Cleavage is thought to proceed via intramolecular alkylation of the phenol group liberated during standard ammonolysis (55 °C, 16 h) by the corresponding alkyl phosphorothioate diester.

Recently, Murata and Wada [275] have designed a phosphate linker based on 3-azidomethyl-4-hydroxybenzyl alcohol (Figure 19.11), which is cleavable by reductive elimination under neutral conditions, and applied it to the synthesis of methyl phosphotriester analogues by the modified H-phosphonate method [143] on macroporous polystyrene (24 µmol g^{-1} loading). After completion of the synthesis, the azido group is reduced to amino by treatment with 0.2 M methyldiphenylphosphine in aqueous dioxane, which triggers intramolecular O→N acyl transfer, fragmentation of the linker and expulsion of the 3′-phosphate. Notably, the authors added mercaptoethanol to trap the quinone methide formed, and were compelled to use glutaric acid ester to attach the linker to the resin because succinylation of the phenol was reversible.

19.3.3
Linkers for the Synthesis of Oligonucleotides 3′-Functionalized with Other Chemical Groups

19.3.3.1 Amino Group

Many approaches have been described for the preparation of oligonucleotides modified with extra amino groups due to the importance of those as the attachment points for other chemical groups, for example, fluorescent dyes, intercalators, biotin and so on [9b, 11, 12b,d] and as bearers of positive charge at physiological pH [276]. Most of the linkers designed for the introduction of amino groups are bifunctional [277] and some may be quite sophisticated [278]. Here we limit the scope to only monofunctional linkers for the modification of the oligonucleotide 3′-end with aminoalkyl groups, for example, via the 3′-phosphodiester linkage.

A very simple and robust linker for the synthesis of 3′-aminohexyl-tailed oligonucleotides based on trimellitic acid (Figure 19.12) has been reported by Petrie et al. [279] 6-Amino-1-hexanol was treated with trimellitic anhydride to form

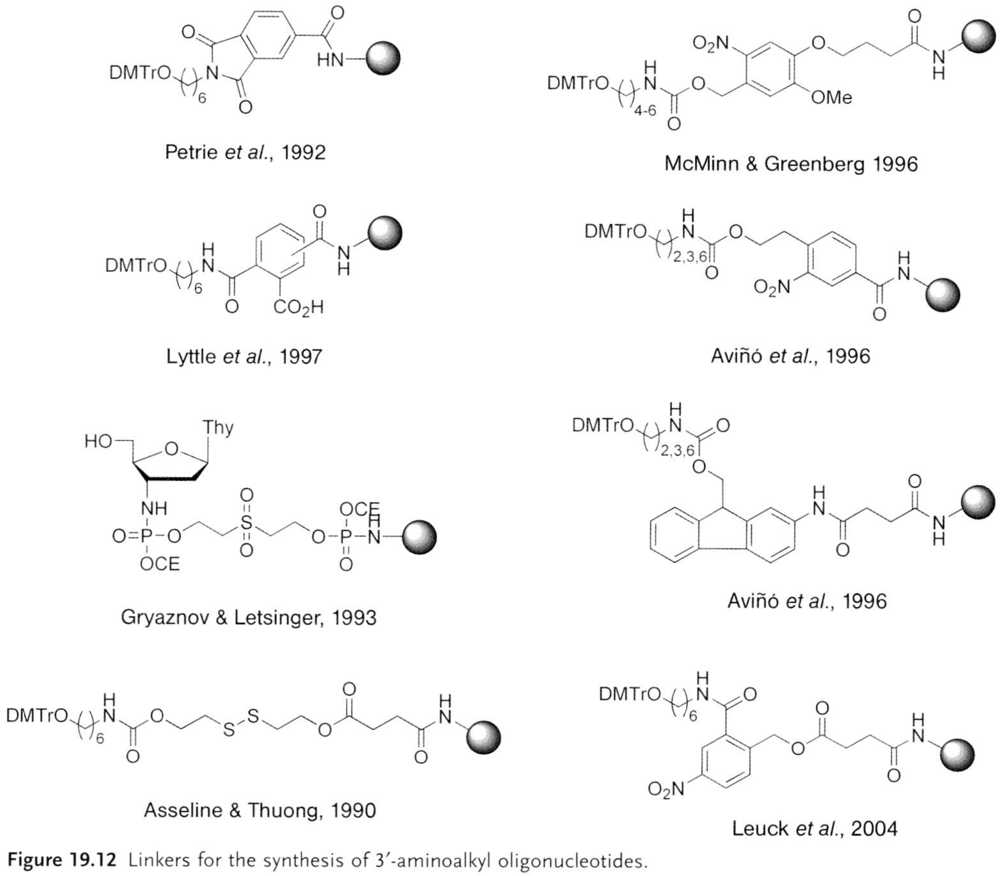

Figure 19.12 Linkers for the synthesis of 3′-aminoalkyl oligonucleotides.

phthalimide, which was in turn dimethoxytritylated and attached to LCAA-CPG via the free carboxyl group (20 μmol g^{-1}). The phthalimide opens up and releases free 3′-amino oligonucleotide during standard ammonia treatment. The authors have found their support superior to the commercially available 3-Fmoc-amino-1,2-propanediol support [280]. Subsequently, Lyttle et al. [281] have added 3′-hydroxy- and 3′-mercaptoalkyl modifications to the repertoire of applicability of the trimellitic anchor (Section 19.3.3.2). A disadvantage of the solid-supported phthalimide is that it requires relatively drastic conditions for the amino group release (80–90% detachment after 17 h at 55 °C) [282].

Gryaznov and Letsinger [283] have used the known 2,2′-sulfonyldiethanol phosphoramidite [257c] for direct attachment to LCAA-CPG support (38 μmol g^{-1} DMTr) followed by selective monophosphitylation with 2-cyanoethyl-N,N,N′,N′-tetraisopropyl phosphordiamidite [284], subsequent tetrazole-catalyzed hydrolysis to the H-phosphonate diester and its oxidative amination [285] with 3′-amino-3′-deoxythymidine (Figure 19.12). The linker obtained was subjected to solid-phase oligonucleotide assembly by the phosphoramidite method, which afforded free 3′-amino-terminated oligomer after ammonolysis due to the apparent instability of the corresponding N-phosphorylated derivative. This anchor is an example of a combination of linkers. Later, a similar combination of a β-sulfonylethyl linker [251] and the phosphoramidate group was exploited by Kumar et al. [286] for the preparation of 3′-aminopentyl oligonucleotides.

Most of the linkers described for the synthesis of 3′-aminoalkyl oligonucleotides employ the carbamate (urethane) linkage to the polymer. A 2,2′-dithiodiethanol-derivatised silica gel support (Section 19.3.2.3) has been transformed by Asseline and Thuong [287] into the N-ω-hydroxyhexylcarbamate (Figure 19.12) via the consecutive treatments with 1,1′-carbonyldiimidazole and 6-amino-1-hexanol followed by the phosphoramidite synthesis of a DNA undecamer. Deprotection by 0.1 M DTT in concentrated aqueous ammonia unmasked the β-mercaptoethyl carbamate, which fragments under basic conditions with the elimination of thiirane to furnish 3′-aminohexyl oligodeoxyribonucleotide. The structure of the latter has been confirmed by the reaction with acridine-9-isothiocyanate.

Aviñó et al. [288] have adapted the NPEOC and Fmoc anchors for the attachment of various O-protected α,ω-aminoalcohols as well as 3′-amino-3′-deoxynucleosides to LCAA-CPG (Figure 19.12). The supports obtained were derived from the known linkers (Section 19.3.1.3 and Section 19.3.2.2) by solid-phase transformation of the hydroxyl linker into the carbamate by the reaction with 1,1′-carbonyldiimidazole followed by aminoalcohol treatment. The authors have found that both anchors, NPEOC and Fmoc, can be cleaved by standard ammonolysis without the need for DBU, but the former gives a cleaner product.

McMinn and Greenberg [289] have modified a photolabile 2-nitroveratryl linker (Section 19.3.1.3) on LCAA-CPG support (49–54 μmol g^{-1}) to suit the 3′-aminoalkyl oligonucleotide synthesis by converting it into the corresponding carbamate with a range of α,ω-aminoalcohols (Figure 19.12). Photolysis at 400 nm for 3 h produced 70–98% yields of a 20-mer while 30–40-mers gave reduced yields. The level of thymidine photodimer formation was estimated to be less than 3%. The carbamate anchor was used to obtain the N-linked 3′-conjugates from fully protected oligonucleotide fragments in solution [290], 3′,5′-bis-conjugates [291] and bis-oligonucleotides ligated internally through the amide bond [292].

Recently, Leuck et al. [282] have described an acyl-type anchor for the fast and efficient preparation of 3′-aminoalkyl oligonucleotides based on the 2-hydroxymethyl-6-nitrobenzoyl group (Figure 19.12). The linker is prepared from 6-nitrophthalide by acylation of the respective aminoalcohol and attachment to LCAA-CPG through the benzylic hydroxyl via the succinyl group. Cleavage is achieved under conditions of basic hydrolysis by the intramolecular attack of the liberated hydroxyl onto the amide carbonyl with the expulsion of the amine. Complete detachment of the oligomer requires 2h at 55 °C in concentrated ammonia or 30 min at 65 °C in 40% aqueous methylamine–concentrated aqueous ammonia (1:1 v/v), or can be accomplished at room temperature within 24 h in concentrated ammonia. The new anchor was shown to be superior to the phthalimido support [279] in terms of product yield and deprotection time.

19.3.3.2 Thiol Group

Oligonucleotides bearing a thiol group represent an important class of chemically functionalized nucleic acids due to the possibility of their alkylation, for example, by maleimides, and ease of oxidation with the formation of a disulfide bond, which furnishes a good linkage for attachment of various other molecules. The main challenge here is to minimize the concurrent oxidation of the free thiol group. This is why the synthesis of thiol oligonucleotides has always been the center of attention for many nucleic acid chemists. This chapter highlights the 3′-thiol modification involving solid supports.

Gupta et al. [293] have prepared a disulfide-linked support (Figure 19.13) by treatment of aminopropyl CPG with an excess of 3,3′-dithiobis(N-succinimidylpropionate) followed by O-dimethoxytritylated aminopentanol (25 µmol g^{-1}). The disulfide can be reduced by 50 mM DTT, pH 8.5 after 6 h at 40 °C. The 3′-mercaptoalkyl oligodeoxyribonucleotides synthesized on this support were detached after exposing it to 50 mM DTT in concentrated ammonia for 16 h at 55 °C. The presence of a thiol group has been confirmed by the reaction with the thiol-specific dye N-(iodoacetamidoethyl)-1-naphthylamine-5-sulfonic acid (1,5-I-AEDANS) and successful immobilization on the 2,2′-dithiobis(5-nitropyridine)-activated mercaptopropyl CPG.

Later, the same group [268] reported an improved version of a disulfide-linked CPG (Figure 19.13) that permitted cleavage of the intact disulfide-containing oligonucleotide during standard ammonia deprotection. Their supports (30 µmol g^{-1}) have been prepared by esterification of succinylaminopropyl CPG with 2,2′-dithiobis(5-nitropyridine)-activated mercaptoethanol followed by disulfide exchange with O-dimethoxytritylated α,ω-mercaptoalcohol (n = 2, 3 or 6). The support with n = 2 was identical to the one used by Asseline and Thuong [287] and furnished 3′-phosphate oligomers after DTT reduction under basic conditions, while the others upon ammonolysis afforded 3′-disulfide protected oligonucleotide

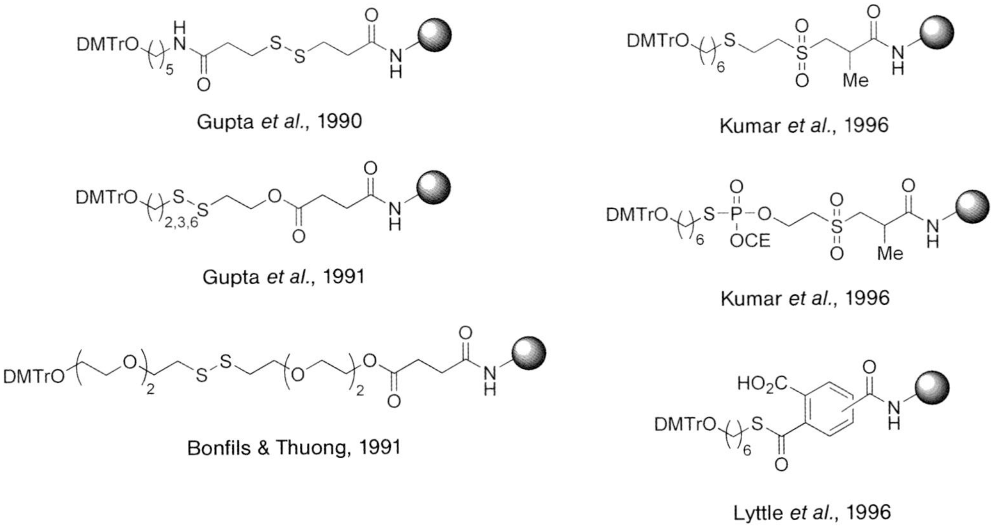

Figure 19.13 Linkers for the synthesis of 3′-mercaptoalkyl oligonucleotides.

precursors that can be isolated without special precautions and converted into the reactive 3′-mercaptoalkyl derivatives by separate reduction step.

A more hydrophilic version of a disulfide support has been developed by Bonfils and Thuong [294] who synthesized a polymer with a ω,ω′-dithiodi-{2-[2-(2-ethoxy)ethoxy]ethyl} linker (Figure 19.13). The support performed similarly to those described above [267, 268, 293], and the heterobifunctional oligonucleotides obtained have been used for the preparation of peptide–oligonucleotide conjugates [296].

Non-disulfide linkers for the synthesis of 3′-mercaptohexyl oligodeoxyribonucleotides have also been published. Kumar et al. [286] has designed a β-elimination-type linker (Figure 19.13). A β-sulfonylethanol support [251] has been sulfonylated with tresyl chloride $CF_3CH_2SO_2Cl$ followed by displacement of the tresylate with O-dimethoxytritylated 6-mercapto-1-hexanol. Alternatively, the polymer [251] after careful phosphitylation with 2-cyanoethyl-N,N,N',N'-tetraisopropyl phosphordiamidite followed by tetrazole-catalyzed hydrolysis has been converted into the H-phosphonate diester [286], which in turn was transformed into the corresponding S-alkyl phosphorothiolate by the reaction with carbon tetrachloride and O-dimethoxytritylated 6-mercapto-1-hexanol. The loadings of 24–25 µmol g^{-1} were obtained, and the 3′-mercaptohexyl oligonucleotides were isolated after the standard ammonolytic deprotection (55 °C, 16 h). The authors have noted that addition of some dithiothreitol (DTT) prevents Michael addition of acrylonitrile formed from the phosphate 2-cyanoethyl groups to the thiol group.

Lyttle et al. [281] used trimellitic anhydride supported on aminopropyl CPG to immobilize the ubiquitous O-dimethoxytritylated 6-mercapto-1-hexanol via the thioester linkage cleavable in ammonia (Figure 19.13). The linkage survived all the conditions of the phosphoramidite synthesis, but it also has been necessary to add some DTT to concentrated aqueous ammonia to prevent spontaneous oxidation of 3′-mercaptohexyl oligonucleotides.

19.3.3.3 Carboxyl Group

Carboxylated oligonucleotides are interesting as reagents for chemoselective conjugation with other molecules with the formation of the amide bond either on solid phase or in solution [295]. Hovinen et al. [297] have described a thymidine 3′-formacetal linker (Figure 19.14) with an internal ester group attached through the succinyl anchor to LCAA-CPG (7–9 µmol g^{-1}). The authors have demonstrated a successful preparation of 3′-carboxyalkyl oligonucleotides via 0.1 M sodium hydroxide followed by concentrated aqueous ammonia deprotection, and 3′-aminoalkyl oligomers via 0.5 M hydrazine hydrate in pyridine–acetic acid (4:1 v/v) followed by 1,3-diaminopropane in ethanol (1:1 v/v).

However, the preparation of the above linker was tedious and multistep, the loading rather low, and so, subsequently, Lönnberg and coworkers [298] came up with a simpler anchor (Figure 19.14) built from two molecules of γ-hydroxybutyric acid linked by an ester linkage and connected to the polymer by an amide bond (22 µmol g^{-1}). The support proved to be versatile and allowed for the synthesis of oligonucleotides modified at the 3′-end with carboxyl, carboxamido, amino and thiol groups. In a separate study [299] the authors have compared several related linkers, including γ-hydroxybutyryloxypropyl succinyl and bis-glycolyl (Figure 19.14), trying to strike the optimal balance between reactivity towards amines, susceptibility to hydrolysis in aqueous amine solution and minimization of cytosine transamination. They have found that the bis-glycolyl anchor on LCAA-CPG and aminomethyl PS is the most reactive (90% cleavage in 2.5 h) compared to the

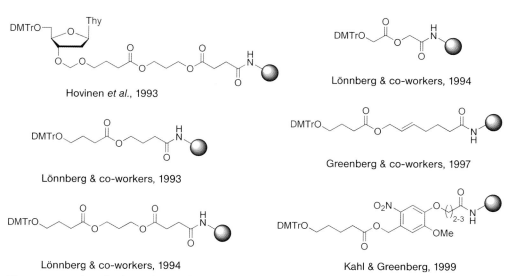

Figure 19.14 Linkers for the synthesis of 3′-carboxyalkyl oligonucleotides.

γ-hydroxybutyryl (8–30 h), the competing hydrolysis slows down when concentration of the diamine is decreased from 50% to 5% v/v, and N4-benzoylcytosine is relatively resistant to transamination in aqueous solution. Presumably, nucleoside derivatives with labile N-protecting groups would be very useful in this case.

Greenberg and coworkers [300] have utilized their allyl linker (Section 19.3.1.3) to prepare 3′-γ-carboxypropyl oligonucleotides after palladium(0)-mediated cleavage and have optimized deprotection conditions (Figure 19.14). Kahl and Greenberg [301] have also exploited photolabile supports (Section 19.3.1.3 and Section 19.3.3.1) for the same purpose and obtained good yields of fully protected 3′-δ-carboxybutyl oligonucleotide fragments, which then have been used for solution-phase conjugation with amines and peptides [301], and for template-free ligation with fully protected 5′-ζ-aminohexyl oligonucleotides obtained from 3′-carbonate photolabile support [292] (Section 19.3.1.3).

19.3.4
Speciality Linkers

An interesting acid-labile linker has been designed for a non-standard RNA synthesis with 5′-Fmoc protection [303] and methyl phosphoramidite chemistry by Palom *et al.* [304]. The authors prepared an acetal anchor for the ribonucleoside 2′,3′-diol group (Figure 19.15) by the reaction of 5′-Fmoc-uridine with 4-formylphenoxyacetic acid, and attached it to LCAA-CPG (30–40 µmol g^{-1}). The nucleoside can be detached from this support in ca. 70% yield by overnight treatment with dilute aqueous HCl, pH 2, conditions now believed to be too drastic for RNA.

Another rare anchoring group was introduced by Pedroso and coworkers [302] for the preparation of cyclic oligonucleotides by the phosphoramidite synthesis and phosphotriester-type cyclization on solid phase. 3-Chloro-4-hydroxyphenylacetic acid 2,4,5-trichlorophenyl ester was phosphitylated with a 2-cyanoethyl phosphoramidite and, after iodine oxidation, used to attach the first nucleoside unit to aminomethyl polystyrene resin or TentaGel (Figure 19.15). The synthesis continued with methyl phosphoramidites until the last coupling, after which the oligomer was detritylated, decyanoethylated (triethylamine–pyridine 1 : 1 v/v) and the

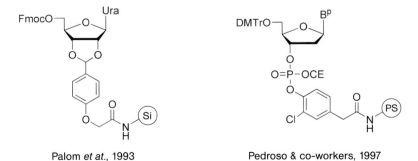

Palom *et at.*, 1993   Pedroso & co-workers, 1997

Figure 19.15 Speciality linkers for unconventional oligonucleotide syntheses.

sole 3′-terminal phosphodiester was then subjected to MSNT-promoted on-resin cyclization with the free 5′-hydroxyl. After completion of the circularization step, the phosphate methyl groups were removed by usual thiophenol treatment, and the cyclic oligonucleotide detached from resin by 0.1 M tetramethylguanidinium *syn*-pyridine-2-aldoximate in aqueous dioxane for 8-16 h at ambient temperature followed by final ammonolysis. Notably, although the yields are variable and dependent on the cycle size (up to 50% for the smallest cycles and around 10% or less for the largest cycles), the purity of the circular product was exceptionally high (>90% by HPLC). This is because only the cyclic product was cleaved from the resin by oximate treatment at its sole phosphotriester residue while the unreacted linear oligomer having only phosphodiester groups remained immobilized.

19.4
Oligonucleotide Synthesis on Soluble Polymer Supports

Very soon after Merrifield introduced the solid phase synthesis concept another polymer-supported technique was conceived [305], namely liquid phase synthesis on a soluble polymer support [306]. The inventors [305] argued that the new method may offer an advantage over the "classical" solid phase synthesis in terms of potentially better reaction kinetics due to a higher diffusion rate in homogeneous solution than inside the solvated gel bead. Immediately after its conception, the liquid phase polymer-supported synthesis was applied for the preparation of oligonucleotides, by the then dominant phosphodiester method, independently by Cramer *et al.* [308] and Hayatsu and Khorana [307]. These groups [307a, 308] demonstrated the successful TpTpT synthesis on a soluble atactic polystyrene with the 5′-thymidine residue joined to the polymer by the methoxytrityl linker (Figure 19.16); this was followed by more dinucleotide syntheses, including A, C and G, in quite respectable stepwise yields (88–96%) [310b], which indeed compared favorably with the contemporary solid-phase syntheses. However, the main handicap was again associated with the phosphodiester method *per se*: the accumulation of polar negatively charged phosphate groups on a hydrophobic polystyrene led to the oligonucleotidyl-polymer acquiring surfactant properties and the associated difficulty of precipitation of the above in solid form out of aqueous solution (V.K. Potapov, personal communication) [307b]. Therefore, the search continued for more hydrophilic soluble polymers that could be precipitated from the non-polar solvents, as well as for the new chemistries of internucleotide bond formation.

The answer to the former problem came once again from peptide synthesis, when Bayer and coworkers [309] reported the first use of poly(ethylene glycol) (PEG) as a soluble support for peptide synthesis. The idea was immediately picked up by Köster, who prepared a PEG-20000 terminating in a dimethoxytrityl linker (Figure 19.16) and demonstrated the attachment of thymidine [310]. Later, Bayer and coworkers [311] assembled up to a pentanucleotide by the phosphodiester method on a PEG-NH$_2$ support with a phosphoramidate linker cleavable by isoamyl

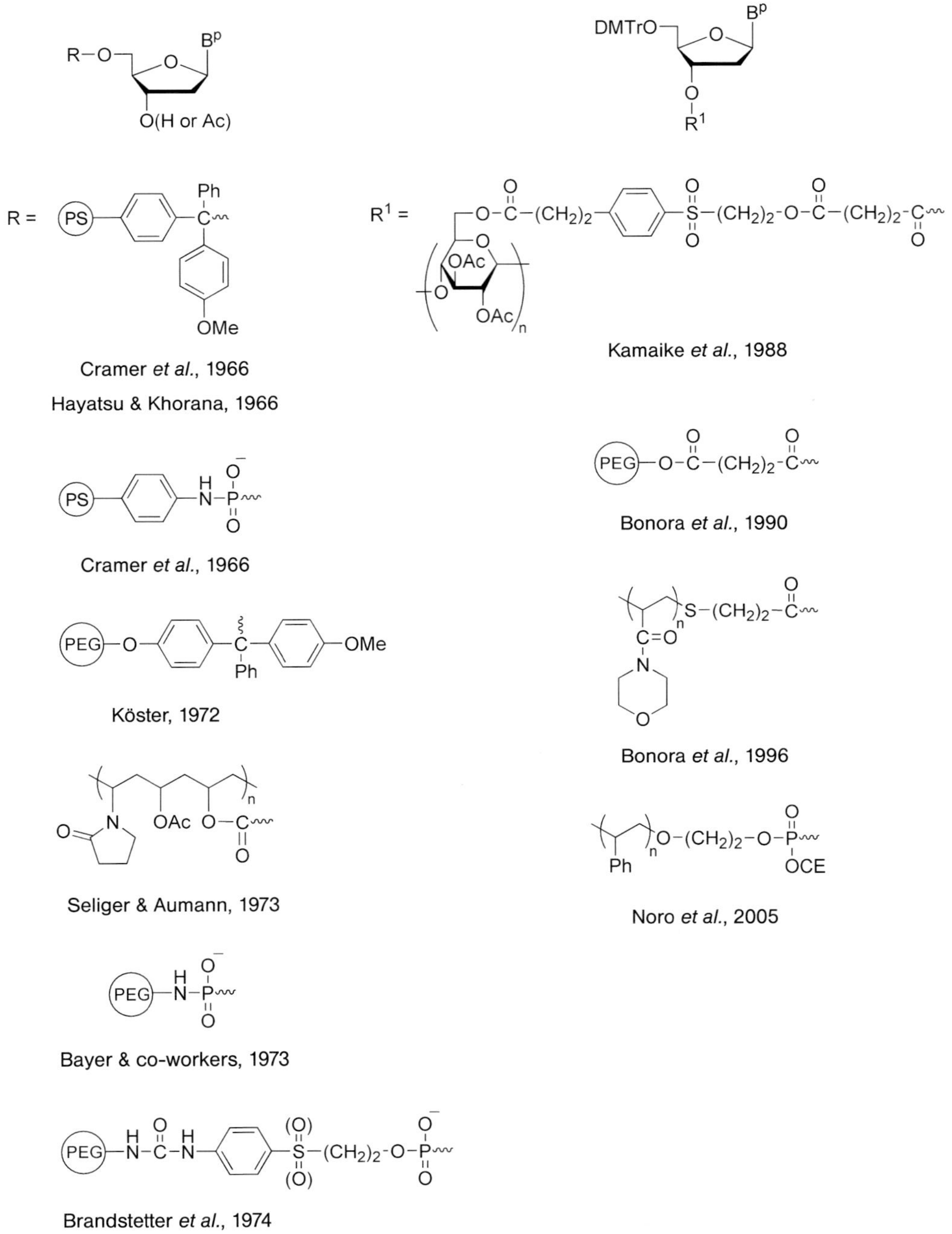

Figure 19.16 Supports and linkers for the soluble polymer supported oligonucleotide synthesis.

nitrite in acetic acid–pyridine mixture (Section 19.3.2.1). Later, Brandstetter *et al.* [258] described the attachment of thymidine 5′-phosphate to a bifunctional PEG equipped with a "safety-catch" linker.

The next candidate polymer explored was poly(vinyl alcohol) (MW ca. 70000), which is soluble in the mixture of hexamethylphosphortriamide and pyridine [312]. It was shown that the support can be loaded with up to $10\,\text{mmol}\,\text{g}^{-1}$ of mononucleotide (T, U or rG) with >90% efficiency and, apart from precipitation from ethanol, may be dialyzed or even chromatographed on a Sephadex G-50 column. No linker attachment or oligonucleotide synthesis data were reported, though. Seliger and Aumann [313] described a polyvinylpyrrolidone and poly(vinyl acetate) bulk copolymer (MW ca. 42000), partially hydrolyzed to the alcohol content of $1\,\text{mmol}\,\text{g}^{-1}$, to which a 3′-protected thymidine was attached through the alkali-labile 5′-carbonate linkage. No other useful synthetic data were obtained at the time, although a similar insoluble polymer (Merckogel) was used later in the large-scale automated solid phase synthesis of oligonucleotides [196].

The true potential of PEG and other soluble polymer-supported liquid-phase oligonucleotide synthesis was uncovered only when the more advanced chemistries, first phosphotriester and then phosphoramidite and H-phosphonate, became available. First, Kamaike *et al.* [314] described a phosphotriester synthesis of DNA and RNA oligomers supported by a cellulose acetate polymer soluble in the condensation solvent pyridine but easily precipitated by ethanol afterwards. The linker employed was a 3′-linked β-eliminable 2-arenesulfonylethyl succinate (Figure 19.16). Then, in the 1990s Bonora *et al.* [315] published the first article in a series on the development of an optimized liquid phase synthesis of oligonucleotides supported by a PEG-5000 monomethyl ether, using a rapid phosphotriester method at the beginning as the most adapted for the large-scale assembly. The authors described the advantages of the method as follows: (i) the synthesis under homogeneous conditions requires lower excess of expensive monomers; (ii) large amounts of oligonucleotides can be obtained from a single synthetic run; and (iii) the efficiency is easily monitored by non-destructive, for example, spectrophotometric, techniques.

Later, a switch was made to the phosphoramidite chemistry as the most high-yielding and rapid [316]. The H-phosphonate method was also shown to be adaptable for PEG-supported variant, albeit with inferior results [317]. Since then the phosphoramidite approach and PEG support have been used for the preparation of phosphorothioate antisense oligonucleotides [318], antisense oligonucleotide-PEG conjugates [319] with various molecular mass poly(ethylene glycol)s [319c], including branched PEGs [319b,d] triplex-forming oligonucleotide-PEG conjugates [320] with anti-gene properties [320b], and peptide–oligonucleotide-PEG conjugates [321].

Finally, the same Italian group studied monocarboxy-functionalized poly-*N*-acryloylmorpholine (PAcM) (MW 6000) as a new soluble polymer support for liquid-phase oligonucleotide synthesis [322] and a possible alternative to PEG, although its advantages over the latter were not immediately apparent. Preparation and antisense properties of oligonucleotide-polyacryloylmorpholine conjugates were also described [323].

Recently, Noro et al. [324] have investigated liquid phase synthesis by the phosphoramidite scheme on linear polystyrene terminated with a hydroxyl group, which was obtained by living anionic polymerization, and have prepared an oligonucleotide–PS conjugates up to a pentanucleotide level with high efficiency, the oligonucleotide being linked to the polymer by a phosphotriester bond. The authors studied the properties of their conjugates by transmission electron microscopy and small-angle X-ray scattering, confirming their cylindrical structure.

19.5
Oligonucleotide Synthesis with Polymer-Supported Reagents

Chemical synthesis of oligonucleotides prompted the application of polymer-supported reagents (Figure 19.17) very early in its history. The first approach described by Rubinstein and Patchornik [325] was developed during the reign of the Khorana's method. The Israeli researchers synthesized a polymer-supported hindered sulfonyl chloride and used it for the preparation of dinucleotides by the phosphodiester method, solving one of the problems of the latter – contamination of the product with the sulfonic acid and the sulfonated nucleoside by-products, which in the case of the polymer-assisted condensation were removed by filtration. The coupling agent was shown to be equally applicable for the phosphotriester synthesis as well. However, this example has not been followed up by others until the last decade.

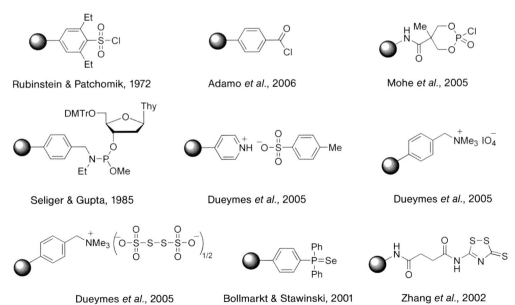

Figure 19.17 Polymer-supported reagents used for oligonucleotide synthesis.

The methodology has been resurrected very recently in the H-phosphonate variant. An attempt by Adamo *et al.* [326] to use a chlorosulfonated polystyrene resin for activation of the nucleoside H-phosphonate monoesters was not particularly impressive, but the corresponding chlorocarbonylated polystyrene performed well, permitting the synthesis of up to a hexanucleotide. Similar results were reported by Mohe *et al.* [327] for the dinucleoside H-phosphonate synthesis promoted by a phosphorochloridate supported on polystyrene.

The use of solid-supported reagents in the phosphoramidite synthesis was pioneered in the 1980s by Seliger and Gupta [328] who prepared an immobilized version of a methoxy phosphoramidite by the reaction of the corresponding 3′-phosphorochloridite with either N-ethylaminomethyl polystyrene (2% DVB) prepared from the chloromethyl resin or silica gel modified with piperazino groups. The viability of the concept has been demonstrated in both manual and machine-assisted solid phase synthesis with a starting nucleoside bound to a standard silica gel support and chain elongation with the thymidine phosphoramidite resin and tetrazole activator. A pentanucleotide was obtained with 78% yield (>95% stepwise yield). The authors noted a remarkable stability of the polymer-supported phosphoramidite (<10% loss of activity after 15 months storage).

Recently, Dueymes *et al.* [329] attempted to carry out a solution phase phosphoramidite synthesis of regular and phosphorothioate oligonucleotides assisted exclusively by polymer-supported reagents. The authors were faced with difficulties at every stage of the assembly, and have been able to demonstrate the preparation of the di- and trinucleotide blocks with activation, oxidation, sulfurization and even removal of the 3′-levulinoyl group performed by the respective reagents. A polymeric activator, polyvinylpyridinium *p*-toluenesulfonate, has been used also for the regioselective phosphitylation of nucleosides [330]. The analogous polyvinylpyridine hydrochloride performed poorly in the nucleoside 3′-phosphitylation reaction [331]. The oxidation was performed using a supported version of a tetraalkylammonium periodate [107], the sulfurization was achieved by the Amberlyst A-26 immobilized tetrathionate, as prompted by a previous observation by Efimov [118]. Finally, the 3′-delevulinoylation was accomplished by hydrazine bound to a sulfonated polystyrene resin. Notably, contrary to the previous successful application of sulfonic acid resins (e.g., the use of Dowex for the 5′-DMTr group removal from oligonucleotides [332] or both 1% DVB or macroporous sulfonated polystyrene for the detritylation of 3′-silylated deoxyribonucleosides [333]), the use of polymer-supported sulfonic acid for the 5′-DMTr ether cleavage in stepwise synthesis failed, and the authors [329] have reverted to the conventional benzenesulfonic acid treatment followed by scavenging of the excess acid with polyvinylpyridine. All the above reactions required considerably more time (hours) for completion than the respective operations during the standard solid phase synthesis cycle (tens of seconds to minutes).

Two other polymer-supported reagents have been described for the conversion of internucleoside phosphite triesters into the corresponding P(V) analogues. Triphenylphosphine selenide on polystyrene (Figure 19.17), obtained by KSeCN treatment of a commercially available polystyryl diphenylphoshine resin, has been employed by Bollmark and Stawinski [334] to effect rapid (ca. 20 min) selenization of the dinucleoside methylphosphites. The corresponding H-phosphonate diesters required prior trimethylsilylation for the efficient selenium transfer. Another reagent was designed by Zhang *et al.* [335] for the sulfurization of the nucleoside phosphite triesters by attachment of 3-amino-1,2,4-dithiazoline-5-thione (ADTT) [122b] via its amino group onto the succinylated Toyopearl HW-55F. The reagent performed well on a model nucleoside phosphite triester (30 min reaction, ratio phosphorothioate/phosphodiester >99:1). However, its applicability for the synthesis of longer oligonucleoside phosphorothioates remains to be elucidated.

19.6
Conclusions

To conclude, it is evident that applications for, modifications of and improvements in, polymer-supported oligonucleotide synthesis are still forthcoming despite the level of perfection achieved already [2b]. New supports, reagents and techniques are constantly emerging out of universities and industrial companies involved in nucleic acid chemistry and chemical biology. New methods for the phosphodiester bond formation on solid phase [336] are being put forward, challenging the almost undisputable superiority of the classical phosphoramidite method [8]. From the latest advances one might cite Reese and coworkers [337] who have recently transformed the solution-phase H-phosphonate method into the one of choice for practical large-scale synthesis of short and medium size phosphorothioate analogues for possible therapeutic purposes [337b], or Ferreira *et al.* [338] who anticipated the merger of two powerful chemistries for solid-phase oligonucleotide synthesis, phosphoramidite and H-phosphonate. Further advances in the preparation of RNA and its analogues are rapidly being made fueled by the discovery of RNA interference and the desire to see the nucleic acid drugs finally enter clinic [10g–i]. Another vigorously evolving area is DNA nanotechnology [339] the high-ranking promises of which include such hot topics as nucleic acid-based nanomaterials, nanocomputing and nanorobotics [340]. One should expect more and more publications coming out in this exciting field of chemical sciences, the chemistry of the nucleic acids.

List of Abbreviations

AcOH	acetic acid
ADTT	3-amino-1,2,4-dithiazoline-5-thione
1,5-I-AEDANS	*N*-(iodoacetaminoethyl)-1-naphthylamine-5-sulfonic acid
AIBN	2,2′-azobis(2-methylpropionitrile)

AMA	40% aqueous methylamine–concentrated aqueous ammonia (1 : 1 v/v)
BSA	N,O-bis(trimethylsilyl)acetamide
BTT	5-benzylthiotetrazole
CE	β-cyanoethyl
CPG	controlled pore glass
dba	dibenzylideneacetone
DBU	1,8-diazabicyclo[5.4.0]undec-7-ene
DCC	N,N'-dicyclohexylcarbodiimide
DCI	4,5-dicyanoimidazole
DDD	diethyldithiocarbonate disulfide
DIPHOS	1,2-bis(diphenylphoshino)ethane
DMF	N,N-dimethylformamide
DMTr	4,4'-dimethoxytrityl
DTT	dithiothreitol
DVB	divinylbenzene
EDITH	3-ethoxy-1,2,4-dithiazoline-5-one
ETT	5-ethylthiotetrazole
Fmoc	9-fluorenylmethoxycarbonyl
HPLC	high performance liquid chromatography
LCAA-CPG	long-chain alkyl amino controlled pore glass
MMTr	4-methoxytrityl
MS-Cl	mesitylenesulfonyl chloride
MSNT	1-mesitylenesulfonyl-3-nitro-1,2,4-triazole
NBS	N-bromosuccinimide
NMR	nuclear magnetic resonance
NPE	2-(4-nitrophenyl)ethyl
NPEOC	2-(4-nitrophenyl)ethoxycarbonyl
NPT	5-nitrophenyltetrazole
PA	polyamide
PAcM	polyacryloylmorpholine
PADS	bis(phenylacetyl) disulfide
PCR	polymerase chain reaction
PEG	poly(ethylene glycol)
PEG-PS	poly(ethylene glycol) grafted polystyrene
PS	polystyrene
PTFE	poly(tetrafluoroethylene)
snRNA	small nuclear ribonucleic acid
TAMRA	tetramethylrhodamine
TBAF	tetrabutylammonium fluoride
TBHP	*tert*-butyl hydroperoxide
TETD	tetraethylthiuram disulfide
THF	tetrahydrofuran
TPS-Cl	2,4,6-triisopropylbenzenesulfonyl chloride

References

1 See, for example, the two very personal accounts of one of the experts in the field: (a) Reese, C.B. (2002) *Tetrahedron*, **58**, 8893. (b) Reese, C.B. (2005) *Organic & Biomolecular Chemistry*, **3**, 3851.
2 (a) Katzhendler, J., Cohen, S., Rahamim, E., et al. (1989) *Tetrahedron*, **45**, 2777. For a more recent discussion see (b) Kozlov, I.A., Dang, M., Sikes, K., et al. (2005) *Nucleosides Nucleotides Nucleic Acids*, **24**, 1037.
3 (a) Zhdanov, R.I. and Zhenodarova, S.M. (1975) *Synthesis*, **4**, 222. (b) Slotin, L.A. (1977) *Synthesis*, **11**, 737. (c) Amarnath, V. and Broom, A.D. (1977) *Chemical Reviews*, **77**, 183. (d) Ikehara, M., Ohtsuka, E. and Markham, A. (1979) *Advances in Carbohydrate Chemistry and Biochemistry*, **36**, 135.
4 (a) Reese, C.B. (1976) *Phosphorus Sulfur*, **1**, 245. (b) Reese, C.B. (1978) *Tetrahedron*, **34**, 3143.
5 (a) Cramer, F. (1961) *Angewandte Chemie*, **73**, 49. (b) Cramer, F. (1966) *Angewandte Chemie (International Edition in English)*, **5**, 173.
6 Khorana, H.G. (1968) *Pure and Applied Chemistry*, **17**, 349.
7 (a) Ohtsuka, E., Ikehara, M. and Söll, D. (1982) *Nucleic Acids Research*, **10**, 6553. (b) Narang, S.A. (1983) *Tetrahedron*, **39**, 3. (c) Crockett, G.C. (1983). *Aldrichimica Acta*, **16**, 47. (d) Itakura, K., Rossi, J.J. and Wallace, R.B. (1984) *Annual Review of Biochemistry*, **53**, 323. (e) Sonveaux, E. (1986) *Bioorganic Chemistry*, **14**, 274.
8 (a) Caruthers, M. (1985) *Science*, **230**, 281. (b) Caruthers, M.H., Barone, A.D., Beaucage, S.L., et al. (1987) *Methods in Enzymology*, **154**, 287. (c) Caruthers, M. H. (1987) *Reactive Polymers, Ion Exchangers, Sorbents*, **6**, 159. (d) Caruthers, M.H. (1991) *Accounts of Chemical Research*, **24**, 278. (e) Caruthers, M.H., Beaton, G., Wu, J.V. and Wiesler, W. (1992) *Methods in Enzymology*, **211**, 3.
9 (a) Beaucage, S.L. and Iyer, R.P. (1992) *Tetrahedron*, **48**, 2223. (b) Beaucage, S.L. and Iyer, R.P. (1993) *Tetrahedron*, **49**, 1925. (c) Beaucage, S.L. and Iyer, R.P. (1993) *Tetrahedron*, **49**, 6123. (d) Beaucage, S.L. and Iyer, R.P. (1993) *Tetrahedron*, **49**, 10441.
10 (a) Uhlmann, E. and Peyman, A. (1990) *Chemical Reviews*, **90**, 543. (b) Englisch, U. and Gauss, D.H. (1991) *Angewandte Chemie (International Edition in English)*, **30**, 613. (c) Milligan, J.F., Matteucci, M.D. and Martin, J.C. (1993) *Journal of Medicinal Chemistry*, **36**, 1923. (d) De Mesmaeker, A., Haener, R., Martin, P. and Moser, H.E. (1995) *Accounts of Chemical Research*, **28**, 366. (e) Nielsen, P. E. (1995) *Annual Review of Biophysics and Biomolecular Structure*, **24**, 167. (f) Micklefield, J. (2001) *Current Medicinal Chemistry*, **8**, 1157. (g) Opalinska, J.B. and Gewirtz, A.M. (2002) *Nature Reviews. Drug Discovery*, **1**, 503–14. (h) Kurreck, J. (2003) *European Journal of Biochemistry*, **270**, 1628. (i) Crooke, S.T. (2004) *Annual Review of Medicine*, **55**, 61.
11 A classical review on the subject: (a) Goodchild, J. (1990) *Bioconjugate Chemistry*, **1**, 165. For a fairly recent compilation see (b) Virta, P., Katajisto, J., Niittymäki, T. and Lönnberg, H. (2003) *Tetrahedron*, **59**, 5137.
12 (a) Gait, M.J. (ed.) (1990) *Oligonucleotide Synthesis: A Practical Approach*, 2nd edn, IRL Press, Oxford. (b) Eckstein, F. (ed.) (1991) *Oligonucleotides and Analogues: A Practical Approach*, Oxford University Press, Oxford. (c) Agrawal, S. (ed.) (1993) Protocols for oligonucleotides and analogs: synthesis and properties, in *Methods in Molecular Biology*, Vol. **20**, Humana Press Inc. (d) Agrawal, S. (ed.) (1993) Protocols for oligonucleotide conjugates: synthesis and analytical techniques, in *Methods in Molecular Biology*, Vol. **26**, Humana Press Inc. (e) Lichtenstein, C. and Nellen, W. (eds) (1997) *Antisense Technology. A Practical Approach*, Oxford University Press. (f) Herdewijn, P. (ed.) (2004) Oligonucleotide synthesis: methods and applications, in *Methods in Molecular Biology*, Vol. **288**, Humana Press Ltd.
13 Michelson, A.M. and Todd, A.R. (1955) *Journal of the Chemical Society*, 2632.

14 Ralph, R.K. and Khorana, H.G. (1961) *Journal of the American Chemical Society*, **83**, 2926.
15 (a) Corby, N.S., Kenner, G.W. and Todd, A.R. (1952) *Journal of the Chemical Society*, 3669. (b) Kenner, G.W., Todd, A.R. and Weymouth, F.J. (1952) *Journal of the Chemical Society*, 3675. (c) Michelson, A.M., Szabo, L. and Todd, A.R. (1956) *Journal of the Chemical Society*, 1546. (d) Hall, R.H., Todd, A. and Webb, R.F. (1957) *Journal of the Chemical Society*, 3291. (e) Schofield, J.A. and Todd, A. (1961) *Journal of the Chemical Society*, 2316.
16 Gilham, P.T. and Khorana, H.G. (1958) *Journal of the American Chemical Society*, **80**, 6212.
17 Ohtsuka, E., Moon, M.W. and Khorana, H.G. (1962) *Journal of the American Chemical Society*, **87**, 2956.
18 Agarwal, K.L., Yamazaki, A., Cashion, P.J. and Khorana, H.G. (1972) *Angewandte Chemie (International Edition in English)*, **11**, 451.
19 Khorana, H.G. (1969) *Angewandte Chemie (International Edition in English)*, **81**, 1027.
20 Agarwal, K.L., Büchi, H., Caruthers, M.H., et al. (1970) *Nature*, **227**, 27.
21 Khorana, H.G., Turner, A.F. and Vizsolyi, J.P. (1961) *Journal of the American Chemical Society*, **83**, 686.
22 Büchi, H. and Khorana, H.G. (1972) *Journal of Molecular Biology*, **72**, 251.
23 Schaller, H., Weimann, G., Lerch, B. and Khorana, H.G. (1963) *Journal of the American Chemical Society*, **85**, 3821.
24 Weber, H. and Khorana, H.G. (1972) *Journal of Molecular Biology*, **72**, 219.
25 Smith, M., Rammler, D.H., Goldberg, I.H. and Khorana, H.G. (1962) *Journal of the American Chemical Society*, **84**, 430.
26 Eckstein, F. and Rizk, I. (1967) *Angewandte Chemie (International Edition in English)*, **6**, 695.
27 Jacob, T.M. and Khorana, H.G. (1964) *Journal of the American Chemical Society*, **86**, 1630.
28 Lohrmann, R. and Khorana, H.G. (1966) *Journal of the American Chemical Society*, **88**, 829.
29 Letsinger, R.L. and Ogilvie, K.K. (1967) *Journal of the American Chemical Society*, **89**, 4801.
30 Letsinger, R.L. and Ogilvie, K.K. (1969) *Journal of the American Chemical Society*, **91**, 3350.
31 Eckstein, F. and Rizk, I. (1969) *Chemische Berichte*, **102**, 2362.
32 Reese, C.B. and Saffhill, R. (1968) *Chemical Communications*, 767.
33 Itakura, K., Katagiri, N., Bahl, C.P., et al. (1975) *Journal of the American Chemical Society*, **97**, 7327.
34 Katagiri, N., Itakura, K. and Narang, S.A. (1975) *Journal of the American Chemical Society*, **97**, 7332.
35 Stawinski, J., Hosumi, T., Narang, S.A., et al. (1977) *Nucleic Acids Research*, **4**, 353.
36 Sood, A.K. and Narang, S.A. (1977) *Nucleic Acids Research*, **4**, 2757.
37 Reese, C.B., Titmas, R.C. and Yau, L. (1978) *Tetrahedron Letters*, **19**, 2727.
38 Reese, C.B. and Zhang, P.-Z. (1993) *Journal of the Chemical Society, Perkin Transactions 1*, 2291.
39 Sekine, M., Matsuzaki, J. and Hata, T. (1981) *Tetrahedron Letters*, **22**, 3209.
40 Sekine, M., Matsuzaki, J.-I. and Hata, T. (1985) *Tetrahedron*, **41**, 5279.
41 Hotoda, H., Wada, T., Sekine, M. and Hata, T. (1989) *Nucleic Acids Research*, **17**, 5291.
42 van der Marel, G., van Boeckel, C.A.A., Wille, G. and van Boom, J.H. (1981) *Tetrahedron Letters*, **22**, 3887.
43 Wreesmann, C.T.J., Fidder, A., van der Marel, G.A. and van Boom, J.H. (1983) *Nucleic Acids Research*, **11**, 8389.
44 Marugg, J.E., Tromp, M., Jhurani, P., et al. (1984) *Tetrahedron*, **40**, 73.
45 Efimov, V.A., Reverdatto, S.V. and Chakhmakcheva, O.G. (1982) *Tetrahedron Letters*, **23**, 961.
46 Efimov, V.A., Reverdatto, S.V. and Chakhmakhcheva, O.G. (1982) *Nucleic Acids Research*, **10**, 6675.
47 Efimov, V.A., Chakhmakhcheva, O.G. and Ovchinnikov, Yu.A. (1985) *Nucleic Acids Research*, **13**, 3651.
48 Efimov, V.A., Buryakova, A.A., Dubey, I.Y., et al. (1986) *Nucleic Acids Research*, **14**, 6525.

49 Van Boom, J.H., Burgers, P.M.J., Owen, G.R., Reese, C.B. and Saffhill, R. (1971) *Journal of the Chemical Society D*, 869.
50 (a) Sekine, M. and Hata, T. (1975) *Tetrahedron Letters*, **16**, 1711. (b) Sekine, M., Hamaoki, K. and Hata, T. (1979) *Journal of Organic Chemistry*, **44**, 2325. For a review of the chemistry see: (c) Sekine, M. and Hata, T. (1986) *Journal of Synthetic Organic Chemistry, Japan*, **44**, 229.
51 (a) Chattopadhyaya, J.B. and Reese, C.B. (1979) *Tetrahedron Letters*, **20**, 5059. (b) Jones, S.S., Rayner, B., Reese, C.B., et al. (1980) *Tetrahedron*, **36**, 3075. (c) Chattopadhyaya, J.B. and Reese, C.B. (1980) *Nucleic Acids Research*, **8**, 2039.
52 Devine, K.G. and Reese, C.B. (1986) *Tetrahedron Letters*, **27**, 5529.
53 Letsinger, R.L. and Mahadevan, V. (1965) *Journal of the American Chemical Society*, **87**, 3526.
54 Letsinger, R.L. and Kornet, M.J. (1963) *Journal of the American Chemical Society*, **85**, 3045.
55 Merrifield, R.B. (1963) *Journal of the American Chemical Society*, **85**, 2149.
56 Letsinger, R.L., Kornet, M.J., Mahadevan, V. and Jerina, D.M. (1964) *Journal of the American Chemical Society*, **86**, 5163.
57 Letsinger, R.L. and Mahadevan, V. (1966) *Journal of the American Chemical Society*, **88**, 5319.
58 Letsinger, R.L., Caruthers, M.H. and Jerina, D.M. (1967) *Biochemistry*, **6**, 1379.
59 Shimidzu, T. and Letsinger, R.L. (1968) *Journal of Organic Chemistry*, **33**, 708.
60 (a) Reese, C.B. and Ubasawa, A. (1980) *Tetrahedron Letters*, **21**, 2265. (b) Reese, C.B. and Richards, K.H. (1985) *Tetrahedron Letters*, **26**, 2245.
61 Use of PCl_3 for the phosphonylation of nucleosides: Phosphorylation of ribonucleosides with phosphorus trichloride: (a) Honjo, M., Marumoto, R., Kobayashi K. and Yoshioka, Y. (1966) *Tetrahedron Letters*, **7**, 3851. (b) Yoshikawa, M., Sakuraba, M. and Kusashio, K. (1970) *Bulletin of the Chemical Society of Japan*, **43**, 456. (c) Kabachnik, M.M., Potapov, V.K. and Shabarova, Z.A. (1970) *Doklady Akademii Nauk SSSR*, **195**, 1107.
62 Nucleoside silylphosphites: Hata, T. and Sekine, M. (1974) *Tetrahedron Letters*, **15**, 3943.
63 (a) Letsinger, R.L., Finnan, J.L., Heavner, G.A. and Lunsford, W.B. (1975) *Journal of the American Chemical Society*, **97**, 3278. (b) Letsinger, R.L. and Lunsford, W.B. (1976) *Journal of the American Chemical Society*, **98**, 3655.
64 Ogilvie, K.K. and Nemer, M.J. (1981) *Tetrahedron Letters*, **22**, 2531.
65 Tanaka, T. and Letsinger, R.L. (1982) *Nucleic Acids Research*, **10**, 3249.
66 Daub, G.W. and van Tamelen, E.E. (1977) *Journal of the American Chemical Society*, **99**, 3526.
67 Smith, D.J.H., Ogilvie, K.K. and Gillen, M.F. (1980) *Tetrahedron Letters*, **21**, 861.
68 Matteucci, M.D. and Caruthers, M.H. (1980) *Tetrahedron Letters*, **21**, 719.
69 Alvarado-Urbina, G., Sathe, G.M., Liu, W.-C., et al. (1981) *Science*, **214**, 270.
70 Chow, F., Kempe, T. and Palm, G. (1981) *Nucleic Acids Research*, **9**, 2807.
71 Ogilvie, K.K., Theriault, N. and Sadana, K.L. (1977) *Journal of the American Chemical Society*, **99**, 7741.
72 Ogilvie, K.K. and Nemer, M.J. (1980) *Tetrahedron Letters*, **21**, 4159.
73 Pon, R.T. and Ogilvie, K.K. (1984) *Tetrahedron Letters*, **25**, 713.
74 Ogilvie, K.K., Nemer, M.J. and Gillen, M.F. (1984) *Tetrahedron Letters*, **25**, 1669.
75 Burgers, P.M.J. and Eckstein, F. (1978) *Tetrahedron Letters*, **19**, 3835.
76 Nemer, M.J. and Ogilvie, K.K. (1980) *Tetrahedron Letters*, **21**, 4149.
77 Nemer, M.J. and Ogilvie, K.K. (1980) *Tetrahedron Letters*, **21**, 4153.
78 Sinha, N.D., Großbruchhaus, V. and Köster, H. (1983) *Tetrahedron Letters*, **24**, 877.
79 Letsinger, R.L., Groody, E.P. and Tanaka, T. (1982) *Journal of the American Chemical Society*, **104**, 6805.
80 Letsinger, R.L., Groody, E.P., Lander, N. and Tanaka, T. (1984) *Tetrahedron*, **40**, 137.
81 Fourrey, J.L. and Shire, D.J. (1981) *Tetrahedron Letters*, **22**, 729.
82 Matteucci, M.D. and Caruthers, M.H. (1981) *Journal of the American Chemical Society*, **103**, 3185.

83 Caruthers, M.H., Beaucage, S.L., Efcavitch, J.W., *et al.* (1980) *Nucleic Acids Research. Symposium Series*, **7**, 215.

84 See, for example, the papers cited in the next reference and in the review article by Nifantiev: Nifantiev, E.E., Grachev, M.K. and Burmistrov, S.Yu. (2000) *Chemical Reviews*, **100**, 3755.

85 Beaucage, S.L. and Caruthers, M.H. (1981) *Tetrahedron Letters*, **22**, 1859.

86 Dorman, M.A., Noble, S.A., McBride, L.J. and Caruthers, M.H. (1984) *Tetrahedron*, **40**, 95.

87 McBride, L.J. and Caruthers, M.H. (1983) *Tetrahedron Letters*, **24**, 245.

88 Adams, S.P., Kavka, K.S., Wykes, E.J., Holder, S.B. and Galluppi, G.R. (1983) *Journal of the American Chemical Society*, **105**, 661.

89 Dörper, T. and Winnacker, E.-L. (1983) *Nucleic Acids Research*, **11**, 2575.

90 Sinha, N.D., Biernat, J. and Köster, H. (1983) *Tetrahedron Letters*, **24**, 5843.

91 Sinha, N.D., Biernat, J., McManus, J. and Köster, H. (1984) *Nucleic Acids Research*, **12**, 4539.

92 Köster, H., Biernat, J., McManus, J., *et al.* (1984) *Tetrahedron*, **40**, 103.

93 Gao, X., Gaffney, B.L., Senior, M., *et al.* (1985) *Nucleic Acids Research*, **13**, 573.

94 Andrus, A. and Beaucage, S.L. (1988) *Tetrahedron Letters*, **29**, 5479.

95 Scaringe, S.A., Wincott, F.E. and Caruthers, M.H. (1998) *Journal of the American Chemical Society*, **120**, 11820.

96 Jäger, A. and Engels, J. (1984) *Tetrahedron Letters*, **25**, 1437.

97 See the Nifantiev review (reference [84]), the references cited therein, and the following papers: (a) Dahl, O. (1983) *Phosphorus Sulfur*, **18**, 201. (b) Dahl, B. H., Nielsen, J. and Dahl, O. (1987) *Nucleic Acids Research*, **15**, 1729. (c) Berner, S., Mühlegger, K. and Seliger, H. (1988) *Nucleosides Nucleotides*, **7**, 763. (d) Berner, S., Mühlegger, K. and Seliger, H. (1989) *Nucleic Acids Research*, **17**, 853.

98 Pon, R.T. (1987) *Tetrahedron Letters*, **28**, 3643.

99 Montserrat, F.X., Grandas, A., Eritja, R. and Pedroso, E. (1994) *Tetrahedron*, **50**, 2617.

100 Rao, M.V., Reese, C.B., Schehimann, V. and Yu, P.S. (1993) *Journal of the Chemical Society, Perkin Transactions 1*, 43.

101 Froehler, B.C. and Matteucci, M.D. (1983) *Tetrahedron Letters*, **24**, 3171.

102 Wolter, A., Biernat, J. and Köster, H. (1986) *Nucleosides Nucleotides*, **5**, 65.

103 (a) Tsou, D., Wright, P., Lloyd, D. and Andrus, A. (1994) Innovations and perspectives in solid phase synthesis. peptides, proteins and nucleic acids, in *Biological and Biomedical Applications* (ed. R. Epton), Mayflower Worldwide Ltd, p. 125. (b) Sproat, B., Colonna, F., Mullah, B., *et al.* (1995) *Nucleosides Nucleotides*, **14**, 255. (c) Wincott, F., DiRenzo, A., Shaffer, C., *et al.* (1995) *Nucleic Acids Research*, **23**, 2677.

104 (a) Pitsch, S., Weiss, P.A., Jenny, L., *et al.* (2001) *Helvetica Chimica Acta*, **84**, 3773. (b) Welz, R. and Müller, S. (2002) *Tetrahedron Letters*, **43**, 795.

105 Vargeese, C., Carter, J., Yegge, J., *et al.* (1998) *Nucleic Acids Research*, **26**, 1046.

106 Hayakawa, Y., Uchiyama, M. and Noyori, R. (1986) *Tetrahedron Letters*, **27**, 4191.

107 Fourrey, J.-L. and Varenne, J. (1985) *Tetrahedron Letters*, **26**, 1217.

108 Hayakawa, Y., Uchiyama, M. and Noyori, R. (1986) *Tetrahedron Letters*, **27**, 4195.

109 Manoharan, M., Lu, Y., Casper, M.D. and Just, G. (2000) *Organic Letters*, **2**, 243.

110 Chappell, M.D. and Halcomb, R.L. (1999) *Tetrahedron Letters*, **40**, 1.

111 Kataoka, M., Hattori, A., Okino, S., *et al.* (2001) *Organic Letters*, **3**, 815.

112 (a) Stec, W.J., Zon, G. and Egan, W. (1984) *Journal of the American Chemical Society*, **106**, 6077. (b) Stein, C.A., Subasinghe, C., Shinozuka, K. and Cohen, J.S. (1988) *Nucleic Acids Research*, **16**, 3209. For a recent improvement see: (c) Krotz, A.H., Hang, A., Gorman, D. and Scozzari, A.N. (2005) *Nucleosides Nucleotides Nucleic Acids*, **24**, 1293.

113 (a) Kamer, P.C.J., Roelen, H.C.P., Elst, H., van der Marel, G.A. and van Boom, J.H. (1989) *Tetrahedron Letters*, **30**, 6757. (b) Roelen, H.C.P., Kamer, P.C.J., Elst, H., *et al.* (1991) *Recueil des Travaux Chimiques des Pays-Bas*, **110**, 325. (c) Cheruvallath, Z.

S., Wheeler, P.D., Cole, D.L. and Ravikumar, V.T. (1999) *Nucleosides Nucleotides*, **18**, 485. (d) Cheruvallath, Z. S., Wheeler, P.D., Cole, D.L. and Ravikumar, V.T. (1999) *Nucleosides Nucleotides*, **18**, 1195. (e) Cheruvallath, Z.S., Carty, R.L., Moore, M.N., et al. (2000) *Organic Process Research & Development*, **4**, 199. (f) Krotz, A.H., Gorman, D., Mataruse, P., et al. (2004) *Organic Process Research & Development*, **8**, 852. (g) Ravikumar, V.T., Andrade, M., Carty, R.L., et al. (2006) *Bioorganic & Medicinal Chemistry Letters*, **16**, 2513.

114 Vu, H. and Hirschbein, B.L. (1991) *Tetrahedron Letters*, **32**, 3005.

115 Rao, M.V., Reese, C.B. and Zhao, Z. (1992) *Tetrahedron Letters*, **33**, 4839.

116 Stec, W.J., Uznanski, B., Wilk, A., Hirschbein, et al. (1993) *Tetrahedron Letters*, **34**, 5317.

117 Rao, M.V. and Macfarlane, K. (1994) *Tetrahedron Letters*, **35**, 6741.

118 Efimov, V.A., Kalinkina, A.L., Chakhmakhcheva, O.G., et al. (1995) *Nucleic Acids Research*, **23**, 4029.

119 (a) Eleuteri, A., Cheruvallath, Z.S., Capaldi, D.C., et al. (1999) *Nucleosides Nucleotides*, **18**, 1803. (b) Cheruvallath, Z.S., Kumar, R.K., Rentel, C., et al. (2003) *Nucleosides Nucleotides Nucleic Acids*, **22**, 461.

120 (a) Iyer, R.P., Egan, W., Regan, J.B. and Beaucage, S.L. (1990) *Journal of the American Chemical Society*, **112**, 1253. (b) Iyer, R.P., Phillips, L.R., Egan, W., et al. (1990) *Journal of Organic Chemistry*, **55**, 4693.

121 (a) Xu, Q., Musier-Forsyth, K., Hammer, R.P. and Barany, G. (1996) *Nucleic Acids Research*, **24**, 1602. (b) Xu, Q., Barany, G., Hammer, R.P. and Musier-Forsyth, K. (1996) *Nucleic Acids Research*, **24**, 3643.

122 (a) Zhang, Z., Nichols, A., Tang, J.X., Han, Y. and Tang, J.Y. (1999) *Tetrahedron Letters*, **40**, 2095. (b) Tang, J.-Y., Han, Y., Tang, J.X. and Zhang, Z. (2000) *Organic Process Research & Development*, **4**, 194.

123 Kabachnik, M.M., Potapov, V.K., Shabarova, Z.A. and Prokofiev, M.A. (1971) *Doklady Akademii Nauk SSSR. Seria Khimiia*, **201**, 858.

124 For conceptually similar approaches see: (a) Jayaraman, K. and McClaugherty, H. (1982) *Tetrahedron Letters*, **23**, 5377. (b) Cao, T.M., Bingham, S.E. and Sung, M.T. (1983) *Tetrahedron Letters*, **24**, 1019. For the most recent one: (c) Mihaichuk, J.C., Hurley, T.B., Vagle, K.E., et al. (2000) *Organic Process Research & Development*, **4**, 214.

125 (a) Garegg, P.J., Regberg, T., Stawinski, J. and Strömberg, R. (1985) *Chemica Scripta*, **25**, 280. (b) Garegg, P.J., Regberg, T., Stawinski, J. and Strömberg, R. (1986) *Chemica Scripta*, **26**, 59.

126 Froehler, B.C. and Matteucci, M.D. (1986) *Tetrahedron Letters*, **27**, 469.

127 Garegg, P.J., Lindh, I., Regberg, T., et al. (1986) *Tetrahedron Letters*, **27**, 4051.

128 Froehler, B.C., Ng, P.G. and Matteucci, M.D. (1986) *Nucleic Acids Research*, **14**, 5399.

129 Garegg, P.J., Lindh, I., Regberg, T., et al. (1986) *Tetrahedron Letters*, **27**, 4055.

130 Tanaka, T., Tamatsukuri, S. and Ikehara, M. (1987) *Nucleic Acids Research*, **15**, 7235.

131 Stawinski, J., Strömberg, R., Thelin, M. and Westman, E. (1988) *Nucleic Acids Research*, **16**, 9285.

132 Sakatsume, O., Yamane, H., Takaku, H. and Yamamoto, N. (1989) *Tetrahedron Letters*, **30**, 6375.

133 Sakatsume, O., Ohtsuki, M., Takaku, H. and Reese, C.B. (1989) *Nucleic Acids Research*, **17**, 3689.

134 Rozners, E., Westman, E. and Strömberg, R. (1994) *Nucleic Acids Research*, **22**, 94.

135 Gaffney, B.L. and Jones, R.A. (1988) *Tetrahedron Letters*, **29**, 2619.

136 Heinonen, P., Winqvist, A., Sanghvi, Y. and Strömberg, R. (2003) *Nucleosides Nucleotides Nucleic Acids*, **22**, 1387.

137 Andrus, A., Efcavitch, J.W., McBride, L.J. and Giusti, B. (1988) *Tetrahedron Letters*, **29**, 861.

138 (a) Garegg, P.J., Regberg, T., Stawinski, J. and Strömberg, R. (1987) *Nucleosides Nucleotides*, **6**, 283. (b) Garegg, P.J., Regberg, T., Stawinski, J. and Strömberg, R. (1987) *Nucleosides Nucleotides*, **6**, 655.

139 Froehler, B.C. (1986) *Tetrahedron Letters*, **27**, 5575.

140 Froehler, B., Ng, P. and Matteucci, M. (1988) *Nucleic Acid Res*, **16**, 4831.

141 Peyrottes, S., Vasseur, J.-J., Imbach, J.-L. and Rayner, B. (1996) *Nucleic Acid Res*, **24**, 1841.

142 See, for example, two review papers of Stawinski and coworkers: (a) Cieslak, J., Sobkowski, M., Jankowska, J., *et al.* (2001) *Acta Biochimica Polonica*, **40**, 429. (b) Stawinski, J. and Kraszewski, A. (2002) *Accounts of Chemical Research*, **35**, 952.

143 Wada, T., Sato, Y., Honda, F., Kawahara, S. and Sekine, M. (1997) *Journal of the American Chemical Society*, **119**, 12710.

144 Köster, H. and Cramer, F. (1972) *Liebigs Annalen der Chemie*, **766**, 6.

145 Seliger, H. and Görtz, H.-H. (1981) *Angewandte Chemie (International Edition in English)*, **20**, 683.

146 Melby, L.R. and Strobach, D.R. (1967) *Journal of the American Chemical Society*, **89**, 450.

147 Blackburn, G.M., Brown, M.J. and Harris, M.R. (1967) *Journal of the Chemical Society, C*, 2438.

148 Cramer, F. and Köster, H. (1968) *Angewandte Chemie (International Edition in English)*, **7**, 473.

149 Freist, W. and Cramer, F. (1970) *Angewandte Chemie (International Edition in English)*, **9**, 368.

150 Ogilvie, K.K. and Kroeker, K. (1972) *Canadian Journal of Chemistry*, **50**, 1211.

151 Yip, K.F. and Tsou, K.C. (1971) *Journal of the American Chemical Society*, **93**, 3272.

152 (a) Miyoshi, K., Arentzen, R., Huang, T. and Itakura, K. (1980) *Nucleic Acids Research*, **8**, 5507. (b) Ito, H., Ike, Y., Ikuta, S. and Itakura, K. (1982) *Nucleic Acids Research*, **10**, 1755.

153 Ike, Y., Ikuta, S., Sato, M., Huang, T. and Itakura, K. (1983) *Nucleic Acids Research*, **11**, 477.

154 Miller, P.S., Agris, C.H., Murakami, A., *et al.* (1983) *Nucleic Acids Research*, **11**, 6225.

155 Belagaje, R. and Brush, C.K. (1982) *Nucleic Acids Research*, **10**, 6295.

156 (a) Bardella, F., Giralt, E. and Pedroso, E. (1990) *Tetrahedron Letters*, **31**, 6231. (b) Montserrat, F.X., Grandas, A., Eritja, R. and Pedroso, E. (1994) *Tetrahedron*, **50**, 2617.

157 Bardella, F., Eritja, R., Pedroso, E. and Giralt, E. (1993) *Bioorganic & Medicinal Chemistry Letters*, **3**, 2193.

158 Köster, H. and Geussenhainer, S. (1972) *Angewandte Chemie (International Edition in English)*, **11**, 713.

159 (a) Köster, H. and Cramer, F. (1974) *Liebigs Annalen der Chemie*, **1974**, 946. (b) Köster, H. and Cramer, F. (1974) *Liebigs Annalen der Chemie*, **1974**, 959.

160 McCollum, C. and Andrus, A. (1991) *Tetrahedron Letters*, **32**, 4069.

161 Sproat, B., Colonna, F., Mullah, B., *et al.* (1995) *Nucleosides Nucleotides*, **14**, 255.

162 (a) Stetsenko, D.A., Malakhov, A.D. and Gait, M.J. (2002) *Organic Letters*, **4**, 3259. (b) Stetsenko, D.A., Malakhov, A.D. and Gait, M.J. (2003) *Nucleosides Nucleotides Nucleic Acids*, **22**, 1379. (c) Sekine, M., Ohkubo, A. and Seio, K. (2003) *Journal of Organic Chemistry*, **68**, 5478. (d) Ohkubo, A., Seio, K. and Sekine, M. (2004) *Tetrahedron Letters*, **45**, 363.

163 Bayer, E. (1991) *Angewandte Chemie (International Edition in English)*, **30**, 113.

164 Gao, H., Gaffney, B.L. and Jones, R.A. (1991) *Tetrahedron Letters*, **32**, 5477.

165 (a) Wright, P., Lloyd, D., Rapp, W. and Andrus, A. (1993) *Tetrahedron Letters*, **34**, 3373. (b) Tsou, D., Hampel, A., Andrus, A. and Vinayak, R. (1995) *Nucleosides Nucleotides*, **14**, 1481.

166 (a) Conte, M.R., Mayol, L., Montesarchio, D., Piccialli, G. and Santacroce, C. (1993) *Nucleosides Nucleotides*, **12**, 351. (b) Hayakawa, Y. and Kataoka, M. (1998) *Journal of the American Chemical Society*, **120**, 12395.

167 Potapov, V.K., Veiko, V.P., Koroleva, O.N. and Shabarova, Z.A. (1979) *Nucleic Acids Research*, **6**, 2041.

168 (a) Birch-Hirschfeld, E., Földes-Papp, Z., Gührs, K.-H. and Seliger, H. (1994) *Nucleic Acids Research*, **22**, 1760. (b) Birch-Hirschfeld, E., Gührs, K.-H., Földes-Papp, Z. and Seliger, H. (1996) *Helvetica Chimica Acta*, **79**, 137.

169 Atherton, E., Clive, D.L.J. and Sheppard, R.C. (1975) *Journal of the American Chemical Society*, **97**, 6584.

170 Gait, M.J. and Sheppard, R.C. (1976) *Journal of the American Chemical Society*, **98**, 8514.

171 Köster, H. and Heidmann, W. (1976) *Angewandte Chemie (International Edition in English)*, **15**, 546.
172 Köster, H. and Heidmann, W. (1976) *Angewandte Chemie (International Edition in English)*, **15**, 547.
173 Gait, M.J. and Sheppard, R.C. (1977) *Nucleic Acids Research*, **4**, 1135.
174 Gait, M.J. and Sheppard, R.C. (1977) *Nucleic Acids Research*, **4**, 4391.
175 Narang, C.K., Brunfeldt, K. and Norris, K.E. (1977) *Tetrahedron Letters*, **18**, 1819.
176 Miyoshi, K. and Itakura, K. (1979) *Tetrahedron Letters*, **20**, 3635.
177 (a) Miyoshi, K., Miyake, T., Hozumi, T. and Itakura, K. (1980) *Nucleic Acids Research*, **8**, 5473. (b) Miyoshi, K., Huang, T. and Itakura, K. (1980) *Nucleic Acids Research*, **8**, 5491.
178 Gait, M.J., Singh, M., Sheppard, R.C., et al. (1980) *Nucleic Acids Research*, **8**, 1081.
179 Markham, A.F., Edge, M.D., Atkinson, T.C., et al. (1980) *Nucleic Acids Research*, **8**, 5193.
180 Duckworth, M.L., Gait, M.J., Goelet, P., et al. (1981) *Nucleic Acids Research*, **9**, 1691.
181 Atherton, E., Brown, E., Sheppard, R.C. and Rosevear, A. (1981) *Journal of the Chemical Society, Chemical Communications*, 1151.
182 (a) Gait, M.J., Matthes, H.W.D., Singh, M. and Titmas, R.C. (1982) *Journal of the Chemical Society, Chemical Communications*, 37. (b) Gait, M.J., Matthes, H.W.D., Singh, M., et al. (1982) *Nucleic Acids Research*, **10**, 6243.
183 Minganti, C., Ganesh, K.N., Sproat, B.S. and Gait, M.J. (1985) *Analytical Biochemistry*, **147**, 63.
184 (a) Barbato, S., De Napoli, L., Mayol, L., et al. (1987) *Tetrahedron Letters*, **28**, 5727. (b) Barbato, S., De Napoli, L., Mayol, L., et al. (1989) *Tetrahedron*, **45**, 4523.
185 Köster, H. (1972) *Tetrahedron Letters*, **13**, 1527.
186 (a) Fritz, J.S. and King, J.N. (1976) *Analytical Chemistry*, **48**, 570. (b) Tundo, P. and Venturello, P. (1979) *Journal of the American Chemical Society*, **101**, 6606.

187 (a) Gough, G.R., Brunden, M.J. and Gilham, P.T. (1981) *Tetrahedron Letters*, **22**, 4177. (b) Kohli, V., Balland, A., Wintzerith, M., et al. (1982) *Nucleic Acids Research*, **10**, 7439. (c) Köster, H., Stumpe, A., and Wolter, A. (1983) *Tetrahedron Letters*, **24**, 747. (d) Sproat, B.S. and Bannwarth, W. (1983) *Tetrahedron Letters*, **24**, 5771.
188 Pon, R.T., Usman, N. and Ogilvie, K.K. (1988) *Biotechniques*, **6**, 768.
189 Pon, R.T. (1993) *Methods in Molecular Biology*, **20**, 465.
190 Groger, G. and Seliger, H. (1988) *Nucleosides Nucleotides*, **7**, 773.
191 Chapman, T.M. and Kleid, D.G. (1973) *Journal of the Chemical Society, Chemical Communications*, **5**, 193.
192 Köster, H. and Heyns, K. (1972) *Tetrahedron Letters*, **13**, 1531.
193 (a) Horn, T., Vasser, M.P., Struble, M.E. and Crea, R. (1980) *Nucleic Acid Research. Symposium Series*, **7**, 225. (b) Crea, R. and Horn, T. (1980) *Nucleic Acids Research*, **8**, 2331.
194 (a) Frank, R., Heikens, W., Heisterberg-Moutsis, G. and Blöcker, H. (1983) *Nucleic Acids Research*, **11**, 4365. (b) Matthes, H.W.D., Zenke, W.M., Grundström, T., et al. (1984) *EMBO Journal*, **3**, 801. (c) Ott, J. and Eckstein, F. (1984) *Nucleic Acids Research*, **12**, 9137.
195 Reddy, M.P., Michael, M.A., Farooqui, F. and Girgis, N.S. (1994) *Tetrahedron Letters*, **35**, 5771.
196 Jaisankar, P., Hinz, M., Happ, E. and Seliger, H. (1998) *Nucleosides Nucleotides*, **17**, 1787.
197 Seliger, H., Kotschi, U., Scharpf, C., et al. (1989) *Journal of Chromatography A*, **476**, 49.
198 (a) Seliger, H., Bader, R., Birch-Hirschfeld, E., et al. (1995) *Reactive Functional Polymers*, **26**, 119. (b) Bader, R., Hinz, M., Schu, B. and Seliger, H. (1997) *Nucleosides Nucleotides*, **16**, 829.
199 Devivar, R.V., Koontz, S.L., Peltier, W.J., et al. (1999) *Bioorganic & Medicinal Chemistry Letters*, **9**, 1239.
200 Kumar, P., Sharma, A.K., Gupta, K.C., et al. (2001) *Helvetica Chimica Acta*, **84**, 345.

201 Pacard, E., Brook, M.A., Ragheb, A.M., et al. (2006) *Colloids and Surfaces. B, Biointerfaces*, **47**, 176.
202 Pless, R.C. and Letsinger, R.L. (1975) *Nucleic Acids Research*, **2**, 773.
203 Letsinger, R.L., Caruthers, M.H., Miller, P.S. and Ogilvie, K.K. (1967) *Journal of the American Chemical Society*, **89**, 7146.
204 Dobrynin, V.N., Chernov, B.K. and Kolosov, M.N. (1980) *Bioorganicheskaia Khimiia (Russia)*, **6**, 138.
205 Köster, H., Hoppe, N., Kohli, V., et al. (1980) *Nucleic Acids Research. Symposium Series*, **7**, 39.
206 (a) Pon, R.T. and Yu, S. (1997) *Tetrahedron Letters*, **38**, 3331. (b) Pon, R.T. and Yu, S. (1999) *Synlett*, **11**, 1778.
207 Damha, M.J., Giannaris, P.A. and Zabarylo, S.V. (1990) *Nucleic Acids Research*, **18**, 3813.
208 Sharma, P., Sharma, A.K., Malhotra, V.P. and Gupta, K.C. (1992) *Nucleic Acids Research*, **20**, 4100.
209 (a) Gupta, K.C. and Kumar, P. (1992) *Bioorganic & Medicinal Chemistry Letters*, **2**, 727. (b) Kumar, P., Ghosh, N., Sadana, K.L., Garg, et al. (1993) *Nucleosides Nucleotides*, **12**, 565.
210 Berner, S., Gröger, G., Mühlegger, K. and Seliger, H. (1989) *Nucleosides Nucleotides*, **8**, 1165.
211 Palom, Y., Grandas, A. and Pedroso, E. (1998) *Nucleosides Nucleotides*, **17**, 1177.
212 Brown, T., Pritchard, C.E., Turner, G. and Salisbury, S.A. (1989) *Journal of the Chemical Society, Chemical Communications*, 891.
213 Stengele, K.-P. and Pfleiderer, W. (1990) *Tetrahedron Letters*, **31**, 2549.
214 (a) Uhlmann, E. and Pfleiderer, W. (1980) *Tetrahedron Letters*, **21**, 1181. (b) Himmelsbach, F., Schulz, B.S., Trichtinger, T., et al. (1984) *Tetrahedron*, **40**, 59.
215 Sekine, M., Tsuruoka, H., Iimura, S., et al. (1996) *Journal of Organic Chemistry*, **61**, 4087.
216 Alul, R.H., Singman, C.N., Zhang, G. and Letsinger, R.L. (1991) *Nucleic Acids Research*, **19**, 1527.
217 Wada, T., Mochizuki, A., Sato, Y. and Sekine, M. (1998) *Tetrahedron Letters*, **39**, 5593.
218 Aviñó, A., García, R.G., Díaz, A., et al. (1996) *Nucleosides Nucleotides*, **15**, 1871.
219 Wada, T., Honda, F., Sato, Y. and Sekine, M. (1999) *Tetrahedron Letters*, **40**, 915.
220 Pon, R.T. and Yu, S. (1997) *Nucleic Acids Research*, **25**, 3629.
221 Guzaev, A. and Lönnberg, H. (1997) *Tetrahedron Letters*, **38**, 3989.
222 (a) Mullah, B. and Andrus, A. (1997) *Tetrahedron Letters*, **38**, 5751. (b) Mullah, B., Livak, K., Andrus, A. and Kenney, P. (1998) *Nucleic Acids Research*, **26**, 1026.
223 Dobrynin, V.N., Filippov, S.A., Bystrov, N.S., et al. (1983) *Bioorganicheskaia Khimiia (Russia)*, **9**, 706.
224 Pon, R.T. and Yu, S. (1997) *Tetrahedron Letters*, **38**, 3327.
225 The Q-linked supports are available from Glen Research, Inc., Sterling, USA.
226 Pon, R.T., Yu, S. and Sanghvi, Y.S. (1999) *Bioconjugate Chemistry*, **10**, 1051.
227 Pon, R.T. and Yu, S. (2001) *Tetrahedron Letters*, **42**, 8943.
228 (a) Pon, R.T. and Yu, S. (2002) *Nucleic Acids Research*, **Suppl. 2**, 1. (b) Pon, R.T., Yu, S. and Sanghvi, Y.S. (2002) *Journal of Organic Chemistry*, **67**, 856.
229 Norris, K.E., Norris, F. and Brunfeldt, K. (1980) *Nucleic Acids Research. Symposium Series*, **7**, 233.
230 (a) Efimov, V.A., Pashkova, I.N., Kalinkina, A.L. and Chakhmakhcheva, O.G. (1993) *Bioorganicheskaia Khimiia (Russia)*, **19**, 800. (b) Efimov, V.A., Kalinkina, A.L. and Chakhmakhcheva, O.G. (1993) *Nucleic Acid Research*, **21**, 5337.
231 Wada, T., Kobori, A., Kawahara, S. and Sekine, M. (1998) *Tetrahedron Letters*, **39**, 6907.
232 (a) Sproat, B.S., Minganti, C. and Gait, M.J. (1985) *Nucleosides Nucleotides*, **4**, 139. (b) Sproat, B.S. and Brown, D.M. (1985) *Nucleic Acids Research*, **13**, 2979.
233 Eritja, R., Robles, J., Fernandez-Forner, D., et al. (1991) *Tetrahedron Letters*, **32**, 1511.
234 (a) Eritja, R., Robles, J., Aviñó, A., Albericio, F. and Pedroso, E. (1992) *Tetrahedron*, **48**, 4171. (b) Aviñó, A., Garcia, R.G., Marquez, V.E. and Eritja, R. (1995) *Bioorganic & Medicinal Chemistry Letters*, **5**, 2331.

235 Aviñó, A. and Eritja, R. (1994) *Nucleosides Nucleotides*, **13**, 2059.
236 de la Torre, B.G., Aviñó, A., Tarrason, G., Piulats, J., Albericio, F. and Eritja, R. (1994) *Tetrahedron Letters*, **35**, 2733.
237 (a) Rabanal, F., Giralt, E. and Albericio, F. (1992) *Tetrahedron Letters*, **33**, 1775. (b) Rabanal, F., Giralt, E. and Albericio, F. (1995) *Tetrahedron*, **51**, 1449. (c) Albericio, F., Cruz, M., Debéthune, L., et al. (2001) *Synthetic Communications*, **31**, 225.
238 (a) Robles, J., Maseda, M., Beltrán, M., et al. (1997) *Bioconjugate Chemistry*, **8**, 785. (b) de la Torre, B.G., Albericio, F., Saison-Behmoaras, E., et al. (1999) *Bioconjugate Chemistry*, **10**, 1005.
239 Greenberg, M.M. and Gilmore, J.L. (1994) *Journal of Organic Chemistry*, **59**, 746; see also the application of the same carbonate photolinker in Reference [219].
240 Venkatesan, H. and Greenberg, M.M. (1996) *Journal of Organic Chemistry*, **61**, 525.
241 Greenberg, M.M., Matray, T.J., Kahl, J.D., et al. (1998) *Journal of Organic Chemistry*, **63**, 4062.
242 Chen, T., Fu, J. and Greenberg, M.M. (2000) *Organic Letters*, **2**, 3691.
243 A concept: (a) Mullen, D.G. and Barany, G. (1987) *Tetrahedron Letters*, **28**, 491. (b) Mullen, D.G. and Barany, G.J. (1988) *Organic Chemistry*, **53**, 5240. For a recent example see (c) Lipshutz, B.H. and Shin, Y.-J. (2001) *Tetrahedron Letters*, **42**, 5629.
244 Routledge, A., Wallis, M.P., Ross, K.C. and Fraser, W. (1995) *Bioorganic & Medicinal Chemistry Letters*, **5**, 2059.
245 Kobori, A., Miyata, K., Ushioda, M., et al. (2002) *Journal of Organic Chemistry*, **67**, 476.
246 (a) Ohkubo, A., Ezawa, Y., Seio, K. and Sekine, M. (2004) *Journal of the American Chemical Society*, **126**, 10884. (b) Ohkubo, A., Sakamoto, K., Miyata, K., et al. (2005) *Organic Letters*, **7**, 5389.
247 Kobori, A., Miyata, K., Ushioda, M., et al. (2002) *Chemistry Letters*, 16.
248 Ohkubo, A., Aoki, K., Seio, K. and Sekine, M. (2004) *Tetrahedron Letters*, **45**, 979.

249 Markiewicz, W.T. (1979) *Journal of Chemical Research (M)*, 0181.
250 Kwiatkowski, M., Nilsson, M. and Landegren, U. (1996) *Nucleic Acids Research*, **24**, 4632.
251 Markiewicz, W.T. and Wyrzykiewicz, T.K. (1989) *Nucleic Acids Research*, **17**, 7149.
252 A 3′-phosphoramidate linked to LCAA-CPG, applicable also for PS and TentaGel: Gryaznov, S.M. and Letsinger, R.L. (1992) *Tetrahedron Letters*, **33**, 4127.
253 Ohtsuka, E., Morioka, S. and Ikehara, M. (1972) *Journal of the American Chemical Society*, **94**, 3229.
254 Kadokura, M., Wada, T., Seio, K., et al. (2001) *Tetrahedron Letters*, **42**, 8853.
255 (a) Sommer, H. and Cramer, F. (1972) *Angewandte Chemie (International Edition in English)*, **11**, 717. (b) Sommer, H. and Cramer, F. (1974) *Chemische Berichte*, **107**, 24.
256 (a) Narang, S.A., Bhanot, O.S., Goodchild, J., et al. (1972) *Journal of the American Chemical Society*, **94**, 6183. (b) Agarwal, K.L., Fridkin, M., Jay, E. and Khorana, H.G. (1973) *Journal of the American Chemical Society*, **95**, 2020. (c) Greene, P.J., Poonian, M.S., Nussbaum, A.L., et al. (1975) *Journal of Molecular Biology*, **99**, 237. (d) Agarwal, K.L., Berlin, Y.A., Fritz, H.J., et al. (1976) *Journal of the American Chemical Society*, **98**, 1065.
257 (a) Claesen, C., Tesser, G.I., Dreef, C.E., et al. (1984) *Tetrahedron Letters*, **25**, 1307. (b) Balgobin, N. and Chattopadhyaya, J. (1985) *Acta Chemica Scandinavica*, **B39**, 883. (c) Horn, T. and Urdea, M.S. (1986) *Tetrahedron Letters*, **27**, 4705.
258 Brandstetter, F., Schott, H. and Bayer, E. (1974) *Tetrahedron Letters*, **15**, 2705.
259 Balgobin, N., Josephson, S. and Chattopadhyaya, J.B. (1981) *Tetrahedron Letters*, **22**, 1915.
260 Efimov, V.A., Buryakova, A.A., Reverdatto, S.V., et al. (1983) *Nucleic Acids Research*, **11**, 8369.
261 Reintamm, T., Müller, U., Oretskaya, T.S., et al. (1990) *Bioorganicheskaia Khimiia (Russia)*, **16**, 524.
262 Felder, E., Schwyzer, R., Charubala, R., Pfleiderer, W. and Schulz, B. (1984) *Tetrahedron Letters*, **25**, 3967.
263 Asseline, U. and Thuong, N.T. (1989) *Tetrahedron Letters*, **30**, 2521.

264 Efimov, V.A., Kalinkina, A.L. and Chakhmakhcheva, O.G. (1995) *Russian Journal Bioorganic Chemistry*, **21**, 527.
265 Asseline, U., Bonfils, E., Kurfürst, R., et al. (1992) *Tetrahedron*, **48**, 1233.
266 Lartia, R. and Asseline, U. (2004) *Tetrahedron Letters*, **45**, 5949.
267 Gupta, K.C., Sharma, P., Kumar, P. and Sathyanarayana, S. (1991) *Nucleic Acids Research*, **19**, 3019.
268 Kumar, P., Bose, N.K. and Gupta, K.C. (1991) *Tetrahedron Letters*, **32**, 967.
269 Zhang, X. and Jones, R.A. (1996) *Tetrahedron Letters*, **37**, 3789.
270 McMinn, D.L., Hirsch, R. and Greenberg, M.M. (1998) *Tetrahedron Letters*, **39**, 4155.
271 Dell'Aquila, C., Imbach, J.-L. and Rayner, B. (1997) *Tetrahedron Letters*, **38**, 5289.
272 Roland, A., Xiao, Y., Jin, Y. and Iyer, R.P. (2001) *Tetrahedron Letters*, **42**, 3669.
273 Langenegger, S.M., Moesch, L., Natt, F., Hall, J. and Häner, R. (2003) *Helvetica Chimica Acta*, **86**, 3476.
274 Cheruvallath, Z.S., Cole, D.L. and Ravikumar, V.T. (2003) *Bioorganic & Medicinal Chemistry Letters*, **13**, 281.
275 Murata, A. and Wada, T. (2006) *Tetrahedron Letters*, **47**, 2147.
276 Asseline, U. (2002) Properties of modified oligo nucleotides involving positive charges: Influence of their number and placement, in *Modified Nucleosides, Synthesis and Applications* (ed. D. Loakes), Research Signpost, p. 33.
277 See Stetsenko, D.A. and Gait, M.J. (2001) *Bioconjugate Chemistry*, **12**, 576 and the references cited therein.
278 See, e.g. (a) Dioubankova, N.N., Malakhov, A.D., Stetsenko, D.A., et al. (2002) *Organic Letters*, **4**, 4607. (b) Dioubankova, N.N., Malakhov, A.D., Stetsenko, D.A., et al. (2006) *Tetrahedron*, **62**, 6762.
279 Petrie, C.R., Reed, M.W., Adams, A.D. and Meyer Jc, R.B. (1992) *Bioconjugate Chemistry*, **3**, 85.
280 Nelson, P.S., Frye, R.A. and Liu, E. (1989) *Nucleic Acids Research*, **17**, 7187.
281 Lyttle, M.H., Adams, H., Hudson, D. and Cook, R.M. (1997) *Bioconjugate Chemistry*, **8**, 193.
282 (a) Leuck, M., Giare, R., Paul, M., et al. (2004) *Tetrahedron Letters*, **45**, 317. (b) Leuck, M., Giare, R., Zien, N., et al. (2005) *Nucleosides Nucleotides Nucleic Acids*, **24**, 989.
283 Gryaznov, S.M. and Letsinger, R.L. (1993) *Tetrahedron Letters*, **34**, 1261.
284 (a) Nielsen, J., Taagard, M., Marugg, J.E., et al. (1986) *Nucleic Acids Research*, **14**, 7391. (b) Nielsen, J. and Dahl, O. (1987) *Nucleic Acids Research*, **15**, 3626.
285 The Atherton–Todd reaction: Atherton, F.R., Openshaw, H.T. and Todd, A.R. (1945) *Journal of the Chemical Society*, 660.
286 Kumar, P., Gupta, K.C., Rosch, R. and Seliger, H. (1996) *Bioorganic & Medicinal Chemistry Letters*, **6**, 2247.
287 Asseline, U. and Thuong, N.T. (1990) *Tetrahedron Letters*, **31**, 81.
288 Aviñó, A., Garcia, R.G., Albericio, F., et al. (1996) *Bioorganic & Medicinal Chemistry*, **4**, 1649.
289 McMinn, D.L. and Greenberg, M.M. (1996) *Tetrahedron*, **52**, 3827.
290 (a) McMinn, D.L., Matray, T.J. and Greenberg, M.M. (1997) *Journal of Organic Chemistry*, **62**, 7074. (b) McMinn, D.L. and Greenberg, M.M. (1998) *Journal of the American Chemical Society*, **120**, 3289.
291 Kahl, J.D., McMinn, D.L. and Greenberg, M.M. (1998) *Journal of Organic Chemistry*, **63**, 4870.
292 Greenberg, M.M. and Kahl, J.D. (2001) *Journal of Organic Chemistry*, **66**, 7151.
293 Gupta, K.C., Sharma, P., Sathyanarayana, S. and Kumar, P. (1990) *Tetrahedron Letters*, **31**, 2471.
294 Bonfils, E. and Thuong, N.T. (1991) *Tetrahedron Letters*, **32**, 3053.
295 (a) Kremsky, J.N., Wooters, J.L., Dougherty, et al. (1987) *Nucleic Acids Research*, **15**, 2891. (b) Gottikh, M., Asseline, U. and Thuong, N.T. (1990) *Tetrahedron Letters*, **31**, 6657. (c) Hovinen, J., Guzaev, A., Azhayev, A. and Lönnberg, H. (1994) *Journal of the Chemical Society, Perkin Transactions 1*, 2745. (d) Guzaev, A., Hovinen, J., Azhayev, A. and Lönnberg, H. (1995) *Nucleosides Nucleotides*, **14**, 833. (e) Kachalova, A.V., Stetsenko, D.A., Romanova, et al. (2002) *Helvetica Chimica Acta*, **85**, 2409. (f) Kachalova, A.V., Stetsenko, D.A., Gait,

M.J. and Oretskaya, T.S. (2004) *Bioorganic & Medicinal Chemistry Letters*, **14**, 801. (g) Kachalova, A., Zubin, E., Stetsenko, D., et al. (2004) *Organic & Biomolecular Chemistry*, **2**, 2793. (h) Lebedev, A.V., Combs, D. and Hogrefe, R.I. (2007) *Bioconjugate Chemistry*, **18**, 1530.

296 Bonfils, E., Depierreux, C., Midoux, P., et al. (1992) *Nucleic Acids Research*, **20**, 4621.

297 Hovinen, J., Gouzaev, A.P., Azhayev, A.V. and Lönnberg, H. (1993) *Tetrahedron Letters*, **34**, 5163.

298 Hovinen, J., Guzaev, A., Azhayev, A. and Lönnberg, H. (1993) *Tetrahedron Letters*, **34** (50), 8169.

299 Hovinen, J., Guzaev, A., Azhayev, A. and Lönnberg, H. (1994) *Tetrahedron*, **50**, 7203.

300 Matray, T.J., Yoo, D.J., McMinn, D.L. and Greenberg, M.M. (1997) *Bioconjugate Chemistry*, **8**, 99.

301 Kahl, J.D. and Greenberg, M.M. (1999) *Journal of Organic Chemistry*, **64**, 507.

302 (a) Alazzouzi, E., Escaja, N., Straub, M., et al. (1997) *Nucleosides Nucleotides*, **16**, 1513. (b) Alazzouzi, E., Escaja, N., Grandas, A. and Pedroso, E. (1997) *Angewandte Chemie (International Edition in English)*, **36**, 1506. (c) Frieden, M., Grandas, A. and Pedroso, E. (1999) *Nucleosides Nucleotides*, **18**, 1181.

303 Lehman, C., Xu, Y., Christodoulou, C., et al. (1989) *Nucleic Acids Research*, **17**, 2379.

304 Palom, Y., Alazzouzi, E., Gordillo, F., et al. (1993) *Tetrahedron Letters*, **34**, 2195.

305 Shemyakin, M.M., Ovchinnikov, Yu.A., Kiryushkin, A.A. and Kozhevnikova, I.V. (1965) *Tetrahedron Letters*, **6**, 2323.

306 For a review see: Gravert, D.J. and Janda, K.D. (1997) *Current Opinion in Chemical Biology*, **1**, 107.

307 (a) Hayatsu, H. and Khorana, H.G. (1966) *Journal of the American Chemical Society*, **88**, 3182. (b) Hayatsu, H. and Khorana, H.G. (1967) *Journal of the American Chemical Society*, **89**, 3880.

308 Cramer, F., Helbig, R., Hettler, H., et al. (1966) *Angewandte Chemie (International Edition in English)*, **5**, 601.

309 Mutter, M., Hagenmaier, H. and Bayer, E. (1971) *Angewandte Chemie (International Edition in English)*, **10**, 811.

310 Köster, H. (1972) *Tetrahedron Letters*, **13**, 1534.

311 Brandstetter, F., Schott, H. and Bayer, E. (1973) *Tetrahedron Letters*, **14**, 2997.

312 Schott, H. (1973) *Angewandte Chemie (International Edition in English)*, **12**, 246.

313 Seliger, H. and Aumann, G. (1973) *Tetrahedron Letters*, **14**, 2911.

314 (a) Kamaike, K., Hasegawa, Y. and Ishido, Y. (1988) *Tetrahedron Letters*, **29**, 647. (b) Kamaike, K., Hasegawa, Y., Masuda, I., et al. (1990) *Tetrahedron*, **46**, 163.

315 (a) Bonora, G.M., Scremin, C.L., Colonna, F.P. and Garbesi, A. (1990) *Nucleic Acids Research*, **18**, 3155. (b) Bonora, G.M., Scremin, C.L., Colonna, F.P. and Garbesi, A. (1991) *Nucleosides Nucleotides*, **10**, 269. (c) Colonna, F.P., Scremin, C.L. and Bonora, G.M. (1991) *Tetrahedron Letters*, **32**, 3251.

316 Bonora, G.M., Biancotto, G., Maffini, M. and Scremin, C.L. (1993) *Nucleic Acids Research*, **21**, 1213.

317 (a) Zaramella, S. and Bonora, G.M. (1995) *Nucleosides Nucleotides*, **14**, 809. For an interesting variation of the H-phosphonate synthesis on a PEG support see: (b) Padiya, K.J. and Salunkhe, M.M. (2000) *Bioorganic & Medicinal Chemistry*, **8**, 337.

318 Tetraethylthiuram disulfide (TETD) sulfurization: (a) Scremin, C.L. and Bonora, G.M. (1993) *Tetrahedron Letters*, **34**, 4663. Dithiodicarbonate disulfide (DDD) sulfurization: (b) Bonora, G.M., Rossin, R., Zaramella, S. et al. (2000) *Organic Process Research & Development*, **4**, 225.

319 (a) Bonora, G.M., Ivanova, E., Zarytova, V., et al. (1997) *Bioconjugate Chemistry*, **8**, 793. (b) Burcovich, B., Veronese, F.M., Zarytova, V. and Bonora, G.M. (1998) *Nucleosides Nucleotides*, **17**, 1567. (c) Bonora, G.M., Ivanova, E., Komarova, N., et al. (1999) *Nucleosides Nucleotides*, **18**, 1723. (d) Vorobjev, P.E., Zarytova, V.F. and Bonora, G.M. (1999) *Nucleosides Nucleotides*, **18**, 2745. (e) Ballico, M., Cogoi, S., Drioli, S. and Bonora, G.M. (2003) *Bioconjugate Chemistry*, **14**, 1038.

320 (a) Ballico, M., Drioli, S., Morvan, F., et al. (2001) *Bioconjugate Chemistry*, **12**, 719. (b) Rapozzi, V., Cogoi, S., Spessotto, P., et al. (2002) *Biochemistry*, **41**, 502.

321 Bonora, G.M., Ballico, M., Campaner, P., *et al.* (2003) *Nucleosides Nucleotides*, **22**, 1255.
322 Bonora, G.M., Baldan, A., Schiavon, O., *et al.* (1996) *Tetrahedron Letters*, **37**, 4761.
323 Bonora, G.M., De Franco, A.M., Rossin, R., *et al.* (2000) *Nucleosides Nucleotides*, **19**, 1281.
324 Noro, A., Nagata, Y., Tsukamoto, M., *et al.* (2005) *Biomacromolecules*, **6**, 2328.
325 (a) Rubinstein, M. and Patchornik, A. (1972) *Tetrahedron Letters*, **13**, 2881. (b) Rubinstein, M. and Patchornik, A. (1975) *Tetrahedron*, **31**, 1517.
326 Adamo, I., Dueymes, C., Schönberger, A., *et al.* (2006) *European Journal of Organic Chemistry*, **2**, 436.
327 Mohe, N., Heinonen, P., Sanghvi, Y.S. and Strömberg, R. (2005) *Nucleosides Nucleotides Nucleic Acids*, **24**, 897.
328 Seliger, H. and Gupta, K.C. (1985) *Angewandte Chemie (International Edition in English)*, **24**, 685.
329 Dueymes, C., Schönberger, A., Adamo, I., *et al.* (2005) *Organic Letters*, **7**, 3485.
330 Zlatev, I., Kato, Y., Meyer, A., *et al.* (2006) *Tetrahedron Letters*, **47**, 8379.
331 Sanghvi, Y.S., Guo, Z., Pfundheller, H.M. and Converso, A. (2000) *Organic Process Research & Development*, **4**, 175.
332 Iyer, R.P., Jiang, Z., Yu, D., *et al.* (1995) *Synthetic Communications*, **25**, 3611.
333 Patil, S.V., Mane, R.B. and Salunkhe, M.M. (1994) *Synthetic Communications*, **24**, 2423.
334 Bollmark, M. and Stawinski, J. (2001) *Chemical Communications*. 771.
335 Zhang, Z., Han, Y., Tang, J.X. and Tang, J.-Y. (2002) *Tetrahedron Letters*, **43**, 4347.
336 Padiya, K.J., Mohe, N.U. and Salunkhe, M.M. (2002) *Synthetic Communications*, **32**, 917.
337 (a) Reese, C.B. and Song, Q. (1999) *Journal of the Chemical Society, Perkin Transactions 1*, 1477. (b) Reese, C.B. and Yan, H. (2002) *Journal of the Chemical Society, Perkin Transactions 1*, 2619.
338 Ferreira, F., Meyer, A., Vasseur, J.-J. and Morvan, F. (2005) *Journal of Organic Chemistry*, **70**, 9198.
339 Wengel, J. (2004) *Organic & Biomolecular Chemistry*, **2**, 277.
340 Ito, Y. and Fukusaki, E. (2004) *Journal of Molecular Catalysis B: Enzymatic*, **28**, 155.

20
Oligosaccharides

Peter H. Seeberger[1] and Harald Wippo[2]

[1]Swiss Federal Institute of Technology (ETH), Laboratory of Organic Chemistry, ZürichWolfgang-Pauli-Str. 10, 8093 Zürich, Switzerland
[2]Senn Chemicals AG, P.O. Box 267, 8157 Dielsdorf, Switzerland

20.1
Introduction

Oligosaccharides are important target structures for bioorganic chemists. The significance of oligosaccharides and glycoconjugates in biological recognition processes is widely recognized as very important [1–11]. The fast and reliable synthesis of useful amounts of pure oligosaccharides remains a demanding task for organic chemists. Whereas oligonucleotides and oligopeptides are routinely prepared by automated synthesisers, the synthesis of oligosaccharides is a far more delicate task. Several difficulties are encountered in oligosaccharide synthesis: the protecting group pattern of the building blocks, the stereochemical control during the coupling reaction that generates a new stereogenic center during each glycosylation and glycosylation reactions are often unpredictable regarding product distribution, side product formation and yield [12].

Considerable effort has been made in recent years to improve the yields of oligosaccharide synthesis. The selection of the appropriate functional resin is, in many cases, crucial to the outcome of oligosaccharide syntheses. Here, we briefly review the functional resins that have been used for oligosaccharide synthesis. The chapter is divided into two main sections: First the most common insoluble and soluble resins applied in oligosaccharide synthesis are reviewed and then the different types of linkers applied in oligosaccharide synthesis are summarized. Two subsections are dedicated to ionic liquids (Section 20.3.3), representing a special case of "soluble resins", and magnetic particles (Section 20.2.3), which are considered a relatively new concept of insoluble solid supports.

We clearly distinguish resins and linkers: resins are considered to be inert during the reactions whereas linkers are basically immobilized protecting groups attached to the resin. Linkers connect the resin with the first building block and play an important role as spacer. As for all protecting groups, the right choice is

The Power of Functional Resins in Organic Synthesis. Judit Tulla-Puche and Fernando Albericio
Copyright © 2008 WILEY-VCH Verlag GmbH & Co. KGaA, Weinheim
ISBN: 978-3-527-31936-7

important for synthesis design and compatibility of all chemical steps during oligosaccharide assembly. In addition, the cleavage of the linker from the target molecule should be high yielding and not cause the formation of side products.

In the chemical literature, linkers are further divided into integral linkers that are part of the resin and non-integral or grafted linkers, where a part of the linker is attached to the core of the resin. The integral linkers exhibit some disadvantages since the reaction takes place in direct proximity to the resin. Steric and electronic effects will exert a direct influence on the outcome of the reaction. An additional problem to be alleviated is the determination of the exact loading of a resin with an integrated linker [13].

20.2
Insoluble Resins

20.2.1
Controlled-Pore Glass (CPG)

Controlled-pore Glass (CPG) solid supports have been used extensively for oligonucleotide assembly, but less frequently for oligosaccharide synthesis. CPG-supports are non-swelling, exhibit only small loading capacities due to the limited functionlized surface and tolerate a broad range of solvents. The mechanical instability due to fracture somewhat hampers the handling of the CPG beads. Unfortunately, CPG-supports are incompatible with silyl ether protecting groups that are commonly employed for temporary protection of hydroxyl groups in carbohydrate chemistry. In 1998, Schmidt *et al.* reported the synthesis of an alpha(1 → 2)-linked trimannoside using a mercapto-functionalized CPG-support [14] and glycosyl trichloroacetimidate-building blocks. The CPG supports are available as mercapto-propyl-functionalized CPG (300 µmol g^{-1} loading) [15], amino-functionalized CPG (20–50 µmol g^{-1}) [16] or glycol-functionalized CPG (~20 µmol g^{-1}) [17]. Hindsgaul *et al.* [18] used the tagged CPG-attached linker **1** for solid phase oligosaccharide tagging (SPOT) (Scheme 20.1).

N-Tetrose (LNT) (**2**) and fucosylated LNT lacto-*N*-fucopentaose (LNFP) (**3**) were used as specific structures for the SPOT experiment (Scheme 20.2). Benzyl hydroxylamine capture groups were attached through an ester function to solid support **4**. Reducing sugar **5** was captured to form oxime ether **6**. After capping with acetic anhydride, the oxime ether was reduced and the intermediate N,O-alkylhydroxylamine **7** was tagged with tetramethylrhodamine isothiocyanate (**8**). Cleavage of **9** with LiOH(aq) released tagged sugar **10** ready for analysis.

20.2.2
Polystyrene Resins

20.2.2.1 Polystyrene Resin Cross-linked with Divinylbenzene
Most oligosaccharide syntheses have relied on Merrifield's resin [polystyrene (PS), cross-linked with 1% divinylbenzene], which was originally a nitrated chloro-

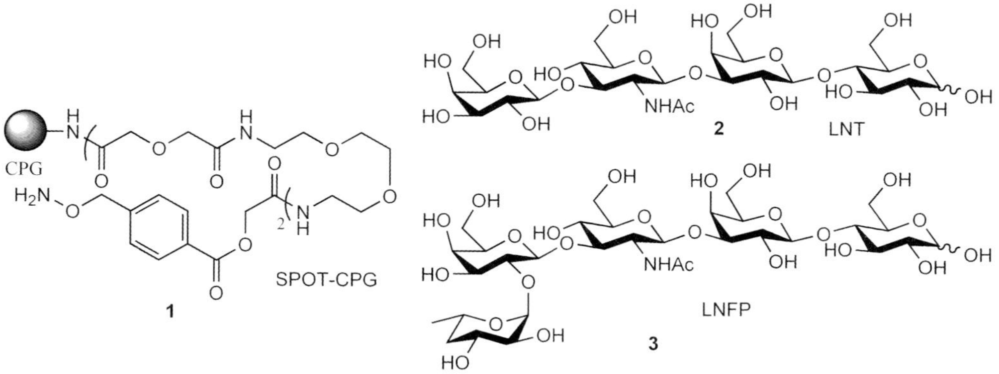

Scheme 20.1 Structures of SPOT-CPG (**1**), LNT (**2**) and LNFP (**3**).

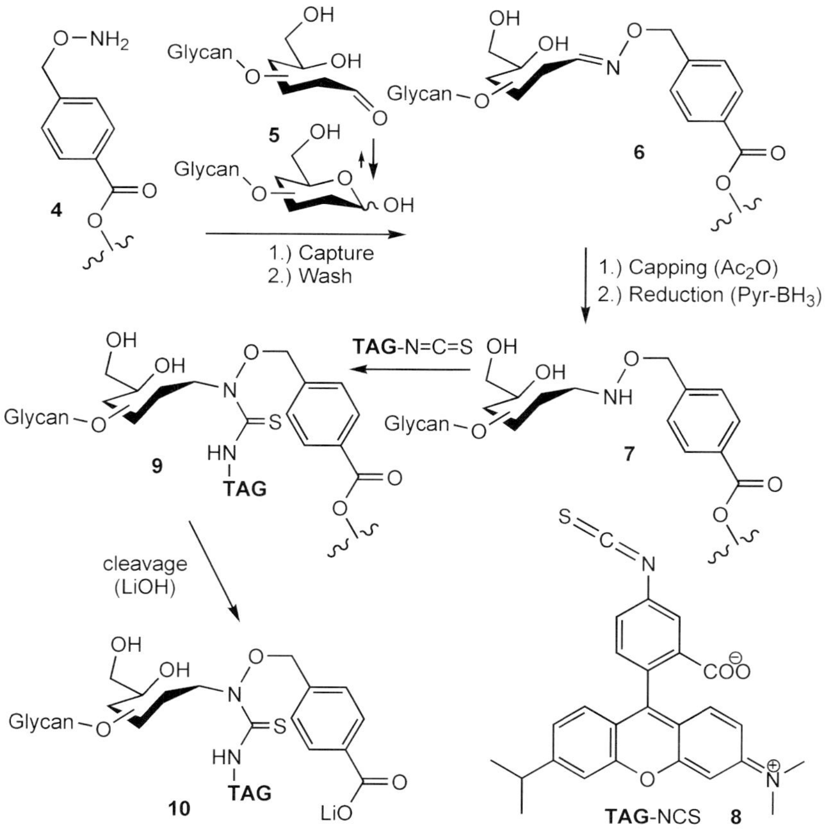

Scheme 20.2 SPOT methodology.

methyl polystyrene resin applied in peptide syntheses [19]. Cross-linked PS is often the resin of choice due to high loading capacities (0.5–2.5 mmol g^{-1}), the broad range of reaction conditions that are tolerated, its mechanical durability and low price. Nevertheless, PS has some drawbacks that must be considered for successful syntheses. Prior to use, the resin needs to be swollen in solvents such as THF, dichloromethane, DMF or dioxane to gain access to all reactive sites. Only solvents that sufficiently swell the resin can be applied [20].

20.2.2.2 PEG Grafted Polystyrenes

The use of more polar solvents is possible when poly(ethylene glycol) (PEG) grafted polystyrene is used. Resins with long PEG-chains are produced by means of anionic graft copolymerization, whereas short PEG-chains can be attached via classical ether synthesis. Graft copolymers with PEG chains of about 2000–3000 dalton are usually employed for best performance. The benzyl ether linkage between the PEG chains and the PS-backbone can be cleaved and may result in PEG-contamination of the product. Resins such as TentaGels [21], a trade mark of Rapp Polymere [22], show remarkable swelling properties even in water, at the expense of lower loading capacities (0.2–0.3 mmol g^{-1}) and higher prices. Argo-Gels, provide higher loading capacities (0.3–0.5 mmol g^{-1}) due to flexible PEG-grafts, generating almost solution-like conditions for the polymer-bound molecules. These gels can be used with a wide range of solvents, including water and contain only a small amount of freely accessible PEG. Table 20.1 provides an overview over commercially available ArgoGels with different functional groups [23].

A range of PEG-based resins are available, including PEGA [24], POEPOP [25] and POEPS [26, 27], that do not stand up to the harsh conditions employed during Friedel-Crafts reactions or acetolysis. To circumvent these drawbacks, the SPOCC that contains only primary ether bonds, secondary and quarternary carbon atoms was introduced. The synthesis of SPOCC resin is summarized in Scheme 20.3 [27].

20.2.3
Magnetic Particles

Magnetic particles have received increasing attention in recent years [28–33]. Functionalized magnetic particles have relatively high loading capacities, similar to those of classical insoluble organic resins, and do not swell in organic solvents. Magnetic particles can be used in capillary reactors to combine the advantages of microreactor technology and solid phase synthesis. An external electromagnetic field allows one to trap and release the magnetic particles (magnetic trap) [34–36]. Interesting applications in fully automated synthesisers may evolve from these beginnings. Shimomura *et al.* have described the use of magnetic particles modified by graft polymerization of acrylic acid (**11**). 3-Aminophenylboronic acid (**12**) was coupled to the free carboxylate groups of polymer **11** (Scheme 20.4). Complexation of polymer **13** with mono-, di- and oligosaccharides is possible [37]. This methodology may facilitate the isolation of sugars from complex natural extracts by magnetically forced sedimentation of the sugar-loaded magnetic particles.

Table 20.1 Selection of commercially available ArgoGels.

Structure	Gel	Functional group
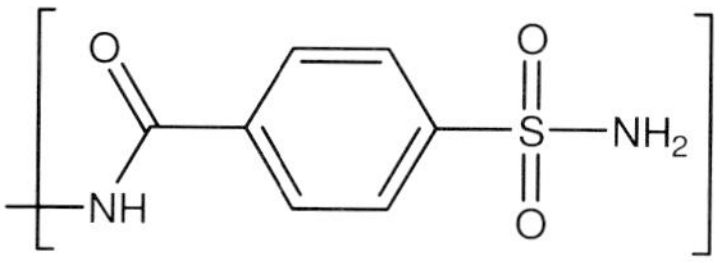 [benzamide-sulfonamide structure]	ArgoGel™-AS-SO$_2$NH$_2$	0.3–0.4 mmol g^{-1}
–[OCH$_2$CH$_2$-Cl]	ArgoGel™-Cl	—
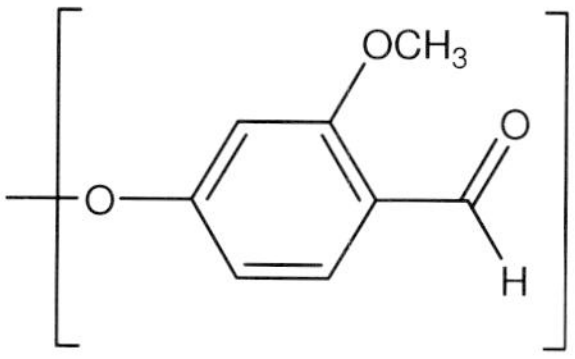 [2-methoxy-4-alkoxy-benzaldehyde structure]	ArgoGel™-MB-CHO	0.35–0.45 mmol-aldehyde g^{-1}
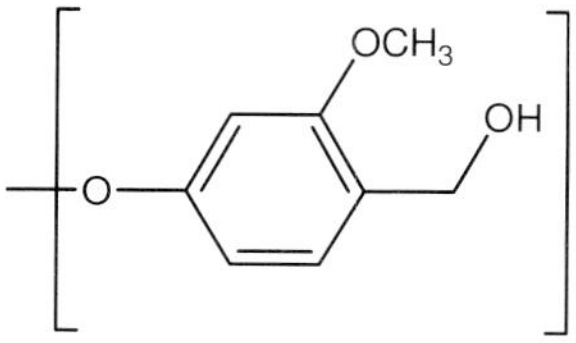 [2-methoxy-4-alkoxy-benzyl alcohol structure]	ArgoGel™-MB-CH$_2$OH	0.35–0.40 mmol-alcohol g^{-1}
–[OCH$_2$CH$_2$-OH]	ArgoGel™-OH	0.40–0.50 mmol-alcohol g^{-1}

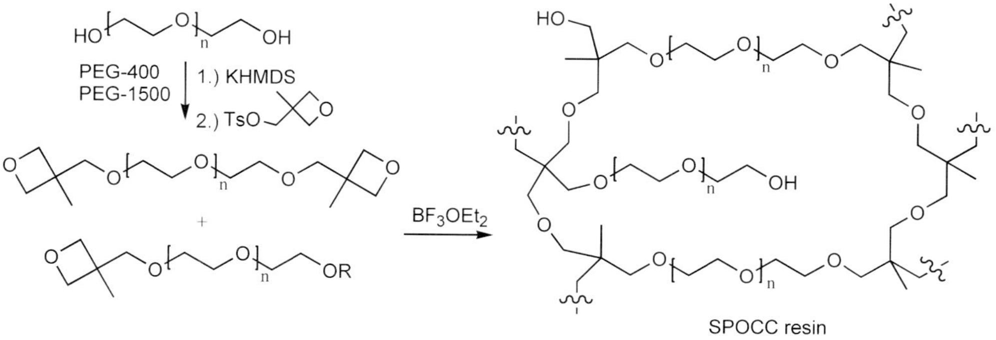

Scheme 20.3 Synthesis of SPOCC resin.

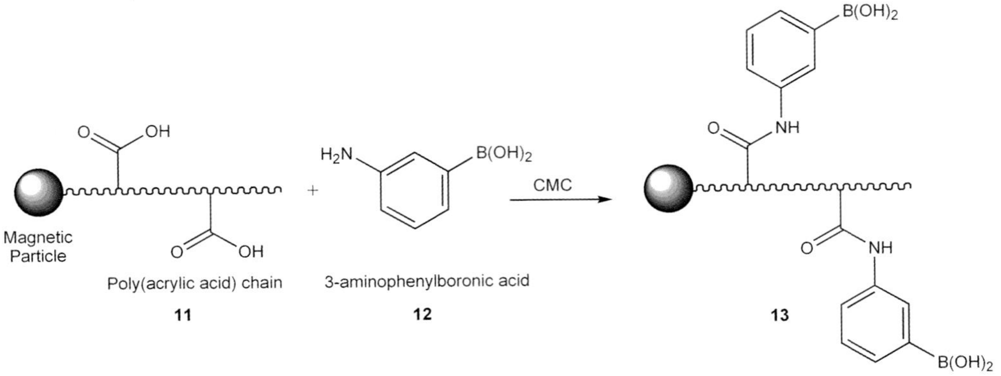

Scheme 20.4 Coupling of 3-aminophenylboronic acid (12) to poly(acrylic acid) (11) attached to a magnetic surface.

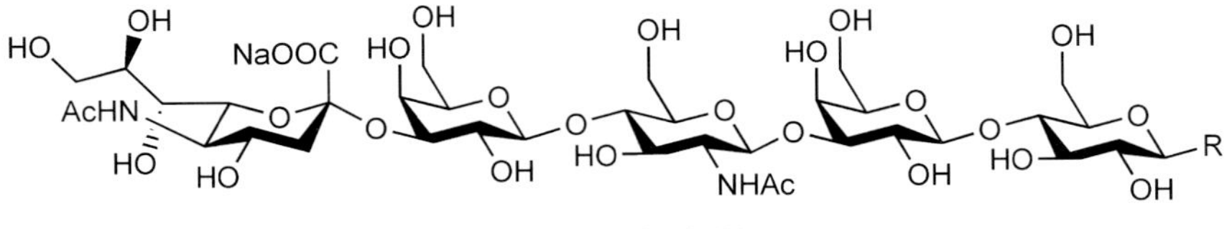

Figure 20.1 Structure of the *Neisseria meningitidis* pentasaccharide (14).

20.3
Soluble Resins

Soluble resins combine the advantages of solution phase synthesis with the simple work-up of solid phase synthesis. The soluble resin and the reagents form a homogeneous system during the reaction. After each coupling step the polymer is precipitated and carefully washed to remove excess reagents. Incomplete precipitation results in a loss of resin after each precipitation step. A further drawback is the limited temperature range, since most soluble resins precipitate below −45 °C.

20.3.1
MPEG Resins

MPEG-type resins have been widely applied in oligosaccharide syntheses [38–40]. Whitfield et al. [41] synthesized the *Neisseria meningitidis* pentasaccharide [Neu5Acα(2→3)-Galβ(1→4)GlcNAcβ(1→3)Galβ(1→4)Glc-R] (14) (Figure 20.1) by using chemoenzymatic and polymer-supported methods.

Protected thioglycoside 15 was attached to the soluble polymer support via a DOX-linker 16 (Scheme 20.5) using the NIS/AgOTf protocol, followed by acidic hydrolysis of the acetal using acetic acid. Disaccharide 17 was coupled with glycosyl

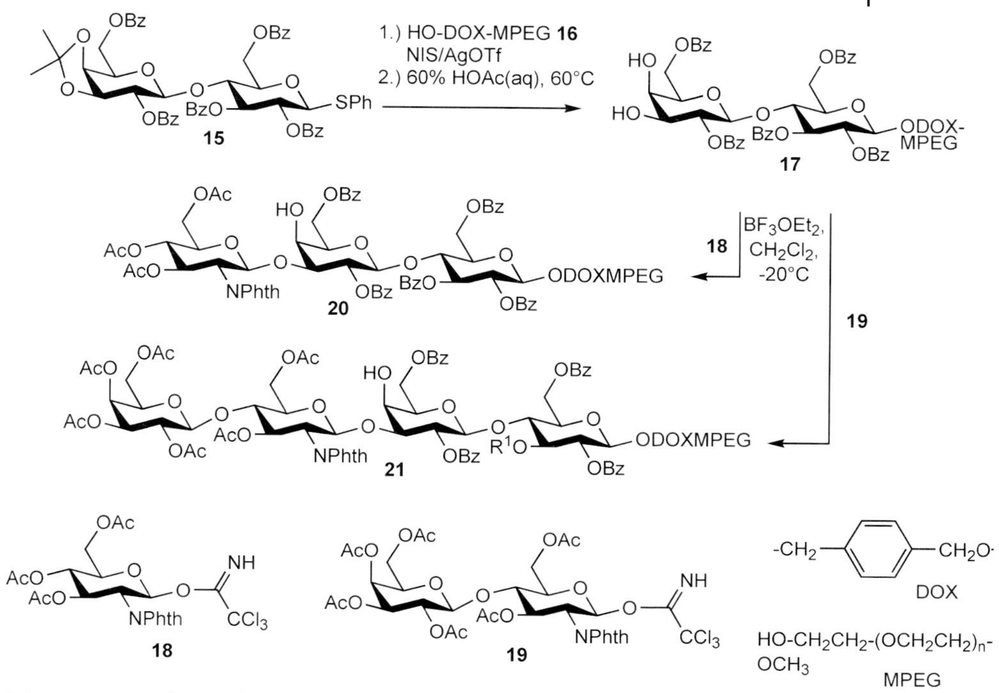

Scheme 20.5 Syntheses of tri- and tetrasaccharides (**20** and **21**) on soluble MPEG-support as precursors in the synthesis of the *Neisseria meningitidis* pentasaccharide (**14**).

trichloroacetimidates **18** and **19**, respectively, to furnish trisaccharide **20** and tetrasaccharide **21** in high yields.

20.3.2
Hyperbranched Soluble Resins

Nonlinear, dendritic or hyperbranched polymers have received considerable attention in recent years. These polymers have seen little use as supports for oligosaccharide synthesis compared to Merrifield or soluble MPEG resins. Nevertheless, some interesting applications of hyperbranched polymers such as hyperbranched polyester and poly(amidoamine) to oligsaccharide synthesis have been reported.

20.3.2.1 Hyperbranched Polyester

A dimannoside was synthesized by Parquette *et al.* using hyperbranched polyester **22** [42] (Figure 20.2). This polyester is soluble in most aprotic solvents and can be almost quantitatively precipitated by methanol or separated by size-exclusion chromatography (SEC).

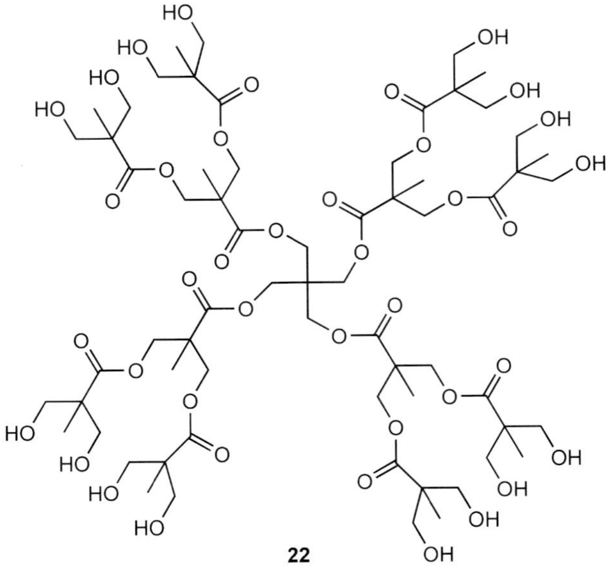

Figure 20.2 Idealized structure of Boltron Hyperbranched Polymer (**22**).

The polymer was first synthesized from pentaerythritol and dimethylolpropionic acid as the repeating monomer [43]. The resulting hyperbranched polyester can be produced in large quantities at low costs and is soluble in a broad range of polar solvents such as THF, diethylether, dichloromethane and acetone, but readily precipitates in high yield in methanol. In aqueous media the polymeric ester can be hydrolyzed, allowing for simple work-up of the reaction and extraction of the product. Connecting the first mannose **23** via photocleavable linker **24** to polymeric support **22** allows for easy product isolation initiated by photolytic cleavage of the linker [43] (Scheme 20.6).

Glycosylation of partially deprotected **26** with the mannoside building block **27** afforded, after size exclusion chromatography and cleavage from the polymer support, exclusively the α-linked dimannoside derived from **28** in 49% yield (Scheme 20.7). Replacing the acetyl protecting group by a benzyl group on C2 of mannoside building block **27** raises the yield of dimannoside to 78%, but results in a complete loss of α,β-selectivity.

20.3.2.2 PAMAM

PAMAMs [poly(amidoamine) dendrimers] are commercially available hyperbranched cores. T-antigen [Thomsen-Friedenreich, β-Gal-(1,3)-α-D-GalNAc-OR, **29**] was successfully attached to PAMAM **30** via a 3-(propyl)mercaptopropionic acid linker to investigate T-antigen–protein interactions [44] (Scheme 20.8).

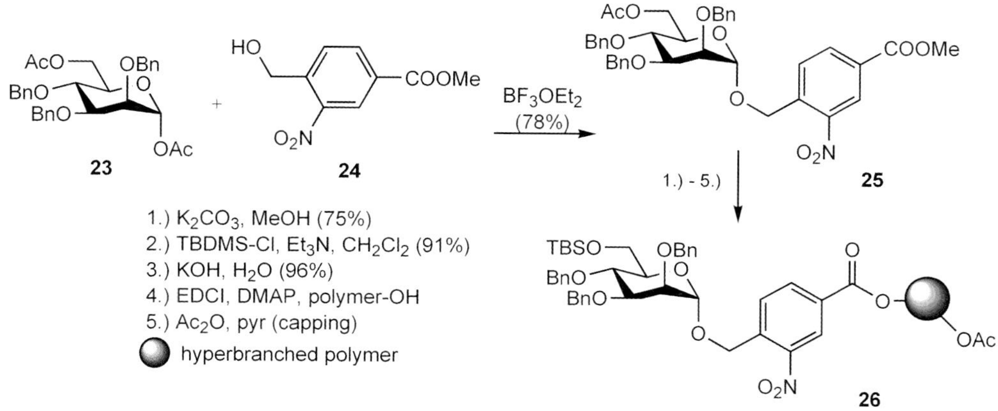
AcO
OBn
BnO
BnO
OAc
23
HO
COOMe
O2N
24
BF3OEt2
(78%)
25
1.) - 5.)
TBSO
O
26
1.) K2CO3, MeOH (75%)
2.) TBDMS-Cl, Et3N, CH2Cl2 (91%)
3.) KOH, H2O (96%)
4.) EDCI, DMAP, polymer-OH
5.) Ac2O, pyr (capping)
hyperbranched polymer

Scheme 20.6 Attachment of the first mannoside building block to a resin-bound photo-cleavable linker.

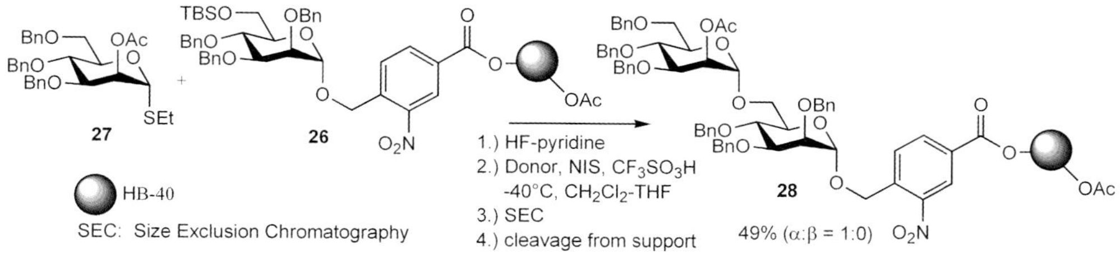
BnO
OAc
SEt
27
TBSO
OBn
26
O2N
HB-40
SEC: Size Exclusion Chromatography
1.) HF-pyridine
2.) Donor, NIS, CF3SO3H
-40°C, CH2Cl2-THF
3.) SEC
4.) cleavage from support
28
49% (α:β = 1:0)

Scheme 20.7 Dimannoside solid phase synthesis.

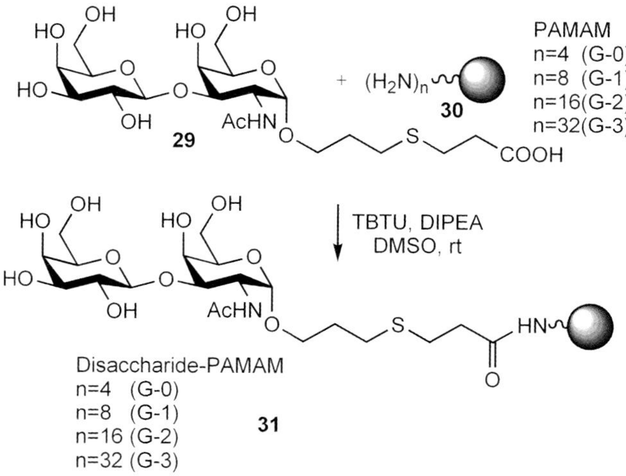
HO
OH
AcHN
29
(H2N)n
30
PAMAM
n=4 (G-0)
n=8 (G-1)
n=16(G-2)
n=32(G-3)
S
COOH
TBTU, DIPEA
DMSO, rt
HN
Disaccharide-PAMAM
n=4 (G-0)
n=8 (G-1)
n=16 (G-2)
n=32 (G-3)
31

Scheme 20.8 Synthesis of T-antigen-linker-PAMAMs (**31**).

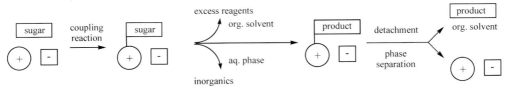

Scheme 20.9 Ionic liquids as tag for oligosaccharide synthesis.

20.3.3
Ionic Liquids

Ionic liquids (IL) are organic salts with melting points below ambient or reaction temperature and are characterized by very low vapor pressures and high thermal stabilities. The polarity of ionic liquids can be tuned to provide hydrophilic as well as hydrophobic liquids. Ionic liquids are a very special case of a "soluble support". In principle, a solid or a soluble support, like a resin or a soluble polymer, is replaced by an ionic liquid having a functional group incorporated for attachment of the first building block. Oligosaccharide syntheses are conducted in homogeneous solution. Polar reagents can be extracted with water when a hydrophobic ionic liquid is used (liquid–liquid extraction). The homogeneous nature of this approach requires fewer building blocks than solid phase synthesis due to the homogeneous system. We illustrate an oligosaccharide synthesis using ionic liquids as tag and not as solvent [45]. The substrate is attached to an ionic liquid of distinct polarity, which should enable isolation of the IL-bound product applying liquid–liquid extraction. After removal of excess reagents and monomeric building blocks, the IL-bound product is used in the next synthesis step. The final step liberates the bound product and the ionic liquid can be reused. Evidently, the nature of the bound product influences the distribution between the solvent and the water phase and the properties will change during the progress of the synthesis (Scheme 20.9).

The protected triglucan **32** was synthesized using [mimBF$_4$] as a tag and did not require purification by column chromatography throughout the entire synthesis (Scheme 20.10).

20.4
Linkers

20.4.1
Acid- and Base-Labile Linkers

20.4.1.1 Acid-Labile Linkers
Acid-labile linkers have been used primarily for peptide chemistry but also for oligosaccharide synthesis. Amino-functionalized Rink resin has served to prepare disaccharide libraries (Scheme 20.11) [46].

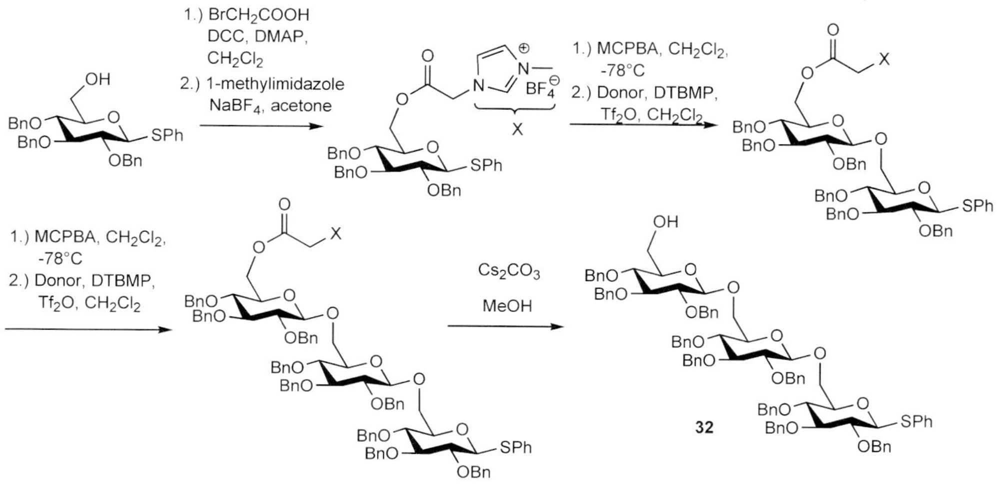

Scheme 20.10 Triglucan synthesis using an ionic liquid tag.

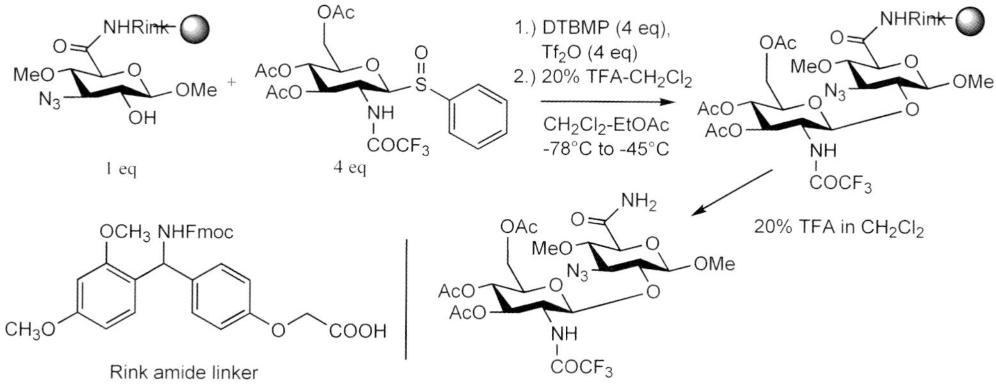

Scheme 20.11 Disaccharide synthesis and acid cleavage of Rink resin.

A tris(alkoxy)benzylamine (BAL) linker was used for the preparation of oligosaccharides containing D-glucosamine as the terminal sugar [47, 48]. D-Glucosamine (**34**) was attached by reductive amination to the resin-bound BAL linker **33** in high yield (Scheme 20.12). Partially protected D-glucosamine acceptor **37**, attached to the BAL-modified aminomethylated polystyrene, was coupled with 2,3,4,6-tetra-*O*-benzoyl-α-D-glucopyranosyl trichloroacetimidate (**36**) to furnish disaccharide **38**. Subsequent N-acylation affords an acid-labile linkage that is cleavable by TFA–H$_2$O (19:1) to furnish disaccharide **39** in 82% yield (five steps from aminomethylated PS, α:β ratio: 1:10) (Scheme 20.13).

Hanessian and coworkers connected benzylidene acetals of carbohydrates to Wang aldehyde resin. This linker was used as a temporary protecting group for

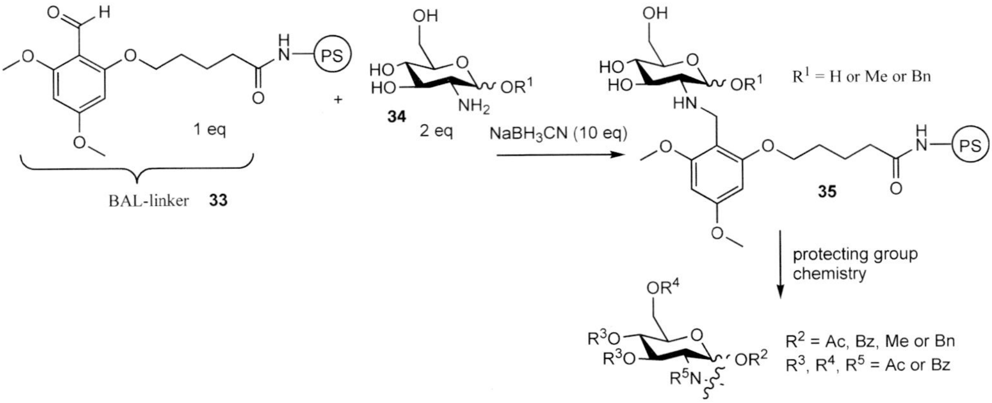

Scheme 20.12 Attachment of the first sugar moiety to BAL-AM-PS by reductive amination.

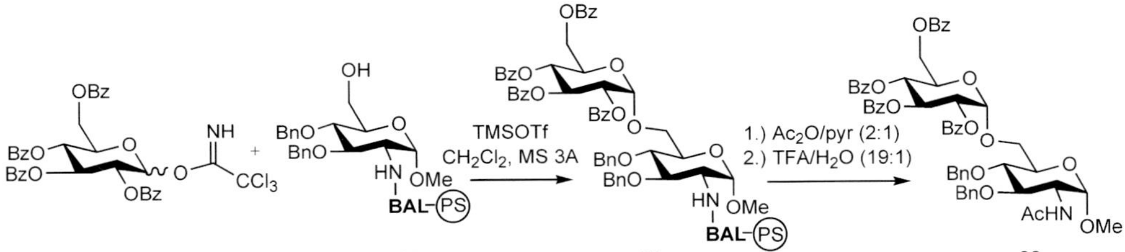

Scheme 20.13 Disaccharide synthesis using the BAL-linker.

the preparation of differentially protected monosaccharides [49]. The amino-functionalized Rink resin (Scheme 20.11) as well as the benzylidene acetal linker are cleaved by trifluoroacetic acid (TFA). Ogawa and Ito [50] have developed a Wang resin linker for the synthesis of polylactosamines that withstood the mildly acidic conditions during the glycosylation step using catalytic amounts of TMSOTf. The oligosaccharide product was cleaved with excess triphenylmethyl-borontetra-fluoride (TrBF$_4$).

20.4.1.2 Base-Labile Linkers

The base-labile succinoyl linker, usually employed in automated DNA synthesis, has also been applied to oligosaccharide synthesis. This linker was used on amino-functionalized TentaGel and PS supports and is cleaved with aqueous ammonia. Heparin-like oligosaccharides were assembled on PEG and MPEG polymeric supports employing succinoyl linkers that were cleaved by hydrazin acetate in THF [51]. The same linker was used for the synthesis of a protected tetrasaccharide related to the mucin oligosaccharides surrounding *Xenopus laevis* oocytes on Tenta-Gel and MPEG polymers [52]. Another base-labile linker was introduced by

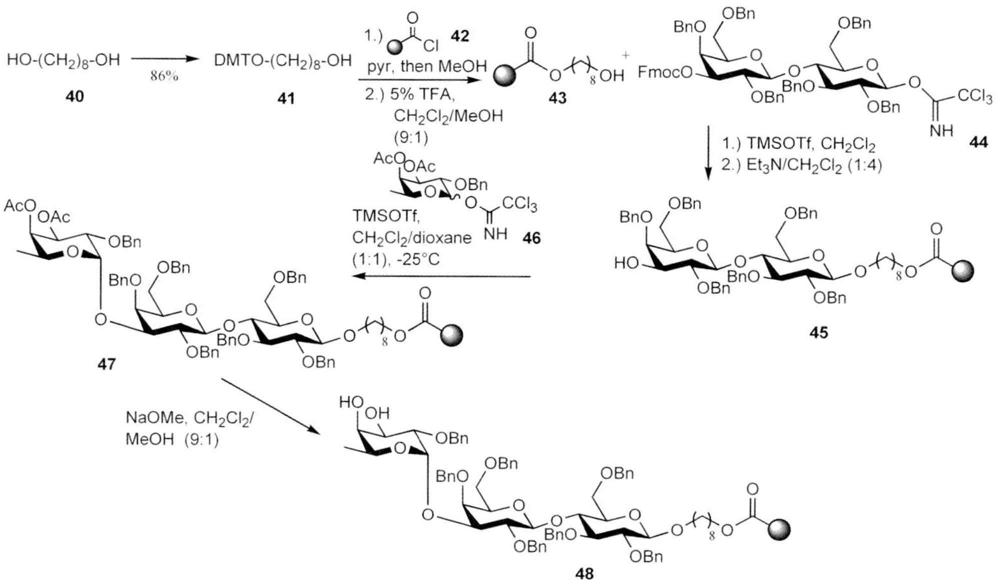

Scheme 20.14 Trisaccharide synthesis employing a base sensitive linker.

Schmidt [53]. Acyl chloride support **42** was reacted with DMT-protected octane-1,8-diol **41** (Scheme 20.14). The first building block (disaccharide trichloroacetimidate **44**) was attached to resin-bound linker **43** in the presence of TMSOTf. Fmoc-cleavage was achieved while the ester linkage remained intact. After one further glycosylation, cleavage with sodium methoxide furnished partially protected trisaccharide **48**.

20.4.2
Linkers Cleaved by Olefin Metathesis

Fully protected tumor-associated carbohydrate antigen Globo-H (**50**), a potential breast and prostate cancer vaccine [54, 55], was assembled by automated solid support synthesis using oct-4-ene-1,8-diol linker **49** and building blocks **51–56** (Scheme 20.15). The linker is stable to acid and base but is cleaved cleanly by olefin metathesis [56]. The octenediol linker performs best on Merrifield's PS resin, affording routinely loadings in a range of 0.1–0.3 mmol g^{-1}.

20.4.3
Photocleavable Linkers

Nicolaou [57] has reported the synthesis of a dodecasaccharide employing a *p*-hydroxybenzoic acid spacer group connected via a photolabile linker [3-(hydroxymethyl)-4-nitrophenol] to the resin (Scheme 20.16). Sugar mono- or oligomers

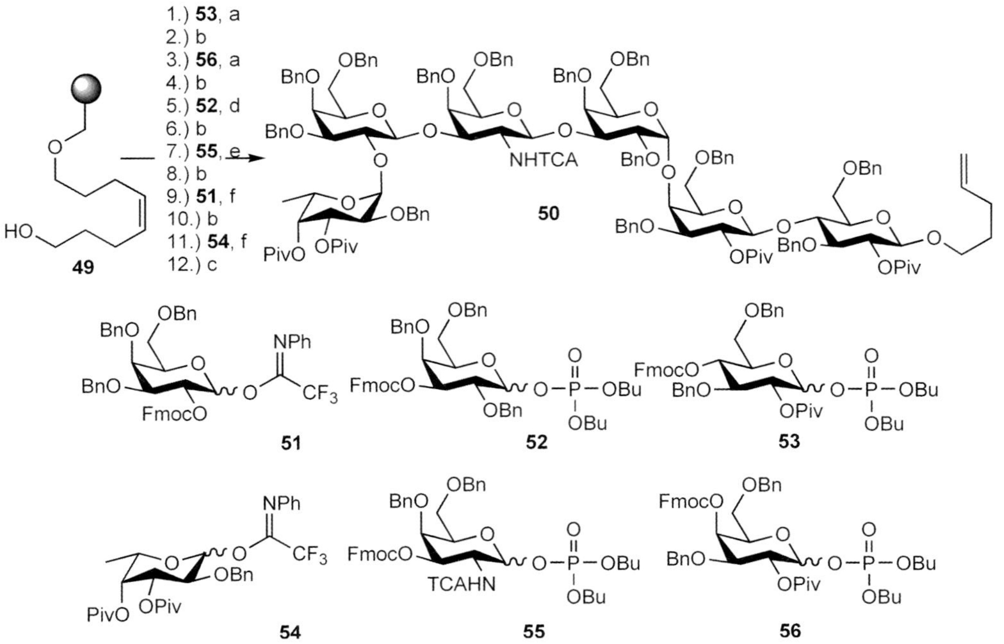

Scheme 20.15 Automated solid phase synthesis of Globo-H (**50**). Conditions: (a) building block (5 equiv), TMSOTf (5 equiv), DCM, −15 °C, repeated once for 45 min each; (b) piperidine (20% in 2 mL of DMF), repeated twice for 5 min each; (c) Grubbs' catalyst (first generation), ethylene atmosphere, DCM, rt, overnight; (d) building block (5 equiv), TMSOTf (5 equiv), Et₂O, DCM, −50 °C, repeated once for 3 h each; (e) building block (3.3 equiv), TMSOTf (3.3 equiv), DCM, −15 °C, repeated twice for 25 min each; (f) building block (5 equiv), TMSOTf (0.5 equiv), DCM, −10 °C, repeated once for 25 min each.

were released as thioglycosides upon activation with trimethyl(phenylthio)silane/zinc iodide. Photolytic cleavage resulted in mono- or oligomers containing the spacer group. This methodology includes a convergent approach for synthesizing complex oligosaccharides with repeated substructures and is applicable to the synthesis of combinatorial libraries of oligosaccharides.

20.4.4
Silyl Ether Linkers

Silyl ethers are widely used as temporary protecting groups for hydroxyl groups in organic chemistry. Selective cleavage is accomplished mostly by treatment with acid such as HCl in methanol or fluoride anions (e.g. HF/pyridine or tetrabutylammonium fluoride). A di-isopropyl aryl silane linker was attached to the C4 hydroxyl group of a partially protected glucal [58] but proved too labile for subsequent couplings. Magnusson et al. [59] introduced a linker that is accessible starting from

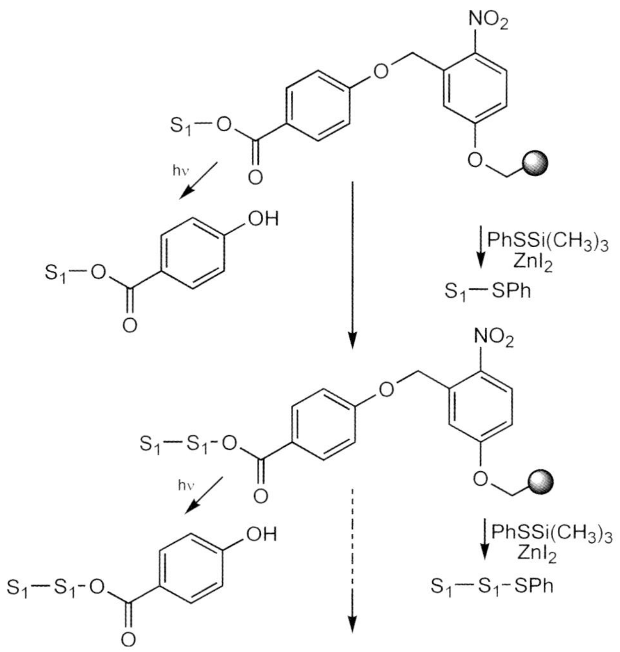

Scheme 20.16 Block-type solid phase strategy employing a photolabile linker and thioglycoside building blocks.

2-(hexenyldimethylsilyl)ethanol (**57**) in four steps. The linker **58** was reacted with galactosyl building block **59** to furnish linker galactoside **60** in 90% yield (Scheme 20.17). Subsequent cleavage of the benzoyl groups and saponification of the methyl ester delivers deprotected galactoside **61**. Coupling of **61** to AM-PS with diisopropylcarbodiimide (DIC) furnished **62**.

Acylation of **62** with propionic anhydride in pyridine, cleavage from the linker using borontrifluoride etherate and acetylation of the anomeric hydroxyl group furnished fully protected galactoside **63**. This galactoside is suitable for selective introduction of protecting groups to deliver a polymer-bound galactosyl building block for solid phase oligosaccharide synthesis.

20.4.5
Boronate Linkers

Boronate linkers are used rarely for oligosaccharide synthesis due to their moisture sensitivity. Synthesis of a polystyrylboronic acid **64** resin allowed for easy diol protection and was used for the solid support synthesis of methyl 2,3-di-*O*-benzoyl-D-galactopyranoside. Relatively low loading (0.2–0.6 mmol functional groups per g) and moisture sensitivity limited the utility of the approach [60]. Polystyryl boronic acid **64** was employed to protect and attach glucopyranoside **65**, with the C2 and C3 hydroxyl groups ready for further couplings [61].

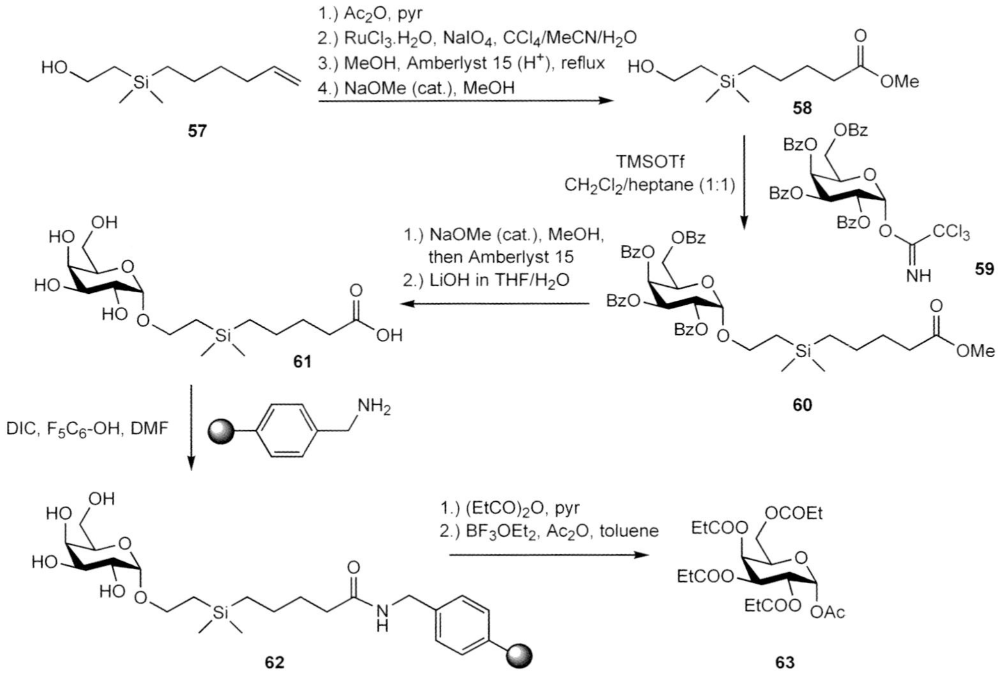

Scheme 20.17 Galactoside building block synthesis using a resin-bound silyl linker.

Resin-bound **66** reacted much faster with incoming building blocks on position 2-O, particularly when those were bulky and the resin was only partially swollen. Position 3-O was reacted with a further building block to yield **68**. Cleavage under mild conditions and coupling to polystyryltrityl chloride allowed for selective protection of the C4 hydroxyl group to give **71**. Cleavage from the resin and introduction of the last building block at the C6 hydroxyl group yielded **73** (Scheme 20.18).

20.4.6
Thiol-Group Containing Linkers

Schmidt et al. [62] have described the synthesis of oligomannosides using an alkyl thiol containing linker connected to Merrifield's resin. Mannosyl trichloroacetimidate was activated with TMSOTf. The fully protected oligomannoside was released from the solid support by oxidation (NBS/DTBP) (Scheme 20.19).

Oligosaccharide–protein binding has been investigated by surface plasmon resonance (SPR) of synthetic oligosaccharide structures covalently immobilized on bovine-serum-albumin-coated glass slides [63]. The hydrophilic thiol linker, based on 2-[2-(2-mercaptoethoxy)ethoxy]ethanol (Figure 20.3), is usually attached to the anomeric center of the mono- or oligosaccharide structures.

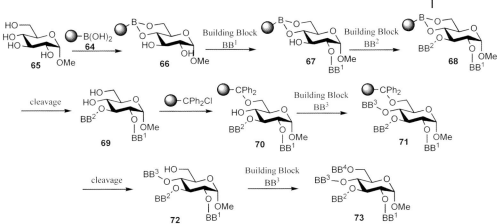

Scheme 20.18 Regio-orientation strategy using boronate linkers attached to PS.

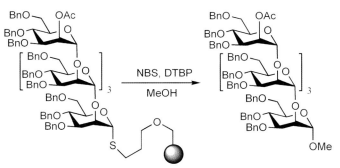

Scheme 20.19 An alkylthiol linker for solid phase oligosaccharide synthesis.

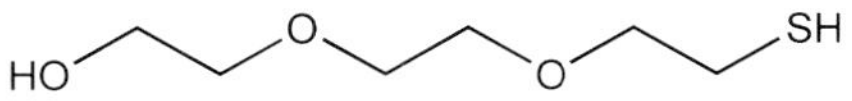

Figure 20.3 Structure of 2-[2-(2-mercaptoethoxy)ethoxy]ethanol.

20.4.7
Linkers Cleaved Under Oxidative Conditions

The most commonly employed, oxidatively removable linker is related to the *p*-methoxybenzyl group (PMB) [64]. Resin-bound PMB can be cleaved by the action of 2,3-dichloro-5,6-dicyano-*p*-benzoquinone (DDQ). The acid lability of PMB can be successfully reduced by attaching an acylamino moiety in the para position of the benzene ring. The PMB-derived linker may be attached to hydroxyl groups other than the anomeric position to facilitate product purification and analysis (Scheme 20.20).

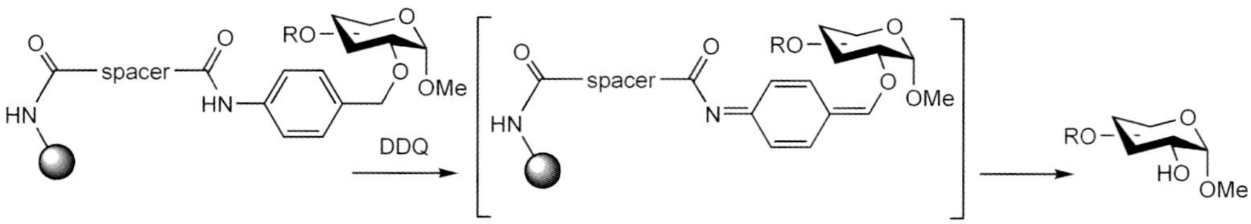

Scheme 20.20 PMB linker cleaved by oxidation.

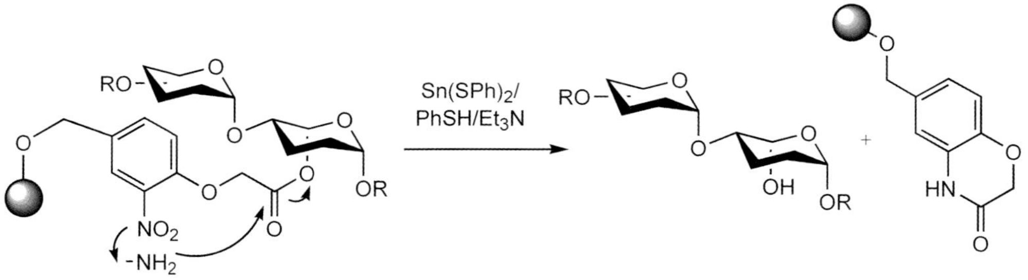

Scheme 20.21 Reductive cleavage of a nitro-Wang linker.

20.4.8
Linkers Cleaved by Reduction

A nitro-introduced Wang resin-type linker for soluble and insoluble resins has been explored [65]. The linker is completely stable during glycosylations that employ TMSOTf or Cp_2HfCl_2-AgOTf as activator. The product was released by reduction of the nitro group, followed by intramolecular cyclization leading to a hydroxamic acid derivative and the target molecule. Reduction was accomplished by $Sn(SPh)_2/PhSH/Et_3N$ (Scheme 20.21) [66].

PEG (average MW 5000) was selected as soluble support, whereas for solid phase synthesis Merrifield's resin and an additional spacer was employed to reduce steric hinderance imposed by the solid phase.

20.4.9
Cleavage by Hydrogenolysis

The α,α'-dioxyxylyl diether linker is, presumably, the most frequently applied linker that is cleavable by hydrogenolysis. This linker was synthesized by means of immobilized *Candida antartica* lipase glucose γ-aminobutyric ester [67]. Acylation of the immobilized glucose building block and GABA (γ-aminobutyric acid) occurs selectively on C6. The synthesis commenced with the union of perbenzoylated glucosyl trichloroacetimidate (**74**) and the α,α'-dioxyxylyl diether linker attached to a MPEG polymer (**75**, Scheme 20.22). Cleavage of the benzoyl groups of **76** followed by enzyme-catalyzed acylation of the C6 hydroxyl group with *N*-Cbz-

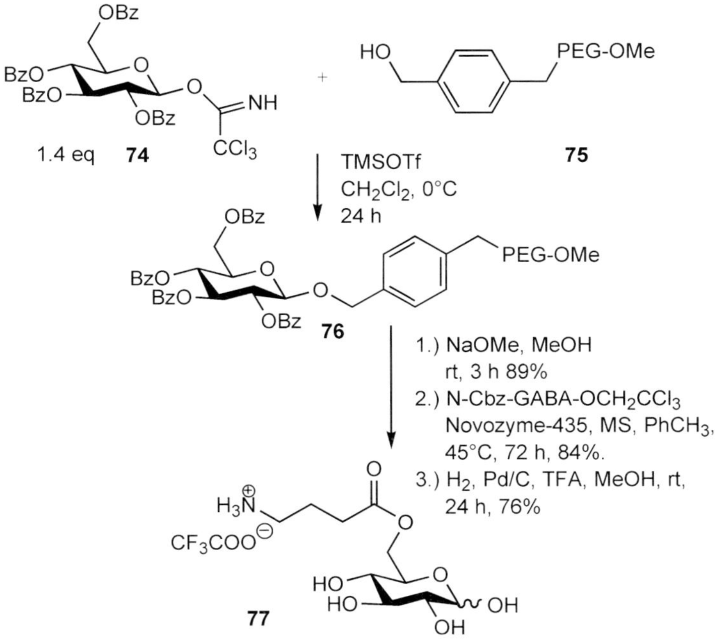

Scheme 20.22 Synthesis of glucose γ-aminobutyric ester.

GABA-OCH$_2$CCl$_3$ furnished the fully deprotected target molecule **77** in good yield after hydrogenolysis. The use of a soluble polymer support is often beneficial in enzyme-controlled reactions.

Compounds such as **77** are candidates for crossing the blood–brain barrier by using the glucose carrier protein GLUT-1 [67]. The reaction of polystyrene acid chloride and monotritylated octan-1,8-diol resulted in an ester-type linker [68]. After detritylation, benzylated Fmoc-protected lactosyl trichloroacetimidate was coupled as the first building block. Following the assembly of a trisaccharide, cleavage from the solid support was accomplished by treatment with sodium methoxide in methanol to furnish 8-hydroxyoctyl derived trisaccharide [68] or via catalytic hydrogenolysis [69].

20.4.10
Enzymatically Cleavable Linkers

We discuss exclusively enzymatically cleavable linker systems that have been applied to oligosaccharide synthesis. An α-chymotrypsin-labile linker containing the phenylalanine moiety, allowing α-chymotrypsin-induced cleavage at the carboxyl group of the amino acid, may serve as a representative example. The synthesis started with the glycosylation of Cbz-Phe-NH-(CH$_2$)$_6$OH (**78**) followed by catalytic hydrogenation (Scheme 20.23). The free α-amino group of **80** was

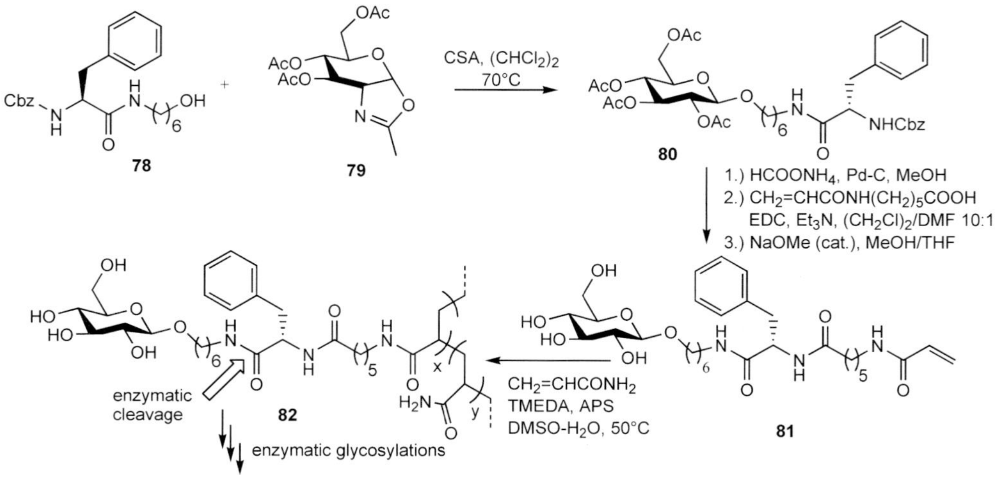

Scheme 20.23 Application of an enzymatically cleavable linker.

condensed with 6-acrylamidocaproic acid and the resulting compound was de-O-acetylated.

The GlcNAc derivative **81** was copolymerized with acrylamide in the presence of ammonium persulfate (APS) and N,N,N',N'-tetramethylethylenediamine (TMEDA) in high yield. The resulting polymer (**82**) was used for enzymatic glycosylations and the final product was released by α-chymotrypsin-catalyzed cleavage, providing the target molecule with an already installed 6-aminohexyl linker on the GlcNAc moiety (Scheme 20.23) [70].

20.5
Capture and Release Techniques

Product isolation and purification is often a difficult and time-consuming task. Capture and release techniques are useful to purify and isolate the products of complex multistep syntheses. The desired products are usually labeled with a tag that specifically binds to resins or a stationary phases applied in column chromatography. Binding can be covalent or by adsorption. By-products are easily removed by washing, followed by controlled release of the desired products.

Two applications of capture and release techniques targeting a trisaccharide are discussed. The synthesis of the protected trisaccharide β-D-Glu-(1→2)-α-D-Man-(1→6)-β-D-Man took advantage of solutionphase polymer-support methodology combined with resin capture–release purification [71]. MPEG (M_w ~ 750 Da) was chosen as a polar tag and was attached to the first building block. The product was isolated simply by passage through a short silica gel column. Non-tagged fractions were eluted first, whereas a more polar solvent mixture eluted the MPEG-bound material. Glycosylation was achieved by N-iodosuccinimide (NIS) and trifluoromethanesulfonic acid (TfOH) followed by capping with acetic anhydride in pyridine. The last sugar building block contained a chloroacetyl anchor, preparing the target structure for capture–release purification.

Chloroacetylated MPEG-bound trisaccharide 83 was reacted with cysteine-conjugated Wang resin (84). The Fmoc-group was removed by treatment with 4-DMAP after purification of intermediate 85 to liberate trisaccharide 86 by intramolecular cyclization (Scheme 20.24). A modified approach was employed for the synthesis of a protected trimannoside using cap and capture–release purfication methods combined with solid phase synthesis, whereas the CPG-group [CPG = p-(5-ethoxycarbonyl)pentyloxybenzyl] served as a unique tag to fish out the desired final product [72]. Mannosyl building block 87 featuring a succinoyl tag at the C6 hydroxyl group was coupled to PEG-NH$_2$ followed by the hydrazinolysis of the levaloyl ester (Scheme 20.25). Acceptor 89 was coupled with the protected mannosyl trichloroacetimidate 90, followed by levaloyl cleavage. Dimannosyl acceptor 92 was reacted with the tagged mannosyl trichloroacetimidate 93 to furnish the fully protected trimannoside 94. Cleavage of the esters and acetylation followed by a coupling step gave resin-bound trimannoside 96. Mild acidic cleavage released protected target structure 97. The described synthesis approach allowed the introduction of a unique group in the last coupling step to facilitate the capture by an amino-group functionalized resin. The side products and reagents remained in solution and release from the resin yielded homogeneous oligosaccharide.

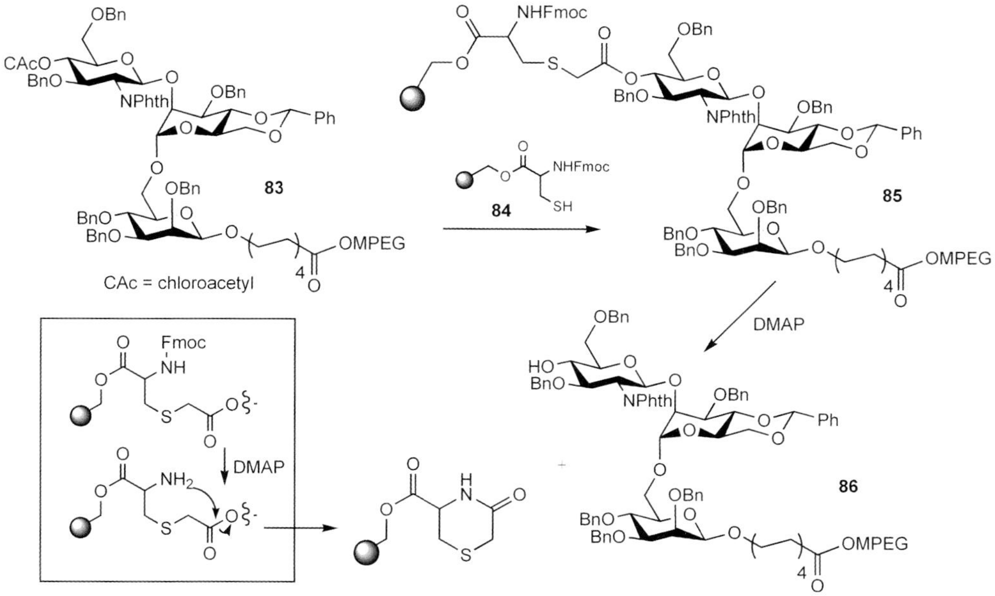

Scheme 20.24 Capture and release technique using resin-bound Fmoc-Cys-OH.

606 | 20 Oligosaccharides

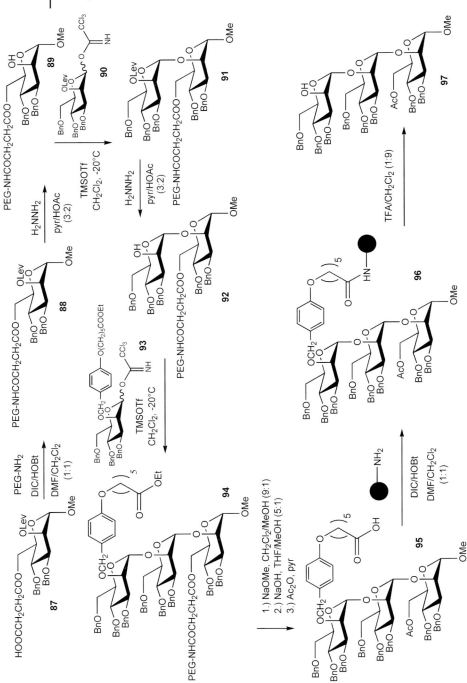

Scheme 20.25 Cap and capture–release techniques applied to a trimannoside synthesis.

20.6
Conclusions

We have reviewed the most commonly used solid and soluble phases as well as the different linker concepts currently employed for oligosaccharide synthesis. Section 20.3.3 (ionic liquids) and Section 20.2.3 (magnetic particles) describe less frequently employed methods of replacing classical polymer phases. Many of the described polymers and linkers are not unique to oligosaccharide synthesis – in fact many of them were first introduced for peptide chemistry. The relevance of the polymers and linkers will depend strongly on applicability to automated oligosaccharide synthesis. Getting access to synthetically pure oligosaccharides for medicinal applications will be of utmost importance in the future.

List of Abbreviations

Ac	acetyl
Ac$_2$O	acetic anhydride
AgOTf	silver trifluoromethane sulfonate
AM-PS	aminomethylated polystyrene
APS	ammonium persulfate
BAL	backbone amide linker
BF$_3$.OEt$_2$	boron trifluoride diethyl etherate
Bn	benzyl
Bu	butyl
Bz	benzoyl
CAc	chloroacetyl
Cbz = Z	benzyloxycarbonyl
CMC	N-cyclohexyl-N'-(2-morpholinoethyl)carbodiimide
Cp$_2$HfCl$_2$	hafnocene dichloride
CPG (as a solid support)	controlled pore glass
CPG (as a protecting group)	[p-(5-ethoxycarbonyl)pentyloxy]benzyl
CSA	camphorsulfonic acid
DCC	N,N'-dicyclohexylcarbodiimide
DCM	dichloromethane
DDQ	2,3-dichloro-5,6-dicyano-1,4-benzoquinone
DIC	N,N'-diisopropylcarbodiimide
DIPEA	N-ethyldiisopropylamine
4-DMAP	4-N,N-(dimethylamino)pyridine
DMF	dimethylformamide
DMSO	dimethyl sulfoxide
DMT	4,4′-dimethoxytrityl
DNA	deoxyribonucleic acid
DOX-linker	α,α′-dioxyxylyl diether linker
DTBMP	2,6-di-*tert*-butyl-4-methyl pyridine

DTBP	2,6-di-*tert*-butylpyridine
EDCI or EDC	1-(3-dimethylaminopropyl)-3-ethylcarbodiimide
EtOAc	ethyl acetate
Fmoc	9-fluorene methyloxycarbonyl
Fmoc-Cys-OH	*N*-Fmoc protected cysteine
GABA	γ-aminobutyric acid
HF	hydrogen fluoride
HOAc	acetic acid
HOBt	1-hydroxybenzotriazole
IL	ionic liquid
KHMDS	potassium hexamethyldisilazane
KOH	potassium hydroxide
Lev	levulinoyl
LNFP	lacto-*N*-fucopentaose
LNT	*N*-tetrose
MCPBA	3-chloroperoxybenzoic acid
MeCN	acetonitrile
[mimBF$_4$]	1-methylimidazolium-tetrafluoroborate
MPEG	monomethyl poly(ethylene glycol)
MS	molecular sieve
NaOMe	sodium methoxide
NBS	*N*-bromosuccinimide
NIS	*N*-iodosuccinimide
PAMAM	polyamidoamine dendrimer
PEG	poly(ethylene glycol)
PEGA	poly(ethylene glycol)/dimethylacrylamide copolymer
Ph	phenyl
Phth	phthaloyl
Piv	pivaloyl
PMB	*p*-methoxybenzyl
POEPOP	polyoxyethylene-polyoxypropylene
POEPS	polyoxyethylene–polystyrene
PS	polystyrene
pyr	pyridine
pyr-BH$_3$	pyridine borane complex
rt	room temperature
SEC	size exclusion chromatography
SPOCC	polyoxyethylene–polyoxetane
SPOT	solid phase oligosaccharide tagging
SPR	surface plasmon resonance
TBS or TBDMS	*tert*-butyldimethylsilyl
TBTU	*O*-(benzotriazol-1-yl)-*N,N,N′,N′*-tetramethyluronium tetrafluoroborate

TCA	trichloroacetyl
TFA	trifluoroacetic acid
Tf$_2$O	trifluoromethanesulfonic anhydride
TfOH	trifluoromethanesulfonic acid
THF	tetrahydrofuran
TMEDA	*N,N,N',N'*-tetramethylethylenediamine
TMSOTf	trimethylsilyl trifluoromethane sulfonate
TrBF$_4$	triphenylmethyl-borontetrafluoride

References

1 Allen, H.J. and Kisailus, E.C. (eds) (1992) *Glycoconjugates: Composition, Structure, and Function*, Dekker, New York, pp. 685.
2 Lee, Y.C. and Lee, R.T. (eds) (1994) *Neoglycoconjugates: Preparations and Applications*, Academic Press, London.
3 Kobata, A. (1993) *Accounts of Chemical Research*, **26** (6), 319–24.
4 Varki, A. (1993) *Glycobiology*, **3** (2), 97–130.
5 Phillips, M.L., Nudelman, E., Gaeta, F.C.A., et al. (1990) *Science*, **250**, 1130–2.
6 Lasky, L.A. (1992) *Science*, **258**, 964–9.
7 Miller, D.J., Macek, M.B. and Schur, B.D. (1992) *Nature*, **357**, 589–93.
8 Schulze, I.T. and Manger, I.D. (1992) *Glycoconjugate Journal*, **9** (2), 63–6.
9 Yaki, T., Hirabayashi, Y., Ishikawa, H., et al. (1986) *Journal of Biological Chemistry*, **261** (7), 3075–8.
10 Spohr, U. and Lemieux, R.U. (1988) *Carbohydrate Research*, **174**, 211–37.
11 Levy, D.E., Tang, P.C. and Musser, J.H. (1994) *Annual Report in Medicinal Chemistry*, Vol. 29 (ed. W.K. Hagmann), Academic Press, San Diego, pp. 215–46.
12 Ruttens, B. and Van der Eycken, J. (2002) *Tetrahedron Letters*, **43** (12), 2215–21.
13 Guillier, F., Orain, D. and Bradley, M. (2000) *Chemical Reviews*, **100** (6), 2093–4.
14 Heckel, A., Mross, E., Jung, K.H., et al. (1998) *Synlett*, **2**, 171–3.
15 Adinolfi, M., Barone, G., De Napoli, L., et al. (1998) *Tetrahedron Letters*, **39** (14), 1953–6.
16 Millipore Corporation (2006) Millipore data sheet DNA Nucleoside Controlled Pore Glass (CPG®) Media, Billerica, MA 01821, USA.
17 Millipore Corporation (2006) See product detail of Millipore Glycerole Controlled-Pore Glass, Billerica, MA 01821, USA.
18 Lohse, A., Martins, R., Jørgensen, M. and Hindsgaul, O. (2006) *Angewandte Chemie – International Edition* **45** (25), 4167–72.
19 Merrifield, R.B. (1963) *Journal of the American Chemical Society*, **85** (14), 2149.
20 Sherrington, D.C. (1998) *Chemical Communications*, **21**, 2275–86.
21 Bayer, E. (1991) *Angewandte Chemie – International Edition in English*, **30** (2), 113–47.
22 Rapp Polymere, D-72072, Tübingen, Germany. http://www.rapp-polymere.com
23 Sigma-Aldrich Co. (2007) Online Catalog-Subdirectory http://www.sigmaaldrich.com/homepage/site_level_pages/CatalogHome.html Keyword: ArgoGel.
24 Ferreras, R.M., Delaisse, J.M., Foged, N.T. and Meldal, M.P. (1998) *Journal of Peptide Science*, **4** (3), 195–210.
25 Schleyer, A., Meldal, M., Manat, R., et al. (1997) *Angewandte Chemie – International Edition in English*, **36** (18), 1976–8.
26 Renil, M. and Meldal, M. (1996) *Tetrahedron Letters*, **37** (34), 6185–8.
27 Grøtli, M., Gotfredsen, C.H., Rademann, J., et al. (2000) *Journal of Combinatorial Chemistry*, **2** (2), 108–19.
28 Gupta, A.K. and Gupta, M. (2005) *Biomaterials*, **26** (18), 3995–4021.
29 Khng, H.P., Cunliffe, D., Davies, S., et al. (1998) *Biotechnology and Bioengineering*, **60** (4), 419–24.

30 Heebøll-Nielsen, A., Dalkiær, M., Hubbuch, J.J. and Thomas, O.R.T. (2004) *Biotechnology and Bioengineering*, **87** (3), 311–23.
31 Bozhinova, D., Galunsky, B., Yueping, G., et al. (2004) *Biotechnology Letters* **26** (4), 343–50.
32 Oster, J., Parker, J. and à Brassard, L. (2001) *Journal of Magnetism and Magnetic Materials*, **225** (1–2), 145–50.
33 Andersson, H., van Der Wijngaart, W., Enoksson, P. and Stemme, G. (2000) *Sensors and Actuators*, **B67**, 203–8.
34 Nakamura, K., Abe, Y. and Kogure, N. (2006) Patent JP2006247492.
35 Ahlers, H., Müller, P.J. and Ozegowski, J.H. (2004) Patent DE10231925.
36 Oder, R.R. (2003) Patent WO03072531.
37 Shimomura, M., Ono, B., Oshima, K. and Miyauchi, S. (2006) *Polymer*, **47** (16), 5785–90.
38 Wentworth, P. Jr and Janda, K.D. (1999) *Chemical Communications*, **19**, 1917–24.
39 Gravert, D.J. and Janda, K.D. (1997) *Chemical Reviews*, **97** (2), 489–509.
40 Krepinsky, J.J. (1996) Advances in Polymer-Supported Solution Synthesis of Oligosaccharides, in *Modern Methods in Carbohydrate Synthesis* (eds S.H. Khan and R.A. O'Neill), Harwood Academic Publishers, Amsterdam, pp. 194–224.
41 Yan, F., Wakarchuk, W.W., Gilbert, M., Richards, J.C. and Whitfield, D.M. (2000) *Carbohydrate Research*, **328** (1), 3–16.
42 Kantchev, A.B. and Parquette, J.R. (1999) *Tetrahedron Letters*, **40** (46), 8049–53.
43 Malmstroem, E., Johansson, M. and Hult, A. (1995) *Macromolecules*, **28** (5), 1698–703.
44 Roy, R. and Baek, M.G. (2003) *Methods in Enzymology*, **362**, 240–9.
45 He, X. and Chan, T.H. (2006) *Synthesis*, **10**, 1645–51.
46 Silva, D.J., Wang, H., Allanson, N.M., et al. (1999) *Journal of Organic Chemistry*, **64** (16), 5926–9.
47 Tolborg, J.F. and Jensen, K.J. (2000) *Chemical Communications*, **2**, 147–8.
48 A peptide synthesis using the BAL-linker is described in Boas,U., Brask, J., Christensen, J.B. and Jensen, K.J. (2002) *Journal of Combinatorial Chemistry*, **4** (3), 223–8.
49 Hanessian, S. and Huynh, H.K. (1999) *Synlett*, **1**, 102–4.
50 Shimizu, H., Ito, Y., Kanie, O. and Ogawa, T. (1996) *Bioorganic & Medicinal Chemistry Letters*, **6** (23), 2841–6.
51 Ojeda, R., Terentí, O., de Paz, J.L. and Martín-Lomas, M. (2004) *Glycoconjugate Journal*, **21** (5), 179–95.
52 Geurtsen, R. and Boons, G. J. (2002) *European Journal of Organic Chemistry*, **9**, 1473–7.
53 Roussel, F., Knerr, L. and Schmidt, R.R. (2001) *European Journal of Organic Chemistry*, **11**, 2067–73.
54 Slovin, S.F., Ragapathi, G., Adluri, S., et al. (1999) *Proceedings of the National Academy of Sciences of the United States of America*, **96** (10), 5710–15.
55 Huang, C.Y., Thayer, D.A., Chang, A.Y., et al. (2006) *Proceedings of the National Academy of Sciences of the United States of America*, **103** (1), 15–20.
56 Chatterjee, A.K. and Grubbs, R.H. (1999) *Organic Letters*, **1** (11), 1751–3.
57 Nicolaou, K.C., Watanabe, N., Li, J., et al. (1998) *Angewandte Chemie International Edition*, **37** (11), 1559–61.
58 Savin, K.A., Woo, J.C.G. and Danishefsky, S.J. (1999) *Journal of Organic Chemistry*, **64** (11), 4183–6.
59 Weigelt, D. and Magnusson, G. (1998) *Tetrahedron Letters*, **39** (18), 2839–42.
60 Seymour, E. and Fréchet, J.M.J. (1976) *Tetrahedron Letters*, **17** (15), 1149.
61 Liao, Y., Li, Z.M. and Wong, H.N.C. (2001) *Chinese Journal of Chemistry*, **19** (11), 1119–29.
62 Rademann, J. and Schmidt, R.R. (1997) *Journal of Organic Chemistry*, **62** (11), 3650–3.
63 Ratner, D.M., Adams, E.W., Su, J., et al. (2004) *ChemBioChem*, **5** (3), 379–83.
64 Fukase, K., Nakai, Y., Egusa, K., et al. (1999) *Synlett*, **7**, 1074–8.
65 Manabe, S. and Ito, Y. (2001) *Chemical and Pharmaceutical Bulletin*, **49** (9), 1234–5.
66 Manabe, S., Nakahara, Y. and Ito, Y. (2000) *Synlett*, **9**, 1241–4.
67 de Torres, C. and Fernández-Mayoralas, A. (2003) *Tetrahedron Letters*, **44** (11), 2383–5.

68 Roussel, F., Knerr, L., Grathwohl, M. and Schmidt, R.R. (2000) *Organic Letters*, **2** (20), 3043–6.

69 Roussel, F., Takhi, M. and Schmidt, R.R. (2001) *Journal of Organic Chemistry*, **66** (25), 8540–8.

70 Reents, R., Jeyaraj, D.A. and Waldmann, H. (2001) *Advanced Synthesis Catalysis*, **343** (6–7), 501–13.

71 Hanashima, S., Inamori, K., Manabe, S., Taniguchi, N. and Ito, Y. (2006) *Chemistry – A European Journal*, **12** (13), 3449–62.

72 Wu, J. and Guo, Z. (2006) *Journal of Organic Chemistry*, **71** (18), 7067–70.

21
High-Throughput Synthesis of Natural Products

Nicolas Winssinger, Sofia Barluenga and Pierre-Yves Dakas

Université Louis Pasteur-CNRS, Organic and Bioorganic Chemistry Laboratory, Institut de Science et Ingénierie Supramoléculaires, 8 allée Gaspard Monge, 67000 Strasbourg, France

21.1
Introduction

The success of solid phase chemistry for peptide and oligonucleotide synthesis, followed by the development of combinatorial methods for the preparation of libraries, inspired chemists in the natural product synthesis arena to explore the potential of these techniques to facilitate access to complex natural products and their analogues. The demonstration that pharmaceutically relevant molecules such as the benzodiazapenes [1, 2] could be prepared by solid phase synthesis provided an encouraging milestone, showing that this technique was applicable beyond the highly repetitive nature of amide bond formation and phosphoramidate couplings of peptides and oligonucleotides synthesis, respectively. The first complex natural products to be fully synthesized on a solid phase were the epothilones in 1997 [3]. Over the past ten years, a tremendous number of new linkers and strategies for solid phase synthesis have been reported with applications to various natural products, ranging from alkaloids to polyketides and terpine derived secondary metabolites. While the facile automation of solid phase chemistry makes this strategy extremely attractive, the last ten years have also confirmed some of the inherent limitations of solid phase synthesis, namely the incompatibility of some reagents with the heterogeneous nature of the solid phase, the slower reaction rates and, most importantly, the fact that complex transformations cannot always be driven to a clean product despite the excess of reagents utilized. This later fact has stimulated the development of two alternative strategies, polymer supported reagents and isolation tags such a poly(ethylene glycol) (PEG) or fluorous tags.

Natural product synthesis has long been a testing ground for new methodologies and an inspiration for the development of new strategies [4]. This chapter is not intended to be comprehensive but rather showcases the state of the art through a series of case studies in high-throughput synthesis using solid phase chemistry,

The Power of Functional Resins in Organic Synthesis. Judit Tulla-Puche and Fernando Albericio
Copyright © 2008 WILEY-VCH Verlag GmbH & Co. KGaA, Weinheim
ISBN: 978-3-527-31936-7

immobilized reagents and isolation tags. While a number of striking solid phase synthesis of depsipetides or other peptidic natural products have been reported in the recent literature [5–9], these will not be covered here due to the similarity of the protocols to regular peptide synthesis.

The inherent complexity of target-oriented synthesis has undoubtedly stimulated an expansion in the arsenal of techniques available in solid phase synthesis. Nevertheless, the primary motivation for most of the reported work still lies in the rich biological activities that are found in several secondary metabolite families. Indeed, these natural products have been selected through evolutionary pressure for specific biological activities by interacting and modulating the function of proteins or RNAs and as such offer a validated starting point in diversity space of biologically active compounds [10–12].

This chapter is divided in five sections. The first one highlights the potential of immobilizing a natural product core structure on solid phase to facilitate its diversification into a library of analogues. The next three sections cover the *de novo* synthesis of natural products using solid phase synthesis, immobilized reagents or isolation tags respectively. Finally, the last section reviews six landmark libraries based on natural products or natural product motifs.

21.2
Solid Phase Elaboration of Natural Product Scaffolds

The complexity of some biologically relevant natural products makes a *de novo* combinatorial synthesis too demanding; however, one may still capitalize on the privileged core structure of natural products by simply modifying some of the appendages or side-chains. An inspiring example is the discovery of Taxotere, where a simple change in the structure of the side-chain led to significant improvements in the therapeutic properties [13]. Such accomplishments unquestionably warrant a more systematic approach to natural product derivatization. Four different examples are presented in Scheme 21.1 to illustrate this approach. The first one was reported by Xiao and coworkers using a semisynthetic intermediate of Taxol [14], which was elaborated via a three-step sequence involving an Fmoc deprotection followed by a selective amide bond formation and a subsequent acylation. The clinical success of Taxol in the treatment of cancers stimulated a widespread search for new pharmacophores with the same mode of action, namely stabilization of polymerized microtubules. Several new natural products emerged in the 1990s as potentially superior to Taxol, including discodermolide, epothilones and sarcodictyins [15]. In the second example, Nicolaou and coworkers borrowed an advanced but still malleable sarcodyctyin intermediate from their total synthesis program [16] to be immobilized on polystyrene for diversification [17, 18]. The attachment to the resin was achieved via a Witting reaction and the scaffold was elaborated with three points of diversity: the secondary alcohol, after acetate deprotection, was derivatized by esterification or reaction with isocyanates; the primary alcohol, after silyl deprotection, was elaborated in three different manners: simply by esterification or

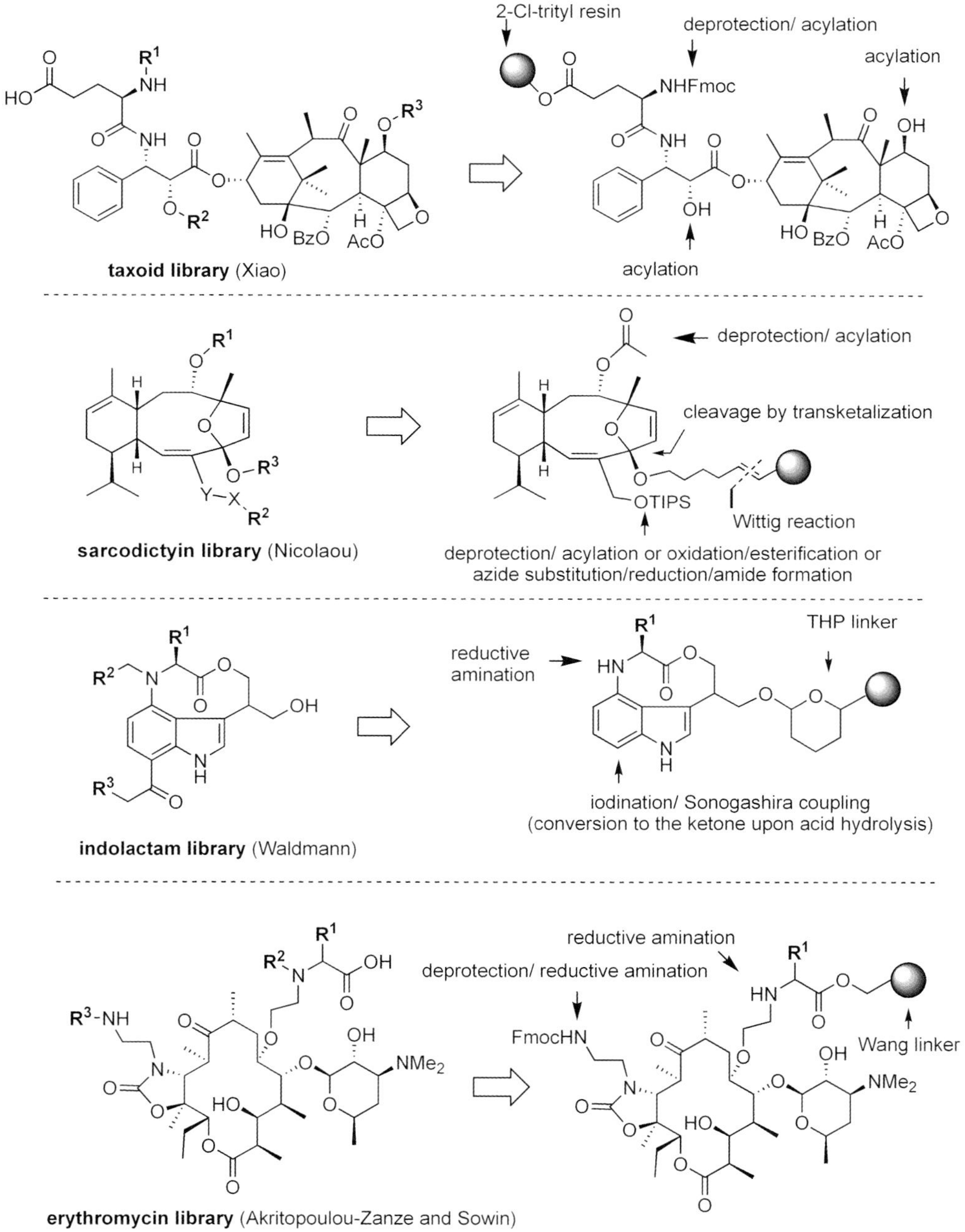

Scheme 21.1 Selected examples of solid phase immobilization of important natural product scaffolds for diversification.

carbonate formation, by oxidation to the carboxylic acid and conversion into esters or amides or last, but not least, by conversion into an azide that was reduced to the corresponding amine and further elaborated into amides. Finally, the library was released from the solid support via a transketalization in the presence of different alcohols, thereby introducing the third point of diversity. Notably, the library was prepared using the radiofrequency encoding [19], which allowed to track each pool of resin in the combinatorial synthesis. The sarcodyctyin library offered a rapid structure–activity profile of this important natural product, thus clearly identifying the areas that could lead to improved activity.

In the third example, Waldmann and coworkers recognized the potential of the telocidin family and in particular indolactam V for their activity against kinases. With the aim of synthesizing a library based on this important structural motif and taking into account that the free hydroxyl group is necessary for biological activity, the core scaffold was loaded via this alcohol to a dihydrofuran resin [20, 21]. The scaffold was then diversified via a four-step sequence involving a reductive amination of the aniline, a regioselective iodination of the aromatic core followed by a Sonogashira coupling with diverse alkynes. Upon TFA treatment to release the compounds from the resin, the alkyne moiety was hydrated to the corresponding ketone.

The fourth example is part of a lead discovery program at Abbott Labs, taking advantage of macrolide cores as a template for combinatorial derivatization. In this example, Akritopoulou-Zanze and Sowin took a semisynthetic derivative of erythromycin, which was coupled via reductive amination to an amino acid loaded on a Wang-type resin [22]. A three-step sequence employing a second reductive amination, Fmoc deprotection and a third reductive amination provided after TFA cleavage a library of over 70 000 compounds with good overall purity. Other notable examples not shown in Scheme 21.1 include the derivatization of the vancomycin's core, leading to important discoveries regarding the mode of action of this important antibiotic [23, 24].

An interesting extension of these principles has been reported by Hoefle and coworkers using chemical degradation of secondary metabolites to obtain novel and unique chiral building blocks [25]. Such building blocks can then be reassembled in a combinatorial fashion. In a proof of principle, the authors used SPOT synthesis (synthesis in a macroarray format) to demonstrate the possible recombination of natural product fragments. More recently, Furlan and coworkers showed that crude natural product extracts could be diversified by targeting specific functional groups [26]. For example, treating crude extracts with hydrazine led to the identification of a bioactive pyrazole that was not present in the original extract.

21.3
Solid Phase Synthesis of Natural Products

Epothilones attracted tremendous interest from the scientific community as they were shown to be potentially superior to taxol [27]. In a landmark publication,

Nicolaou and coworkers showed that epothilone could be accessed using solid phase synthesis (Scheme 21.2) [3, 28]. The key step relied on a ring closing metathesis to form the macrocycle with concomitant release from the resin. The required polymer-bound diene was obtained from three fairly simple fragments using a Wittig olefination, an aldol coupling and an esterification. A particularly attractive feature of cyclorelease reactions is that they can function as a self-purification method if the sequence is designed such that the last building block is required for the release of the resin. In such cases, only intermediates that have gone through the whole synthetic sequence can be released. This later fact has inspired the development of other cyclorelease reactions from the same group base on a Stille reaction that was used to prepare zearalenone [29] and on a Wadsworth–Horner–Emmons reaction to prepare muscone [30].

The prostaglandins are important actors in cellular communication and play a central role in various physiological processes, notably in the pain response. Aside from inspiring the creativity of synthetic chemists, prostaglandins have several

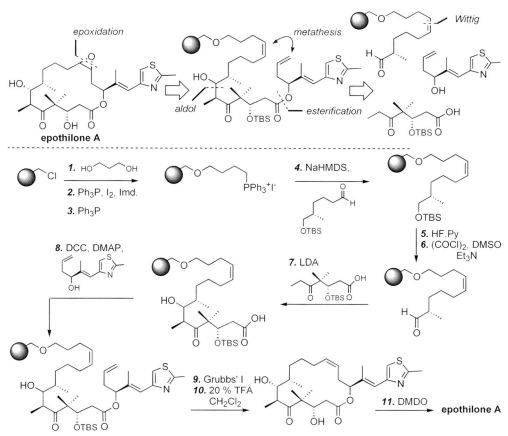

Scheme 21.2 Solid phase synthesis of epothilone (Nicolaou et al.).

therapeutical applications, including child birth delivery, erectile dysfunction and glaucoma [31]. Ellman and coworkers developed an elegant synthesis [32, 33] of this important family using a palladium-mediated sp^2–sp^3 coupling (Suzuki reaction) followed by a diastereoselective conjugate addition with higher order cuprates (Scheme 21.3). The two newly formed stereogenic centers were controlled by the stereochemistry of the ether bond to the resin.

Based on their standing interest in natural products as validated starting points for ligand discovery [6], Waldmann and coworkers synthesized a library of potential phosphatase inhibitors based on the dysidiolide motif (Scheme 21.4) [11, 34]. The terminal olefin was ingeniously exploited as an attachment point to the resin by incorporating in the linker a second olefin that upon treatment with a metathesis catalyst would participate in a fast five-membered ring cyclization, thus releasing the compound from the resin as a RCM byproduct. It has indeed been shown that such reactions are faster than simple cross metathesis. The key decalin system was assembled using an asymmetric Diels–Alder reaction while the γ-hydroxybutenolide was obtained by oxidation of a furan with singlet oxygen. Other notable features of the synthesis are two Wittig olefinations and a lithiated furan addition to a polymer-bound aldehyde. The significance of this synthesis lies not only in the importance of the products that were generated but also in the breadth of synthetic transformations used.

The macrosphelides were isolated in the mid-1990s and found to inhibit cell-adhesion [35], bringing these natural products into the chemical biology spotlight. Takahashi and coworkers reported a combinatorial synthesis of 122 macrosphelides, including the natural ones based on two palladium-catalyzed chemoselective carbonylations of vinyl halide, first in an intermolecular fashion and then in a

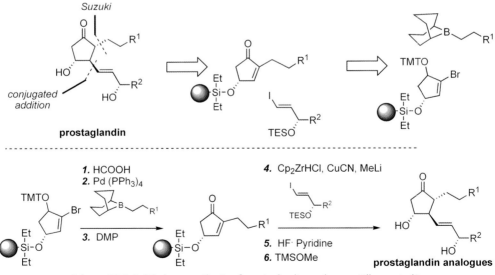

Scheme 21.3 Solid phase synthesis of prostaglandin analogues (Ellman et al.).

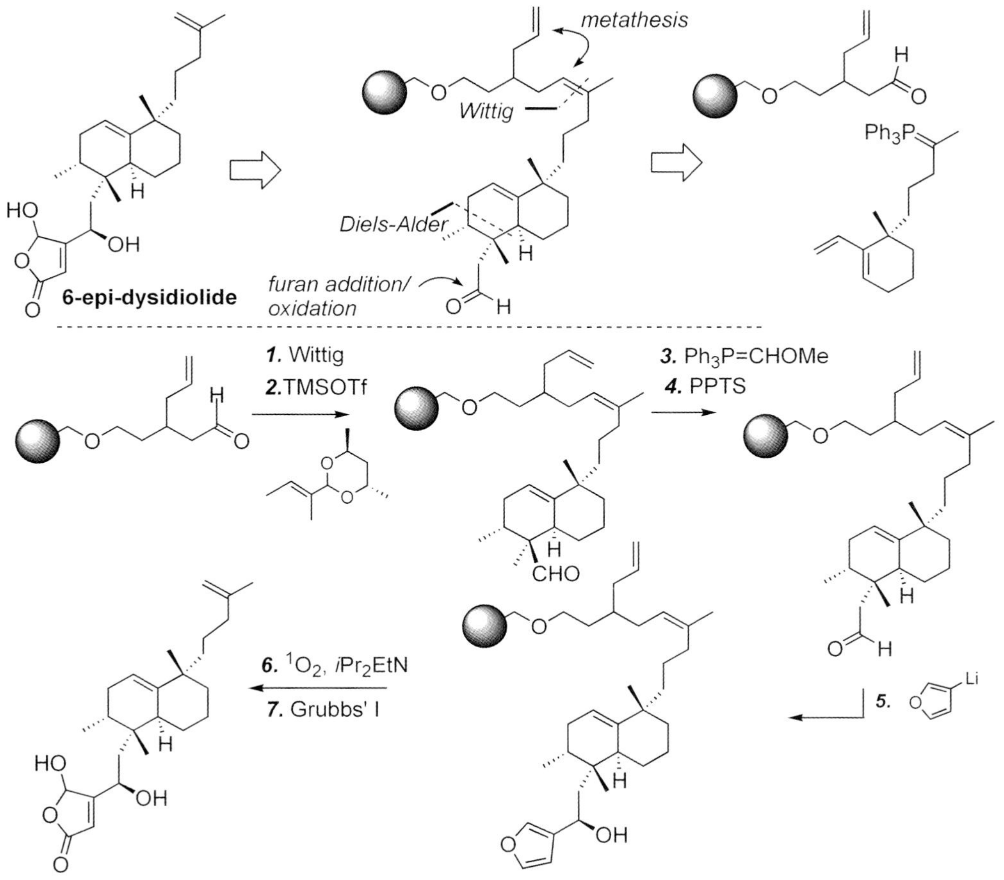

Scheme 21.4 Solid phase synthesis of dysidiolide (Waldmann et al.).

macrocyclization. For this purpose, the authors exploited the difference of reactivity between the vinyl bromide and vinyl iodide to achieve the sequential couplings (Scheme 21.5) [36]. Remarkably, this chemistry was found to be sufficiently general to obtain 122 compounds in good yield and purity out of 128 examples. The library was prepared using radiofrequency encoding split and mix combinatorial synthesis, which is significant as the exigence of CO insertion reactions would most likely not be compatible with microtiter plate technologies.

The bleomycins are a family of glycopeptides-derived antitumor antibiotics that can induce DNA or RNA cleavage in sequence or shape selective manner. Hecht an coworkers developed a solid phase synthesis (Scheme 21.6) which was used to assemble a library of 108 analogues thereby providing a thorough evaluation of the contribution of individual substituents on the efficiency of cleavage [37, 38]. Interestingly, two analogues exhibited superior potencies compared to the natural deglycobleomycin A_6.

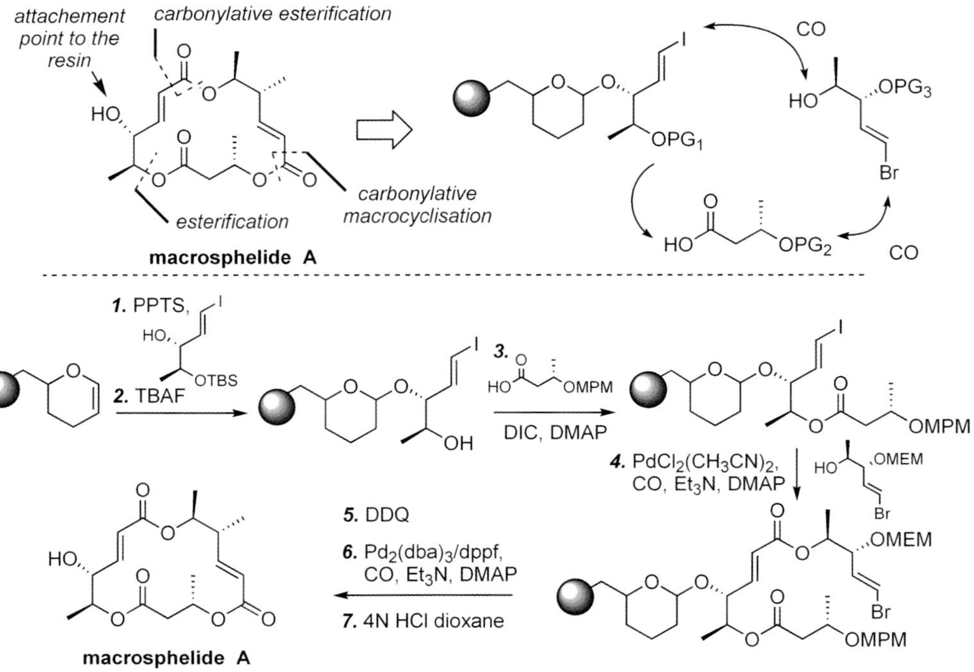

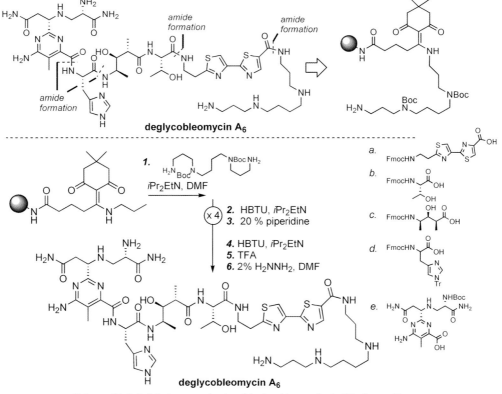

Scheme 21.5 Solid phase synthesis of macrosphelides (Takahashi et al.).

Scheme 21.6 Solid phase synthesis of deglycobleomycin A₆ (Hecht et al.).

The lamellarins alkaloids encompass to date approximately 35 compounds isolated from marine organisms. This family shows an interesting range of pharmacological activities, including antitumor and anti-HIV properties. Alvarez and coworkers devised a very efficient strategy, constructing the highly substituted pyrrole ring using a [3 + 2] dipolar cycloaddition (Scheme 21.7) [39]. The key steps on solid phase include a Sonogashira coupling followed by Baeyer–Villiger reaction to convert an aromatic aldehyde into a phenol and the cycloaddition. Cleavage of the final product with $AlCl_3$ led to complete deprotection of the iPr ether and partial deprotection of the methyl ether, thus yielding both lamellarin U and L.

Saframycin belongs to a family of complex polycyclic alkaloids endowed with spectacular antiproliferative activity. The successful demonstration that analogues of the natural product are clinically effective for the treatment of solid tumors has stimulated intense efforts towards the total synthesis and medicinal chemistry of these alkaloids. Myers and Lanman adapted their solution phase synthesis of saframycin to a ten-step solid phase synthesis, leading to the preparation of 16 analogues [40]. The key steps are two Pictet–Spengler reactions, which are used

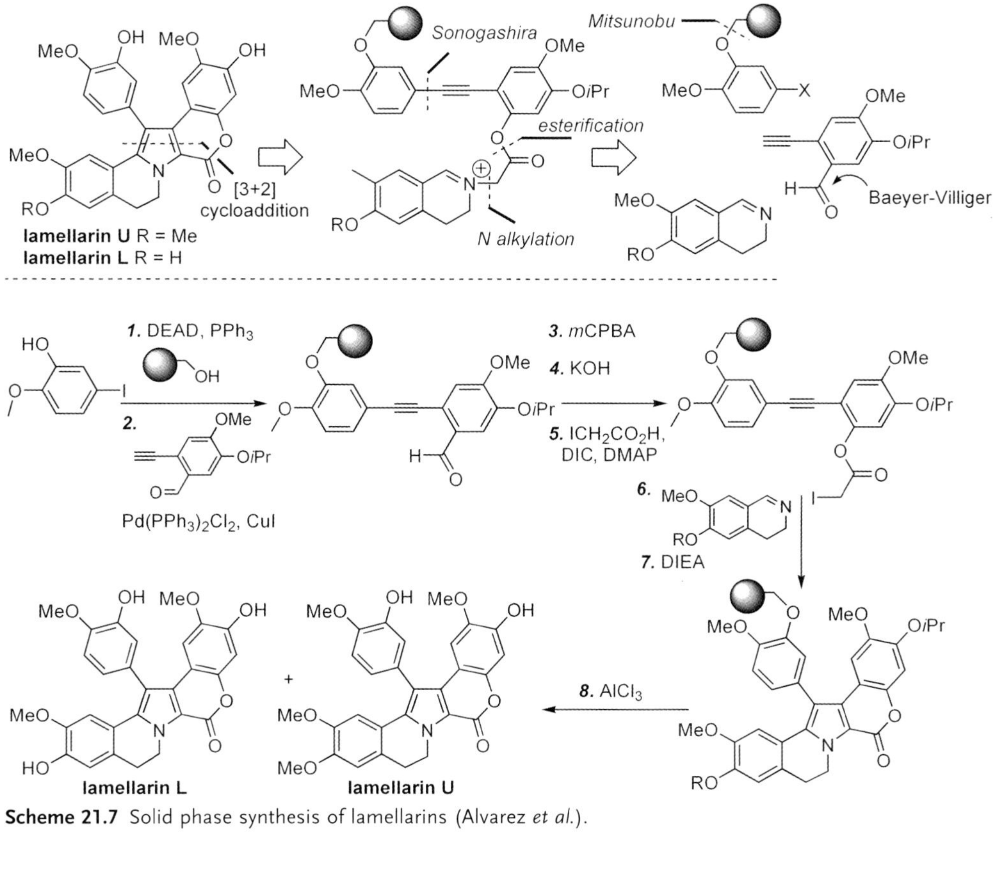

Scheme 21.7 Solid phase synthesis of lamellarins (Alvarez et al.).

to form the bicyclic system (Scheme 21.8). The cleavage is induced by treatment with ZnCl$_2$, which coordinates to the cyano group, assisting its departure and engendering the last cyclization with concomitant release from the polymer, thus affording an iminium ion in solution which is trapped by the cyanide. This cascade reaction afforded the complex molecule in remarkably good overall yield.

Aigialomycins belong to the family of resorcylic macrolides, which contains a high number of ATPase and kinases inhibitors. Specifically, aigialomycin D was found to have antimalarial activity. Despite the lack of structural similarities to purines, several of these compounds have been found to be competitive with ATP for kinases and ATPases. Based on the premise that this motif may be a privileged scaffold for the inhibition of these two important families of enzymes [41], Winssinger and coworkers developed a solid phase synthesis of aigialomycin D and related analogues [42]. The key feature of the synthesis is the use of a thioether linker, which facilitates alkylation of the benzylic position and offers two modes

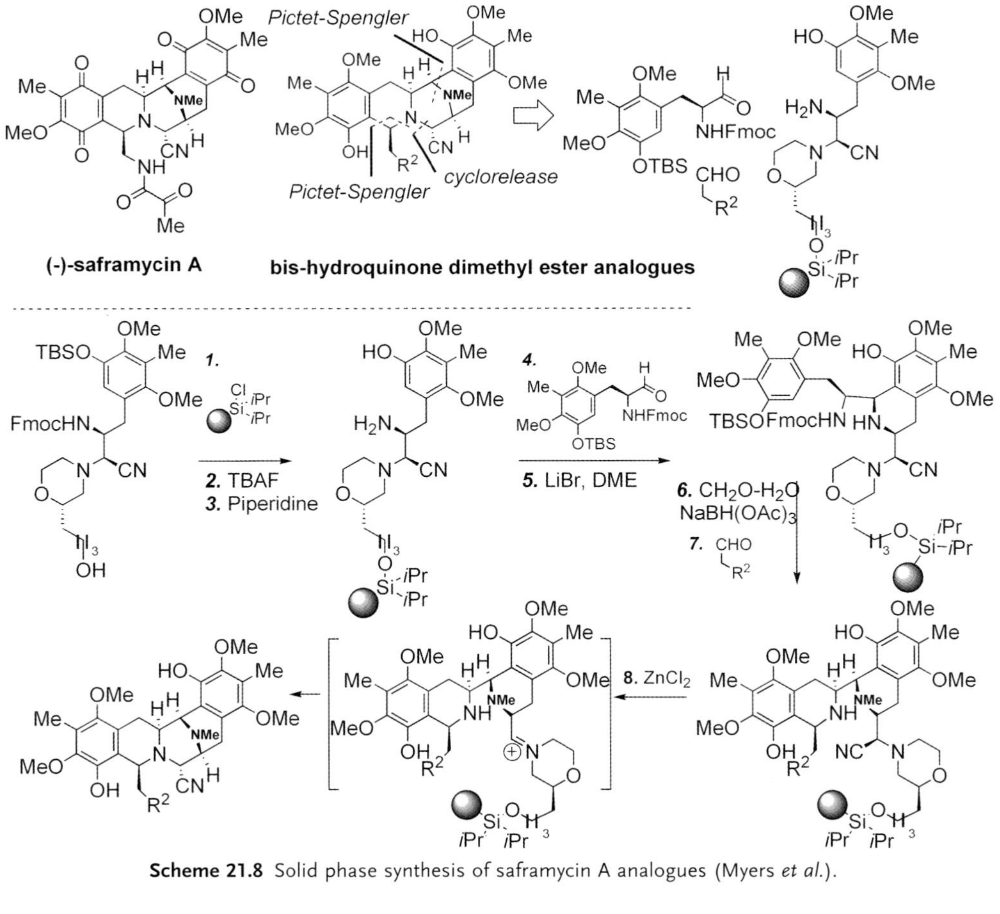

Scheme 21.8 Solid phase synthesis of saframycin A analogues (Myers *et al.*).

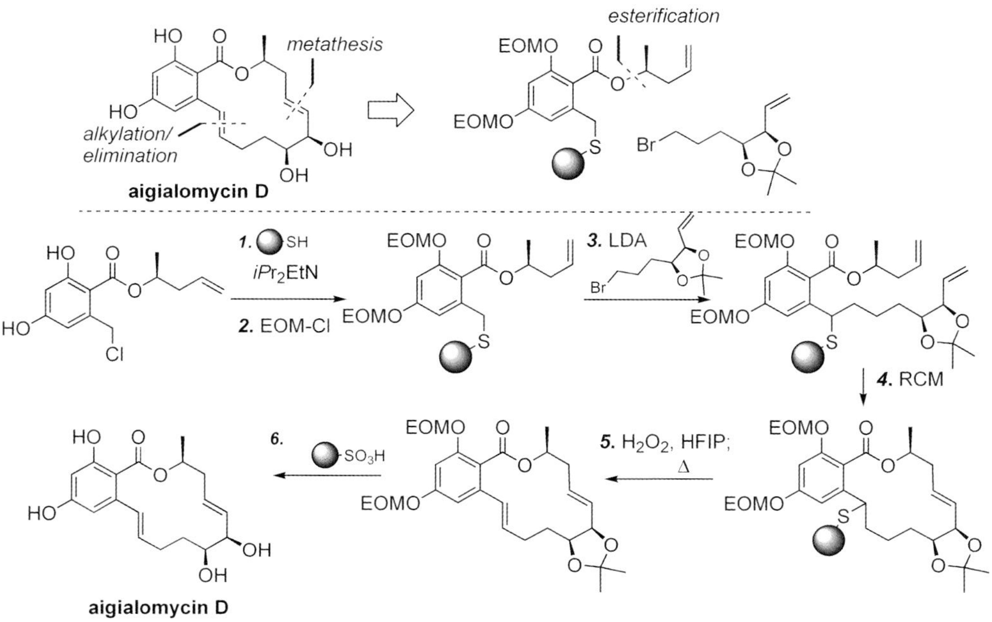

Scheme 21.9 Solid phase synthesis of aigialomycin D (Winssinger et al.).

of cleavage, namely reductive cleavage affording the alkane product or oxidative cleavage affording the alkene as found in aigialomycin D (Scheme 21.9). The macrocycle is formed by a ring-closing metathesis, affording the desired compound in high yield and purity. Importantly, replacement of the allylic diol by an allylic epoxide could be used as a handle for diversification by acid-mediated opening in the presence of different nucleophiles. The authors also showed that aigialomycin D is a CDK inhibitor, thus supporting the privileged nature of this scaffold for kinase inhibition.

The examples cited highlight the state of the art in the solid phase synthesis of natural products ranging from alkaloids to polyketides and terpene derived structures. While the well optimized protocols of peptide and DNA synthesis allow for over 50 steps to be carried out linearly on solid phase, the efficacy of solid phase transformations for the larger repertoire of reactions (albeit less optimized) generally limits the length of a synthesis to a range of 10–12 steps with the current technologies.

21.4
Synthesis of Natural Products Using Immobilized Reagents

As seen in the previous section, tremendous progress has been achieved in solid phase synthesis over the past ten years, enabling the synthesis of complex molecules. However, there remain severe drawbacks to this technique, which can be generally summarized as follows: reactions are difficult to monitor, reactions tend to be slower than their solution counterpart, extra steps are required to attach and release the product from the resin, long linear sequences are difficult to achieve. An alternative strategy that has been developed concurrently is the use of immobilized reagents and scavengers [43, 44]. Three examples of natural product syntheses that did not require traditional workup and chromatography are presented below.

Plicamine belongs to the group of amaryllidaceae alkaloids, which are endowed with rich pharmacology. Furthermore, its polycyclic architecture presents an intricate challenge in synthesis. Ley and coworkers recognized that the fused tricyclic structure could arise from an oxidative phenolic coupling followed by a Michael addition to the resulting conjugate system (Scheme 21.10) [45, 46]. The multistep sequence, involving a reductive amination, acylation, hypervalent iodine oxidative coupling, ketone reduction, methylation, N-alkylation and benzylic oxidation was carried out using solid-supported reagents. In this way, work-up requires only a

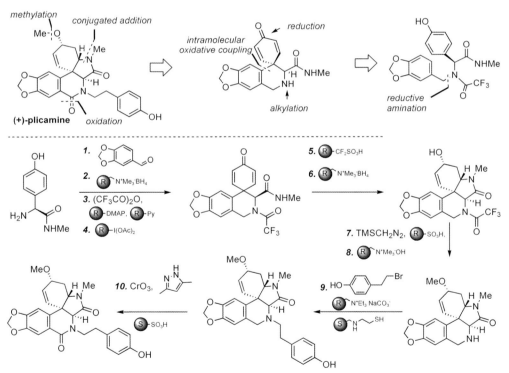

Scheme 21.10 Synthesis of plicamine using polymer-supported reagents (Ley et al.). The letters R and S on the polymer matrix denote a polymer bound reagent and scavenger, respectively.

simple filtration and concentration, operations that can be readily automated. In cases where one of the reagents had to be in solution, such as the bromide in the alkylation, a scavenger resin was used to sequester the excess reagent. The final product was obtained in 43% yield using ten different polymer-bound reagents, which is truly remarkable for a synthesis of this complexity. For the last step, involving the oxidation of the benzylic amine to the amide, a quantitative transformation could not be achieved; however, the unoxidized material was readily removed by the addition of an acidic resin, which captures the remaining starting material in the form of a polymer-bound salt.

In another impressive display of the virtues of polymer bound reagents, Ley and coworkers synthesized epothilone C (Scheme 21.11) [47, 48]. As previously discussed, the antitumor activity of the epothilones has provoked tremendous interest in this family of molecules. Based on previous syntheses, the molecule was disconnected in three fragments that could be assembled in a convergent fashion whilst incorporating exciting and newly devised immobilized reagent methodologies. The final product was obtained with over 90% purity without a single conventional workup or column purification, using a total of 27 transformations (17 steps for the longest linear sequence). The success of this synthesis hinges not only on the use of polymer-bound reagents to achieve a protecting group manipulation, an oxidation or the use of polymer-bound acids to neutralize reactions performed under basic conditions and capture the base (diisopropyl amine or hexamethydisilazane) but also on the creative use of the catch and release strategy. For example, the thiazole fragment A was further purified from impurities lacking the alkyl iodide functionality upon loading onto the polymer-bound triphenylphosphine. The clean fragment was then released upon Wittig olefination in the next step. Another example is the purification by catch and release after the macrocyclization and silyl deprotection. Using sulfonic acid resin, the thioazole is protonated and epothilone remains immobilized on the resin in the form of a salt. The resin is then washed with ammonia in MeOH, thus releasing epothilone C free of the byproducts from the Yamaguchi reagent and silyl protecting groups.

The third example is the synthesis of pochonin D, which was identified as a potential HSP90 by computational analysis. The synthesis of pochonin D was achieved in six steps for the longest linear sequence, providing the final compound in 31% overall yield (Scheme 21.12) [49]. Pochonind D was confirmed to be a nearly as potent an HSP90 inhibitor as radicicol. The significance of this work is that the developed chemistry was amenable to semi-automated parallel synthesis of libraries based on this scaffold [50]. Thus, it was shown that a diversity of core structures could be obtained and further diversified using polymer-bound reagents to obtain a library of over 100 compounds. Screening of this library against a panel of 24 kinases led to the identification of several new inhibitors of therapeutically relevant kinases, thus demonstrating the original hypothesis that the resorcylide scaffold may be a privileged structure for such function.

Other notable syntheses of natural products using polymer-bound reagents include epibatidine [51] and carpanone [52]. These techniques have also been used to prepare pharmaceutically relevant products such as sildenafil [53], and can be

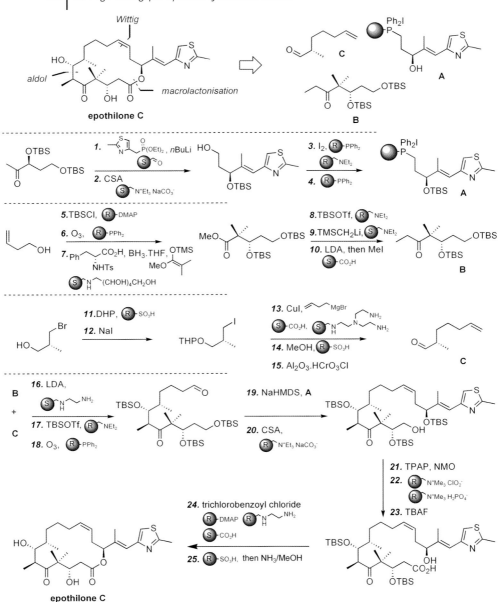

Scheme 21.11 Synthesis of epothilone C using polymer-supported reagents (Ley *et al.*). The letters R and S on the polymer matrix denote a polymer bound reagent and scavenger, respectively.

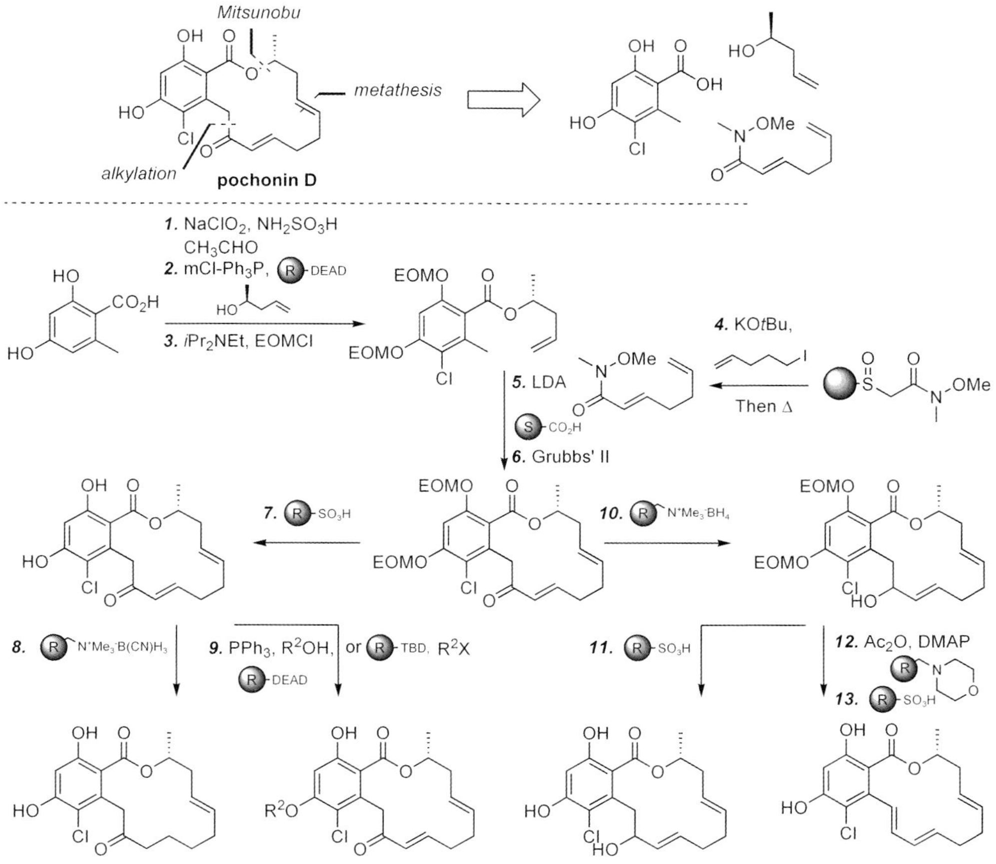

Scheme 21.12 Synthesis of pochonin D and analogues using polymer-supported reagents (Winssinger et al.) The letters R and S on the polymer matrix denote a polymer bound reagent and scavenger, respectively.

used in continuous flow mode [54], as has been demonstrated by Ley and coworkers with the development of a mesofluidic flow reactor for gram-scale production. The reactor incorporates on-chip mixing and columns of solid supported reagents. These examples demonstrate the scope and utility of supported reagents and scavenging techniques and should thus encourage the use of polymer-bound reagents alongside conventional methods in total synthesis or library synthesis programs.

21.5
Synthesis of Natural Products Using Isolation Tags

An alternative method to overcome the limitations of solid phase synthesis is the use of isolation tags. For example, long polyethylene glycol (PEG) chains are particularly insoluble in ethers while being soluble in most other organic solvents. Thus, a molecule derivatized with such a chain can be easily precipitated in Et_2O to recover the desired product. This fact had been originally explored in the context of peptide synthesis [55–57]; however, it was mostly abandoned in favor of solid phase synthesis, which is more easily automated. Janda and coworkers elegantly applied the use of a soluble polymer to the synthesis of prostaglandins [58]. A key feature of this synthesis is the Lindlar reduction of the alkyne to obtain the cis-alkene, which would be impossible on solid phase due to the insoluble nature of the catalyst making it impossible to penetrate within the matrix of polystyrene. The choice of soluble polymer tag proved to be essential as the synthetic scheme required low temperature reactions in THF, which were not compatible with PEG tags' solubility at such temperatures. The authors found that the noncross-linked polystyrene (NCPS) worked admirably well for their purpose (Scheme 21.13).

In the mid 1990s, Curran and coworkers began to explore the possibility of using a fluorous phase [59] to selectively isolate compounds tagged with a perfluoroalkyl, applying the principle to both liquid–liquid extraction (the fluorous phase is not miscible with either the aqueous or organic phase) or solid-phase extraction using fluorous silica gel (the tagged compound adheres to the column until a fluorophilic solvent is used) [60]. This effectively provides an efficient and automatable method to isolate the product from a crude mixture. Curran and coworkers extended the utility of this concept by demonstrating that a mixture of compounds tagged with

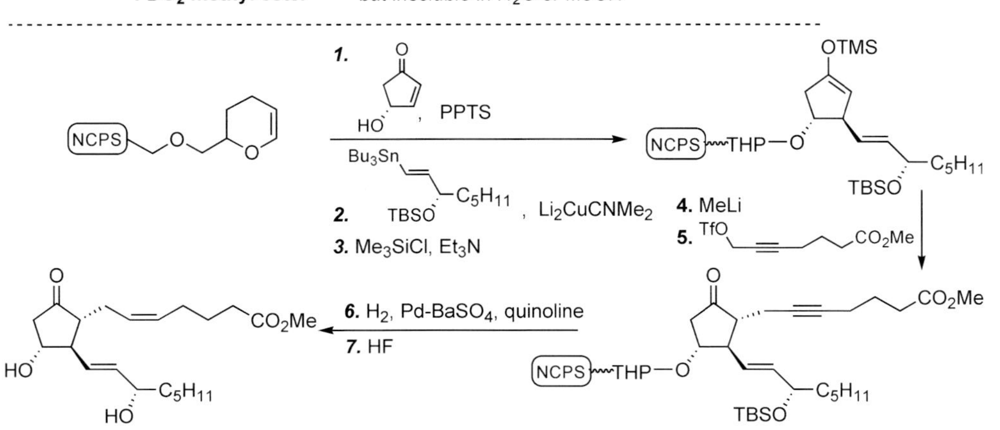

Scheme 21.13 Synthesis of prostaglandin using isolation tags (Janda et al.). NCPS: noncross-linked polystyrene.

different lengths of fluorinated alkyl chain could be resolved on fluorinated silica gel, thus offering the possibility to carry out a synthesis using a mixture of compounds, each labeled with a unique tag length, that is ultimately resolved into its individual components [61–63]. This was elegantly applied to the synthesis of several natural products and their stereoisomers, including the sawfly sex pheromone [64], passifloricins [65], lagunapyrone [66] and discodermolide [61]. The synthesis of a library of stereoisomers of murisolins perhaps best illustrates the power of this concept (Scheme 21.14) [67, 68]. Murisolins were originally reported to be exceptionally cytotoxic; however, their exact stereochemistry could not be easily assigned unmistakably as these lipophilic compounds are especially reluctant to crystallize. Curran and coworkers synthesized a library of 28 stereoisomers using a convergent coupling of two fragments, each of them being a mixture of four stereoisomers tagged with different length of ethylene glycol or fluorinated alkyl chains. This library allowed to unequivocally assigning the stereochemistry of murisolin. Interestingly, several diastereoisomers had identical proton NMRs.

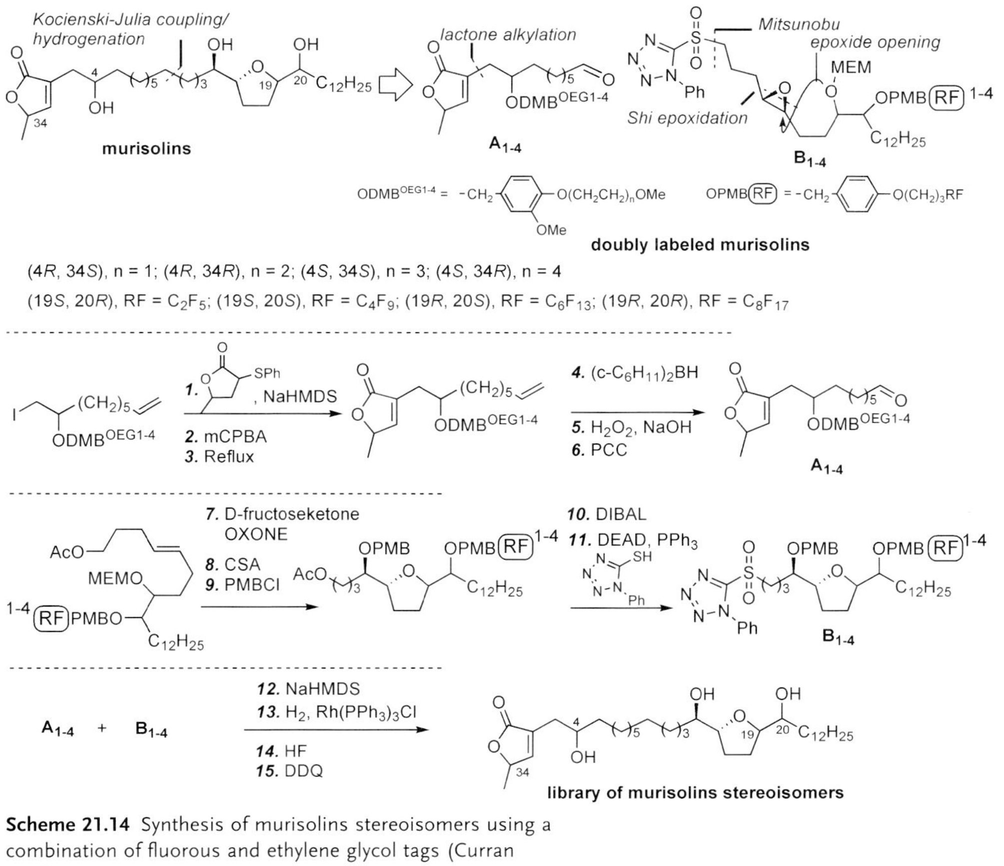

Scheme 21.14 Synthesis of murisolins stereoisomers using a combination of fluorous and ethylene glycol tags (Curran et al.).

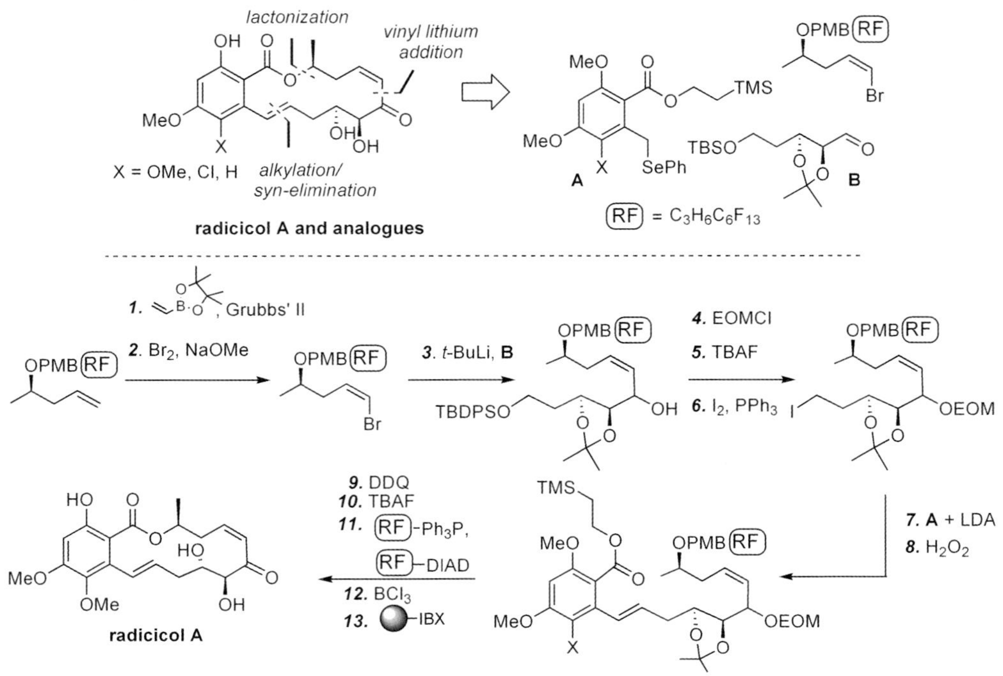

Scheme 21.15 Synthesis of radicicol A and analogues using fluorous isolation tags (Winssinger et al.).

The union of fluorous tags with ethylene glycol tags synthesis provided the possibility of deconvoluting 16 compounds in a mixture.

Radiciol A was reported to accelerate the degradation of select mRNA sequences and is closely related to other resorcylic macrolides containing a cis-enone, which have been shown to irreversibly inactivate select kinases. Based on the interest in covalent inhibitors for chemical biology, Winssinger and coworkers developed a convergent synthesis towards this family of natural product, relying on both fluorous tags and immobilized reagents (Scheme 21.15) [69]. They showed that this strategy could be used to access diverse resorcylides bearing the essential cis-enone moiety in a semi-automated fashion, enabling the authors to establish a thorough profile of the kinase selectivity for these irreversible inhibitors. Significantly, these compounds inactivated only two to three kinases out of 127 tested at low nanomolar concentration.

21.6
Combinatorial Synthesis of Libraries Based on Important Natural Product Motifs

The demand for bioactive molecules arising from progress in both genomics and proteomics has created a tremendous pressure on chemical synthesis to increase

21.6.1
Structure-Based Libraries Targeting Kinases and Other Purine-Dependent Enzymes

The purine scaffold is a key structural element of the substrates and ligands in many biosynthetic, regulatory and signal transduction proteins, including G-coupled proteins, kinases, motor proteins and polymerases. It follows that libraries based on this scaffold may provide versatile leads to probe signaling and metabolic pathways. Olomoucine (Scheme 21.16), a natural derivative of purine that selectively inhibits CDK2, indicated that, despite the ubiquitous role of purines, selective inhibition was possible. Schultz and coworkers reported an initial library of 348 purine derivatives in 1998 this has been corrected, which led to the discovery of an improved and selective CDK2 inhibitor named purvalanol B (6 nM, 1000-fold improvement over olomoucine) [70]. Beyond kinases, this library also afforded inhibitors of carbohydrate sulfotransferases, which is not surprising considering such enzymes derive the sulfate group from 3′-phosphoadenosine 5′-phosphosulfate (PAPS) [71], and estrogen sulfotransferase inhibitors [72]. A screen against the malaria parasite *Plasmodium falciparum* identified two subfamilies of purines with good activities (nanomolar) against the parasite but modest CDK activities, raising the possibility to achieve a good therapeutic window [73]. Purine libraries were also screened in a phenotypic assay for myotube fission (myotubes are multinucleated cells that eventually develop into mature muscle fibers and have the capacity to regenerate wounded muscles), leading to the discovery of a new purine analog named myoseverin [74]. Myoseverin reverted terminal muscle-differentiated cells to a state that was responsive to environmental cues [75]. In an extension of this study, a library of purines was screened in a phenotypic assay for perturbation of mitotic spindle assembly in Xenopus egg extracts. Out of a collection of 1561 compounds, 15 compounds destabilized microtubules without targeting tubulin directly. Affinity chromatography with one compound, named diminutol, suggested NQO1, an NADP-dependent oxidoreductase, as the target, which was confirmed through immunodepletion studies [76]. The chemistry developed to access libraries of purines was extended to other heterocycles, including pyrimidines, quinazolines, pyrazines, phthalazines, pyridazines and quinoxalines. A library of over 45 000 compounds was prepared using a combinatorial scaffold approach. For a library of this size, a split–pool synthesis was necessary; however, the authors used the optical encoding system [77], which directs the path of small reactors harboring 5–10 mg of resin through the split and pool cycles and allows a reformatting of the final library into a spatially addressable format for the cleavage of the final library [78]. This larger library and focused libraries derived from library hits have yielded an impressive collection of biological probes and potential therapeutics, including several compounds that control stem cell fate [79–82] or even reverted differentiated cells [83]. More

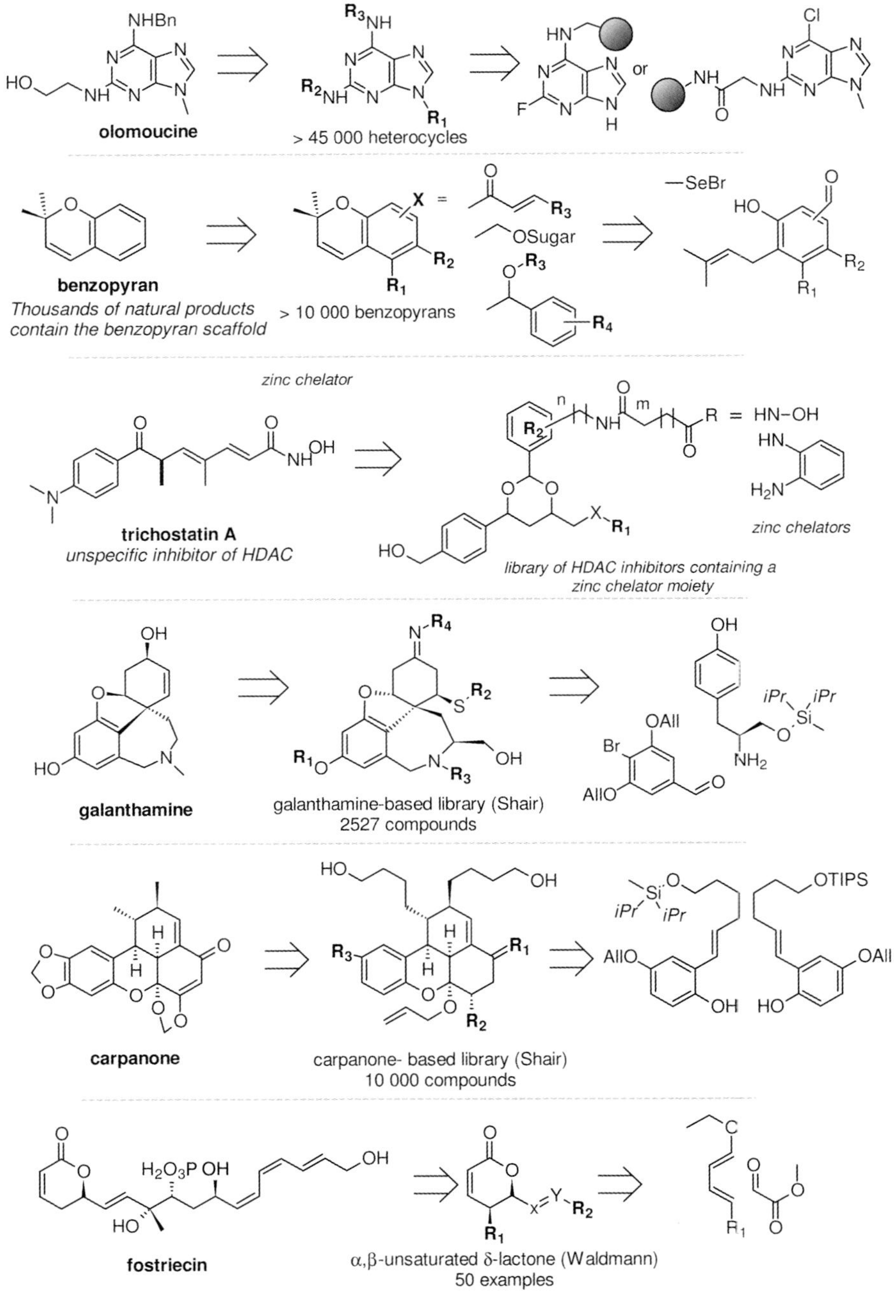

21 High-Throughput Synthesis of Natural Products

olomoucine → > 45 000 heterocycles

benzopyran
Thousands of natural products contain the benzopyran scaffold → > 10 000 benzopyrans

trichostatin A
unspecific inhibitor of HDAC

library of HDAC inhibitors containing a zinc chelator moiety

zinc chelators

galanthamine → galanthamine-based library (Shair) 2527 compounds

carpanone → carpanone-based library (Shair) 10 000 compounds

fostriecin → α,β-unsaturated δ-lactone (Waldmann) 50 examples

A number of natural products (fostriecin, callystatin A, goniothalamin) bear the α,β-unsaturated δ-lactone core structure

Scheme 21.16 Selected examples of libraries inspired by prominent natural product motifs.

recently, several non-ATP-competitive inhibitors of Bcr-Abl kinase were identified that bind to an allosteric site distant from the ATP-binding site and are synergic with imatinib, the Bcr-abl inhibitor currently used for the treatment of chronic myelogenous leukemia [84].

21.6.2
Libraries Based on a Privileged Scaffold – Discovery of Fexaramine

The benzopyran motif is found in more than 4000 natural products and designed analogues including many biologically active compounds. Based on this observation, Nicolaou and coworkers developed a highly efficient and divergent synthesis based on the benzopyran core motif, which was formed by cycloaddition reaction upon loading on a selenium-based resin (Scheme 21.16). This chemistry was used to prepare over 10 000 compounds using split–pool synthesis and optical encoding with directed sorting [77, 85]. The automated sorting was essential in this library synthesis as not all intermediates followed the same synthetic paths. The final library was further diversified by reaction on the pyran moiety [86]. The library was first screened for antibiotic activity, leading to the identification of several benzopyrans with activity against methycillin-resistant bacteria at a comparable level to vancomycin [87]. In collaboration with Evans and coworkers, the library was screened for FXR agonists [88], a nuclear receptor involved in the regulation of cholesterol biosynthesis. The efficient regulation of cholesterol metabolism is essential for mammals as its misregulation leads to arteriosclerosis and heart diseases. It is controlled through a complex feedback loop consisting of LXR and FXR nuclear receptors. The LXR (liver X receptors) are activated by oxysterols (early cholesterol metabolites), leading to up-regulation of CYP7A1, the enzyme catalyzing the rate-limiting step in the conversion of cholesterol into bile acids. The later ones (e.g. chenodeoxycholic acid CDCA) are, however, ligands for FXR (farnesoid X receptors) whose activation downregulates the CYP7A1, and closes the feedback circuit. Additionally, both LXR and FXR are involved in the regulation of several gene products responsible for cholesterol absorption, metabolism and transport. Selective small-molecule FXR agonists would be powerful tools to dissect FXR's physiological function and may provide lead structures for potential therapeutics. The benzopyran library was screened using a cell-based reporter assay in which a FXR responsive promoter was linked to a luciferase reporter. The initial hits were further optimized in terms of potency and pharmacological properties, affording a FXR ligand, fexaramine, with 25 nM affinity. Expression profiling experiments were then carried out to evaluate the effects of fexaramine-induced FXR activation on a genomic scale [89]. Notably, fexaramine did not agonize other nuclear hormone receptors, showing high specificity in contrast to bile acid. Fexaramine is thus a useful tool to dissect the FXR genetic network from the bile acid network. A co-crystal structure of FXR-fexaramine complex was obtained, providing insightful structural information and suggesting a mechanism for the initial steps of bile acid signaling.

21.6.3
Inhibitors of Histone Deacetylases (HDAC)

Histone deacetylases (HDACs) are zinc hydrolases that remove the lysine's ε-N-acetyl group from proteins, thereby modulating their function. With histones, this deacetylation restores their cationic character and increases their affinity for DNA, thus down-regulating DNA transcription. Nine human HDACs have been identified thus far and, while several inhibitors are known, none are selective for individual HDAC and, thus, have limited utility to deconvolute the function of individual HDAC. Interestingly, inhibition of HDACs results in hyperacetylation of α-tubulin, suggesting that HDACs modulate the function of proteins beyond the histones. Inspired by the structure of a natural product, trichostatin A, containing a hydroxamic acid known to chelate to zinc, Schreiber and coworkers designed a library based on the previously made 1,3-dioxanes, targeting the HDAC by incorporating a hydroxamic acid or a 2-aminocarboxyamide functionality (Scheme 21.16). A library of 7200 compounds was prepared on the aforementioned polystyrene macrobeads using split–pool synthesis and chemical encoding [90]. After the synthesis, the macrobeads were separated into individual wells of a microtiter plate and the compounds were cleaved from the resin to obtain a stock solution that was used in biological assays (one bead-one stock solution) [91, 92]. Upon identification of an active compound, the chemical tag remaining on the bead was oxidatively cleaved and analyzed with gas chromatography to determine the structure of the lead. A cell-based cytoblot assay to measure histone and tubulin acetylation level in the presence of inhibitors followed by secondary fluorescence assays led to the identification of tubacin, which selectively induces α-tubulin hyperacetylation (EC_{50} = 2.9 μM), and histacin – a compound that selectively induces histone hyperacetylation (EC_{50} = 34 μM) [93]. Tubacin, in contrast to trichostatin A, had no effect on gene expression and did not affect cell cycle progression. Interestingly, tubacin only inhibited one of the HDAC6's two deacteylase domains. While α-tubulin hyperacetylation had no impact on microtubule dynamics, it was shown to reduce cellular motility, which is consistent with the fact that overexpression of HDAC6 had previously been shown to increase cell motility [94]. Further experiments highlighted the role of α-tubulin acetylation in mediating the localization of microtubule-associated proteins such as p58, a protein that mediates binding of microtubules to the Golgis [95].

21.6.4
Secramine

Galanthamine is a natural alkaloid and a potent acetylcholinesterase inhibitor. Recognizing the potential to access this rigid polycyclic core through a biomimetic oxidative coupling/Michael addition, Shair and coworkers developed a highly efficient diversity-oriented synthesis based on this scaffold (Scheme 21.16) [96]. A library of 2527 compounds was prepared on the aforementioned macrobeads by split–pool synthesis and distributed in microtiter plates for cleavage. The structure of individual compounds was obtained by mass spectroscopic analysis of the library. The library was screened in a phenotypic assay for protein trafficking – translocation of proteins from the endoplasmatic reticulum via the Golgi apparatus to the plasma membrane – using a fluorescent fusion protein between VSVG, which is targeted toward the cell membrane, and GFP, leading to the identification of an inhibitor named secramine. It blocked protein trafficking out of the Golgi at 2 µM concentration, whereas galanthamine has no such effect up to 100 µM. Further experiments showed that secramine inhibited the activation of Rho GTPase Cdc42 and thus inhibited Cdc42-dependent functions. As this inhibition is independent of the prenylation state of Cdc42, secramine is fast acting compared to prenylation inhibitors [97, 98].

21.6.5
Carpanone

Based in the biomimetic synthesis of carpanone reported by Chapman, Shair and coworkers developed the split–pool synthesis of a 10 000 member library. Carpanone as a core structure from a natural product has several features that makes it attractive as a scaffold for libraries: it is a complex rigid polycycle yet readily accessible via a cascade of two complexity-building reactions – a phenol oxidative heterodimerization as previously described by Chapman, but using modified conditions [(diacetoxy)iodobenzene], and a subsequent intramolecular inverse electron-demand Diels–Alder cycloaddition. The library was screened in a fluorescence-based phenotypic assay for vesicular traffic, yielding inhibitors in the micromolar range for the Golgi mediate exocytosis. As for the galanthamine library, the significance of the identified compounds lies in the opportunities they provide in uncovering new biological mechanism.

21.6.6
Natural-Product Inspired Synthesis of αβ-Unsaturated-δ-Lactones

The identification of key structural motifs that may bias a library towards biologically relevant diversity-space can dramatically increase the outcome of screens. Waldmann and coworkers recognized αβ-unsaturated-δ-lactones as such an important motif as it is present in several biologically active and useful natural products such as fostriecin (phosphatase inhibitor), callystatin A (nuclear export inhibitor) and pironetin (immunosuppressant) [99]. In terms of chemistry, this important motif can be easily accessed with high enantiomeric purity by an oxa-Diels–Alder reaction catalyzed by a chiral titanium complex. The synthesized library was screened in phenotypic cell-based assays, yielding new inhibitors of viral entry and modulators of cell cycle progression. Notably, several biologically interesting lead compounds were identified from a library of only 50 members, supporting the hypothesis that a biologically validated scaffold enhances the hit rate in screens.

21.7
Conclusion

The past ten years have brought tremendous progress in the scope of the chemistry that can be carried in a high-throughput fashion using either automated equipment or simply by facilitating the parallel processing of reactions. While solid phase synthesis is the simplest format to automate and lends itself to the powerful split and pool synthetic scheme for library purposes, it suffers from the heaviest limitation in terms of reaction scope. The number of complex natural products that have been synthesized with either polymer-bound reagents or the fluorous isolation tags technology testifies to the power of these two alternative approaches. It can be anticipated that the increasing commercial accessibility to these latter two formats will lead researchers from many disciplines to embrace them. A cursory look at the number of commercial small molecule libraries and the expansion of libraries within pharmaceutical industries testifies to the impact that high-throughput synthesis is having. However, more compounds does not always translate to more hits and the past ten years have also shown that interesting biological activity is usually clustered on small islands within diversity space. Natural products typically reside on these small islands and as such will continue to play an important role in providing lead structures and important structural motifs in addition to the new challenges. While many facets surrounding natural product isolation, synthesis and derivatization remain an art form, predominantly due to the heterogeneity of the challenges that each of these substance present, systematization of the chemical methods to access and elaborate natural products is of primordial importance.

List of Abbreviations

BOC	butoxycarbonyl
CSA	camphorsulfonic acid
DCC	N,N-dicyclohexylcarbodiimide
DEAD	diethyl azodicarboxylate
DIBAL	diisobutylaluminium hydride
DIC	N,N-diisopropylcarbodiimide
DDQ	2,3-dichloro-5,6-dicyano-1,4-benzoquinone
DHP	dihydropyran
DIEA	diisopropylethylamine
DMAP	4-dimethylaminopyridine
DMDO	dimethyldioxirane
DMF	dimethylformamide
EOM	ethoxymethyl
Grubbs' II	Grubb's second generation catalyst: (ruthenium[1,3-bis(2,4,6-trimethylphenyl)-2-imidazolidinylidene]dichloro(phenylmethylene)(tricyclohexylphosphane)

HBTU	O-benzotriazole-N,N,N',N'-tetramethyluronium hexafluorophosphate
HFIP	hexafluoroisopropanol
IBX	2-iodoxybenzoic acid
LDA	lithium diisopropylamide
mCPBA	meta-chloroperoxybenzoic acid
NaHMDS	sodium hexamethyldisilazide
NMO	N-methylmorpholine-N-oxide
PCC	pyridinium chlorochromate
PMB	para-methoxybenzyl
PPTS	pyridinium para-toluenesulfonate
TBAF	tetra-n-butylammonium fluoride
RCM	Ring closing metathesis
PS-TBD	polystyrene-(1,5,7)-triaza-bicyclo[4.4.0]dodeca-5-ene-7-methyl
TBDPS	t-butyldiphenylsilyl
TBS	t-butyldimethylsilyl
TES	triethylsilyl
TFA	trifluoroacetic acid
THP	tetrahydropyran
TIPS	triisopropylsilyl
TPAP	tetrapropylammonium perruthenate
TMS	trimethylsilyl
TMT	trimethoxytrityl

References

1 Bunin, B.A., Plunkett, M.J. and Ellman, J.A. (1994) *Proceedings of the National Academy of Sciences of the United States of America*, **91**, 4708–12.

2 DeWitt, S.H., Kiely, J.S., Stankovic, C.J., et al. (1993) *Proceedings of the National Academy of Sciences of the United States of America*, **90**, 6909–13.

3 Nicolaou, K.C., Winssinger, N., Pastor, J., et al. (1997) *Nature*, **387**, 268–72.

4 Nicolaou, K.C., Vourloumis, D., Winssinger, N. and Baran, P.S. (2000) *Angewandte Chemie – International Edition in English*, **39**, 44–122.

5 Yurek-George, A., Habens, F., Brimmell, M., et al. (2004) *Journal of the American Chemical Society*, **126**, 1030–1.

6 Bourel-Bonnet, L., Rao, K.V., Hamann, M.T. and Ganesan, A. (2005) *Journal of Medicinal Chemistry*, **48**, 1330–5.

7 Cruz, L.J., Insua, M., Baz, J., Trujillo, M., et al. (2006) *Journal of Organic Chemistry*, **71**, 3335–8.

8 Krishnamoorthy, R., Vazquez-Serrano, L.D., Turk, J.A., et al. (2006) *Journal of the American Chemical Society*, **128**, 15392–3.

9 Shaginian, A., Rosen, M.C., Binkowski, B.F. and Belshaw, P.J. (2004) *Chemistry – A European Journal*, **10**, 4334–40.

10 Breinbauer, R., Vetter, I.R. and Waldmann, H. (2002) *Angewandte Chemie – International Edition in English*, **41**, 2879–90.

11 Brohm, D., Metzger, S., Bhargava, A., et al. (2002) *Angewandte Chemie – International Edition in English*, Vol. **41**, pp. 307–11.

12 Dekker, F.J., Koch, M.A. and Waldmann, H. (2005) *Current Opinion in Chemical Biology*, **9**, 232–9.

13 Nicolaou, K.C., Guy, R.K. and Potier, P. (1996) *Scientific American*, **274**, 94–8.

14 Xiao, X.Y., Parandoosh, Z. and Nova, M.P. (1997) *Journal of Organic Chemistry*, **62**, 6029–33.

15 Stachel, S.J., Biswas, K. and Danishefsky, S.J. (2001) *Current Pharmaceutical Design*, **7**, 1277–90.

16 Nicolaou, K.C., Xu, J.Y., Kim, S., et al. (1997) *Journal of the American Chemical Society*, **119**, 11353–4.
17 Nicolaou, K.C., Winssinger, N., Vourloumis, D., et al. (1998) *Journal of the American Chemical Society*, **120**, 10814–26.
18 Nicolaou, K.C., Pfefferkorn, J., Xu, J., et al. (1999) *Chemical and Pharmaceutical Bulletin*, **47**, 1199–213.
19 Nicolaou, K.C., Xiao, X.Y., Parandoosh, Z., et al. (1995) *Angewandte Chemie-International Edition in English*, **34**, 2289–91.
20 Meseguer, B., Alonso-Diaz, D., Griebenow, N., et al. (1999) *Angewandte Chemie-International Edition in English*, **38**, 2902–6.
21 Meseguer, B., Alonso-Diaz, D., Griebenow, N., et al. (2000) *Chemistry-A European Journal*, **6**, 3943–57.
22 Akritopoulou-Zanze, I. and Sowin, T.J. (2001) *Journal of Combinatorial Chemistry*, **3**, 301–11.
23 Nicolaou, K.C., Winssinger, N., Hughes, R., et al. (2000) *Angewandte Chemie-International Edition in English*, **39**, 1084–8.
24 Xu, R., Greiveldinger, G., Marenus, L.E., et al. (1999) *Journal of the American Chemical Society*, **121**, 4898–9.
25 Niggemann, J., Michaelis, K., Frank, R., et al. (2002) *Journal of the Chemical Society, Perkin Transactions 1*, 2490–503.
26 Lopez, S.N., Ramallo, I.A., Sierra, M.G., et al. (2007) *Proceedings of the National Academy of Sciences of the United States of America*, **104**, 441–4.
27 Nicolaou, K.C., Roschangar, F. and Vourloumis, D. (1998) *Angewandte Chemie-International Edition in English*, **37**, 2014–45
28 Nicolaou, K.C., Vourloumis, D., Li, T., et al. (1997) *Angewandte Chemie-International Edition in English*, **36**, 2097–103.
29 Nicolaou, K.C., Winssinger, N., Pastor, J. and Murphy, F. (1998) *Angewandte Chemie-International Edition in English*, **37**, 2534–7.
30 Nicolaou, K.C., Pastor, J., Winssinger, N. and Murphy, F. (1998) *Journal of the American Chemical Society*, **120**, 5132–3.
31 Das, S., Chandrasekhar, S., Yadav, J.S. and Gree, R. (2007) *Chemical Reviews*, **107** (7) 3286–3337.
32 Thompson, L.A., Moore, F.L., Moon, Y.C. and Ellman, J.A. (1998) *Journal of Organic Chemistry*, **63**, 2066–7.
33 Dragoli, D.R., Thompson, L.A., O'Brien, J. and Ellman, J.A. (1999) *Journal of Organic Chemistry*, **1**, 534–9.
34 Brohm, D., Philippe, N., Metzger, S., et al. (2002) *Journal of the American Chemical Society*, **124**, 13171–8.
35 Hayashi, M., Kim, Y.P., Hiraoka, H., et al. (1995) *Journal of Antibiotics*, **48**, 1435–9.
36 Takahashi, T., Kusaka, S., Doi, T., et al. (2003) *Angewandte Chemie-International Edition in English*, **42**, 5230–4.
37 Cagir, A., Tao, Z.F., Sucheck, S.J. and Hecht, S.M. (2003) *Bioorganic and Medicinal Chemistry*, **11**, 5179–87.
38 Leitheiser, C.J., Smith, K.L., Rishel, M.J., et al. (2003) *Journal of the American Chemical Society*, **125**, 8218–27.
39 Cironi, P., Manzanares, I., Albericio, F. and Alvarez, M. (2003) *Organic Letters*, **5**, 2959–62.
40 Myers, A.G. and Lanman, B.A. (2002) *Journal of the American Chemical Society*, **124**, 12969–71.
41 Winssinger, N. and Barluenga, S. (2007) *Chemical Communications*, 22–36.
42 Barluenga, S., Dakas, P.Y., Ferandin, Y., et al. (2006) *Angewandte Chemie-International Edition in English*, **45**, 3951–4.
43 Ley, S.V. and Baxendale, I.R. (2002) *Nature Reviews Drug Discovery*, **1**, 573–86.
44 Ley, S.V., Baxendale, I.R., Bream, R.N., et al. (2000) *Journal of the Chemical Society, Perkin Transactions 1*, **23**, 3815–4195.
45 Baxendale, I.R., Ley, S.V., Nessi, M. and Piutti, C. (2002) *Tetrahedron*, **58**, 6285–304.
46 Baxendale, I.R., Ley, S.V. and Piutti, C. (2002) *Angewandte Chemie-International Edition in English*, **41**, 2194–7.
47 Storer, R.I., Takemoto, T., Jackson, P.S., et al. (2004) *Chemistry-A European Journal*, **10**, 2529–47.
48 Storer, R.I., Takemoto, T., Jackson, P.S. and Ley, S.V. (2003) *Angewandte Chemie-International Edition in English*, **42**, 2521–5.
49 Moulin, E., Zoete, V., Barluenga, S., et al. (2005) *Journal of the American Chemical Society*, **127**, 6999–7004.

50 Moulin, E., Barluenga, S., Totzke, F. and Winssinger, N. (2006) *Chemistry – A European Journal*, **12**, 8819–34.
51 Habermann, J., Ley, S. and Scott, J. (1999) *Journal of the Chemical Society, Perkin Transactions 1*, 1253–5.
52 Baxendale, I., Lee, A. and Ley, S. (2002) *Journal of the Chemical Society, Perkin Transactions 1*, 1850–7.
53 Baxendale, I.R. and Ley, S.V. (2000) *Bioorganic and Medicinal Chemistry Letters*, **10**, 1983–6.
54 Baumann, M., Baxendale, I.R., Ley, S.V., et al. (2006) *Organic Letters*, **8**, 5231–4.
55 Mutter, M. and Bayer, E. (1980) *Peptides*, **2**, 285–332.
56 Bayer, E. and Mutter, M. (1972) *Nature*, **237**, 512–13.
57 Mutter, M., Hagenmaier, H. and Bayer, E. (1971) *Angewandte Chemie – International Edition in English*, **10**, 811–12.
58 Chen, S. and Janda, K.D. (1997) *Journal of the American Chemical Society*, **119**, 8724–5.
59 Studer, A., Hadida, S., Ferritto, R., et al. (1997) *Science*, **275**, 823–6.
60 Zhang, W. and Curran, D.P. (2006) *Tetrahedron*, **62**, 11837–65.
61 Luo, Z., Zhang, Q., Oderaotoshi, Y. and Curran, D.P. (2001) *Science*, **291**, 1766–9.
62 Zhang, Q. and Curran, D.P. (2005) *Chemistry – A European Journal*, **11**, 4866–80.
63 Curran, D.P. (2004) *Handbook of Fluorous Chemistry*, pp. 101–27, Wiley-VCH.
64 Dandapani, S., Jeske, M. and Curran, D.P. (2004) *Proceedings of the National Academy of Sciences of the United States of America*, **101**, 12008–12.
65 Curran, D.P., Moura-Letts, G. and Pohlman, M. (2006) *Angewandte Chemie – International Edition in English*, **45**, 2423–6.
66 Yang, F., Newsome, J.J. and Curran, D.P. (2006) *Journal of the American Chemical Society*, **128**, 14200–5.
67 Curran, D.P., Zhang, Q., Richard, C., et al. (2006) *Journal of the American Chemical Society*, **128**, 9561–73.
68 Zhang, Q., Lu, H., Richard, C. and Curran, D.P. (2004) *Journal of the American Chemical Society*, **126**, 36–7.
69 Dakas, P.Y., Barluenga, S., Totzke, F., et al. (2007) *Angewandte Chemie – International Edition in English* 2007, 6899–6902.
70 Gray, N.S., Wodicka, L., Thunnissen, A.M., et al. (1998) *Science*, **281**, 533–8.
71 Armstrong, J.I., Portley, A.R., Chang, Y.T., et al. (2000) *Angewandte Chemie – International Edition in English*, **39**, 1303–6.
72 Verdugo, D.E., Cancilla, M.T., Ge, X., et al. (2001) *Journal of Medicinal Chemistry*, **44**, 2683–6.
73 Harmse, L., van Zyl, R., Gray, N., et al. (2001) *Biochemical Pharmacology*, **62**, 341–8.
74 Rosania, G.R., Chang, Y.T., Perez, O., et al. (2000) *Nature Biotechnology*, **18**, 304–8.
75 Perez, O.D., Chang, Y.T., Rosania, G., et al. (2002) *Chemistry and Biology*, **9**, 475–83.
76 Wignall, S.M., Gray, N.S., Chang, Y.T., et al. (2004) *Chemistry and Biology*, **11**, 135–46.
77 Nicolaou, K.C., Pfefferkorn, J.A., Mitchell, H.J., et al. (2000) *Journal of the American Chemical Society*, **122**, 9954–67.
78 Ding, S., Gray, N.S., Wu, X., et al. (2002) *Journal of the American Chemical Society*, **124**, 1594–6.
79 Wu, X., Ding, S., Ding, Q., et al. (2002) *Journal of the American Chemical Society*, **124**, 14520–1.
80 Ding, S., Wu, T.Y.H., Brinker, A., et al. (2003) *Proceedings of the National Academy of Sciences of the United States of America*, **100**, 7632–7.
81 Wu, X., Ding, S., Ding, Q., et al. (2004) *Journal of the American Chemical Society*, **126**, 1590–1.
82 Warashina, M., Min, K.H., Kuwabara, T., et al. (2006) *Angewandte Chemie – International Edition in English*, **45**, 591–3.
83 Chen, S., Zhang, Q., Wu, X., et al. (2004) *Journal of the American Chemical Society*, **126**, 410–11.
84 Adrian, F.J., Ding, Q., Sim, T., et al. (2006) *Nature Chemical Biology*, **2**, 95–102.
85 Nicolaou, K., Pfefferkorn, J., Roecker, A., et al. (2000) *Journal of the American Chemical Society*, **122**, 9939–53.
86 Nicolaou, K., Pfefferkorn, J., Barluenga, S., et al. (2000) *Journal of the American Chemical Society*, **122**, 9968–76.
87 Nicolaou, K.C., Roecker, A.J., Barluenga, S., et al. (2001) *ChemBioChem*, **2**, 460–5.

88 Nicolaou, K.C., Evans, R.M., Roecker, A.J., *et al.* (2003) *Organic and Biomolecular Chemistry*, **1**, 908–920.
89 Downes, M., Verdecia, M.A., Roecker, A.J., *et al.* (2003) *Molecular Cell*, **11**, 1079–92.
90 Sternson, S.M., Wong, J.C., Grozinger, C.M. and Schreiber, S.L. (2001) *Organic Letters*, **3**, 4239–42.
91 Blackwell, H.E., Perez, L., Stavenger, R.A., *et al.* (2001) *Chemistry and Biology*, **8**, 1167–82.
92 Clemons, P.A., Koehler, A.N., Wagner, B.K., *et al.* (2001) *Chemistry and Biology*, **8**, 1183–95.
93 Haggarty, S.J., Koeller, K.M., Wong, J.C., *et al.* (2003) *Chemistry and Biology*, **10**, 383–96.
94 Hubbert, C., Guardiola, A., Shao, R., *et al.* (2002) *Nature*, **417**, 455–8.
95 Haggarty, S.J., Koeller, K.M., Wong, J.C., *et al.* (2003) *Proceedings of the National Academy of Sciences of the United States of America*, **100**, 4389–94.
96 Pelish, H.E., Westwood, N.J., Feng, Y., Kirchhausen, T. and Shair, M.D. (2001) *Journal of the American Chemical Society*, **123**, 6740–1.
97 Pelish, H.E., Peterson, J.R., Salvarezza, S.B., *et al.* (2006) *Nature Chemical Biology*, **2**, 39–46.
98 Pelish, H.E., Ciesla, W., Tanaka, N., *et al.* (2006) *Biochemical Pharmacology*, **71**, 1720–6.
99 Lessmann, T., Leuenberger, M.G., Menninger, S., *et al.* (2007) *Chemistry and Biology*, **14**, 443–51.

Index

A

α-chymotrypsin 603
Aβ (β-amyloid (1-42)) synthesis 504, 507, 508
Accurel MP 1000 373
acetal-protected carbonyls 450
acetal type-linkers 404, 405, 564, 595, 596
acetic acid
– catalytic effect in aminolysis 422
– 2-(4-hydroxy-3-methoxyphenoxy) 391
acetylation, enantioselective 158, 376
acetylcholinesterase inhibitors 67, 634
acid- and base-catalyzed one-pot reactions 114
acid and base mixed ion exchange resins 263
acid chlorides
– as electrophilic scavengers 188
– fluorous tagged 197
– ologomeric bis- 203, 204
– scavenging of 217, 221, 222
acid-labile linkers *see* linkers, acid-labile
acid-labile protecting groups 532
acid reagents, on silica gel 108–113
acid stability
– of linkers 5, 383, 387
– of PEG resins 9
acidic and basic scavengers 209–211
acidic organocatalysts
– ion exchange resins as 250–256
– polymers with acidic sites 260, 261
acidity
– cleavage of safety-catch linkers 447
– of leaving groups 418
acrylamide, from acrylonitrile 368, 377, 378
acrylic acid, α-hydroxy 422
activation, safety-catch linkers 438–442
active esters (activated esters)
– electrophilicity of 160, 172
– enzyme coupling with 374

– on N-hydroxybenzotriazole polymers 168–174
– on N-hydroxysuccinimide (polymers) 146, 163–167, 422
– on phenol-derived polymers 159–163
– on poly-HBTU 148
active sites, PTC 275
acyl linkers 546–549
acylation
– carboxyl activation for 142
– catalyzed by perfluorinated sulfonic acid polymers 257, 258
– enzyme catalyzed 367
1-adamantanecarbonyl chloride 538
ADDP (1,1′–(azodicarbonyl)dipiperidine) 125
adsorbents, alternatives to functional resins 230, 231
adsorption, enzyme immobilization by 373, 374
ADTT (3-amino-1,2,4-dithiazoline-5-thione) 570
1,5-I-AEDANS (*N*-(iodoacetamidoethyl)-1-naphthylamine-5-sulfonic acid) 561
aerobic oxidation 86, 89, 91
affinity chromatography 9, 374
aggregation
– colorimetric studies of 50, 55
– enzyme cross-linking and 371
– in molecular catalysis 247
– silica gel nanoparticles on PEG 544
– in stepwise SPPS 500
AIBN (2,2′-azobis(isobutyronitrile)) 30
aigialomycins 622, *623*
alcohol dehydrogenase 368
alcohols
– acid catalyzed reactions 256
– acid-labile linkers 409, 411

– acidity of, as leaving groups 418
– alkylation by supported reagents 101, 339
– α,ω-amino- 560
– benzoylation 286
– as chiral auxiliaries 332–336
– dehydration, catalytic 319
– enantiopreferential effects of lipases 370
– hindered 534
– hydrogenation 322
– immobilization with acid-labile linkers 404
– kinetic resolution of secondary 377
– nucleophile-labile linkers 418, 420
– α,ω-mercapto- 561, 562
– oxidation by supported reagents 84–90
– photolabile linkers 484
– safety-catch linkers 445, 449
– scavenging of, by nucleophiles 193, 203
aldehydes
– chiral oxazolidine 348, 349
– fluorous tagged 197, 198, 207
– hydrogenation of citral 318
– acid-labile linkers 409
– safety-catch linkers 445
– resin-supported 188, 189
– scavengers 188, 189, 197, 198
– scavenging of, by amines 207, 213, 214
– scavenging of, by hydrazines 209, 210, 214
– SPS using diol-based linkers 404
aldol condensations
– anion exchange resins and 262, 270
– asymmetric solid-phase 344–348, 347
– catalyzed by lipase 367
– catalyzed by peptides 292
– catalyzed by proline 286–289
– catalyzed by transition metals 320, 321
– enantiospecific 423
– perfluorinated sulfonic resins and 259
– retro-aldol reaction 557
– self-aldolization 322
Aldrich Chemicals (Sigma Aldrich) 232, 233, 366, 367, 374
– *see also* Sigma-Aldrich
alkaloids
– amaryllidaceae 213, 624
– cinchona 270
– lamellarin and saframycin 621
– oxomaritidine 91
alkanes
– alkyl chains as isolation tags 629
– traceless linkers 451, 452, 454–456

alkene functionalized linkers 488
alkenes
– E/Z ratio 125, 126
– epoxidation of 375
– hydrogenation of dienes 318, 323
– olefin isomerization 125, 268, 319, 320, 323
– safety-catch linkers 445
– traceless linkers 451–454, 456, 457
p-alkoxybenzyl linkers 88, 390, 488, 489
alkylation
– asymmetric α-alkylation 350–352, *353*
– catalyzed by perfluorinated sulfonic acid polymers 258
– chiral alcohols and ketones 339
– cyclohexanones 338
alkynes
– hydrogenation to alkenes 319
– propargyl linkers 449
– traceless linkers 457
Alloc (allyloxycarbonyl) group 164, 448, 497, 498
Altromycin B glycoside 87
Alzheimer's disease 54, 60
Amano PS-C 366, 367
amaryllidaceae alkaloids 624
amide bonds
– covalent molecular imprinting 19
– non-integral linkers/handles 4, 6
amide-forming coupling processes 142–144, 154, 157
amides
– prepared from nucleophile-labile linkers 421
– safety-catch linkers 443, 447
aminal type-linkers 404, 405
amines
– asymmetric α-alkylation 350, 351
– chiral auxiliaries derived from 336–341
– enantiopreferential effects of lipases 369
– interactions with acid-labile linkers 391, 400, 411
– linkers for 3′-functionalized oligonucleotides 559–561
– polymeric, as organocatalysts 260, 261, 263–270
– quaternary, in anion exchange 127, 128
– scavenging by electrophiles 186–192, 196–198, *201, 203*
– secondary, as nucleophiles 121, 122, 129
– supported, as nucleophilic scavengers 186–192, 196–198, *201, 203*, 213
aminium coupling reagents 146–150, 498

amino acids *see also* peptides
– acid cleavage   407
– alkylation   277
– β-homoarlyglycines   335
– catalytic imidazolidinones   289, 290
– choice of C-terminal   500, 502
– epoxidation with   283, 284
– glutamic acid recognition   66
– hydantoins from   427
– hydrogenation of dehydro-   229
– (2S, 4R)-4-hydroxyornithine   88
– metal affinity of   241
– N-protected   147–149, 159–164, 168–170, 390, 391
– organocatalysts   285–293
– peptide coupling reactions   142
– photocleavage efficiencies   473
– proline derivatives   336, 340
– sarcosine   547
– sterically hindered   150, 151, 500
– tiopronin   53
– unnatural   517, 520
aminopropylsiloxane-grafted CPG   475
ammonium hydroxides, quaternary   105
ammonium salts in PTC   275, 278, 280
amylin   505
amyloid plaques   54, 60
anchors and linkers, oligonucleotide synthesis   545–565
anhydrides as electrophilic scavengers
– fluorous-tagged   196, 197
– resin-supported   188
– ROMP-gel   202–205
anhydrolactols   355
anion exchange resins
– catalysis by   261–263, 276
– enzyme adsorption   373
– nucleophilic reagents   127, 128
– Pd-loaded   325
ANP (3-amino-3-(2-nitrophenyl)propionic acid) linkers   475
Anteunis' racemization test   147–149, 164
anthracenes, polymer-supported   209
antibacterial silver nanoparticles   62
antibiotics
– bleomycins   619
– erythromycin and vancomycin   616
– surface imprinting   21
antibody-conjugated nanoshells   55, 56
antisense research   529, 540, 567
antisickling agents   170
antitumor activity   625
antiviral activity   74
6-APA (6-aminopenicillanic acid)   377, 378

aprotic solvents   30
aptamers; DNA   48, 49, 55, 71
arenes
– alkylation   258
– direct release   403–405
– from traceless linkers   451–454, 456–459
argadin   516–519
ArgoGels   588, *589*
arsonic acids   282
arylation reactions   103
arylsulfonate resins   129
aspartame   377
asymmetric synthesis *see* chirality; stereoselectivity
ATRP (atom radical transfer polymerization)   188
aza-Baylis-Hillman reactions   265, 267, 268
aza-Wittig reaction   127
azadirachtin   87
azalactones   188
azides   132

B

backbone linkers   6
Baeyer-Villiger oxidation   282, 283, 367
BAL linkers   393–396, *510*, 512, 516, 595, 596
Barton-McCombie deoxygenation   134, 135
Barton's base   126
base-labile linkers   429–432, 546–549, 554–556
base-labile resins   429–432
base reagents, solid supported
– potassium fluoride on alumina (KF/Al2O3)   101–104
– silica-gel bound organic bases   104–106
basic organocatalysts
– chiral   270–273
– non-chiral   263–270
bathochromic and hyperchromic shifts   73, 476, 277, 479
BAX-sulfate   221
Baylis-Hillman reactions   106, 265, 267
beads   22
benzaldehydes, nitroso-   471–273
benzaldehydes, polymer-supported   189
benzhydryl linkers   398–400, 474
benzhydrylamine resins   397, 398, 469
benzidine phosphoramidate   553
benzimidazoles   214, *215*
benzodiazepines   613
– 1,5-benzodiazepines   254
– 1,4-benzodiazepinones   428, 456, *457*

benzoic acid
- 4-(3-hydroxy-4,4-dimethyl-2-oxopyrrolidin-1-yl) 332, *334*
- 4-bromomethyl-3-nitro 469, 470
- *p*-hydroxy- 597
benzoin 355, 477, 478
benzopyran-4-one, 6-carboxy 422
benzopyrans 254, *632*, 633
benzothiazoles 434
benzotriazole, N-hydroxy- *see* HOBt
benzoyl protecting groups 532
benzyl alcohol 3-azidomethly-4-hydroxy- 558
benzyl hydrazide linkers 447, 457, *458*
benzyl hydroxylamine 586
benzyl protecting groups 497
benzyl sulfide 485, *486*
benzylic linkers and resins *see also* nitroaromatic compounds
- *p*-alkoxybenzyl linkers 488, *489*
- hydroxybenzylic resins 418
- substituent effects 383, 469
benzylidine acetal linkers 595, 596
β-amyloid (1-42) synthesis 504, 507, 508
BHA (benzhydrylamine) resins 397, 398, 469
- *see also* MBHA
bifunctional catalysts 115, 319
Biginelli reaction 112, 209, 260
binding isotherms 32–34
binding properties of MIPS 31–34
biocatalysts *see also* enzymes
- chemocatalysts distinguished from 247, 248
- current status 365, 377
biomedical applications of nanoparticles 45, 56, 57, 69, 73
biomolecules *see also* oligonucleotides; oligosaccharides; peptides
- application of solid-supported chemistry 4
- ChemMatrix resin 9
- molecular recognition 16
- nanoparticle size 46
biorecognition 56, 59
biotin 54, 56, 64, 94, 166
biphenyl derivatives 455, 485
bipyridyl scavengers 212
bis-glycolyl linker 563
bisacetoxybromate 87, *88*
bleomycins 619, *620*
blood-brain barrier 74, 603
Boc (*tert*-butoxycarbonyl) *see also* peptide synthesis
- linkers for thioesters *510*

- protecting group for amines 11, 497
- protecting group for piperazine 121
- scavenging excess 218, 221
BOP (benzotriazol-1-yl-N-oxytris(dimethylamino)phosphonium hexafluorophosphate) 150
borohydrides 95
boronate linkers 599, 600, *601*
boronic acids 18, 93, 94, 588, *589*
bovine serum albumin 600
bradykinin 159, 421, 422
bromination, electrophilic 133
BSA (N,O-bis(trimethylsilyl)acetamide) 547
BTC (bischlorotrimethyl carbonate) 167
BTC (phosgene equivalent) 175, 176
bulk polymerization and surface imprinting 21
butenolides *452*, 456, 618
buyrolactones 340, *452*, 456
byproducts *see* side reactions

C

C-O coupling reactions 321, 322
C-terminal amino acids 500, 502, 506, 508
CALA/CALB (*Candida antarctica* A/B-lipase) 375
callystatin *632*, 635
CALNN pentapeptide 54
cancer treatment 73, 74, 91
capture and release *see* catch-and-release techniques
carbamates
- anticholinesterase activity 67
- enzyme linkages 374
- intermediates 105, 106, 427, 444
- linkers 391, 468, 474, 549, 560, 561
- nucleophile-labile resins 422, 423
- synthesis of 167, *392*
carbazates 391, *392*, 471
carbazoles 259
carbenes 294
carbodiimides *see also* DCC; DIC; DIPCDI
- acid attachment to linkers 390, 546
- HOSu derived polymers 166, 170
- supported as coupling reagents 142–146, 174, 498
carbohydrates
- chemoselective esterification 375
- chiral auxiliaries derived from 330–332
- covalent molecular imprinting and 18
- enzyme covalent bonding to 374
carbon nanotubes 67, *68*
carbon tetrahalides 123, 151

carbonic acid derivatives
- linkers 549–551
- nucleophile-labile resins 422
carbonyl group cleavage, nucleophile-labile resins 412
carboxy-based linkers and resins 424–426, 563, 564
carboxylic acids
- acid-based linkers 425, 548, 549
- enantiopreferential effects of lipases 370
- peroxycarboxylic acids 375
- acid-labile linkers 408
- photolabile linkers 468, 479, 484
- safety-catch linkers 443, 447
- synthesis of chiral α-substituted 331, *334*, 344, *345*, 357–360
carpanone derivatives 213, 625, 632, 635
carriers, enzyme immobilization on 372–374
catalyst-substrate co-immobilization 292
catalysts *see also* biocatalysts; enzymes; metal-catalyzed reactions; organocatalysts; PTC
- effects of attachment to supports 249
- leaching of 317
- molecular and surface catalysis 247
- photocatalysis 294, 295
catalytic promiscuity 367
catch-and-release techniques 86, 212, 340, 604–606, 625
cations
- acidolysis leaving groups 383, 385, 402, 403
- aminium coupling reagents 498
CBS (Corey, Bakshi and Shibata) catalyst 93, 94
CCTU (*S*-phenyltetramethylisothiouronium chloride) 511
Cdc42 635
CDI (carbonyldiimidazole) 341
CDK (cyclin dependent kinase) inhibitors 623, 631
cell adhesion 59
cellular magnetic-linked immunosorbent assay (CMALISA) 59
cellulose 10, 543
cellulose acetate 556
ceruloplasmin 59, 60
cesium fluoride on alumina 106, 107
chalcones 277, 283, 284, 290

chelating agents
- metal ion mediated molecular imprinting 19, 31
- metal scavenging resins 207, 237, 240
- PEG-GNP heterodimers 56
chemical ligation, SPPS 508–516
ChemMatrix resin 9, 497, 504–506
chemocatalysts and biocatalysts 247, 248, 365
chemometrical approach 30, 31
ChemRoutes 231
Chimassorb 944 282
chiral auxiliaries 329
- alcohols 332–336
- amine-derived 336–341
- enone 352–355
- carbohydrate-derived 330–332
- cyclohexanone as 352
- hydrobenzoin-derived 355, 360, *361*
- oxazolidinones, oxazolidines and oxazolines 341–348
- stereoinductive potential 358
- sulfoxides, sulfinamides and sulfoximes as 348–351
chiral basic catalysts 270–273
chiral building blocks 616
chiral column chromatography 230
chirality *see also* stereoselective synthesis
- asymmetric dihydroxylation 92
- asymmetric metal catalysis 228, 229, 240
- asymmetric synthesis of β–phosphono-malonates 107
- chiral boranes 92–94
- chiral oxidation catalysts 283–285
- chiral PTC 276–279
- chiral sulfonamides 93, 158
- chiral sulfonates, alkylation 130
- chiral titanium complexes 635
- enantioselective epoxidation 88, 89
- MIP templates and stationary phases 35
- prochiral substrates 376
chitosan composite (CMC) 67, 270
cholesterol
- FXR agonists 633
- imprinting 19, *20*, 36
chromenes 254
chromic acid, polymer-supported 90
chromium arene complexes 485
chromium carbonyl 336, *337*, 459
cinchona alkaloids 270
citral 318
CLEA/CLEC (cross-linked enzyme aggregates/crystals) 371
CLEAR family resins 8, 9

cleavage mechanisms
– acid labile linkers   405–408
– in aigialomycin synthesis   623
– olefin metathesis   597
– safety-catch linkers   438–442, 450
click chemistry   286, 294, 374, 520, 521
CMALISA (cellular magnetic-linked immunosorbent assay)   59
co-oxidants, solid supported   86–88
cobalt carbonyl   449
cocaine   48
cofactors   367
colloids
– colloidal metals   46, 47, 55, 62, 313, 318
– PTC   276
– synthetic use   69, 75, 544
colorimetry see also optical sensors
– aggregation studies   50, 55
– detection of lead and mercury ions   50, 51
– DNA functionalized GNPs   47–52
– enzyme assay   49, 50
combinatorial synthesis   545
– carboxylic acid libraries   126
– Merrifield resins   425, 426
– MIP libraries   30
– natural product libraries   613, 614, 616, 618, 619, 630–635
– oligosaccharides   471, 598
– peptides   495, 496
– solid phase   290, 291
– techniques applicable   185, 189, 196, 213
computation in molecular modeling   31
condensing agents, oligonucleotide synthesis   532, 533, 538, 546
confocal microscopy   58
conjugate addition reactions   344, 348
continuous fixed-bed catalysis   258
continuous flow synthesis   215, 216, 627
control polymers (CP)   31, 32
convergent approaches, SPPS   499, 500, 598
cooperative catalytic systems   113–115
coordination complexes, transition metal   310, 311
core-shell strategies   24, 69, 71, 72
coupling reagents   142
– aminium and uronium reagents   146–150
– carbodiimides and isoureas   142–146, 172
– DKP suppression   471
– oligonucleotide synthesis   530, 538, 568
covalent enzyme immobilization   374
covalent molecular imprinting   17–19
covalent supporting, transition metals   310
m-CPBA (chloroperbenzoic acid)   438, 439, 444, 452, 455

CPG (controlled pore glass) resin
– aminopropyl-   553, 561, 562
– LCAA-CPG   541, 548, 556, 560, 561, 563, 564
– 3-mercaptopropyl-   556, 586
– NPE-linked   549–551
– oligonucleotide synthesis   540–542, 544
– oligosaccharide synthesis   586, 587
– safety-catch linkers   440
– siloxane grafted   475, 552
CPG (p-(5-ethoxycarbonyl)pentyloxybenzyl) tag   605
cross-linking
– Apt-GNP interparticle   55
– CBS catalysts   94
– ChemMatrix resin   9
– chitosan composite (CMC)   67
– enzyme immobilization   371, 372
– macroporous and microreticular resins   251
– molecular imprinting   16, 19
– polymeric catalysts   267
– polystyrene (PS) resins   7, 8, 84, 538
– reactant penetration   315
– transition metal catalysts supports and   310
cross-linking monomers   28, 29
cross-reactivity and selectivity, MIPs   33
crown ethers   275, 276, 280, 389, 420
2-CTC (2-chlorotrityl) resins   500–502, 507, 508, 513, 517, 521
2-cyanoethyl protecting group   532, 535, 548
cyclative cleavage   459, 460
cyclic alkaloids   624
cyclic amides   448
cyclic linkers, acid labile   402, 403
cyclic oligonucleotides   565
cyclic peptides   394, 395, 426, 427, 514, 516–519
cyclization reactions   123, 124, 127, 136, 170
– see also RCM
– intramolecular nucleophilic displacement   426–428
cycloaddition reactions see also Diels-Alder reactions
– 1,3-dipolar   344, 346, 520, 521
– facial selectivity   335, 336
– MCM reagents   105, 106
– photocycloadditions   352–355
cyclohexanone   338, 352
cyclorelease reactions   617
cysteine protease inhibitors   186

D

DBU (diaza(1,3)bicyclo[5.4.0]undecane) 390, 509
DCC (dicyclohexylcarbodiimide)
– active esters from 170
– analog 143, 144
– a coupling agent 159, 163, 164, 532, 549
– peptide acylation 498, 519
DCHA (dicyclohexyl amine) salt 509
DCI (4,5-dicyanoimidazole) 535
DCT (dichlorotriazine) 193, 194, *195*, 198, *200, 201*
DDD (diethyldithiocarbonate disulfide) 536
Dde1-(4,4-dimethyl-2,6-dioxocyclohexylidene) ethyl 424
DDQ (2,3-dichloro-5,6-dicyano-1,4-benzoquinone) 488, 601
DEAD (diethyl azodicarboxylate) 125
DEAE (diethylaminoethanol) 373
decalins 618
decarboxylation as a traceless method 458, 459
Degussa 235, 374
dehydration plus hydrogenation (hydrogenolysis) 323
Deloxan 235
dendritic structures 286, 294, 296, 591–593
deoxygenation of water 323, 325, 326
depsipeptide methodology 503, 506
depsipeptides 519, 520, 614
DHP (dihydropyran) linker 404
diastereoselective reactions 107, 286, 330, 331, 340
– see also enantioselective reactions
diazenes 425, 447
diazo transfer reactions 132, 133
diazomethane, diphenyl- 397
diazonium compounds 457
diazonium salts 191
dibenzocycloheptadienes 402, 403
DIC (1,3-diisopropylcarbodiimide) 162, 172
DIEA (N,N-diisopropylethylamine) 504, 506, 511, 512, 519, 521
Diels-Alder reactions
– asymmetric, decalins 618
– δ-lactone preparation 635
– diazene preparation 425
– intramolecular 635
– scavenging dienophiles 209, 220
– stereoselective 331, *335*, 336, *337, 338, 344, 347*
dienophile scavenging 209, 220
diglycolyl 548

dihalocarbenes 102
diketones as scavengers 189
dimannosides 591, 592, *593*
dimedone 434
diminutol 631
dioxiranes 283
DIPCDI (N,N'-diisopropylcarbodiimide) 498, 503, 504, 508
diphenyl phosphorochloridate 530, 537
diphenyldiazomethane resin 397
DIPHOS (1,2-bis(diphenylphoshino)ethane) 551
diptericin 511
discodermolide 614, 629
disiloxanediyl linkers 551, 552
displacement curves, MIP binding 33
disulfide-based linkers 556, 557, 561, 562
1,3-dithiane linkers 487, 488
DKP (diketopiperazine)
– alkaloid preparation 427
– avoidance using various coupling agents 470, 471, 500
– peptide synthesis byproduct 402, 426, 482, 507, 512, 519
DMAP (4-(dimethylamino)pyridine)
– acid attachment to linkers 390
– active esters from 146, 172
– analogs 263–265, 267, 272
– catalysis by free and immobilized 105, 106, 146
– polymer-bound reagents 131, 144
– scavenging 210, *211*
DMB (3',5'-dimethoxybenzoin) 479
Dmbb (N-4,5-dimethoxy-2-mercaptobenzyl) 513
DMC (2-chloro-1,3-dimethylimidazolinium chloride) 149
DMP (2,5-dimethylphenacyl) 482, 483
DMTr (4,4'-dimethoxytrityl) group 532, 546
DNA see nucleic acids
DNA aptamers 48, 49, 55, 71
Dowex resins 252, 256, 263
drug delivery with nanoparticles 45, 57, 59–61, 64, 71, 73, 75
drug discovery see also pharmaceuticals
– click chemistry 521
– library synthesis 214
– small molecule combinatorial chemistry 3
– solution-phase multi-step parallel sequences 183
– use of metal ions 227–230

DSA (4-(4-methoxyphenyl-aminomethyl)-3-methoxyphenylsulfinyl-6-hexanoic acid) linker *439*, 446
DSB (4-(2,5-dimethyl-4-methylsulfinylphenyl)-4-hydroxybutanoic acid) linker *439*, 446
DSC (differential scanning calorimetry) 320
DTT (dithiothreitol) 556, 557, 560–563, 570
duloxetine 93
dye attachment techniques 266, 292, 561
dysidiolides 618, *619*

E

E/Z ratio (alkenes) 125, 126
EA[H]Q (2-ethylanthra[hydro]quinone) 324
EC (Enzyme Commission) system 366
EC_{50} values 33
E1cb reactions 429
EDAC (ethyl 3-(dimethylamino) propylcarbodiimide) 166
EDITH (3-ethoxy-1,2,4-dithiazoline-5-one) 536
EDT (1,2-ethanedithiol) 504
ee (enantiomeric excess) *see* enantioselective reactions
EEDQ (2-ethoxy-1-ethoxycarbonyl-1,2-dihydroquinoline) 152, 153
EGDMA (ethylene glycol dimethacrylate) 28, 261
electrochemical sensors 36
electron-withdrawing groups
– base-labile linkers and resins 429, 430
– nitroaromatic resins 470
– side reactions 470
electroorganic synthesis 106
electrophile scavengers *see* nucleophilic scavengers
electrophilic reagents, solid supported
– acylating and sulfonylating reagents 131, 132
– alkylating reagents 129, 130
– brominating reagents 133
– nitrogen electrophiles 132, 133
– silylating reagents 130
electrophilic scavengers
– fluorous tagged 194–200
– PEG supported 157, 193, 194, *195*
– PS-resin supported 185–192
– ROMP and ROMP-gel derived 200–205
– silica-supported 194, *195*
– tabulated 185
electrostatic binding of enzymes 373

elimination reactions
– β-eliminations 554–558, 562, 567
– PTC 276
– unimolecular 429
ELISA (enzyme-linked immunosorbent assay) 59
emulsion polymerization 22–24
enantiopreference prediction 369, 370
enantioselective reactions *see also* chirality; diastereoselective reactions; stereoselective synthesis
– acetylation 158, 376
– alkylation 278
– catalysis 248, 270, 271
– epoxidation 88, 89
– hydrogenation 229
– oxidoreductases 367
– polymer reuse 331
– reduction 93, 94
– synthesis of grossamide 216
endonucleases, colorimetric assay 49, 50
enfuvirtide 500–502
enkephalin 422
enones 352–355, 630
entrapment immobilization 372
enzyme-cleavable linkers 603, 604
enzyme mimetics, MIPs as 36
enzymes
– colorimetric assay 49
– targeting purine dependent 631–633
enzymes, immobilized
– classification and availability 366
– enantiopreference 369, 370
– immobilization approaches 371–374
– popular enzymes in each EC division 366–368
– solvents 365, 367–369
ephedrine as an auxiliary 339, 340
epibatidine 625
epimerization 147, 149
– *see also* racemization
epothilones
– anticancer potential 614, 616
– historic solid-phase syntheses 125, 613, 614, 617, 618
– total synthesis of epothilone C 86, 91, 625, *626*
epoxides 105, 111, 260, 375, 623
– fluorous tagged 197
erythromycins *615,* 616
ESR (electron spin resonance) spectroscopy 315
esterification
– alcohol-polymer attachment through 332

– chemoselective, of carbohydrates   375
– hindered components   149
– interesterification   377
– involving organocatalysis   252, 256, 261, 271
– nucleoside-polymer attachment through   546
– transesterification   269, 367, 420, 421
esters *see also* active esters
– acid-labile linkers   390
– linkers for oligonucleotide synthesis   532
– nucleophile-labile resin attachment   417, 418, 420
– phenacyl   479–481
– stereopreferential enzymatic hydrolysis   369
ethanol
– 2-(4-carboxyphenylthio)-   554
– 2,2′-dithiodi-   556, 560
– β-sulfonyl-   562
ethers
– linkages, acid-labile linkers   384, 386
– preparation by organocatalysis   252, 258, 259
Eupergit C   374
expressed protein ligation   514

F

ferrofluids   57
fexaramine   633
films   24
FITC (fluorescein isothiocyanate)   58, 61
flow-through *see* continuous flow synthesis
9*H*-fluorene, 9-methylene-   207, *208*
fluorene-based linkers   429
fluorescence imaging   58, 63, 73
fluorescent labels   166
fluoroboric acid on silica gel   111
α-fluorotropinone   285
fluorous isolation tags   613, 628, 630
fluorous tagged electrophilic scavengers   194–200
fluorous tagged nucleophilic scavengers   219, 220
fluoxetine   93, 95
Fmoc (9-fluorenylmethoxycarbonyl)-protected amines
– in SPPS   497, 498, *510*
– product removal   207
FMPB (4-(4-formyl-2-methoxyphenoxy)butyric acid) resin   396
formyl (indole) linkers   396
fostriecin   *632*, 635
free-radical initiators   30

Friedel-Crafts reactions   588
– acylation   160, 174, 257, 261, 474, 480
– alkylation   159, 168, 258
– asymmetric   290
– benzoylation   112, 113
– intramolecular   258
frontal chromatography   33
frontalin   358, *359*
functional groups, acid-labile linkers   408–412
functional monomers in MIP   25–27, 30
functional sites
– distribution in gel-type supports   10
– effects of attachment to supports   249
– number of, and catalytic activity   255, 256
furanopyrazines   396
FXR agonists   633

G

GA (glutaraldehyde)   67, 371, 372, 374
gadolinium   228
galactosides   598, 599
galactosylamines   330, 331
galanthamine derivatives   *632*, 634, 635
gel matrices   372
gel-type supports   7–10, 238, 251, 539, 540
– *see also* microporous resins; silica gel
gene expression detection   51–53
gene therapy   74
germanium compounds   135, 457
Globo-H antigen   597
GLUT-1 protein   603
glutamic acid recognition   66
glutaraldehyde (GA)   67, 371, 372, 374
glycosylation   592
gold nanoparticles (GNPs)
– functionalization   47–53
– in H_2O_2 biosensors   67
– synthesis and properties   46, 47
gold nanoshells   55
Golgi apparatus   635
graft copolymers   588
grafted linkers   586
grossamide   216
gylcerol   323

H

H-phosphonate method   530, 531, 534–538
– oligonucleotide synthesis   540, 542, 548, 549, 567, 569, 570
HAL linker   393
halide electrophilic scavengers   192

handles *see* linkers
HATU (1-[bis(dimethylamino)methylene]-1H-1,2,3-triazolo-[4,5-b]pyridinium hexafluorophosphate 3-oxide)   498, 505, 506, 512
HBTU (2-(1H-Benzotriazole-1-yl)-1,1,3,3-tetramethyluronium hexafluorophosphate)   147, 148, 498, 501, 504, 505
HCTU (1-[bis(dimethylamino)methylene]-6-chloro-1H-benzotriazolium hexafluorophosphate 3-oxide)   498
HDAC (histone deacetylase) inhibitors   634
hemoglobin   67, 75
heparin-ilke compounds   596
heterocyclic structures
– combinatorial synthesis   631
– intramolecular nucleophilic attack   418, 428
– polymer catalyzed synthesis   254, 263
heterodimers, metal   55–57
heterogeneous catalysts   101, 113
heterogeneous polymerization   22, 23
heterolytic cleavage   451
heteropoly acid   112
high loading resins   90
high-throughput synthesis   613
highly cross-linked PS *see* macroporous resins
hindered acids   149–151, 500
hindered alcohols   534
hirudin   502, 503
histacin   634
HIV (human immunodeficiency virus)   74, 144
– HIV-1 protease   515, 516
HMB/HMBA (4-hydroxymethylbenzoic) linkers/resins   421, 511, 521
HMFS (N-[9-(hydroxymethyl)-2-fluorenyl] succinamic acid) linker   429, 430
HMPA (4-hydroxymethylphenoxyacetic acid)   386, 390
HMPB (4-(4-hydroxymethylphenoxy)butyric acid) linker   500
HMPPA (3-(4-hydroxymethylphenylsulfanyl) propanoic acid) linker   439, 446
HOAt (1-hydroxy-7-azabenzotriazole)   144, 145, 498, 504
HOBt (N-hydroxybenzotriazole)
– active ester formation   143–145, 158, 162, 168–174
– aminium and iminium salts   147, 148
– as an additive   144, 162, 519

– as coupling reagent   487, 498, 501, 503, 504, 508, 509
– polymers   150, 153, 168–174
– polymers, active ester formation   168–174
homogeneous polymerization   22
HOSu (N-hydroxysuccinimide)   148, 158, 163–168, 170, 422
HSTU (2-succinimido-1,1,3,3-tetramethyluronium hexafluorophosphate)   148
Human IAPP (islet amyloid polypeptide)   505
Hünig's base   153, 155, 172
Hunsdiecker reaction   135
HWE (Horner-Wadsworth-Emmons) reaction   126
hyaluronic acid (HA)   61
hybrid approaches, SPPS   499–502
hydantoins   427, 428, 449
hydrazine hydrate   95
hydrazines
– aldehyde scavenging   209, 210
– chiral   340, *342*
– immobilization to acid-labile linkers   391
– nucleophilic cleavage of linkers   424, 425
– polymer-supported   122
– safety-catch linkers   512
m-hydrobenzoin auxiliaries   355–360, *361*
hydrocyanation of imines   290, 291
hydrogen peroxide (H_2O_2)
– biosensors   67
– epoxidation   277
– oxidizing resin regeneration   281
– peroxy acids   375
– supported catalysis   282–284
– synthesis   324
– thioether activation   444, 445, 554
hydrogenation reactions
– of alkenes and alkynes   95
– using transition metal catalysts   316–319, 324, 325
hydrogenolysis   323, 602, 603
hydrolase enantiopreference   369, 370
hydrophilic PEG-based resins   8–10
hydroquinone-O,O'-diacetic acid   548
hydrosilylation reactions   320
hydrosols   69
hydroxamic acids, supported   176
α-hydroxyacetone oxime   558
N-hydroxybenzotriazole *see* HOBt
hydroxyl groups
– attachment   330, 332
– chemoselective esterification   375
– covalent molecular imprinting   18

– functionalization of resins and linkers 420–422
4-hydroxyproline 286, 287
hyperbranched soluble resins 591–593
hyperchromic shifts 476–479
hypervalent iodine reagents 84, 85

I

IBX (o-iodoxybenzoic acid) 84, 85
ICP-MS (inductive coupled plasma mass spectrometry) 230
IIDQ (2-isobutoxy-1-isobutoxycarbonyl-1,2-dihydroquinoline) 153
imatinib 633
imidazole
– 1,1′-carbonyldi- 550
– imidazole-based liquids 221
– 1,1′-oxalyldi- 548
imidazolidinones 289, 290
imidazolium organocatalysts 293, 294
imines, diastereoselective allylation 340, *343*
iminium ions 622
iminophosphoranes 269
immobilized reagents *see* solid-supported reagents
immunofluorescence detection 62, 63
immunoglobulins 56
imprinting *see* binding; MIP; molecular imprinting
imprinting factors 32
incarceration of catalysts 89
indolactams *615*, 616
inductively coupled plasma optical emission spectrometry (ICP-OES) 58
iniferter (initiator-transfer agent-terminator) approach 24
inorganic chemistry, medicinal 228
integral linkers 4, 5, 384, 586
internal reference amino acids (IRaas) 6, 7
intramolecular processes
– cyclization 449, 482, 516
– Diels-Alder reactions 635
– N to S acyl shift 513
– nucleophilic 418, 426–428
iodobenzene derivatives 103, 104
ion exchange resins *see also* anion exchange
– acidic organocatalysts 250–256
– enzyme immobilization using 373
– mixed acid and basic 263
– perfluorinated 311
ionic gels 238, 239
ionic liquids 220–222, 594
IRaas (internal reference amino acids) 6, 7

iron oxide nanoparticles (IONPs) 57–61
irregular shaped particles
– molecular imprinting 21
– organic nanoparticles 69, 70, *71*
– silica scavenger supports 237
ISEC (inverse steric exclusion chromatography) 315, 320
isobutyryl protecting groups 532
isocyanate electrophilic scavengers
– fluorous-tagged 196, 197
– resin-supported 185–188, 204
– silica-supported 194
isocyanates, scavenging of 212, 217
isolation tags 605, 613, 627–630
– *see also* fluorous isolation tags; magnetic labeling; SPOT
isonitrile-functionalized polymer supports 317
isotherm models 32–34
isourea coupling reagents 142–146
isoxazole and isoxazolines 425
isoxazolinoisoquinolines 432

J

Johnson Matthey Chemicals 236

K

Kaiser resin 174, 175
Kazlauskas' rule 369, 370, 376
KCL (kinetically controlled ligation) 514
ketones
– α-alkylation 339
– diketones as scavengers 189
– enantioselective reduction 367
– methyl isobutyl (MIBK) 322, 323
– released from acid-labile linkers 411
– SPS using diol-based linkers 404
– traceless linkers 453
kinase inhibitors 623, 631–633
kinetic resolution 367, 369, 376, 677
– *see also* enantioselective reactions
Knoevenagel reactions 104, 105, 107, 263, 269, 270
Knorr linkers 398

L

lactams
– β-lactams 271, 348, *349*
– ε-lactams 460
– medium ring 170, 171
lactoferrin 59, 60
δ-lactones 632, 635
lagunapyrone 629
lamellarin family 621

Langmuir binding isotherms 32
lanthanide catalysis 320, 321
laser ablation 69
LCAA (long-chain alkyl amide)-CPG 541, 548, 556, 560, 561, 563, 564
leaching of catalysts 317, 321, 374
lead (Pb^{2+}), colorimetric detection 50, 51
leukemia 633
library syntheses
– amide and sulfonamide 162, 172, 186, 194, 197, 422
– amino acids and ureas 196, 197
– application of scavengers 184, 187–190, 199, 203, 214–216
– benzodiazepines 456, 457
– carbodiimides 145, 146
– DCC resins 144, 186
– enzymes and other natural product families 616, 618, 619, 625, 627, 630–635
– MIP libraries 30
– nucleotides 558
– oligosaccharides 594, 598
– peptides and related substances 85, 291, 427, 450, 489
– Taxol analogs 614–616
– the Wittig reaction 126, 614
ligand exchange reactions 311
linalool 319
linderol A 92
linker combinations 560
linkers 4–7
– see also chiral auxiliaries; cross-linking
– 1,3-dithiane 487, 488
– alkene functionalized 488
– cleavage mechanisms 597, 602–604
– distinguished from supports/resins 4, 585
– disulfide-based reduction 556, 557
– GNP functionalization 47, 48
– integral and grafted 586
– MNP functionalization 55, 56
– oligonucleotides 559–564, 566
– p-alkoxybenzyl 88, 488
– preformed 384
– protecting groups or reagents 387
– sub-linkers 356, 360
linkers, acid-labile 501
– benzyl and alcoxybenzyl 383–387, 390, 391, 394
– cleavage mechanisms 405–408
– cyclic 402, 403, 511
– design 383–387
– integral vs preformed linkers 384
– leaving group stability 384–386, 418

– oligonucleotides 545, 546, 553–558
– oligosaccharide synthesis 594–596
– peptide thioester formation 512
– RNA synthesis 564
– safety-catch (SCAL) 439, 446
– types of linker 387–405
– use for common functional groups 408–412
linkers, base-labile 429–432, 546–549, 554–556, 596, 597
linkers, nucleophile-labile 418
linkers, oxidatively cleaved 601, 602
linkers, photolabile see also nitroveratryl linkers; photocleavage
– 3'-carboxylalkyl nucleosides 564
– functionalized phenacyl linkers 477–483
– miscellaneous 485–487
– nitrobenzyl-derived linkers 467–477
– nucleoside 3'-carbonates 551
– oligosaccharide synthesis 597, 598
– pivaloyl linkers 484, 485
linkers, reductively cleaved 602
linkers, safety-catch 437–450
– see also safety-catch principle
– overview of linker types 438–442
– PEG-supported oligonucleotide synthesis 567
– peptide thioester preparation 510, 511
– photolabile safety-catch 482, 483
linkers, traceless 450–460
lipases 366, 367
liquid chromatography and MIPs 34, 35
liquid-phase synthesis see soluble polymer supports
LNT (N-tetrose) 586, 587
loading and enzyme activity 372, 373
low cross-linked polystyrene 538, 539
lyophilized enzyme powder 370

M

macrolactonization reactions 144
macroporous (maroreticular) resins
– catalytic application 253, 254, 315, 316
– distinguished from microreticular 251
– macroporous silica 541
– metal scavenging 238
– oligonucleotide synthesis 540
macrosphelides 618, 620
magnetism see also MRI; NMR spectroscopy
– magnetic labeling 57
– magnetic particles 588, 589
– nanoparticles 61, 211, 212
– superparamagnetic behavior 57
maleimide, poly(ethylene-co-N-hydroxy- 163

maleimide electrophilic scavengers 189–191
malonyl anchors and linkers 548, 557
MAMP (Merrifield alpha methoxy phenyl) linker 398, 400
mandelic acids 356, 358, 520
Mannich-Michael reactions 331
Maraviroc 144
mass sensitive sensors 36
mass spectroscopy 230, 635
MBHA (p-methylbenzhydrylamine) resins 4, 5, 397, *419*, 508, 509
MCM-41 (Mobil's Composition of Matter-41) 105, 266
medicinal inorganic chemistry 228
membranes *see also* cellulose
– molecular imprinting 24
– polyolefinic 11
α-mercaptobenzhydryl linker 400
mercaptopropionic linker 508
mercury (Hg^{2+}) colorimetric detection 50, 51
Merrifield resin (chloromethyl polystyrene) 389–391, 393, 394, 586
"meso trick," 376
mesofluidic flow reactors 627
metal-based therapies 227, 228
metal-catalyzed reactions *see also* transition metal catalysts
– KF/Al_2O_3, 102, 103
– pharmaceutical use 228, 229
– removal of residues 206–209
metal colloids 313
metal heterodimers 55–57
metal ion mediated molecular imprinting 19, *20*
metal nanoparticles *see* MNPs
metal scavengers
– producers and products 231–236
– resin type 237–239
metal sensors containing DNAzymes 51
metalloproteinase inhibitors 215
methacrylic acid monomer 17, 25
methoxylation reactions 106
MIA (molecularly imprinted sorbent assays) 35
MIBK (methyl isobutyl ketone) 322, 323
Michael acceptors 189, 265
Michael additions
– anion exchange catalyzed 263
– enantioselectivity 270
– iminophosphoranes 269, 270
– lipase catalyzed 367
– pyrroles with ketones 253

Michael reactions
– aza-Michael-Knoevenagel 105
– phospho-Michael reaction 107
– thia-Michael reaction 108, 110
"microgels," 212, 313
microporous (microreticular) resins
– distinguished from macroporous 251, 315
– as transition metal catalyst supports 313, 314
milling 69
MIP-QCM sensors 36
MIPs (molecularly imprinted polymers)
– applications 34–37
– characterization 31–34
– chemical and physical properties 34
– design 25–31
– films and membranes 24, 66
– formats 19–24
– homogeneous and heterogeneous polymerization 22, 23
– imprinting factor 32
MISPE (molecularly imprinted polymer solid-phase extraction) 35
Mitsunobu reactions 125, 131, 132, 444, 513
MNPs (metal nanoparticles)
– gold nanoparticles 46–55, 67
– iron oxide NPs 57–61
– nanoshells and metal heterodimers 55–57
– nanowires 65–67
– oxidizing agents 89
– quantum dots 62–65
– reducing agents 95
– silver nanoparticles 61, 62
– transition metal catalysts based 311, 314, 318, 325
modified surface type supports 10–12
molecular catalysts 247
molecular imprinting 15–19
– *see also* MIPs
– films and membranes 24
– nanowires 66
– optimization of conditions 30, 31
– organocatalysis using 294–296
– sensors based on 35, 36
molecular modeling and MIP design 31
molecular recognition 15, 46, 54
molecularly imprinted polymer solid-phase extraction (MISPE) 35
molecularly imprinted polymers *see* MIPs
molecularly imprinted sorbent assays (MIA) 35

monomers in MIP design 25–28
Morita-Baylis-Hillman reaction 265
morpholine and its derivatives 430, 535, 541
Mosher acids/amides 145, 165
MPA (3-mercaptopropionic acid) 514
MPEG (monomethyl polyethyleneglycol) resins 590
MRI (magnetic resonance imaging) 45, 46, 56–60, 227
mRNA 52, 53
MS-Cl (mesitylenesulfonyl chloride) 532
MSNT (1-(mesitylene-2-sulfonyl)-3-nitro-1H-1,2,4-triazole) 390, 532, 565
Mukaiyama reagent 154, 155
multi-step total syntheses 115, 183
multi-supported reagents 88, 89
"Multipin concept," 11
murisolins 629
muscone 617
myoseverin 631

N

N-O bonds, hydrogenation 324, 325
n→π* absorbances 476, 477
N-terminal amino acids 506, 508
N to S acyl shift 510, 513, 314
NAD (nicotinamide adenine dinucleotide) 367
Nafion resins 256–261, 311, 321, 322
nanoparticles (NPs) see also MNPs
– biomedical applications 45
– carbon nanoparticles 67, 68
– magnetically tagged 211, 212
– organic nanoparticles 68–75
– silica nanospheres 113, 231, 544
nanoshells and metal heterodimers 55–57
nanowires 65–67
native chemical ligation (NCL), SPPS 508–516
natural products
– combinatorial synthesis 630–635
– immobilizing core structures 614–616
– immobilizing reagents 623–627
– solid-phase synthesis 616–623
– using isolation tags 627–630
NBB (3-nitrobenzamidobenzyl) resins 418–421, 469
NBH (2-nitrobenzhydryl) resins 474
NBOC (2-nitrobenzyloxycarbonyl) group 468
NBS (N-bromosuccinimide) 511
NCL (native chemical ligation), SPPS 508–516

NCPS (noncross-linked polystyrene) 628
Neisseria meningitidis 590, 591
nickel nanoparticles 95
NIPA (N-(2-iodylphenyl)-acylamide) resin 85
NIS (N-iodosuccinimide) 605
nitrile-converting enzymes 368, 377
nitrite, polymer-bound 132, 133
nitroalkenes 104, *106*, 125
nitroaromatic compounds 112, 168, 169, 324
– 3-nitro-4-chlorobenzyl alcohol 169
– *p*-nitrophenyl hydrogenation 324
– photolabile 467–477, 551, 597
o-nitrobenzyl linkers 468, 476, 477
– 1-(2-nitrophenyl)-1,3-propanediol 558
– 5-hydroxy-2-nitrobenzyl 471, 474
– 2-hydroxymethly-6-nitrobenzyl 561
– α-substituted 474, 475
nitrogen BET adsorption 320
nitrogen electrophiles 132, 133
nitrogen oxidation 447
nitroveratryl linkers 471–474, 484, 485, 551, 558, 560
– NVOC 468, 484, 485
NMM (N-methyl morpholine) 151, 155–157
NMO (N-methylmorpholine N-oxide) 86, 87
NMP (N-methylpyrrolidone) 511
NMR spectroscopy 315, 540
N,N-dimethylacrylamide 311
non-chiral oxidation catalysts 280–283
non-covalent molecular imprinting 16, 17, 25–27
non-integral linkers 4–6
non-specific binding 31
norbornenes 200, 202, *222*, 347
norzooanemonin 89
Novozym 435, 366, 367, 373, 375–377
NPE (2-(4-nitrophenyl)ethyl) linker 429, 430, 547, 549–551, 556
NPEOC (2-(4-nitrophenyl)ethoxycarbonyl) 547, 549, 560
nuclear localization sequences 54
nucleic acids see also oligonucleotides
– GNP functionalization 47–53
– mRNA 52, 53
– RNA interference 570
– synthesis involving safety-catch linkers 448
nucleophile-labile resins and linkers 417–428

– intermolecular nucleophilic displacement 420–425
– intramolecular nucleophilic displacement 426–428
nucleophile scavengers *see* electrophilic scavengers
nucleophilic organocatalysts   293
nucleophilic reagents, solid supported 121–128
– anchored to anion exchange resins   127, 128
– phosphanes   123–125
– secondary amines   121, 122
– Wittig reagents   126, 127
nucleophilic scavengers   206–216
– ionic liquids   220–222
– silica-supported   217–219
nucleoside 3′-(hemi)succinates   541, 543, 544, 546
NVOC (6-nitroveratryloxycarbonyl) group 468, 484, 485, 497, 498

O

O-acyl isopeptide method   506–508
O to S acyl shift   510, 512, 513
OBAC (oligomeric bis-acid chloride)   203, 204
obliquine   213
OCMC (O-carboxymethylchitosan) polymer   60
O'Donnell-Corey-Lygo methodology   277, 279
olefin isomerization   125, 268, 319, 320, 323
olefin metathesis   597
– *see also* RCM
olefins *see* alkenes
oligomannosides   600
oligonucleotides *see also* biomolecules; nucleic acids
– H-phosphonate method   534–538
– linkers for 3′-functionalized   557–564
– phosphite triester and phosphoramidite methods   534–536
– phosphodiester and phosphotriester methods   530–533
– photorelease   482
– polymer-supported reagents   568–570
– quantum dots and   63
– soluble polymer supports   565–568
– suitable linkers and anchor groups 545–565
– suitable solid supports for synthesis 538–544

– synthesis   430, 450, *531*, 559–564
oligosaccharide synthesis
– capture and release techniques   604–606
– linkers   594–604
– magnetic particle isolation   588
– using DMAP   425
– using insoluble resins   586–589
– using nitroveratryl linkers   471, 472
– using *p*-acylaminobenzyl linkers   489
– using soluble resins   590–594
olomoucine   631, *632*
one-pot reactions
– Bayliss-Hillman reaction   267
– using catalytic resins   254, 263
– using mixed acid and base   114–116
– yielding imides   108
– yielding ONPs   70
– yielding opioids   319
OPC (oligomeric phosphonyl chloride)   205, *206*
opsomization   74
optical encoding   633
– *see also* radiofrequency encoding
optical properties of nanoparticles   47, 73
optical sensors   35, 36
– *see also* colorimetry
optically active compounds *see* chirality
optimization studies, molecular imprinting 30, 31
organic bases, silica-gel bound   104–106
organic nanoparticles (ONPs)
– functionalization   70–72
– synthesis and properties   68–70
– types and applications   73–75
organocatalysts
– amino acids as multifunctional   285
– attempted definition of   248
– imidazolium, thiazolium and related structures   293–294
– phase transfer catalysis   273–280
– polymer supported acidic   250–261
– polymer supported amino acids and peptides   285–293
– polymer supported basic   261–273
– polymer supported oxidation   280–285
organometallic catalysts *see* transition metal catalysts
OSC (oligomeric sulfonyl chloride)   204
osmium tetroxide   86, 91, 92
oxalyl anchors   548
oxazolidines   341–348, 504
oxazolidinones   341–348, 404, *406*, 423, 425, 513
oxazolines   214, *215*, 341–348

oxidatively cleaved linkers   601, *602*
oxidizing agents, solid supported
– aerobic oxidation   86, 89, 91
– functional resin applications   90–92
– multi-supported reagents   88, 89
– newly-developed   84–90
oxidoreductase biocatalysts   367, 368
oxime, α-hydroxyacetone   558
oxime (*p*-nitrobenzophenone oxime) resin   422, 423, 426, 427
oxomaritidine   91
Oxone   281, 283, 285
oxyacyl resins   420
oxytocin   421
ozonolysis   125, 488

P

P- *see abbreviation for monomer*
p-nitromandelic acid   520
PAcM (poly-*N*-acryloylmorpholine)   567
PADS (bis(phenylacetyl) disulfide)   536
PAL-PEG-PS resin   505, 521
PAL (peptide amide linker) linker   393
palladium catalysis
– allylation   443
– cleavage   458
– commercial applications   322–325
– cyclizations   425
– incorporation in a resin matrix   311–313, 325
– resin-supported   315–319
– Suzuki reaction   618
palladium scavengers   206–208, 211, 212, 218, 231, 239, 240
PAM (4-(hydroxymethyl)phenylacetamidomethyl) resins   420, 421, 508, 509, 511, 519
PAM linker   390, *419*
PAMAMs (poly(amidoamine) dendrimers)   592, 593
PAN (polyaniline)   324
paroxetine   376
particle shape and size
– molecular imprinting   21
– optical properties   73
– size distribution   34, 69, 70, 237
passifloricins   629
patents   496
pathogen detection   63
PDI (para-diisocyanate)   186
PDT (photodynamic therapy)   227, 228
PEG (polyethylene glycol) resins *see also* ChemMatrix; SPOCC
PEG-bound triphenylphosphane   124
PEG-HOBt polymer   173

PEG isolation tags   613, 627, 628
PEG-NP conjugates   54, 56, 73, 74
PEG-PS (poly(ethylene glycol)-polystyrene) resins   8
– *see also* TentaGel
– hydrophilic PEG-based resins   8–10
– oligonucleotide synthesis   540, 544
– oligosaccharide synthesis   588
– peptide synthesis   497
PEG-supported chlorotriazines   156, 157
PEG-supported electrophilic scavengers   193, 194
PEG-supported nucleophilic scavengers   212
PEG-supported oligonucleotide synthesis   565
PEG-supported proline   288
PEG-supported PTC catalysts   278
PEGA resin   9, 588
PEI (polyethyleneimine)   373
pellicular solid supports   11, 12
penicillins   377, *378*
pentaerythritol 2-mercaptoalkyl sulfides   241
peptide aldehydes   85, 488, 489
peptide coupling reactions
– carbodiimide reagents for   142, 145
– chlorinated triazines   156
– DCC   163, 164
– DMC   149
– EEDQ   153
– HOBt   169, 173
– HOSu   163
– P-BOP   150, 151
– suppressing DKP formation   471
– using active esters   158, 159
peptide homodimers   515, 516
peptide-oligonucleotide conjugates   540, 562
peptide synthesizer use   285, 504, 541
peptides *see also* biomolecules; SPPS
– acid-labile linkers   387, 389–391, 393–398, 400, 402, 404–407
– nanoparticle functionalization   53–55, 57–61
– nucleophile-labile resins   420–422, 424–427, 430, 431
– pharmaceutical significance   495
– photolabile linkers   467–471, 475, 479–481, 483–485
– safety-catch resins   444, 446, 447, 449, 450
– scavenger, in synthesis   407, 408
– supported, as organocatalysts   285, 290–293

peptidotriazoles 521
perbenzoic acid, *m*-chloro *438*, *439*, 444, *452*, 455, 535
perchloric acid on silica-gel 110, 111
perfluorinated ion exchange polymers 311
– see also Nafion resins
perfluorinated sulfonic acid polymers 256–260
periodate resins 84, 90
PGA (poly(γ-glutamic acid)) 62
"pH memory" 369
pharmaceuticals see also drug delivery; drug discovery
– MISPE analysis 35
– nanoparticle use 46
– peptides as 495
phase-switch purification 212
phase transfer catalysts 389
phenacyl linkers 477, 479–483
phenols
– deprotecting phenolic linkers 448
– released from acid-labile linkers 409
phenylselenilic acid 283
phosgenes 124, 131, 175, 176, 423, 428
phosphatase inhibitors 618
phosphines (phosphanes) see also triphenylphosphines
– Mitsunobu reaction 131, 132
– supported, as nucleophilic reagents 123–127
– supported, as nucleophilic scavengers 207, 208
– supported, as organocatalysts 263–270
phosphite triester method 534–536
phosphitilation 537, 569
phosphodiester method 530–533, 541
phosphodiesterase inhibitors 173
phosphonium ions, acetonyl 261
phosphonium salts
– PTC 275, *278*, 280
– SPPS coupling reagents 498
– traceless linkers 458
phosphoramidate and phosphorothiolate linkers 553, 554
phosphoramidite method, oligonucleotides 534–536
– applications 549, 551, 556, 558, 560, 564, 567
– compared to alternatives 533, 538, 569, 570
– suitable supports for 540–544
– trityl linkers 544–546
phosphorochloridates 530, 537, 569
phosphorothioates 570

phosphorus derivatives, as coupling reagents 150, 152
phosphotriester method 532, 533
photocatalysis 294, 295
photocleavage 450, 455, 459, 592, *593*
– see also linkers, photolabile
photocylizations/photocycloadditions 352–355, 483
photoenolization 482, 483
"phoxime" resin 423, 428
phthalimide linkers 560, 561
phthaloyl linkers 549
physical properties of MIPs 34
physicochemical methods, pore structure analysis 315, 316
Pictet-Spengler reactions 421, 427, 621, 622
Pinnick type oxidations 89
piperazines see also DKP
– piperazine-2-carboxamides 422
– polymer bound 121, 569
piperidines
– stereoselective synthesis 330, *333*
– thioester lability 509–514
π→π* absorbances 476, 477
pironetin 635
π-shielding 335
pivaloyl chloride 537, 538
pivaloyl linkers 484, 485
pK_{aH} values 418
plasmons see SPR detection
platelet-derived growth factors (PDGFs) 55
platinum 228, 240, 318
plicamine 624
PMMA (poly(methyl methacrylate)) 176
Pnm (*p*-nitromandelic acid) 520
pochonin D 625, *627*
POEPS (polyoxyethylelne-polystyrene) and POEPOP (polyoxyethylene-polyoxypropylene) resins 9, 588
poly(4-iodostyrene) 283
poly-L-lysine 543
polyacrylamide resins 541, 542, 554
polycationic ultra-borohydride 95, 96
polyesters, hyperbranched 591, 592
polyethylene, macroporous 544
poly(ethylene-*co*-*N*-hydroxymaleimide) 163
poly(ethyleneglycol) see PEG
polylactosamines 596
polymer beads 22, 23
polymer functionalization with IONP 61
polymer incarcerated metals 89, 95
Polymer Laboratories 233–235
polymer-supported organocatalysts
– acidic 250–261

- amino acid 285–293
- basic 261–273
polymer-supported oxidation catalysts
- chiral 283–285
- non-chiral 280–283
polymer-supported reagents *see* solid supported
polymerization
- initiation 30
- outcome 34
- pre-polymerization mixtures 21, 22
poly(*N*-isopropylacrylamide) 256, 318
poly(N-vinyl pyrrolidone) 325
polyolefinic membranes 11
polyoxyethylene *see* POEPS
polypeptides *see* peptides, proteins
polypropylene, surface modified 544
polypyrrole 66
polysaccharides 543
- *see also* oligosaccharide synthesis
- disaccharide libraries 594
- dodecasaccharides 597
- tetrasaccharides 590, 591, 596
- trisaccharides 590, 591, 603, 604
polysialic acids (PSAs) 56, 57
polysphorin 96
polystyrene (PS) resins 7, 8
- *see also* PEG-PS
- aminomethyl- 540, 564
- chloromethyl- 165, 384, 387, 569
- conventional sulfonated resins 250–254
- grafted onto PTFE 540, 541, 545
- hydroxyethyl- 386
- low cross-linked polystyrene 538, 539
- microporous, as redox reagent support 84
- modified sulfonated resins 254–260
- noncross-linked (NCPS) 628
- oligonucleotide synthesis 538–541
- oligosaccharide synthesis 586–588
- polystyrene-thiophenol 166
- "popcorn" polystyrene 532, 538, 545
- sulfonated, s acid catalysts 250–260
- thermal stability 326
polystyrene supported reagents
- carbodiimide coupling reagents 142, 143
polystyrene supported scavengers
- electrophilic 185–192
- metal 237–239
- nucleophilic 206–216
poly(vinyl acetate) 543, 567
poly(vinyl alcohol) 189, *190*, 543, 567
polyvinyl pyridinium sulfonate 569
polyvinyl pyrrolidone 567

"popcorn" polystyrene 532, 538, 545
pore structure in microporous resins 315, 316
porogens
- macroporous and microreticular resins 251, 315
- MIP design 28–30
- pre-polymerization mixtures 21
potassium fluoride on alumina (KF/Al$_2$O$_3$) 101–104
potassium hydroxide 107
PPA/SiO$_2$ (polyphosphoric acid on silica) 108, 113, 114
PPOA ((4-(2-bromopropionyl)phenoxy)acetic acid) 481
PPTS (pyridinium *p*-toluenesulfonate) 404
pre-polymerization mixture 21, 22
precipitation of metals within a polymer 311–313, *314*
preformed linkers 384
pregalbin 229
prochiral substrates 376
progesterone derivatives 166
proline and its derivatives *see also* pseudoprolines
- CBS catalyst 93, 94
- DMAP analog 272
- organocatalysts 286–290
- prolinol-derived chiral auxiliaries 336, 340
1,2-propanediol 323
1,3-propanediol 376
- 1-(2-nitrophenyl)- 558
propargyl alcohol 449
2-propene-2-ol hydrogenation 317
propionic acid
- 3-(4-hydroxymethylphenoxy)- *501*
- 3-(propyl)mercapto- 592, *593*
prostaglandins 617, 618, 628
protecting groups
- linker activation by removal 448
- nucleosides 532
- peptide synthesis 497, 498
protective group strategies 437
protein microarray technologies 64
protein trafficking 635
proteins *see also* peptides
- GNP functionalization with 53–55
prot(i)odesilylation 403, 456
proton acceptors 498
PS-EDC (1-ethyl-3-(3-dimethylamino-propyl)carbodiimide hydrochloride 145, 146

PSDIB (polymer-supported diacetoxyiodosobenzene) 88
pseudoephedrine 339
pseudolatexes 69
pseudoprolines (ΨPro) 503–506, 508
PSP (polymer-supported perruthenate) 90, 91
PTC (phase transfer catalysis) 270
– chiral PTC 276–278
– non-chiral PTC 273–276
– oxidation catalysis 281
– soluble supports 278–280
purification *see also* scavengers
– adsorbents 230, 231
– application of solid-supported chemistry 4
purvalanol 631
PVP (poly(vinyl pyridine)) 260
– poly(2-vinylpyridine) 316
– poly(4-vinylpyridine) 324
PyAOP (7-azabenzotriazol-1-yl-oxytris (pyrrolidino)phosphonium hexafluorophosphate) 498
PyBOP (benzotriazol-1-yl-oxytris(pyrrolidino) phosphonium hexafluorophosphate) 498, 511, 517
pycnometry 320
pyrazolo-pyrimidine carboxamides 186, *187*
pyrazolo-pyrimidines, 4-amino 216
pyridinium salts 154, 260
pyridyl N-oxide 272
pyrimidines 175, 444, 453
pyrrolidines, cycloaddition 190
pyrrolo[2,1-*c*][1,4]benzodiazepines (PDBs) 91

Q
Q linker 548, 549
QCM (quartz crystal microbalance) sensors 36
quantum dots 62–65
QuardaPure 233
quaternary amine functionality
– anion exchange resins 127, 128
– cinchona alkaloids 276, 278
quaternary ammonium salts 211, 220, 221, 262, 351
quinine 270, 272
quinoline
– 2-isobutoxy-1-isobutoxycarbonyl-1,2-dihydro- 153
quinolines
– isoquinolines and 431
– N-acyl-dihydro 447

R
racemization *see also* epimerization; stereoselective synthesis
– avoiding 500, 502, 507, 511
– tests 147–150, 164
radical-mediated reactions 134–136, 479, 484
radicicols 625, 630
radiofrequency encoding 616, 619
RANTES synthesis 505, 506
RCM (ring closure metathesis) 460, 617, 618, 623
reagents, excess 4, 184
Reaxa 233
recognition *see* biorecognition; molecular recognition
reducing agents, solid supported 92–96
– *see also* oxidizing agents
– metal nanoparticle precipitation 312, 313
reductively cleaved linkers 602
regioselective phosphitilation 569
REM (regenerative Michael acceptors) resin 430, 431, *439*, 445
reporter molecules 35, 64, 65, 633
resins *see* solid supports
resorcylic macrolides 622, 630
reversed micelles 370
rhodium catalysis 229, 231, 320, 336
ring closure *see* cyclization reactions; RCM
Rink acid resin 398, *399*, 594, 596
Rink amide 387, *388*, 393, 398
Rink amino methyl resin 507
RNA interference 570
ROMP (ring-opening metathesis polymerization) 126, 146, 149, 164, 165
– electrophilic scavengers 200–205
– nucleophilic scavengers 222
Rose Bengal 295
rosiglitazone 90
RPMA (reverse-phase protein microarrays) 64, 65
Ru-TsDPEN 94, 95
ruthenium catalysis 229, 323, 377, 460
ruthenium derivatives (PSP) 90, 91
ruthenium scavenging 208, 209, 218, 240, 241

S
sacrificial spacer approach 19, *20*
safety-catch principle 431, 455, 479, 554
– *see also* linkers, safety-catch
saframycin 621, *622*
sarcodictyins 614–616

sarcosine 547
Sasrin linkers 384, 386, 393, 500
Sasrin resin *419*, 421
sawfly sex pheromones 629
SAXS (small angle x-ray scattering) 320, 568
SCAL (safety-catch acid-labile) linkers 439, 446
scavengers *see also* purification
– current status 183–185
– dienophiles 209
– electrophilic scavengers 157, 185–205
– excess reagent 625–627
– metal scavengers 231–236
– natural product synthesis 624
– nitrosobenzaldehyde side products 471
– nucleophilic scavengers 205–222
– osmium tetroxide 86
– peptide synthesis 407, 408
– phase-switch purification 212
– reactive cations 406
– tagged scavengers 184
– water 407
scavengers of electrophiles *see* nucleophilic scavengers
scavengers of nucleophiles *see* electrophilic scavengers
SEC (size-exclusion chromatography) 55, 591, 592
secondary metabolites 614, 616
secramine 634, 635
seed particle polymerization 24
selectivity, hydrogenation over Pd 317
selectivity, MIPs 33, 35
selenilic acid, phenyl- 283
selenium-based resins 633
selenium linkers 445, 456
selenization 570
SEM (scanning electron micrography) 21, 23, 66
semi-covalent molecular imprinting 19, 20
semiconducting nanomaterials 62, 63, 65–57
semiconductors
– quantum dots 62, 63
sensors
– CNTs 67
– DNAzymes 51
– MIPs 35, 36
sequestration *see* scavengers
side-chain anchoring 512, 516–519
side reactions
– silane-based scavengers 407
– trityl-based linkers 400

– nitroaromatic resins 470–474
– oligonucleotide synthesis 531, 533, 546, 547
– SPPS 507
– trimannoside synthesis 605
– Wang resins 5, 6
Sieber amide 402, 500
Sigma-Aldrich (Aldrich Chemicals) 232, 233, 366, 367, 374
silanes
– cation scavengers 407
sildenafil 173, *174*, 625
silica composite beads 24
silica gel-bound organic bases 104–106
silica gel-supported reagents 109
– fluoroboric acid 111
– heteropoly acid 108, 112
– metal scavengers 231–233, 236, 237
– oligonuceotide synthesis 540, 541, 544, 545, 569
– perchloric acid 110, 111
– polyphosphoric acid 108, 113, 114
silica nanoparticles 55
"silica sulfuric acid" (SSA) 108–110
silica-supported reagents
– electrophilic scavengers 194, 195
– nucleophilic scavengers 217–219
– *p*-toluenesulfonic acid 210
– P_2O_5 112
– polytrifluoromethanesulfosiloxane 112
Silicycle 231, 232, 236
siloxane derivatives 112, 475, 551, 552
silver nanoparticles 61, 62
silyl-based linkers 403, 404, 456, 457, 551–553
silyl ether linkers 598, 599
silyl ether protecting groups 586
silylation of succinyl anchors 547
size-exclusion chromatography (SEC) 55, 591, 592
Smopex 236
S_N1-type mechanisms 383, 400, 405
S_N2-type mechanisms 407
sodium carbonate 107, 114
sol-gel process 372
solid-phase binding assays 35
solid-phase chemistry *see* solid supports
solid-phase extraction (SPE) 35
solid-supported reagents
– acid reagents 108–113
– base reagents 101–107
– co-oxidants 86–88
– coupling reagents 141–176
– electrophilic reagents 129–133
– natural product synthesis using 623–627

- nucleophilic reagents   121–128
- oxidizing agents   84–92
- radical-mediated reactions   134–136
- reducing agents   92–96
solid-supported transition metals *see* transition metal catalysts
solid supports
- base-labile resins   429–432
- distinguished from linkers   4, 585
- effect on organocatalysts   249, 250
- effect on transition metal catalysts   313–316, 326
- history of solid supports   3, 4
- hydrophilicity/lipophilicity   324
- nature of, for redox reagents   84
- nucleophile-labile resins   417–428
- oligonucleotide synthesis   538–544
- peptide synthesis   497
- requirements, source and classification of   7
- temperature stability   317, 326
soluble polymer supports   313, 565–568, 590–594
solution phase syntheses   183, 569, 570
- *see also* soluble polymer supports
solvent effects   104, 255, 274, 312
solvent polarity   7, 30
solvents *see also* porogens
- for immobilized enzymes   365, 367–369
Sonogashira reactions   102, 103
spacers
- acid-labile linkers   386, 387, 389
- photocleavable linkers   597, 598
spectroscopy *see also* ESR; NMR; XPS
- MIP complex formation   30
SPIONs (superparamagnetic iron oxide nanoparticles)   58, 60
split-pool synthesis   631, 633, 635, 636
SPOCC (polyoxyethylene-polyoxetane) resin   9, 497, 504, 588, *589*
SPOT (solid phase oligosaccharide tagging) procedure   10, 11, 586, *587*
SPOT synthesis   616
SPPS (solid-phase peptide synthesis) *see also* peptides
- acid-labile resins   390, 397–400, 407, 408
- base-labile resins   421
- click chemistry   520, 521
- coupling reagents   498
- cyclic peptides   516–519
- depsipeptides   519, 520
- early work   8, 407, 408
- methodological tools   496–498

- native chemical ligation   508–516
- publications and patents   496
- stepwise approaches   499, 503–508
- synthesis of long peptides   499–508
SPR (surface plasmon resonance) detection   35, 47, 50, 600
- plasmon resonance extinction   55
SSA ("silica sulfuric acid")   108–110
stannanes   134, 135
star polymers   267
statistical (chemometrical) techniques   30, 31
Staudinger reaction   124, 127, 271
stereoinductive potential   358
stereoselective syntheses   88, 92, 108, *110*, 111, 115, 351–353, 355–361
- acid reagents   108, *110*, 111
- chiral auxiliaries   351–353, 355–361
- thiochromans   115
stereoselectivity *see also* chirality; enantioselective reactions
- alkene epoxidation   88
- carbonyl reduction by enzymes   367
- dihydroxylation   92
- esterification   294
- hydrogenation   318
- hydrosilylation   320
- isomerization   268
StratoSpheres   233
Strecker reaction   290, 291
streptavidin   64
styrene oxide   111
sub-linkers   356, 360
substituent effects
- benzoin linkers   478
- benzylic linkers and resins   383, 469
- HOBt (*N*-hydroxybenzotriazole)   171
substitution reactions, with PTC   276
succinoyl linkers/tags   596, 605
succinyl anchors and linkers   546–548, 558, 567
sugars *see* carbohydrates; oligosaccharides; polysaccharides
sulfides, oxidation of   89, 90, 443–445
sulfinamide chiral auxiliaries   348–351
sulfonamide linkers   438, 443
sulfonamide substituted HOBt   171, 172
sulfonamides
- chiral   93
- N-acyl   443
- released from acid-labile linkers   411
sulfonate scavengers   210, 211
sulfonated polystyrene resins   250–260
sulfone traceless linkers   456
sulfones, hydroxy   351, *352*

sulfonic acid scavengers   210
sulfonic-based linkers and resins
   424–426
sulfonyl chlorides
– fluorous tagged   197
– oligomeric   204, *205*
– polymer supported   131, 568
– scavengers   217
β-sulfonylethyl groups   554, 556, 560
sulfotransferase inhibitors   631
sulfoxides   348–351, 446
sulfoxime chiral auxiliaries   348–351
sulfur-based traceless linkers   455
sulfurization reagents   536, 569
super acids   255
superparamagnetic behavior   57
supported reagents *see* polymer-supported; solid supported
supramolecular assemblies   16
surface area, pellicular supports   11
surface catalysis   247
surface imprinting and bulk polymerization   21
suspension polymerization   23
Sweetzyme IT   371
swelling efficiency   469
"switch" peptide transitions   506
"synthesis machines"   215
synthetic enzyme development   36

T

T-20 (enfuvirtide)   500–502
T-antigen   592, *593*
tablets, scavengers in   211
tagging *see* isolation tags
TAMA-Cl   221
TAME (*tert*-amyl methyl ether)   322, 323
TAMPAL (trityl-associated mercaptopropionic acid leucine) resin   509
TAMRA (tetramethylrhodamine)   548
tandem ligation   514
targeted drug delivery   59–61
Tat peptides   53, 57, 58, 60
taxoids   614–616
TBAB (tetrabutylammonium bromide)   351
TBD (1,5,7-triazabicyclo[4.4.0]dec-5-ene)   105, 106
TBHP (*tert*-butylhydroperoxide)   536
TBTU (*N*-[(1*H*-benzotriazol-1-yl)(dimethylamino) methylene]-*N*-methylmethanaminium tetrafluoroborate   146, 147, 498
TCFP (trichickenfootphos)   229
telocydins   616
TEM (transmission electron microscopy)
– GNP-peptide conjugates   53, 54
– nanowires   66
– oligonucleotides   568
temperature stability *see* thermal stability
templates
– molecular imprinting   15, 17, 18, 21, 25, 33, 35
– transition states in catalysis   296
TEMPO (2,2,6,6-tetramethyl-1-piperidinyloxy) radical   85–89, 281, 282, 367
TentaGel resins   10, 540, 557, 558, 564, 588
TETD (tetraethylthiuram disulfide)   536
tetrabutylammonium salts   211, 351
tetrazole and tetrazolides   534, 535, 569
TFA cleavage   5
TFFH (tetramethylfluorochloroformamidinium hexafluorophosphate)   149, 498
TFMSA (trifluoromethanesulfonic acid)   397, 405
thermal stability
– gel-type resins   251
– polymer supports   317, 318
thermoresponsive solubility   280
thiazolium organocatalysts   293, 294
thioacetals   487, 488
thioacids   400
thiocholine   67
thiochromans   114, 115
thioesters   *395*, 422, 443, 508–514
thioethers   444, 454, 554–556
thiohydroxamic acid/esters   459, 485, *486*
thiols
– acid-labile linkers   412
– linkers for oligonucleotide synthesis   561–563
– linkers for oligosaccharide synthesis   600
– NP functionalization   47, 56
– scavengers of cations   406
– scavenging of, by nucleophiles   190, *191*, 198, *201*, 203
– supported, as metal scavengers   241
– supported, as nucleophilic scavengers   207, 218, 219
thiophene derivatives   423
thrombin colorimetric detection   49
thymosin   170
tin compounds   134, 135, 240, 241, 310
tiopronin   53
titanium, chiral complexes   635

TMEDA (N,N,N′,N′-
  tetramethylethylenediamine) 604
TMT (trimercaptotriazine) 240
TPAP oxidations 86, 90, 91
TPS-Cl (2,4,6-triisopropylbenzenesulfonyl
  chloride) 532
traceless linkers 450–460
– cleavage mechanism and product
  451–454
– definition 450, 451
TRAM (traceless release of acrylamides)
  linker *439*, 445
transesterification 269, 367, 420, 421
transition metal catalysts 239, 309
– *see also individual metals*
– based on MNPs 311–314, 318, 325
– commercial and industrial uses 322–326
– reactions using polymer-supported
  316–322
– synthesis of polymer-supported 309–315
transition metal photolabile complexes 485,
  486
transition states, templating 296
transketalization 616
triazenes 129, 130, 457, 487
triazines
– chlorinated-, as coupling reagents 155–
  157 (*see also* DCT)
– trimercapto- (TMT) as scavengers 240
triazolides 548
triazolo-pyrimidines 186, *187*
trichostatin derivatives *632*, 634
triglucans 594
triglycerides 377
trimannosides 586, 605
triphase catalysis *see* PTC
triphenylphosphines 123–125, 267
– ditriflate 151
– oxide 192, 194
– ROMP-derived 222
trityl-based linkers 400–402, 511, 540, 541,
  545, 546
– *see also* 2-CTC
trityl protecting groups 532
TsDPEN ((1*R*, 2*R*)-*N*-(*p*-tolylsulfonyl)-1,2-
  diphenylethylenediamine) 94, 95
TSTU (2-succinimido-1,1,3,3-
  tetramethyluronium tetrafluoroborate)
  148
tubulin and tubacin 634
12-tungstophosphoric acid 112
two-step swelling polymerization 24
2-CTC (2-chlorotrityl) resins 500–502,
  507, 508, 513, 517, 521

U
Ugi condensations 331, *332*
ultraresins 84, 90
unnatural amino acids 517, 520
uronium coupling reagents 146–150
USPIOs (ultrasmall superparamagnetic iron
  oxide NPs) 57–59

V
vancomycin 445, 616, 633
Varian Inc. 233–235
VAZ (2-vinyl-4,4-dimethyl-5-oxazolone)
  188
vinyl esters
– enzyme catalyzed acylation with 367
– enzyme kinetic resolution with 376
vinyl ethers 336
vinyl functionalized cross-linkers 30
vinyl halides 618, 619
vinyl sulfones 430

W
Wang aldehyde resin 595, 596
Wang linkers 384, 386, 390–392, 402
Wang resins
– non-integral linkers/handles 4–6
– nucleophile-labile resins *419*, 421–423,
  428
– *p*-alkoxybenzyl linkers 488
– radical reagents on 135
water
– deoxygenation 323, 325, 326
– enzyme stability and activity 369
– scavenger 407
WAXS (wide angle x-ray scattering)
  320
Wittig olefinations 126, 127, 134, 432, 614,
  617, 618, 625

X
XAL (3-methoxyxanthine) linkers 387,
  402
XPS (X-ray photoelectron spectroscopy)
  324
XRD (X-ray diffraction) 324
XRMA (X-ray microprobe analysis) 315
xylyl linkers (α,α′-dioxyxylyl diether) 602

Y
Young's racemization test 147–149
ytterbium 209

Z
zearalenone 617